THE MOLECULAR IMMUNOLOGY OF COMPLEX CARBOHYDRATES—2

ADVANCES IN EXPERIMENTAL MEDICINE AND BIOLOGY

Recent Volumes in this Series

Volume 483
TAURINE 4: Taurine and Excitable Tissues
Edited by Laura Della Corte, Ryan J. Huxtable, Giampietro Sgaragli, and Keith F. Tipton

Volume 484
PHYLOGENETIC PERSPECTIVES ON THE VERTEBRATE IMMUNE SYSTEM
Edited by Gregory Beck, Manickam Sugumaran, and Edwin L. Cooper

Volume 485
GENES AND PROTEINS UNDERLYING MICROBIAL URINARY TRACT VIRULENCE: Basic Aspects and Applications
Edited by Levente Emődy, Tibor Pál, Jörg Hacker, and Gabriele Blum-Oehler

Volume 486
PURINE AND PYRIMIDINE METABOLISM IN MAN X
Edited by Esther Zoref-Shani and Oded Sperling

Volume 487
NEUROPATHOLOGY AND GENETICS OF DEMENTIA
Edited by Markus Tolnay and Alphonse Probst

Volume 488
HEADSPACE ANALYSIS OF FOODS AND FLAVORS: Theory and Practice
Edited by Russel L. Rous

Volume 489
HEMOPHILIA CARE IN THE NEW MILLENNIUM
Edited by Dougald M. Monroe, Ulla Hedner, Maureane R. Hoffman, Claude Negrier, Geoffrey F. Savidge, and Gilbert C. White II

Volume 490
MECHANISMS OF LYMPHOCYTE ACTIVATION AND IMMUNE REGULATION VIII
Edited by Sudhir Gupta

Volume 491
THE MOLECULAR IMMUNOLOGY OF COMPLEX CARBOHYDRATES—2
Edited by Albert M. Wu

Volume 492
NEUROIMMUNE CIRCUITS, DRUGS OF ABUSE, AND INFECTIOUS DISEASES
Edited by Herman Friedman, Thomas W. Klein, and John J. Madden

A Continuation Order Plan is available for this series. A continuation order will bring delivery of each new volume immediately upon publication. Volumes are billed only upon actual shipment. For further information please contact the publisher.

THE MOLECULAR IMMUNOLOGY OF COMPLEX CARBOHYDRATES—2

Edited by

Albert M. Wu

Glyco Research Laboratory
Institutes of Basic Medical Sciences and Natural Products
College of Medicine
Chang-Gung University
Kwei-san, Tay-yuan, Taiwan

Springer Science+Business Media, LLC

Library of Congress Cataloging-in-Publication Data

The molecular immunology of complex carbohydrates-2/edited by Albert M. Wu.
p. ; cm.—(Advances in experimental medicine and biology; 491)
"Satellite meeting of the 15th International Glycoconjugate Conference held August 22–27 in Tokyo, Japan"—Pref.
Includes bibliographical references and index.
ISBN 978-1-4613-5469-7 ISBN 978-1-4615-1267-7 (eBook)
DOI 10.1007/978-1-4615-1267-7
1. Carbohydrates—Immunology—Congresses. 2. Lectins—Congresses. 3. Glycoconjugates—Immunology—Congresses. I. Wu, Albert M. II. International Symposium on Glycoconjugates (15th: 1999: Tokyo, Japan) III. Advances in experimental medicine and biology; v. 491.
[DNLM: 1. Carbohydrates—immunology—Congresses.2. Glycoconjugates—Congresses. 3. Lectins—Congresses. QU 75 M718 2001]
QR186.6.C37 M65 2001
616.07'9—dc21

2001018619

Proceedings of the Molecular Immunology of Complex Carbohydrates—2, held August 28–September 2, 1999, in Taiwan.

ISBN 978-1-4613-5469-7

Originally published by Kluwer Academic / Plenum Publishers, New York in 2001
Softcover reprint of the hardcover 1st edition 2001

10 9 8 7 6 5 4 3 2 1

A C.I.P. record for this book is available from the Library of Congress

To Elvin Abraham Kabat (September 1, 1914–June 16, 2000), one of the pioneers in quantitative immunochemistry and my postdoctoral training adviser, as well as to colleagues and friends, who have sincerely supported me in my scientific career. Their valuable assistance has made my research possible.

Elvin Abraham Kabat

September 1, 1914 – June 16, 2000

This book is dedicated to the memory of Elvin A. Kabat, Ph.D., one of the great immunologists of the 20th century, and a pioneer in the immunology of carbohydrates. Dr. Kabat graduated from the City College of New York at the age of 18 in 1932, and obtained a Ph.D. in Biochemistry from Columbia University College of Physicians and Surgeons in 1937. His mentor at Columbia was Dr. Michael Heidelberger, who played a major role in the evolution of immunology from a descriptive field into a quantitative, chemical discipline. After graduating from Columbia, Dr. Kabat spent a postdoctoral year as a Rockefeller Foundation fellow at Uppsala University in Sweden. During that time he performed the first studies of the electrophoretic mobility of purified antibodies.

It is not possible, in this brief memorial, to do justice to the scope and variety of Dr. Kabat's scientific accomplishments during his career of almost seven decades. Instead, in keeping with the topic of this conference, I will emphasize his studies of the immunology of carbohydrates, and how this studies contributed to the development of fundamental concepts in immunochemistry. During the late 1940s he initiated studies of the chemical nature of the ABO and Lewis blood group substances. Over several decades his laboratory, and that of Morgan and Watkins in England, identified the antigenic determinants and the complex structures of the oligosaccharide chains that displayed these important antigens. Knowledge of the structural basis of these antigens made possible later studies of the enzymatic and genetic basis of this important hereditary polymorphism.

His studies of dextrans originated in the 1950s in the analysis of allergic reactions to dextrans used as blood expanders. After demonstrating that the hypersensitivity reactions were indeed directed at dextran carbohydrate, and not, as previously assumed, at protein contaminants, he realized that dextrans provided a model system for analysis of antigenic determinants and antibody specificity. His studies led him to conclude that antibodies against dextran could recognize structures as small as di- and trisaccharides, and as large as hexasaccharides. He also suggested, based on studies of linear and branching dextrans, that some antibody combining sites were cavities that recognized the terminal nonreducing ends of carbohydrate chains, and other antibody binding sites were grooves, which permitted them to bind to internal structures of extended carbohydrate structures. These predictions, based on studies of heterogeneous polyclonal antibodies, were verified by his subsequent studies of monoclonal antibodies, and by crystallographic studies performed by other scientists several decades later. In parallel to his analyses of antibodies against carbohydrates, Dr. Kabat and his collaborators also established the carbohydrate-binding specificity of many lectins.

Dr. Elvin Abraham Kabat, pioneer in the field of Quantitative Immunochemistry, receiving the National Medal of Science from President Bush in 1991.

Editor Wu meets with Dr. Elvin Abraham Kabat at College of Physicians and Surgeons Columbia University, New York, N.Y., USA, January 27, 1995.

From left: W. T. J. Morgan, E. A. Kabat, W. M. Watkins, K. O. Lloyd and A. M. Wu.
Cyst Mucin Group, Seattle, WA, August 24, 1995

Honoring Dr. E.A. Kabat, Seattle, WA, August 24, 1995

MICC Taiwan

Molecular Immunology of Complex Carbohydrates II

國際 醣類分子免疫學 學術研討會

Aug 28 -Sept 2, 1999

HONORING
DR. ELVIN KABAT

Aug. 28 – Sept. 2, 1999

NISHIMURA

吳
長庚醫學院
明道

MICC

Taiwan

Molecular Immunology of Complex Carbohydrates II

國際 醣類分子免疫學 學術研討會

Aug 28 -Sept 2, 1999

HONORING

Dr. Winifred M. Watkins

Aug. 28 – Sept. 2, 1999

Participants of the INTERATIONAL SYMPOSIUM on Molecular Immunology of Complex Carbohydrates-II, and the Taiwan-Canada Glycobiology Workshop

國際 醣類分子免疫學 學術會議　台灣 · 加拿大 醣類生物學 學術研討會

Institute of Biological Chemistry, Academia Sinica, Taipei, Taiwan. Aug.28-Sept.2, 1999

Scientists and students came from all over the world to work with Dr. Kabat, and many people who trained with him became leaders in glycobiology and immunology. His passion for science, integrity and high standards made him a demanding taskmaster, and his critiques of experimental data could be unsparing. His former trainees enjoyed getting together at international meetings to reminisce about their experiences in his laboratory. The serious atmosphere of the laboratory was leavened by his wonderful sense of humor, and his talent as a raconteur. Scientists trained in his laboratory carried with them a model of how science should be performed, and his trainees maintained enduring personal and professional relationships with him.

His scientific accomplishments were recognized by awards and honorary degrees too numerous to list. Among these awards were election to the National Academy of Sciences, the Louisa Gross Horwitz Prize from Columbia University, which he shared with Dr. Heidelberger and Dr. Henry Kunkel, and the United States' highest award for scientific achievement, the National Medal of Science. The scientific community has lost one of its greatest and most committed members. On behalf of all his former associates and colleagues, I would like to extend our sympathy to his family, and to celebrate with them a life of remarkable achievement.

Donald M. Marcus
Dept. of Medicine
Baylor College of Medicine
Houston, Texas 77030

CONTRIBUTORS

Annick Barre (**I-3**), Institut de Pharmacologie et Biologie Structurale, UPR CNRS 9062, 205 Route de Narbonne, 31077 Toulouse Cedex, France

Paul V. Beum (**III-3**), Department of Biochemistry and Molecular Biology, University of Nebraska Medical Center, Omaha, Nebraska 68198-4525, USA.

V. P. Bhavanandan (**I-7**), Department of Biochemistry & Molecular Biology, The Pennsylvania State University College of Medicine, Hershey, Pennsylvania 17033, USA.

C. Fred Brewer (**I-2**), Department of Molecular Pharmacology and Microbiology and Immunology, Albert Einstein College of Medicine, Bronx, New York 10641, USA

Jean-Robert Brisson (**V-9**), Institute for Biological Sciences, National Research Council of Canada, Ottawa, Ontario, K1A 0R6 Canada.

Pi-Wan Cheng (**III-3**), Department of Biochemistry and Molecular Biology, University of Nebraska Medical Center, Omaha, Nebraska, 68198-4525, USA.

Teh-Ying Chou (**IV-3**), Department of Pathology & Institute of Biochemistry, National Yang-Ming University, Taipei, 112, Taiwan.

Cheng-Keng Chuang (**IV-4**), Graduate Institute of Clinical Medicine, c/o Chang Gung Memorial Hospital, Building 12J, Kweishan, Tao-yuan 333, Taiwan.

Julia L. Clarke (**III-1**), Department of Haematology, Imperial College School of Medicine, Hammersmith Hospital, London W12 ONN, United Kingdom.

Andrew Cox (**V-7**), Institute for Biological Sciences, National Research Council of Canada, Ottawa, Ontario, da, K1A 0R6, Canada.

Els J.M. Van Damme (**I-3**), Laboratory for Phytopathology and Plant Protection, Katholieke Universiteit Leuven, Willem de Croylaan 42, 3001 Leuven, Belgium.

Anne Dell (**II-5**), Department of Biochemistry, Imperial College, South Kensington, London, SW7 2AZ, United Kingdom

Maria Duk (**I-9** and **II-2**), Department of Immunochemistry, Ludwik Hirszfeld institute of Immunology and Experimental Therapy, Polish Academy of Sciences, UL. Czerska 12, 53-114, Wroclaw, Poland.

Ten Feizi (**I-5**), The Glycosciences Laboratory, Imperial college school of medicine Northwick Park Campus, Harrow, HA1 3UJ, United Kingdom.

Kiyoshi Furukawa (**III-4**), Department of Biosignal Research, Tokyo Metropolitan Institute of Gerontology, Itabashi-ku, Tokyo 173-0015, Japan.

Hans-Joachim Gabius (**I-6**), Institute of Physiological Chemistry, Faculty of Veterinary Medicine, Ludwig-Maximilians-University, Veterinärstr. 13, D-80539 München, Germany.

Nechama Gilboa-Garber (**I-8**), Faculty of Life Sciences, Bar-Ilan University, 52 900, Ramat-Gan, Israel.

Xiaoxuan Guo (**I-7**), Department of Biochemistry & Molecular Biology, The Pennsylvania State University College of Medicine, Hershey, Pennsylvania 17033, USA.

Senitiroh Hakomori (**IV-1**), Pacific N.W. Research Institute, University of Washington, 720 Broadway, Seattle, WA 98122 USA.

Yoichiro Hama (**II-4, III-7**), Department of Applied Biological Sciences, Faculty of Agriculture, Saga University, Saga 840-8502, Japan.

Gerald W. Hart (**IV-3**), Department of Biological Chemistry, The John Hopkins University School of Medicine, Baltimore, MD 21205, U.S.A.

Anthony Herp (**A-I**), Glyco-Immunochemistry Research Laboratory, College of Medicine, Chang-Gung University, Tao-yuan, 333, Taiwan.

Sen Hiraizumi (**III-4**), Department of Surgery, Sendai City Hospital, Wakabayashi-ku, Sendai 984-0075, Japan.

Derek W. Hood (**V-7**), Molecular Infectious Diseases Group, University of Oxford, Department of Paediatrics, John Radcliffe Hospital, Headington, Oxford, 0X3 3DU, United Kingdom.

Kuo-Yuan Hwa (**II-6**), Institute of Biological Chemistry, Academia Sinica, Nankang, Taipei 115, Taiwan.

Tatsuro Irimura (**IV-2**), Laboratory of Cancer Biology and Molecular Immunology, Graduate School of Pharmaceutical Sciences, The University of Tokyo, 7-3-1 Hongo, Bunkyo-ku, Tokyo 113-0033, Japan.

Matthias Iwersen (**III-5**), Biochemisches Institut, Christian-Albrechts-Universität, Olshausenstra_e 40, D-24098 Kiel, Germany.

Harold J. Jennings (**V-4** and **V-9**), Institute for Biological Sciences, National Research Council of Canada, Ottawa, Ontario, K1A 0R6, Canada.

Weiping Jiang (**I-7**), Department of Biochemistry & Molecular Biology, The Pennsylvania State University College of Medicine, Hershey, Pennsylvania 17033, USA.

Herbert Kaltner (**I-6**), Institute of Physiological Chemistry, Faculty of Veterinary Medicine, Ludwig-Maximilians-University, Veterinärstr, 13, D-80539 München, Germany.

Reiji Kannagi (**III-2** and **A-II**), Program of Molecular Pathology, Aichi Cancer Center, Research Institute, 1-1, Kanoka-den, Chikusa-ku, Nagoya, 464-8681, Japan.

Karl-Anders Karlsson (**V-1**), Institute of Medical Biochemistry, Göteborg University, P.O. BOX 440, SE-405 30 Göteborg, Sweden.

Yutaka Kawamura (**IV-2**), Laboratory of Cancer Biology and Molecular Immunology, Graduate School of Pharmaceutical Sciences, The University of Tokyo, 7-3-1 Hongo, Bunkyo-ku, Tokyo 113-0033, Japan.

Kay-hooi Khoo (**II-5**), Institute of Biological Chemistry, Academia Sinica, Taipei, 115, Taiwan.

Noriaki Kitamura (**III-4**), Faculty of science and Technology, Science University of Tokyo, Noda, Chiba 278-0022, Japan.

Guido Kohla (**III-5**), Biochemisches Institut, Christian-Albrechts-Universität, Olshausenstra_e 40, D-24098 Kiel, Germany.

Mamoru Kyogashima (**III-6**), Seikagaku Corporation Tokyo Research Institute 1253, Tateno 3-chome Higashiyamato-shi, Tokyo 207-0021, Japan.

Chin-Jen Lee (**V-3** and **V-6**), Center For Biologics Evaluation and Research,FDA, Bethesda, MD 20817, USA.

Lucia H. Lee (**V-3**), Center For Biologics Evaluation and Research, FDA, Bethesda MD 20817, USA.

Su-Chen Li (**II-4** and **III-7**), Department of Biochemistry SL43 Tulane University School of Medicine, 1430 Tulane Avenue, New Orleans, LA 70112, USA.

Yu-Ten Li (**II-4** and **III-7**), Department of Biochemistry SL43 Tulane University School of Medicine, 1430 Tulane Avenue, New Orleans, LA 70112, USA.

Zhong-Ming Li (**V-6**), Center For Biologics Evaluation and Research, FDA, Bethesda, MD 20817, USA.

Shuen-Kuei Liao (**IV-4**), Graduate Institute of Clinical Medicine, c/o Chang Gung Memorial Hospital, Building 12J, Kweishan, Tao-yuan 333, Taiwan.

Chien-Cheng Lin (**IV-4**), Graduate Institute of Basic Medical Sciences, College of Medicine, Chang Gung University, Tao-yuan, 333, Taiwan.

Chun-Cheng Lin (**II-7**), Institute of Chemistry, Academia Sinica, Nankang, Taipei 115, Taiwan.

Chen-Hung Lin (**II-7**), Institute Biological Chemistry, Academia Sinica, Nankang, Taipei 115, Taiwan.

Halina Lis (**I-1**), Department of Biological Chemistry, The Weizmann Institute of Science, Rehovot 76100, Israel.

Elwira Lisowska (**I-9**, **II-2** and **II-3**), Department of Immunochemistry, Ludwik Hirszfeld institute of Immunology and Experimental Therapy, Polish Academy of Sciences, UL. Czerska 12, 53-114, Wroclaw, Poland.

Cheng-shiung Lu (**V-3**), Center for Disease Control, Taipei, 100, Taiwan.

Adele Martin (**V-7**), Institute for Biological Sciences, National Research Council, Ottawa, Ontaria, da K1A 0R6, Canada.

E. Richard Moxon (**V-7**), Molecular infectious Diseases Group, University of Oxford, Department of Paediatrics, John Radcliffe Hospital, Headington, Oxford, 0x3 3DU, United Kingdom.

Hiroki Nakagawa (**II-4**), Department of Applied Biological Sciences, Faculty of Agriculture Saga University, Saga 840-8502, Japan.

Yoko Nemoto-Sasaki (**IV-2**), Laboratory of Cancer Biology and Molecular Immunology, Graduate School of Pharmaceutical Sciences, The University of Tokyo, Tokyo 113-0033, Japan.

Masayuki Ota (**IV-2**), Laboratory of Cancer Biology and Molecular Immunology, Graduate School of Pharmaceutical Sciences, The University of Tokyo, Tokyo 113-0033, Japan.

Willy J. Peumans (**I-3**), Laboratory for Phytopathology and Plant Protection, Katholieke Universiteit Leuven, Willem de Croylaan 42, 3001 Leuven, Belgium.

Jörg Pommerencke (**III-5**), Biochemisches Institut, Christian-Albrechts-Universität Olshausenstra_e 40, D-24098 Kiel, Germany.

Sabine Puch (**I-7**), Department of Biochemistry & Molecular Biology, The Pennsylvania State University College of Medicine, Hershey, Pennsylvania 17033, USA.

James C. Richards (**V-7**), Institute for Biological Sciences, National Research Council, Ottawa, Ontaria, da K1A 0R6, Canada.

Pierre Rouge (**I-3**), Institut de Pharmacologie et Biologie Structurale, UPR CNRS 9062, 205 Route de Narbonne, 31077 Toulouse Cedex, France

Takeshi Sato (**III-4**), Department of Biosignal Research, Tokyo Metropolitan Institute of Gerontology, Itabashi-ku, Tokyo 173-0015, Japan.

Roland Schauer (**III-5**), Biochemisches Institut, Christian-Albrechts-Universität, Olshausenstraβe 40, D-24098, Kiel, Germany.

Hiltrud Schmid (**III-5**), Biochemisches Institut, Christian-Albrechts-Universität Olshausenstraβe 40, D-24098 Kiel, Germany.

Elke K.H. Schweda (**V-7**), University College of South Stockholm, Clinical Research Centre, NOVUM, S-141 86, Huddinge, Sweden.

Nathan Sharon (**I-1**), Department of Biological Chemistry, The Weizmann Institute of Science, Rehovot 76100, Israel.

Yung-Chi Shen (**IV-4**), Graduate Institute of Clinical Medicine, c/o Chang Gung Memorial Hospital, Building 12J, Kweishan, Tao-yuan 333, Taiwan.

Shuh-Chyung Song (**A-I**), Department of Pathology and Lab. Medicine, University of Rochester Medical Center, 601, Elmwood Avenue, Box 626 Rochester, New York, 14642, USA.

Toshihisa Sumi (**II-4**), Department of Applied Biological Sciences, Faculty of Agriculture, Saga University, Saga 840-8502, Japan.

Yasuo Suzuki (**V-2**), Department of Biochemistry, University of Shizuoka School of Pharmaceutical Sciences, 52-1 Yada, Shizuoka, 422-8526, Japan.

Chao-Ming Tsai (**V-8**), Division of Bacterial Products, Center for Biologics Evaluation and Research, FDA, Bethesda, MD 20892, USA.

Ming-Sung Tsai (**A-I**), Glyco-Immunochemistry Research Laboratory, College of Medicine, Chang-Gung University, Tao-yuan, 333, Taiwan.

Johnnes F.G. Vliegenthart (**II-1**), Bijvoet Center, Department Bio-Organic Chemistry, Utrecht University, P.O.Box 80.075 NL-3508 TB Utrecht, The Netherlands.

Denong Wang (**V-5**), Immunology Laboratory, Columbia genome center, College of Physicians and Surgeons, Columbia University, New York, New York, 10032, USA.

Winifred Watkins (**III-1**), Department of Haematology, Imperial College School of Medicine, Hammersmith Hospital, London, W12 0NN, United Kingdom.

Albert M. Wu (**I-4, I-8, I-9, V-3** and **A-I**), Glyco-Immunochemistry Research Laboratory, College of Medicine, Chang-Gung University, Tao-yuan, 333, Taiwan.

Wei Zou (**V-4**), Institute for Biological Sciences, National Research Council of Canada, Ottawa, Ontario, K1A 0R6, Canada.

PREFACE

I organized this symposium, "Molecular Immunology of Complex Carbohydrates-2 (MICC-2)," at the Institute of Biological Chemistry, Academia Sinica, Taipei, Taiwan between August 28-September 2, 1999, as a satellite meeting of the 15th International Glycoconjugate Conference (held August 22–27 in Tokyo, Japan). I also held a Taiwan-Canada Glycobiology Workshop after this meeting at the Institutes of Biological Chemistry and Chemistry. To promote glycobiology in Taiwan, I offered a Complexs Carbohydrate & Medicine-2 workshop at the Glyco- Research Laboratory, College of Medicine, Chang-Gung University, Kwei-san, Tao-yuan, Taiwan before the MICC-2 symposium. The lecture and poster materials of these three meetings are collected in these proceedings, which are divided into five Sections and two Appendixes.

Section I, entitled "Protein-Carbohydrate Interactions of Plant and Animal Lectins," provides current concepts of lectin-carbohydrate interactions; classification of lectins, based on amino acid sequences, molecular structures, and lectin affinity for carbohydrates. However, the relationships between amino acid sequences and carbohydrate affinity of lectins have to be more thoroughly characterized. The reviews on animal lectins in this Section explore new areas of lectins. Section II, "Aspects of Structure and Antigenicity of Glycoconjugates," provides important information on structural concepts of glyco-immunology. "Glycotope Expression (Glycosylation), Metabolism and Functions," which play important roles in life processes, are discussed in Section III. Four articles on advances in knowledge on structural roles of glycans and treatment of cancer are discussed in Section IV. Complex carbohydrates as vaccines is one of the promising directions pursued in glycobiology and ten reviews related to complex carbohydrates in microbial and viral infections and vaccine design are selected for inclusion in Section V. Two guides, "A Guide to the Carbohydrate Specificities of Applied Lecins-II (Updated in 2000)" and "A Guide to Monoclonal Antibodies Directed to Glycotopes," are included at the end of the proceedings. These two guides are likely to be two of the most comprehensive handbooks detailing carbohydrate specificities of applied lectins and anti-carbohydrate monoclonal antibodies. It is obvious that the new methodologies in structural analyses and lectin-carbohydrate binding assays are greatly contributing to the progress in glycosciences. These proceedings provide information on glycotopes required for essential basic concepts and applications. Furthermore, not only previous chapters have been updated, but many new advances in glycotopes have been included. MICC-2 represents the glycan connection between Taiwan biologists and colleagues from other parts of the world.

The meeting organizers are grateful to: Taiwan National Science Council, Chang-Gung complex (Hospital and College of Medicine), Tao-yuan, Taiwan; Institutes of Biological Chemistry (Dr. Shih-Hsiung Wu) and Chemistry (Dr. Sunney Chan), Academia Sinica, Taipei; and Development Center for Biotechnology (Dr. Michael M.G. Yet), Taipei, for supporting these meetings. I also want to thank the committee of Glyco XV, especially Drs. A. Kobata and K. Furukawa, Tokyo, for encouraging the MICC-2 meeting in Taiwan; the members and staff of Institutes of Biololgical Chemistry and Chemistry, particularly Drs. K.-H. Khoo, Chun-Hung Lin, and Chun-Cheng Lin, for their help and Dr. A. Herp in providing his editing experience.

Albert M. Wu
Glyco Research Laboratory
College of Medicine
Chang-Gung University
Kwei-san, Tao-yuan
Taiwan

CONTENTS

SECTION I

Carbohydrate-Protein Interactions: Plant and Animal Lectins

SECTION II

Structure and Antigenicity of Glycoconjugates

SECTION III

Glycotopes Expression, Metabolism, and Functions

SECTION IV

Complex Carbohydrates in Development and Treatment of Cancer

SECTION V

Complex Carbohydrates in Microbial and Viral Infections and Vaccine Design

Appendix

THE STRUCTURAL BASIS FOR CARBOHYDRATE RECOGNITION BY LECTINS

Nathan Sharon and Halina Lis

Department of Biological Chemistry, The Weizmann Institute of Science, Rehovot 76100, Israel

Lectins (Sharon and Lis, 1989; Lis and Sharon, 1998), a class of carbohydrate-specific proteins ubiquitous in Nature, bind mono- and oligosaccharides of eukaryote glycoconjugates reversibly, but are devoid of catalytic activity, and in contrast to antibodies, are not products of an immune response. Each lectin molecule contains typically two or more carbohydrate combining sites, i.e., they are di- or polyvalent. Therefore, when they react with cells, for example erythrocytes, they will combine with the cell surface sugars and may cause crosslinking of the cells, as well as their subsequent precipitation, a phenomenon referred to as cell agglutination. The erythrocyte agglutinating, or hemagglutinating, activity of lectins is a major attribute of these proteins and is used routinely for their detection and characterization. Lectins also form crosslinks between polysaccharide or glycoprotein molecules in solution and induce their precipitation. Both the agglutination and precipitation reactions of lectins are inhibited by the sugars for which the lectins are specific.

Although lectins have been first described at the turn of the century, their study started to gain momentum only in the 1960's (Sharon and Lis, 1998). They were then shown to be invaluable tools for the structural and functional investigation of complex carbohydrates, especially glycoproteins, and for the examination of changes that occur on cell surfaces during physiological and pathological processes, from cell differentiation to cancer. At present, they are the focus of intense attention because of the realization that they act as recognition determinants in diverse biological processes. These include clearance of glycoproteins from the circulatory system, control of intracellular traffic of glycoproteins, adhesion of infectious agents to host cells, recruitment of leukocytes to inflammatory sites, as well as cell interactions in the immune system, in malignancy and metastasis. In depth understanding at the molecular level of how lectins bind carbohydrates is of theoretical interest and may pave the way for the design and production of drugs for fighting a wide range of diseases, from infection to inflammation to cancer.

During the last decade, there has been remarkable progress in elucidating the features of lectins that are important for carbohydrate binding (Lis and Sharon, 1998). In particular, high res-

The Molecular Immunology of Complex Carbohydrates–2.
Edited by Albert M. Wu, Kluwer Academic/Plenum Publishers, 2001.

olution X-ray crystallography of lectins in complex with their ligands allowed the identification of the chemical groups on the protein and on the carbohydrate that interact with each other and of the types of bond formed. Further information on the contribution of individual amino acids to the activity of lectins has been obtained by site-directed mutagenesis and by molecular modeling. Of special interest are the studies of lectin-oligosaccharide complexes, since they provide a basis for the understanding of how lectins recognize their natural ligands.

In this article we focus on what has been learned on this subject from the investigation of legume lectins, and also draw conclusions from a comparison of data on the combining sites of these lectins with those of lectins from other sources.

LEGUME LECTINS

Legume lectins represent the largest and most thoroughly studied family of proteins of this class, some 100 members of which have been characterized, almost all obtained from seeds of the plants (Sharon and Lis, 1990; Loris et al., 1997; Hamerlyck et al.,1998; Van Damme et al., 1998). Well known examples are concanavalin A, phytohemagglutinin (PHA), soybean agglutinin (SBA), peanut agglutinin (PNA) and coral tree or *Erythrina corallodendron* lectin (ECorL). The legume lectins are extremely useful as a model system for the understanding of the molecular basis of carbohydrate-protein interactions because they are easy to obtain in purified form and exhibit an amazing variety of sugar specificities. They often have the ability to distinguish between very similar carbohydrates, such as glucose and galactose, that differ in the configuration of a single hydroxyl (4-OH) or in the position of a glycosidic linkage [NeuAc(α2-3)Gal vs NeuAc(α2-6)Gal]. Fundamental insights obtained from the study of these lectins have been widely applied to lectins of other families such as the pharmaceutically important C-type animal lectins. Indeed, well established concepts like subsite multivalency were first formulated, or the formation of crosslinked lattices was first observed, during studies of legume lectins and proved to be valid outside the legume lectin family. An important additional reason for the interest in legume lectins is their structural similarity to lectins from other sources, whether of animals or microorganisms (Table 1).

Typically, the members of this family consist of two or four subunits (or protomers), either identical or slightly different, each with a single, small carbohydrate combining site with the same specificity. They also contain one atom each of tightly attached calcium and manganese ions per subunit that are required for carbohydrate binding. In addition to their carbohydrate combining sites, several of the legume lectins possess hydrophobic sites that bind specifically non-polar compounds such as adenine, indoleacetic acid or porphyrins.

The subunits of the legume lectins are of 25-30 kDa and are commonly composed of single polypeptide chains of about 250 amino acids. In some (e.g. those from pea and lentil) the polypeptides are fragmented into a light (α, 5-7 kDa) and heavy (β, ca. 20 kDa) chain. Most, but not all, are glycosylated and carry one or two asparagine-linked oligosaccharides per subunit. The oligosaccharides are either of the oligomannose type, identical with those found in animal

Table 1.
Structurally similar lectins

Family/Group	Source	Specificity	Function
Legume lectins	Plants	Diverse	Binding of Rhizobia, defense
Galectins	Animals	Galactose	Modulation of cell adhesion
MR60/ERGIC-53, VIP-36	Animals	Mannose	Intracellular sorting of glycoproteins
Sialidase wings	Bacteria	Sialic acid	Unknown

glycoproteins, or contain in addition fucose and xylose, forming a class of N-linked glycans that are plant specific. In the latter glycans, the xylose is linked β1-2 to the β-mannose of the pentasaccharide core Man(α1-3)[Man(α1-6)]Man(β1-4)GlcNAc(β1-4)GlcNAc, and the fucose is linked α 1-3 to the asparagine-linked *N*-acetylglucosamine.

I. PRIMARY STRUCTURES

The primary structures of the lectins of this family are remarkably homologous, with about 20% of identical amino acids, and close to 20% of similar ones. The conserved amino acids include several of those that participate in hydrogen bonding or hydrophobic interactions with the monosaccharide held in the combining site, and almost all residues that coordinate the metal ions. Lectins from plants belonging to the Diocleae tribe (that include the genera *Canavalia* and *Dioclea*), e.g., concanavalin A and *Dioclea grandiflora* lectin, occupy a special position, since they display an unusual homology, referred to as "circular homology", with the other legume lectins. For concanavalin A, this homology is obtained by aligning residue 123 with the amino terminal residue of the other lectins, proceeding to the carboxyl end of concanavalin A and continuing along its amino terminal region. It is the result of an unique rearrangement of the peptide chain that occurs in the last step of the synthesis of the lectin. Quite surprisingly, two mannose-specific animal lectins (MR60/ERGIC-53 and VIP36) are homologous with those of the legumes, and contain also two of the conserved residues that hydrogen bond with the monosaccharide in the combining site of the legume lectins.

II. TERTIARY AND QUATERNARY STRUCTURES

At the time of writing, over 80 three dimensional structures of twenty legume lectins, without as well as with various carbohydrate ligands, have been elucidated by high resolution X-ray crystallography (for information, see the 3D lectin data base: *www.cermav.cnrs.fr/databank.lectine*). In a few other cases, molecular models have been constructed. The legume lectin subunits, all of which are almost superimposable (Figure 1) are in the shape of a dome formed largely of three β sheets: an antiparallel 6-stranded back sheet, a 7-stranded front sheet and a smaller, 5-stranded sheet that plays a major role in holding the two large sheets together. The majority of residues not included in the β structures are in loops and β bends that connect the strands of the β sheets. The six-stranded sheet is almost flat, while the seven-stranded one is concave. This topology is related to the jellyroll fold and is known as "the lectin fold" (Srinivasan et al., 1996). The same fold has been discovered in a number of other proteins from diverse sources, including bacteria, fungi and mammals, most of which bind carbohydrates. They include the jellyroll wings of *Vibrio cholerae* sialidase; isopenicillin N-synthase of *Aspergillus nidulans*; mammalian galectins and human C-reactive protein.

In the legume lectins, the combining sites of the carbohydrate and of the metal ions are located mainly in the β folds of the seven-stranded sheet, the concavity of which provides a shallow carbohydrate binding site. This site is located at the top of each protomer that is easily accessible not just to monosacharides, but to oligo- and polysaccharides as well. The Ca^{2+} and Mn^{2+} (which can be substituted by another transition metal ion such as zinc, nickel or cobalt) are 4.25 Å apart and are in close proximity (9-13 Å) to the carbohydrate binding site.

In spite of the great similarity of the primary and tertiary structures of the legume lectins, there are marked differences in their quaternary structure, whether they are in the form of dimers or of tetramers (Hamelryck et al, 1999; Prabu et al., 1999; Vijayan and Chandra 1999) (Figure 2). These may result from the presence of covalently linked carbohydrate in the contact region between the subunits, as is the case for ECorL, or small variations in amino acids in these regions, for instance as in PNA or *Griffonia simplicifolia* lectin IV (GSIV), that differ thus from concanavalin A which has a so called canonical quaternary structure.

III. COMBINING SITES

As mentioned, the carbohydrate combining sites of legume lectins, similarly to those of other proteins of this class, are in the form of shallow depressions on the surface of the protein.

Fig. 1. Tertiary structures of ECorL in complex with galactose (1LTE, thin line) and of *Lens culinaris* lectin in complex with mannose (1LEM, thick line)

Typically, only one or two edges or faces of the carbohydrate ligand are bound to the protein. This is in contrast to carbohydrate-binding bacterial periplasmic receptors, specific for e.g., glucose or galactose, in which the ligand binding site is buried in the interior of the protein. In general, the combining sites appear to be preformed, since few conformational changes occur upon carbohydrate binding.

1. For monosaccharides

Over the last years, our laboratory has focused on the combining site of ECorL. The range of sugars recognized by this lectin is listed in Figure 3. Based on the high resolution three dimensional structure of the ECorL-lactose complex, in which originally only the galactose part of the ligand was clearly seen (Shaanan et al., 1991), on site-directed mutagenesis studies of the lectin (Adar and Sharon, 1996), and on data from related lectins, we have concluded that legume lectins, irrespective of their specificity, bind monosaccharides in their primary combining sites through the side chains of a constellation of four invariant residues, three of which are an aspartic acid, an asparagine and an aromatic amino acid (Sharon, 1993; Adar and Sharon, 1996) The only exceptions to this generalization are *Dolichos biflorus* lectin (DBL), to be discussed later, (Hamelryck et al., 1999) and PHA-L (Hamelryck et al., 1996), in the combining site of which the invariant aromatic acid is replaced by leucine. The fourth conserved amino acid is glycine present in all legume lectins except for concanavalin A, where it is replaced by arginine. However, the glycine and arginine form H-bonds with the monosaccharide in the primary site via their main chain amide, and not via their side chains, as the aspartic acid and asparagine referred to earlier.

In the ECorL-lactose complex, the hydrophobic surface formed by the C3-H, C4-H and

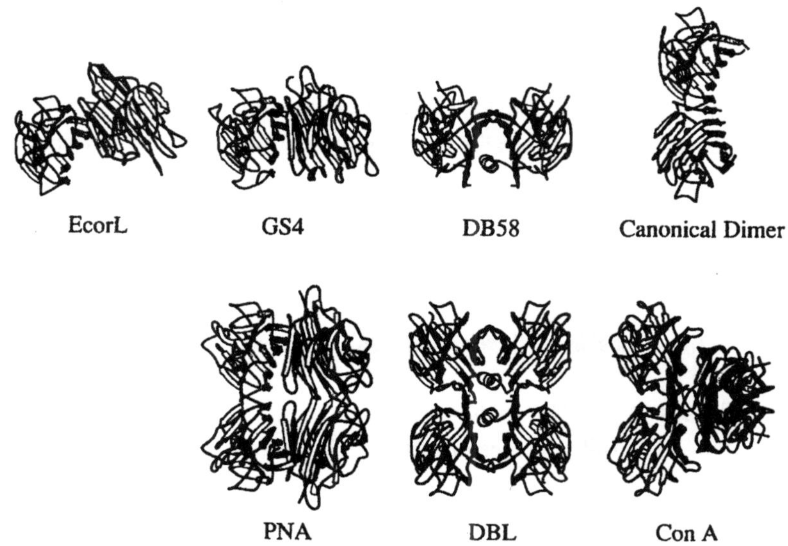

Fig. 2. The different lectin oligomeric types, including the novel DB58 dimer type (from Hamelryck et al., 1999). The canonical dimer is represented by lentil lectin. Each tetramer type can be considered as a dimer of dimers. In the figure, the dimers and the corresponding tetramers are vertically aligned. One of the subunits is in the same orientation in all oligomers (upper left monomer fit the three tetramers). β Strands are shown as arrows.

C5-H of galactose is stacked over the aromatic ring of Phe131. Stacking of the hydrophobic face of sugar rings on aromatic side chains is a common feature of the complexes of legume lectins with carbohydrates, irrespective of their specificity, since positions equivalent to Phe133 in ECorL are almost always occupied by aromatic residues. Similar stacking is also found in sugar complexes of other lectins, for instance the galectins, as well as of certain bacterial toxins, such as *E.coli* lytic toxin (Rini, 1995). Replacement of the aromatic amino acid which stacks against the carbohydrate ring with a smaller, non-aromatic residue [e.g. Phe131 in ECorL with valine or alanine (Adar and Sharon, 1996) or Tyr124 in *Griffonia simplicifolia* lectin II (GSII) with glycine (Zhu et al., 1996)] afforded an inactive protein, illustrating the essential role of the this kind of interaction for ligand binding.

Coming back to the key H-bonding amino acid side chains, it was found that substitution, by site-directed mutagenesis, of Asp89 or Asn133 in ECorL by alanine resulted in loss of sugar binding ability (Adar and Sharon, 1996). Since the X-ray data showed that Asp89 forms H-bonds with the 3- and 4-OH of the galactose in the combining site and Asn133 is H-bonded with the 4-OH of the ligand, this is in line with the inability of the lectin to bind 3- or 4-fluorogalactose. Mutation of asparagine and/or aspartic acid located at equivalent positions in other legume lectins, such as PHA-L (Mirkov and Chrispeels, 1993), pea (*Pisum sativum*) lectin (PSL) (van Eijsden et al., 1994) or GSII (Zhu et al., 1996) abolished acivity. The key role of these amino acids in ligand binding has been similarly demonstrated for the homologous, mannose-specific animal lectin MR60/ERGIC-53 (Roche and Monsigny, 1996).

The asparagine just mentioned also coordinates the Ca2+ present in all legume lectins, which thus helps to position this residue in the correct place for ligand binding. The Ca2+ and the adjacent Mn2+ are both linked by coordination bonds to common residues (Asp129 and Asp136 in ECorL). The two metal ions are further stabilized in their positions by four conserved water molecules, two for each metal ion (Loris et al., 1994a; Adar et al., 1998). One of the water ligands of the calcium ion forms a bridge with the carbonyl group of the carbohydrate-binding aspartic acid residue (e.g. Asp208 in concanavalin A or Asp89 in ECorL). Another characteristic of the combining site of legume lectins is the presence of a rare *cis*-peptide bond between this key binding aspartic acid and the preceding amino acid, which is almost always alanine. This bond is sta-

Gal[a]<GalNAc<Lac<Fuc(α1-2)Lac<LacNAc<
Fuc(α1-2)LacNAc<<MeαGalNDns<MeβGalNDns[b]

[a] $K_a=1.6x10^3$ (microcalorimetry)
[b] $K_a=3.5x10^5$ (microcalorimetry); $7.3x10^5$ (spectrofluorimetry)

No binding of 3 or 4 fluorogalactose, nor of GlcNDns.

*Based on data of Kinzy et al., 1992, Adar and Sharon, 1996 and Moreno et al, 1997

Fig. 3. Specificity of *Erythrina corallodendron* lectin*

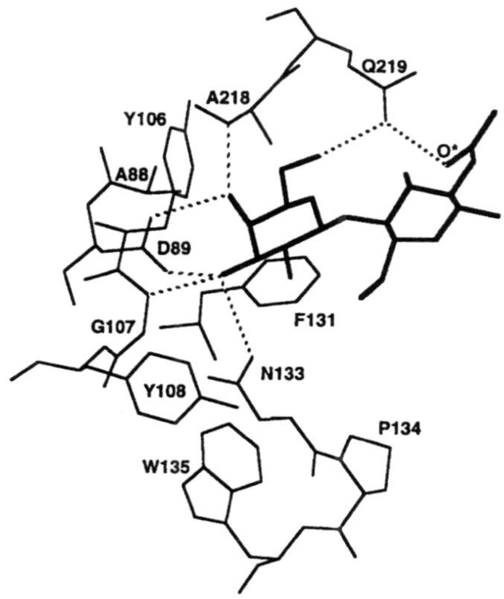

Fig. 4. *N*-Acetyllactosamine in the combining site of *Erythrina corallodendron* lectin (From Adar and Sharon, 1996).

bilized by the metal ions and it too is needed for the proper orientation of the aspartic acid in the combining site.

Water molecules serve not only in the ligating of the metal ions to the legume lectins, but they also mediate the interactions of the lectins with sugars. In the ECorL-lactose complex (Shaanan et al, 1991), as well as in the complexes of the lectin with *N*-acetylgalactosamine and with *N*-acetyllactosamine (Elgavish and Shaanan, 1998), a water molecule bridges between the 6-OH of galactose and the main chain oxygen of Leu86. Water has been shown to play such a role in other lectin-carbohydrate complexes too.

The fact that the key amino acids involved in the binding of the carbohydrate are highly conserved in all legume lectins and have an identical spatial disposition raises the puzzling question how discrimination between very similar monosaccharides, such as mannose (or glucose) on the one hand and galactose on the other, is achieved. This is possible because the monosaccharide in the primary combining site of the Man/Glc specific lectins is oriented differently than in the galactose specific ones (Figure 5). For instance, concanavalin A (Naismith et al., 1994; Bradbrook et al., 1998) and the lectins from *Lathyrus ochrus* (Bourne et al., 1990) and from lentil (Loris et al., 1994b) bind glucose and mannose such that the OD1 and OD2 of aspartic acid are hydrogen bonded to 6-OH and 4-OH of these monosaccharides, whereas the ND2 of asparagine forms such a bond to their 4-OH. In addition, the main chain amide of glycine (or of arginine in the case of concanavalin A and of the other Diocleae lectins) hydrogen bonds with the 3-OH of the monosaccharide. On the other hand, in lectins specific for galactose, such as ECorL (Shaanan et al., 1991), SBA (Dessen et al., 1995) and PNA (Banerjee et al., 1996), the OD1 and OD2 of aspartic acid are H-bonded to the 4-OH and 3-OH, respectively, while the ND2 of asparagine and the NH of glycine are both bonded to the 3-OH (Figure 5). In this way, the same constellation of four highly conserved residues in legume lectins, correctly positioned by the two

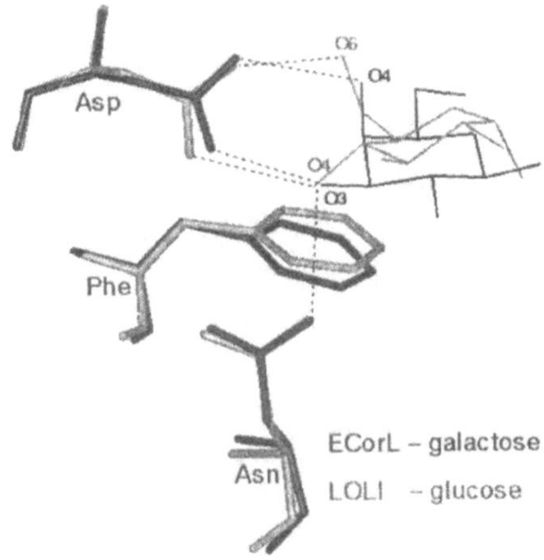

Fig. 5. Superposition of galactose in the combining site of ECorL on mannose in that of LOLI.

metal ions, either directly or via water molecules, provides the framework required for binding of diverse monosaccharides.

As pointed out by Sharma and Surolia (1997), the amino acids that form the monosaccharide, or primary, combining sites of legume lectins are derived from distinct regions of the polypeptide chain that form loops designated A, B, C and D, connecting the β strands of the protein. The conserved aspartic acid and glycine belong to loops A and B, respectively, whereas the conserved asparagine and hydrophobic residue are in loop C. Loop D consists of amino acids that line parts of the binding pocket, and they form hydrogen bonds with the ligated carbohydrate mainly by their backbone NH and CO groups, one example being Ala218 in ECorL (Figure 4). There is a good correlation between the length of the latter loop and the monosaccharide specificity of the legume lectins. Thus, in all Man/Glc specific legume lectins this loop is identical in size, consisting of a variable sequence of 18 amino acids, and it is longer than in lectins specific for Gal/GalNAc or for oligosaccharides only. The conformation of loop D is also an important determinant for discriminating between mannose and galactose. For example, in the lectins of lentil, pea and *Lathyrus ochrus* that are Man/Glc specific, loop D contains the sequence from Thr28 to Ala33 in the β chain that makes it sterically impossible for other sugars to position themselves favorably relative to the four key carbohydrate binding residues. The suggestion has therefore been made that loop D should be considered as the monosaccharide specificity loop (Loris et al., 1998).

Within the Gal/GalNAc specific lectins, the ratio of affinities for the two sugars varies widely. Extreme examples are PNA that does not bind *N*-acetylgalactosamine at all and DBL that has little if any ability to interact with galactose. In between there are lectins such as ECorL that binds *N*-acetylgalactosamine only about two times better than galactose, and SBA that binds the former monosaccharide some 30 times better than the latter. Both ECorL and the very similar *Erythrina cristagalli* lectin (ECL or ECA), as well as SBA, exhibit exceptionally high affinities for *N*-dansygalactosamine, about 250 times higher than for *N*-acetylgalactosamine in the case of the former lectins (De Boeck et al., 1984a; Arango et al., 1993) and 20 times higher in the case of the latter (De Boeck et al., 1984b), whereas PNA does not bind *N*-dansygalactosamine. The high affinity of the dansyl derivatives to ECorL and SBA is strictly sugar specific, since these lectins do not bind *N*-dansylglucosamine (Streicher, private communication). Examination of the three dimensional structure of ECorL led to the conclusion that its ability to accommodate galactose derivatives with bulky substituents at C-2 depends on the presence of a cavity of appropriate size facing this part of the galactopyranose ring (Arango et al, 1993) (Figure 6). Later work has shown that such a cavity is also present in SBA (Dessen et al, 1994), but that in PNA it is partially blocked by the side chain of a glutamic acid (Sharma et al., 1998). Evidence for this role of the glutamic acid was obtained in our laboratory before the three dimensional structure of the PNA-carbohydrate complex was solved. Comparison of the primary structure of PNA with that of ECorL, combined with the crystallographic data on the structure of the ECorL-lactose complex, showed that the former lectin has two extra amino acids, a glutamic acid and an asparagine, not found in the region of the cavity of the latter lectin (Arango et al. 1993). Introduction of these two residues into recombinant ECorL, although decreasing only to a small extent (about 4 times) its affinity for *N*-acetylgalactosamine, diminished it markedly (up to 43 times) for the much bulkier *N*-dansylgalactosamine.

Molecular modeling shows that in the complex of ECorL with Meα-*N*-dansylgalactosaminide the dansyl moiety of the ligand fits snugly and lies close to, and in parallel with, the indole group of Trp135, resulting in the formation of a charge-transfer complex between the two. This may account for the high affinity of the *N*-dansyl derivatives for the lectin (Adar et al., 1998). However, other factors may also play a role, since replacement of Trp135 by alanine decreased this affinity no more than tenfold (Table 2) and the resultant mutant still bound Meα-*N*-dansylgalactosaminide 70 times better than galactose. Replacement by alanine or threonine of Tyr108, the other aromatic residue lining the cavity, affected only marginally this affinity, ruling out the possibility that the tyrosine interacts to any significant extent with the dansyl moiety.

The cavity in ECorL also accomodates a fucose substituted on the 2-OH of galactose in an

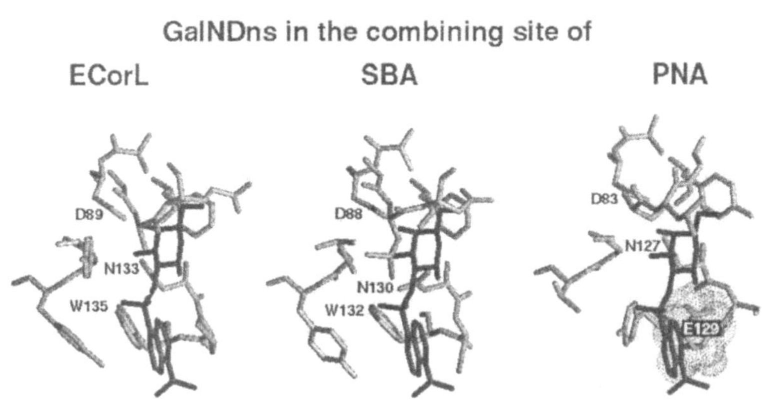

Fig. 6. Cavity facing the 2-OH in galactose-specific lectins.

Table 2.
Contribution of 2-substituents of galactose to the affinity of ligands for *Erythrina corallodendron* lectin and its W135A mutant* (Data from Adar et al., 1996, Moreno et al. 1997, and unpublished).

Ligand	*2- Substituent*	*Contribution to affinity for*	
		ECorL	*W135A EcorL*
Fuc(α1-2)Gal(β1-4)Glc	Fucose	2-5x	2x
Fuc(α1-2)Gal(β1-4)GlcNAc	Fucose	2-5x	n.t.
MeαGalNDns	*N*-Dansyl	250x	35x

*Calculated from data of Moreno et al., 1997 and Adar et al., 1998.

α-linkage, which explains why the H-type 2 trisaccharide Fuc(α1-2)Gal(β1-4)GlcNAc is the best ligand for this lectin (see later). Similar cavities, responsible for the high affinity binding of substituted monosaccharides, are found in other legume lectins as well. Thus, in DBL the cavity provides bonds to the acetamide of *N*-acetylgalactosamine (Hamelryck et al., 1999); in the lentil lectin it harbors the nitrophenyl group of 3-O-nitrophenylmannose (Loris et al., 1994a) and in GSIV it accomodates the α1-2 linked fucose of the Le^b oligosaccharide (Delbaere et al., 1993).

The unusually strict specificity of DBL for *N*-acetylgalactosamine and its inability to bind galactose is believed to be due to the substitution of the conserved aromatic residue of the binding site by leucine (Leu127), thereby eliminating the possibility of aromatic stacking against the surface of the ligand (cf. above). This conclusion is based on the finding that a mutant of DBL in which Leu127 was replaced by phenylalanine acquired the ability to combine with galactose (Hamelryck et al., 1999). In the binding of *N*-acetylgalactosamine to DBL, the absence of aromatic stacking is compensated by a hydrogen bond between the oxygen of its acetamide and the NH of Gly103, and by hydrophobic interactions between the methyl group of the acetamide and Trp132 and Tyr104. A similar compensating mechanism for the specific recognition of acetamido sugars was recently observed in the complex of tachylectin 2, a lectin from the Japanese horseshoe crab, *Tachypleus tridentatus* and *N*-acetylglucosamine (Beisel et al., 1999).

2. For oligosaccharides

As is the case with lectins in general, whether from legumes or from other sources, ECorL combines with suitable oligosaccharides better than with monosaccharides, suggesting the occurrence of additional interactions with the parts of the ligand located outside the primary binding site (Figure 3). The amino acids involved in these interactions are not necessarily the same as the key residues of the primary site discussed above. The slightly higher affinity of ECorL for lactose (by a factor of about 2) appears to be due to an H-bond between the amide of Gln219 and the 3-OH of the glucose moiety of the disaccharide. This was suggested on the basis of site-directed mutagenesis of ECorL (Adar and Sharon, 1996) and confirmed by improved X-ray cystallographic analysis of the ECorL-lactose complex (Elgavish and Shaanan, 1998). The above studies also led to the conclusion that an H-bond between the same glutamine and the carbonyl of the acetamide of *N*-acetylactosamine may account for the 4-8 times higher affinity of the latter disaccharide to the lectin as compared to lactose.

Among all oligosaccharides tested, Fuc(α1-2)Gal(β1-4)GlcNAc is the best ligand for

ECorL (2-5 times better than *N*-acetyllactosamine). No crystal structure of the complex of the trisaccharide with ECorL is available, but molecular modeling revealed a perfect fit of the fucose in the cavity extending from the primary binding site of the lectin (Moreno et al., 1997). The amide of Asn133 which H-bonds with the 3-OH of galactose forms such bonds also with the 2-OH of the fucose and with the glycosidic oxygen of the latter sugar. This fucose hydroxyl may in addition H-bond with the backbone NH of Gly107. The modeling also indicates the presence of a fourth such bond between the 3-OH of the fucose and the hydroxyl of Tyr108. Also, Pro134 and Trp135 lining the cavity make hydrophobic contacts with the CH_3 of this monosaccharide, as well as with its C2- and C3-hydrogens, and together with Tyr108 they close the binding site. The extensive network of bonds notwithstanding, the fucose makes only a small contribution to the affinity of the trisaccharide to the lectin. This contrasts with the much larger contribution of the dansyl group of *N*-dansylgalactosamine, which in the complex of the latter compound with the lectin occupies the same site as the fucose (Table 2). Here we have another illustration of the fact that affinities (or energetics) of protein-ligand complexes cannot be deduced from structural considerations (Janin, 1997).

In the complex of DBL with the blood group A trisaccharide, GalNAc(α1-3)[Fuc-(α1-2)]Gal, only the *N*-acetylgalactosamine, bound in the primary combining site, is seen (Hamelryck et al., 1999). It is hydrogen bonded with the conserved sugar binding residues Asp85, Gly103 and Asn129 found in all legume lectins. The backbone amide of Gly103, in addition to the conserved hydrogen bond to the 3-OH of the *N*-acetylgalactosamine, forms a similar bond with the oxygen of the acetamide. Additional hydrogen bonds are formed between the *N*-acetylgalactosamine of the trisaccharide and two amino acids belonging to the specificity loop, namely between the 6-OH and the side chain of Ser215 and between the 4-OH and the backbone nitrogen of Leu214. The latter amino acid is responsible for the α-anomeric preference of the lectin, since binding of the β anomer of *N*-acetylgalactosamine would bring the anomeric oxygen too close to its side chain. In addition to hydrogen bonds, hydrophobic interactions play and important role in ligand binding to DBL. The side chains of Tyr104 and Trp132 form a hydrophobic pocket in the vicinity of the methyl group of *N*-acetylgalactosamine and Tyr218 makes a favorable contact with the C-6 of the monosaccharide. As already mentioned, the aromatic amino acid that is conserved in the combining site of legume lectins is in DBL replaced by the aliphatic leucine, but the loss of the commonly occuring stacking interaction is compensated by contacts with the acetamide. In the complex of DBL with the Forssman disaccharide, GalNAc(α1-3) GalNAc, both monosaccharide constituents are clearly seen. The non-reducing *N*-acetylgalactosamine is bound in the primary combining site in exactly the same manner as the *N*-acetylgalactosamine of the blood group A trisaccharide, whereas the reducing moiety extends from the monosacharide binding site and interacts with the protein only through its acetamide group.

An unusual case, in which water bridges are responsible for generating specificity in protein-carbohydrate interactions, was revealed by comparing the crystal structure of the complexes of PNA with lactose and with Gal(β1-3)GalNAc (Ravishankar et al., 1997). The two disaccharides differ markedly in their affinity for the lectin, which is 20 times higher for the latter disaccharide than for the former one. In both lectin-disaccharide complexes the hydrogen bonding interactions with the galactose moiety are the same, but they are different for the reducing moieties, in spite of the fact that the same two amino acids are involved, i.e. Ser211 via its side chain OH, and Gly213 via its main chain amide. In the case of lactose, the two bonds are with the 3-OH of the glucose, whereas in the case of Gal(β1-3)GalNAc they are with the 4-OH of *N*-acetylgalactosamine. Because of the differences in the glycosidic linkages (β1-4 versus β1-3) and in the conformation of the two disaccharides, these two hydroxyls occupy the same position in the complex. The number and nature of hydrophobic protein-sugar contacts are also nearly the same in both complexes. The only additional interactions found in the PNA-Gal(β1-3)GalNAc complex as compared to the PNA-lactose complex are via two water bridges involving the carbonyl oxygen of the acetamide of *N*-acetygalactosamine. One water molecule connects the carbohydrate to the main chain oxygen of Ile101 and the other connects it to the side chain of Asn41 and the main chain amide of Leu212. These water molecules are present in all four subunits of the PNA-lactose complex as well, but there is no sugar atom in their immediate vicinity and they interact only with the protein. Thus, the high affinity of PNA for Gal(β1-3)GalNAc is due essentially to the above water bridges. Replacement by site-directed mutagenesis of Leu212 in PNA by alanine led to a 100-fold decrease in the affinity of the lectin for Gal(β1-3)GalNAc, while having no effect on its affinity for lactose (Sharma et al., 1996).

In the complex of GSIV with the methyl glycoside of the Leb tetrasaccharide, Fuc(α1-2)Gal(β1-3)[Fuc(α1-4)]GlcNAcβOMe, the ligand is bound with its galactose positioned in the primary combining site, forming the same bonds with the four conserved amino acids as does galactose in the combining site of ECorL (Delbaere et al., 1993). However, in contrast to the specificity loop of the latter lectin (Thr216-Glu224) that hydrogen bonds to the galactose in the combining site, in GSIVthe corresponding loop (Val221-Glu225) interacts with the α1-4 linked fucose due a rather unusual conformation of the loop. The inability of galactose in the primary combining site of GSIV to form hydrogen bonds with the specificity loop may account for the inability of this monosaccharide to bind to the lectin on its own.

IV. COMPARISON WITH OTHER LECTINS

Lectins other than those of the legumes also combine with carbohydrates primarily by a network of hydrogen bonds and hydrophobic interactions, However, while in the legume lectins only a limited repertoire of amino acid side chains functions in hydrogen bonding, in other lectins different amino acids are often employed for the same purpose, even when combining with the same ligand (Table 3).

For instance, in the crystal structures of galectin-2 from human spleen in complex with lactose (Lobsanov et al., 1993) and of galectin-1 from bovine spleen in complex with *N*-acetyllactosamine (Liao et al., 1994) the side chains of four amino acids (asparagine arginine, glutamic acid and histidine), are hydrogen bonded to the 4-OH of the terminal, non-reducing

Table 3.
Different lectins combine with the same monosaccharide by different H-bonding side chains

Ligand	Lectin*	H-Bonding residues	References
Mannose	Con A	Asn, Asp	see text
	GNL	Asn, Asp, Gln, Tyr	Hester et al., 1995
	MBP-A	Asn, Glu	Weis et al., 1992
Galactose	ACA	Asn, His,	Transue et al., 1997
	ECorL	Asn, Asp	see text
	Galectin-1	Arg, Asn, Glu, His	Liao et al., 1994

*ConA, concanavalin A; GNL, *Galanthus nivalis* lectin; MBP-A, mannose binding protein; ACA, *Amaranthus caudatus* aglutinin; ECorL, *Erythtrina corallodendron* lectin

galactose, and a likewise conserved tryptophan is stacked against the sugar ring; the 6-OH is also hydrogen bonded with the protein, but neither is the 2-OH nor the 3-OH. In the galectin-lactose complex the glucose moiety contributes to binding by interacting with the protein via its 2-OH and 3-OH, while in the galectin-*N*-acetyllactosamine complex, only the 3-OH of *N*-acetylglucosamine is hydrogen bonded to the protein.

Examination of the primary sites of available three dimensional structures of lectin-carbohydrate complexes reveals that any amino acid capable of forming hydrogen bonds with sugar hydroxyls or with common sugar substituents, for example acetamide, does perform such a function in one lectin or another (Table 4). This is also true for the reducing moieties of disaccharides in the few lectin-disaccharide complexes examined, as illustrated by the examples in Table 5. In addition to the diversity of amino acids taking part in the complexation of carbohydrates, metal ions may in rare cases play a direct role in ligand binding, as found in the selectins (Rini, 1995; Weis and Drickamer, 1996). Since most saccharides are uncharged, ionic (charge-charge) interactions are not often found in lectin-saccharide complexes.

Table 4.
Hydrogen bonding residues in the primary sites of lectins

Source & family	Arg	Asn	Asp	Gln	Glu	His	Ser	Tyr	Me^{2+}
Plants									
Legumes		+	+						+[a]
Cereals					+		+	+	
Amaramthaceae		+				+			
Amaryllidaceae		+	+	+				+	
Moraceae[b]			+						
Animals									
Collectins		+			+				+ [c]
Galectins	+	+			+	+			
Selectins		+			+	+			+[c]
Siglecs	+								
Viruses									
Influenza							+	+	
Murine polyoma	+	+				+		+	

[a] Structural.; [b]Amine of N-terminal glycine also hydrogen bonds with the ligand; [c] Binds the ligand

Table 5.
H-Bonding residues with ring oxygens of the reducing monosaccharide in complexes of lectins with lactose and *N*-acetylactosamine

Lectin	Amino acid
ECorL	Gln
PNA	Ser
SBA	Asp
Galectin-1	Arg, Glu

The above survey clearly demonstrates the great diversity of the carbohydrate binding sites of lectins, and shows that Nature solves in a variety of different ways the problem of constructing combining sites for carbohydrates.

SUMMARY

1. Different carbohydrate-specific proteins, such as lectins, may combine with the same monosaccharide or oligosaccharide by different H-bonding and hydrophobic side chains.

2. Homologous proteins with distinct specificities may bind different monosaccharides (e.g., for glucose and galactose that differ in the configuration of a single hydroxyl) by the same set of invariant residues that are identically positioned in their tertiary structures.

3. The energetics of protein-carbohydrate interactions cannot be derived from structural information.

4. Nature solves in a variety of different ways the problem of constructing combining sites for carbohydrates, just as it provides diverse solutions for other functions of proteins.

REFERENCES

1. Adar, R. and Sharon, N., 1996, Mutational studies of the combining site residues of *Erythrina corallodendron* lectin, *Eur. J. Biochem.* 239, 668-674.
2. Adar, R., Ågström, J. Moreno, E., Karlsson, K.A., Streicher, H. and Sharon, N., 1998, Structural studies of the combining site of *Erythrina collarodendron* lectin. Role of tryptophan 135, *Prot. Sci.* 7, 52-63.
3. Arango, R., Rodriguez-Arango, E., Adar, R., Belenky, D., Loontiens, F., Rozenblatt, S. and Sharon, N., 1993, Modification by site-directed mutagenesis of the specificity of *Erythrina corallodendron* lectin for galactose derivatives with bulky constituents, *FEBS Lett.* 330, 133-136.
4. Banerjee, R., Das, K., Ravishankar, R., Suguna, K., Surolia, A. and Vijayan, M., 1996, Conformation, protein-carbohydrate interactions and a novel subunit association in the refined structure of the peanut lectin-lactose complex, *J. Mol. Biol.* 259, 281-296.
5. Beisel, H.G., Kawabata, S., Iwanaga, S., Huber, R. and Bode, W., 1999, Tachylectin 2: crystal structure of a specific GlcNAc/GalNAc-binding protein involved in the innate immunity host defense of the Japanese horseshoe crab *Tachypleus tridentatus*, *EMBO J.* 18, 2313-2322.
6. Bourne, Y., Roussel, A., Frey, M., Rougé, P., Fonntecilla-Camps, J.-C. and Cambillau, C., 1990, Three-dimensional structure of complexes of *Lathyrus isolectin* I with glucose and mannoose: fine specificity of the monosacharide -binding site, *Proteins Struct. Funct. Genet.* 8, 365-376.

7. Bradbrook, G.M., Gleichmann, T., Harrop, S.J., Habash, J., Raftery, J., Kalb (Gilboa), J., Yariv, J., Hillier, I.H. and Helliwell J.R., 1998, X-Ray and molecular dynamic studies of concanavalin-A glucoside and mannoside complexes, *J. Chem.Soc. Faraday Trans.* 94, 1603-1611.
8. De Boeck, H., Loontiens, F.G., Lis, H. and Sharon, N., 1984a, Binding of simple carbohydrates and some *N*-acetyllactosamine-containing oligosaccharides to *Erythrina cristagalli* agglutinin as followed with a fluorescent indicator ligand, *Arch. Biochem. Biophys.* 234, 297-304.
9. DeBoeck, H., Lis, H., van Tilbeurgh, H., Sharon, N.and Loontiens, F.G., 1984b, Binding of simple carbohydrates and some chromophoric derivatives of soybean agglutinin as followed by titrimetric procedures and stopped-flow kinetics, *J.Biol. Chem.* 259, 7067-7074.
10. Delbaere, L.T.J., Vandonselaar, M., Prasad, L., Wilson, K.S.and Dauter, Z., 1993, Structure of the lectin IV of *Griffonia simplicifolia* and its complex with Lewis b human blood group determinant at 2.0 Å resolution, *J. Mol. Biol.* 230, 950-965.
11. Dessen, A., Gupta, D., Sabesan, S., Brewer, C.F. and Sacchettini, J.C., 1995, X-Ray crystal structure of soybean agglutinin cross-linked with a biantennary analog of the blood group I carbohydrate antigen, *Biochemistry* 34, 4933-4942.
12. Elgavish, S. and Shaanan, B., 1998, Structures of the *Erythrina corallodendron* lectin and its complexes with mono- and disaccharides, *J. Mol. Biol.* 277, 917-932.
13. Fiedler, K.and Simons, K., 1994, A putative novel class of animal lectins in the secretory pathway homologous to leguminous lectins, *Cell* 77, 625-626.
14. Hamelryck, T.W., Dao-Thi, M.-H., Poortmans, F., Chrispeels, M.J., Wyns, L. and Loris, R., 1996, The crystallographic structure of phytohemagglutinin-L, *J. Biol. Chem.* 271, 20479-20485.
15. Hamelryck, T.W., Loris, R., Bouckaert, J. and Wyns, L., 1998, Structural features of the legume lectins. *Trends Glycosci. Glycobiol.* 10, 349-360.
16. Hamelryck, T.W., Loris, R., Bouckaert, J., Dao-Thi, M.-H., Strecker, G., Imberty, A., Fernandez, E., Wyns, L. and Etzler, M.E., 1999, Carbohydrate binding, quaternary structure and a novel hydrophobic binding site in two legume lectin oligomers
17. from Dolichos biflorus. J. Mol. Biol. 286, 1161-1177.
18. Hester, G., Kaku, H., Goldstein, I.J. and Wright, C.S., 1995, Structure of mannose-specific snowdrop (*Galanthus nivalis*) lectin is representative of a new plant lectin family, *Nature Struct. Biol.* 2, 472-479.
19. Janin, J.,1997, Ågströms and calories, *Structure* 3, 473-479.
20. Kinzy, W., Belenky, D.M., Loontiens, F.G. and Sharon, N. 1992 Synthesis of methyl α- and β-*N*-dansylgalactosaminides, probes for the combining sites of GalNAc-specific lectins. *Glycoconjugate J.* 9, 225-227.
20. Liao, D.-I., Kapadia, G., Ahmed, H., Vasta, G.R. and Herzberg, O., 1994, Structure of S-lectin, a developmentally regulated vertebrate β-galactoside-binding protein, *Proc. Natl. Acad. Sci. USA* 91, 1428-1432.
21. Lis, H. and Sharon, N. 1998, Lectins: carbohydrate-specific proteins that mediate cellular recognition, *Chem. Revs.* 99, 637-674
22. Lobsanov, Y.D., Gitt, M.A., Leffler, H., Barondes, S.H. and Rini, J.M., 1994, X-Ray crystal structureof the human dimeric S-lac lectin L-14-II, in complex with lactose at 2 Å resolution, *J. Biol. Chem.* 268, 27034-27038.
23. Loris, R., Stas, P.P.G. and Wyns, L., 1994a, Conserved waters in legume lectin crystal structures. The importance of bound water for the sequence-structure relationship within the legume family, *J. Biol Chem.* 269, 26722-26733.
24. Loris. R., Casset, F., Bouckaert, J., Pletinckx, J., Dao-Thi, M.-H., Poortmans, F., Imberty, A., Perez, S. and Wyns, L., 1994b, The monosaccharide binding site of lentil lectin: an X-ray and molecular modeling, *Glycoconjugate J.* 11, 507-517.
25. Loris, R., Hamelryck, T., Bouckaert, J. and Wyns, L. 1998, Legume lectin structure, *Biochim. Biophys. Acta* 1383, 9-36.
26. Mirkov, T.E. and Chrispeels, M.J., 1993, Mutatioin of Asn128 to Asp of *Phaseolus vulgaris* leucoagglutinisn (PHA-L) eliminates carbohydrate-binding and biological activity, *Glycobiology* 3, 381-387.

27. Moreno, E., Teneberg, S., Adar, R., Sharon, N., Karlsson, K.-A. and Ågström, J., 1997, Re-definition of the carbohydrate specificity of *Erythrina collarodendron* lectin based on solid phase binding assays and molecular modeling of native and recombinant forms obtained by site-directed mutagenesis, *Biochemistry* 36, 4429-4437
28. Naismith, J.H., Emmerich, C., Habash, J., Harop, S.J., Helliwell, J.R., Hunter, W.N., Raftery, J., Kalb (Gilboa), A.J. and Yariv, J., 1994, Refined structure of concanavalin A complexed with methyl α-D-mannopyranoside at 2.0 Å resolution and comparison with the saccharide-free structure, *Acta Cryst.* D50, 847-858.
29. Prabu, M.M., Sugana, K. and Vijayan, M., 1999, Variability in quaternary association of proteins with the same tertiary fold: a case study and rationalization involving legume letins, *Proteins Struct. Funct. Genetics* 35, 58-69.
30. Ravishankar, R., Ravindran, N., Suguna, A., Surolia, A. and Vijayan, M., 1997, Crystal structure of the peanut lectin-T-antigen complex. Carbohydrate specificity generated by water bridges. *Curr. Science* 72, 855.
31. Rini, J., 1995, Lectin structure, *Annu. Rev. Biophys. Biomol. Struct* 24, 551-577.
32. Roche, A.-C. and Monsigny, M., 1996, Trafficking of endogenous glycoproteins mediated by intracellular lectins: facts and hypotheses. *Chemtracts Biochem. Mol. Biol.* 6, 188-201.
33. Shaanan, B., Lis, H. and Sharon, N., 1991, Structure of a lectin with ordered carbohydrate, in complex with lactose, *Science* 254, 862-866.
34. Sharma, V. and Surolia, A., 1997, Analyses of carbohydrate recognition by legume lectins: size of the combining site loops and their primary specificity. *J. Mol. Biol.* 267, 433-445.
35. Sharma, V., Vijayan, M. and Surolia, A., 1996, Imparting exquisite specificity to peanut agglutinin for the tumor-associated Thomsen-Friedenreich antigen by redesign of its combining site, *J. Biol. Chem.* 271, 21209-21213.
36. Sharon, N., 1993, Lectin-carbohydrate complexes of plants and animals: an atomic view, *Trends Biochem. Sci.* 18, 221-226.
37. Sharon, N. and Lis, H., 1989, *Lectins*, Chapman and Hall, London.
38. Sharon, N. and Lis, H., 1990, Legume lectins - a large family of homologous proteins, *FASEB J.* 4 3198-3208.
39. Sharon, N. and Lis, H., 1998, 110 Years of lectin research, *Carbohydrates in Europe* No. 23, 12-17.
40. Srinivasan, N., Rufino, S.D., Pepys, M.B., Wood, S. and Blundell, T.L. 1996, A superfamily of proteins with the lectin fold, *Chemtracts Biochem. Mol. Biol.* 6, 149-164.
41. Surolia, A., Sharon, N. and Schwarz, F.P., 1996, Thermodynamics of monosaccharide and disaccharide binding to *Erythrina corallodendron* lectin, *J. Biol. Chem.* 271, 17697-17703.
42. Transue, T.R., Smith, A.K., Mo, H., Golsdtein, I.J. and Saper, M.A., 1997, Structure of benzyl T-antigen disaccharide bound to *Amaranthus caudatus* agglutinin, *Nature Struct. Biol.* 4, 779- 783.
43. Van Damme, E.J.M., Peumans, W.J., Pusztai, A. and Bardocz, S., *Handbook of plant lectins: properties and biomedical applications* (1997) 452 pp., John Wiley & Sons, Chichester, England.
44. Van Eijsden, R.R., De Pater, B.S. and Kijne, J.W.. 1994, Mutational analysis of the sugar binding site of pea lectin, *Glycoconjugate J.* 11, 375-380.
45. Vijayan, M. and Chandra, N., 1999, Lectins. *Curr. Opin. Struct. Biol.* 9, 707-714.
46. Weis, W.I., Drickamer, K. and Hendrickson, W.A., 1992, Structure of a C-type mannose-binding protein complexed with an oligosaccharide. *Nature* 360, 127-134.
47. Weis W. and Drickamer K., 1996, Structural basis of lectin-carbohydrate recognition, *Annu. Rev. Biochem* . 65, 441-473.

LECTIN CROSS-LINKING INTERACTIONS WITH MULTIVALENT CARBOHYDRATES[1,2]

C. Fred Brewer
Departments of Molecular Pharmacology, and Microbiology and Immunology
Albert Einstein College of Medicine, Bronx, New York 10641, USA

Lectins are carbohydrate binding proteins that are widely conserved in nature, including in plants, animals, and microorganisms (Varki et al., 1999). Lectins with similar folding patterns and carbohydrate binding activities have been found in nematodes and vertebrates, suggesting the importance of their biological functions (Kasai and Hirabayashi, 1996). The biological roles of many animal lectins are now known. For example, a number of mammalian lectins are involved in receptor mediated endocytosis of glycoproteins (cf. Drickamer and Taylor, 1993), and others including the selectins shown to be involved in cellular recognition and adhesion (Drickamer and Taylor, 1993). Certain members of the galectin family have been implicated in metastasis (Konstantinov et al., 1996; Lotan and Raz, 1988), while galectin-1 has been shown to be involved in the induction of apoptosis in activated human T-cells (Pace et al., 1999; Perillo et al., 1995).

A common feature of lectins is their multivalent binding activities. The x-ray crystal structures of many plant (Loris et al., 1998) and animal lectins (Rini, 1995) demonstrate their oligomeric quaternary structures. As a consequence, lectin binding to cells often leads to cross-linking and aggregation

---

[1] This work was supported by Grant CA-16054 from the National Cancer Institute, Department of Health, Education and Welfare,and Core Grant P30 CA-13330 from the same agency (C.F.B.). Tel. No. (718) 430-2227.

[2] The abbreviations used are: SBA, lectin from *Glycine max* (soybean agglutinin); GalNAc, 2-acetamido-2-deoxy-D-galactopyranose; Gal, D-galactopyranose; Glc, D-glucopyranose; GlcNAc, 2-acetamido-2-deoxy-D-glucopyranose; LacNAc, Galβ(1-4)Glc.

The Molecular Immunology of Complex Carbohydrates–2.
Edited by Albert M. Wu, Kluwer Academic/Plenum Publishers, 2001.

of specific glycoproteins and glycolipids receptors. In many cases, these cross-linking interactions are associated with signal transduction effects including the mitogenic activities of lectins (Nicolson, 1976), the induction of mating reactions in fungi (Kooijman et al., 1989), and, as mentioned, the induction of apoptosis in activated human T-cells (Pace et al., 1999; Perillo et al., 1995).

Recent structural studies of multivalent lectin-carbohydrate cross-linked complexes have provided new insights into the specificity of these interactions. In general, two types of multivalent complexes are observed, designated type 1 and type 2 (Brewer, 1996). In type 1

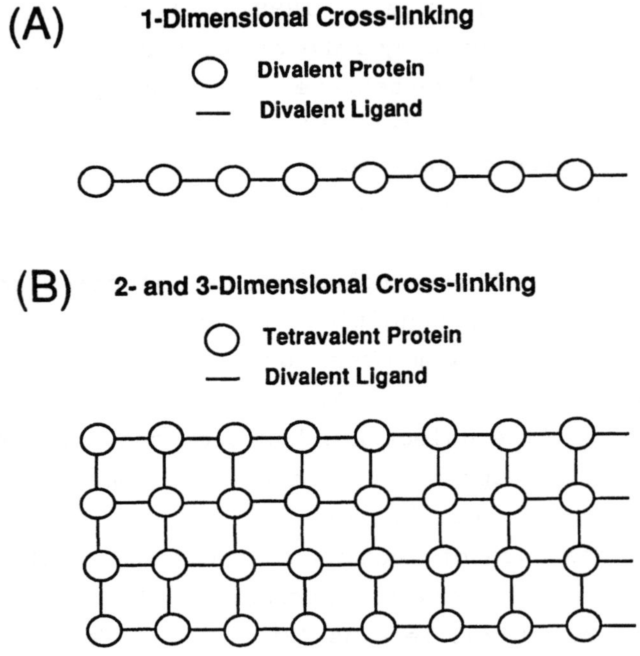

Fig. 1. Fig. 1A shows a schematic diagram of a type 1 cross-linked complex between a divalent lectin and divalent carbohydrate. Fig. 1B shows a schematic diagram of a type 2 cross-linked complex between a tetravalent lectin and divalent carbohydrate. In both schemes, the lectin is represented by a circle and the carbohydrate by a line. (From Olsen et al. (1997) with permission from the American Chemical Society.)

complexes, cross-linking occurs between a divalent carbohydrate and divalent lectin. Fig. 1A shows a schematic diagram of type 1 interactions, which may result in the formation of linear (1-dimensional) cross-linked complexes. These complexes are often soluble (cf. Bhattacharyya et al., 1988; Mandal and Brewer, 1993), and possess a high degree of flexibility and polymorphism. In addition, such complexes in theory can accommodate more than one type of bivalent carbohydrate or divalent lectin in the cross-linked complex. Examples include the x-ray crystal structures of the dimeric 14 kDa Gal-specific lectin (galectin-1) from bovine heart muscle cross-linked with a bivalent N-linked biantennary complex glycopeptide (Bourne et al., 1994). Three different crystal structures are observed for the galectin cross-linked with three different conformations of the glycopeptide. These findings thus emphasize the structural flexibility of type 1 interactions.

In type 2 complexes, cross-linking occurs between lectins and carbohydrates in which one of the two molecules possesses a valency greater than two. An example is a cross-linked complex between a tetravalent lectin and a divalent carbohydrate as shown schematically in Fig. 1B. Such interactions lead to the formation of 2- and 3-dimensional cross-linked complexes, which may be insoluble and precipitate from solution (Bhattacharyya et al., 1990; Mandal and Brewer, 1992a; Mandal and Brewer, 1992b). Type 2 complexes may also exists as ordered cross-linked lattices, as illustrated in Fig. 1B. These interactions have been termed homogeneous cross-linking, since a given lattice consists of a single carbohydrate and a single lectin, even in the presence of multiple carbohydrates and lectins in solution (Brewer, 1996).

Until recently, the molecular basis for the formation of these ordered carbohydrate-lectin type 2 lattices was poorly understood. However, the x-ray crystal structures of a single tetravalent lectin cross-linked with four different bivalent carbohydrates have recently been reported (Dessen et al., 1995; Olsen et al., 1997). These results have provided insight into the molecular basis for the formation of type 2 homogeneous carbohydrate-lectin cross-linked complexes. A summary of these findings is presented below.

X-ray Structures of SBA Cross-linked with Four Isomeric Bivalent Pentasaccharides

The soybean agglutinin (SBA) is a tetrameric lectin with binding specificity for GalNAc and Gal residues. Previous quantitative precipitation experiments and electron micrograph studies (Gupta et al., 1994) suggested that SBA formed unique type 2 cross-linked complexes with the four bivalent pentasaccharides shown in Fig. 2. The x-ray structure of SBA cross-linked with the 2,6-pentasaccharide was subsequently reported (Dessen et al., 1995). The structure revealed that lattice formation was promoted uniquely by the bridging action of the bivalent pentasaccharide. The structure of the complex was determined by molecular replacement techniques and refined

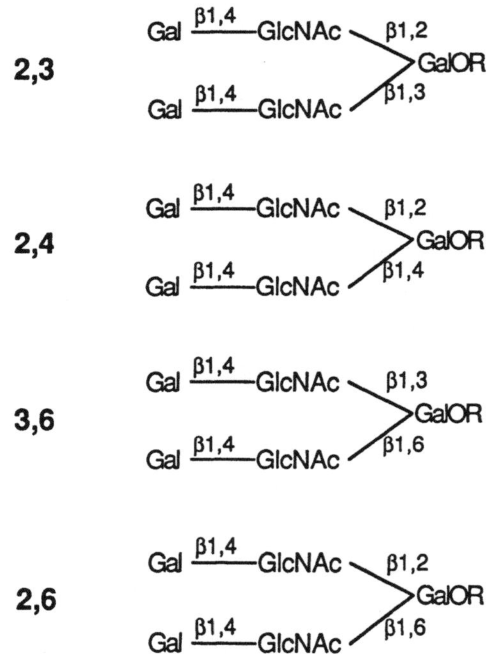

Fig. 2. Structures of the 2,3-, 2,4-, 3,6- and 2,6-pentasaccharides. The aglycon moiety R is $-(CH_2)_5COOCH_3$. The core galactose residue is in the β-anomeric configuration. (From Olsen et al. (1997) with permission from the American Chemical Society.)

at 2.6 Å resolution to an R value of 20.1%. The crystals were hexagonal with a $P6_422$ space group, which differs significantly from that of crystals of the free protein (space group C2). Analysis of the crystal lattice allowed identification of a sixfold axis of symmetry which forms the basis of the structural network. The lattice is generated with pentasaccharide molecules cross-linking neighboring SBA subunits. Association between SBA tetramers involved binding of the terminal galactose residues of the pentasaccharide at identical sites in each monomer, with the oligosaccharide cross-linking to a symmetry related neighbor molecule.

More recently, the x-ray crystal structures of SBA cross-linked with the 2,3-, 2,4- and 3,6-pentasaccharides in Fig. 2 were reported (Olsen et al., 1997). The structure of a portion of the 3,6-pentsaccharide complex with SBA is shown in Fig. 3. Examination of the Cα coordinates of SBA in the cross-linked complexes with the 2,3-, 2,4- and 3,6-pentasaccharides are similar to one another as well as to the structure of the lectin in the 2,6-pentasaccharide complex (Dessen et al., 1995).

TABLE 1. Space Group and Unit Cell Parameters[a] of SBA Cross-linked with the 2,3-, 2,4-, 3,6- and 2,6-Pentasaccharides in Fig. 1.

Complex	Space Group	a(Å)	b(Å)	c(Å)	α(°)	β(°)	γ(°)
SBA-2,3 penta.	$I4_122$	122.6	122.6	90.6	90	90	90
SBA-2,4 penta.	$P6_422$	144.6	144.6	107.2	90	90	120
SBA-3,6 penta.	$P6_422$	143.3	143.3	107.8	90	90	120
SBA-2,6 penta.	$P6_422$	144.9	144.9	109.4	90	90	120

[a] From Olsen et al. (1997) with permission from the American Chemical Society.

Table 1 shows the space group and unit cell dimensions of all four cross-linked complexes. While the 2,4-, 3,6- and 2,6-pentasaccharide/SBA lattices belong to the same space group ($P6_422$), they have distinctly different unit cell dimensions.

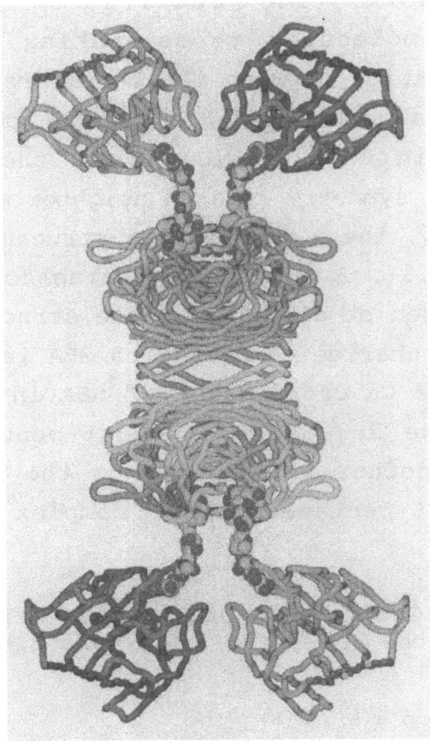

Fig. 3. X-ray crystallographic structure of a portion of the cross-linked lattice formed by the tetrameric soybean agglutinin (SBA) and the 3,6-pentasaccharide (Fig. 1). The view is in the z plane perpendicular to a 2-fold axis of the tetramer. One SBA tetramer (shown as yellow, magenta, green and blue trace models) in the middle is cross-linked by four pentasaccharide molecules (CPK models shown in red, yellow and dark blue) to four SBA monomers (shown in aqua). The structure is from Olsen et al. (1997). The figure was made by Jon Christopher of the Department of Biochemistry and Biophysics, Texas A & M University, using the program SPOCK <http://quorum.tamu.edu/spock>.

Figure 3. X-ray crystallographic structure of a portion of the cross-linked lattice formed by the tetrameric soybean agglutinin (SBA) and the 3, 6-pentasaccharide (Fig. 1). The view is in the z plane perpendicular to a 2-fold axis of the tetramer. One SBA tetramer (shown as yellow, magenta, green, and blue trace models) in the middle is cross-linked by four pentasaccharide molecules (CPK models shown in red, yellow, and dark blue) to four SBA monomers (shown in aqua). The structure is from Olsen *et al.* (1997). The figure was made by Jon Christopher of the Department of Biochemistry and Biophysics, Texas A & M University, using the program SPOCK <http://quorum.tamu.edu/spock>.

The 2,4- and 2,6-pentasaccharide lattices exhibit small differences (0.3 Å) along their a and b axes, with a larger difference (2.2 Å) along the c axes. The 3,6-pentasaccharide lattice differs from the other two lattices along the a and b axes (1.3 - 1.6 Å). The lattice of the 3,6-pentasaccharide/SBA complex also differs along the c axis by 0.6 Å relative to the lattice of the 2,4-pentasaccharide/SBA complex. Differences in the c axes of all three lattices can be related to the distances separating the C-6 carbon atoms of the terminal Gal residues of each cross-linking pentasaccharide in symmetry related SBA monomers. The distances of 19.2 Å, 19.7 Å and 20.6 Å that separate the C-6 atoms of the terminal Gal residues of the 2,4-, 3,6- and 2,6-pentasaccharides, respectively, correlate with the order of the distances along the c axes for all three complexes of 107.2 Å, 107.8 Å and 109.4 Å (Table 1), respectively. Thus, the c axes of the three complexes reflect differences in the lengths of the three oligosaccharides.

In the case of the 2,3-pentasaccharide, the distance between the C-6 atoms of the terminal Gal residues is 15.0 Å, which is considerably shorter than the respective distances of the other three carbohydrates. More importantly, the angle between the β-LacNAc arms is sufficiently acute to cause the lattice to adopt a different space group ($I4_122$). The crystal lattices of the 2,4-, 3,6- and 2,6-pentasaccharide/SBA complexes possess a common 6-fold axis of symmetry. A salient feature of the lattice(s) is the absence of a high degree of strong protein-protein contacts, and formation of large, solvent-filled cavities (~70%). By comparison, the 2,3-pentasaccharide complex shows a 4-fold axis of symmetry. The 4-fold axis of symmetry results in a cavity between lectin molecules with a diameter of ~45 Å, as compared to ~100 Å for the other three lattices.

Since the structure of SBA itself does not significantly change in the four different complexes, differences in the respective lattice structures of the lectin are due the cross-linking carbohydrates. These results also demonstrate that the molecular basis for the formation of homogeneous type 2 carbohydrate-protein cross-linked complexes between SBA and the four carbohydrates in Fig. 1 is crystal packing interactions in each complex.

SUMMARY

The present findings provide a molecular basis for a new paradigm of specificity in multivalent carbohydrate-lectin interactions, namely the formation of type 2 homogeneous cross-linked lattices between multivalent carbohydrates and lectins. The present x-ray data demonstrate that the cross-linked complexes formed between a series of structurally related divalent carbohydrates and a single tetravalent lectin (SBA) are distinct and due to crystal packing interactions. These results thus provide a molecular basis for the formation of homogeneous type 2 cross-linked complexes between lectins and multivalent carbohydrates and glycoconjugates.

These findings are also relevant to the observations that lectin-carbohydrate cross-linking interactions are involved in cellular recognition and signal transduction processes. For example, activated human T-cells undergo apoptosis due to binding and cross-linking of specific glycoprotein receptors by galectin-1 (Pace et al., 1999). Confocal microscopy shows that the galectin cross-linked glycoprotein receptors form homogeneous aggregates from a population of previously dispersed molecules on the surface of the cells. The crystal structures of the four SBA/pentasaccharide complexes thus repesent models for lectin-carbohydrate clustering *in vivo*.

REFERENCES

Bhattacharyya, L., Fant, J., Lonn, H. and Brewer, C. F., 1990, Binding and precipitating activities of *lotus tetragonolobus* isolectins with L-fucosyl oligosaccharides. Formation of unique homogeneous cross-linked lattices observed by electron microscopy. *Biochemistry* 29:7523.

Bhattacharyya, L., Haraldsson, M. and Brewer, C. F., 1988, Precipitation of galactose-specific lectins by complex-type oligosaccharides and glycopeptides: studies with lectins from *ricinus communis* (sgglutinin I), *erythrina indica*, *erythrina arborescens*, *abrus precatorius* (agglutinin), and *glycine max* (soybean). *Biochemistry* 27:1034.

Bourne, Y., Bolgiano, B., Liao, D.-l., Strecker, G., Cantau, P., Herzberg, O., Feizi, T. and Cambillau, C., 1994, Crosslinking of mammalian lectin (galactin-1) by complex biantennary saccharides. *Nature Struct. Biol.* 1:863.

Brewer, C. F., 1996, Multivalent lectin-carbohydrate cross-linking interactions. *Chemtracts - Biochem. Mol. Biol.* 6:165.

Dessen, A., Gupta, D., Sabesan, S., Brewer, C. F. and Sacchettini, J. C., 1995, X-ray crystal structure of the soybean agglutinin cross-linked with a biantennary analog of the blood group I carbohydrate antigen. *Biochemistry* 34:4933.

Drickamer, K. and Taylor, M. E., 1993, Biology of animal lectins. Ann. Rev. Cell Biol. 9:237.

Gupta, D., Bhattacharyya, L., Fant, J., Macaluso, F., Sabesan, S.and Brewer, C. F., 1994, Observation of unique cross-linked lattices between multiantennary carbohydrates and the soybean lectin. Presence of pseudo twofold axes of symmetry in complex type carbohydrates. *Biochemistry* 33:7495.

Kasai, K.-i. and Hirabayashi, J., 1996, Galectins: a family of animal lectins that decipher glycocodes. *J. Biochem.* 119:1.

Konstantinov, K. N., Robbins, B. A. and Liu, F.-T., 1996, Galectin-3, a β-galactoside-binding animal lectin, is a marker of anaplastic large-cell lymphoma. *Amer. J. Path.* 148:25.

Kooijman, R., de Wildt, P., van der Bliet, G., Homan, W., Kalshoven, H., Musgarave, A. and van der Ende, H., 1989, Wheat germ agglutinin induces mating reactions in chlamydomonas engametos by cross-linking agglutinin-associated glycoproteins in the flagellar membrane. *J. Cell.Biol.* 109:1677.

Loris, R., Hamelryck, T., Bouckaert, J. and Wyns, L., 1998, Legume lectin structure. *Biochim. Biophys. Acta* 1383:9.

Lotan, R. and Raz, A.,1988, Lectins in cancer cells. *Ann. N. Y. Acad. Sci.* 551:385.

Mandal, D. K. and Brewer, C. F., 1992a, Cross-linking activity of the 14-kilodalton β-galactose-specific vertebrate lectin with asialofetuin: Comparison with several galactose-specific plant lectins. *Biochemistry* 31:8465.

Mandal, D. K. and Brewer, C. F., 1992b, Interactions of concanavalin A with glycoproteins: formation of homogeneous glycoprotein-lectin cross-linked complexes in mixed precipitation systems. *Biochemistry* 31:12602.

Mandal, D. K. and Brewer, C. F., 1993, Differences in the binding affinities of dimeric concanavalin A (including acetyl and succinyl derivatives) and tetrameric concanavalin A with large oligomannose-type glycopeptides. *Biochemistry* 32:5116.

Nicolson, G. L., 1976, Transmembrane control of the receptors on normal and tumor cells. I. Cytoplasmic influence over cell surface components. *Biochim. Biophys. Acta* 457:57.

Olsen, L. R., Dessen, A., Gupta, D., Sabesan, S., Sacchettini, J.C. and Brewer, C. F., 1997, X-ray crystallographic studies of unique cross-linked lattices between four isomeric biantennary oligosaccharides and soybean agglutinin. Biochemistry 36:15073.

Pace, K. E., Lee, C., Stewart, P. L. and Baum, L. G., 1999, Restricted receptor segregation into membrane microdomains occurs on human T cells during apoptosis induced by galectin-1. *J. Immunol.* 163:3801.

Perillo, N. L., Pace, K. E., Seilhamer, J. J. and Baum, L. G., 1995, Apoptosis of T cells mediated by galectin-1. Nature 378:736.

Rini, J. M., 1995, Lectin structure. *Annu. Rev. Biophys. Biomol. Struct.* 24:551.

Varki, A., Cummings, R., Esko, J., Freeze, H., Hart, G. and Marth, J., 1999, *Essentials of Glycobiology*, Cold Spring Harbor Laboratory Press, Cold Spring Harbor.

CLASSIFICATION OF PLANT LECTINS IN FAMILIES OF STRUCTURALLY AND EVOLUTIONARY RELATED PROTEINS

Willy J. Peumans[1], Els J. M. Van Damme[1], Annick Barre[2] and Pierre Rougé[2]

[1]*Laboratory for Phytopathology and Plant Protection, Katholieke Universiteit Leuven, Willem de Croylaan 42, 3001 Leuven, Belgium;*
[2]*Institut de Pharmacologie et Biologie Structurale, UPR CNRS 9062, 205 Route de Narbonne, 31077 Toulouse Cedex, France*

Many plants contain one or more proteins that bind specifically to well-defined mono- or oligosaccharides. According to the currently available information, these carbohydrate-binding plant proteins, which are better known as lectins, agglutinins or hemagglutinins are a very extended group of proteins. Recent compilations have shown that roughly 500 different plant lectins have been isolated and characterized in some detail (1). Until a few years ago the interrelationships between the different members of this rather large group of plant proteins remained unclear. As a result, two major misconceptions arose with respect to the group of plant lectins as a whole. First, many scientists have the idea that all plant lectins are alike. Second, others regard plant lectins as a very complex and heterogeneous class of proteins because of the obvious differences in molecular structure, biochemical properties and carbohydrate-binding specificity. As a matter of fact the truth is in between these two extreme conceptions. The rapid progress in the structural analysis of lectins and molecular cloning of lectin genes allowed to trace the evolutionary relationships within this extended class of plant proteins and enabled to classify the great majority of al known plant lectins in seven families of structurally and evolutionary related proteins (2). This contribution aims to give a concise overview of each of these seven families. Thereby, attention is given to the occurrence, molecular structure, three-dimensional structure and carbohydrate-binding specificity of the different lectin families. Before discussing each of the seven families (in an alphabetic order) it is clearly stated which plant proteins are considered as lectins. In addition, some comments are given with regard to a few relevant aspects of the overall structure of plant lectins.

I. Which plant proteins are considered as lectins?

The definition of the term 'lectin' is one of the controversial issues in lectinology. Since

The Molecular Immunology of Complex Carbohydrates–2.
Edited by Albert M. Wu, Kluwer Academic/Plenum Publishers, 2001.

this issue has been addressed in detail a few years ago, the definition proposed by Peumans and Van Damme (3) will be adopted here. According to this definition all plant proteins possessing at least one non-catalytic domain that binds reversibly to a specific mono- or oligosaccharide are considered as lectins. Proteins, which are structurally and evolutionary related to lectins but lack active carbohydrate-binding sites do not meet this criterion and hence are not regarded as lectins but as lectin-related proteins.

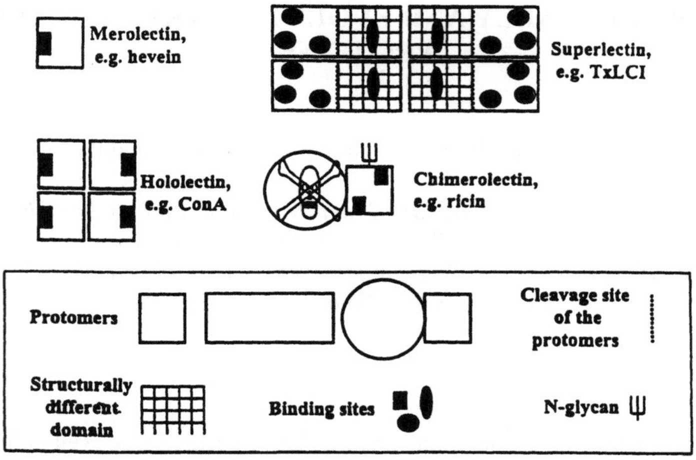

Fig. 1. Schematic representation of the overall structure of the different subclasses of plant lectins. TxLCI refers to the mannose/GalNAc-binding tulip lectin. All symbols shown in the boxed area apply also to Figures 2 and 4-6.

II. Overall structure of plant lectins

As will be discussed below in detail, native plant lectins exhibit a wide variety of molecular structures. These differences are not only due to differences in the structure of the composing subunits but may also result from differences in the association of two or more subunits. In addition, post-translational modification(s) may create additional differences. At present, there is considerable confusion for what concerns the description of the structure of native lectins and lectin subunits. To avoid any further ambiguity all structures are described here in terms of 'protomers'. Thereby, a protomer is defined as the mature polypeptide or the sum of all mature polypeptides derived from the primary translation products of the relevant lectin mRNA. Depending on the number of composing protomers a lectin is called a monomer, dimer, trimer, tetramer etc. In those cases where a protomer consists of two polypeptides, a monomeric lectin contains two different subunits.

Plant lectins are further subdivided on the basis of their overall structure in merolectins, hololectins, chimerolectins and superlectins (Figure 1). Proteins that consist exclusively of a single carbohydrate-binding domain are defined as merolectins. Examples are the small chitin-binding proteins composed of a single so-called hevein domain (see description of chitin-binding lectins). Merolectins are by definition monovalent and hence are incapable of precipitating glycoconjugates or agglutinating cells. Hololectins also consist exclusively of carbohydrate-binding domains but contain at least two identical or very similar domains. Since hololectins are di- or multivalent they can agglutinate cells and/or precipitate glycoconjugates. The great majority of all (agglutinating) plant lectins are hololectins. Superlectins are a special class of hololectins. They consist of at least two carbohydrate-binding domains, which are not identical or similar but recognize structurally different sugars. The tulip bulb lectin TxLCI, which is built up of two dissimilar tandemly arrayed carbohydrate-binding domains recognizing mannose and GalNAc, respectively, is one of the few examples of the group of superlectins (4). Unlike the mero-, holo- and superlectins, chimerolectins do not exclusively consist of carbohydrate-domains but are 'chimeric' proteins built up of a carbohydrate-binding domain tandemly arrayed with an unrelated domain with a well defined catalytic (or another biological) activity that acts independently of the carbohydrate-binding domain. Depending on the number of sugar-binding sites chimerolectins behave as merolectins or hololectins. For example, multivalent type 2 ribosome-inactivating proteins (RIP) (5) readily agglutinate cells whereas monovalent class I plant chitinases (with a single chitin-binding domain) (6) do not.

III. Families of structurally and evolutionary related plant lectins

During the last two decades an enormous progress has been made in the sequencing of lectins and lectin genes as well as in the structural analysis of lectins by X-ray crystallography and (to a far lesser extent) NMR. As a result, it became possible to trace with a fairly high precision the evolutionary relationships between different plant lectins. According to the available information the great majority of al known plant lectins can be classified into seven families of structurally and evolutionary related proteins. These families are (in an alphabetic order): (i) the amaranthin family (amaranthins), (ii) the chitin-binding lectins composed of hevein domains, (iii) the Cucurbitaceae phloem lectins, (iv) the jacalin-related lectins, (v) the legume lectins, (vi) the monocot mannose-binding lectins (MMBL) and (vii) the type 2 ribosome-inactivating proteins (type 2 RIP). Each of these seven families is described below in detail. The description starts with a brief historical note followed by an overview of the occurrence of the lectin family. After these introductory data, the molecular structure of the protomers and native lectins is highlighted, followed by (if available) a general description of the 3-D structure and the carbohydrate-binding specificity.

1. The amaranthin family (amaranthins)

'Amaranthins' are a small family of lectins found in seeds of *Amaranthus* species. The family is named after 'amaranthin', which is the trivial name of the first member of this family isolated from *Amaranthus caudatus* seeds.

History

Although the occurrence of amaranthins was reported for the first time in 1954 (7), they were purified and characterized from *A. leucocarpus* and *A. cruentus* only in 1988 (8-9). In 1997, both the primary structure and the three-dimensional structure of the *Amaranthus caudatus* lectin have been reported by Transue et al. (10).

Occurrence

Amaranthins have been found exclusively in seeds of species of the genus *Amaranthus* (family Amaranthaceae). Documented examples are *A. caudatus*, *A. spinosus*, *A. leucocarpus* and *A. cruentus*.

Molecular structure

All amaranthins studied hitherto are homodimers composed of two identical unglycosylated subunits of about 33 kDa (Figure 2; 11). The protomers of amaranthin, which has been completely sequenced by chemical methods, comprises 300 amino acid residues and consists of two homologous domains (10). None of the amaranthin genes has been cloned. However, a previously cloned seed-specific protein from *A. hypochondriacus* has virtually the same sequence as amaranthin (12). According to the deduced sequence of the presumed *A. hypochondriacus* lectin, the amaranthins are synthesized in the cytoplasm and posttranslationally processed by the removal of a *C*-terminal propeptide.

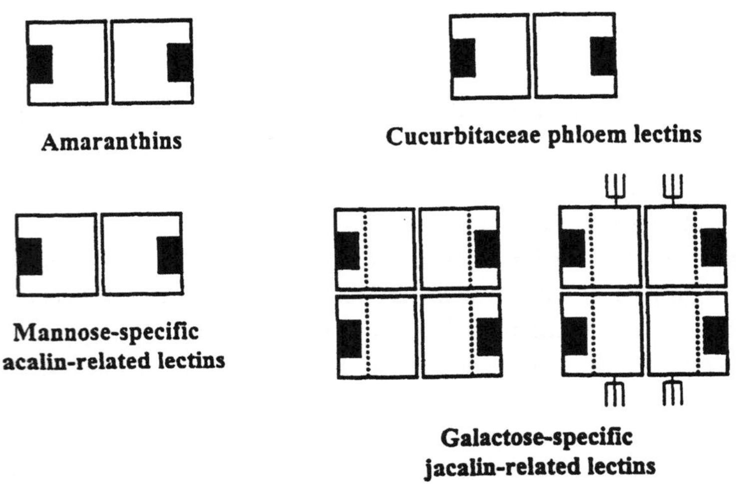

Fig. 2. Molecular structure of the amaranthins, Cucurbitaceae phloem lectins and jacalin-related lectins.

Three-dimensional structure

The protomers of the *Amaranthus caudatus* agglutinin consist of two domains (called *N*- and *C*-domain) linked by a short helix (10). Both domains have a typical β-trefoil structure consisting of six strands of antiparallel β-sheet capped by three β-hairpins into a β-barrel. Two protomers associate head to tail by extensive non-covalent contacts between both monomers giving rise to the native homodimer. Two shallow depressions formed at the interface between the *N*- and *C*-domains of the two facing monomers form the two surface-exposed carbohydrate-binding sites (Figure 3A). The binding of benzylated T-antigen disaccharide (Gal-β1,3-GalNAc-α-*O*-benzyl) to amaranthin, depends on a complex hydrogen bonding pattern with amino acid residues that are mainly located on surface-exposed hairpins and turns.

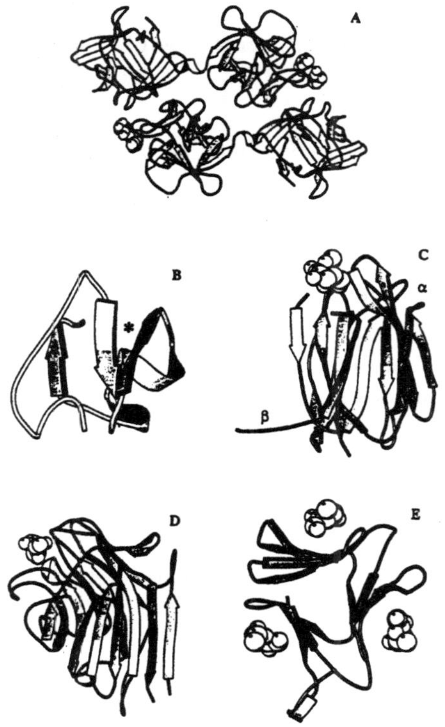

Fig. 3. A: Dimeric structure of amaranthin showing the two occupied T-antigen (cpk view) binding sites located at the interface of the two monomers. B: Structure of hevein showing the putative location of the carbohydrate binding site (*). C: Structure of the monomer of jacalin resulting from the non-covalent association of a light (β) and a heavy (α) subunit, with lactose (cpk view) inserted in the carbohydrate-binding site. D: Monomer of *Lathyrus ochrus* isolectin I (LoLI) with D-mannose (cpk view) occupying the monosaccharide binding site. E: Structure of the monomer of GNA showing the three mannose-binding sites occupied by three D-mannose molecules (cpk view). Arrows correspond to strands of β-sheet and short α-helices are represented by flat ribbons. Cartoons were rendered with Molscript (122).

Specificity

Amaranthin is usually considered as a GalNAc-specific lectin but has a much higher affinity for the T-antigen disaccharide Galß(1,3)GalNAc (11). This particular specificity suggests that the amaranthins are targeted against typical animal glycoconjugates.

2. The Cucurbitaceae phloem lectins

'Cucurbitaceae phloem lectins' are a small family of lectins, which are apparently confined to the Cucurbitaceae family. The name of the family refers to their (exclusive) occurrence in the phloem exudate of Cucurbitaceae species.

History

In 1968 Hossaini discovered an agglutinating factor in pumpkin (*Cucurbita maxima*) (13). Ten years later, Sabnis and Hart (14) showed that lectins are predominant proteins in the phloem exudate of several Cucurbitaceae species, and described for the first time the isolation and partial characterization of the pumpkin lectin. Determination of the carbohydrate-binding specificity and molecular structure of the Cucurbitaceae phloem lectins was accomplished in the following years (15-16). The first member of this lectin family was cloned in 1992 (17).

Occurrence

Hitherto, Cucurbitaceae phloem lectins have been found exclusively in the phloem exudate of Cucurbitaceae species more precisely in representatives of the genera *Citrullus, Coccinia, Cucumis, Cucurbita, Luffa* and *Sechium.*

Molecular structure

All Cucurbitaceae phloem lectins which have been characterized hitherto are dimers built up of two (identical) unglycosylated subunits of about 25 kDa (Figure 2; 18). No complete amino acid sequences are available except the deduced sequences of the *Cucurbita maxima* and *Cucurbita pepo* phloem lectins (17, 19). According to these deduced sequences the *Cucurbita maxima* and *Cucurbita pepo* phloem lectins are synthesized in the cytoplasm. It is not known whether the primary translation products undergo posttranslational processing.

Three-dimensional structure

At present, no 3-D structure has been determined yet of any of the Cucurbitaceae phloem lectins.

Specificity

All Cucurbitaceae phloem lectins clearly exhibit specificity towards oligomers of GlcNAc but not to the monomer (15, 18). The inhibitory potency of the GlcNAc-oligomers increases with chain length up to 5 residues (18). It has been shown that the *Luffa acutangula* and the *Cucurbita maxima* lectins are capable of binding internal di-*N*-acetylchitobiosyl sequences of *N*-linked glycan chains from fetuin, ovalbumin and soybean lectin (18).

3. The chitin-binding lectins composed of hevein domains

The chitin-binding lectins composed of hevein domains are a very extended and heterogeneous family that comprises all proteins possessing at least one so-called hevein domain. The name hevein refers to a small 43 amino-acid protein with a fully active chitin-binding site that was originally discovered in the latex of the rubber tree (*Hevea brasiliensis*) (20). Accordingly, all structural units of about 40 amino acid residues which exhibit sequence similarity to hevein and possess chitin-binding activity are considered hevein domains.

History

The history of the chitin-binding lectins starts with the discovery of an agglutinating factor from potato tubers by Marcusson-Begun in 1926 (21). Another important milestone was the observation made by Aub and collaborators (22) that a lipase-containing extract from wheat germ agglutinated several transformed cell lines but did not cause agglutination of their untransformed counterparts. Later, Burger and Goldberg (23) succeeded in separating an agglutinating factor called 'wheat germ agglutinin' (WGA) from the lipase activity. WGA was the first chitin-binding lectin to be isolated and characterized (24-25), crystallized (26) and cloned (27). The concept of the hevein domain as a structural chitin-binding unit is of a relatively recent date because the chitin-binding activity of hevein was not recognized until 1991 (28) even though the protein was already isolated and sequenced in 1975 (20).

Occurrence

Chitin-binding lectins with hevein domains are very widespread in the plant kingdom. For example, class I chitinases, which are considered (chimero)lectins because of the presence of an *N*-terminal chitin-binding hevein domain are an ubiquitous group of plant proteins (6, 29). Apart from class I chitinases, chitin-binding lectins have been found in many species of different taxonomic groups. Single hevein domain proteins were purified from *Hevea brasiliensis* (rubber tree, Euphorbiaceae), *Amaranthus caudatus* (grain amaranth, Amaranthaceae) and elderberry (*Sambucus nigra,* Caprifoliaceae). Multi-hevein domain chitin-binding lectins have been found in several taxonomic groups. They are widespread within the Gramineae and Solanaceae, and have been isolated from several *Phytolacca* species (pokeweed, Phytolaccaceae), *Urtica dioica* (stinging nettle, Urticaceae), *Chelidonium majus* (greater celandine, Papaveraceae), and *Viscum album* (mistletoe, Viscaceae).

Molecular structure

The family of chitin-binding lectins is not uniform with respect to the molecular structure of the native lectins because it comprises besides hololectins also merolectins and different types of chimerolectins. A schematic overview of the currently known structures is shown in Figure 4. Merolectins composed of a single hevein domain are the simplest chitin-binding lectins. Some of these merolectins like hevein and hevein-like proteins (e.g. from elderberry) consist of a complete hevein domain of about 43 amino acid residues with 4 intrachain disulfide bonds. Others like the chitin-binding antimicrobial peptides from *Amaranthus caudatus* (AcAMP) consist of a single truncated hevein domain of 30 amino acid residues containing 3 disulfide bridges (30). Hololectins have been identified which are built up of protomers consisting of two, three, four or seven tandemly arrayed hevein domains. Two domain protomers occur in the monomeric *Urtica dioica* (stinging nettle) agglutinin (UDA) and the dimeric chitin-binding lectin from *Viscum album* (31-32) as well as in the *Phytolacca americana* lectin Pa-5/PL-D (33). Three-domain protomers have been found exclusively in *Phytolacca americana* root lectins Pa-4/PL-C (34). Numerous lectins composed of four hevein domain protomers have been identified in Gramineae species. All these Gramineae lectins (e.g. WGA, rice lectin) are dimers. It should be mentioned, however, that in rice (*Oryza sativa)* the protomers of the lectins are post-translationally cleaved into two smaller fragments of two hevein domains each. In one documented example, namely *Phytolacca americana* lectin PL-B the protomers consist of seven tandemly arrayed hevein domains. The latter lectin is also the only example of an *N*-glycosylated chitin-binding hololectin (35). Besides mero- and hololectins, the family of chitin-binding lectins comprises at least two different types of chimerolectins. A first group are the Solanaceae lectins (e.g. the potato or *Solanum tuberosum* agglutinin), which are dimers built up of protomers consisting of an *N*-terminal chitin-binding domain of three hevein repeats linked to an *O*-glycosylated serine-hydroxyproline-rich (extensin-like) domain (36-37). The second group is the class I chitinases, which contain a single *N*-terminal hevein domain linked through a glycine/proline-rich hinge domain to a catalytically active chitinase domain (6,38).

The availability of detailed sequence information (obtained from both protein and DNA sequencing) strongly facilitated the determination of the molecular structure of the chitin-binding lectins and made also a crucial contribution to the unraveling of the biosynthesis and post-translational processing of numerous members of this lectin family. All chitin-binding lectins studied hitherto are synthesized as pre(pro)proteins and undergo co-translational and post-translational modifications, which in turn indicates that they follow the secretory pathway. There are, however, some important differences with respect to the post-translational processing. Most chitin-binding lectins (e.g. AcAMP, Gramineae lectins and class I chitinases) are synthesized as preproproteins with a relatively small *C*-terminal propeptide. Accordingly, post-translational processing removes only a minor part of the lectin precursor. In some other cases, however, the proproteins contain an extended propeptide so that the post-translational processing results in the removal of a major part of the precursor. For example, hevein (43 AA) and UDA (89 AA) are the remnants of much larger precursor molecules.

Three-dimensional structure

The three-dimensional structure of the carbohydrate-binding domains of the chitin-binding lectins is determined primarily by the structure of the hevein unit(s) (Figure 3B). According to ^{1}H-NMR spectrometry the most striking conformational feature of hevein consists of a stretch of amino acid residues located at the *N*-terminal end of the polypeptide chain which forms two strands of antiparallel β-sheet followed by an α-helix (39). The folding of the *N*-terminal region is stabilized by the four intra-chain disulfide bonds. Both strands of antiparallel β-sheet contain several residues (namely, Ser^{19}, Trp^{21} and Trp^{23}), which are involved in the binding of GlcNAc-containing oligosaccharides (40). Tyr^{30}, which is located in the first α-helix also participates in the binding of GlcNAc-oligomers. According to modeling studies based on the NMR-coordinates of hevein the 3-D structure of UDA is

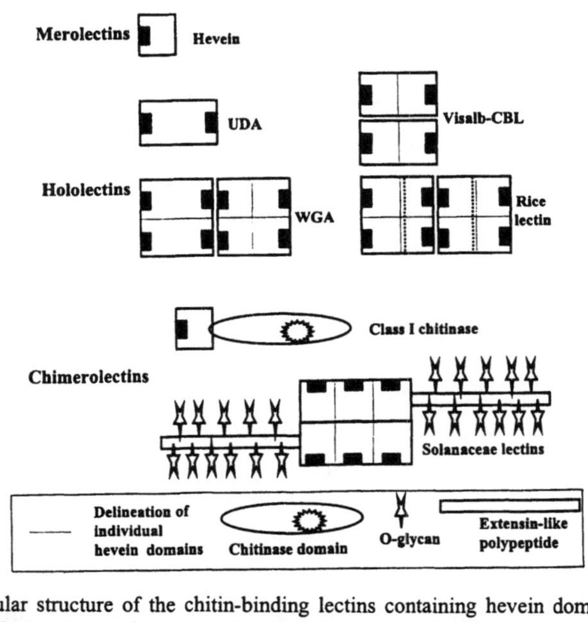

Fig. 4. Molecular structure of the chitin-binding lectins containing hevein domains. UDA and Visalb-CBL refer to the *Urtica dioica* agglutinin and the *Viscum album* chitin-binding lectin, respectively.

quite similar to that of hevein except that UDA lacks the second α-helix of hevein. The 30 amino acid antimicrobial peptide Ac-AMP2 from *Amaranthus caudatus* seeds exhibits a three-dimensional conformation similar to that of hevein (41). However, Ac-AMP2 lacks the α-helix and the strand of β-sheet located at the *C*-terminus of hevein. Since most of the residues of hevein involved in the binding of GlcNAc-oligomers are conserved in Ac-AMP2 this protein is fully capable of binding chito-oligosaccharides.

The 3-D structure of WGA is far more complex than that of hevein and UDA because the native lectin is a dimer of two non-covalently bound protomers consisting of four hevein domains. Each hevein domain of WGA is apart from a very short α-helix of only five residues devoid of regular secondary structures and accordingly consists mainly of coil and turn structures. The four disulfide bonds of each domain (32 in total) stabilize the overall folding of the dimer, in which the two dimers associate in a "head-to-tail" fashion. Each of the four individual domains of the WGA monomer possesses a carbohydrate-binding site.

Carbohydrate-binding specificity

The sugar-binding activity and specificity of the chitin-binding lectins apparently relies exclusively on their hevein domains. Since all chitin-binding lectins (including class I chitinases) are retained on crude chitin, all hevein domains apparently bind native chitin. Detailed specificity studies done with WGA, UDA and the potato and thorn apple lectins leave no doubt that the carbohydrate-binding sites of the chitin-binding lectins are complex and accommodate preferentially GlcNAc-trimers or tetramers. This explains why the relative inhibitory potency of the di-, tri- and tetramers of GlcNAc on WGA is 131-, 3700- and 4500-fold higher, respectively, than that of GlcNAc (42). Moreover, the occurrence of complex sugar binding sites also explains why apart from the Gramineae lectins none of the chitin-binding lectins is effectively inhibited by GlcNAc. It is also noteworthy that the multiple binding sites of some chitin-binding lectins are not identical as was shown for WGA and UDA, which contain both high- and low-affinity sites.

4. The jacalin-related lectins

The jacalin-related lectins are a relatively novel family of lectins. Until a few years ago only a few members were known. However, evidence is accumulating that this lectin family is very widespread among higher plants. The term jacalin-related lectins is derived from 'jacalin', which is the trivial name for the lectin from the seeds of jack fruit (*Artocarpus integrifolia*). Two subgroups are distinguished on the basis of the specificity of the lectins namely the small subfamily of GalNAc-specific Moraceae lectins and the extended subfamily of mannose-specific jacalin-related lectins.

History

The report of Jones et al. (43) that seeds of *Maclura pomifera* contain a potent hemagglutinin is the first documented evidence for the occurrence of a lectin of the jacalin family. However, the purification of the *Maclura pomifera* seed lectin was reported only in 1977 (44) and that of jacalin in 1981 (45). The importance of the discovery of these novel lectins became evident when it turned out that both novel lectins recognize specifically Galß(1,3)GalNAc-residues (46-47). As soon as the specific IgA-binding activity and mitogenicity of the jack fruit lectin were recognized (48), jacalin was widely used as a powerful tool in immunology. Further interest in jacalin and related lectins followed when the potent anti-HIV (49) activity and insecticidal properties (50) were discovered. In the meantime, jacalin and the *Maclura pomifera* lectin were completely sequenced by chemical methods (51). Jacalin was the first member of this family to be cloned (52) and analyzed by X-ray crystallography (53). The discovery of the mannose-specific subfamily dates back to 1996 when cloning of a rhizome lectin from *Calystegia sepium* revealed that this mannose/maltose-specific lectin shared a high sequence similarity with the Moraceae lectins (54).

Occurrence

The subfamily of galactose-specific jacalin-related lectins is apparently confined to a few genera of the family Moraceae. Hitherto, these lectins have exclusively been found, indeed, in seeds and some vegetative tissues of *Artocarpus* species and *Maclura pomifera*. In contrast, the mannose-specific subfamily is widely distributed among higher plants. Documented examples of purified lectins are the rhizome lectin of *Calystegia sepium* (54), the tuber lectin of the Jerusalem artichoke (*Helianthus tuberosus,* Asteraceae) (55) and a lectin called KM+ from jackfruit seeds (56). In addition, sequences of putative mannose-binding jacalin-related lectins have been identified in rice (*Oryza sativa*), barley (*Hordeum vulgare*) and wheat (*Triticum aestivum*).

Molecular structure

Though all jacalin-related lectins are undoubtedly related the galactose- and mannose-specific subfamilies markedly differ with respect to their molecular structure (Figure 2). All galactose-specific Moraceae lectins are built up of four identical protomers, which consist of a large α-chain and a small ß-chain, and harbor a single sugar-binding site. Both chains are derived from a single preprolectin that undergoes a complex co- and post-translational processing involving the removal of a signal peptide, cleavage of an *N*-terminal propeptide and excision of a 4-amino acid residue linker between the α-chain and ß-chain. Some, but not all isoforms of these lectins are *N*-glycosylated. Most probably, the synthesis and intracellular transport of the galactose-specific jacalin-related lectins follow the secretory pathway.

All known mannose-specific jacalin-related lectins are built up of very similar (uncleaved) protomers of about 150 amino acid residues. Apart from the tetrameric *Helianthus tuberosus* lectin, the mannose-specific jacalin-related lectins are dimers. Unlike their galactose-specific counterparts the mannose-specific jacalin-related lectins are not co- or post-translationally processed. It appears, indeed, that the mature lectin protomers comprise the entire open reading frame of the corresponding genes. Accordingly, one can reasonably expect that the mannose-specific jacalin-related lectins are synthesized and located in the cytoplasm.

Three-dimensional structure

Hitherto only a few members of the galactose-binding subfamily have been studied by X-ray diffraction analysis. The resolution of the 3-D structure of jacalin (53) demonstrated that each protomer consists of a three-fold symmetric β-prism made of three four-stranded β-sheets (Figure 3C). Of the twelve strands eleven are formed by the α-chain whereas the β-chain forms the twelfth strand. Four protomers associate by non-covalent interactions into a tetrameric structure. Thereby, the β-chains, which occupy the central region of the tetramer, play a crucial role. Each protomer harbors a single carbohydrate-binding site. This site is made up by residues Gly[1], Tyr[122], Trp[123] and Asp[125] of the α-chain, which form a network of 9 hydrogen bonds with O3, O4, O5 and O6 of methyl-α-D-galactose. It should be emphasized here that the generation of Gly[1] of the α-chain requires a correct post-translational processing of projacalin into an α and β chain.

The 3-D structure of the *Maclura pomifera* lectin and its carbohydrate-binding site are virtually identical to the corresponding structures of jacalin. No structure has been reported yet of any of the mannose-binding jacalin-related lectins. Although molecular modeling studies indicate that these mannose-specific lectins have the same overall β-prism fold as their galactose-binding counterparts, it is still unclear whether the difference in specificity between the two subfamilies can be ascribed to differences in the 3-D structure of the binding sites. It is important to realize in this respect that the protomers of the mannose-binding lectins are not cleaved, which implies that no free *N*-terminal Gly residue is generated at the position corresponding to Gly[1] of the α-chain of jacalin.

Carbohydrate-binding specificity

There is some heterogeneity within the family of jacalin-related lectins for what concerns the carbohydrate-binding specificity. First, the family clearly comprises a galactose-specific and a mannose-specific subgroup. According to detailed specificity studies jacalin and the *Maclura pomifera* lectin react with terminal non-reducing α-D-galactosyl residues and have a high affinity for Galß(1,3)GalNAc (46-47). In contrast, the mannose-specific jacalin-related lectins from *Calystegia sepium* and *Helianthus tuberosus* do not recognize galactose but bind mannose and maltose and have a high affinity for oligomannosides (55, 57). Second, although the sugar-specificity of the mannose-specific jacalin-related lectins has not yet been studied in great detail, it appears that different members of this subfamily definitely differ with respect to their fine specificity. Further studies with other mannose-specific jacalin-related lectins will be required to get a better overview of the variation in specificity within this group of lectins.

5. The legume lectins

Legume lectins are a large but homogeneous group of carbohydrate-binding proteins, which are exclusively found in legumes. The name 'legume lectin' is self-evident and emphasizes the exclusive occurrence within the legume family. Several legume lectins made an important contribution to the development of lectin research and are valuable tools in biological and biomedical research. Legume lectins are also the only plant lectins that are metalloproteins.

History

The report on the isolation of a presumed toxic lectin called 'robin' from the bark of the legume tree black locust (*Robinia pseudoacacia*) by Power and Cambier in 1890 can be regarded as the onset of the history of the legume lectins (21). In an era where lectins were still associated with toxicity, the discovery by Landsteiner and Raubitschek (58) of non-toxic lectins in the legumes *Phaseolus vulgaris* (bean), *Pisum sativum* (pea), *Lens culinaris* (lentil) and *Vicia sativa* (vetch) provided ample evidence that lectins are not necessarily toxins. From then on lectins were discovered in many legume seeds and a great deal of lectin research dealt with legume seed lectins. As a result all breakthroughs in plant lectin research have been achieved with legume lectins. The inhibition of the agglutination activity of ConA by cane sugar was the basis of the concept of lectins as carbohydrate-binding proteins (59). Similarly, the observation that extracts from some legume seeds specifically agglutinate blood cells of one particular blood group within the ABO system led to the discovery of blood group specificity (60-61). Furthermore, the triggering by the *Phaseolus vulgaris* seed lectin of quiescent, non-dividing lymphocytes into a derepressed state of active growth and proliferation evidenced that lectins are bioactive proteins (62). Finally, legume lectins played a crucial role in the pioneering work in the field of biochemistry, physiology and molecular biology of plant lectins. ConA was the first plant lectin to be purified and crystallized (59), and analyzed by X-ray diffraction (63-64). The first plant lectin to be cloned was also a legume lectin, namely the soybean seed lectin (65).

Occurrence

Hitherto, legume lectins have exclusively been found in legumes (Fabaceae). According to recent compilations more than a hundred legume lectins have been isolated from seeds or vegetative tissues of over 70 different species scattered over various taxonomic groups. Many legume lectins are typical seed lectins. Others are confined to vegetative tissues like leaves and bark. Several legumes contain two or more different lectins in their seeds or vegetative tissues.

Molecular structure

All legume lectins are built up of protomers of 30 kDa which are all derived from similar primary translation products of 250-300 amino acid residues. The mature protomers of most legume lectins consist of a single polypeptide chain of approximately 250 amino acid residues. Accordingly these lectins are usually called 'one-chain' legume lectins. In some cases, the protomers are either completely or partly cleaved by a proteolytic post-translational processing into two smaller polypeptides. Depending on the position of the cleavage site along the protomer the resulting polypeptides are similar or dissimilar in size. Due to the apparent occurence of two different polypeptides legume lectins built up of cleaved protomers are called 'two chain' legume lectins.

Many but not all legume lectins are *N*-glycosylated. In general, glycosylated lectin protomers possess one or two glycan chains. Both high-mannose type and complex type glycans have been found in legume lectins. In some cases, high-mannose and complex type *N*-glycans occur on a single lectin protomer (for example in the *Phaseolus vulgaris* hemagglutinin) (66-67). Differences in glycosylation (in terms of the number and/or structure of the *N*-glycans) often lead to the formation of glycoforms which can mistakenly be considered as isolectins.

Of all plant lectins the legume lectins are the only metalloproteins. Each protomer contains, indeed, a tightly bound Mn^{2+} and Ca^{2+} ion, which are both essential for the carbohydrate-binding activity of the lectin. It is also noteworthy that the amino acid residues, which are required for the binding of the divalent cations are markedly conserved in all legume lectins.

The legume lectin family exhibits some heterogeneity with respect to the molecular structure of the native proteins. Basically, all legume lectins consist of either two or four non-covalently bound protomers. Taking into consideration that both cleaved and uncleaved protomers occur, which in some cases are glycosylated, at least eight different molecular forms of legume lectins can be distinguished (Figure 5). Additional heterogeneity can be introduced by the formation of heterotetrameric lectins when two different lectin polypeptides are synthesized as is illustrated by the classical example of the PHA isolectins. The E and L subunits synthesized in the cells of bean (*Phaseolus vulgaris*) cotyledons associate randomly resulting in the homotetrameric E_4 and L_4 isoforms and the hetero-tetrameric E_1L_3, E_2L_2 and E_3L_1 isolectins (68).

All legume lectins are synthesized on the endoplasmic reticulum (ER) as pre(pro)proteins and subsequently enter the secretory pathway to their final destination in the storage protein vacuoles. Co-translational processing involves the removal of a signal peptide and can be followed by *N*-glycosylation. Further processing of both the polypeptide backbone and *N*-linked glycan chains takes place during or after intracellular transport from the ER via the Golgi apparatus to the storage protein vacuoles (69). Documented post-translational modifications of the polypeptide backbone include (i) removal of an *N*-terminal or *C*-terminal propeptide, (ii) proteolytic cleavage at a single site and (iii) excision of a (short) internal peptide by a double proteolytic cleavage. Depending on each particular lectin, one or more of these post-translational modifications take place. Lectins from species of the genera *Canavalia* and *Dioclea* undergo a unique mode of polypeptide processing which was first discovered for ConA (70-71).

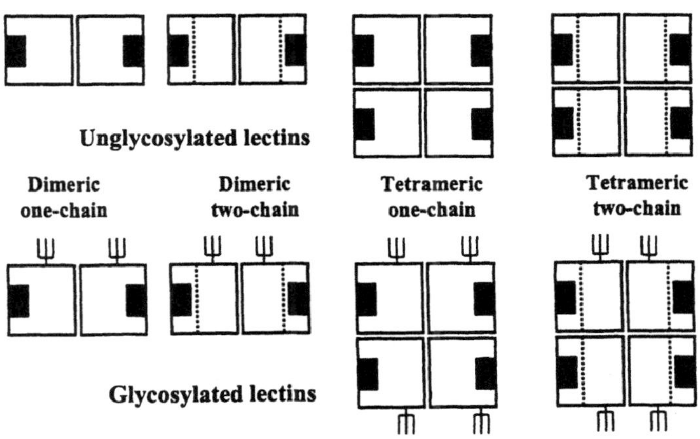

Fig. 5. Molecular structure of the legume lectins.

Three-dimensional structure

Numerous legume lectins have been crystallized and their 3-D structure solved at atomic resolution (Figure 3D). The available data leave no doubt that all the legume lectin protomers (irrespective whether they are single- or two-chain lectins) have a very similar three-dimensional conformation. Basically, the protomers form a flattened dome-shaped structure, which is built from a curved seven-stranded β-sheet (front face) and a flat six-stranded β-sheet (back face) interconnected by turns and loops. Since the orientation of the antiparallel strands of β-sheet changes from one face to the other, a very rigid and strong structure is created (which explains why legume lectins are fairly resistant towards many proteolytic enzymes). The monosaccharide-binding site (i.e. the site which is responsible for the binding of simple sugars) is formed by four loops located at the upper part of the dome. An interesting feature is the occurrence of a *cis*-peptide bond involving an aspartic residue (e.g. Asp^{81} in LoLI) in the peptide chain which orients the side chain of this residue in such a way that it can interact with a Ca^{2+} ion. A second divalent cation, Mn^{2+}, is also bound to the polypeptide. Both divalent cations are required to stabilize the functional conformation of the monosaccharide binding site (72). Besides a carbohydrate-binding site, the legume lectin protomers also contain a hydrophobic cavity, which is mainly formed by conserved hydrophobic residues.

Most legume lectins are built up of two protomers, which interact with each other in a two-fold symmetry plane. As a result, a twelve-stranded β-sandwich is formed in which the two facing monomers associate by their flat bottoms and the loops forming the monosaccharide-binding sites are located at both ends of the dimer. This very rigid structure contains a front wall of fourteen antiparallel strands of β-sheet associated to a back wall of twelve antiparallel strands of β-sheet. It should be mentioned here that the formation of this canonical divalent structure can be prevented by glycan chains as is illustrated by the *Erythrina corallodendron* lectin in which the two lectin monomers are associated differently (73).

Many native legume lectins consist of four protomers. Basically, a non-covalent association of two dimers into tetramers can be achieved in different ways. Tetrameric lectins possess - like most other homotetrameric proteins, a 222 (D2) symmetry except PNA which displays an unusual quaternary arrangement exhibiting neither a 222 (D2) nor a fourfold (C4) symmetry (74). For example, in PHA-L two canonical dimers become associated through their back walls in such a way that two monosaccharide-binding sites occur at both sides of the tetramer (75). Most probably, the central cleft delineated by this type of association enables the binding of adenine and adenine-related plant hormones by both PHA-L and PHA-E. The glycan chains present on the surface of the PHA protomers cannot prevent the association of the dimers because they are directed towards the solvent.

Legume lectins can interact with both simple and complex sugars (76-77). The interaction with simple sugars and sugar derivatives apparently depends on the presence of a monosaccharide-binding site at the surface of the lectin protomer. A few amino acid residues belonging to four loops located at the top of the dome-shaped protomer form a monosaccharide-binding site which accounts for this specific recognition. For example in LoLI residues Asp^{81}, Gly^{99}, Asn^{125}, Gly^{208}, Ala^{209} and Glu^{210} create a network of seven hydrogen bonds with O3, O4, O5 and O6 of the simple sugar. In addition, the aromatic residue Phe^{123} in LoLI enables a hydrophobic interaction with the pyranose ring of the sugar. Legume lectins also interact with more complex glycans such as *N*- and *O*-linked oligosaccharide side chains of glycoproteins. This interaction, which determines the fine carbohydrate-binding specificity of legume lectins, is far more specific and allows to subdivide lectins with similar monosaccharide-binding specificities (78-79).

Carbohydrate-binding specificity

The family of legume lectins is very heterogeneous with respect to the carbohydrate-binding specificity of the different lectins. As a matter of fact, legume lectins cover a much broader range of specificities than any other lectin family. Virtually all the classical specificity groups are represented in the legume lectin family. Only mannose-binding and mannose/ maltose-specific lectins have not been identified (yet) in legumes. Moreover, fucose-specific and mannose/glucose-binding lectins have hitherto exclusively been found in legumes. It is also noteworthy that almost all lectins with a so-called 'complex' specificity are legume lectins. Although the broad specificity range of the legume lectins is on a first sight difficult to reconcile with both the high sequence identities and the highly conserved 3-D structure of the protomers, there is a logic explanation for this apparent contradiction. Substitutions of a few amino acids involved in sugar binding activity and variations in the length of a particular loop profoundly change the structure of the binding site without affecting the overall 3-D structure of the protomer (80-81). This particularity suggests that the evolution of the legume lectins was driven towards a diversity of carbohydrate-binding specificity.

6. The monocot mannose-binding lectins (MMBL)

The collective name 'monocot mannose-binding lectins' (MMBL) was given to a superfamily of mannose-binding lectins, which are apparently confined to a relatively small taxonomic group of the monocotyledonous plants. MMBL are neither structurally nor evolutionary related to mannose-binding legume or jacalin-related lectins.

History

Although the discovery of lectin activity in garlic in 1980 by Nachbar and Oppenheim can be regarded as the first documented evidence for the occurrence of the MMBL (82), the MMBL were only recognized as a novel lectin family after the isolation and cloning of a lectin with an exclusive specificity towards mannose from snowdrop bulbs (83-84) convincingly demonstrated that the snowdrop lectin (called *Galanthus nivalis* agglutinin or GNA) differed from all previously described plant lectins with respect to its specificity and primary structure. The subsequent isolation and cloning of numerous mannose-specific lectins from other Amaryllidaceae species (85), and from several Alliaceae (86-87) and Orchidaceae species (88) soon revealed that all these lectins are closely related. Once it appeared that similar lectins occur also in several other monocot families, the term 'monocot mannose-binding lectins' was introduced as a collective name for this novel lectin family.

Occurrence

MMBL are particularly widespread in the Amaryllidaceae, Alliaceae and Orchidaceae species (89), and are probably also very common in the Araceae family (90). MMBL have been found in several genera of the family Liliaceae (e.g. *Tulipa*, *Polygonatum, Scilla* and *Aloe*) but are certainly not ubiquitous in this family. Hitherto, only one MMBL has been isolated from the family Bromeliaceae (namely *Neoregelia flandria)* (91).

Molecular structure

The superfamily of MMBL exhibits some heterogeneity with respect to the molecular structure of the native proteins. Basically, this lectin family comprises lectins consisting of 12 kDa protomers and lectins built up of 30 kDa protomers consisting of two tandemly arrayed lectin domains. MMBL built up of 30 kDa protomers are further subdivided in two subgroups depending whether the composing lectin domains are almost identical or dissimilar.

The majority of the currently known MMBL consist of one, two or four one-domain protomers of about 110 amino acid residues (Figure 6). MMBL consisting of a single one-domain protomer have been identified in the orchids *Listera ovata, Epipactis helleborine* and *Gastrodia elata* (88, 92). These monomeric proteins definitely bind mannose and are considered merolectins because they possess a single active carbohydrate-binding site.

Most MMBL are hololectins composed of two or four identical (non-covalently linked) one-domain protomers. For example, all Amaryllidaceae and Bromeliaceae lectins, most of the Alliaceae and Orchidaceae lectins, and some Liliaceae lectins are homodimers or homotetramers. Hitherto, only one heterodimeric MMBL has been identified in bulbs of ramsons (*Allium ursinum*), which consists of two slightly different one-domain protomers that are encoded by different genes (93).

Molecular cloning provided conclusive evidence that several MMBL are built up of protomers consisting of two tandemly arrayed lectin domains (called two-domain protomers). Most of these lectins are hololectins because both lectin domains of each protomer recognize mannose. However, there are also a few examples of superlectins built up of protomers consisting of two lectin domains with a different specificity.

Cloning of the *Allium sativum* bulb lectin I (ASA-I) demonstrated that a single precursor consisting of two highly similar tandemly arrayed lectin domains (separated from each other by a short linker sequence) is processed into two slightly different polypeptides of 11.5 and 12.5 kDa, which remain bound to each other by non-covalent interactions. As a result, native ASA-I behaves as a heterodimer while it consists in fact of a single protomer (94). A similar mechanism holds true for the Araceae lectins (e.g. from *Arum maculatum* and *Colocasia esculenta*) (90). These lectins also are derived from two-domain precursors, which are post-translationally processed into two polypeptides of similar size. Since these mature lectins are composed of two protomers, they behave as heterotetramers. In a few cases, the two-domain protomers are not or only partially cleaved into two separate polypeptides. For example, the tulip lectin TxLC-I is built up of four identical subunits of 28 kDa, which are only partly cleaved into two different polypeptides of 14 kDa. Since in this particular case the *N*- and *C*-terminal bind mannose and GalNAc, respectively, TxLC-I is considered a superlectin (95).

All MMBP are synthesized as preproproteins. In some cases the *C*-terminal propeptide(s) contain one or two *N*-linked glycans. Co-translational processing involves the removal of a signal peptide and if applicable *N*-glycosylation. During or after transport of the lectins from the ER into their final destination (presumably vacuoles) the C-terminal and in some cases an internal propeptide are proteolytically removed.

Three-dimensional structure

Detailed information about the 3-D structure of the MMBL was obtained from X-ray diffraction studies of GNA and some related lectins (96-97). Native GNA is built up of four identical non-covalently bound protomers of 109 residues (12 kDa). These protomers consist of three tandemly arrayed subdomains (called subdomain I, II and III) each of which corresponds to a four-stranded β-sheet. Interestingly, the four-stranded bundle forming subdomain I is completed by a strand of β-sheet located in the extended *C*-terminal end of the polypeptide chain. Subdomains I, II and III are connected by loops and form a twelve-stranded β-barrel containing three mannose-binding sites, which are located in the clefts

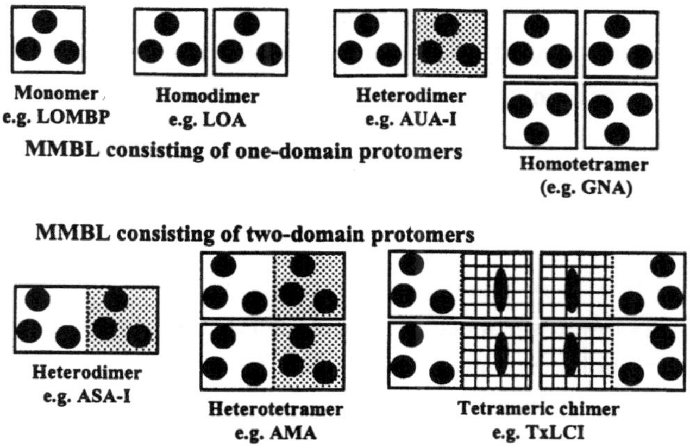

Fig. 6. Molecular structure of the MMBL. Abbreviations refer to the following lectins: LOMBP=the monomeric (non-agglutinating) *Listera ovata* (twayblade) lectin; LOA=the dimeric (agglutinating) *Listera ovata* lectin; AUA-I= the *Allium ursinum* (ramsons) bulb lectin I; ASA-I= the *Allium sativum* (garlic) bulb lectin I; GNA=*Galanthus nivalis* (snowdrop) agglutinin; AMA=*Arum maculatum* (lords and ladies) agglutinin; TxLCI=the mannose/GalNAc-binding tulip lectin.

formed by the three bundles of β-sheet (Figure 3E). A single disulfide bond between cysteine residues Cys^{29} and Cys^{52} occurs in the protomer. Four protomers (A, B, C and D) associate by non-covalently interactions into the GNA homotetramer, which has the shape of a flattened crown with a central 16 Å wide solvent channel. The four protomers associate pairwise (i.e. A and D, and B and C) by hydrogen-bonds into tight dimers which are stabilized by *C*-terminal strand exchange. Further association of the A-D and B-C dimers into tetramers involves mainly hydrophobic interactions. Since each protomer contains three mannose-binding sites native GNA tetramers possess 12 functional mannose-binding sites.

Molecular modeling of numerous MMBL from the families Alliaceae, Araceae, Orchidaceae and Liliaceae indicated that all the lectins exhibit an overall three-dimensional structure very similar to that of GNA. However, according to docking experiments with D-mannose the protomers of some lectins contain only two or one functional site(s) because the replacement of some crucial amino acids in the carbohydrate-binding site by more extended or hydrophobic residues prevents binding of D-mannose (98).

The resolution of the structure of GNA revealed that each of the three mannose-binding sites is made up of the same four amino acid residues, namely Gln, Asp, Asn and Tyr. These four residues bind O2 (Asp, Asn), O3 (Gln) and O4 (Tyr) of mannose through a network of four hydrogen bonds. In addition, a valine residue interacts with C3 and C4 of mannose through hydrophobic interactions.

Since GNA and other related MMBL bind with a high affinity to mannose-containing glycoproteins it could be expected that these lectins possess also extended carbohydrate-binding sites. Structural analysis of GNA complexed with di-, tri- and pentamannosides (99) confirmed the presence of an extended binding site and, in addition, revealed two different types of interactions with complex oligomannosides. A first extended site, which comprises the binding sites of two adjacent subdomains III and the surrounding regions belonging to two different dimers (A and B or C and D), enables GNA to interact with the two D-mannose units of $\alpha(1,3)$-linked oligomannosides. The second extended site is formed at the interface between two GNA tetramers. This extended site comprises the mannose-binding site of subdomain III of monomer (B) of a first GNA tetramer and the mannose-binding site II of monomer (A) of a second GNA tetramer. It is questionable, however, whether the second extended site is functional in solution because it is only formed upon packing of the tetramers in the crystals.

Carbohydrate-binding specificity

Except for a few two-domain (super)lectins all MMBL studied hitherto exhibit an exclusive specificity towards mannose. However, since the mannose concentrations required for an efficient inhibition are high the MMBL apparently have a low affinity for the monosaccharide. In general oligomannosides are much more potent inhibitors of the MMBL, which suggests that their carbohydrate-binding sites accommodate preferentially oligomannosyl-residues. It should also be emphasized that the MMBL definitely differ from each other for what concerns their fine specificity. For example, according to detailed hapten inhibition assays, GNA has the highest affinity for terminal Manα(1,3)Man, whereas the daffodil lectin and *Listera ovata* agglutinin preferentially bind Manα(1,6)Man and Manα(1,3)Manα(1,3)Man, respectively (100-102).

There are also a few MMBL, which are not inhibited by mannose. For example, the two-domain tulip lectin TxLC-I, and the two-domain lectins from *Scilla campanulata* and *Arum maculatum* can not be inhibited by any simple sugar. The apparent insensitivity of TxLC-I to mannose is due to the fact that the protomers consist of an *N*-terminal mannose-binding domain and a GalNAc-specific *C*-terminal domain. As result, the lectin is not inhibited by mannose or GalNAc but only by a combination of both monosaccharides.

7. The type 2 ribosome-inactivating proteins (type 2 RIP) and related lectins

Ribosome-inactivating proteins (RIP) are a class of enzymes that catalytically inactivate eukaryotic ribosomes (5). This activity is based on the enzymatic removal of a specific adenine residue from a highly conserved loop (A4324 of rat rRNA) in the large subunit of the ribosomal RNA. Due a conformational change in the affected loop elongation factor EF2 can no longer bind to the ribosome and protein synthesis is arrested. RIP consisting of a single catalytically active polypeptide of about 30 kDa are called type 1 RIP to distinguish them from type 2 RIP, which are chimeric proteins built up of an enzymatically active A chain covalently linked to a B chain with lectin activity. Most type 2 RIP are capable of penetrating eukaryotic cells and subsequently inactivating the ribosomes. However, type 2 RIP markedly differ in cytotoxicity ranging from virtually non-toxic to extremely toxic.

History

The first documented description of a plant lectin dates back to 1888 when Stillmark (103) isolated 'ricin' from castor bean seeds and attributed the toxicity of the protein to its agglutinating activity. In the following decades similar proteins were isolated from seeds of *Abrus precatorius* and *Croton tiglium* (called abrin and crotin, respectively), and from roots of *Adenia digitata* (modeccin). Type 2 RIP received a renewed interest after the discovery that these toxins were more toxic to Ehrlich ascites cells than to normal cells (104). The intensified research led to the elucidation of the so-called two-chain structure of the type 2

RIP and the discovery of their potent inhibitory activity on protein synthesis, the mechanism of which was elucidated after discovery of the RNA *N*-glycosidase activity of the A chain of the toxins (105).

Research on type 2 RIP was for a long time focussed on the potent toxins ricin and abrin. This explains why ricin was the first type 2 RIP to be sequenced and cloned. The first 3-D structure of a type 2 RIP was determined for ricin.

Occurrence

In spite of the very long history of type 2 RIP only a relatively small number of these lectins have been isolated and characterized. Hitherto, genuine type 2 RIP have been found in the families Cucurbitaceae (e.g. *Momordica charantia*), Euphorbiaceae (e.g. *Ricinus communis*, *Croton* sp.), Iridaceae (e.g. *Iris* sp.), Lauraceae (e.g. *Cinnamonum camphora*), Leguminoseae (e.g. *Abrus precatorius*), Passifloraceae (e.g. *Adenia* sp.), Ranunculaceae (e.g. *Eranthis hyemalis*), Sambucaceae (*Sambucus* sp.), and Viscaceae (e.g. *Viscum album*). Type 2 RIP occur in seeds and vegetative tissues.

Molecular structure

The family of type 2 RIP/lectins is somewhat heterogeneous for what concerns the molecular structure of the native proteins because it comprises besides the genuine chimeric type 2 RIP, also hololectins and merolectins (Figure 7).

All type 2 RIP characterized hitherto consist of very similar protomers. These protomers are typical chimeric proteins composed of two functionally and structurally different A and B chains harboring an enzymatic and a carbohydrate-binding activity, respectively. The *N*-terminal A chain (25-30 kDa), which is evolutionary related to the single chain type 1 RIP possesses *N*-glycosidase activity whereas the *C*-terminal B chain (30-35 kDa) contains at least two distinct carbohydrate-binding sites. It should be emphasized here that the A and B chains are derived from a single precursor of about 60-65 kDa, from which a linker is post-translationally excised between the *N*- and the *C*-terminal domains. Both chains remain together after processing through a disulfide bond between cysteine residues located at the *C*-terminus of the A domain and the *N*-terminus of the B chain. The mature protomers (further referred to as [A-s-s-B]-pairs) possess enzymatic and carbohydrate-binding activity. All type 2 RIP are glycoproteins that contain two to four glycan chains.

Native type 2 RIP consist of one, two or four identical [A-s-s-B]-pairs (or protomers). Accordingly, they are considered as monomeric, dimeric and tetrameric proteins, respectively. Examples of monomeric type 2 RIP are the classical toxins ricin, abrin, modeccin and volkensin, and the cytotoxic lectins from *Eranthis hyemalis* and *Iris* sp. ML-II and ML-III from mistletoe are also monomeric type 2 RIP, whereas ML-I forms dimers at concentrations above 20 μg/ml. The agglutinins from *Abrus precatorius* and *Ricinus communis* are dimeric type 2 RIP. Dimeric type 2 RIP have also been found in *Momordica charantia* and *Sambucus* sp. Hitherto, the Neu5Acα(2,6)Gal/GalNAc-binding type 2 RIP from elderberry bark (SNA-I) is the only documented example of a tetrameric type 2 RIP (106). In this lectin, the four protomers are pairwise linked through an intermolecular disulfide bridge between the B chains of two adjacent protomers.

Hololectins consisting of subunits homologous to the B chain of type 2 RIP have exclusively been found in *Sambucus* sp. Two types of hololectins occur. First, an alternative processing of a genuine type 2 precursor whereby the first 8 residues (including the cysteine residue involved in the disulfide bond linking the A and B chains) are removed yields a slightly truncated free B-chain. Association of two such truncated B-chains results in the homodimeric hololectin *Sambucus nigra* agglutinin II (SNA-II) (107). Second, elderberry expresses a fruit-specific type 2 RIP gene which has a major deletion in the A chain. Post-translational processing of the precursor yields a protein that is highly similar to the B-chain of the genuine elderberry type 2 RIP. Association of the mature protomers derived from the truncated type 2 RIP gene into dimers gives a homodimeric hololectin called *Sambucus nigra* agglutinin IV (SNA-IV) (108). Both SNA-II and SNA-IV are glycoproteins containing two *N*-linked glycans.

Fruits and bark of elderberry contain also a small fetuin-binding protein consisting of a single 22 kDa polypeptide chain, which corresponds to the *C*-terminal part of the B chain of the NeuAcα(2,6)Gal/GalNAc-binding type 2 RIP from fruits (SNA-If) and bark (SNA-I), respectively. Since these highly truncated SNA-If and SNA-I (called TrSNA-If and TrSNA-I, respectively) possess only one carbohydrate-binding site per molecule they are merolectins (109).

Three-dimensional structure

Most of information on the 3-D structures of type 2 RIP was obtained by X-ray crystallographic analysis of ricin (110-112). Since the A chain is not relevant for the lectin activity of type 2 RIP, the structure of the *N*-glycosidase domain is only briefly discussed here. Three distinct domains are distinguished in the ricin A chain, which contains eight α-

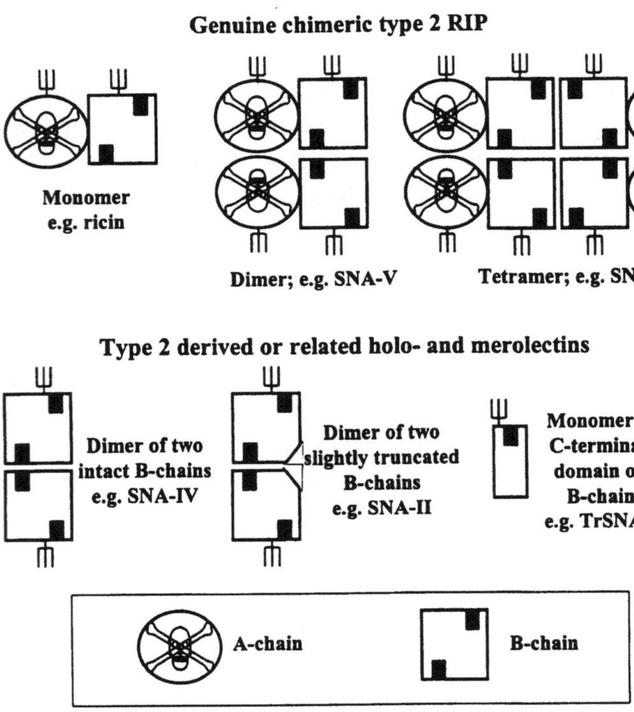

Fig.7. Molecular structure of type 2 RIP and related lectins. Abbreviations refer to the following lectins: SNA-I, -II, -IV and -V = *Sambucus nigra* (elderberry) agglutinins I, II, IV and V, respectively; TrSNA-I = truncated B chain of SNA-I.

helices and six strands of β-sheet. The catalytically active site consists of residues Tyr^{80}, Tyr^{123}, Glu^{177}, Arg^{180} and Trp^{211}. Several other residues located in the vicinity of this site (namely Asn^{78}, Arg^{134}, Gln^{173}, Ala^{178}, Glu^{208} and Asn^{209}) are probably required to maintain its catalytic conformation (111, 113-114).

The ricin B chain has a rather complex structure characterized by the occurrence of internal repeats. First, each B chain consists of two similar tandemly arrayed domains called domain 1 and domain 2. Second, each domain consists of four subdomains. These subdomains are called 1λ, 1α, 1β and 1γ, and 2λ, 2α, 2β and 2γ for domain 1 and 2, respectively. Interestingly, subdomains α, β and γ share a high residual similarity. Apart from a few short strands of β-sheet the B chain of ricin is devoid of extended regular secondary structures like α-helices or β-sheets and consists mainly of coil structures linked by turns and loops. The folding of the B chain is stabilized by four disulfide bridges. Domain 1 and 2 have a very similar overall folding characterized by a typical β-trefoil structure (115).

X-ray crystallographic studies of abrin and the mistletoe lectin ML-I confirmed that these type 2 RIP have the same overall three-dimensional conformation as ricin (116). In addition, molecular modelling indicated that also the B chain of type 2 RIP and related hololectins from elderberry (*Sambucus nigra*) have the same overall structure and conformation as the B chain of ricin.

The resolution of the structure of ricin yielded also detailed information about the 3-D structure of the sugar-binding sites. Ricin harbors at least two carbohydrate-binding sites, which are located at the *N*-terminal subdomain 1α and the *C*-terminal subdomain 2γ. Both sites comprise five critical amino acid residues (Asp^{22}, Gln^{35}, Trp^{37}, Asn^{46}, Gln^{47} for site 1, and Asp^{234}, Ile^{246}, Tyr^{248}, Asn^{255} and Gln^{256} for site 2). Both sites bind the Gal residue of lactose through a network of four (domain 1) or three (domain 2) hydrogen bonds. However, the binding site of domain 1 is a low affinity site whereas that of domain 2 is a high affinity site for lactose (117-118). Hydrophobic interactions between the pyranose ring of Gal and the aromatic (Trp^{37}, Tyr^{248}) or hydrophobic (Ile^{246}) residues of the binding sites complete the binding of lactose. Site 2 of ricin binds exclusively GalNAc. This specificity apparently results from a rotation (of about 15°) of GalNAc inside the site, so that there is no steric clash between the *N*-acetyl group of this sugar and the side chain of the Asp^{250} (112). Besides site 1 and 2, the B-chain of ricin possibly contains a third carbohydrate-binding (119-121).

Carbohydrate-binding specificity

Most type 2 RIP are considered Gal-, GalNAc- or Gal/GalNAc-specific lectins. However, several type 2 RIP have a clear preference for di- or oligosaccharides. In addition, a few type 2 RIP are neither Gal- nor GalNAc-specific but react only with NeuAcα(2,6)-Gal/GalNAc-containing glycans (107). Hololectins and merolectins derived from type 2 RIP exhibit the same specificity as the parent type 2 RIP.

Type 2 RIP and the subunits of the related hololectins usually possess two (or possibly even three) carbohydrate-binding sites with a (slightly) different specificity and/or affinity. This simultaneous occurrence of different binding sites render the results of specificity studies less clear-cut because the overall specificity of the type 2 RIP is predominantly determined by the specificity of the binding site with the highest affinity for the sugar. As a result, the specificity of the low-affinity site can not be determined properly.

SUMMARY

The majority of plant lectins can be classified in seven families of structurally and evolutionary related proteins. Within a given lectin family most but not necessarily all members are built up of protomers with a similar primary structure and overall 3-D fold. The overall structure of the native lectins is not only determined by the structure of the protomers but depends also on the degree of oligomerization and in some cases on the post-translational processing of the lectin precursors.

In general, lectin families are fairly homogeneous for what concerns the overall specificity of the individual lectins, which illustrates that the 3-D structure of the binding site has been conserved during evolution. In the case of the jacalin-related lectins the occurrence of a mannose- and galactose-binding subfamily can be explained by the fact that a post-translational cleavage of the protomers (of the galactose-binding subfamily) yields a slightly altered binding site. Unlike the other families, the legume lectins display a wide range of specificites, which is clearly reflected in the occurrence of sugar-binding sites with a different 3-D structure.

REFERENCES

1. Van Damme, E.J.M., Peumans, W.J., Pusztai, A., and Bardocz, S. (1998) Handbook of Plant Lectins: Properties and Biomedical Applications. John Wiley & Sons, Chichester, UK.
2. Van Damme, E.J.M., Peumans, W.J., Barre, A., and Rougé, P. (1998) Plant lectins: a composite of several distinct families of structurally and evolutionary related proteins with diverse biological roles. Crit. Rev. Plt. Sci., 17: 575-692.
3. Peumans, W.J., and Van Damme, E.J.M. (1995) The role of lectins in plant defense. Histochem. J., 27: 253-271.
4. Van Damme, E.J.M., Barre, A., Rougé, P., Van Leuven, F., and Peumans, W.J. (1996) Isolation and molecular cloning of a novel type 2 ribosome-inactivating protein with an inactive B chain from elderberry (*Sambucus nigra*) bark. J. Biol. Chem., 272: 8353-8360.
5. Barbieri, L., Batelli, G.B., and Stirpe, F. (1993) Ribosome-inactivating proteins from plants. Biochim. Biophys. Acta, 1154: 237-282.
6. Collinge, D.B., Kragh, K.M., Mikkelsen, J.D., Nielsen, K.K., Rasmussen, U., and Vad, K. (1993) Plant chitinases. Plant J., 3: 31-40.
7. Bird, G.W.G. (1954) Observations on the interactions of the erythrocytes of various species with certain seed agglutinins. Br. J. Exp. Pathol., 35: 252.
8. Koeppe S.J., and Rupnov, J.H. (1988) Purification and characterization of a lectin from the seeds of amaranth (*Amaranthus cruentus*). J. Food Sc., 53: 1412-1417.
9. Zenteno, E., and Ochoa, J.-L. (1988) Purification of a lectin from *Amaranthus leucocarpus* by affinity chromatography. Phytochem., 27: 313-317.
10. Transue, T.R., Smith, A.K., Mo, H., Goldstein, I.J., and Saper, M.A. (1997) Structure of benzyl T-antigen disaccharide bound to *Amaranthus caudatus* lectin. Nature Struct. Biol., 10: 779-783.
11. Rinderle, S.J., Goldstein, I.J., Matta, K.L., and Ratcliffe, R.M. (1989) Isolation and characterization of amaranthin, a lectin present in the seeds of *Amaranthus caudatus*, that recognizes the T- (or cryptic T)-antigen. J. Biol. Chem., 264: 16123-16131.
12. Raina, A., and Datta, A. (1992) Molecular cloning of a gene encoding a seed-specific protein

with nutritionally balanced amino acid composition from *Amaranthus*. Proc. Natl. Acad. Sci. USA, 89: 11774-11778.

13. Hossaini, A. (1968) Hemolytic and hemagglutinating activities of 222 plants. Vox Sang., 15: 410-417.
14. Sabnis, D.D., and Hart, J.W. (1978) The isolation and some properties of a lectin (haemagglutinin) from *Cucurbita* phloem exudate. Planta, 142: 97-101.
15. Allen, A.K. (1979) A lectin from the exudate of the fruit of the vegetable marrow (*Cucurbita pepo*) that has a specificity for β-1,4-linked *N*-acetylglucosamine oligosaccharides. Biochem. J., 183: 133-137.
16. Read, S.M., and Northcote, D.H. (1983) Subunit structure and interactions of the phloem proteins of *Cucurbita maxima* (pumpkin). Eur. J. Biochem., 134: 561-569.
17. Bostwick, D.E., Dannehofer, J.M., Skaggs, M.I., Lister, R.M., Larkins, B.A., and Thompson, G.A. (1992) Pumpkin phloem lectin genes are specifically expressed in companion cells. Plant Cell, 4: 1539-1548.
18. Anantharam V., Patanjali, S.R., Swamy, M.J., Sanadi, A.R., Goldstein, I.J., and Surolia, A. (1986) Isolation, macromolecular properties, and combining site of a chito-oligosaccharide-specific lectin from the exudate of ridge gourd (*Luffa acutangula*). J. Biol. Chem., 261: 14621-14627.
19. Wang, M.-B., Boulter, D., and Gatehouse, J.A. (1994) Characterization and sequencing of cDNA clone encoding the phloem protein PP2 of *Cucurbita pepo*. Plant Mol. Biol., 24: 159-170.
20. Waljuno, K., Scholma, R. A., Beintema, J., Mariono, A., and Hahn, A. M. (1975) Amino acid sequence of hevein. Proc. Int. Rubber Conf., Kuala Lumpur, 2: 518-531.
21. Kocourek, J. (1986) Historical background. *In* The Lectins, Properties, Functions, and Applications in Biology and Medicine (Liener, I.E., Sharon, N. and Goldstein, I.J., eds.), pp 1-32 Academic Press, Orlando, USA.
22. Aub, J.C., Tieslau, C., and Lankester, A. (1963) Reactions of normal and tumor cell surfaces to enzymes. I. Wheat-germ lipase and associated mucopolysaccharides. Proc. Natl. Acad. Sci. USA, 50: 613-619.
23. Burger, M.M., and Goldberg, A.R. (1967) Identification of a tumor-specific determinant on neoplastic cell surfaces. Proc. Natl. Acad. Sci. USA, 57: 359-366.
24. Nagata, Y., and Burger, M.M. (1972) Wheat germ agglutinin. Isolation and crystallization. J. Biol. Chem., 247: 2248-2250.
25. LeVine, D., Kaplan, M.J., and Greenaway, P.J. (1972) The purification and characterization of wheat-germ agglutinin. Biochem. J., 129: 847-856.
26. Wright, C.S. (1977) The crystal structure of wheat germ agglutinin at 2.2 Å resolution. J. Mol. Biol., 111: 439-457.
27. Raikhel, N.V., and Wilkins, T.A. (1987) Isolation and characterization of a cDNA clone encoding wheat germ agglutinin. Proc. Natl. Acad. Sci. USA, 84: 6745-6749.
28. Van Parijs, J., Broekaert, W.F., Goldstein, I.J., and Peumans, W.J. (1991) Hevein: an antifungal protein from rubber-tree (*Hevea brasiliensis*) latex. Planta, 183: 258-262.
29. Linthorst, H.J.M. (1991) Pathogenesis-related proteins of plants. Crit. Rev. Plant Sci., 10: 123-150.
30. Broekaert, W.F., Marien, W., Terras, F.R.G., De Bolle, M.F.C., Proost, P., Van Damme, J., Dillen, L., Claeys, M., Rees, S.B., Vanderleyden, J., and Cammue, B.P.A. (1992) Antimicrobial peptides from *Amaranthus caudatus* seeds with sequence homology to the cysteine/glycine-rich domain of chitin-binding proteins. Biochemistry, 31: 4308-4314.
31. Beintema, J.J., and Peumans, W.J. (1992) The primary structure of stinging nettle (*Urtica dioica*) agglutinin: a two-domain member of the hevein family. FEBS Lett., 299: 131-134.
32. Peumans, W.J., Verhaert, P., Pfüller, U., and Van Damme, E.J.M. (1996) Isolation and partial characterization of a small chitin-binding lectin from mistletoe (*Viscum album*). FEBS Lett., 396: 261-265.

33. Yamaguchi, K.I., Mori, A., and Funatsu, G. (1996) Amino acid sequence and some properties of lectin-D from the roots of pokeweed (*Phytolacca americana*). Biosci. Biotechnol. Biochem., 60: 1380-1382.
34. Yamaguchi, K.I., Mori, A., and Funatsu, G. (1995) The complete amino acid sequence of lectin-C from the roots of pokeweed (*Phytolacca americana*). Biosci. Biotechnol. Biochem., 59: 1384-1385.
35. Yamaguchi, K.I., Yurino, N., Kino, M., Ishiguro, M., and Funatsu, G. (1997) The amino acid sequence of mitogenic lectin-B from the roots of pokeweed (*Phytolacca americana*). Biosci. Biotechnol. Biochem., 61: 690-698.
36. Kieliszewski, M.J., Showalter, A.M., and Leykam, J.F. (1994) Potato lectin: a modular protein sharing sequence similarities with the extensin family, the hevein lectin family, and snake venom disintegrins (platelet aggregation inhibitors). Plant J., 5: 849-861.
37. Allen, A.K., Bolwell, G.P., Brown, D.S., Sidebottom, C., and Slabas, A.R. (1996) Potato lectin: a three-domain glycoprotein with novel hydroxyproline-containing sequences and sequence similarities to wheat-germ agglutinin. Int. J. Biochem. Cell Biol., 28: 1285-1291.
38. Beintema, J.J. (1994) Structural features of plant chitinases and chitin-binding proteins. FEBS Lett., 350: 159-163.
39. Andersen, N.H., Cao, B., Rodriguez-Romero, A., and Arreguin, B. (1993) Hevein: NMR assignment and assessment of solution-state folding for the agglutinin-toxin motif. Biochemistry, 32: 1407-1422.
40. Asensio, J.L., Cañada, F.J., Bruix, M., Rodriguez-Romero, A., and Jimenez-Barbero, J. (1995) The interaction of hevein with *N*-acetylglucosamine-containing oligosaccharides. Solution structure of hevein complexed to chitobiose. Eur. J. Biochem., 230: 621-633.
41. Martins, J.C., Maes, D., Loris, R., Pepermans, H.A.M., Wyns, L., Willem, R., and Verheyden, P. (1996) ^{1}H NMR study of the solution structure of Ac-AMP2, a sugar binding antimicrobial protein isolated from *Amaranthus caudatus*. J. Mol. Biol., 258: 322-333.
42. Goldstein, I.J., Hughes, R.C., Monsigny, M., Osawa, T., and Sharon, N. (1980) What should be called a lectin? Nature, 285: 66.
43. Jones, J. M., Cawley, L.P., and Teresa, G.W. (1967) Hemagglutinins (lectins) extracted from *Maclura pomifera.* Vox Sang., 12: 211-214.
44. Bausch, J.N., and Poretz, R.D. (1977) Purification and properties of the hemagglutinin from *Maclura pomifera* seeds. Biochemistry, 16: 5790-5794.
45. Moreira, R.A., and Ainouz, I.L. (1981) Lectins from seeds of jack fruit (*Artocarpus integrifolia* L.): Isolation and purification of two isolectins from the albumin fraction. Bio. Plant (Praha), 23: 186-192.
46. Sarkar, M., Wu, A.M., and Kabat, E.A. (1981) Immunochemical studies on the carbohydrate specificity of *Maclura pomifera* lectin. Arch. Biochem. Biophys., 209: 204-218.
47. Sastry, M.V.K., Banerjee, P., Patanjali, S.R., Swamy, M.J., Swarnalatha, G.V., and Surolia, A. (1986) Analysis of the saccharide binding to *Artocarpus integrifolia* lectin reveals specific recognition of T-antigen (β-D-Gal(1,3)D-GalNAc). J. Biol. Chem., 261: 11726-11733.
48. Roque-Bareira, M.C., and Campos-Neto, A. (1985) Jacalin: an IgA-binding lectin. J. Immunol., 134: 1740-1743.
49. Kabir, S., and Daar, A.S. (1994) The composition and properties of jacalin, a lectin of diverse applications obtained from the jackfruit (*Artocarpus heterophyllus*)seeds. Immunol. Invest., 23: 167-188.
50. Czapla, T.H., and Lang, B.A. (1990) Effect of plant lectins on the larval development of European corn borer (Lepidoptera: Pyralidae) and southern corn rootworm (Coleoptera:Chrysomelidae). J. Econ. Entomol., 83: 2480-2485.
51. Young, N.M., Johnston, R.A.Z., and Watson, D.C. (1991) The amino acid sequences of jacalin and the *Maclura pomifera* agglutinin. FEBS Lett., 282: 382-384.

52. Yang, H., and Czapla, T.H. (1993) Isolation and characterization of cDNA clones encoding jacalin isolectins. J. Biol. Chem., 268: 5905-5910.
53. Sankaranarayanan, R., Sekar, K., Banerjee, R., Sharma, V., Surolia, A., and Vijayan, M. (1996) A novel mode of carbohydrate recognition in jacalin, a Moraceae plant lectin with a β-prism fold. Nature Struct. Biol., 3: 596-603.
54. Van Damme, E.J.M., Barre, A., Verhaert, P., Rougé, P., and Peumans, W.J. (1996) Molecular cloning of the mitogenic mannose/maltose-specific rhizome lectin from *Calystegia sepium*. FEBS Lett., 397: 352-356.
55. Van Damme, E.J.M., Barre, A., Mazard, A.-M., Verhaert, P., Horman, A., Debray, H., Rougé, P., and Peumans, W.J., (1999) Characterization and molecular cloning of the lectin from *Helianthus tuberosus*. Eur. J. Biochem., 199: 135-142.
56. Rosa, J.C., De Oliveira, P.S.L., Garratt, R., Beltramini, L., Resing, K., Roque-Barreira, M.-C., and Greene, L.J. (1999) KM+, a mannose-binding lectin from *Artocarpus integrifolia*: amino acid sequence, predicted tertiary structure, carbohydrate recognition, and analysis of the β-prism fold. Protein Sci., 8: 13-24
57. Peumans, W.J., Winter, H.C., Bemer, V., Van Leuven, F., Goldstein, I.J., Truffa-Bachi, P., and, Van Damme, E.J.M. (1997) Isolation of a novel plant lectin with an unusual specificity from *Calystegia sepium*. Glycoconjugate J., 14: 259-265.
58. Landsteiner, K., and Raubitschek, H. (1907) Beobachtungen über Hämolyse und Hämagglutination. Zentralbl. Bakteriol. Parasitenk. Infektionskr. Hyg. Abt., 1: Orig., 45: 660-667.
59. Summer, J.B., and Howell, S.F. (1936) The identification of the hemagglutinin of the Jack bean with concanavalin A. J. Bacteriol., 32: 227-237.
60. Renkonen, K.O. (1948) Studies on hemagglutinins present in seeds of some representatives of Leguminoseae. Ann. Med. Exp. Biol. Fenn., 26: 66-72.
61. Boyd, W.C., and Reguera, R.M. (1949) Studies on haemagglutinins present in seeds of some representatives of the family Leguminoseae. J. Immunol., 62: 333-339.
62. Nowell, P.C. (1960) Phytohemagglutinin: an initiator of mitosis in cultures of normal human leukocytes. Cancer Res., 20: 462-466.
63. Edelman, G.M., Cunningham, B.A., Reeke, G.N.Jr., Becker, J.W., Waxdal, M.J., and Wang, J.L. (1972) The covalent and three-dimensional structure of concanavalin A. Proc. Natl. Acad. Sci. U.S.A., 69: 2580-2584.
64. Hardman, K.D., and Ainsworth, C.F. (1972) Structure of concanavalin A at 2.4 Å resolution. Biochemistry, 11: 4910-4919.
65. Vodkin, L.O., Rhodes, P.R., and Goldberg, R.B. (1983) A lectin gene insertion has the structural features of a transposable element. Cell, 34: 1023-1031.
66. Sturm, A., and Chrispeels, M.J. (1986) The high mannose oligosaccharide of phytohemagglutinin is attached to Asn^{12} and the modified oligosaccharide to Asn^{60}. Plant Physiol., 81: 320-322.
67. Sturm, A., Bergwerff, A.A., and Vliegenthart, F.G. (1992) ^{1}H-NMR structural determination of the *N*-linked carbohydrate chains on glycopeptides obtained from the bean lectin phytohemagglutinin. Eur. J. Biochem., 204: 313-316.
68. Feldsted, R.L., Egorin, M.J., Leavitt, R.D., and Bachur, N.R. (1977) Recombinations of subunits of *Phaseolus vulgaris* isolectins. J. Biol. Chem., 252: 2967-2971.
69. Van Driessche, E. (1988) Structure and function of Leguminosae lectins. *In* Advances in Lectin Research. Vol. 1, pp. 73-134, Franz, H., Ed., VEB Verlag Volk und Gesundheit, Berlin, Germany.
70. Carrington, D.M., Auffret, A., and Hanke, D.A. (1985) Polypeptide ligation occurs during post-translational modification of concanavalin A. Nature, 313: 64-67.
71. Bowles, D.J., Marcus, S.E., Pappin, D.J.C., Findlay, J.B.C., Eliopoulos, E., Maycox, P. R., and Burgess, J. (1986) Posttranslational processing of concanavalin A precursors in jackbean cotyledons. J. Cell Biol., 102: 1284-1297.

72. Agrawal, B.B.L., and Goldstein, I.J. (1968) Protein-carbohydrate interaction. VI. Isolation of concanavalin A by specific adsorption on cross-linked dextran gels. Biochim. Biophys. Acta, 147: 262-271.
73. Shaanan, B., Lis, H., and Sharon, N. (1991) Structure of a legume lectin with an ordered *N*-linked carbohydrate in complex with lactose. Science 254: 862-866.
74. Banerjee, R., Mande, S. C., Ganesh, V., Das, K., Dhanaraj, V., Mahanta, S. K., Suguna, K., Surolia, A., and Vijayan, M. (1994) Crystal structure of peanut lectin, a protein with an unusual quaternary structure. Proc. Natl. Acad. Sci. U.S.A., 91: 227-231.
75. Hamelryck, T.W., Dao, M.-H., Poortmans, F., Chrispeels, M.J., Wyns, L., and Loris, R. (1996) The crystallographic structure of phytohemagglutinin-L. J. Biol. Chem., 271: 20479-20485.
76. Drickamer, K. (1995) Multiplicity of carbohydrate interactions. Nature Struct. Biol., 2: 437-439.
77. Weis, W.I., and Drickamer, K. (1996) Structural basis of lectin-carbohydrate recognition. Annu. Rev. Biochem., 65: 441-473.
78. Debray, H., Decout, D., Strecker, G., Spik, G., and Montreuil, J. (1981) Specificity of twelve lectins towards oligosaccharides and glycopeptides related to *N*-glycosylproteins. Eur. J. Biochem., 117: 41-55.
79. Kornfeld, K., Reitman, M.L., and Kornfeld, R. (1981) The carbohydrate-binding specificity of pea and lentil lectins. Fucose is an important determinant. J. Biol. Chem., 256: 6633-6640.
80. Young, N.M., and Oomen, R.P. (1992) Analysis of sequence variation among legume lectins. A ring of hypervariable residues forms the perimeter of the carbohydrate-binding site. J. Mol. Biol., 228: 924-934.
81. Sharma, V., and Surolia, A. (1997) Analyses of carbohydrate recognition by legume lectins: size of the combining site loops and their primary specificity. J. Mol. Biol., 267: 433-445.
82. Nachbar, M.S., and Oppenheim, J.D. (1980) Lectins in the United States diet: a survey of lectins in commonly consumed foods and a review of the literature. Am. J. Clin. Nutr., 33: 2338-2345.
83. Van Damme, E.J.M., Allen, A.K., and Peumans, W.J. (1987) Isolation and characterization of a lectin with exclusive specificity towards mannose from snowdrop (*Galanthus nivalis*) bulbs. FEBS Lett., 215: 140-144.
84. Van Damme, E.J.M., Kaku, H., Perini, F., Goldstein, I.J., Peeters, B., Yagi, F., Decock, B., and Peumans, W.J. (1991) Biosynthesis, primary structure and molecular cloning of snowdrop (*Galanthus nivalis* L.) lectin. Eur. J. Biochem., 202: 23-30.
85. Van Damme, E.J.M., Allen, A.K., and Peumans, W.J. (1988) Related mannose-specific lectins from different species of the family Amaryllidaceae. Physiol. Plant., 73: 52-57.
86. Van Damme, E.J.M., Goldstein, I.J., and Peumans, W.J. (1991) A comparative study of related mannose-binding lectins from the Amaryllidaceae and Alliaceae. Phytochem., 30: 509-514.
87. Van Damme, E.J.M., Smeets, K., Balzarini, J., Pusztai, A., Van Leuven, F., Goldstein, I. J., and Peumans, W.J. (1993) Cloning and characterization of the lectin cDNA clones from onion, shallot and leek. Plant Mol. Biol., 23: 365-376.
88. Van Damme, E.J.M., Balzarini, J., Smeets, K., Van Leuven, F., and Peumans, W.J. (1994) The monomeric and dimeric mannose-binding proteins from the Orchidaceae species *Listera ovata* and *Epipactis helleborine*: sequence homologies and differences in biological activities. Glycoconjugate J., 11: 321-332.
89. Van Damme, E.J.M., Smeets, K., and Peumans, W.J. (1995) The mannose-binding monocot lectins and their genes. *In* Lectins: Biomedical Perspectives (Pusztai, A. and Bardocz, S., eds.), pp. 59-80. Taylor and Francis, London, UK.
90. Van Damme, E.J.M., Goossens, K., Smeets, K., Van Leuven, F., Verhaert, P., and Peumans, W.J. (1995) The major tuber storage protein of Araceae species is a lectin: Characterization

and molecular cloning of the lectin from *Arum maculatum* L. Plant Physiol., 107: 1147-1158.

91. Yagi, F., Hidaka, M., Minami, Y., and Tadera, K. (1996) A lectin from leaves of *Neoregelia flandria* recognizes D-glucose, D-mannose and *N*-acetylglucosamine, differing from the mannose-specific lectins from other monocotyledonous species. Plant Cell Physiol., 37: 1007-1012.
92. Xu, Q., Liu, Y., Wang, X., Gu, H., and Chen, Z. (1998) Purification and characterization of a novel anti-fungal protein from *Gastrodia elata*. Plant Physiol. Biochem., 36: 899-905.
93. Van Damme, E.J.M., Smeets, K., Torrekens S., Van Leuven, F., and Peumans, W.J. (1993) The mannose-specific lectins from ramsons (*Allium ursinum* L.) are encoded by three sets of genes. Eur. J. Biochem., 217: 123-129.
94. Van Damme, E.J.M., Smeets, K., Torrekens, S., Van Leuven, F., Goldstein, I.J., and Peumans, W.J. (1992) The closely related homomeric and heterodimeric mannose-binding lectins from garlic are encoded by one-domain and two-domain lectin genes, respectively. Eur. J. Biochem., 206: 413-420.
95. Van Damme, E.J.M., Briké, F., Winter, H. C., Van Leuven, F., Goldstein, I.J., and Peumans, W.J. (1996) Molecular cloning of two different mannose-binding lectins from tulip bulbs. Eur. J. Biochem., 236: 419-427.
96. Wright, C.S., Kaku, H., and Goldstein, I.J. (1990) Crystallization and preliminary X-ray diffraction results of snowdrop (*Galanthus nivalis*) lectin. J. Biol. Chem., 265: 1676-1677.
97. Hester, G., Kaku, H., Goldstein, I.J., and Wright, C.S. (1995) Structure of mannose-specific snowdrop (*Galanthus nivalis*) lectin is representative of a new plant lectin family. Nature Struct. Biol., 2: 472-479.
98. Barre, A., Van Damme, E.J.M., Peumans, W.J., and Rougé, P. (1996) Structure-function relationship of monocot mannose-binding lectins. Plant Physiol., 112: 1531-1540.
99. Wright, C.S., and Hester, G. (1996) The 2.0 Å structure of a cross-linked complex between snowdrop lectin and a branched mannopentaose: evidence for two unique binding modes. Structure, 11: 1339-1352.
100. Shibuya, N., Goldstein, I.J., Van Damme, E.J.M., and Peumans, W.J. (1988) Binding properties of a mannose-specific lectin from the snowdrop (*Galanthus nivalis*) bulb. J. Biol. Chem., 263: 728-734.
101. Kaku, H., Van Damme, E.J.M., Peumans, W.J., and Goldstein, I.J. (1990) Carbohydrate-binding specificity of the daffodil (*Narcissus pseudonarcissus*) and amaryllis (*Hippeastrum* hybr.) bulb lectins. Arch. Biochem. Biophys., 279: 298-304.
102. Saito, K., Komae, A., Kakuta, M., Van Damme, E.J.M., Peumans, W.J., Goldstein, I.J., and Misaki, A. (1993) The alpha-mannosyl-binding lectin from leaves of the orchid twayblade (*Listera ovata*). Application to separation of α-D-mannans from α-D-glucans. Eur. J. Biochem., 217: 677-681.
103. Stillmark, H. (1888) Über Ricin ein giftiges Ferment aus den Samen von *Ricinus communis* L. und einige anderen Euphorbiaceen. Inaugural Dissertation Dorpat, Tartu.
104. Lin, J.-Y., Tserng, K.-Y., Chen, C.-C., Lin, L.-T., and Tung, T.-C. (1970) Abrin and ricin: new anti-tumour substances. Nature, 227: 292-293.
105. Endo, Y., Mitsui, K., Motizuki, M., and Tsurugi, K. (1987) The mechanism of action of ricin and related toxic lectins on eukaryotic ribosomes. The site and the characteristics of the modification in 28S ribosomal RNA caused by the toxins. J. Biol. Chem., 262: 5908-5912.
106. Van Damme, E.J.M., Barre, A., Rougé, P., Van Leuven, F., and Peumans, W.J. (1996) The NeuAc (α-2,6)-Gal/GalNAc binding lectin from elderberry (*Sambucus nigra*) bark, a type 2 ribosome inactivating protein with an unusual specificity and structure. Eur. J. Biochem., 235: 128-137.
107. Van Damme, E.J.M., Barre, A., Rougé, P., Van Leuven, F., and Peumans, W.J. (1996) Characterization and molecular cloning of SNAV (nigrin b), a GalNAc-specific type 2 ribosome-inactivating protein from the bark of elderberry (*Sambucus nigra*). Eur. J. Biochem., 237: 505-513.

108. Van Damme, E.J.M., Roy, S., Barre, A., Rougé, P., Van Leuven, F., and Peumans, W.J., (1997) The major elderberry (*Sambucus nigra*) fruit protein is a lectin derived from a truncated type 2 ribosome-inactivating protein. Plant J., 12: 1251-1260.
109. Peumans, W.J., Roy, S., Barre, A., Rougé, P., Van Leuven, F., and Van Damme, E.J.M., (1998) Elderberry (*Sambucus nigra*) contains truncated Neu5Ac(α-2,6)Gal/GalNAc-binding type 2 ribosome-inactivating proteins. FEBS Lett., 425: 35-39.
110. Rutenber, E., Katzin, B.J., Collins, E.J., Mlsna, D., Ready, M.P., and Robertus, J.D. (1991) Crystallographic refinement of Ricin to 2.5 Å. Proteins Struct. Funct. Genet., 10: 240-250.
111. Katzin, B.J., Collins, E.J., and Robertus, J.D. (1991) Structure of Ricin A-chain at 2.5 Å. Proteins Struct. Funct. Genet., 10: 251-259.
112. Rutenber, E., and Robertus, J.D. (1991) Structure of Ricin B-chain at 2.5 Å resolution. Proteins Struct. Funct. Genet., 10: 260-269.
113. Kim, Y., and Robertus, J.D. (1992) Analysis of several key active site residues of ricin A chain by mutagenesis and X-ray crystallography. Protein Engin., 5: 775-779.
114. Chaddock, J.A., and Roberts, L.M. (1993) Mutagenesis and kinetic analysis of the active site Glu177 of ricin A-chain. Protein Engin., 6: 425-431.
115. Murzin, A.G., Lesk, A.M., and Chothia, C. (1992) β-trefoil fold. Patterns of structure and sequence in the Kunitz inhibitors interleukins-1β and 1α and fibroblast growth factors. J. Mol. Biol., 223: 531-543.
116. Tahirov, T.H., Lu. T.H., Liaw, Y.C., Chen, Y.L., and Lin, J.Y. (1995) Crystal structure of abrin-a at 2.14 Å. J. Mol. Biol., 250: 354-367.
117. Yamasaki, N., Hatakeyama, T., and Funatsu, G. (1985) Ricin D-saccharide interaction as studied by ultraviolet difference spectroscopy. J. Biochem., 98: 1555-1560.
118. Hatakeyama, T., Yamasaki, N., and Funatsu, G. (1986) Identification of the tryptophan residue located at the low-affinity saccharide binding site of ricin D. J. Biochem., 100: 781-788.
119. Fu, T., Burbage, C., Tagge, E.P., Brothers, T., Willingham, M.C., and Frankel, A.E. (1996) Ricin toxin contains three lectin sites which contribute to its in vivo toxicity. Int. J. Immunopharmacol., 18: 685-692.
120. Frankel, A.E., Burbage, C., Fu, T., Tagge, E., Chandler, J., and Willingham, M.C. (1996) Ricin toxin contains at least three galactose-binding sites located in B chain subdomains 1 alpha, 1 beta and 2 gamma. Biochemistry, 35: 14749-14756.
121. Venkatesh, Y.P., and Lambert, J.M. (1997) Galactose-induced dimerization of blocked ricin at acidic pH: evidence for a third galactose-binding site in ricin B-chain. Glycobiology, 7: 329-335.
122. Kraulis, P.J. (1991) Molscript: a program to produce both detailed and schematic plots of protein structures. J. Appl. Cryst., 24: 946-950.

EXPRESSION OF BINDING PROPERTIES OF Gal/GalNAc REACTIVE LECTINS BY MAMMALIAN GLYCOTOPES[1,2]

(AN UPDATED REPORT)

Albert M. Wu

Glyco-Immunochemistry Research Laboratory., Institute of Molecular and Cellular Biology, College of Medicine Chang-Gung University, Kwei-san, Tao-yuan, 333, Taiwan

By the early eighties, the carbohydrate specificities of many lectins were grouped by the ability of monosaccharides or their derivatives to inhibit lectin-induced hemagglutination (1-7). But, it was found that lectins of the same apparent monosaccharide affinity were demonstrating different reactivities toward different oligosaccharide chains, and different affinities to animal cells and glycoproteins, which implies that they have their own binding specificity extending beyond the monosaccharide (5-16). At the same time, the concept of the combining sites of many lectins extending to five or more sugars, and many multi-branched oligosaccharides exhibiting a significant increment in their lectin-binding reactivities as compared to their linear counterparts, was accepted. For example, *Datura stramonium* (TAL, thorn apple) shows over a 500- fold higher specificity for a biantennary penta- 2,6 – saccharide, which contains two branches of Galβ1→4GlcNAcβ1→6Man and Galβ1→4GlcNAcβ1→2Man at the nonreducing ends, than for one of its branched trisaccharides (17). Furthermore, some lectins having broad spectra of binding properties possess dual or multiple affinities for various disaccharides. Thus, *Wistaria floribunda* (WFA,18) is reactive for GalNAcα1→3GalNAc (**F**), GalNAcα1→3Gal (**A**), and GalNAcα1→Ser(Thr) (**Tn**) residues: *Abrus precatorius* agglutinin (APA,13), for Galβ1→3GalNAc (**T**), Galβ1→3/4GlcNAc (**I/II**); *Artocarpus integrifolia* (Jacalin, AIL,19), for Galβ1→3GalNAc (**T**) and GalNAcα1→Ser/Thr (**Tn**); *Bauhinia purpure alba*, for (20)

---

[1]This paper is dedicated to Dr. A. Herp, who has assisted my research for over two decades.
[2]This work was supported by Grants from Chang-Gung Medical Research Project (CMRP No. 676), Kwei-san, Tao-yuan, Taiwan and the National Science Council (NSC 87-2316-b-182-005), Taipei, Taiwan.
[3]Abbreviations as shown in Tables I to III.

Galβ1→3GalNAc (**T**), GalNAcα1→Ser(Thr) (**Tn**) and Galβ1→(3)4GlcNAc (**I/II**) residues, and *Datura stramonium* (17), for *N*-acetyllactosamine (Galβ1→4GlcNAc, **II**) and *N*-acetylchitobiose (GlcNAcβ1→4GlcNAc, **C**). In a previous review (5-7), we reported that organizing and grouping binding properties of lectins, based on disaccharide determinants and GalNAcα1→Ser(Thr) sequence rather than monosaccharide inhibition pattern, should facilitate the selection of lectins as structural probes for glycans as well as the interpretation of the distribution and binding properties of the glycotopes on the cell surface. From the information provided by inhibition profiles of lectin-glycan binding (5-7, 11-31), the lectins selected in this report can be classified according to a single or a combination of two or more of the mammalian structural units. They are – 1. **F**, GalNAcα1→3GalNAc; 2. **A**, GalNAcα1→3Gal; 3. **T**, Galβ1→3GalNAc; 4. **I**, Galβ1→3GlcNAc; 5. **II**, Galβ1→4GlcNAc; 6. **B**, Galα1→3Gal; 7. **E**, Galα1→4Gal; 8. **L**, Galβ1→4Glc; 9. **P**, GalNAcβ1→3Gal; 10. **S**, GalNAcβ1→4Gal, and 11. **Tn**, GalNAcα1→Ser (Thr) of the peptide chain. Examples of the binding properties of these lectins can be demonstrated by *Ricinus communis* agglutinin (RCA_I), grouped as **II** specific lectin and its binding property is **II** > **I** > **B** > **T**; *Abrus precatorius* agglutinin (APA), classified as **T** and its carbohydrate specificity is **T** > **I/II** > **E** > **B** > **Tn**; *Artocarpus integrifolia* (jacalin, AIL), as **T/Tn** specific and its binding reactivity is **T** > **Tn** >> **I (II)**, sialyl **T** and sialyl **Tn,** and *Geodia cydonium* (GCL), as **F/A** specific, and with affinity for **F** > $\mathbf{A_h}$ [GalNAcα1→3(LFucα1→2)Gal] >> **I** > **L**. These structural units (lectin determinants) in mammalian glycoconjugates (32-40) are shown in Fig. 1 and Table I. The lectin $\mathbf{T_\beta}$ and $\mathbf{F_\beta}$ determinants are expressed in mammalian glycosphingolipids. $\mathbf{T_\alpha}$, **Tn** and $\mathbf{F_\alpha}$ occur in glycoproteins, and the other determinants are present in many glycoconjugates, such as in the human blood group A, B, H, Le^a, Le^b, and Ii active glycoproteins prepared from human ovarian cyst fluid (35-38), salivary mucins (21-23, 40), gastric mucins (13) and Tamm Horsfall glycoprotein (THGP) (24,25), etc. According to the strength of affinity of lectins for these structural units (Table I), the binding properties of these lectins, as shown in Table II, can be expressed by a decreasing order of affinity of lectin determinants.

The reactivities of lectin determinants (lectin active disaccharides) represent a combination of the binding power of two individual sugars. As listed in Table II, *Abrus precatorius* agglutinin (APA) (13), peanut (PNA) (26), *Geodia cydonium* (GCL) (27), *Maclura pomifera* (MPL) (28) and *Bauhinia purpurea alba* (BPL) (20) are all Galβ1→3GlcNAc specific (7). However, the inhibitory profile of the monosaccharides with these lectins is quite different; GalNAc > Gal in MPL and *Artocarpus integrifolia* (jacalin), and Gal >> GalNAc (inactive) in PNA and APA. Thus, it is expected that the proportional contribution of their monosaccharide residues should be different (Table III).

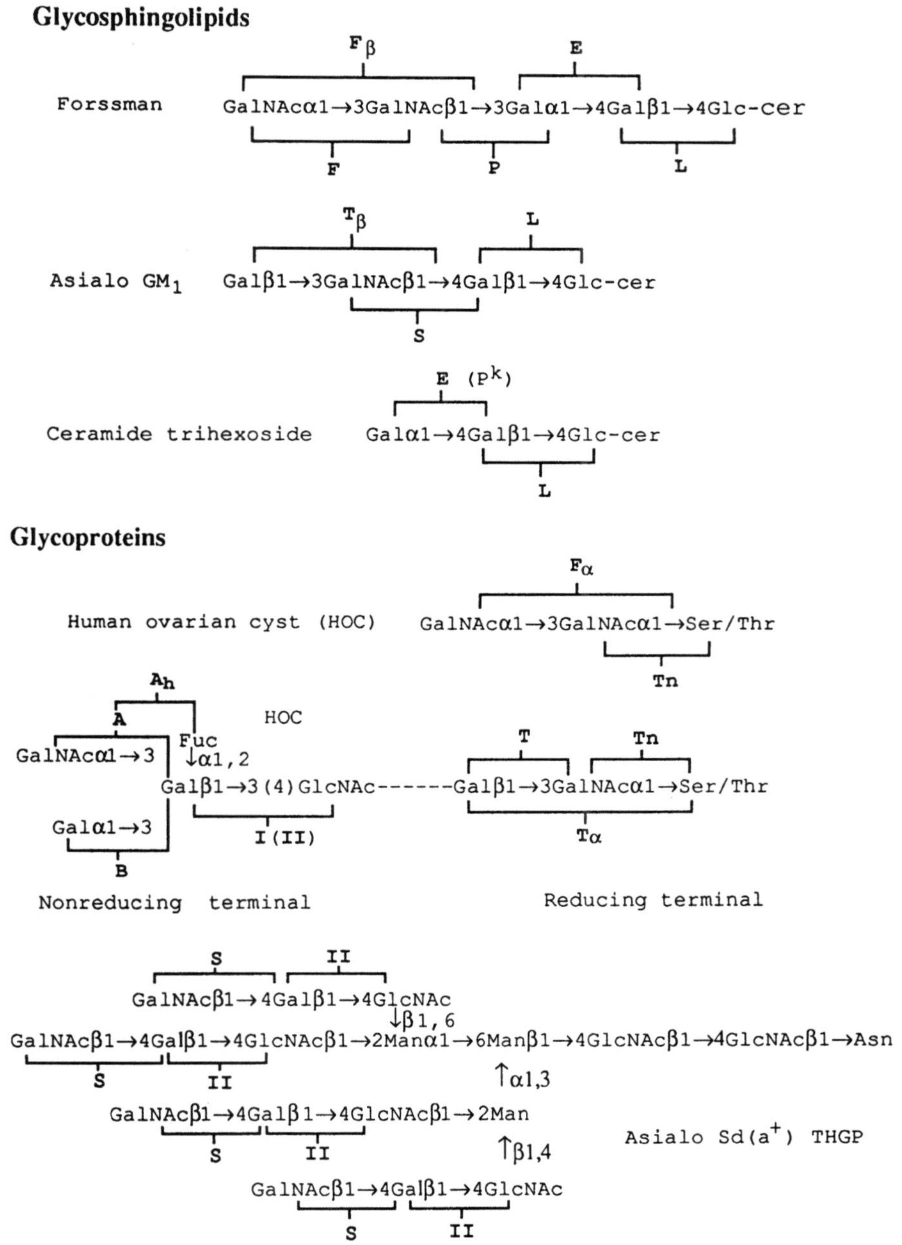

Fig. 1. Proposed mammalian glycoconjugate units (lectin determinants) used to express binding properties of Gal- and GalNAc-specific lectins. Lectin **F** determinant is found in glycosphinglipid (the principal glycolipid of mammalian tissues) of the tissues of the guinea pig, horse, cat, and chicken. It can also be found on the surface of some bacteria, viruses, human gastric carcinoma, and colon tumors (33-35,37). All **(F, A/A$_h$, B, I/II, T, Tn)** determinants are found in human blood groups A, B, H, Le^a, Le^b, and Ii active glycoproteins, prepared form human ovarian cyst fluid (HOC, 35-38). The Gal β1→3GalNAc group at the nonreducing end of asialo-GM$_1$ is also considered as a lectin **T$_β$** determinant. Determinants of lectin **I** and **II** discussed in this article are equivalent to human blood group type I (Lacto-*N*-biose, βGal1→3GlcNAc) and type II (*N*-acetyllactosamine) carbohydrate sequences (38). **S** determinant is found in human Tamm-Horfall glycoprotein (THGP) (39) and GM$_2$.

Table I. Carbohydrate structural Units in Mammalian Glycoconjugates

	Codes[a]	Structural units	Sources
1	**Fα**	GalNAcα1→3GalNAcα1→	In *O*-linked glycoproteins
	Fβ	GalNAcα1→3GalNAcβ1→	Mainly in glycosphingolipids - Forssman pentasaccharide
2	**A**	GalNAcα1→3Gal	Human blood group A active disaccharide
	A_h	GalNAcα1→3[LFucα1→2]Gal	Human blood group A active trisaccharide
3	**Tn**	GalNAcα1→Ser/Thr of protein core	Tn antigen
4	**Tα**	Galβ1→3GalNAcα1→Ser/Thr of protein core	The mucine-type sugar sequence on the human erythrocyte membrane
	Tβ	Galβ1→3GalNAcβ1→...ceramide	Gangliosides such as asialo-GM_1 disaccharide
5	**I**	Galβ1→3GlcNAc	Human blood group type I and II carbohydrate sequences. Most of the lectins reactive with **II** are also reactive with **I**. Lectin **I(II)** determinants can be found at the nonreducing end of the carbohydrate chains derived from *N*-and *O*-glycans
6	**II**	Galβ1→4GlcNAc	
	Tri-II	Triantennary Galβ1→4GlcNAc	
	mII	Multivalent Galβ1→4GlcNAc	
7	**B**	Galα1→3Gal	Human blood group B specific disaccharide
8	**E**	Galα1→4Gal	Blood group P^K active disaccharide. Sheep hydatid cyst glycoproteins, salivary glycoproteins of the Chinese swiftlet, glycosphingolipids in human erythrocytes, and small intestine
9	**L**	Galβ1→4Glc	In lactosyl ceramides
10	**P**	GalNAcβ1→3Galα1→	Blood group P related disaccharide
11	**S**	GalNAcβ1→4Galβ1→	Asialo GM_2 disaccharide; human blood group Sd(a+) related disaccharide in most human secretions, such as urine Tamm-Horsfall glycoprotein etc.

[a]α, β: anomer of sugars; m: multivalent

Table II. Grouping and Binding Properties of Gal/GalNAc-Specific Lectins

Codes	Lectins (5-7)	Carbohydrate specificity
F/A	*Codium fragile* subspecies *tomentosoides* (CFT, 14,15)	**Fp**, and **Tα** >**A_h** >> **I/II**
	Dolichos biflorus (DBL, 18)	**F**>**A_h**[a]>**A**>**Tn**
	Helix pomatia (HPL, 18)	**F**>**A** (>**A_h**[b])≥**Tn**, **T**
	Hog peanut (ABL, *Amphicarpaea bracteata*)	**F**>**A**>**Tn**
	Wistaria floribunda (WFL, 18)	**A** (>**A_h**[b]), **F**>**Tn**, **I** (**II**)
	Geodia cydonium (GCL, 41)	**Fβ**, **A_h** > **I** > **II**
	Griffonia (Bandeiraea) simplicifolia-A_4 (GSI-A_4,29)	**Fp** > **A_h** > GalNAc > **E** > **B**
F/II	*Caragana arborescens* agglutinin (CAA, 42)	**Fp** > **II**, m**Tn** > sialyl **Tn**
	Wistaria sinensis agglutinin (WSA, 43)	**Fp** ≥ m**II** > **II**, **Tn**, **I** and **A_h**
A	Lima bean (LBL)	Hexa-**A_h**[a]>**A_h**[b]>>**B**
	Soybean (SBL, *Glycine max*)	**A** (>**A_h**[b]), **Tn** and **I** (**II**)
	Vicia villosa (VVL) (a mixture of A_4, A_2B_2 and B_4)	**A** (>**A_h**[b]) and **Tn** mainly
Tn	*Codium fragile* subspecies *tomentosoides* (CFT, 14,15)	**Fp** and **Tα** >**Tn** clusters>**A_h**>> **I/II**
	Vicia villosa B_4 (VVL-B_4)	Two **Tn**>>one **Tn**>> one or two **T**
	Salvia sclarea (SSL)	Two **Tn** > single or three sequential **Tn** structures
T	Peanut (PNA, *Arachis hypogaea*, 26)	**T**>>**I** (**II**) >> **Tn**
T/Tn	*Codium fragile* subspecies *tomentosoides* (CFT, 14,15)	**Fα**, **A**, **Tα** and **Tn**
	Maclura pomifera (MPL, 28)	**T**>**Tn**
	Artocarpus integrifolia (jacalin, AIL, 19,44,45)	**T**>**Tn**>>>**I** (**II**)
T, Tn/II	*Bauhinia purpurea alba* (BPL, 20)	**T**>**I** (**II**) and **Tn**
T/II	*Ricinus communis* toxin (ricin, RCA_2)	**T**>**I** (**II**) and **Tn**
	Abrus precatorius agglutinin (APA, 13)	**T**>**I/II**>**E**>**B**>**Tn**
	Sophora japonica (SJL, 46)[c]	**T** and **I** (**II**)
I (II)	*Ricinus communis* agglutinin (RCA_1, 47)	**II**>**I**>**B**>**T**
	Datura stramonium (TAL, thorn apple, 17)	Biantennary **I** (**II**) (penta-2,6)>>**C**[d]
	Erythrina cristagalli (coral tree, ECL, 48)	Multiple antennary **I/II**
	Geodia cydonium (GCL, 27)	**A** & m**II** clusters, **F** > **A** > >> **II**
B	*Griffonia (Bandeiraea) simplicifolia*-B_4 (GSI-B_4,49)	**B**>**E**>**A**
E	*Abrus precatorius* toxin-a (Abrin-a, 50)	**E**,**II**
	Mistletoe lectin-I (ML-I 50, 51)	**E**,**I/II**

[a]Substitution of Fucα1→2 to subterminal Gal is important for binding. [b]Substitution of Fucα1→2 to subterminal Gal blocks binding. [c]Some helping factor is required for precipitation of the lectin. [d]C, chitin oligosaccharide. m: multivalent.

Table III. Differential binding contribution of monosaccharide residues in Galβ1→3GalNAc (T) active lectins.

Lectins	Individual contribution[a]		Enhancement of inhibitory potency after combination of Galβ1→ and GalNAcα→[b]	
	GalNAcα→	Galβ→		
[1] *Maclura pomifera* (MPL, 28)	++++	±	T/GalNAcα1→	1.0
[2] *Bauhinia purpurea alba* (BPA, 20)	++++	±	Tα/GalNAcα1→	2.0
[3] *Codium fragile subspecies tomentosoides* (CFT 14,15)	+++++	– (Inactive)	T/GalNAcα1→	12.5
[4] *Amaranthus caudatus* agglutinin (ACA, 52)	+++++	±	Tα/GalNAcα1→	80.0
[5] *Artocarpus integrifolia* (Jacalin, AIL) (19, 44, 45[c])	++++	±	T/GalNAcα1→	40.0
[6] *Abrus precatorius* agglutinin (APA, 13)	±	+++++	T/Galβ1→	2.0
[7] *Arachis hypogaea* agglutinin (Peanut, PNA 26)	– (Inactive)	+++++	T/Galβ1→	3.6
[8] *Ricinus communis* toxin (Ricin, RCA_2)[d]	+	++++	T/Galβ1→	1.3

a, Based on inhibitory potency of Galβ1→/GalNAcα→ or GalNAcα→/Galβ→; +++++, only GalNAc or Gal active; ++++, 4/5; ±, < 1/5; –, inactive.
b, [2] BPA, Tα/Galβ1→ = 11.0; [6] APA, T/GalNAc = 27; [8] RCA_2 (Ricin), T/GalNAc = 6.0. c, T_α > T >> T_β (19); d, unpublished data.

To obtain a complete description of the carbohydrate specificities of the applied lectins, the following information is suggested to be given– 1. Monosaccharide specificity (Gal, GalNAc, GlcNAc, and/or Man).; 2. Expression of reactivities toward mammalian disaccharide structural units (lectin determinants in Table I) by a decreasing order; 3. the most active ligand; 4. simple multivalent or cluster effect such as glycopeptides and multi antennary glycotopes to inhibit binding; 5. complex multivalent or cluster effects present in macromolecules with known glycotopes.

In this report, we illustrate only, as shown in Table II, the abilities of mammalian disaccharide structural units (lectin determinants) to inhibit lectin-glycan binding. Several cases of the potency of multivalent determinants and determinant clusters are also shown. The inhibitory profile of the complex multivalent determinants or complex clusters is not included as this is only in the early stage of our studies (27, 42).

Due to the multiple reactivity of lectins toward mammalian glycotopes (Table II), the possible existence of different combining sites or subsites in the same molecule has to be examined, and the differential binding properties of these combining sites (if any) have to be characterized.

Recent advances in the characterization, cloning and structural analysis allowed to classify plant lectins into seven families of structurally and evolutionary related proteins. Within each lectin family the overall profile and structure of the carbohydrate-binding site(s) are conserved. A closer examination of the carbohydrate specificity further indicates that most plant lectins are not targeted against plant carbohydrates but preferentially bind foreign glycans (53,54). To establish the relationship among the amino acid sequences of the combining sites of plant lectins and mammalian glycotopes should be an important direction to be addressed in lectinology.

SUMMARY

Expression of the binding properties of Gal/GalNAc specific lectins, based on the affinity of decreasing order of mammalian glycotopes (determinants) rather than monosaccharide inhibition pattern, is probably one of the best ways to express carbohydrate specifity and should facilitate the selection of lectins as structural probes for studying mammalian glycobiology.

Eleven mammalian structural units have been selected to express the binding domain of applied lectins. They are: 1. **F**, GalNAcα1→3GalNAc; 2. **A**, GalNAcα1→3Gal; 3. **T**, Galβ1→3GalNAc; 4. **I**, Galβ1→3GlcNAc; 5. **II**, Galβ1→4GlcNAc; 6. **B**, Galα1→3Gal; 7. **E**, Galα1→4Gal; 8. **L**, Galβ1→4Glc; 9. **P**, GalNAcβ1→3Gal; 10. **S**, GalNAcβ1→4Gal and 11. **Tn**, GalNAcα1→Ser (Thr) of the peptide chain. Thus, the carbohydrate specificity of Gal/GalNAc reactive lectins can be divided into classes according to their highest affinity for the above disaccharides and/or **Tn** residue. Examples of the binding properties of these lectins can be demonstrated by *Ricinus communis* agglutinin (RCA_1), grouped as **II** specific lectin and its binding property is **II** > **I** > **B** > **T**; *Abrus precatorius* agglutinin (APA), classified as **T** and its carbohydrate specificity is **T** > **I/II** > **E** > **B** > **Tn**; *Artocarpus integrifolia* (jacalin, AIL), as **T/Tn** specific and its binding reactivity is **T** > **Tn** >> **I (II)** and *Geodia cydonium* (GCL), as **F/A** specific, and with affinity for **F** > $\mathbf{A_h}$ [GalNAcα1→3(LFucα1→2)Gal] >> **I** > **L**.

Due to the multiple reactivity of lectins toward mammalian glycotopes, the possible existence of different combining sites or subsites in the same molecule has to be examined, and the differential binding properties of these combining sites (if any) have to be characterized. To establish the relationship among the amino acid sequences of the combining sites of plant lectins and mammalian glycotopes should be an important direction to be addressed in lectinology.

REFERENCES

1 Mäkelä, O. (1957) Studies in hemagglutinins of leguminosae seeds. *Ann. Med. Exp. Biol. Fenn. Suppl.*, 11: 1-156.

2 Gallagher, J. T. (1984) Carbohydrate-binding properties of lectins: a possible approach to lectin nomenclature and classification. *Biosci. Rep.* 4: 621-632.

3 Goldstein, I.J. and Poretz, R.D. (1986) Isolation, physiochemical characterization, and carbohydrate binding specificity of lectins. In: " The lectins, properties, function and applications in biology and medicine." (Liener,. I.E., Sharon, N., and Goldstein, I.J. eds.), Academic Press, New York, 33-247.

4 Wu, A. M. (1984) Differential binding characteristics and applications of DGalβ1→3DGalNAc specific lectins. *Mol. Cell. Biochem.* 61: 131-141.

5 Wu, A. M. and Sugii, S. (1988) Differential binding properties of GalNAc and/or Gal specific lectins. *Adv. Exp. Med. Biol.*, 228: 205-263.

6 Wu, A. M. and Sugii, S. (1991) Coding and classification of D-galactose, *N*-acetyl-D-galactosamine, and β-D-Gal*p*-[1→3(4)-β-D-Glc*p*NAc, specificities of applied lectins. *Carbohydr. Res.*, 213: 127-143.

7 Wu, A. M., Song, S. C., Sugii, S. and Herp, A. (1997) Differential binding properties of Gal/GalNAc specific lectins available for characterization of glycoreceptors. *Ind. J. Biochem. Biophys.* 34: 61-71.

8 Nicolson, G.L. (1974) The interaction of lectins with animal cell surfaces. *Int. Rev. Cytol.*, 39: 89-190.

9 Lis, H. and Sharon, N. (1974) Lectins: their chemistry and application to immunology. In: " The antigen", (Sela, M. ed.), Academic Press, New York, 4: 35-67.

10 Lis, H., and Sharon, N. (1986) Lectin as molecules and as tools. *Ann. Rev. Biochem.* 55, 35-67.

11 Wu, A.M., Sugii, S., and Herp, A. (1988) A guide for carbohydrate specificities of lectins in: "The molecular immunology of complex carbohydrates." (Wu, A.M. ed.) *Adv. Exp. Med. Biol.*, 228:819-848.

12 Wu, A. M., Sugii, S., Gruezo, F. G. and Kabat, E. A. (1988) Immunochemical studies on the *N*-acetyllactosamine β(1→6)-linked trisaccharide specificity of *Ricinus communis* agglutinin. *Carbohydr. Res.* 178: 243-257.

13 Wu, A. M., Lin, S. R., Chin, L. K., Chow, L. P. and Lin, J. Y. (1992) Defining the carbohydrate specificities of *Abrus precatorius* agglutinin as T (Galβ1→3GalNAc) > I/II (Galβ1→3/4GlcNAc). *J. Biol. Chem.*, 267: 19130-19139.

14 Wu, A. M., Song, S. C., Hwang, P. Y., Wu, J. H. and Chang, K. S. S. (1995) Binding studies on the combining site of a GalNAcα1→ specific lectin with Thomsen-Friedenreich activity prepared from green marine algae *Codium fragile* subspecies *tomentosoides*. *Eur. J. Biochem.* 233: 145-151.

15 Wu, A. M., Song, S. C., Chang, S. C., Wu, J. H., Chang, K. S. S. and Kabat, E. A. (1997) Further characterization of the binding properties of a GalNAc specific lectin from *Codium fragile* subspecies *tomentosoides*. *Glycobiology* 7: 1061-1066.

16 Chen, C. P., Song, S. C., Gilbo-Garbar, N., Chang, K. S. S. and Wu, A. M. (1998) studies on the binding site of the galactose-specific agglutinin PA-IL from *Pseudomonas aeruginosa*. *Glycobiology* 8: 7-16.

17 Crowley, J.F., Goldstein, I.J., Arnarp, J. and Lonngren, J. (1984) Carbohydrate binding studies on the lectin from Datura stramonium seeds. *Arch. Biochem. Biophys.*, 231: 524-533.

18 Baker, D.A., Sugii, S., Kabat, E.A., Ratcliffe, R.M., Hermentin, P. and Lemieux, R.U. (1983) Immunochemical studies on the combining sites of Forssman hapten reactive hemagglutinins from *Dolichos biflorus*, *Helix pomatia* and *Wistaria floribunda*. *Biochemistry*, 22: 2741-2750.

19 Mahanta, S.K., Sastry, M.V.K. and Surolia, A. (1990) Topography of the combining region of a Thomsen-Friedenreich-antigen lectin jacalin (*Artocarpus integrifolia* agglutinin). *Biochem. J.* 265: 831-840.

20 Wu, A.M., Kabat, E.A., Gruezo, F.G. and Allen, H.J. (1980) Immunochemical studies on the combining site of the D-galactopyranose and 2-acetamido-2-deoxy-D-galactopyranose specific lectin isolated from *Bauhinia purpurea alba* seeds. *Arch. Biochem. Biophys.*, 204: 622-639.

21 Wu, A.M., Wu, J.H. and Shen, F.-S. (1994) Interaction of a novel **Tn** (GalNAcα1→Ser/Thr) glycoprotein with Gal, GalNAc and GlcNAc specific lectins. *Biochem. Biophys. Res. Commun.*, 198: 251-256.

22 Wu, A.M., Shen, F.-S., Herp, A. and Wu, J.H. (1994) Interaction of hamster submaxillary sialyl-Tn and Tn glycoproteins with Gal, GalNAc, specific lectins. *Mol. Immunol.*, 31: 485- 490.

23 Wu, A.M., Herp, A., Song, S.-C., Wu, J.H. and Chang, Kenneth, S.S. (1995) Interaction of native and asialo rat sublingual glycoproteins with lectins. *Life Sciences*, 57: 1841-1852.

24 Wu, A.M., Watkins, W.M., Song, S.-C., Herp, A. and Wu, J.H. (1995) Native and asialo-Tamm – Horsfall glycoproteins as important ligands for the detection of GalNAcβ1→ and Galβ1→4GlcNAc active lectins. *Biochem. Biophys. Res. Commun.*, 209: 103-110.

25 Wu, J.H., Wu, A.M., Watkins, W.M., Chen, C.-P. and Song, S.-C. (1996) Interaction of a human blood group sd (a-) Tamm-Horsfall glycoprotein with applied lectins. *FEBS Letters*, 384: 231-234.

26 Pereira, M.E.A., Kabat, E.A., Lotan, R. and Sharon, N. (1976) Immunochemical studies on the specificity of the peanut (*Arachis hypogaea*) agglutinin. *Carbohydr. Res.*, 51: 107-118.

27 Wu, J.H., Song, S.C., Chen, Y.Y., Tsai, M.C., Kabat, E.A. and Wu, A.M. (1998) Multi-antennary Galβ1→4GlcNAc and Galβ1→3GalNAc clusters as important ligands for a lectin isolated from the sponge. *Geodia cydonium. FEBS Lett.*, 427: 134-138.

28 Sarkar, M., Wu, A.M., and Kabat, E.A. (1981) Immunochemical studies on the carbohydrate specificity of *Maclura pomifera* lectin. *Arch. Biochem. Biophys.*, 209: 204-218.

29 Wu, A.M., Wu, J.H., Chen, Y.-Y., Song, S.-C., and Kabat, E.A. (1999) Further characterization of the combining sites of *Bandeiraea* (*Griffonia*) *simplicifolia* lectin-I, isolectin A_4 (GSI-A_4)., *Glycobiology*, 9: 1161-1170.

30 Song, S.-C., Chen, Y.-Y., Herp, A., Tsai, M.-S. and Wu, A.M. (1999) Lectinochemical studies on the combining site of a GalNAcα or β→ and Galβ1→4GlcNAc specific lectin prepared from *Wistaria sinensis* seeds., *Eur. J. Biol.*, 266: 778-788.

31 Wu, A. M., Song, Shuh-Chyung, Ming-Sung Tsai, Chen Yuen-Yuen and Nechama Gilboa-Garber : Defining the carbohydrate specificitie of *Aplysia* gonad lectin (AGL) exhibiting a peculiar D-galacturonic acid >>> galactose affinity. *J. Biol. Chem.*, 275, 14017-14024, 2000.

32 Springer, G.F. (1971) Blood group and Forssman antigenic determinants shared between microbes and mammalian cells. *Prog. Allergy*, 15: 9-77.

33 Makita, A., Suzuki, C. and Yohizawa, Z. (1966) Chemical and immunological characterization of the Forssman hapten isolated from equine organs. *J. Biochem.*, 60: 502-513.

34 Hakomori, S.-I. and Kannagi, R. (1983) Glycosphingolipids as tumor-associated and differentiation markers. *J. Nat. Cancer Inst.*, 71: 231-351.

35 Wu, A.M. (1988) Structural concepts of the blood group A, B, H, Le^a, Le^b, I, and i active glycoproteins purified from human ovarian cyst fluid in: "The molecular immunology of complex carbohydrates." (Wu, A.M. ed.). *Adv. Exp. Med. Biol.*, 228: 351-392.

36 Wu, A.M., Wu, J.H., Watkins, W.M.. Chen, C.P. and Tsai, M.C. (1996) Binding properties of a blood group Le(a+) active sialoglycoprotein, purified from human ovarian cyst, with applied lectins. *Biochim. Biophys. Acta*, 1316: 139-144.

37 Maisonrouge-McAuliffe, F. and Kabat, E.A. (1976) Immunochemical studies on blood groups. Heterogeneity of oligosaccharides liberated by degradation with alkaline borohydride of two human ovarian cyst fractions differing in B, I and i activities and in reactivity toward *Concanavalin* A. *Arch. Biochem. Biophys.*, 175: 81-89.

38 Watkins W.M. (1980) Biochemistry and genetics of the ABO, Lewis, and P blood group systems., *Adv. Hum. Genet.*, 10: 1-136.

39 Hard, K., Van Zadelhoff, G., Moonen, P., Kamerling, J.P. and Vliegenthart, J.F.G. (1992) The Asn-linked carbohydrate chains of human Tamm-Horsfall glycoprotein of one male. *Eur. J. Biochem.*, 209: 895-915.

40 Herp, A., Borelli, B., and Wu, A.M. (1988) Biochemistry and lectin binding properties of mammalian salivary mucus glycoprotein. in: "The molecular immunology of complex carbohydrates." (Wu, A.M. ed.). *Adv. Exp. Med. Biol.*, 228: 395-435.

41 Hanisch, F.-G., Baldus, S.-E., and Kummel, T.A. (1996) Forssman disaccharide is the specific ligand of a galectin from the sponge *Geodia cydonium* but does not mediate its binding to nuclear protein np56., *Glycobiology*, 6, 321-336.

42 Wu, A.M., Wu, J.H., Chen, Y.-Y., Tsai, M.-S. and Herp, A. (1999) Forssman pentasaccharide and polyvalent Galβ1→4GlcNAc as major ligands with affinity for *Caragana arborescens* agglutinin. *FEBS Letters.*, 463, 225-230.

43 Song, S.-C., Chen, Y.-Y., Herp, A., Tsai, M.-S. and Wu, A.M. (1999) Lectinochemical characterization of a GalNAc and multi-Galβ1→4GlcNAc reactive lectin from *Wistaria sinensis* seeds. *Eur. J. Biochem.*, 266, 1-16.

44 Sastry, M.V.K., Banarjee, S.R., Swamy, M.J., Swarnalatha, G.V. and Surolia, A. (1986) Analysis of saccharide binding to *Artocarpus integrifolia* lectin reveals specific recognition of T- antigen (β-D-Gal(1-3)D-GalNAc)., *J. Biol. Chem.*, 261: 11726-11733.

45 Ahmed, H. and Chatterjee, B.P. (1989) Further characterization and immunochemical studies on the carbohydrate specificity of Jackfruit (Artocarpus integrifolia) lectin., *J. Biol. Chem.*, 264: 9365-9372.

46 Wu, A.M., Kabat, E.A., Gruezo, F.G. and Poretz, R.D. (1981) Immunochemical studies on the reactivities and combining sites of the D-galactopyranose and 2-acetamido-2-deoxy -D-galactopyranose specific lectin purified from *Sophora japonica* seeds. *Arch. Biochem. Biophys.*, 209, 191-203.

47 Wu, A.M., Herp, A. and Wu, J.H. (1993) Defining carbohydrate specificity of *Ricinus communis* agglutinin as Galβ1→4GlcNAc (**II**) > Galβ1→3GlcNAc (**I**) > Galα1→3Gal (**B**) > Galβ1→3GalNAc (**T**). *Mol. Immunol.*, 30, 333-339.

48 Kaladas, P.M., Kabat, E.A., Iglesias, J.L., Lis, H. and Sharon, N. (1982) Immunochemical studies on the combining site of the D-galactose/*N*-acetyl-D-galactosamine specific lectin from *Erythrina cristagalli* seeds. *Arch. Biochem. Biophys.*, 217, 624-637.

49 Wu, A.M., Song, S.-C., Wu, J.H. and Kabat, E.A. (1995) Affinity of *Bandeiraea (Griffonia) simplicifolia* lectin-I, isolectin B_4 (BSI-B_4) for Galα1→4Gal ligand. *Biochem. Biophys. Res. Commun.*, 216, 814-820.

50 Wu, A.M., Song, S.-C. , Wu, J.H., Pfuller, U., Chow, L.P. and Lin, J.Y. (1995) A sheep hydatid cyst glycoprotein as receptors for three toxic lectins, as well as *Abrus precatorius* and *Ricinus communis* agglutinins. *Biochim. Biophys. Acta.*, 1243, 124-128.

51 Wu, A.M., Chin, L.K., Franz, H., Pfuller, U. and Herp, A. (1992) Carbohydrate specificities of the receptor sites of Mistletoe toxic lectin-I. *Biochim. Biophys. Acta.*, 1117, 232-234.

52 Rinderle, S.J., Goldstein, I.J., Matta, K.I. and Ratcliffe, M. (1989) Isolation and characterization of Amaraathin, a lectin present in the seeds of *Amaranthus caudatus*, that recognizes the T-(or Cryptic T)-antigen. *J. Biol. Chem.*, 261, 16123-16131.

53 Peumans, W.J., Barre, A., Hao, Q., Rouge, P. and Van Damme, E.J.M. (2000) Higher plants developed structurally different motifs to recognize foreign glycans. *Trends in Glycoscience and Glycotechology*, 12: 83-101.

54 Peumans, W.J., Van Damme, E.J.M., Barre, A. and Rouge, P. (2001) Classification of plant lectins in families of structurally and evolutionary related proteins. (I-3). In: "The molecular immunology of complex carbohydrates-2."(Wu, A.M. ed.). *Adv. Exp. Med. Biol.* 491: 27-54.

'GLYCO-EPITOPE' ASSIGNMENTS FOR THE SELECTINS: ADVANCES ENABLED BY THE NEOGLYCOLIPID (NGL) TECHNOLOGY IN CONJUNCTION WITH SYNTHETIC CARBOHYDRATE CHEMISTRY[1]

Ten Feizi

The Glycosciences Laboratory, Imperial College School of Medicine
Northwick Park Campus, Harrow, HA1 3UJ, UK

The three selectins (E-, L- and P-selectins) are membrane-associated cell adhesion molecules with crucial roles in the leukocyte-endothelium interactions that enable lymphocyte recirculation (L-selectin) and leukocyte immigration into sites of inflammation and tissue injury (all three selectins) (1). The discovery of lectin-type domains at the tips of the selectins stimulated intense research into their carbohydrate ligands (2). Knowledge that the E- and P-selectins on the inflamed or activated endothelium bind to granulocytes and monocytes served to focus research into the Lewis^{x} (Le^{x}) and sialyl-Le^{x} sequences (as in Fig.1, structures 1 and 2) which were already established as being carbohydrate differentiation antigens on these leukocyte types (3-6). Indeed, it was readily shown that all three selectins recognize the 3′-sialyl-Le^{x} sequence (and also the isomeric sequence 3′-sialyl-Le^{a}, as in structure 4, Fig.1); and that the E- and P-selectins can also bind the non-acidic Le^{x} and Le^{a} sequences (1, 7-9), as in structures 1 and 3, Fig.1. A picture has been emerging, however, of subtle differences in the binding specificities of the selectins, such that variant carbohydrate sequences related to the sialyl-Le^{x} and –Le^{a} sequences are preferentially bound by one or other of the selectins.

This review is largely concerned with binding specificities of the E- and L-selectins towards structurally defined oligosaccharides. I highlight the neoglycolipid (NGL) technology (10), and applications of this technology, which, in conjunction with liquid secondary ion mass spectrometry (11), has been a powerful means of examining details of the binding specificities of the selectins, and has led to the discovery novel ligands among naturally occurring and chemically synthesized oligosaccharides. Also discussed here are some observations on the interactions of the selectins with biotinylated oligosaccharide ligands.

I. Neoglycolipids (NGLs): Powerful Probes for Ligand Discovery for Carbohydrate-Binding Receptors

NGLs, which are oligosaccharide probes derived by chemical conjugation of oligosaccharides to lipid, were introduced (12) and developed in order to address the need for a micro-procedure for direct binding studies with glycans released from glycoproteins and glycolipids (10, 13), see also (14). NGLs are generated by conjugation of oligosaccharides to an aminolipid such as 1,2-dihexadecyl-*sn*-glycero-3-phosphoethanolamine (DHPE) by reductive amination. As with natural glycolipids, the hydrophobic lipid enables the oligo-

[1] The author is supported by the UK Medical Research Council

saccharides to be coated onto matrices for solid phase binding experiments. Through clustering of the lipid moieties, the oligosaccharides are presented in an oligomeric state, which generates the avidities required for readily detectable binding. NGLs have the advantage that they contain a single lipid moiety, contrasting with the heterogeneous lipids of natural glycolipids. Following the conjugation, each oligosaccharide remains a discrete entity, rather than a population of oligosaccharides being conjugated to a macromolecular carrier. Thus, mixtures of NGLs are amenable to resolution by TLC for binding experiments on chromatograms. The excellent ionization property of NGLs has enabled a powerful mass spectrometric strategy to be developed whereby components resolved on

Galβ1-4GlcNAcβ1-3Galβ1-4Glc — ***Structure 1***
| 1,3
Fucα

Galβ1-4GlcNAcβ1-3Galβ1-4Glc — ***Structure 2***
| 2,3 | 1,3
NeuAcα Fucα

Galβ1-3GlcNAcβ1-3Galβ1-4Glc — ***Structure 3***
| 1,4
Fucα

Galβ1-3GlcNAcβ1-3Galβ1-4Glc — ***Structure 4***
| 2,3 | 1,4
NeuAcα Fucα

Galβ1-3/4GlcNAcβ1-3Gal — ***Structure 5***
| 3 | 1,4/3
HSO_3 Fucα

Galβ1-4GlcNAcβ1-3Galβ1-4Glc — ***Structure 6***
| 3 | 1,3
HSO_3 Fucα

Galβ1-3GlcNAcβ1-3Galβ1-4Glc — ***Structure 7***
| 3 | 1,4
HSO_3 Fucα

Galβ-GlcNAc-Gal — ***Structure 8***
| |
NeuAc Fuc

Gal β1-3\
GalNAcβ1-3Galα1-4Galβ1-4Glc — ***Structure 9***
Galβ1-4GlcNAcβ1-6/
| 1,3
Fucα

GlcUAβ1-3Galβ1-3GlcNAcβ1-3Galβ1-4Glc — ***Structure 10***
| 3
HSO_3

Fig. 1. The Le^a and Le^x-related sequences which occur naturally on the carbohydrate chains of glycoproteins or glycolipids, and have been assigned as carbohydrate differentiation antigens of leukocytes or as ligands for the selectins as described in the text. The term (sialyl or sulfated) *pentasaccharide* is used here for structures 2, 4, 6 and 7, to denote the fucosylated tetrasaccharide backbone that they share.

chromatograms can be sensitively analyzed, *in situ,* by liquid secondary ion mass spectrometry (15, 16). By this means, molecular mass, monosaccharide sequence, branching pattern and the presence of other groups such as sulphate or phosphate can be detected on bioactive oligosaccharide components within mixtures released from glycoproteins (10, 13, 17).

The key feature of the NGL technology is that it enables the pinpointing of the bioactive oligosaccharides within highly heterogeneous mixtures derived from natural glycoconjugates. Thus, aliquots of total oligosaccharide populations (or of sub-populations fractionated on the basis of charge or size) derived from desired glycoproteins, glycolipids or whole cells are converted into NGLs, resolved by TLC, and ligand-bearing oligosaccharides detected by chromatogram-binding experiments. Oligosaccharide fractions containing the desired ligand-bearing components are subjected to successive liquid chromatographies, with monitoring of the bioactivities within aliquots converted into NGLs, until the desired components are isolated. The technology has been powerful in the elucidation of oligosaccharide-binding specificities of antibodies and endogenous carbohydrate-binding proteins, and in studies of the roles of specific oligosaccharides in cell signaling and microbial adhesion studies and as substrates for glycosyltransferases, as reviewed in ref. (13). Notable examples have included the discovery of a unique oligosaccharide antigen on the adhesive proteoglycan of the sponge, *Microciona prolifera* (18), a family of linear and branched oligosaccharides of the lacto- and neolacto-series capped with sulphated-Le^{a} and sulphated-Le^{x} sequences which are potent ligands for E-selectin (17, 19) (these are elaborated further below), and assignments of the sequences of novel *O*-mannosyl glycans on mammalian brain glycoproteins (20, 21).

A limitation of DHPE, shared with the naturally occurring ceramides, is that it does not contain a chromophore, and the derived NGLs have only the UV absorbing property of the parent oligosaccharide for detection by HPLC. Detection and quantitation of NGLs has depended on staining for lipid, hexose or sialic acid residues using primulin, orcinol or resorcinol, respectively (10). HPLC of NGLs with UV detection has not been convenient because of the UV-absorbing property of the solvents required for separation. There was a need, therefore, for a new lipid reagent, which contains a UV-absorbent or fluorescent chromophore, and this has now been addressed as follows.

We have now created a second generation of lipid-linked oligosaccharide probes, fluorescent NGLs, for ligand discovery within highly complex mixtures of oligosaccharides (22). DHPE, has been modified to incorporate a fluorescent label, anthracene. This new lipid reagent, *N*-aminoacetyl-*N*-(9-anthracenylmethyl)-1,2-dihexadecyl-*sn*-glycero-3-phosphoethanol-amine (ADHP), synthesized from anthracenaldehyde and DHPE, gives an intense fluorescence under UV light. Fluorescent NGLs derived from a variety of neutral and acidic oligosaccharides by conjugation to ADHP, by reductive amination, can be detected and quantified by spectrophotometry and scanning densitometry, and resolved by TLC and HPLC with sub-picomol detection. Antigenicities of the ADH-NGLs are well retained, and picomol levels can be detected using monoclonal carbohydrate sequence-specific antibodies. Among *O*-glycans from an ovarian cystadenoma mucin, isomeric oligosaccharide sequences, sialyl-Le^{a}- and sialyl-Le^{x}-active, could be resolved by HPLC as fluorescent NGLs, and sequenced by liquid secondary ion mass spectrometry (22). Thus the NGL technology now uniquely combines high sensitivity of immuno-detection with a comparable sensitivity of chemical detection. Principles are thus established for a streamlined technology whereby an oligosaccharide population is carried through ligand detection and ligand isolation steps, and sequence determination by mass spectrometry, enzymatic sequencing and other state-of-the-art carbohydrate analyses.

II. Importance of the Density of Selectin Expression and the Density of Oligosaccharide Ligands in the Magnitude of the Binding Signals

Co-operative effects of clustering both of E-selectin and its ligands were shown to be important for mediating adhesion. For example, a soluble, monomeric form of E-selectin gave no detectable binding signal with the immobilized clustered ligands such as sialyl-Le^a pentasaccharide, in the form of NGLs, but could be induced to give binding signals when oligomerized in the form of a soluble complex using a non-neutralizing anti-E-selectin antibody and staphylococcal protein A [RA Childs & T Feizi, unpublished] cited in ref. (23). With Chinese hamster ovary (CHO) cells transfected to express E-selectin, adhesion intensity and even specificity were found to dependent on the density of surface expression of the membrane-associated adhesion molecule (24). There was a threshold of density of E-selectin required for binding to the lipid-linked acidic fuco-oligosaccharide ligands (sialyl-Le^a and -Le^x pentasaccharides, structures 2 and 4, Fig.1), and three types of adhesive specificity were observed (24): (a) transfected cells with low levels of E-selectin expression showed no detectable carbohydrate-mediated adhesion; (b) cells with the highest density of E-selectin expression adhered not only to the lipid-linked acidic sequences, but also to the non-acidic Le^a, and to a lesser extent Le^x-active analogues, although ten-fold or higher levels of these latter NGLs were required to give binding intensities equivalent to those observed with the acidic oligosaccharides; (c) cells with intermediate levels of E-selectin expression adhered only to the acidic oligosaccharides.

The *in vivo* significance of binding to the non-acidic Le^a and Le^x type sequences observed under static adhesion assay conditions, discussed above, is uncertain as there is considerable evidence for the importance of sialic acid in many *in vivo* and *in vitro* experiments investigating metastatic potential and cell adhesion. Interestingly, some evidence was obtained that, under conditions of flow *in vitro*, Le^x on the myeloid cell surface may play a greater role in E-selectin-mediated adhesion than sialyl-Le^x (25). Amounts of Le^a antigen on human erythrocytes are too low to elicit E-selectin binding (26) but the enormously dense expression of the Le^a and Le^x antigens on many adenocarcinoma cells may well be able to mediate adhesion. For these reasons, it is interesting to note that, while the presence of the blood group H fucose α1-2 linked to galactose of the sequence of Le^a (as in the Le^b antigen), does not hinder E-selectin adhesion, the additional presence of the blood group A or B monosaccharides, *N*-acetylgalactosamine α1-3-linked as in ALe^b heptasaccharide (24) or galactose α1-3- linked to galactose, as in BLe^b heptasaccharide, hinders E-selectin binding [T Feizi, E Breimer & J O'Brien, unpublished, and cited in ref. (23). In adenocarcinomas of the distal colon, the blood group A, B, H , Le^b, ALe^b and BLe^b antigens are expressed as cancer-associated antigens of the distal colon, and we have raised the possibility (24) that the variable display of these on the tumour cells in different individuals may be among factors that influence metastatic potential.

III. The Efficacy of Lipid-linked Oligosaccharides in Supporting Selectin-Mediated Cell Interactions

Although there have been numerous reports of selectin interactions, *in vitro*, with lipid-linked oligosaccharides, there is still much to be learnt about the degree of involvement of glycolipids as physiological counter-receptors for these adhesion molecules. When lipid-linked oligosaccharide ligands (NGLs) were immobilized, *in vitro*, they were found to effectively support rolling and tethering of cells that express E- or L-selectins at their surface (27). These findings suggest that glycolipids at the cell surface of leukocytes may have roles as counter-receptors for E-selectin. Moreover it is conceivable that glycolipids are among L-selectin counter-receptors at the surface of high endothelial cells.

IV. Discovery of Sulphated Oligosaccharides of Le^a/Le^x Type as Selectin Ligands on Epithelial Glycoproteins

A novel class of acidic selectin ligand, sulphate-containing, was identified by screening for E-selectin binding to NGLs generated from the carbohydrate chains (*O*-glycans) of a mucin-type glycoprotein that had been isolated from a human ovarian cystadenoma (17). An E-selectin-binding oligosaccharide was identified consisting of an equimolar mixture of 3'-sulphated tetrasaccharides of Le^a and Le^x type, structure 5, Fig.1. The binding activity of the mixture was at least equal to that of the 3'-sialyl-Le^x analogue (17). This was corroborated (28) by subsequent work with the individual chemically synthesized tetrasaccharides and also the pentasaccharide analogues, structures 6 and 7, Fig.1. Further studies of NGLs derived from the epithelial glycoprotein, have revealed a substantial family of oligosaccharides of sulpho-Le^a/Le^x type with various branched and linear backbone sequences (19). Thus, in the E-selectin system, now also found with the L- and P-selectins (9) sulphate 3-linked to the terminal galactose can substitute for the carboxyl group of 3-linked sialic acid (29). This suggests that the primary contribution of the sialic acid is the negatively-charged group, and is in accord with evidence (30) that other molecular features of sialic acid have no effect on E-selectin recognition.

V. Carbohydrate Ligands for E-selectin Detected on Murine Leukocytes and Kidney

Murine leukocytes, in sharp contrast with those of the human, do not express serologically detectable Le^x and sialyl-Le^x antigens (31) nor the Le^a and sialyl-Le^a antigens (32); and yet, murine leukocytes are bound by E- and P-selectins as are human leukocytes. We have made two approaches to identifying carbohydrate ligands for E-selectin in the mouse (14) by exploiting the NGL technology (10). Having found evidence for the presence of poly-*N*-acetyllactosamine type sequences among the E-selectin-binding glycolipid population extracted from the murine neutrophilic cell line, 32D cl3, we used endo-β-galactosidase to release susceptible oligosaccharides of this series from the surface of these cells. Oligosaccharide fractions were converted into NGLs, and the main E-selectin-binding component was identified, by TLC-liquid secondary ion mass spectrometry, as a sialo-fuco-oligosaccharide of sialyl-Le^a/Le^x type (structure 8, Fig.1) and was bound by monoclonal anti-sialyl-Le^a (2D3) but not by CSLEX-1. We (J Topping and T. Feizi, unpublished) have subsequently found that CSLEX-1 does not give a binding signal with the sialyl-Le^x sequence based on the trisaccharide backbone Galβ1-4GlcNAcβ1-3Gal that was generated as result of the endo-β-galactosidase digestion (rather a tetrasaccharide backbone or longer is required for binding), whereas antibody 2D3 gives strong binding signals with the sialyl-Le^a analogue based on the trisaccharide backbone Galβ1-3GlcNAcβ1-3Gal. We now interpret the results of Osanai et al. as indicating that any sialyl-Le^x sequence within the fraction investigated (the predicted product of Fuc-TVII (32) would not have been detected with the anti-sialyl-Le^x antibody, CSLEX-1. It is not known however, why the sialyl-Le^a antigen is not detectable serologically, using 2D3 antibody, on the murine leukocytes and also on the intact cells in the neutrophilic cell line investigated. A possible explanation is that the display of the saccharide epitope on these cells is suboptimal for antibody binding, possibly crowded out by other selectin ligands that are not recognized by the antibody. It is moreover, possible that the products of another fucosyltransferase act in concert with those of FucT-VII to generate ligand densities above the threshold needed for selectin-mediated adhesion, but the densities of the individual sialyl-Le^x and Le^a determinants recognized by the antibodies are relatively low.

We identified a second class of E-selectin ligand among glycolipids extracted from the kidney of the BALB/c mouse (14). By chromatogram binding experiments and *in-situ*

LSIMS using NGLs derived from oligosaccharides released with endoglycoceramidase, in conjunction with compositional and linkage analyses, we identified the oligosaccharide moiety as the Le^x-active, extended, branched globo sequence, structure 9, Fig.1, which had been described previously (33). We obtained mass spectrometric evidence for the presence of novel sialyl analogues of the branched globo sequence, raising the possibility that fucosyl sialyl-Le^x analogues may also be present, though not detected. It will be interesting to determine the cellular distribution of the globo-Le^x type sequence, especially among epithelia and malignant cells derived from them, as its presence in the kidney is genetically determined (33), and it may therefore contribute to the metastatic potential of epithelial tumours in different inbred strains of mice.

VI. Explorations of the Ligands for L-selectin

Much evidence for a carbohydrate-mediated adhesive specificity for L-selectin, with involvement of sialic acid, fucose and sulphate was obtained before that for the other two selectins, and before the discovery of a lectin-like protein motif on these proteins see (34) for a review of the elegant experiments that led to those conclusions). Research on the sequences of oligosaccharides recognized by L-selectin were boosted with the realization that the sialyl-Le^x sequence is recognized by the two related proteins E- and P-selectins. *In vitro* binding experiments using normal mouse lymphocytes (35) or a mouse L-selectin IgG chimera (36) or mouse L1-2 pre-B cells transfected with human L-selectin cDNA (35), showed that both the murine and the human L-selectins can bind to 3'-sialyl-Le^x and -Le^a sequences (structures 2 and 4, Fig.1), linked to protein or to lipid. These features, and the lack of recognition of the carboxyl group rather than the polyol tail of the sialic acid (37) are shared with human E-selectin. It was apparent that the adhesive specificities of the two selectins are different. First, it was observed, that L-selectin, but not E-selectin, adheres to the sulphate-containing glycosphingolipid sulphatide (36, 38-42). Second, L-selectin was shown to bind with a greater intensity than E-selectin to a glycoprotein 'PNAd' isolated from peripheral lymph nodes (35). Third, E-selectin but not L-selectin

```
HSO3
 | 6
Galβ1-4GlcNAcβ1-
 | 2,3     | 1,3
NeuAcα   Fucα
```
6'-sulpho-sialyl-Le^x

```
          HSO3
           | 6
Galβ1-4GlcNAcβ1-
 | 2,3     | 1,3
NeuAcα   Fucα
```
6-sulpho-sialyl-Le^x

```
HSO3      HSO3
 | 6       | 6
Galβ1-4GlcNAcβ1-
 | 2,3     | 1,3
NeuAcα   Fucα
```
6',6-sulpho-sialyl-Le^x

Fig. 2. Sulphated sialyl-Le^x-type capping sequences assigned by Hemmerich and colleagues (46) to O-linked oligosaccharides released from GlyCAM-1.

bound to glycoproteins expressing 'CLA', an antigen on cutaneous lymphocytes (35).

Work with structurally defined oligosaccharides of the Le^a, Le^x and sulpho-glucuronyl series (HNK-1, structure 10, Fig. 1) corroborated the importance of sulphate as a recognition element on oligosaccharide ligands for L-selectin (23). From binding and inhibition experiments with the recombinant soluble protein, and lipid-linked oligosaccharides, it became clear that there is an overall preference of the sulpho-Le^a and -Le^x sequences over the sialyl analogues (39, 43-45), to the extent that the non-fucosylated sulphated backbones are bound (albeit with lower intensities), but not the sialyl analogues (39).

VII. Demonstration that the Sulphated Form of the Sialyl-Le^x Sequence with Sulphate at Position 6 of the Penultimate *N*-acetylglucosamine is a Potent Ligand for L-selectin

Considerable information is accumulating on sulphated oligosaccharide sequences of the type found on high endothelial venules, the key attachment sites for L-selectin. Among O-glycans released from the endothelial glycoprotein GlyCAM-1, evidence has been found for the occurrence of sulphated forms of sialyl-Le^x (Fig.2): in the first, designated 6'-sulpho-sialyl-Le^x, the sulphate is at position 6 of the terminal galactose; in the second, 6-sulpho-sialyl-Le^x, sulphate is at position 6 of the penultimate *N*-acetylglucosamine, and in the third, 6',6-sulpho-sialyl-Le^x, both positions are sulphated (46). However, with the limited amounts of oligosaccharide material derived from GlyCAM-1, the potencies of these oligosaccharides relative to the non-sulphated sialyl-Le^x could not be determined. Chemical synthesis approaches have been extremely rewarding here, for they have

SIALYL-LE^X SERIES

Galβ1-4GlcNAcβ1-3Galβ1-4Glc
 | 2,3 | 1,3
NeuAcα Fucα
Sialyl-Le^x

HSO_3
 | 6
Galβ1-4GlcNAcβ1-3Galβ1-4Glc
 | 2,3 | 1,3
NeuAcα Fucα
6'-sulpho-sialyl-Le^x

HSO_3
 | 6
Galβ1-4GlcNAcβ1-3Galβ1-4Glc
 | 2,3 | 1,3
NeuAcα Fucα
6-sulpho-sialyl-Le^x

HSO_3 HSO_3
 | 6 | 6
Galβ1-4GlcNAcβ1-3Galβ1-4Glc
 | 2, 3 | 1,3
NeuAcα Fucα
6',6-sulpho-sialyl-Le^x

SULPHO-LE^X SERIES

Galβ1-4GlcNAcβ1-3Galβ1-4Glc
 | 3 | 1,3
HSO_3 Fucα
3'-sulpho-Le^x

HSO_3
 | 6
Galβ1-4GlcNAcβ1-3Galβ1-4Glc
 | 1,3
Fucα
6'-sulpho-Le^x

HSO_3
 | 6
Galβ1-4GlcNAcβ1-3Galβ1-4Glc
 | 3 | 1, 3
HSO_3 Fucα
6',3'-sulpho-Le^x

HSO_3
 | 6
Galβ1-4GlcNAcβ1-3Galβ1-4Glc
 | 3 | 1, 3
HSO_3 Fucα
6 ,3'-sulpho-Le^x

Fig. 3. Chemically synthesized sequences based on the Le^x pentasaccharide. The term pentasaccharide is used here to denote the fucosylated tetrasaccharide backbone that these sequences share. The sialyl-Le^x-containing sequences were synthesized as glycosylceramides (47, 71) and the non-sialylated sulfo-Le^x series as the free oligosaccharides (56).

enabled (a) the clear assignment of the position of sulphation on sialyl-Le^x that elicits the strongest L-selectin binding signals, (b) the identification of novel more potent ligands, and (c) the discovery of a processing pathway of the sialyl-ligands that may be important for regulation of the expression of selectin ligand activity, *in vivo*. These several developments are highlighted below.

In the first of the synthetic approaches aimed at assessing the L-selectin binding signals elicited by 6'-, or the 6-sulphated sialyl-Le^x, the three sulphated variants of the sialyl-Le^x sequence (Fig.3) were synthesized as pentaglycosylceramides (47). In initial experiments, the ability of these three sulphated compounds to support L-selectin binding appeared comparable to that of the non-sulphated sialyl-Le^x (48). However, this was soon found to be a reflection of the presence of some impurities in each of the sulphated compounds. Once these were removed, the 6-sulpho-sialyl-Le^x pentasaccharide clearly stood out as the preferred ligand for L-selectin, whereas the 6'-sulpho-sialyl-Le^x analogue was inactive, and the 6',6-disulpho-sialyl-Le^x had intermediate activity (49). The 6-sulpho-sialyl-Le^x sequence is immunochemically detectable on high endothelial venules (50, 51). Cell lines that lack L-selectin ligands, when transfected with α1-3 fucosyltransferase (Fuc-TVII) and *N*-acetylglucosamineβ:6 sulphotransferase have now been shown to acquire the ability to support L-selectin-mediated cell adhesion (52-55). Collectively these results establish the 6-sulpho-sialyl-Le^x sequence as a functional ligand for L-selectin.

The second chemical synthesis approach (56) was undertaken with the knowledge that among epithelium-derived oligosaccharides that support E- and L-selectin binding, there are Le^x (and also Le^a) sequences that have sulphate at position 3 of the terminal galactose instead of a 3-linked sialic acid (17, 39). As the 3'-sulpho-Le^x and 3'-sulpho-Le^a are, overall, more potent ligands for L-selectin than the sialyl analogues (39, 43-45), the pentasaccharides 6',3'-sulpho-Le^x, and 6,3'-sulpho-Le^x, and also the monosulphated 6'-sulpho-Le^x (Fig.3) were synthasized (56). In the form of NGLs, these were compared with 3'-sulpho-Le^x pentasaccharide and also with the sulphated sialyl-Le^x analogues for their ability to support L-selectin binding (45). As with the sialyl-Le^x series, the 6-sulpho form of the 3'-sulpho-Le^x pentasaccharide gave the greatest binding signal, the monosulphated, 6'-sulpho-Le^x analogue gave no detectable binding signal with L-selectin and 6',3'-sulpho-Le^x was consistently less potent than the 3'-sulpho-Le^x. Thus on oligosaccharides of this series, sulphate at position 6 of the *N*-acetylglucosamine enhances the L-selectin binding signal elicited both by the 3'-sialyl- and the 3'-sulpho-Le^x sequences, whereas sulphate at position 6 of the terminal galactose hinders the binding. There is, however, an apparently conflicting, and as yet unexplained finding in structural terms. This is the observation that cells transfected with a 6'-sulphotransferase (which transfers sulphate to the galactose) in addition to two other glycosyltransferases (Fuc-TVII and *O*-glycan core 2 enzymes) acquire the ability to support L-selectin binding (53, 55); and in one cell line, the occurrence of enhanced L-selectin binding was observed to cells which were additionally transfected with the 6-sulphotransferases. This issue is considered below (under IX).

VIII. The Finding with Synthetic Oligosaccharides that de *N*-acetylation of the Sialic Acid in the 6-sulpho-3'-sialyl-Le^x Enhances the Binding Signals with L-selectin, Whereas Subsequent Modification of the Carboxyl Group Virtually Abolishes the Signals, has Provided Clues to the Existence of Novel Biosynthetic Pathways for Selectin Ligands

Work with the impurities (synthetic by-products) in the pentaglycosylceramides discussed above (49), has opened up some exciting research directions. These were minor components, which nevertheless had initially concealed the striking differences between the 6'- and the 6-sulphated forms of sialyl-Le^x. By mass spectrometry, we showed that these by-products were analogues of the intended compounds, and that each contained a modified sialic acid. In one of the components which was 'superactive', the sialic acid is de-*N*-acetylated, whereas in the second component which had weak selectin binding activity, there is, in addition, a modification of the carboxyl group (49). Clearly, these are by-products that arose at the final stages of preparation (deprotection) of the intended compounds. Strong alkaline conditions are used to stabilize the sulphate group during the deprotection step (49) and under these conditions, some loss can occur of the *N*-acetyl group of the sialic acid. Moreover, this can proceed to the modification of the carboxyl group of the sialic acid by the reaction of the amino group at C-5 with the carboxyl group at C-2 of the six-membered ring to give an intermolecular amide bond. The stronger L-selectin binding signal with the de-*N*-acetylated compound was corroborated by examining the intentionally synthesized, de-*N*-acetyl-sialyl compound (57).

The striking differences in bioactivity relative to that of the *N*-acetylated compound raised the possibility (49) that such a modification of the selectin ligand may occur *in vivo*. The occurrence of de-*N*-acetyl sialic acid has been reported among sialoglycolipids (GM_3 and GD_3) in certain cell lines and tumour tissues (58-60) and it has been proposed (59) that a reversible de-*N*-acetylation and re-*N*-acetylation occurs *in vivo*. Evidence is now forthcoming that supports the concept of such a post-biosynthetic processing of *N*-acetyl neuraminic acid on selectin ligands. A monoclonal antibody (61) produced by immunizing mice with the unfractionated 6-sulpho-sialyl-Le^x pentaglycosylceramide (47, 49) has proven to be a powerful tool in this respect. This antibody happens to be directed to the minor impurity 6-sulpho-sialyl-Le^x with the modified sialic acid carboxyl group (61). Immunocytochemical studies with this antibody have shown that this form of the 6-sulpho-sialyl-Le^x exists on leukocytes (information is awaited regarding its occurrence on endothelium). Moreover, enzymatic studies using sonicates of leukocytes as enzyme source, and the chemically synthesized de-*N*-acetyl form of 6-sulpho-Le^x pentasaccharide as substrate, together with immunochemical studies with the above mentioned antibody (61), have provided evidence for the presence of an enzyme which converts the substrate to the analogue with the modified carboxyl group. This has been tentatively designated 'cyclic' sialic acid (61). It will be interesting to investigate if as proposed (61), sialyl-Le^x analogues containing this inactive 'cyclic' sialic acid form act as a dormant pool of selectin ligands, which, upon appropriate stimulation, are hydrolyzed by cellular enzymes and become active in selectin-dependent cell adhesion.

IX. A Second Class of Sulphated L-selectin Ligand, and Possible Co-operation with Ligands of the Le^a/Le^x Series

Much of the foregoing discussion has been concerned with the specificity of L-selectin binding to several immobilized acidic oligosaccharide analogues based on the Le^x pentasaccharide sequence. This is a class of oligosaccharide ligand to which L-selectin binds in a calcium-dependent manner. A second class of ligand of which sulphatide (3'-sulphated galactosyl-cermide) is the prototype encompasses various sulphated mono- and disaccharides in which the position of the sulphate on galactose is not critical, and the binding is partly or wholly calcium-independent (40, 43). The second class of L-selectin ligand includes ganglio series sequences with a terminal galactose that is variously sulphated 3-O, 4-O, 6-O, 3,4-O-di- or 3,6-O-di-sulphated [ref. (23) and references therein and (62)]. There is increasing evidence that L-selectin binds also to glycosaminoglycans. When presented as NGLs, glycosaminoglycan disaccharides of keratan sulphate, heparin and chondroitin sulphate types are bound (43), leading to the conclusion that in contrast to the tri- and longer oligosaccharides, clustered short oligosaccharides with 6-O sulphation of *N*-acetylgalactosamine, *N*-acetylglucosamine or glucosamine, 4-O sulphation of *N*-acetylgalactosamine, 2-O-sulphation of uronic acid, N-sulphation of glucosamine, and to a lesser extent the non-sulphated uronic acid-containing disaccharides, can all support L-selectin binding. Endothelial cells contain heparan sulphate glycosaminoglycans with various sulphation patterns, also including non-sulphated glucosamine, to which immobilized L-selectin binds with high avidity (63). All this suggests that endothelial glycosaminoglycans may be among natural ligands for this selectin. We have proposed (43) that such interactions of L-selectin may provide a link between the selectin-mediated and the integrin-mediated adhesion systems in leucocyte extravasation cascades, for glycosaminoglycans serve as reservoirs for inflammatory chemokines; these are short-range stimulators of lymphocyte migration which trigger integrin activation (64).

There is much to be learnt about the interplay and possible co-operativity of these two classes of L-selectin ligand in the natural setting. CHO cells transfected to express the recently cloned galactoseβ:6'-sulphotransferase concomitantly with CD34, fucosyltransferase VII and O-glycan core 2 enzyme were shown to be able to support L-selectin binding, as were the cells transfected with a *N*-acetylglucosamine-6-sulfotransferase (53). When co-transfected, the two sulphotransferases were found to synergize and elicit enhanced L-selectin binding (53). We have suggested (65) that short saccharides 6'-sulphated at galactose may be among products of the galactoseβ:6'-sulphotransferase. The sequences of the 6'-sulphated oligosaccharides in the transfected cell line have not yet been characterized. Definition of these will be crucial to understanding the biochemical basis of the synergistic property of the products of this enzyme.

X. Influence of the Mode of Oligosaccharide Presentation on the Binding Strengths of the Selectins to their Oligosaccharide Ligands – Observations with Biotinylated Forms of the Oligosaccharide Ligands Presented on a Streptavidin Matrix

Biotinylated oligosaccharides are alternative oligosaccharide probes that are potentially applicable for carbohydrate recognition studies where the aim is to work with a set of defined oligosaccharides, rather than to discover the ligands among heterogeneous mixtures of oligosaccharides. In the course of exploring biotinylated oligosaccharides presented on a streptavidin matrix as probes for studies of carbohydrate-protein interactions, we observed differential effects of three biotinylated tags on the interactions of *N*-glycans with certain plant lectins (66). These were the UV absorbent and fluorescent molecules biotinyl-L-3-(2-naphthyl)-alanine hydrazide (BNAH), and 2-amino,6-amidobiotinyl-pyridine (BAP), and the non-fluorescent, 6-(biotinyl)-aminocaproyl hydrazide (BACH) which has a relatively long spacer. BNAH and BACH allow coupling of oligosaccharides under non-reductive conditions thus preserving the pyranose ring of the reducing-end monosaccharide. Coupling to BAP is carried out with reducing reagent yielding a 'ring-opened' reduced sugar. With the plant lectins Concanavalin A and *Ricinus Communis* agglutinin which recognize outer monosaccharides on *N*-glycans, equivalent binding signals were detected using the three biotinylated tags, whereas with *Lens Culinaris* agglutinin (LCA) and *Pisum Sativum* agglutinin (PSA) which recognize the trimannosyl core as well as the fucosylated core with a non-reduced *N*-acetylglucosamine, the BNAH and BACH derivatives (but not BAP) were bound, and the binding signal with BNAH was stronger. For high affinity binding, PSA requires the asparagine at the *N*-glycan attachment site (67). Thus the BNAH derivative which contains an alanine hydrazide is a better mimic of the β-glycosyl asparaginyl core than the BACH derivative, and would be the tag of choice for studies of the interactions of *N*-glycans with plant lectins of this type. However, BNAH is no longer readily available commercially, whereas BACH is.

We have also investigated the efficacy of presentation of biotinylated oligosaccharides on streptavidin-coated microwells, for interactions with three monoclonal antibodies directed at sialyl-Le^a or sulfo-Le^a-related sequences, and the E-, L- and P-selectins which recognize both oligosaccharide series (68). With the antibodies it was observed that if the biotinylated oligosaccharide incorporated the entire antigenic determinant, and additional saccharide length was not included, the length of spacer on the biotinyl tag was a critical factor in the strength of the binding signal. If on the other hand oligosaccharide chain beyond the determinant was included, the length of the spacer on the biotinyl tag was less important. The E-selectin binding data with the biotinylated sialyl- and sulfo-oligosaccharides were in overall accord with previous knowledge. With the L and P-selectins, however, unexpectedly low binding signals were elicited by biotinyl sulpho-Le^a sequences relative to those with the sialyl-analogues. This suppression was more pronounced with the rodent (rat and mouse) than the human L-selectin. The mechanism of this suppression has not yet been explained. We propose that such differential availabilities of oligosaccharides displayed on streptavidin may relate to biological situations, for example, of a given oligosaccharide ligand presented on different carrier proteins (69), or on different *O*-glycan cores on mucin-type glycoproteins (70).

SUMMARY

The neoglycolipid (NGL) technology involving the preparation of lipid-linked oligosaccharide probes for binding experiments with carbohydrate-recognizing proteins, and their analysis by mass spectrometry, is a unique and powerful means of discovering oligosaccharide ligands for carbohydrate-binding proteins, and assigning details of their specificities. The key feature is that it enables the pinpointing and sequence determination of bioactive oligosaccharides within highly heterogeneous mixtures derived from natural glycoconjugates. A new generation of NGLs incorporating a fluorescent label now establishes the principles for a streamlined technology whereby oligosaccharide populations are carried through ligand detection and isolation steps, and sequence determination.

Advances in selectin research made through applications of the NGL technology include (i) demonstration of the importance of density of selectin expression, and of oligosaccharide ligands, in the magnitude and the specificity of the binding signals; (ii) demonstration of the efficacy of lipid-linked oligosaccharides in supporting selectin-mediated cell interactions; (iii) the discovery of 3-sulphated Le^a/Le^x as selectin ligands; (iv) the isolation and sequencing of carbohydrate ligands for E-selectin on murine myeloid cells and kidney; (v) the finding that sulphation at position 6 of the penultimate *N*-acetylglucosamine confers superior L-selectin binding signals not only to 3-sialyl-Le^x but also to 3'-sulpho-Le^x; and (vi) the finding that sialic acid de-*N*-acetylation, or further modification with formation of an intra-molecular amide bond in the carboxyl group, enhances or virtually abolishes, respectively, the potency of the 6'-sulfo-sialyl-Le^x ligand.

Working with biotinylated forms of the oligosaccharide ligands, we have observed that their presentation on a streptavidin matrix influences differentially the efficacy of interactions of the L- and P-selectins (but not E-selectin) with the sialylated and sulphated ligands.

REFERENCES

1. M.P.Bevilacqua and R.M.Nelson, Selectins., J.Clin.Invest. 91:379 (1993).
2. B.K.Brandley, S.J.Swiedler, and P.W.Robbins, Carbohydrate ligands of the LEC cell adhesion molecules., Cell 63:861 (1990).
3. H.C.Gooi, S.J.Thorpe, E.F.Hounsell, H.Rumpold, D.Kraft, O.Forster, and T.Feizi, Marker of peripheral blood granulocytes and monocytes of man recognized by two monoclonal antibodies VEP8 and VEP9 involves the trisaccharide 3-fucosyl-N-acetyllactosamine, Eur.J.Immunol. 13:306 (1983).
4. S.J.Thorpe and T.Feizi, Species differences in the expression of carbohydrate differentiation antigens on mammalian blood cells revealed by immunofluorescence with monoclonal antibodies, Biosci.Reps. 4:673 (1984).
5. T.Feizi, Demonstration by monoclonal antibodies that carbohydrate structures of glycoproteins and glycolipids are onco-developmental antigens, Nature 314:53 (1985).

6. M.Fukuda, N.Hiraoka, and J.C.Yeh, C-type lectins and sialyl Lewis X oligosaccharides. Versatile roles in cell-cell interaction, J.Cell.Biol. 147:467 (1999).
7. R.P.McEver, K.L.Moore, and R.D.Cummings, Leukocyte trafficking mediated by selectin-carbohydrate interactions, J.Biol.Chem. 270:11025 (1995).
8. S.D.Rosen and C.R.Bertozzi, Leukocyte adhesion: Two selectins converge on sulphate, Current Biology 6:261 (1996).
9. P.R.Crocker and T.Feizi, Carbohydrate recognition systems: functional triads in cell-cell interactions, Curr.Opin.Struct.Biol. 6:679 (1996).
10. T.Feizi, M.S.Stoll, W.Chai, and A.M.Lawson, Neoglycolipids: probes of oligosaccharide structure, antigenicity and function, Methods Enzymol. 230:484 (1994).
11. A.M.Lawson, W.Chai, C.-T.Yuen, and T.Feizi, Mass spectrometry in carbohydrate recognition studies, *in:* "Mass Spectrometry in Biological Sciences", A.L.Burlingame and S.A.Carr, eds., Humana Press, Totowa, New Jersey (1996).
12. P.W.Tang, H.C.Gooi, M.Hardy, Y.C.Lee, and T.Feizi, Novel approach to the study of the antigenicities and receptor functions of carbohydrate chains of glycoproteins., Biochem.Biophys.Res. Commun. 132:474 (1985).
13. T.Feizi and R.A.Childs, Neoglycolipids: probes in structure/function assignments to oligosaccharides, Methods Enzymol. 242:205 (1994).
14. T.Osanai, T.Feizi, W.Chai, A.M.Lawson, M.L.Gustavsson, K.Sudo, M.Araki, K.Araki, and C.-T.Yuen, Two families of murine carbohydrate ligands for E-selectin, Biochem.Biophys.Res.Commun. 218:610 (1996).
15. A.M.Lawson, W.Chai, G.C.Cashmore, M.S.Stoll, E.F.Hounsell, and T.Feizi, High-sensitivity structural analyses of oligosaccharide probes (neoglycolipids) by liquid-secondary-ion mass spectrometry, Carbohydr.Res. 200:47 (1990).
16. W.Chai, G.C.Cashmore, R.A.Carruthers, M.S.Stoll, and A.M.Lawson, Optimal procedure for combined high-performance thin-layer chromatography/high-sensitivity liquid secondary ion mass spectrometry, Biol.Mass Spectrom. 20:169 (1991).
17. C.-T.Yuen, A.M.Lawson, W.Chai, M.Larkin, M.S.Stoll, A.C.Stuart, F.X.Sullivan, T.J.Ahern, and T.Feizi, Novel sulfated ligands for the cell adhesion molecule E-selectin revealed by the neoglycolipid technology among O-linked oligosaccharides on an ovarian cystadenoma glycoprotein, Biochemistry 31:9126 (1992).
18. D.Spillmann, K.Hard, O.J.Thomas, J.F.G.Vliegenthart, G.Misevic, M.M.Burger, and J.Finne, Characterization of a novel pyruvylated carbohydrate unit implicated in the cell aggregation of the marine sponge Microciona prolifera, J.Biol.Chem. 268:13378 (1993).
19. W.Chai, T.Feizi, C.-T.Yuen, and A.M.Lawson, Nonreductive release of *O*-linked oligosaccharides from mucin glycoproteins for structure/function assignments as neoglycolipids: application in the detection of novel ligands for E-selectin, Glycobiology 7:861 (1997).
20. C.-T.Yuen, W.Chai, R.W.Loveless, A.M.Lawson, R.U.Margolis, and T.Feizi, Brain contains HNK-1 immunoreactive *O*-glycans of the sulfoglucuronyl lactosamine series that terminate in 2-linked or 2,6-linked hexose (mannose), J.Biol.Chem. 272:8924 (1997).
21. W.Chai, C.T.Yuen, H.Kogelberg, R.A.Carruthers, R.U.Margolis, T.Feizi, and A.M.Lawson, High prevalence of 2-mono- and 2,6-di-substituted Manol-terminating sequences among O-glycans released from brain glycopeptides by reductive alkaline hydrolysis, Eur.J.Biochem. 263:879 (1999).
22. Stoll MS, Feizi T, R.W.Loveless, W.Chai, A.M.Lawson, and C.-T.Yuen, Fluorescent neoglycolipids: improved probes for oligosaccharide ligand discovery, Eur.J.Biochem. 267:1795 (2000).
23. T.Feizi, Oligosaccharides that mediate mammalian cell-cell adhesion: Curr.Opin.Struct.Biol. 3:701 (1993).
24. M.Larkin, T.J.Ahern, M.S.Stoll, M.Shaffer, D.Sako, J.O'Brien, A.M.Lawson, R.A.Childs, K.M.Barone, P.R.Langer-Safer, A.Hasegawa, M.Kiso, G.R.Larsen, and T.Feizi, Spectrum of sialylated and non-sialylated fuco-oligosaccharides bound by the endothelial-leukocyte adhesion molecule E-selectin. Dependence of the carbohydrate binding activity on E-selectin density, J.Biol.Chem. 267:13661 (1992).
25. N.Kojima, K.Handa, W.Newman, and S.-I.Hakomori, Multi-recognition capability of E-selectin in a dynamic flow system, as evidenced by differential effects of sialidases and anti-carbohydrate antibodies on selectin-mediated cell adhesion at low vs hgigh wall shear stress: A preliminary note., Biochem.Biophy.Res.Commun. 189:1686 (1992).
26. T.Feizi, Blood group-related oligosaccharides are ligands in cell-adhesion events, Biochem.Soc.Trans. 20:274 (1992).
27. R.Alon, T.Feizi, C.-T.Yuen, R.C.Fuhlbrigge, and T.A.Springer, Glycolipid ligands for selectins support leukocyte tethering and rolling under physiologic flow conditions, J.Immunol. 154:5356 (1995).
28. C.-T.Yuen, K.Bezouska, J.O'Brien, M.S.Stoll, R.Lemoine, A.Lubineau, M.Kiso, A.Hasegawa, N.J.Bockovich, K.C.Nicolaou, and T.Feizi, Sulfated blood group Lewis^a A superior oligosaccharide ligand for human E-selectin, J.Biol.Chem. 269:1595 (1994).
29. H.Kogelberg, T.A.Frenkiel, S.W.Homans, A.Lubineau, and T.Feizi, Conformational studies on the selectin and natural killer cell receptor ligands sulfo- and sialyl-lacto-*N*-fucopentaoses (SuLNFPII and

SLNFPII) using NMR spectroscopy and molecular dynamics simulations. Comparisons with the nonacidic parent molecule LNFPII, Biochemistry 35:1954 (1996).
30. D.Tyrrell, P.James, N.Rao, C.Foxall, S.Abbas, F.Dasgupta, M.Nashed, A.Hasegawa, M.Kiso, D.Asa, and B.K.Brandley, Structural requirements for the carbohydrate ligand of E-selectin, Proc.Nat.Acad.Sci.USA 88:10372 (1991).
31. P.L.Smith, K.M.Gersten, B.Petryniak, R.J.Kelly, C.Rogers, Y.Natsuka, J.A.Alford III, E.P.Scheidegger, S.Natsuka, and J.B.Lowe, Expression of the α(1-3)fucosyltransferase Fuc-TVII in lymphoid aggregate high endothelial venules correlates with expression of L-selectin ligands, J.Biol.Chem. 271:8250 (1996).
32. P.Maly, A.D.Thall, B.Petryniak, C.E.Rogers, P.L.Smith, R.M.Marks, R.J.Kelly, K.M.Gersten, G.Cheng, T.L.Saunders, S.A.Camper, R.T.Camphausen, F.X.Sullivan, Y.Isogai, O.Hindsgaul, U.H.von Andrian, and J.B.Lowe, The alpha(1,3)fucosyltransferase Fuc-TVII controls leukocyte trafficking through an essential role in L-, E-, and P-selectin ligand biosynthesis, Cell 86:643 (1996).
33. M.Sekine, K.Nakamura, M.Suzuki, F.Inagaki, T.Yamakawa, and A.Suzuki, A single autosomal gene controlling the expression of the extended globoglycolipid carrying SSEA-1 determinant is responsible for the expression of two extended globogangliosides, .J.Biochem.Tokyo. 103:722 (1988).
34. S.D.Rosen, L-Selectin and its biological ligands., Histochemistry 100:185 (1993).
35. E.L.Berg, J.Magnani, R.A.Warnock, M.K.Robinson, and E.C.Butcher, Comparison of L-selectin and E-selectin ligand specificities:the L-selectin can bind the E-selectin ligands sialyl-Le^x and sialyl Le^a, Biochemica Biophysica Research Communications 184:1048 (1992).
36. C.Foxall, S.R.Watson, D.Dowbenko, C.Fennie, L.A.Lasky, M.Kiso, A.Hasegawa, D.Asa, and B.K.Brandley, The three members of the selectin receptor family recognize a common carbohydrate epitope, the sialyl Lewisx oligosaccharide., J.Cell Biol. 117:895 (1992).
37. K.E.Norgard, H.Han, L.Powell, M.Kriegler, A.Varki, and N.M.Varki, Enhanced interaction of L-selectin with the high enthelium venule ligand via selectively oxidized sialic acids, Proc.Natl.Acad.Sci.USA 90:1068 (1993).
38. A.Aruffo, W.Kolanus, G.Walz, F.Fredman, and B.Seed, CD62/P-Selectin recognition of myeloid and tumor cell sulfatides, Cell 67:35 (1991).
39. P.J.Green, T.Tamatani, T.Watanabe, M.Miyasaka, A.Hasegawa, M.Kiso, M.S.Stoll, and T.Feizi, High affinity binding of the leucocyte adhesion molecule L-selectin to 3'-sulfated-Le^a and -Le^x oligosaccharides and the predominance of sulphate in this interaction demonstrated by binding studies with a series of lipid-linked oligosaccharides, Biochem.Biophys.Res.Commun. 188:244 (1992).
40. Y.Suzuki, Y.Toda, T.Tamatani, T.Watanabe, T.Suzuki, T.Nakao, K.Murase, M.Kiso, A.Hasegawa, K.Tanado-Aritomi, I.Ishizuka, and M.Miyasaka, Glycolipids are ligands for a lymphocyte homing receptor, L-selectin (LECAM-1), binding epitope in sulfated sugar chain, Biochem.Biophys.Res.Commun. 190:426 (1993).
41. R.M.Nelson, S.Dolich, A.Aruffo, O.Cecconi, and M.P.Bevilacqua, Higher-affinity oligosaccharide ligands for E-selectin, J.Clin.Invest. 91:1157 (1993).
42. L.K.Needham and R.L.Schnaar, The HNK-1 reactivity Sulfoglucuronyl Glycolipids are ligands for L-selectin and P-selectin but not E-selectin, Proc.Natl.Acad.Sci.USA 90:1359 (1993).
43. P.J.Green, C.-T.Yuen, R.A.Childs, W.Chai, M.Miyasaka, R.Lemoine, A.Lubineau, B.Smith, H.Ueno, K.C.Nicolaou, and T.Feizi, Further studies of the binding specificity of the leukocyte adhesion molecule, L-selectin, towards sulfated oligosaccharides - Suggestion of a link between the selectin- and the integrin-mediated lymphocyte adhesion systems, Glycobiology 5:29 (1995).
44. C.Galustian, R.A.Childs, C.-T.Yuen, A.Hasegawa, M.Kiso, A.Lubineau, G.Shaw, and T.Feizi, Valency dependent patterns of reactivity of human L-selectin towards sialyl and sulfated oligosaccharides of Le^a and Le^x types: Relevance to anti-adhesion therapeutics, Biochemistry 36:5260 (1997).
45. C.Galustian, A.Lubineau, C.le Narvor, M.Kiso, G.Brown, and T.Feizi, L-selectin interactions with novel mono- and multisulfated Lewis^x sequences in comparison with the potent ligand 3'-sulfated Lewis^a, J Biol Chem 274:18213 (1999).
46. S.Hemmerich, H.Leffler, and S.D.Rosen, Structure of the *O*-glycans in GlyCAM-1, an endothelial-derived ligand for L-selectin, J.Biol.Chem. 270:12035 (1995).
47. S.Komba, H.Ishida, M.Kiso, and A.Hasegawa, Synthesis and biological activities of three sulfated sialyl Le^x ganglioside analogues for clarifying the real carbohydrate ligand structure of L-selectin, Bioorg.Med.Chem. 4:1833 (1996).
48. Y.Yoshino, H.Ohmoto, N.Kondo, H.Tsujishita, Y.Hiramatsu, Y.Inoue, H.Kondo, H.Ishida, M.Kiso, and A.Hasegawa, Studies on selectin blockers. 4. Structure-function relationships of sulfated sialyl Lewis X hexasaccharide ceramides toward E-P, P-, and L-selectin binding, J.Med.Chem. 40:455 (1997).
49. C.Galustian, A.M.Lawson, S.Komba, H.Ishida, M.Kiso, and T.Feizi, Sialyl-Lewis^x sequence 6-*O*-sulfated at *N*-acetylglucosamine rather than at galactose is the preferred ligand for L-selectin and de-*N*-acetylation of the sialic acid enhances the binding strength, Biochem.Biophys.Res.Comm 240:748 (1997).
50. C.Mitsuoka, K.N.Kawakami, S.M.Kasugai, N.Hiraiwa, K.Toda, H.Ishida, M.Kiso, A.Hasegawa, and R.Kannagi, Sulfated sialyl Lewis X, the putative L-selectin ligand, detected on endothelial cells of high

endothelial venules by a distinct set of anti-sialyl Lewis X antibodies [published erratum appears in Biochem Biophys Res Commun 1997 Apr 17;233(2):576], Biochem.Biophys.Res.Commun. 230:546 (1997).
51. C.Mitsuoka, K.M.Sawada, F.K.Ando, M.Izawa, H.Nakanishi, S.Nakamura, H.Ishida, M.Kiso, and R.Kannagi, Identification of a major carbohydrate capping group of the L-selectin ligand on high endothelial venules in human lymph nodes as 6-sulfo sialyl Lewis X, J.Biol.Chem. 273:11225 (1998).
52. N.Kimura, C.Mitsuoka, A.Kanamori, N.Hiraiwa, K.Uchimura, T.Muramatsu, T.Tamatani, G.S.Kansas, and R.Kannagi, Reconstruction of functional L-selectin ligands on a cultured human endothelial cell line by cotransfection of $\alpha 1 \rightarrow 3$ fucosyltransferase VII and newly cloned GlcNAcβ:6–sulfotransferase cDNA, Proc.Nat.Acad.Sci.USA 96:4530 (1999).
53. A.Bistrup, S.Bhakta, J.K.Lee, Y.Y.Belov, M.D.Gunn, Z.Feng-Rong, C.C.Huang, R.Kannagi, S.D.Rosen, and S.Hemmerich, Sulfotransferases of two specificities function in the reconstitution of high endothelial cell ligands for L-selectin, J.Cell.Biol 145:899 (1999).
54. N.Hiraoka, B.Petryniak, J.Nakayama, S.Tsuboi, M.Suzuki, J.C.Yeh, D.Izawa, T.Tanaka, M.Miyasaka, J.B.Lowe, and M.Fukuda, A novel, high endothelial venule-specific sulfotransferase expresses 6- sulfo sialyl Lewis(x), an L-selectin ligand displayed by CD34, Immunity. 11:79 (1999).
55. K.Tangemann, A.Bistrup, S.Hemmerich, and S.D.Rosen, Sulfation of a high endothelial venule-expressed ligand for L-selectin. Effects on tethering and rolling of lymphocytes, J.Exp.Med. 190:935 (1999).
56. C.Auge, F.Dagron, R.Lemoine, C.Le Narvor, and A.Lubineau, Syntheses of sulfated derivatives as sialyl Lewis^a and sialyl Lewis^x analogues, *in:* "Carbohydrate Mimics: Concepts and Methods", Y.Chapleur, ed., Verlag Chemie, Weinheim (1997).
57. S.Komba, C.Galustian, H.Ishida, T.Feizi, R.Kannagi, and M.Kiso, The first total synthesis of 6-sulfo-de-*N*-acetylsialyl Lewis^x ganglioside: A superior ligand for human L-selectin, Angew.Chem.Int.Ed. 38:1131 (1999).
58. N.Hanai, T.Dohi, G.A.Nores, and S.Hakomori, A novel ganglioside, de-N-acetyl-GM3 (II3NeuNH2LacCer), acting as a strong promoter for epidermal growth factor receptor kinase and as a stimulator for cell growth, J.Biol.Chem. 263:6296 (1988).
59. A.E.Manzi, E.R.Sjoberg, S.Diaz, and A.Varki, Biosynthesis and turnover of *O*-acetyl and *N*-acetyl groups in the gangliosides of human melanoma cells, J.Biol.Chem. 265:13091 (1990).
60. E.R.Sjoberg, R.Chammas, H.Ozawa, I.Kawashima, K.H.Khoo, H.R.Morris, A.Dell, T.Tai, and A.Varki, Expression of de-N-acetyl-gangliosides in human melanoma cells is induced by genistein or nocodazole, J.Biol.Chem. 270:2921 (1995).
61. C.Mitsuoka, K.Ohmori, N.Kimura, A.Kanamori, S.Komba, H.Ishida, M.Kiso, and R.Kannagi, Regulation of selectin binding activity by cyclization of sialic acid moiety of carbohydrate ligands on human leukocytes, Proceedings of the National Academy of Sciences, U.S.A. 96:1597 (1999).
62. C.R.Bertozzi, S.Fukuda, and S.D.Rosen, Sulfated disaccharide inhibitors of L-selectin: Deriving structural leads from a physiological selectin ligand, Biochemistry 34:14271 (1995).
63. K.Norgard-Sumnicht and A.Varki, Endothelial heparan sulfate proteoglycans that bind to L-selectin have glucosamine residues with unsubstituted amino groups, J.Biol.Chem. 270:12012 (1995).
64. Y.Tanaka, D.H.Adams, and S.Shaw, Proteoglycans on endothelial cells present adhesion-inducing cytokines to leukocytes, Immunol.Today 14:111 (1993).
65. T.Feizi and C.Galustian, Novel oligosaccharide ligands and ligand-processing pathways for the selectins, Trends in Biochemical Sciences 24:369 (1999).
66. C.Leteux, R.A.Childs, W.Chai, M.S.Stoll, H.Kogelberg, and T.Feizi, Biotinyl-L-3-(2-naphthyl)-alanine hydrazide derivatives of *N*-glycans: versatile solid-phase probes for carbohydrate-recognition studies, Glycobiology 8:227 (1998).
67. K.Yamamoto, T.Tsuji, and T.Osawa, Requirement of the core structure of a complex-type glycopeptide for the binding to immobilized lectins, Carbohydr.Res. 110:283 (1982).
68. C.Leteux, M.S.Stoll, R.A.Childs, W.Chai, M.Vorozhaikina, and T.Feizi, Influence of oligosaccharide presentation on the interactions of carbohydrate sequence-specific antibodies and the selectins. Observations with biotinylated oligosaccharides, Journal of Immunological Methods 227:109 (1999).
69. D.Solis, T.Feizi, C.T.Yuen, A.M.Lawson, R.A.Harrison, and R.W.Loveless, Differential recognition by conglutinin and mannan-binding protein of *N*-glycans presented on neoglycolipids and glycoproteins with special reference to complement glycoprotein C3 and ribonuclease B, J.Biol.Chem. 269:11555 (1994).
70. F.Li, P.P.Wilkins, S.Crawley, J.Weinstein, R.D.Cummings, and R.P.McEver, Post-translational modifications of recombinant P-selectin glycoprotein ligand-1 required for binding to P- and E-selectin, J.Biol.Chem. 271:3255 (1996).
71. A.Kameyama, H.Ishida, M.Kiso, and A.Hasegawa, Synthetic studies on sialoglycoconjugates 22: total synthesis of tumor-associated ganglioside, sialyl Lewis X, J.Carbohydr.Chem. 10:549 (1991).

ANIMAL LECTINS: FROM INITIAL DESCRIPTION TO ELABORATED STRUCTURAL AND FUNCTIONAL CLASSIFICATION

Herbert Kaltner and Hans-J. Gabius

Institute of Physiological Chemistry, Faculty of Veterinary Medicine, Ludwig-Maximilians-University, Veterinärstr. 13, D-80539 München, Germany

The title of this book highlights the inherent complexity of glycan chains attached to protein or lipid carriers. On cell surfaces, such complex carbohydrates protrude prominently into the environment as antennae-like structures. Intuitively, one readily dotes on the assumption that this strategic positioning is not merely a coincidence without functional implications. Conversely, glycan determinants capture the imagination with their promise of serving as information-storing units. Indeed, this concept is strongly advocated by theoretical reasoning on coding capacity by biomolecules. On the level of a hexamer, amino acids used as letters of an alphabet can form 6.4×10^7 different structures (six-letter words). When starting again with a panel of twenty different monomers (letters), the astounding number of 1.44×10^{15} hexasaccharides can be established[1]. In terms of coding capacity, glycans as hardware surpass peptides by more than seven orders of magnitude on the level of hexamers. Unquestionably, this complexity impinges on the level of sophistication of glycan analysis methodology[2,3].

Equipped with refined chromatographic, electrophoretic and spectroscopic techniques, it is no longer a source of frustration to embark on glycan analysis. From the realm of this elegant work, it is especially noteworthy that end groups of glycan antennae or side branches with spatial accessibility on glycoproteins can expose rather uncommon saccharide variants relative to the core portions of glycans such as sulfated hexoses/hexosamines in N- and O-glycans[3-6]. Immediately, the idea of their involvement in information transfer comes to

The Molecular Immunology of Complex Carbohydrates–2.
Edited by Albert M. Wu, Kluwer Academic/Plenum Publishers, 2001.

mind. Corroborating this reasoning, defects in the generation of distinct glycan epitopes are attributed etiologically to clinical syndromes. How a glycan assembly dysfunction drastically impairs the normal phenotype is evident in the leukocyte adhesion deficiency type II syndrome (defect in the production of the donor for fucosylation, GDP-Fuc) or the I (inclusion)-cell disease (mucolipidosis II) and its milder variant (Pseudo-Hurler polydystrophy; defect in introducing 6′-phosphate moieties into the high-mannose-type N-glycans of lysosome-destined hydrolases)[6,7]. In the words of Winterburn and Phelps[8] „carbohydrates are ideal for generating compact units with explicit informational properties, since the permutations on linkages are larger than can be achieved by amino acids, and, uniquely in biological polymers, branching is possible". It is no treading on thin ice to follow the authors in their conclusion that „the significance of the glycosyl residues is to impart a discrete recognitional role on the protein" [8]. Via a molecular rendezvous with carbohydrate-utilizing enzymes (e.g. glycosyltransferases, glycosidases and enzymes in carbohydrate metabolism), sugar-specific immunoglobulins and especially members of the large family of carbohydrate-binding proteins which are neither enzymes or antibodies, referred to as lectins, the message can be decoded. This can for example lead to elicitation of intracellular signaling pathways for an array of distinct responses[9]. In this sense, the glycan display can be thought of as a characteristic feature like a signature whose documentation and verification by observers - lectins - triggers various activities in cell sociology. This consideration of the ligand capacity of carbohydrate chains automatically guides us to the question whether lectins can be classified as neatly as their complementary partners. Here, mapping of the structures has rendered main principles of glycan assembly and inherent variability readily discernible[6].

I. GENEALOGY OF ANIMAL LECTINS

The very early phases of research on lectins were dominated by the focus on agglutinating activities in plants such as castor bean or jequirity[10,11]. Likewise, the hemolymph of crustaceans and snake venoms were observed to be a source for erythroagglutinating substances. From mammals, a protein with capacity for „conglutination" of alexinated (activated complement-coated) red blood cells has been described in 1906, which was later referred to as the collectin conglutinin[12]. Like concanavalin A or ricin, whose crosslinking of erythrocytes or glycogen/mucoproteins by protein-sugar recognition is the basis for agglutination, as observed by Landsteiner and Raubitschek (1909) and Sumner and Howell in

1936[10], animal lectins have initially been placed into categories established by the type of monosaccharide inhibiting this activity. As additional characteristic, the dependence on the presence of Ca^{2+}-ions for lectin activity served to furnish a refinement, which has been useful in the monitoring of lectins from tumors[13,14]. Once sequence information for animal lectins became available, it was instrumental to usher in alignment by cluster analysis, a powerful tool in genealogical lectin research[15]. With 14 identical sequence positions and 18 residues conserved in character, a domain of 115-134 amino acids has been unveiled to be characteristic for a family of Ca^{2+}-dependent (C-type) lectins. When the traced similarity is graphically translated into an evolutionary dendrogram, seven subfamilies for C-type lectins can presently be discerned[16]. Remarkably, the sugar specificity of individual sugar receptors even in a subfamily can vary, with galactose, N-acetylgalactosamine or mannose, L-fucose, and N-acetylglucosamine as key target epitopes. In a case, where a crystal structure of the complex is available, the unusual feature of coordination bonds between a Ca^{2+}-ion and the equatorial vicinal hydroxyl groups 3 and 4 of mannose is observed, namely for the mannan-binding lectin, a member of the collectin subfamily[17]. Overall, the C-type-like sequence stretch present in the family members′ structure and its folding pattern is as distinguishing as a fingerprint to assign any newly described protein to this family. Therefore, the structural motif supersedes monosaccharide specificity as decisive factor for lectin classification, as unequivocally illustrated in Table 1.

Currently, five distinct families are conclusively defined[18]. The steadily increasing number of lectin sequences from animals of various branches of the evolutionary tree has also made possible to take stock of a hallmark of structural organization of diverse lectins mentioned in this table. As shown in Fig. 1 for the selectins, a subfamily of C-type lectins, the carbohydrate recognition domain (CRD) is often linked to other well-defined protein modules. In this case and also in I-type lectins of the sialoadhesin subgroup, the modular arrangement with the CRD prominently facing outwards from the cell surface makes sure that the CRD can efficiently function as accessible sensor for any ligand on opposing cell surfaces. A graphic example of the time scale for intrafamily diversification in a genealogical tree is provided by the two avian galectins CG-14 and CG-16 which are assumed to exhibit overlapping but also distinct binding properties and functions[19,20]. The disparate developmental regulation with organ-typical patterns intimates operativeness of intricate control mechanisms[20,21]. Since the time of origin of the two genes after duplication of the putative ancestral genetic coding unit has been assigned to a period close to the divergence of birds and mammals approximately 3 x 10^8 years ago[22], the length of this period should have clearly been sufficient for these changes to be imprinted in the genome.

Table 1. Main families of animal lectins[18]

Family	Structural motif	Carbohydrate ligand	Modular arrangement
C-type	conserved CRD	variable (mannose, galactose, fucose, heparin tetrasaccharide)	yes
I-type	immunoglobulin-like CRD	variable (man6glcNAc2, HNK-1 epitope, hyaluronic acid, α2,3/α2,6-sialyllactose)	yes
galectins	conserved CRD	Galβ1,3(4)GlcNAc core structures with species-and galectin type-dependent differences in affinity for extensions to blood group A,B or H epitopes, internal stretches of poly(N-acetyllactosamine) chains	variable
pentraxins	pentameric subunit arrangement	4,6-cyclic acetal of β-galactose, galactose, sulfated and phosphorylated monosaccharides	yes
P-type	conserved CRD	mannose-6-phosphate-containing glycoproteins	yes

In mammals, so far up to nine members of the galectin family have been registered (Table 2). It is remarkable that besides fine-structural differences in the CRD the topological display of the CRD differs encompassing monomeric galectins (galectins -5/-7), the non-covalently linked homodimeric galectins-1/-2, galectin-3 with propensity for non-covalent self-association via its N-/C-domains and the tandem-repeat proteins[18]. As readily envisaged, such an orientation is ideally suited for bridging and crosslinking purposes.

Table 2. Members of the galectin family of mammalian lectins[18]

Name	Occurrence	Structural features
galectin-1 (galaptin, L-14)	many cell types	homodimer; one CRD per subunit (12-16 kDa);proto type
galectin-2	lower small intestine; clone from human hepatoma	homodimer; one CRD per subunit (43% sequence identity to galectin-1; 14 kDa): proto type
galectin-3 (CBP35, Mac-2, IgE-binding protein, L-29, L-34)	many cell types	monomer with one CRD (oligomer formation in solution and on surfaces); pro-, tyr-, and gly-rich repeats in N-terminal section (29-37 kDa): chimera type
galectin-4	colon, small intestine, stomach, oral epithelium, esophagus	monomer with two partially homologous but distinct CRDs, connected by a linker region (36 kDa), proteolysis generates truncated proto-type-like products: tandem-repeat type
galectin-5	reticulocytes, erythrocytes	monomer with one CRD (17 kDa): proto type
galectin-6	small intestine, colon	tandem-repeat arrangement of two CRDs (33 kDa)
galectin-7	keratinocytes, rat mammary carcinomas	one CRD (12.7 kDa): proto type
galectin-8	lung, liver, kidney, muscle, spleen, brain	homologous to galectins-4 and -6 (tandem-repeat arrangement of two CRDs with unique link peptide; 34 kDa)
galectin-9	small intestine, liver, lung, kidney, thymus (rat/mouse; small intestinal isoform with 31/32 amino acid extension of link peptide); lymphatic tissue and B cells, T cells and macrophages, liver, lung, pancreas, colon carcinoma cells (human)	homologous to galectins-4, -6 and -8 (tandem-repeat arrangement of two CRDs with unique link peptide; 36 kDa)
Charcot-Leyden crystal protein	major autocrystallizing constituent of eosinophils and basophils	one CRD and lysophospholipase activity (16.5 kDa)

Since new information is accrued step by step, the classification of Table 1 should not at all be considered final. The molecular chaperones calnexin and calreticulin, the lectins VIP-36 and ERGIC-53 involved in intracellular trafficking of glycoproteins in the secretory pathway, glycosaminoglycan-binding proteins, the putative carbohydrate-binding capacity of the fibrinogen-like domain in ficolins and β1,3-glucan binding by a member of the hexamerin family of proteins testify that additions to this compilation will soon be required to keep abreast of the knowledge in this field[18,23,24]. However, the stringency of the criteria for classification must not be compromised by premature extension of the presented table without fully meeting its quality criteria. For the lectins mentioned therein and for those, whose proper classification is largely a matter of time, an impressive wealth of information has been gathered on the versatility of functional implications presented in condensed form in the next paragraph.

II. A KALEIDOSCOPE OF LECTIN FUNCTIONS

Recalling the bridging/crosslinking capacity of galectins and the strategic positioning of the ligand contact site of selectins (Fig. 1), a role of lectins in cell-cell and cell-matrix adhesion immediately enters our mind. Theoretically, an array of glycobiological means exist to the same end (Fig. 2).

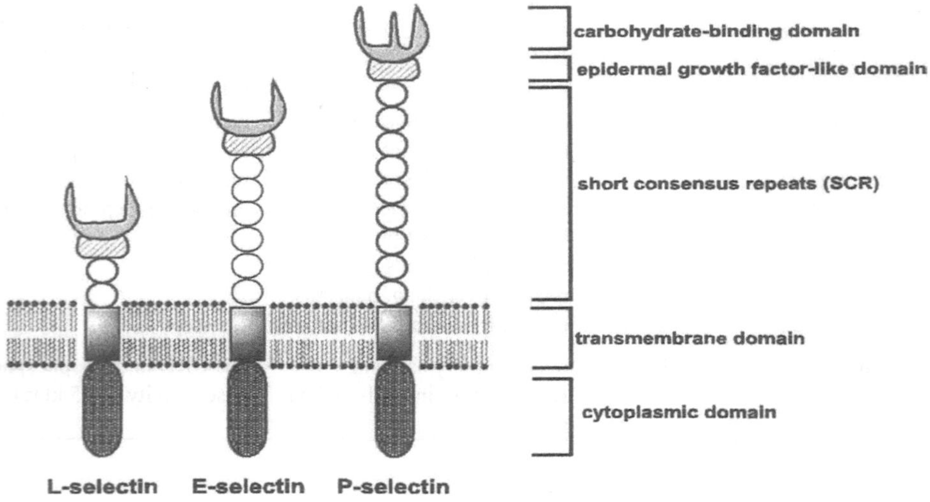

Figure 1. Structural characteristics of selectins. The modular arrangement of the three family members starts with an extracellular N-terminal lectin domain and an epidermal growth factor (EGF)-like region. Two (L-selectin), six (E-selectin), and nine (P-selectin) consensus repeats homologous to a complement-binding domain, a transmembrane region and a cytoplasmic C-terminal domain complete the modular arrangement of this type of transmembrane proteins.

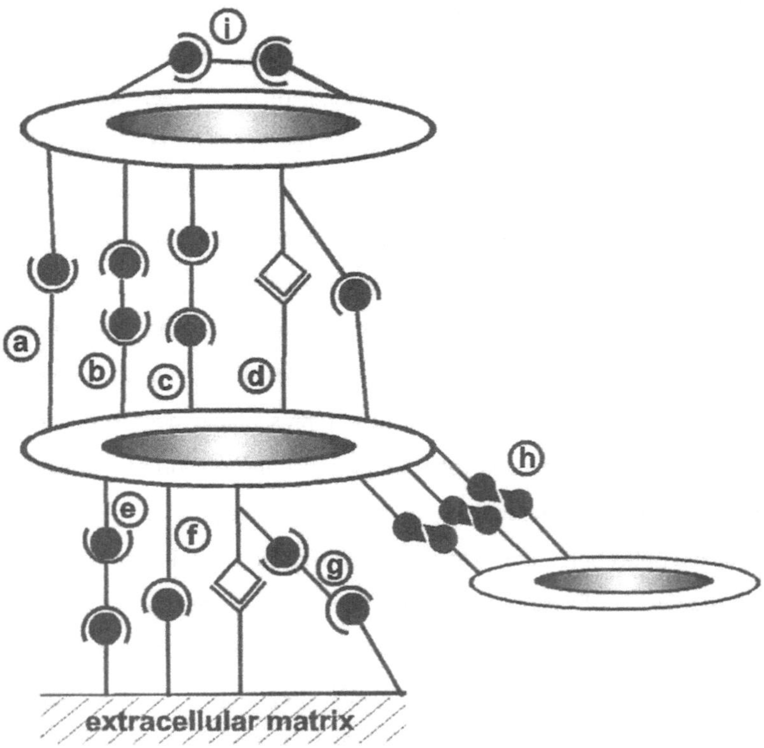

Figure 2 : Schematic drawing of theoretical possibilities for cell surface interactions involving oligosaccharides and lectins. Two cells can associate by presentation of lectin-glycoligand pairs on apposing surfaces such as direct binding between a lectin and its appropriate carbohydrate ligand (**a**). A further binding mode either engages a bi- or multivalent lectin which is capable to bridge carbohydrate ligands on the cell surface of neighboring cells (**b**) or soluble bi- or multivalent carbohydrate ligands (**c**) which can crosslink lectins which are non-covalently associated with the cell surface or anchored with a transmembrane domain in the cell membrane. A peculiar mode of action is given in (**d**). Besides being engaged in protein-carbohydrate interaction the lectin can also bind to a protein ligand on the apposing cell membrane, establishing a two-point association favored by different modules in the lectin sequence. Glycobiological interactions with molecules of the basal lamina and the extracellular matrix are represented by types (**e**)-(**g**). Either the bridging mode (**e**) or the direct binding mode (**f**) result in the adhesion of cells to constituents of the basal laminia (e.g. laminin) and/or the extracellular matrix (e.g. fibronectin). The situation in (**g**) reflects the possibility for independent protein-protein and protein-carbohydrate interactions involving the principal binding mode (**e**), i.e. the binding of a dimeric lectin crosslinking the carbohydrate chains of cell surface proteins, which are also involved in protein-protein interactions with oligosaccharide chains of matrix proteins. Thus, the strength of the interaction can readily be enhanced to effect tight binding. If clusters of complementary carbohydrate surfaces are suitably located, adhesion can also be mediated by carbohydrate-carbohydrate interactions (**h**). Bridging is not restricted to different cell surfaces, but can also occur between ligands on the cell surface of the same cell (**i**). As currently known, the binding type in situation (**a**) is preferred e.g. by C-type lectins such as selectins, several galectins and I-type lectins. Galectins of the proto- and tandem-repeat type such as galectin-1 or galectin-8, respectively, appear capable to mediate their effects via the modes illustrated in (**b**), (**e**), (**f**), (**g**) and (**i**).

Indeed, this recognitive mode is encountered physiologically, turning cell type-selective glycosignatures into a productive molecular rendezvous[18,25,26]. In these processes, lectins from various families, i.e. C-type lectins, I-type lectins and galectins, can cooperate, as illustrated exemplarily for host and tumor cells during the metastatic cascade (Fig. 3). However, it would be a misconception to restrict the range of lectin activities to intercellular communication. Tracking down foreign glycosignatures in innate (non-adaptive) immunity, targeting enzyme activity via a lectin module or employing a lectin as delivery destination for biologically active glycoproteins, e.g. in hepatic clearance of pituitary glycoprotein hormones[5], are counted among the documented functions (Table 3).

Rightfully, while the initial steps of description of lectins in animals has lagged behind the research activities on plant lectins for decades, the given list of animal functions indicates that this area of analysis has reached a presentable degree of maturity, when compared to delineating endogenous roles of plant lectins[27]. Moreover, deciphering these functional implications is a prerequisite to turn these insights from basic science into promising medical applications. For example, combining glycan tailoring by chemoenzymatic synthesis with the design of pharmacologically appropriate carriers as scaffold for immobilization of lectin-directed determinants bodes well for production of neoglycoconjugates as tools for lectin detection and for lectin-mediated targeting[28-31]. These approaches take advantage of the non-uniform expression profile and functional pattern of lectins in diverse cell types. To enhance the target selectivity of these custom-made probes in search of lectins, it is pertinent to delineate the structural details of lectin-ligand interaction and its thermodynamic course.

Table 3. Functions of animal lectins (for collection of recent reviews, see [18, 39-41])

Activity	Example of lectin
ligand-selective molecular chaperones in endoplasmatic reticulum	calnexin, calreticulin
intracellular routing of glycoproteins and vesicles	ERGIC-53, VIP-36, P-type lectins, comitin
intracellular transport and extracellular assembly	non-integrin 67 kDa elastin/laminin-binding protein
cell type-specific endocytosis	hepatic asialoglycoprotein receptor, macrophage C-type lectins, hepatic endothelial cell receptor for GalNAc-4-SO_4-bearing glycoproteins
recognition of foreign glycans (β1,3-glucans, LPS)	CR3 (CD11b/CD18), *Limulus* coagulation factors C and G
recognition of foreign or aberrant glycosignatures on cells (incl. endocytosis or initiation of opsonization or complement activation)	collectins, pentraxins (CRP, limulin), C-type macrophage receptors, L-ficolin
targeting of enzymatic activity in multimodular proteins	acrosin
bridging of molecules	homodimeric and tandem-repeat galectins, cytokines (e.g. IL-2:IL-2R and CD3 of TCR), cerebellar soluble lectin
effector release (H_2O_2, cytokines etc.)	galectins, selectins, CD23
cell growth control and apoptosis	galectins, C-type lectins, amphoterin-like protein, cerebellar soluble lectin
cell routing	selectins, I-type lectins, galectins
cell-cell interactions	selectins and other C-type lectins, galectins, I-type lectins
cell-matrix interactions	galectins, heparin- and hyaluronic acid-binding lectins
matrix network assembly	proteoglycan core proteins (C-type CRD), galectins, non-integrin 67 kDa elastin/laminin-binding protein

Figure 3. Illustration of possible interactions involving endogenous lectins and appropriate glycoconjugates in the metastatic cascade. They can occur within the formation of tumor emboli, in the process of tumor cell adhesion to the endothelial lining and during the extravasation of tumor cells into the underlying tissue. Platelets can be activated by contact to tumor cells (**1**) resulting both in cell membrane presentation of P-selectin and in the release of platelet factors (**2**) which boost the expression of endothelial selectins and ICAM-molecules in conjunction with tumor cell-derived factors such as TGFβ or TNFα (**3**). In resemblance to the leukocyte-endothelial interaction cascade, tumor cell-endothelial cell interaction eventually leads to extravasation of tumor cells (**4**). Enhanced expression of selectins (adhesion molecules) on endothelial cells provokes an increased tethering and rolling of leukocytes, which may have a bearing on homing (**5**). While the selectin/galectin-mediated thromboemboli formation between tumor cells and tumor cells and platelets is supposed to protect viable tumor cells from shear forces in the bloodstream, tumor cells in the aggregates are also thought to escape the surveillance mechanisms of the host immune system, advancing the transport of cancer cells to ectopic sites. Galectins (**6 – 12**) are implicated in tumor metastasis [14,25,42-50]. Several common principles, as outlined in figure 1, are considered to participate in tumor cell aggregation and in adhesion to endothelial cells and to matrix components. Galectin-1 is present on the surface of endothelial cells (**7**)[49]extending the repertoire of molecules supporting tumor cell -endothelial cell adhesion. On adjacent tumor cells the bridging mode of homodimeric galectin-1 (*see* figure 1) can support tumor cell aggregation, exemplified by lysosomal membrane glycoprotein 1/2 (LAMP-1/2)[51] as ligands. Direct binding of galectin-3 to poly-N-acetyllactosamine chains of e.g. carcinoembryonic antigen (**9**), as known from colon carcinoma[52], and homotypic adhesion of tumor cells probably by cross-linking of cell surface galectin-3 mediated by a soluble mucin (**10**)[53] are further possibilities to strengthen the stability of tumor cell aggregates. Matrix interactions can proceed via binding typical glycoprotein constituents such as laminin or fibronectin[54]. I-type lectins (**11**) may also be involved in this complex process. It is reasonable to assume that they may recognize cancer-associated mucins or N-linked glycans owing to their content of α2,6- or α2,3-linked sialic acids in I-type lectin-reactive oligosaccharides in the peripheral part of their sugar antennae.

Inset: During the metastatic process, tumor cells must cross basement membranes to invade organ parenchyma. An initial and crucial step in this process is tumor cell attachment to the basement membrane via cell-surface receptors for e.g. laminin. Both galectin-1 and -3 bind to N-acetyllactosamine epitopes on oligosaccharide chains on laminin (**12**) and fibronectin (**13**), justifying their definition as non-integrin laminin/fibronectin-binding proteins, as a prerequisite for promoting spreading of tumor cells. Presence of galectins can thus modulate the adhesive properties of different tumor cell types[54].

III. HOW TO TAILOR AN OPTIMAL LECTIN LIGAND

A therapeutic aim, i.e. the design of potent carbohydrate ligands or mimetics thereof for anti-adhesive therapy, is depicted in Fig. 4. For this purpose, it would be of great benefit to precisely be aware of the contact points between receptor and ligand for the enthalpic gain, of any changes in the conformation of the ligand and the receptor and any reorganization of the solvent structure in the course of complexation. Translated into experimental approaches, the crystal structure and the properties of all involved reactants by calorimetric and NMR spectroscopic analyses, by molecular mechanics and dynamics simulations and by application of chemically engineered ligands are to be assessed to define reliable guidelines for rational drug design[32].

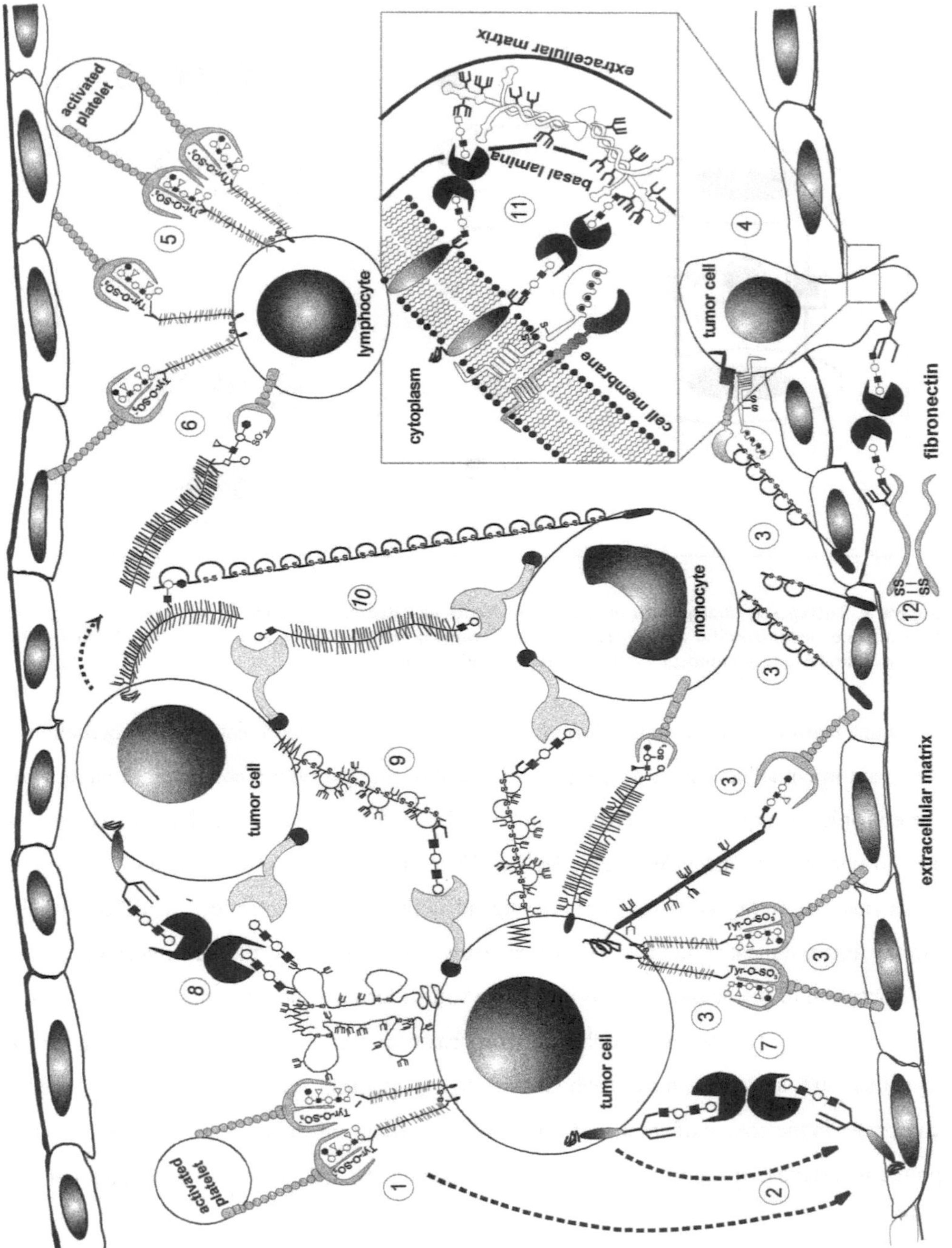
activated platelet
activated platelet
tumor cell
tumor cell
tumor cell
lymphocyte
monocyte
cytoplasm
cell membrane
basal lamina
extracellular matrix
extracellular matrix
fibronectin
1
2
3
4
5
6
7
8
9
10
11
12

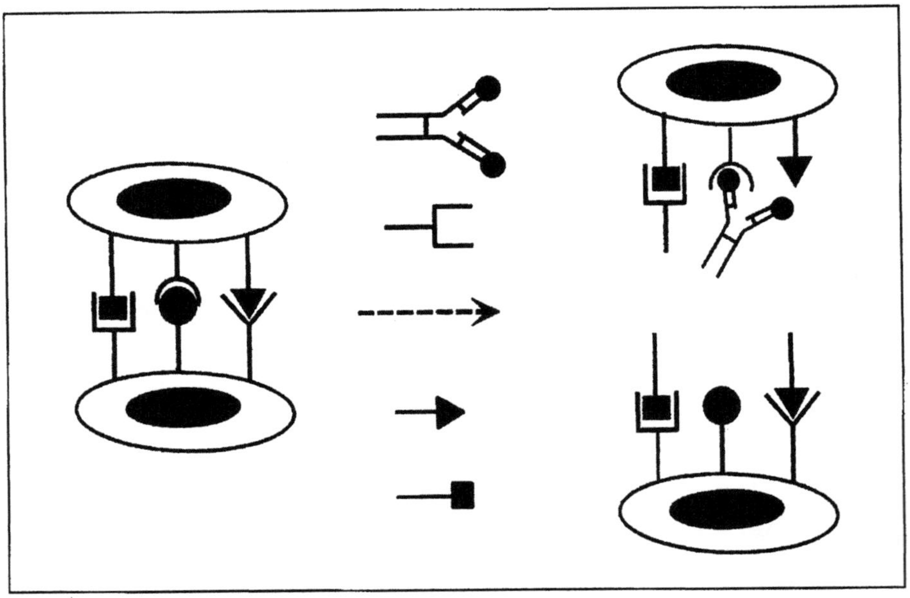

Figure 4. Therapeutic strategies to interfere with lectin-mediated cell adhesion by application of blocking reagents, i.e. lectin-specific antibodies, a ligand-occupying soluble lectin and an excess of tailor-made oligosaccharides or glycomimetics which saturate ligand-binding sites.

A salient feature of the complex formation is the inherent enthalpy-entropy compensation[33]. Any gain in enthalpic driving force is thus inevitably coupled with a concomitant increase of the entropic penalty and vice versa, as shown in Fig. 5 for data sets of binding of galactosides to two unrelated galactoside-specific lectins. Since drug research obviously cannot deal with modifying the endogenous receptor, the ligand is the focus to introduce modifications for affinity improvements. In view of intramolecular ligand mobility the determination of the conformation of the bound ligand is a valuable source for clues how to limit the entropic cost by synthetically preforming the bound conformation. Indeed, animal lectins (i.e. avian galectins and selectins) have been shown to accomplish selection of one low-energy conformer in the course of complex formation[34-37]. A route towards affinity enhancement thus points to establish a rigid binding partner maintaining the full profile of favourable contact points.

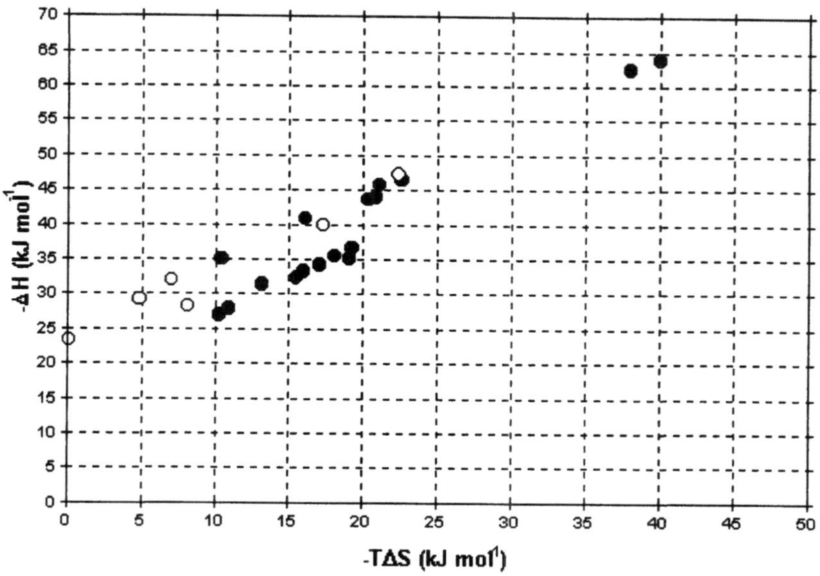

Figure 5. Enthalpy-entropy compensation plot for titration calorimetry-derived data sets of binding of galactosides to *Viscum album* L. agglutinin (●) and the galectin from chicken liver, CG-16 (O)[55].

As long as the the crucial epitopes are present, chemical ingenuity can leave carbohydrates as basic structure class and turn to any substance category. The objective of this work is to spawn derivatives (glycomimetics) with a snugly fit in the CRD of the lectin under scrutiny. Together with the cluster effect of the polymer[29] this synthetic engineering can turn combined interdisciplinary basic research into clinical progress, as predicted for the field of tumor lectinology more than a decade ago[38].

SUMMARY

The genetic code connects the two biochemical dimensions of nucleic acids and proteins. Theoretical calculations on coding capacity reveal that oligosaccharides as hardware surpass peptides by more than seven orders of magnitude based on hexamer synthesis. Thus, the sugar code establishes the third dimension of biological information transfer. Using carbohydrate-binding proteins (lectins, enzymes and antibodies) the information content of such epitopes is decoded. Currently, five families of animal lectins are defined in structural terms, i.e. the C-type, I-type and P-type groups, the galectins and the pentraxins. They are involved in intra- and intercellular glycan routing using oligosaccharides as postal-code equivalents and acting as defense molecules homing in on foreign or aberrant glycosignatures, as crosslinking agent in biosignaling and as coordinator of transient or firm cell-cell/cell-matrix contacts. By delineating the driving forces toward complex formation, knowledge about the causes for specificity can be turned into design of custom-made high-affinity ligands for clinical application, e.g. in anti-adhesion therapy, drug targeting or diagnostic histopathology.

REFERENCES

1. R. A. Laine, The information-storing potential of the sugar code, in: *Glycosciences: Status and Perspectives*, H.-J. Gabius, S. Gabius, eds., pp. 1-14, Chapman & Hall, London-Weinheim (1997).
2. E. F. Hounsell, Methods of glycoconjugate analysis, in: *Glycosciences: Status and Perspectives*, H.-J. Gabius, S. Gabius, eds., pp. 15-30, Chapman & Hall, London-Weinheim (1997).
3. H. Geyer and R. Geyer, Strategies for glycoconjugate analysis, *Acta Anat.* 161: 18-35 (1998).
4. A. Varki, "Unusual" modifications and variations of vertebrate oligosaccharides: are we missing the flowers for the trees ? *Glycobiology* 6: 707-710 (1996).
5. L.V. Hooper,S.M. Manzella, and J. U. Baenziger, The biology of sulfated oligosaccharides, in: *Glycosciences: Status and Perspectives*, H.-J. Gabius, S. Gabius, eds., pp. 261-276, Chapman & Hall, London-Weinheim (1997).
6. G. Reuter and H.-J. Gabius, Eukaryotic glycosylation: whim of nature or multipurpose tool ? *Cell. Mol. Life Sci.* 55: 368-422 (1999).
7. I. Brockhausen, J. Schutzbach, and W. Kuhns, Glycoproteins and their relationship to human disease, *Acta Anat.* 161: 36-78 (1998).
8. P. J. Winterburn and C.F. Phelps, The significance of glycosylated proteins, *Nature* 236: 145-151 (1972).
9. A. Villalobo and H.-J. Gabius, Signaling pathways for transduction of the initial message of the glycocode into cellular responses, *Acta Anat.* 161:110-129 (1998).
10. I. Kocourek, Historical background. In: *The Lectins. Properties, Functions, and Applications in Biology and Medicine*, I. E. Liener, N. Sharon, and I. J. Goldstein, eds., pp. 1-32, Academic Press, Orlando (1986).
11. H.-J. Gabius, Non-carbohydrate binding partners/domains of animal lectins, *Int. J. Biochem.* 26: 469-477 (1994).
12. H.-J. Bordet and F. P. Gay, Sur les relations des sensibilisatrices avec l'alexine, *Ann. Inst. Pasteur* 20: 467-498 (1906).
13. H.-J. Gabius, R. Engelhardt, S. Rehm, and F. Cramer, Biochemical characterization of endogenous carbohydrate-binding proteins from spontaneous murine rhabdomyosarcoma, mammary adenocarcinoma, and ovarian teratoma, *J. Natl. Cancer Inst.* 73: 1349-1357 (1984).
14. H.-J. Gabius, Endogenous lectins in tumors and the immune system, *Cancer Investig.* 5: 39-46 (1987).
15. K. Drickamer, Two distinct classes of carbohydrate-recognition domains in animal lectins, *J. Biol. Chem.* 263: 9557-9560 (1988).
16. K. Drickamer, Evolution of Ca^{2+}-dependent animal lectins, *Progr. Nucleic Acid Res. Mol. Biol.* 45: 207-233 (1993).
17. W. I. Weis and K. Drickamer, Structural basis of lectin-carbohydrate interaction, *Annu. Rev. Biochem.* 65: 441-473 (1996).
18. H.-J. Gabius, Animal lectins, *Eur. J. Biochem.* 243: 543-576 (1997).
19. D. Solís, A. Romero, H. Kaltner, H.-J. Gabius, and T. Diaz-Mauriño, Different architecture of the combining site of the two chicken galectins revealed by chemical mapping studies with synthetic ligand derivatives, *J. Biol. Chem.* 271: 12744-12748 (1996).
20. K.S. Lips, H. Kaltner, G. Reuter, B. Stierstorfer, F. Sinowatz, and H.-J. Gabius, Correspondence of gradual developmental increases of expression of galectin-reactive glycoconjugates with alterations of the total contents of the two differentially regulated galectins in chicken intestine and liver as indications for overlapping functions, *Histol. Histopathol.* 14: 743-760 (1999).
21. E.C. Beyer and S.H. Barondes, Quantitation of two endogenous lactose-inhibitable lectins in embryonic and adult chicken tissues, *J. Cell Biol.* 92: 23-27 (1982).

22. Y. Sakakura, J. Hirabayashi, Y. Oda, Y. Okyama, and K. Kasai, Structure of chicken 16 kDa β-galactoside-binding lectin, *J. Biol. Chem.* 265: 21573-21579 (1990).
23. J. Lu, and Y. Le, Ficolins and the fibrinogen-like domain, *Immunobiology* 199: 190-199 (1998).
24. C. Chen, A. F. Rowley, R.P. Newton, and N.A. Ratcliffe, Identification, purification and properties of a β-1,3-glucan-specific lectin from the serum of the cockroach, *Blaberus discoidalis*, which is implicated in immune defence reactions. *Comp. Biochem. Physiol. B* 122: 309-319 (1999).
25. H.-J. Gabius, Detection and functions of mammalian lectins – with emphasis on membrane lectins, *Biochim. Biophys. Acta* 1071: 1-18 (1991).
26. H. Kaltner and B. Stierstorfer, Animal lectins as cell adhesion molecules, *Acta Anat.* 161: 162-179 (1998).
27. H. Rüdiger, Structure and function of plant lectins, in: *Glycosciences: Status and Perspectives*, H.-J. Gabius, S. Gabius, eds., pp. 415-438, Chapman & Hall, London-Weinheim (1997).
28. H.-J. Gabius and S. Gabius, Chemical and biochemical strategies for the preparation of glycohistochemical tools and their application in lectinology, *Adv. Lectin Res.* 5: 123-157 (1992).
29. Y.C. Lee and R.T. Lee, eds., *Neoglycoconjugates: Preparation and Applications*, Academic Press, San Diego (1994).
30. N.V. Bovin and H.-J. Gabius, Polymer-immobilized carbohydrate ligands: versatile chemical tools for biochemistry and medical sciences, *Chem. Soc. Rev.* 24: 413-421 (1995).
31. S. Gabius, K. Kayser, N.V. Bovin, N. Yamazaki, S. Kojima, H. Kaltner, and H.-J. Gabius, Endogenous lectins and neoglycoconjugates: a sweet approach to tumor diagnosis and targeted drug delivery, *Eur. J. Pharm. Biopharm.* 42:250-261 (1996).
32. H.-J. Gabius, The how and why of protein-carbohydrate interaction: a primer to the theoretical concept and a guide to application in drug design, *Pharm. Res.* 15:23-30 (1998).
33. E.J. Toone, Structure and energetics of protein-carbohydrate complexes, *Curr. Opin. Struct. Biol.* 4:719-728 (1994).
34. H.-C. Siebert, M. Gilleron, H. Kaltner, C.-W. von der Lieth, T. Kozár, N.V. Bovin, E. Y. Korchagina, J.F.G. Vliegenthart, and H.-J. Gabius, NMR-based, molecular dynamics- and random walk molecular mechanics-supported study of conformational aspects of a carbohydrate ligand (Galβ1-2Galβ1-R) for an animal galectin in the free and in the bound state, *Biochem. Biophys. Res. Commun.* 219: 205-212 (1996).
35. L. Poppe, G.S. Brown, J.S. Philo, P.V. Nikrad, and B.H. Shah, Conformation of sLe^x tetrasaccharide, free in solution and bound to E-, P-, and L-selectin, *J. Am. Chem. Soc.* 119:1727-1736 (1997).
36. C.-W. von der Lieth, H.-C. Siebert, T. Kozár, M. Burchert, M. Frank, M. Gilleron, H. Kaltner, G. Kayser, E. Tajkhorshid, N.V. Bovin, J.F.G. Vliegenthart, and H.-J. Gabius, Lectin ligands: new insights into their conformations and their dynamic behavior and the discovery of conformer selection by lectins, *Acta Anat.* 161:91-109 (1998).
37. R. Harris, G.R. Kiddle, R.A. Field, M.J. Milton, B. Ernst, J.L. Magnani, and S.W. Homans, Stable-isotope-assisted NMR studies on ^{13}C-enriched sialyl $Lewis^x$ in solution and bound to E-selectin, *J. Am. Chem. Soc.* 121:2546-2551 (1999).
38. H.-J. Gabius, Tumor lectinology: at the intersection of carbohydrate chemistry, biochemistry, cell biology and oncology, Angew. Chem. Int. Ed. 27:1267-1276 (1988).
39. H.-J. Gabius and S. Gabius (eds.), *Lectins and Glycobiology*, Springer Verlag, Heidelberg-New York (1993).
40. H.-J. Gabius and S. Gabius (eds.), *Glycosciences: Status and Perspectives*, Chapman & Hall, London-Weinheim (1997).
41. H.-J. Gabius and F. Sinowatz (eds.), Special issue on glycosciences, *Acta Anat.* 161: 1-276 (1998).
42. H.-J. Gabius, Endogene Lektine in Tumoren und ihre mögliche Bedeutung für Diagnose und Therapie von Krebserkrankungen, *Onkologie 12*: 175-181 (1989).
43. H.-J. Gabius, Concepts of tumor lectinology, *Cancer Investig.* 15: 454-464 (1997).
44. H.-J. Gabius, R. Engelhardt, F. Cramer, Endogenous tumor lectins: a new class of tumor markers and targets for therapy? *Med. Hypothesis* 18: 47-50 (1985).
45. S. Gabius, V. Schirrmacher, H. Franz, S.S. Joshi, and H.-J. Gabius, Analysis of cell-surface sugar receptor expression by neoglycoenzyme binding and adhesion to plastic-immobilized neoglycoproteins for related weakly and strongly metastatic cell lines of murine tumor model systems, *Int. J. Cancer* 46: 500-507 (1990).
46. A. Raz and R. Lotan Endogenous galactoside-binding lectins: a new class of functional tumor cell surface molecules related to metastasis, *Cancer Metastasis Rev.* 6: 433-452 (1987).
47. T.Irimura, Y. Matsushita, R.C. Sutton, D. Carralero, D.W. Ohannesian, K.R. Cleary, D.M. Ota, G.L. Nicolson, and R. Lotan, Increased content of an endogenous lactose-binding lectin in human colorectal carcinoma progressed to metastatic stages, *Cancer Res. 51*: 387-393 (1991).

48. R. Lotan, P.M. Belloni, R.J. Tressler, D. Lotan, X.C. Xu, and D.K. Meijer (1994) Expression of galectins on microvessel endothelial cells and their involvement in tumor cell adhesion, Glycoconjugate J *11*: 462-468.

49. D. W. Ohannesian and R. Lotan, Galectins in tumor cells; in: *Glycosciences: Status and Perspectives*, H.-J. Gabius, S. Gabius, eds., pp. 459-469, Chapman & Hall, London-Weinheim (1997).

50. M. Fukuda, Lysosomal membrane glycoproteins: structure, biosynthesis, and intracellular trafficking, *J. Biol. Chem.* 266: 21327-21330 (1991).

51. D. W. Ohannesian, D. Lotan, P. Thomas, J.M. Jessup, M. Fukuda, H.-J. Gabius, and R. Lotan, Carcino-embryonic antigen and other glycoconjugates act as ligands for galectin-3 in human colon carcinoma cells, *Cancer Res.* 55: 2191-2199 (1995).

52. H. Inohara and A. Raz, Functional evidence that cell surface galectin-3 mediates homotypic cell adhesion, *Cancer Res.* 55: 3267-3271 (1995).

53. R.S. Bresalier, J.C. Byrd, L. Wang, and A. Raz, Colon cancer mucin: a new ligand for the β-galactoside-binding protein galectin-3, *Cancer Res.* 56: 4354-4357 (1996).

54. S. André, S. Kojima, N. Yamazaki, C. Fink, H. Kaltner, K. Kayser, and H.-J. Gabius, Galectins-1 and –3 and their ligands in tumor biology. Non-uniform properties in cell surface presentation and modulation of adhesion to matrix glycoproteins for various tumor cell lines, in biodistribution of free and liposome-bound galectins and in their expression by breast and colorectal carcinomas with/ without metastatic propensity, *J. Cancer Res. Clin. Oncol.125*: 461-474 (1999).

55.S. Bharadwaj, H. Kaltner, E.Y. Korchagina, N.V. Bovin, H.-J. Gabius, and A. Surolia, Microcalorimetric indications for ligand binding as a function of the protein for galactoside-specific plant and avian lectins, *Biochim. Biophys. Acta 1472:* 191-196 (1999).

GALECTINS AND OTHER ENDOGENOUS CARBOHYDRATE-BINDING PROTEINS OF ANIMAL BLADDER

V.P.Bhavanandan, Sabine Puch, Xiaoxuan Guo and Weiping Jiang
Department of Biochemistry & Molecular Biology,
The Pennsylvania State University College of Medicine, Hershey, Pennsylvania 17033
U.S.A

Interest in bladder glycoconjugates and lectins has recently increased primarily due to the popular hypothesis that bladder epithelial glycoconjugates are deficient in Interstitial Cystitis (IC). Bladder glycoconjugates and lectins also are important in the pathophysiology of urinary tract infections and bladder carcinoma. Interstitial cystitis is a disorder of the bladder that presents with a variety of symptoms including urinary urgency, frequency, nocturia and bladder pain (Messing, 1992; Thompson and Christmas, 1996). The pathology and etiology of IC are still not understood and the disease is considered chronic and incurable (Elbadawi, 1997). A recent population based study concluded that the prevalence of IC in the United States is approximately 60 cases per 100,000 (Curhan et al., 1999). While IC is not fatal, it is a major women's health issue (~90% of the patients are females), and IC patients scored lower quality of life measurements than women on chronic dialysis (Held et al. 1990). Glycosaminoglycans (GAGs) were originally suggested to be the major glycoconjugate component of the urothelium (Parsons et al., 1980, 1990). However, our recent studies show that GAGs are less abundant in either rabbit or human bladder epithelium, but substantial quantities of hyaluronan and chondroitin-4-sulfate were detected in the subepithelial layers (Buckley et. al., 1996; Bhavanandan and Erickson, 1997). In contrast, the epithelium was found to be rich in glycoproteins, specifically mucin

[1] This work was supported in part by US PHS grant DK 57266 from the National Institute of Diabetes, Digestive and Kidney Diseases and a Dean's Feasibility Grant from the Pennsylvania State University College of Medicine.

glycoproteins (Bhavanandan et al., 1998; Buckley et. al., 2000; Higuchi et al., 2000). These results are in agreement with those of others showing only minimal GAGs (Rubben et al., 1983; Grist and Chakraborty, 1994), but significant levels of glycoproteins, in bladder epithelium (Walker et al., 1988; Stefanelli et al., 1990).

A popular hypothesis for IC is that bladder epithelial permeability is increased due to defects in the bladder glycoconjugates allowing noxious urine components to penetrate and irritate the bladder (Lilly and Parsons, 1990). Even though this hypothesis has not been rigorously tested, certain treatments are based on it. For example, a common treatment approach is intravesical instillation of polyanionic polysaccharides such as, heparin and hyaluronan or oral administration of sodium pentosan polysulfate (Parsons and Mulholland, 1987; Parsons et al., 1994; Hanno, 1997; Jepson et al., 1998). It is believed that that these polysaccharides benefit the IC patient by coating the bladder surface and thereby repairing the defect in the mucosa (Parsons, 1997). However, the nature of the structures on the bladder surface that may be responsible for the interaction with the negatively charged exogenous polysaccharides is unclear. The glycocalyx of epithelial surfaces are not known to carry positively charged residues but are typically highly negatively charged as evidenced by the intense staining with reagents such as Ruthenium Red or colloidal iron. Therefore, it is very likely that the exogenously added sulfated polysaccharides are only temporarily held on the surface of the bladder by salt bridges.

An objective of our studies is the development of a method to improve the efficacy of intravesical treatments by modifying the polysaccharides with sugar ligands, so that they will bind to endogenous lectins in the bladder (Gabius, 1997). Compared to the ionic interactions between acidic polysaccharides and bladder components, an appropriately selected lectin-ligand binding is expected to be much more stable because concentrations of inhibitory saccharides high enough to disrupt the binding are unlikely in urine. However, there is no information on identity of the lectins associated with mammalian bladder that may interact with instilled glycoconjugates. The known mammalian lectins fall into four major classes: C-type Ca^{++}-dependent lectins (including selectins), S-type thiol-dependent lectins (the galectins), P-type lectins, and pentraxins (Barondes et al., 1994). This study was undertaken to identify the endogenous carbohydrate-binding activities in rabbit and human bladder and to biochemically characterize the major lectin(s) identified.

I. Histochemical examination of rabbit bladder and human bladder/ureter sections using biotinylated neoglycoconjugates

Paraffin-embedded rabbit bladder sections and paraffin-embedded normal human bladder and ureter surgical sections were obtained from the Department of Pathology of the Milton S. Hershey Medical Center (Hershey, PA). Paraffin-embedded human bladder sections of biopsies of IC patients and normal looking segments of bladder cancer patients were from the Cooperative Human Tissue Network (Eastern Division, University of Pennsylvania Medical Center, Philadelphia, PA). The paraffin sections were heated at 58-60 °C for 20 minutes, deparaffinized using xylene and rehydrated through graded ethanol baths (100-70%) and finally Tris-buffered saline (TBS). Endogenous peroxidase activity was quenched with 1.8% hydrogen peroxide in methanol. After washing with TBS and blocking nonspecific binding sites with 1% BSA, the tissue sections were incubated for 3 hours with the biotinylated neoglycoconjugates at 50 and 100 µg/ml in 0.1% BSA-TBS (Gabius et al., 1992; Danguy et al., 1995). Biotinylated neoglycoconjugates were from Glycotech Corporation (Rockville, MD) and consist of the carbohydrate (20% mol) attached to biotin (5% mol) through PAA which is a polyacrylamide polymer. The slides were then washed with TBS and incubated with the Vectastain ABC reagent (Avidin and biotinylated horseradish peroxidase complex) for 30 min. After further TBS washes, the slides were incubated for 7 min with the enzyme substrate solution containing equal volumes of 0.1% diaminobenzidine tetrahydrochloride and 0.02% hydrogen peroxide. The slides were washed in distilled water, dehydrated through graded ethanol baths and xylene, and mounted using Permount. The specificity of the protein-carbohydrate interaction was established by competitive inhibition experiments in which the sections were pre-incubated with the homologous non-biotinylated PAA-neoglycoconjugate (20 µg/ml) for one hour. The solution was removed and the sections were incubated with a mixture of the neoglycoconjugate-PAA-Biotin in the presence of a 20- to 100-fold excess of the ligand monosaccharide for 2 hours (Akif et al., 1995). Biotinylated saccharide-free carrier for use as a control was prepared by incubating glucose-PAA-Biotin for one hour at room temperature with 0.1 M sodium periodate to oxidize glucose followed by treatment with 0.1 M sodium borohydride in 0.01N sodium hydroxide for 30 min. to reduce the generated aldehyde groups.

The carbohydrate-binding pattern in the mucosa of the rabbit bladder, human bladder, and human ureter is summarized in Table 1. ß-D-galactose neoglycoconjugate, at concentrations of 50 μg/ml and higher, showed the strongest staining in all three tissues. The neoglycoconjugates ß-N-acetylglucosamine and α-D-mannose also showed strong staining, particularly at a concentration of 100 μg/ml. In contrast, α-L-fucose, α-glucose, α-*N*-acetylneuraminic acid, and α-*N*-acetylgalactosamine neoglycoconjugates showed either no or a very weak reaction (Table 1). Two neoglycoconjugates containing β-galactose-terminated disaccharides, Galβ1- 3GalNAcα- and Galβ1-4GlcNAcβ-, were also tested to determine the influence of the penultimate sugar on the binding. There was no noticeable difference in the staining between the two; however, the staining intensity in both cases was lower than with the monosaccharide probe.

Table I. Binding intensities of neoglycoconjugates (100μg/ml) to the mucosa of normal rabbit and human bladders and human ureter [1]

Neoglycoconjugate	Rabbit Bladder	Human Bladder	Human Ureter
α-L-Fuc	0	0	+
β-D-Gal	++++	+++	+++
α-GalNAc	++	+	++
α-D-Glc	+	+	+
β-D-GlcNAc	++++	+++	++
α-D-Man	+++	++	++
α-Neu5Ac	0	0	0
Galβ1-3GalNAcα	++	+	++
Galβ1-4GlcNAcβ	++	++	++
Controls			
0.1% BSA-TBS	0	0	0
PAA-Biotin	0	+	0
β-Gal-PAA-Biotin/ Gal 1 : 20	0	++	++
β-Gal-PAA-Biotin/ Gal 1 : 50	+	+	+
β-GlcNAc-PAA-Biotin/ GlcNAc 1 : 20	+	+	+
β-GlcNAc-PAA-Biotin/ GlcNAc 1 : 50	+	+	+

[1] The intensity of the staining of the epithelium is scored on an arbitrary scale of 0 (no staining) to ++++ (maximum intensity).

The carbohydrate-binding activity of some of the neoglycoconjugates and controls are illustrated in Figure 1. The biotinylated neoglycoconjugate in the presence of the unlabeled homologous monosaccharide showed significantly reduced staining at a 20-fold excess and invariably weak staining at a 50-fold excess. The presence of 5 mM Ca^{2+} in the washing and incubation buffer did not enhance the staining significantly. Further, preincubations of the sections with ethylenediaminetetraacetic acid did not alter the staining with the galactose-PAA. These observations indicate that the lectin activity is not calcium-dependent and therefore does not belong to the family of calcium- dependent (C-type) lectins (Leffler, 1997). The carbohydrate-free carrier, biotinylated glucose neoglycoconjugate pre-treated with periodate to destroy the glucose residue, did not show any staining and thus served as an important negative control. Additional controls ascertained the specificity of the binding of the carbohydrate ligands to lectins in these tissues.

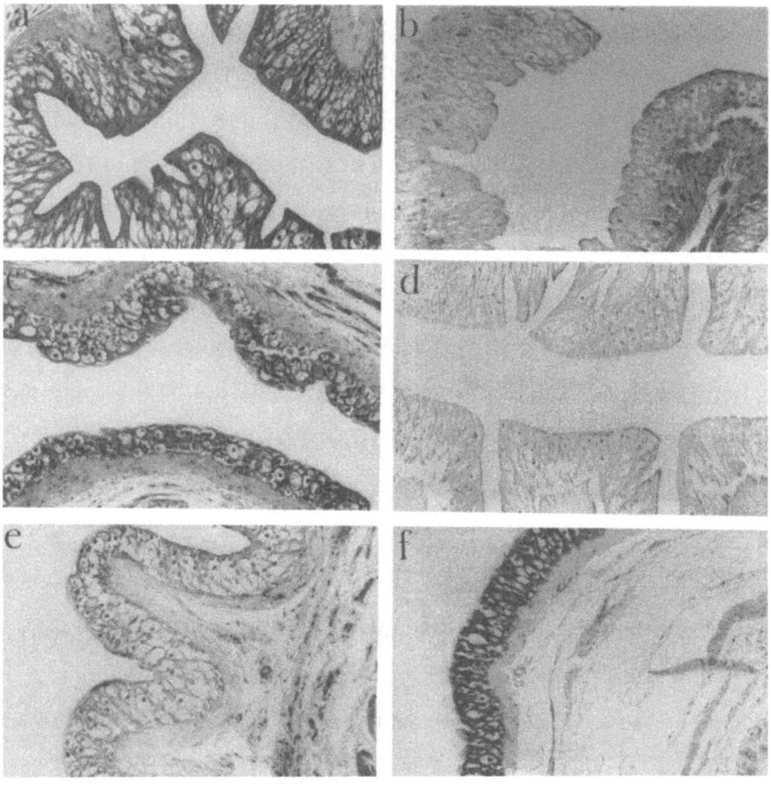

Figure 1. Interaction of rabbit bladder sections with neoglycoconjugates of a) ß-D-galactose at 100 µg/ml, b) as (a) but in the presence of an 20-fold excess of D-galactose, c) ß-*N*-acetyl-D-glucosamine at 100 µg/ml, d) as (c) but in the presence of an 20-fold excess of *N*-acetylglucosamine, e) α-D-mannose at 100 µg/ml, f) Galβ1-4GlcNAcβ- at 100 µg/ml.

The staining by the ß-D-galactose and ß-*N*-acetylglucosamine probes is not limited to the transitional epithelium but strong staining is also found in the underlying *lamina propria* of the rabbit bladder and human ureter sections. The ß-D-galactose neoglycoconjugate also showed strong staining (+ + +) with bladder sections from IC patients (not illustrated) suggesting that the disease does not affect the lectin activity of the bladder epithelium. These observations could be important for the aim of improved treatment since the epithelium of IC patients often seems disrupted, and, therefore, the modified drug would be able to bind to the exposed *lamina propria.*

II. Isolation of the β-galactose-binding protein from rabbit bladder mucosa

1. Rabbit bladder organ cultures: Initial experiments were done with mucosa from rabbit bladders that were metabolically radiolabeled as explant cultures. Fresh normal bladders were obtained from rabbits that were sacrificed for experiments on other organs at the Animal Facility at the Milton S. Hershey Medical Center (Hershey, PA). The bladders, removed immediately after sacrifice of rabbits, were rinsed with ice-cold phosphate-buffered saline (PBS), and established in culture within about 30 min (Buckley et al., 1996). One to two bladders were placed mucosal side up in a scored Nunc culture dish (15 x 60 mm). CMRL 1066 media (4 -5 ml), supplemented with 10 % fetal bovine serum, penicillin-streptomycin, L-glutamine, fungizone, insulin, transferrin, retinol acetate, and hydrocortisone, prewarmed to 37 °C, was added so that the mucosa was not submerged. After incubation in a 5 % CO_2 incubator at 37 °C for 24 hours, the spent media was replaced with fresh media containing 15 μCi/ml of [^{14}C]-L-amino acid mixture and the incubation continued for up to 48 hours to obtain equilibrium radiolabeling of bladder proteins.

2. Protein extraction from bladder epithelium and affinity chromatography: At the end of the incubation, the media was removed, and the bladder epithelium was carefully separated using a blunt instrument (Teflon spatula) and collected into extraction buffer (50 mM Tris, pH 7.5-0.1 % NP-40, 100 mM NaCl, 2 mM $CaCl_2$, 2 mM DTT, 5 mM N-ethylmaleimide, Leupeptin [0.5 μg/ml], Pepstatin [0.7 μg/ml], Pefabloc SC [100 μg/ml]). The suspension was homogenized and extracted by repeated passage through a syringe needle, followed by stirring for 30 min at 4 °C. The homogenate was then centrifuged at

10,000 rpm for 20 min at 4 °C. Aliquots of the supernatant were tested for saccharide-binding protein using four different affinity gels namely, immobilized asialofetuin, lactose, *N*-acetylglucosamine, and aminophenylthiogalactopyranoside (Den and Malinzak, 1977). The binding of protein to these gels was determined by eluting with either 0.2 M lactose or *N*-acetylglucosamine. The immobilized lactose gel was chosen for purification since this gel showed the maximum binding of labeled protein(s). In preparative experiments, the lactose-gel column (5 ml gel) was equilibrated with extraction buffer, and the [^{14}C]-amino acid labeled bladder mucosa extract applied. The column was washed with 10 bed volumes of the extraction buffer to remove unbound material and then eluted with 10 bed volumes of extraction buffer without NP-40 to remove the detergent. The bound [^{14}C]-labeled protein was eluted with 0.2 M lactose in 50 mM Tris/HCl buffer, pH 7.5.

To isolate the unlabeled protein, batches of fifty frozen rabbit bladders, frozen in liquid nitrogen, were purchased from Pel-Freez (Rogers, AR). The bladders were thawed, the mucosa scraped, extracted, and the extract subjected to affinity chromatography as above but on a larger column (25 ml) of immobilized lactose gel.

3. SDS- and non-denaturing PAGE and gel filtration analysis: After removal of lactose by dialysis against 50 mM Tris/HCl buffer, pH 7.5, the purified [^{14}C]-amino acid labeled protein was applied to a SDS-polyacrylamide gel, and electrophoresis carried out at 100 V for 1.5 hours in SDS-running buffer. The gel was fixed, treated with Amplify solution (Amersham), dried, and exposed to an X-ray film at -70 °C for one week. SDS-PAGE on a 10% gradient gel revealed a major band of approximately 30 kDa and a minor band of about 21 kDa (Puch and Bhavanandan, 1999). Electrophoresis under non-denaturing conditions was also carried out with the [^{14}C]-labeled and unlabeled proteins isolated from rabbit bladder. The ^{14}C -labeled protein showed a very diffuse band in the 95-120 kDa range on a 10% gel under the non-denaturing conditions (Figure 2). In order to further purify the protein and to determine its native size, gel filtration was attempted on various gel matrices such as Bio Gel P150 and Sephacryl S200 and using different buffers. However, the recovery in every case was extremely low (< 10 %), possibly due to strong non-specific binding of the protein to the matrices and or insolubility due to aggregation of the protein.

Affinity chromatography of the extract of epithelium from fifty rabbit bladders on a preparative column of immobilized lactose yielded about 15 μg of the protein. SDS-PAGE followed by silver staining revealed a strong band of approximately 18 kDa and a weaker band of about 30kDa. The reproducibility of the above results with the unlabeled protein confirmed that the 18 kDa band is not an artifact. The 18 kDa protein could be a degradation product of the 30 kDa protein due to protease activities even in the presence of protease inhibitors. For instance, galectin-4 was first isolated as a protein of molecular weight ~17,000, but subsequent cDNA cloning led to its identification as a 36,000 protein. Similarly, the N-terminal proline-glycine-rich domain of galectin-3, another ß-galactoside binding protein with a molecular weight of about 30 kDa, was reported to be highly sensitive to collagenase-like enzymes (Raz et al., 1989). Since the link region of galectin-4 is homologous to this domain, it is possible that it is also sensitive to the same enzymes. Because of the difficulties with gel filtration mentioned above, the purified unlabeled protein was subjected to non-denaturing gel electrophoresis. A protein band at about 60 kDa was detectable, but other bands of larger size probably representing higher aggregates were also visible. Thus, it seems likely that the 30 kDa protein forms a dimer under native conditions. The most common quaternary structures for galectins in mammals are monomers and dimers. However, in the presence of glycoconjugate ligands such as immobilized lactose, multivalent aggregates are reported to be formed by some of the monomeric galectins even at micromolar concentrations (Leffler, 1997).

4. Further characterization of the rabbit bladder galactose-binding lectin: The strong band at 30 kDa on SDS-PAGE of the [^{14}C]-labeled protein purified on a lactose affinity column was interesting since five of the known mammalian galectins (galectins-3, -4, -6, -8, -9) are reported to have molecular weights in the 30 and 36 kDa range. Further, galectin-3 and galectin-4 are known to be expressed in epithelial cells and therefore, most likely also in the bladder (Barondes et al., 1994; Gitt et al.,1997)

Other investigators have reported difficulties in purifying galectins (Oda et al., 1993; Raz et al., 1989), which is in agreement with the degradation problems noted in our experiments (discussed above) to isolate the lactose-binding protein from rabbit bladder. In general, the galectins appear to be extremely labile and exhibit non-specific binding characteristics which makes the full biochemical characterization difficult. It must be pointed out that the vast majority of the galectins reported in the literature have been characterized by molecular biological approaches. Of the ten different galectins found so far in mammalian tissues, only galectin-1 to galectin-4 were discovered as proteins isolated by affinity chromatography on immobilized ß-galactosides. None of galectin-6 to galectin-10 was initially discovered based on ß-galactoside binding activity. These observations further suggested that the calcium-independent, galactose-binding proteins in bladder could be galectins.

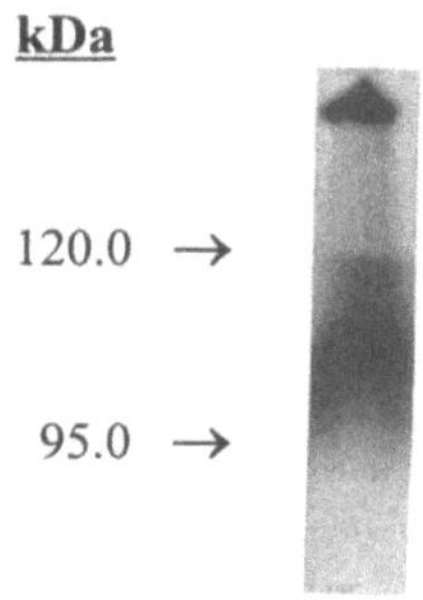

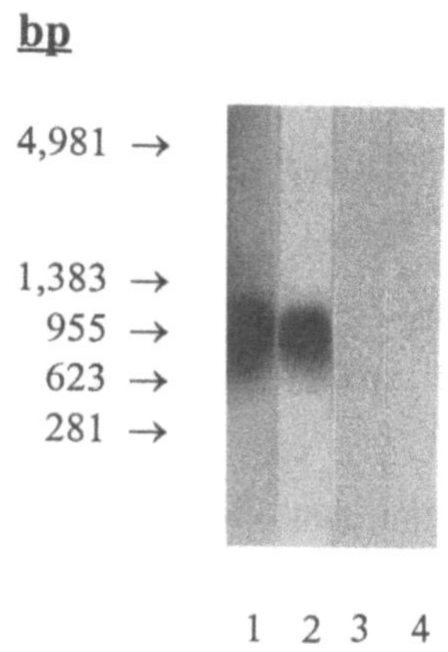

Figure 2 (left panel). Polyacrylamide gel electrophoresis on a 10% non-denaturing gel of the ^{14}C -labeled lactose-binding protein purified from rabbit bladder. The labeled protein band was visualized by autoradiography. Gel filtration standards were run on parallel lanes as marker proteins, the positions of the 120 k and 95 k proteins are indicated.

Figure 3 (right panel). Northern blot analysis of human and rabbit bladder epithelial RNA. Total RNA was subjected to electrophoresis on a 1% agarose gel, transferred to Nylon membrane and hybridized with [^{32}P]- labeled galectin DNA probes as follows: lane 1, human bladder RNA with galectin-3 DNA; lane 2, rabbit bladder RNA with galectin-3 DNA; lane 3, human bladder RNA with galectin-4 DNA; lane 4, rabbit epithelial RNA with galectin-4 DNA. Positions of the RNA molecular length markers run on the same gel are indicated.

III. Molecular Biological Analysis of Bladder Galectins

Because of the reasons discussed above as well as the very limited availability of human bladder tissue for research, we resorted to molecular biological techniques for the characterization of galectins in bladder. Since the available information suggested galectin-3 and galectin-4 as the most likely candidates in bladder, northern blot analysis and polymerase chain reaction (PCR) approaches were used to examine expression of these two galectins in rabbit and human bladders.

1. Isolation of RNA and construction of probes: Total RNA was isolated from mucosa of fresh rabbit bladders using Ultraspec-II RNA kit. Poly(A) RNA was purified from total RNA with Rapid mRNA Purification kit (Ameresco, Solon, Ohio). RNA was also similarly isolated from a fresh human bladder surgical specimen and from RT-4 human bladder carcinoma cells. cDNA was synthesized using the Marathon ™ cDNA Amplification Kit. cDNA of rabbit bladder epithelium was screened for expressions of galectin-3 and galectin-4 using polymerase chain reaction. A primer for galectin-4 [TTGCCTTCCACTTCAATCCT /CCG] was designed based on the alignment of known mammalian species including human (Oda et al., 1993; Chiu et al., 1994; Gitt et al., 1998; Rechreche et al., 1997; Huflejt et al. 1997). The primer for galectin-3 [TTTCTGCCCCTTTCAGATTAT] was synthesized according to Gaudin et al., (1995).

2. Northern blot analysis of rabbit and human bladder RNA: Rabbit and human bladder epithelial RNA preparations (15 μg each) were separated on a 1% agarose gel containing 3-[N-morpholino]propane sulfonic acid, 3.7% formaldehyde and ethidium bromide, blotted onto a nylon membrane and prehybridized with hybridization buffer containing 50 % formamide, 1 x Denhardt, 5 x SSC, 1% SDS, and 0.1 mg/mL herring sperm DNA for 2- 4 hours at 42 °C. Hybridization with [^{32}P]-labeled galectin-3 and galectin-4 rabbit DNA in hybridization buffer was then carried out at 42 °C for 20 hours (Feinberg and Vogelstein, 1983). After several washes with decreasing salt concentration and a final wash with 0.2 x SSC and 1% SDS at 60 °C for 15 min, autoradiography was performed.

The northern blots of RNA hybridized with [^{32}P]- labeled rabbit galectin-3 cDNA probe showed a clear signal at approximately 900 bp with both the rabbit and the human RNA. However, neither the rabbit nor the human RNA gave a detectable signal when hybridized with the labeled galectin-4 cDNA (Figure 3). The blots contained the same amounts of RNA, were hybridized with probes of similar quantity and specific activities, and washed under identical conditions. Thus, the intensity of the signals observed suggest that galectin-3 is expressed at much higher levels in bladder epithelium than galectin-4. The results of PCR, discussed below, are also in agreement with the low expression in rabbit and human bladder of galectin-4 compared to galectin-3. The same results were also obtained using RNA isolated from the RT-4 bladder cancer cells (not illustrated), indicating that the expression of galectins is not altered in malignancy. These data indicate that galectin-3 is expressed in both rabbit and human bladder epithelium. The level of expression of galectin-4 in rabbit is lower than that of galectin-3, but no conclusion can be drawn yet about the galectin-4 expression in human urothelium.

3. Polymerase chain reaction: PCR was performed with the Marathon ™cDNA Amplification Kit using rabbit bladder epithelial cDNA. The PCR-products were separated on a 0.8 % agarose gel and the appropriate DNA-bands eluted with GENECLEAN II Kit (BIO 101 Inc., La Jolla, CA). The purified PCR-products were ligated into pGEM® -T Easy Vector, and the Subcloning Efficiency DH5a Competent cells were transformed with the ligation reactions and selected on LB/Amp/xGal-plates. White colonies were grown in TB-medium, containing ampicillin and glucose, with shaking at 300 rpm. Plasmids were prepared using the QIAprep Spin Plasmid Kit (Qiagen, Santa Clarita, CA). The DNA preparations were sequenced using ABI Prism automated DNA sequencer in the Macromolecular Core Facility of the Hershey Medical Center. Databases were searched for sequence similarities using the World Wide Web page (http://www.ncbi.nlm.nih.gov/cgi-bin/BLAST) of the National Center for Biotechnology Information (NCBI).

PCR products were detected with the galectin-3 and galectin-4 primers, although the signals corresponding to galectin-3 were much stronger than those of galectin-4 in agreement with the results of the northern blot analysis. The PCR-products were cloned in pGEM® -T Easy and subjected to automated sequencing. The full-length sequence of galectin-3 and a partial sequence of galectin-4 were obtained. The search of DNA and protein databases with the resulting sequences revealed that an identical galectin-3 sequence has been reported for New Zealand white rabbit vascular smooth muscle cells (Gaudin et al., 1995). This confirmed that galectin-3 was indeed expressed in the bladder, a tissue which had not been previously examined. Sequences for rabbit galectin-4 have not been documented previously; however, galectin-4 sequences from other species were found to be highly homologous to the determined rabbit sequence suggesting that this PCR product was derived from rabbit galectin-4. Additional PCR was performed with new primers in order to deduce the amino acid sequence of the entire coding region. The determined cDNA sequence contained 1200 base pairs and encoded an amino acid sequence of 328 amino acids (Jiang et al., 1999). The deduced amino acid sequence was compared with known galectin-4 sequences in the database. The rabbit galectin-4 was found to be more similar to human and pig than to the rat and mouse counterparts. Further, the following four distinct regions were present in all galectin-4 sequences: a 19 amino acid N-terminal region, two carbohydrate recognition domains of 130 amino acids each, and a linker region of 49 amino acid residues. The sequence alignment also revealed two peptide motifs common to all cloned galectin-4 and galectin-6. We have proposed these motifs, M-A-F/Y-V-P-A-P-G-Y-Q-P-T-Y-N-P-T-L-P-Y in the N-terminal segment and A-F-H-F-N-P-R-F-D-G-W-D-K-V-V-F in the first carbohydrate recognition domain, as signature sequences to identify related galectins (Jiang et al., 1999).

SUMMARY

Defects in the glycocalyx of the bladder epithelium may be related to the development of bladder diseases including interstitial cystitis which is a chronic bladder disease of unknown etiology. Indirect evidence has implicated alterations in the bladder epithelial glycoconjugates in interstitial cystitis and vesicaler instillation of glycosaminoglycans is promoted as treatments. However, information on the nature of the glycoconjugates of the bladder epithelium and lectins that may interact with the exogenous instilled glycoconjugates is very limited. We have examined the endogenous lectin associated with bladder epithelium by immunohistochemistry using biotinylated neoglycoconjugates. The strong calcium-independent binding of β-D-galactose probe suggested the presence of galectins in rabbit and human bladder. Extracts of rabbit bladder organ cultures metabolically labeled with [14_C]-amino acids were subjected to affinity chromatography on immobilized lactose and the specifically bound material eluted with 0.2 M lactose. SDS-PAGE of the recovered proteins revealed a major band of approximately 30 kDa and a minor band of 21 kDa. Polymerase chain reaction and northern blot analysis showed that both galectin-3 and galectin-4 are expressed in rabbit bladder. Since galectin-3 from rabbit had been previously cloned, we cloned and sequenced galectin-4 from rabbit bladder. The deduced full length sequence of 328 amino acids revealed four distinct regions; a N-terminal peptide of 19 residues, two carbohydrate recognition domains of 130 residues each, and a linker region of 49 residues. Comparison of the rabbit galectin-4 sequence with those of human, pig, rat, and mouse revealed two invariant peptide motifs that are proposed as signature sequences for identifying related galectins.

REFERENCES

Akif F., Gabius H-J., and Danguy A., (1995) Estrous cycle-related alterations in the expression of glycoconjugates and lectins in the mouse endometrium shown histochemically. *Tissue and Cell* 27: 197-206.

Barondes, S.H., Cooper, D.N.W., Gitt, M.A., and Leffler H., (1994) Galectins: Structure and function of a large family of animal lectins. *J. Biol. Chem.* 269: 20807-20810.

Bhavanandan, V.P., and Erickson D.R., (1997) An investigation of the nature of bladder mucosal glycoconjugates and their role in interstitial cystitis. *Indian J. Biochem. Biophys.* 34: 205-211.

Bhavanandan, V.P., Zhu, Q., Yamakami, K., DiIulio, N.A., Nair, S., Capon, C., Lemoine J., and Fournet B., (1998) Purification and characterization of the MUC1 mucin-type glycoprotein, epitectin, from human urine: structures of the major oligosaccharide alditols. *Glycoconjugate J.*, 15: 37-49.

Buckley, M.S., Washington, S., Laurent, C., Erickson, D.R., and Bhavanandan V.P., (1996) Characterization and immunohistochemical localization of the glycoconjugates of the rabbit bladder mucosa. *Arch. Biochem. Biophys.* 330: 163-173.

Buckley, M.S., Xin, P., Washington, S., Herb, N., Erickson, D.R., and Bhavanandan V.P., (2000). Lectin histochemical examination of rabbit bladder glycoproteins and characterization of a mucin isolated from the bladder mucosa. *Arch. Biochem. Biophys.* 375: in press.

Chiu, M., Parry, D.A.D., Feldman, S.R., Klapper, D.G., and O'Keefe, E.J., (1994) An adherans junction protein is a member of the family of lactose-binding lectins. *J. Biol. Chem.* 269: 31770-31776.

Curhan, G.C., Speizer, F.E. Hunter, D.J ., Curhan S.G., and Stampfer M.J., (1999) Epidemiology of interstitial cystitis: A population based study. *J Urol* 161: 549-552.

Danguy, A., Kayser, K., Bovin, N.V., and Gabius H-J., (1995) The relevance of neoglycoconjugates for histology and pathology. *Trends in Glycosci. Glycotech.* 7: 261-275.

Den, H., and Malinzak D.A., (1977) Isolation and properties of beta-D-Galactoside-specific lectin from chick embryo thigh muscle. *J. Biol. Chem.* 252: 5444-5448.

Elbadawi, A., (1997) Interstitial cystitis: A critique of current concepts with a new proposal for pathologic diagnosis and pathogenesis. *Urology* 49: 14-40.

Feinberg, A.P., and Vogelstein B., (1983) A technique for radiolabeling DNA restriction endonuclease fragments to high specific activity. *Analyt. Biochem.* 132: 6-13.

Gabius, H-J., (1997) Animal lectins. *Eur. J. Biochem.* 243: 543-576.

Gabius, H-J., Bahn,H., Holzhausen, H-J., Knolle, J., and Stiller D., (1992) Neoglycoprotein binding to normal urothelium and grade-dependent changes in bladder lesions. *Anticancer Res.* 12: 987-992.

Gaudin J-Ch., Monsigny M., and Legrand A., (1995). Cloning of the cDNA encoding rabbit galectin-3. Gene 16: 249-252

Grist, M., and Chakraborty J., (1994) Identification of a mucin layer in the urinary bladder. *Urology* 44: 26-33.

Gitt, M.A., Jordan, E.T., and Leffler, H., (1997) Galectin-2, galectins-5 and -9, and galectins-4 and -6. *Trends. Glycosci. Glycotechnol.* 9: 87-93.

Gitt, M. A., Colnot, C., Poirier, F., Nani, K.J., Barondes, S.H., and Leffler, H., (1998) Galectin-4 and galectin-6 are two closely related lectins expressed in mouse gastrointestinal tract. *J. Biol. Chem.* 273: 2954-2960.

Hanno, P.M.., (1997) Analysis of long-term Elmiron therapy for interstitial cystitis. *Urology* 49: 93-99.
Held, P.J., Hanno,P.M., Wein, A.J., Pauly, M.V., and Cahn M.A., (1990). Epidemiology of Interstitial Cystitis. in: *Interstitial Cystitis*, P.M. Hanno, D.R. Staskin, R.J. Krane, and A.J.Wein, eds., Springer-Verlag, New York, pp. 29-48.
Higuchi, T., Xin, P., Buckley, M.S., Erickson, D.R., and Bhavanandan V.P., (2000) Characterization of the rabbit homolog of MUC1 glycoprotein isolated from bladder by affinity chromatography on immobilized jacalin. Glycobiology in press
Huflejt, M.E., Jordan, E.T., Gitt, M.A., Barondes, S.H., and Leffler, H., (1997) Strikingly different localization of galectin-3 and galectin-4 in human colon adenocarcinoma T84 cells: Galectin-4 is localized at sites of cell adhesion. *J. Biol. Chem.* 272: 14294-14303.
Jepsen, J.V., Sall,M., Rhodes,P.R., Schmidt, D., Messing, E., and Bruskewitz R.C., (1998) Long-term experience with pentosanpolysulfate in interstitial cystitis. *Urology* 51: 381-387.
Jiang, W., Puch, S., Guo, X., and Bhavanandan, V.P., (1999) Signature sequences for the galectin-4 subfamily *IUBMB Life* 48: 1-5.
Leffler, H., (l997) Introduction to galectins (Review) *Trends in Glycosci. and Glycotechnol.* 9: 9-19.
Lilly, J.D., and Parsons C.L., (1990) Bladder surface glycosaminoglycans is a human epithelial permeability barrier. Surg. Gynecol. Obstet. 171: 493-496.
Messing, E.M. (1992) Interstitial cystitis and related syndromes. in: Campbell's Urology, P.C. Walsh, A.B. Retik, T.A. Stamey, and E.D. Vaughan, Jr. eds. W.B. Saunders Co., Philadelphia pp 982-1005.
Oda, Y., Herrmann, J., Gitt, M.A., Turck, C. W., Burlingame, A.L. Barondes, S.H., and Leffler, H., (1993) Soluble lactose-binding lectin from rat intestine with two different carbohydrate-binding domains in the same peptide chain. *J. Biol. Chem.* 268: 5929-5939
Parsons, C.L., (1997) Epithelial coating techniques in the treatment of interstitial cystitis. *Urology* 49: 100-104.
Parsons, C.L., and Mulholland S.G., (1987) Successful therapy of interstitial cystitis with pentosan polysulfate. *J. Urol.* 138: 513-516.
Parsons, C.L., Boychuk, D., Jones, S., Hurst, R., and Callahan, H., (1990) Bladder surface glycosaminoglycans: An epithelial permeability barrier. *J. Urol.* 143: 139-142.
Parsons, C.L., Housley, T., Schmidt, J.D., and Lebow, D., (1994) Treatment of interstitial cystitis with intravesical heparin. *Brit. J. Urol.* 73: 504-507.
Parsons, C.L., Stauffer, C., and Schmidt J.D., (1980) Bladder-surface glycosaminoglycans: An efficient mechanism of environmental adaptation. *Science* 208: 605-607.
Puch, S., and Bhavanandan, V.P., (1999) Endogenous carbohydrate-binding proteins of rabbit and human bladder *Urology* 53: 848-852.
Raz, A, Pazerini, G., and Carmi, P., (1989) Identification of the metastasis-associated, galactoside-binding lectin as a chimeric gene product with homology to an IgE-binding protein. Cancer Res. 49: 3489-3493.
Rechreche, H., Mallo, G.V., Montalto, G., Dagorn, J.-C., and Iovanna, J.L., (1997) Cloning and expression of the mRNA of human galectin-4, an S-type lectin down-regulated in colorectal cancer. *Eur. J. Biochem.* 248: 225-230.
Rubben, H., Friedrichs, R., Stuhlsatz, H.W., Cosma, S., and Lotzeyer W., (1983) Quantitative analysis of glycosaminoglycans in urothelium and bladder wall of calf. *Urology* 22: 655-657.
Stefanelli, J., Callahan, H.J., Byrne, D.S., and Mulholland S.G., (1990) Antisera to a rabbit urinary tract antigen also react with human bladder and kidney tissue. *J. Urol.* 143: 414-417.
Thompson, A.C., and Christmas, T.J., (1996) Interstitial cystitis — an update. *Br. J. Urology* 78: 813-820.
Walker, S.R., Callahan, H.J., Fritz R., and Mulholland S.G., (1988). Distribution of rabbit mucosal glycoprotein throughout urinary tract. *Urology* 33: 127-130.

BINDING PROPERTIES AND APPLICATIONS OF *APLYSIA* GONAD LECTIN[1,2]

Nechama Gilboa-Garber[a] and Albert M. Wu[b]

[a]*Faculty of Life Sciences, Bar-Ilan University, Ramat Gan, Israel*
[b]*Glyco-Immunochemistry Research Laboratory, Institute of Molecular and Cellular Biology, School of Medicine, Chang-Gung University, Kwei-san 333, Taiwan*

Aplysia, a giant marine snail (sea hare, Fig. 1) widely used for neurobiological investigations, produces several lectins exhibiting diverse sugar specificities (Gilboa-Garber *et al.*, 1984; Ozeki, 1998). Highest lectin activity, *Aplysia* gonad lectin (AGL), resides in the gonad of the adult mollusk (Gilboa-Garber *et al.*, 1985). It is specific to galacturonic acid (GalUA) and also binds galactose and lactose (Gilboa-Garber *et al.*, 1985) but not *N*-acetylgalactosamine, which is most attractive for the lectin of *Helix pomatia* albumin gland (Hammarström and Kabat, 1964). There are additional prominent differences between these two snail lectins: AGL is Ca^{2+}-dependent, while the *Helix pomatia* lectin (HPA) is not dependent on divalent cations (Hammarström and Kabat, 1964). The recently discovered galactose-binding *Aplysia* skin lectin (Ozeki, 1998) resembles HPA in this respect. AGL is composed of two 32-33 kDa subunits (Gilboa-Garber *et al.*, 1985) whereas HPA is a hexamer of 13 kDa subunits (Hammarström *et al.*, 1972). AGL stimulates mitosis in human peripheral lymphocytes (Zipris and Gilboa-Garber, 1987), while HPA has not been reported to be associated with such an activity. AGL, which is produced in the *Aplysia* gonad, accompanies the fertilized eggs

1 This review summarizes research work which involved several coworkers and collaborators over an 18-year-period. Among them: Prof. A. Susswein and Ms. S. Markovici, who kindly supplied us the *Aplysia* gonads and eggs, Ms. L. Mizrahi, Dr. D. Zipris, Dr. D. Avichezer, Dr. D. Sudakevitz, Dr. C. Levene, Ms. H. Mymon, Prof. A. Oren, Ms. B. Lerrer and Ms. E. Lesman-Movshovich, all of them from Israel, as well as Prof. N. Benhamou *et al.*, from Quebec, Canada. Thanks to all of them and to Ms. S. Victor and Ms. A. Goldreich, for the skillful preparation of this manuscript, and Ms. E. Gindi and Ms. T. Anker, for the graphic and photographic presentation of this manuscript. The research was supported by Bar-Ilan cancer research funds. This work was also supported by Grant 676 from the Chang-Gung Medical Research Project, Kwei-san, Taiwan, and Grant NSC 89-2320-B-182-003 from the National Science Council, Taipei, Taiwan.

2 The abbreviations used are: AGL, *Aplysia* gonad lectin; Con A, Concanavalin A; *E. coli*, *Escherichia coli*; Gal, D-galactopyranose; GalNAc, *N*-acetyl-D-galactosamine; GalUA, galacturonic acid; Glc, glucose; GlcNAc, N-acetyl-D-glucosamine; GlcUA, glucuronic acid; Met-α/β Gal, methyl derivative of α/β-Gal; LacNAc, N-acetyllactosamine; MPL, *Maclura pomifera* lectin; PHA, phytohemagglutinin (from *Phaseolus vulgaris* seeds); PNA, peanut agglutinin.

to the seawater (Gilboa-Garber *et al.*, 1985) where it is found associated with egg strips (Kamiya and Shimizu, 1981; Gilboa-Garber *et al.*, 1985). This lectin was not detected in any other organ of the adult animal (Gilboa-Garber *et al.*, 1985; Wilson *et al.*, 1992).

AGL was shown to be relatively resistant to heating up to 70°C and to high pH, but to exhibit sensitivity to low pH and proteolysis (Gilboa-Garber *et al.*, 1985). Its purification was therefore attained by heating to 65°C (for removal of foreign heat-sensitive proteins), precipitation by ammonium sulfate and affinity chromatography on Sepharose 4B to which this lectin readily adsorbs (Gilboa-Garber *et al.*, 1985).

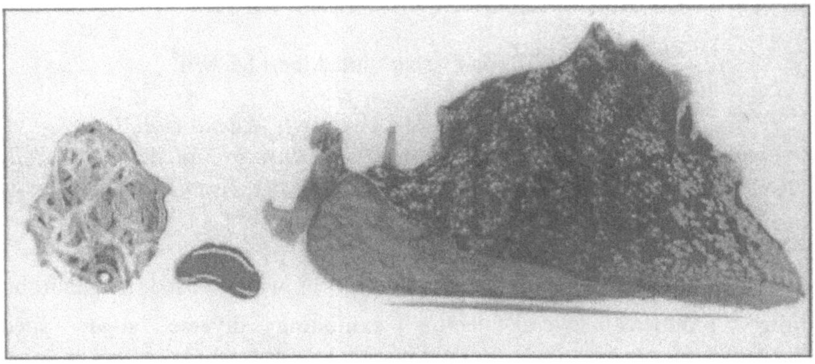

Figure 1. *Aplysia* with its gonad and fertilized egg strips (left)

I. AGL CARBOHYDRATE SPECIFICITY

The basic knowledge of AGL specificity, indicating GalUA (pectins) >> Met β Gal > Gal > Met α Gal,but not GlcUA nor GalNAc, (Gilboa-Garber *et al.*, 1985), was further characterized by enzyme-linked lectino-sorbent assay (ELLSA) and by inhibition of AGL-glycan interaction (Wu *et al.*, 2000). Examination of the lectin binding with 34 glycans revealed that it reacted strongly with the following glycoforms (Fig. 2) : most human blood group precursor (equivalent) glycoproteins (gps) (**structure 1**), two Gal α 1→4Gal-containing gps (asialo bird nest gp, gp #8, **structure 2** and sheep hydatid cyst gp, gp #15), and two GalUA-containing polysaccharides (pectins from apple and citrus fruits, gp # 4 and gp #5), but poorly with most human blood group A and H active and sialylated gps. The abilities of various glycans to inhibit the binding of AGL with Cyst Beach P-1 glycoprotein by ELLSA are shown in Fig. 3. The decreasing order of the reactivity of these glycoforms is Cyst OG 10% 2x PPT (one of the human blood group precursor gps, gp #1 in Fig. 3, **structure 1**) and sheep hydatid cyst gp (blood group P_1 active gp, gp #2 in Fig. 3) > asialo bird nest gp (gp #3), five human blood group precursor gps (gps 4, 5, 8, 9 and 10, **structure 1**) and two GalUA containing gps (pectin-A and pectin-C, gps # 6 and 7) > a blood group B active gp (Cyst Beach, gp #11); mild acid hydrolyzed hog gastric mucin #21 (**II**, gp#12), asialo rat sublingual gp (**II,** gp #13 **structure 3**); pneumoccus type 14

Structure 1

Proposed carbohydrate side chains of blood group active glycoproteins, prepared from human ovarian cyst fluid.

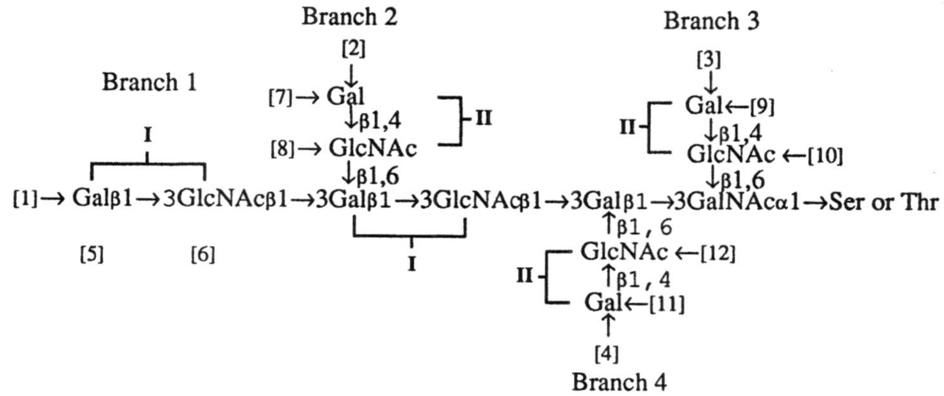

The blood group substances were purified from human ovarian cyst fluid by digestion with pepsin and precipitation with ethanol: the dried ethanol precipitates were extracted with 90% phenol, the insoluble fraction being given after the name of the blood group substance (e.g., Cyst Beach phenol insoluble). The supernatant was fractionally precipitated by addition of 50% ethanol in 95% phenol to the indicated concentrations. The designation 10 or 20% (ppt) denotes a fraction precipitated from phenol at an enthanol concentration of 10 or 20%; 2x signifies that a second phenol extraction from ethanol precipitation was carried out (e.g., Cyst OG 20% 2x).

The four-branched structure (branch 1 to branch 4) shown above represents the internal portion of the carbohydrate moiety of blood group substances, to which the residues responsible for A, B, H, Le^a, Le^b, Le^x and Le^y activities are attached (Lloyd and Kabat, 1968; Rovis *et al.*, 1973; Wu, 1988). This structure represents precursor blood group active glycoproteins (#1, Cyst OG 10% 2x ppt; #9, Cyst Tij 20% of 2nd of 10% in Fig. 3) and the precursor equivalent gp can be prepared by Smith degradation of cyst A, B, H active glycoproteins (#10, Cyst MSS 1st Smith degraded gp; in Fig. 3, Wu, 1988) or mild acid hydrolysis (P-1 gps – #4, Cyst Beach P-1; #5, Cyst Mcdon P-1; #8, Cyst Tighe P-1 in Fig. 3). Numbers in brackets [1] to [12] indicate the site of attachment for the human blood group A, B, H, Le^a, and Le^b determinants (#11, Cyst Beach; #12, Cyst 19; Cyst MSS 10% 2x; Cyst Mcdon; Cyst Tighe phenol insoluble and Cyst JS phenol insoluble in Fig. 3). These determinants as well as the structural units at the nonreducing end are the sources of lectin **A/A_h, B, I/II, T,** and **Tn** determinants (Wu and Sugii, 1988 and 1991; Wu, *et al.*, 1997) . A megalo-saccharide of twenty-four sugars has not been isolated. However, most of the carbohydrate chains isolated are parts of this structure.

Structure 2

The mucus glycoproteins, the so-called nest-cementing substances, isolated from the salivary gland of Chinese swiftlets (genus *Collocalia*). These Galα1→4Gal (**E**) and GalNAcα1→3GalNAcα1→Ser/Thr (**F**α) containing glycoproteins are mainly constituted of sialic acid-rich *O*-glycosylproteins (Wieruszeski *et al.*, 1987 and Wu *et al.*, 1996). The most complex representatives of the monosialyl fraction from *Collocalia* mucin are:

```
        E                  II
Galα1→4Galβ1→4Galβ1→4GlcNAc
                        ↓β1,6
      NeuAcα2→3Galβ1→3GalNAcα1→Ser/Thr
                      Tα
```

and

```
      E                 II
Galα1→4Galβ1→4Galβ1→4GlcNAc                 Tn
                       ↓β1,6
     NeuAcα2→3Galβ1→3GalNAcα1→3GalNAcα1→Ser/Thr
                    T                Fα
```

The other compounds identified are partial structures thereof.

Structure 3

Established structure of the carbohydrate moiety of asialo rat sublingual glycoprotein
This mucin contains the following of chains at three different lengths (Slomiany and Slomiany, 1978).

```
                      II                  II                  II
         1/2 II
One      GlcNAcβ1→3Galβ1→4GlcNAcβ1→3Galβ1→4GlcNAcβ1→3Galβ1→4GlcNAc
         β1→3Galβ1→4GlcNAcβ1→3GalNAcα1→ Ser/Thr ,   4 1/2 II →Tn
               II              Tn
```

```
                      II                  II                  II
         1/2 II
Two      GlcNAcβ1→3Galβ1→4GlcNAcβ1→3Galβ1→4GlcNAcβ1→3Galβ1→4GlcNAc
         β1→3GalNAcα1→Ser/Thr ,   3 1/2 II→Tn
               Tn
```

```
                      II                  II                  Tn
         1/2 II
Two      GlcNAcβ1→3Galβ1→4GlcNAcβ1→3Galβ1→4GlcNAcβ1→3GalNAcα1→Ser/Thr
                              2 1/2 II →Tn
```

The codes used are: **II**, Galβ1→4GlcNAcβ1→3; 1/2 **II**, GlcNAcβ1→3; **Tn**, GalNAcα1→Ser/Thr at nonreducing end. Most of the carbohydrate chains are parts of this structure, variation in this ratio and in chain lengths are expected to occur in different mucin preparations (Wu, *et al.*, 1995).

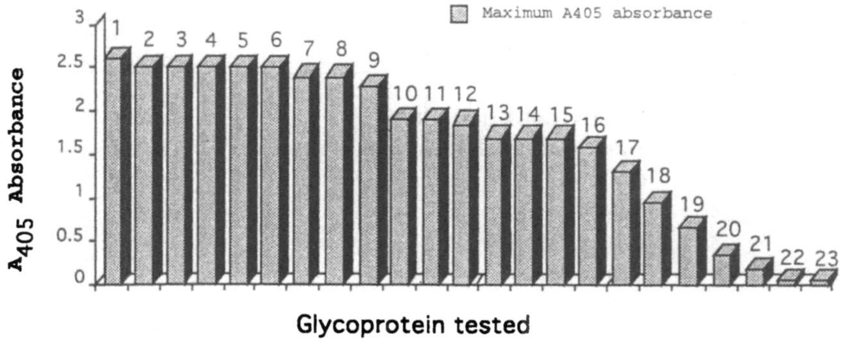

Figure 2. AGL-glycan interaction profile. Enzyme linked lectin sorbent assay (ELLSA) was used (Wu *et al.*, 2000). Binding of AGL to microtiter plates coated with serially diluted human blood group A, B, H, P_1, and Ii active glycoproteins and **T**, **Tn** and **F** containing gps, sialo- and asialo gps and polysaccharides. The lectin was used at a constant amount of 5 ng per well. Total volume50 µl. Maximal A_{405} absorbance was recorded after 2 h incubation. The order of reactivity is: 1. Cyst MSS 1st Smith degradred (**T, Tn, I, II**); 2. Cyst Beach P-1 (**T, Tn, I, II**); 3. Cyst Mcdon P-1 (**T, Tn, I, II**); 4. Pectin A (from apple); 5. Pectin C (from citrus); 6. Asialo rat sublingual gp (RSL, **II**); 7. Cyst Tighe P-1; 8. Asialo bird nest gp (**II, E, T, F**); 9. Cyst OG 10% 2x ppt (**B, I, II,** [Ii]); 10. Hog gastric mucin #21 (**I, II**); 11. Asialo human □1-acid (**II**); 12. Cyst Tij 20% of 2nd 10% (**B**); 13. Hog gastric mucin #14 (**A, I, II**); 14. Asialo porcine salivary gp (PSM, **A, A_h, T, Tn**); 15. Sheep hydatid cyst gp (**E** [P_1]); 16. Pneumococcus type 14 polysaccharide (**II**); 17. Bird nest gp (sialyl **II, E, T, F**); 18. Asialo fetuin; 19. Cyst Beach (**B**); 20. Hog gastric mucin #9 (**A_h, H**); 21. Cyst Mcdon (**A_h**); 22. Cyst JS (**H**); 23. Hog gastric mucin #4 (**A_h, H**).

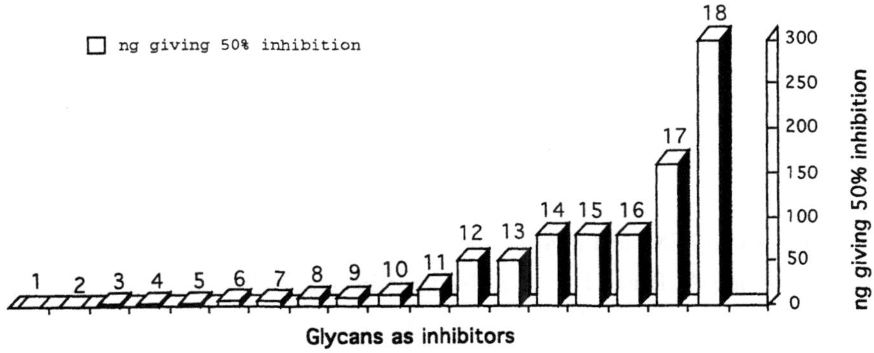

Figure 3. Inhibition profile of AGL-glycoform interaction by various glycans. Inhibition of the emzyme linked lectin sorbent assay was used. The quantity of Cyst Beach P-1 in the coating solution was 10 ng per well; AGL, 5ng/well; total volume, 50 µl. A_{405} was recorded after 2 h incubation. Based on the 50% inhibition (I_{50}), the order of reactivity is: 1. Cyst OG 10% 2x ppt; 2. GalUA and Sheep hydatid (P_1); 3. Asialo bird nest gp; 4. Cyst Beach P-1; 5. Cyst Mcdon P-1; 6. Pectin A; 7. Pectin C; 8. Cyst Tighe P-1; 9. Cyst Tij 20% of 2nd of 10%; 10. Cyst MSS 1st Smith degraded; 11. Cyst Beach (**B**); 12. Hog gastric mucin #21; 13. Asialo RSL (**II**); 14, Pneumococcus type 14 PS; 15. Asialo PSM (**A, H, T, Tn**); 16. Hog gastric mucin #14; 17. Cyst 19 (**B**); 18. Native bird nest gp. In this assay, the I_{50} of Galα1→4Gal **(E)** was 850 nanogram and D-Gal, 7.0 x10^3 nanogram, while Hog gastric mucin #4; Hog gastric mucin #9; Cyst MSS 10% 2x; Cyst Mcdon; PSM; Cyst JS phenol insoluble; Cyst Tighe phenol insoluble; RSL; Fetuin; Asialo fetuin; Bovine salivary gp (BSM, sialyl T, Tn); Asialo BSM; Ovine salivary gp (OSM, sialyl Tn); Asialo OSM did not reach 50% inhibition (Wu *et al.*, 2000).

Table 1. Carbohydrate specificity of AGL[a]

Activity order	Saccharide	Relative inhibitiory potency[b]
1	DiGalUA (GalUAα1→4GalUA)	8.5×10^4
2	TriGalUA (GalUAα1→4GalUAα1→4GalUA)	6.5×10^4
3	GalUA	3.7×10^4
4	Galα1→4Gal (**E**)	55.0
5	Galβ1→3GlcNAc (**I**)	9.2
6	Galβ1→4Man	3.7
7	Galβ1→3DAra	2.9
8	Galβ1→3GalNAc (**T**)	2.0
9	Phenyl-β-Gal	1.8
10	α-Melibiose(Galα1→6Glc)	1.6
11	D-Raffinose (Galα1→6Glcβ1→2D-Fru_f)	1.6
12	MethylβGal	1.6
13	p-NO_2-phenylαGal	1.6
14	p-NO_2-phenylβGal	1.6
15	Galα1→3GalαMethyl	1.4
16	p-NO_2-phenylβGalNAc	1.4
17	Galβ1→3GalNAc αbenzyl ($\mathbf{T}_\alpha$)	1.1
18	Gal	1.0
19	Galβ1→3GlcNAc1→3Galβ1→4Glc	0.8
20	MethylαGal	0.8
21	Galα1→3Gal (**B**)	0.7
22	Galβ1→4Glc (**L**)	0.6
23	Galβ1→4GlcNAc (**II**)	0.5
24	Stachyose (Galα1→6Galα1→6Glcβ1→2DFru_f)	0.4

[a]Data taken from Wu *et al.* (2000)
[b]Gal is normalized to 1.0 (55 nano moles required for 50% inhibition).

polysaccharide, asialo PSM, and mild acid hydrolyzed hog gastric mucin #14 (gps #14 to #16) » blood group A, B and H active glycoproteins and sialylated glycoproteins (gps #17 and #18 etc). With several exceptions, the inhibitory reactivities of glycoforms towards AGL agree with the maximum absorbance values recorded in the binding assay (Fig. 2).

The weak or negative reactivity of AGL with A and H active gps and most sialylated gps (Cyst Mcdon and Cyst MSS, Cyst Beach phenol insoluble, Cyst Tighe phenol insoluble, human α1-acid gp, RSL, and PSM) could be ascribed to the masking effects of LFucα1→, GalNAcα1→ and sialic acid at the terminal Galβ1→ and/or to poor adsorbance of these glycoforms onto microwell plate.

Among the GalUA and mammalian saccharides tested for inhibition of AGL-glycan binding (Table I), GalUA mono- to trisaccharides were the most potent ones. They were 8.5×10^4 times more active than Gal and about 1.5×10^3 more active than the human blood P^k active disaccharide (**E**, Galα1→4Gal). This disaccharide was 6, 28 and 120 times more efficient than Gal β 1→3GlcNAc (**I**), Gal β 1→3GalNAc (**T**), and Gal β 1→4GlcNAc (**II**), respectively, and 35 and 80 times more active than melibiose (Gal α 1→6Glc) and human blood group B active disaccharide (Gal α 1→3Gal), respectively, showing that the decreasing order of the lectin affinity toward α-anomers of Gal is: α 1→4 > α 1→6 > α 1 →3. From the data provided, the carbohydrate specificity of AGL can be defined as GalUA α 1→4 trisaccharides to mono GalUA > branched or cluster forms of **E**, **I** and **II** >> monomeric **E**, **I** and **II**, while GalNAc is inactive.

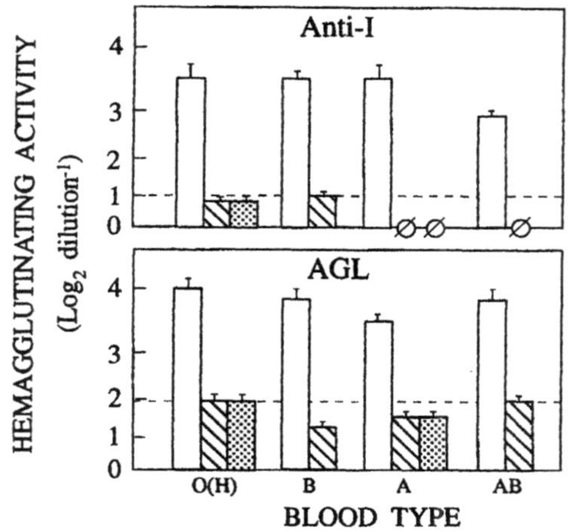

Figure 4. Agglutination of human adult (I-positive, empty bars) and I-negative fetal (striated bars) or rare adult (dotted bars) erythrocytes of different ABO types by human anti-I serum (using 3 sera) and AGL (1-5 μg/ml when papain-treated cells are used). The data represent means of results obtained in 5-15 experiments ± SEM.

II. AGL BLOOD GROUP SPECIFICITY

AGL strongly agglutinates human erythrocytes. Their agglutination is much stronger following enzyme (sialidase > papain) treatment (Gilboa-Garber *et al.*, 1985). Using different human blood group samples revealed that AGL mimics human anti-I antibodies in preferential agglutination of adult I-positive erythrocytes as compared to the I-negative **i**-positive fetal (cord) or rare adult cells (Fig. 4).

The lectin favors branched poly *N*-acetyllactosamine (LacNAc) of the adult I antigen over its linear precursor i antigen of fetal erythrocytes (Fig. 5).

AGL preferential interaction with I vs i was confirmed in adsorption test as compared to I – insensitive *Maclura pomifera* lectin (Fig. 6). In this property, AGL also differs from

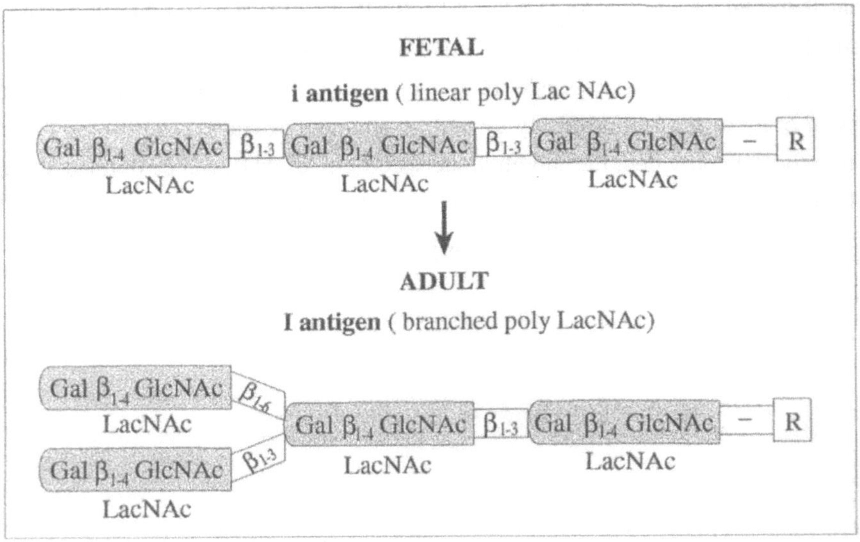

Figure 5. A simplified presentation of the structural difference between adult I and fetal i antigens.

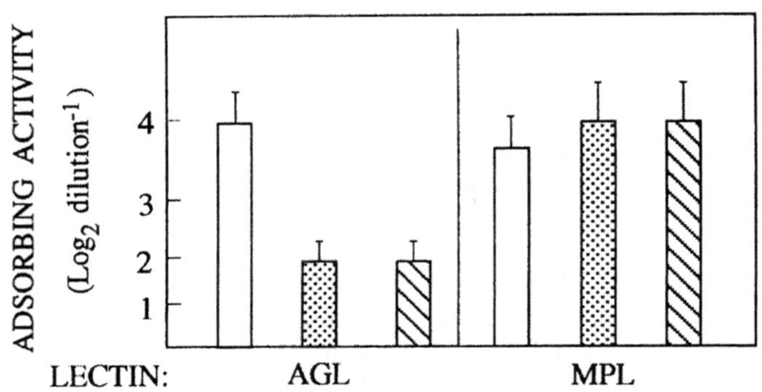

Figure 6. Comparison of adsorption of AGL and *Maclura pomifera* lectin (MPL, which is insensitive to I antigen) upon I-positive (empty bars) and I-negative (adult – dotted bars; fetal–striated bars) erythrocytes. The adsorption results are presented as decrease in the hemagglutinating activities from the original ones. The data represent means of 5 experimental results ± SEM.

galectin-1 (Gilboa-Garber *et al.*, 1999) which interacts with the Ii system antigens without preference of the I antigen.

In contrast to certain plant and microbial lectins shown to exhibit some I-affinity, AGL affinity for the I antigen is not dependent on the presence of ABH/P system antigens (Gilboa-Garber *et al.*, 1999). It agglutinates O_h erythrocytes (Bombay type) more strongly than the H-positive ones, and P-negative (p) and P^k erythrocytes ($p>P^k$) more strongly than P-positive cells. In addition, it also exhibits T antigen affinity–reacting with sialidase-treated erythrocytes more strongly than with papain-treated ones and much more

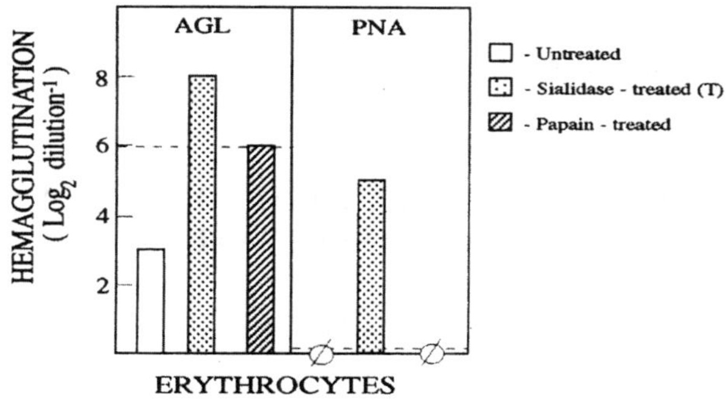

Figure 7. Comparison of the agglutination of untreated, sialidase-treated **(T)** and papain-treated erythrocytes (empty, dotted and striated bars, respectively) by AGL and the peanut lectin PNA. The data represent means of more than 10 experimental results.

strongly than with untreated cells (Fig. 7). A prominent hemagglutination by AGL was also observed with a single Tn blood sample examined.

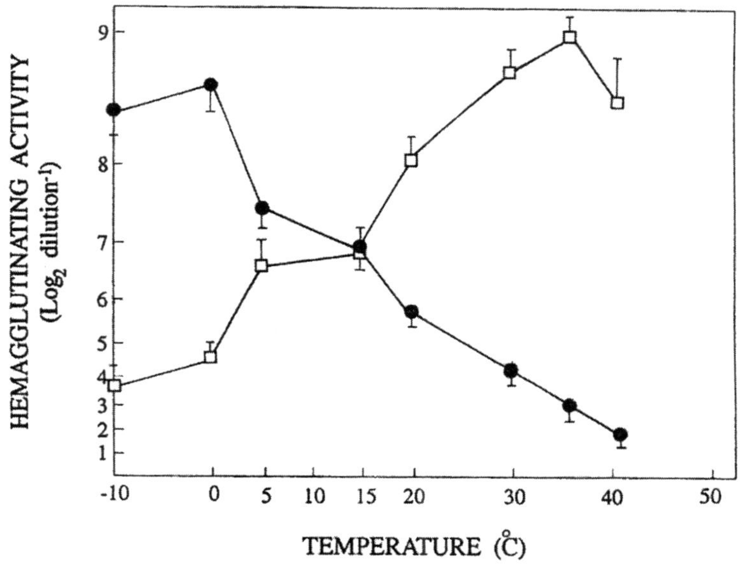

Figure 8. Temperature-dependence of the agglutinating activities of AGL (•) and Con A (□) towards human erythrocytes.

An interesting feature unveiled during the above described hemagglutination studies was the enhancement of AGL interactions with untreated or papain-treated erythrocytes at 4°C – a property shared by all the I-detecting agglutinins, including the anti-I antibodies and galectins (Gilboa-Garber *et al.,* 1999), in contrast to other antibodies and lectins such as anti-A and Con A (Fig. 8). However, when the AGL binding site interacted with the T antigen following its unmasking by sialidase treatment of RBCs, the hemagglutination temperature profile changed and no enhancement was observed at 4°C. The dual antigen (Ii and T) specificity of the same binding site of AGL provides a special occasion for a study of the temperature effect on the interactions of lectin binding sites with their different receptor saccharides (Sudakevitz and Gilboa-Garber, 2000).

III. AGL INTERACTION WITH SOLUBLE ANIMAL GLYCOCONJUGATES

Comparison of the avidity of AGL for a series of glycoproteins and asialoglycoproteins as well as polysaccharides showed a very high avidity of AGL for GalUA residues of pectins and for terminal galactosyl of animal substances exhibiting combined **T**, **Tn**, **I**, **II**, **E**, **T** and **F** glycotopes (lectin determinants, Wu and Sugii, 1988), as well as P^k, but not for Tn, A or H antigens alone (Wu *et al.*, 2000). The latter results may indicate that the strong hemagglutination observed in the single Tn case examined by us may be due to either a combination of this antigen with I antigen or reduced sialylation.

1. Human Body Fluids

Studies on interactions of AGL with human body fluids revealed only slight inhibition of the AGL hemagglutinating activity by human sera, variable inhibitions by human salivas

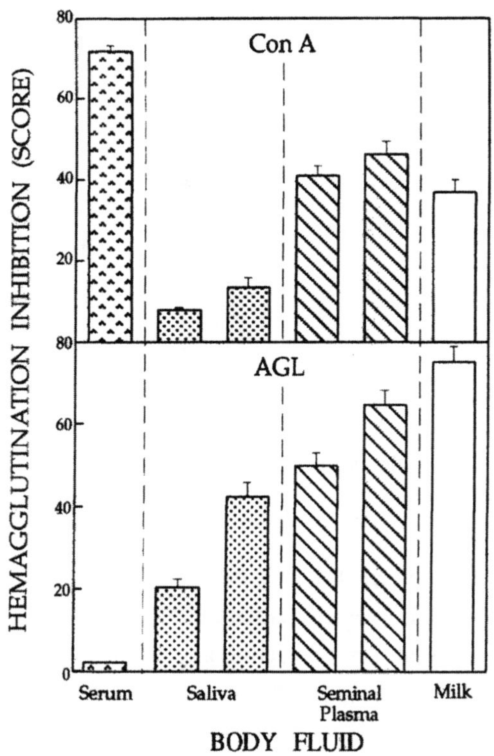

Figure 9. Comparative inhibition of Con A and AGL hemagglutinating activities by human sera, saliva,seminal plasma and milk samples (B. Lerrer's and E. Lesman-Movshovich's, Ph.D. theses, directed by N. Gilboa-Garber).

and seminal fluids (Lerrer and Gilboa-Garber, 1998) and highest degree of inhibition by human milk (Fig. 9) (Lesman-Movshovich *et al.*, 1998).

The results presented in Fig. 9 indicate that AGL and Con A significantly differ in their reactivity with the body fluid glycoconjugates. Human serum contains more of those inhibiting Con A while milk is richer in the AGL-inhibiting ones. The ability of AGL to bind galactosylated molecules in human body fluids might be valuable for purification of these components which may be useful for inhibition of microbial and neoplastic cell adhesion (Gilboa-Garber and Lerrer, 1997).

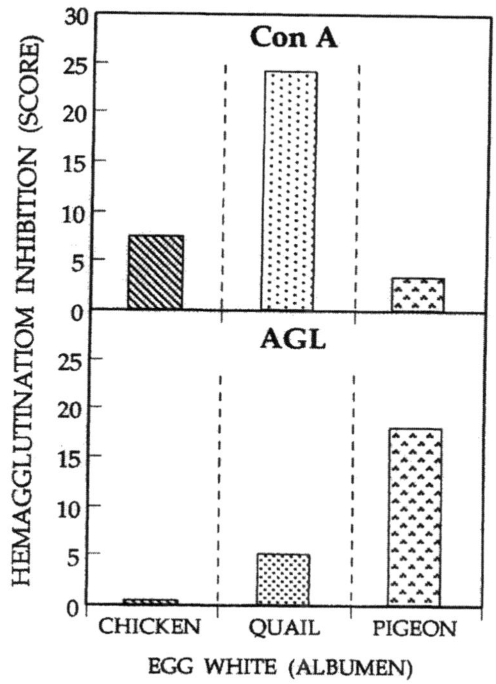

Figure 10. Comparative inhibition of Con A and AGL hemagglutinating activities (scores) by the egg white of chicken, quail and pigeon. (B. Lerrer's, Ph.D. thesis, directed by N. Gilboa-Garber).

2. Fowl Egg White (Albumen) Samples

Very interesting results were also obtained studying the inhibition of AGL by egg white samples of chicken, quail and pigeon. The results presented in Figure 10 indicate that AGL and Con A were differently inhibited by these albumens. The chicken white exhibited fair inhibition of Con A, mainly due to the presence of glycosylated (Glc/GlcNAc) avidin and also conalbumen and ovomucoid trypsin inhibitors, but not of AGL. The quail albumen very strongly inhibited Con A but only slightly AGL. The pigeon white showed reciprocal behavior: very strong inhibition of inhibiting AGL and only weakly of Con A.

IV. CELL GROWTH STIMULATION BY AGL

AGL stimulation of mitosis in human peripheral lymphocytes was observed microscopically and by ^{3}H-thymidine incorporation (Table 2) as well as by determination of interleukin 2 (IL-2) formation (Zipris and Gilboa-Garber, 1987). AGL was also shown to enhance neurite outgrowth in cultured *Aplysia* neurons and to increase their viability (as a neurotrophic agent) in cultures (Wilson *et al.*, 1992).

Table 2. Stimulation of human T and B lymphocytes by AGL, as compared to PHA, assayed by ^{3}H-thymidine incorporation (Zipris and Gilboa-Garber, 1987)

Lectin added	"T" cells	"B" cells
None	1,534 ± 512	1,004 ± 122
AGL (62 μg/ml)	47,505 ± 7,091	16,578 ± 687
PHA (125 μg/ml)	80,178 ± 12,852	20,416 ± 4,924

All the above growth stimulating activities were shown to be specifically inhibited by the respective saccharides.

V. AGGLUTINATION OF CANCER CELLS AND REDUCTION OF THEIR TUMORIGENICITY BY AGL

AGL was shown to strongly agglutinate cancer cells (Leibovici *et al.*, 1986; Avichezer *et al.*, 1987). This agglutination was inhibitable by GalUA. Cell viability under these conditions was only slightly inhibited. Subsequent inoculation of the lectin-coated cells to mice revealed a significant reduction in their tumorigenicity as compared to the lectin-free cells. Thirty μg of AGL totally abolished tumor appearance and 5 μg/2×10^5 cells led to a 60% reduction in tumorigenicity (Leibovici *et al.*, 1986; Avichezer *et al.*, 1987). The DNA synthesis (followed by ^{3}H-thymidine incorporation) in the tumor cells was unaffected by the lectin, although the cell membrane was probably affected and the coated cells were eliminated (Avichezer *et al.*, 1987). The reason for the increased interaction of AGL with cancer cells might be related to the findings of higher exposure of poly LacNAc residues and also stronger interaction of such cells with galectins, which detect the carcinoembryonic antigens (Ohannesian *et al.*, 1994). It may likewise be due to cancer-associated T and Tn antigens (Springer, 1997).

VI. APPLICATION OF AGL FOR HISTOCHEMICAL/ CYTOCHEMICAL STUDIES OF PLANTS AND FUNGI

Benhamou *et al.* (1988) tagged AGL with colloidal gold particles and used it for labeling carbohydrates in resin-embedded sections of various plant and fungal tissues fixed with glutaraldehyde and osmium tetroxide. The patterns of the AGL binding sites were compared to those obtained with the Gal-specific lectin Ricin. The two lectins differed in their carbohydrate binding. The AGL labeling was lost if the pectin was specifically removed by digestion with pectinase and washings or if the lectin was specifically inhibited by GalUA.

The staining indicated that AGL interacted with the pectic components known to occur in higher plant walls (up to 35% of the wall mass) and in fungal (e.g., ascomycetes) cell walls. The introduction of this lectin into cytochemistry was found to endorse results hitherto obtained using gold-complexed pectinase (Benhamou *et al.*, 1988). Usage of the gold-labeled lectin for the ultrastructural study of galacturonic acid distribution in fungi which are pathogenic for plants revealed heterogeneity in the cell wall composition of both their mycelia and spores (Benhamou, 1988) and also enabled their typing and identification.

VII. APPLICATION OF AGL FOR STUDY OF PROKARYOTIC CELL SURFACE COMPOSITION AND MICROBIAL TYPING

Marine bacteria, isolated from the *Aplysia* surroundings, were generally agglutinated by *Aplysia* egg (Kamiya and Shimizu, 1981) and gonad (Zipris *et al.*, 1986) lectins. AGL was also found to differentially interact with nosocomial microorganisms. Among thirty isolates of *Escherichia coli* examined, about three were found to be agglutinated by it. Since AGL may select bacteria which bear carcinoembryonic-like antigens, Thomsen-Friedenreich (T) and human blood group I active ligand as described above, such bacteria might be useful for cancer diagnosis and treatment according to Springer's strategy (Springer, 1997).

Preliminary results (Gilboa-Garber, Sudakevitz and Cohen) showing stronger agglutination of AGL-positive *E. coli* strains by sera of healthy subjects as compared to those of cancer patients, support this idea. Exciting results were also obtained using AGL for the study of the S (surface) glycoproteins of halophilic Archaea and for their typing (Gilboa-Garber *et al.*, 1998). There were serious difficulties in the investigation of the structure of these microorganisms due to the dependence of their structural integrity upon highly hypertonic environment. The salinity required hampered usage of antibodies, lectins, and several enzymes for their characterization, since these proteins do not function in the

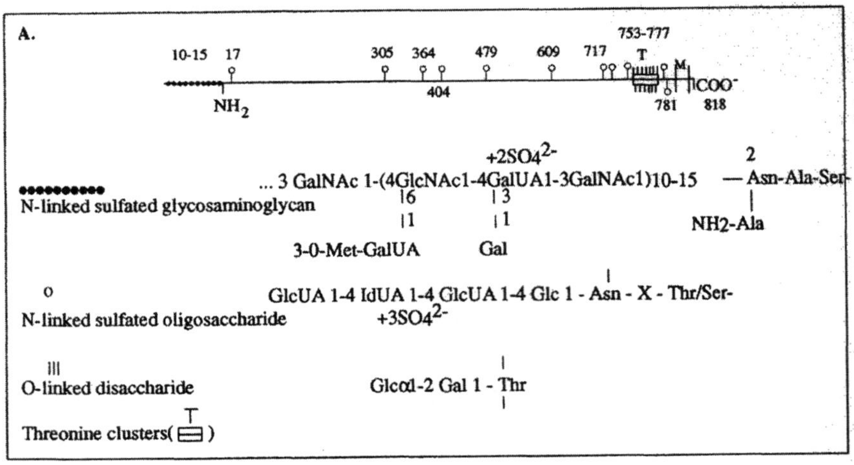

Figure 11. Model of the S-layer glycoprotein of *Halobacterium salinarum* and its glycosylation sites according to Lechner and Wieland (1989).

presence of high salt concentrations. As a result, there were significant discrepancies between the structures described by different groups of investigators and even by the same groups during the research period. The first complete structural description of archaeal envelope glycoprotein saccharides was that of *Halobacterium salinarum* (Lechner and Wieland, 1989) followed by that of *Haloferax volcanii* (Sumper *et al.*, 1990). The latest model of *Hb. Salinarum* S-layer gp (Mr about 120 kDa) described is composed of a core protein (about 87 kDa), rich in acidic amino acids, bearing many acidic and neutral saccharide chains which are attached to it in a manner resembling animal proteoglycans (Fig. 11).

As may be seen in this model, the core protein bears three types of complex saccharides: one of them, a major (around 10 kDa) acidic *N*-linked sulfated glycosaminoglycan – at the sub-N-terminal domain of the core protein, as a single copy. This polysaccharide is composed of 10-15 repeats of branched sulfated pentasaccharide containing terminal 3-*0*-Met-GalUA. This saccharide complex is considered as the main shape – maintaining component of the S-layer. The second type is *N*-linked sulfated linear tetrasaccharide composed of 2-3 sulfated hexuronic acid residues (GlcUA, IdUA, but not GalUA). This low molecular mass tetrasaccharide is distributed along the protein core in about 10 separate copies. The third type is a collagen-like *O*-linked Glcα1→2Gal bound in about 12-15 copies per molecule via threonines, which are clustered close to the C-terminal domain of the core. In order to use lectins for the study of the above saccharides, we defined a salinity threshold high enough for survival of the Archaea while sufficiently low for the lectin binding. Since no clear agglutination could be observed, we examined the adsorption of the lectins to the Archaea using hemagglutination inhibition assay.

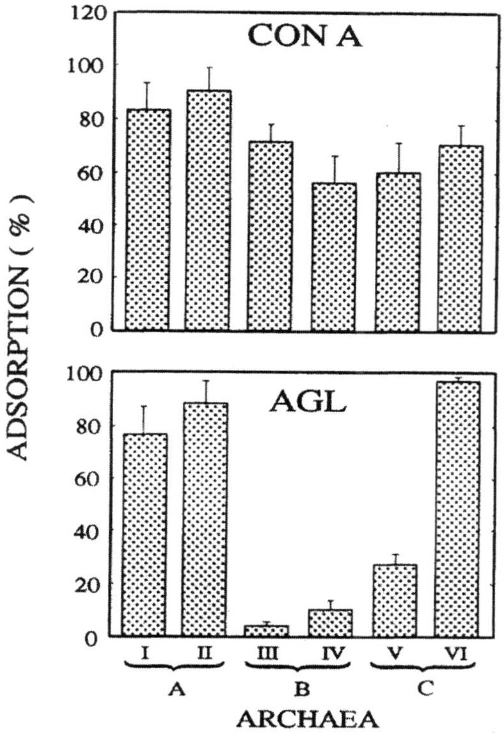

Figure 12. Adsorption of Con A and AGL onto the halophilic Archaea (followed by hemagglutination inhibition test). The data represent means ± SEM obtained in 9-15 tests. The Archaea examined: A. *Halobacterium salinarum* strains "*halobium*" (I) and "*salinarium*" (II); B. *Haloferax* sps. *volcanii* (III) and *Mediterranei* (IV) and; C. *Haloarcula* sps. *marismortui* (V) and *vallismortis* (VI).

The presence of many copies of terminal glucose in *Hb. salinarum* (Fig. 11) was definitely confirmed by Con A, which exhibited highest reactivity (Fig. 12: A I, II). The presence of 3-0-Met GalUA in its first major saccharide (in 10-15 repeats) entreated AGL usage. Indeed as may be seen in Fig. 12, both strains of *Hb. salinarum* (A) adsorbed AGL almost as strongly as Con A.

Haloferax volcanii (BIII) also nicely adsorbed Con A (Fig. 12), although slightly less than *Halobacterium salinarum*. This finding is in good agreement with the model of its structure (Fig. 13) (Sumper *et al.*, 1990).

As seen in Fig. 13, *Hf. volcanii* possesses somewhat fewer saccharides than *Hb. salinarum*: 7 instead of 12 N-glycosylation sites and also fewer O-linked Glcα1$\rightarrow$2Gal residues. Therefore, its slightly weaker reaction with Con A (Fig. 12: B III), as compared to that of the latter, agrees with the models. The reaction of the second *Haloferax* (*mediterranei*, Fig. 12: B IV) with Con A indicates a similar trend. The same conclusion also applies to the two *Haloarcula* sps.

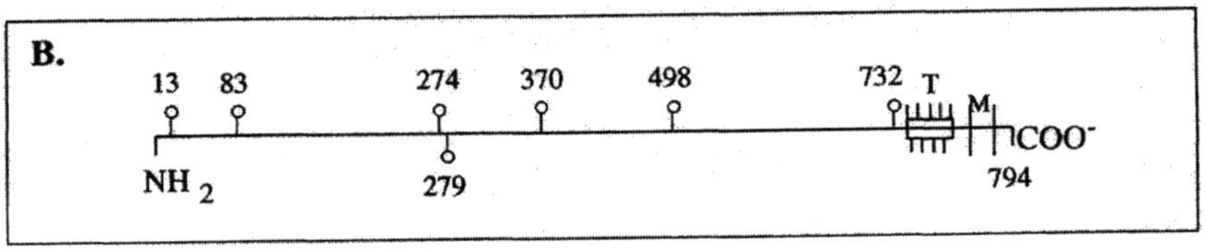

Figure 13. Model of the S-layer glycoprotein of *Haloferax volcanii*, according to Sumper *et al.* (1990). Details as in Fig. 8.

The results with AGL, as represented in Figure 12 are even more significant. The fact that AGL does not interact with *Hf. volcanii* (in contrast to Con A) confirms the model presented in Figure 13, namely – absence of the major GalUA-bearing acidic *N*-linked (at the sub-*N*-terminus) sulfated glycosaminoglycan (the first saccharide type) which was supposed to absorb AGL. A similar result with *Hf. mediterranei* (Fig. 12: B IV) indicates that the two *Haloferax* sps are similar in structure. Confronting this example are the two *Haloarcula* sps (Fig. 12: C V & VI). Examination of their adsorption of Con A and AGL revealed that both strongly adsorbed Con A but profoundly differed in regard to AGL: while *Ha. Marismortui* exhibited a weak to moderate AGL adsorption, *Ha. vallismortis* very strongly adsorbed this lectin, even more so than *Hb. salinarum*. It would be interesting to compare these results when the structure of those of Archaea will be described. The results presented here and those obtained with the Archaea data previously described (Gilboa-Garber *et al.*, 1998) demonstrate the value of this lectin for archaeal cell surface study as well as for their typing.

VIII. POSSIBLE BIOLOGICAL FUNCTIONS OF AGL

There are several possible functions which could be ascribed to AGL, in analogy to lectin functions in other animals, plants and microorganisms and based on the lectin biological effects.

In the gonad, the lectin function might be as follows:

1. Protection of gametes and fertilized eggs.
2. Organization of the oocytes and sperm cells in the gonad.
3. Prevention of autologous fertilization in the hermaphroditic gonad.

In the fertilized eggs, the lectin contributions might be as follows:

1. Protection of the egg embryos.
2. Organization of the eggs in the strip matrix.
3. Stimulation of cell division in the developing embryos.
4. Adhesion of the floating egg strips to algae (e.g., *Ulva*) which ensures the nutrition of the hatching offsprings.

SUMMARY

Adult *Aplysia* gonad contains high levels of a galactophilic lectin (MW around 65 kDa; composed of 2 subunits of apparent single species). It binds galactose and various α/β-galactosides (but not *N*-acetylgalactosamine), in addition to an outstanding high affinity for galacturonic acid. This lectin is relatively resistant to heating up to 70°C and to alkaline pH, but sensitive to proteolysis and low pH. It resembles galectins in binding to poly LacNAc (preferentially branched) complexes at low temperatures (0°-4°C) more avidly than at room temperature or at 37°C, but differs from them in being Ca^{2+}-dependent. It agglutinates papain/sialidase-treated erythrocytes more strongly than untreated cells and stimulates mitosis in peripheral human lymphocytes (inducing IL-2 formation). This lectin also enhances neurite outgrowth and increases their viability, while suppressing cell tumorigenicity. It is useful for histochemical/ cytochemical studies of galacturonic acid in plant tissues and fungi and for the study of cell surface composition of various prokaryotic (including halophilic Archaea) and eukaryotic cells and for their typing. It is useful as a reagent for I-antigen detection in adult human erythrocytes (anti-I), exhibiting strongest agglutination of O(h) Bombay-type erythrocytes and also exhibits sensitivity to the T antigen. It binds galactosylated molecules in human body fluids (shown by hemagglutination - inhibition tests), including saliva, seminal fluid and milk (detecting individual divergence) and in fowl egg albumens (exhibiting highest affinity for that of pigeon). Therefore, it might be valuable as a probe and fishhook for fishing compounds exhibiting anti-bacterial/neoplastic cell adhesion activities.

REFERENCES

Avichezer, D., Leibovici, J., Gilboa-Garber, N., and Michowitz, M., 1987, New galactophilic lectins reduce tumorigenicity and preserve immunogenicity of Lewis lung carcinoma cells, in: *Lectures and Symposia 14th Int. Cancer Cong., Budapest, 1986,* K. Lapis and S. Ackhardt, eds., Karger, Basel, pp. 79-86.

Benhamou, N., 1988, Ultrastructural study of galacturonic acid distribution in some pathogenic fungi using gold-complexed *Aplysia depilans* gonad lectin. *Can. J. Microbiol.* 35: 349-358.

Benhamou, N., Gilboa-Garber, N., Trudel, J., and Asselin, A., 1988, A new lectin-gold complex for ultrastructural localization of galacturonic acids. *J. Histochem. Cytochem.* 36: 1403-1411.

Gilboa-Garber, N., and Lerrer, B., 1997, Potent cell-cell interaction blocking factors in human seminal plasma as shown by means of bacterial, plant and animal lectins. *Proc. of the Annual Meeting of the Israel Soc. Microbiol., Ramat Gan,* p. 69 (Abstract).

Gilboa-Garber, N., Mizrahi, L., and Susswein, A.J., 1984, Detection of lectins in the reproductive system and hemolymph from species of the sea hare, *Aplysia. Mar. Biol. Lett.* 5: 105-114.

Gilboa-Garber, N., Mymon, H., and Oren, A., 1998, Typing of halophilic Archaea and characterization of their cell surface carbohydrates by use of lectins. *FEMS Microbiol. Lett.* 163: 91-97.

Gilboa-Garber, N., Sudakevitz, D., and Levene, C., 1999, The *Aplysia* lectin anti-I specificity as compared to that of human anti-I and several other I-detecting lectins. *Transfusion.* 39: 1060-1064.

Gilboa-Garber, N., Susswein, A.J., Mizrahi, L., and Avichezer, D., 1985, Purification and characterization of the gonad lectin of *Aplysia depilans*. *FEBS Lett.* 181: 267-270.

Hammarström, S., and Kabat, E.A., 1964, Purification and characterization of a blood group reactive hemagglutinin from the snail *Helix pomatia* and a study of its combining site. *Biochemistry* 8: 2696.

Hammarström, S., Westoo, A., and Bjork, I., 1972, Subunit structure of *Helix pomatia* A hemagglutinin. *Scand. J. Immunol.* 1: 295-309.

Kamiya, H., and Shimizu, Y., 1981, A natural agglutinin inhibitable by D-galacturonic acid in the sea hare *Aplysia* eggs: characterization and purification. *Bull. Jap. Soc. Sci. Fish* 47: 255.

Lechner, J., and Wieland, F., 1989, Structure and biosynthesis of prokaryotic glycoproteins. *Ann. Rev. Biochem.* 58: 173-194.

Leibovici, J., Avichezer, D., and Gilboa-Garber, N., 1986, Effect of lectins on tumorigenicity of AKR lymphoma cells of varying malignancy. *Anticancer Res.* 6: 1411-1415.

Lerrer, B., and Gilboa-Garber, N., 1998, Estimation of anticancer potential of glycoconjugates from human milk and other body fluids examined by inhibition of galactophilic animal and related lectins. *5th IUBMB Conference on the Biochemistry of Health and Diseases, Jerusalem,* p. 74 (Abstract).

Lesman-Movshovich, E., Lerrer, B., and Gilboa-Garber, N., 1998, The anti-infection potential of human milk glycoconjugates as shown by inhibition of microbial related lectins. *Proc. 5th IUBMB Conference on the Biochemistry of Health and Diseases, Jerusalem,* p. 117 (Abstract).

Lloyd, K.O. and Kabat, E.A., 1968, Immunochemical studies on blood groups. XLI. Proposed structures for the carbohydrate portion of blood group A, B, H, Lewis^a and Lewis^b substances. Proc. Nat. Acad. Sci., 61: 1470-1477.

Ohannesian, D.W., Lotan, D., and Lotan, R., 1994, Concomitant increases in galectin-1 and its glycoconjugate ligands (carcinoembryonic antigen, lamp-1, and lamp-2) in cultured human colon carcinoma cells by sodium butyrate. *Cancer Res.* 54: 5992-6000.

Ozeki, Y., 1998, Purification and cell attachment activity of a D-galactose-binding lectin from the skin of sea hare, *Aplysia kuordai. Biochem. Mol. Biol. Int.* 45: 989-995.

Rovis, L., Anderson, B., Kabat, E.A., Gruezo, F., and Liao, J., 1973, Structures of oligosaccharides produced by based-borohydride degradation of human ovarian cyst blood group H, Le^b, Le^a active glycoproteins., Biochemistry, 12: 5340-5354.

Slomiany, A. and Slomiany B.L., 1978, Structures of the *O*-glycosidically linked carbohydrate units of fetuin. *J. Biol. Chem.* 253: 7301-7306.

Springer, G.F., 1997, Immunoreactive T and Tn epitopes in cancer diagnosis, prognosis, and immunotherapy. *J. Mol. Med.* 75: 594-602.

Sudakevitz, D., and Gilboa-Garber, N., 2000, Different temperature profiles are exhibited in *Aplysia* gonad lectin (AGL) binding to the saccharide determinants of the human erythrocyte I and T antigens. Proc. Ann. Meeting, Israel Soc. Biochem. Mol. Biol., Haifa, p. 18.

Sumper, M., Berg, E., Mengele, R., and Strobel, I., 1990, Primary structure and glycosylation of the S-layer protein of *Haloferax volcanii. J. Bacteriol.* 172: 7111-7118.

Wieruszeski, J.M., Michalski, J.C., Montreuil, J., Strecker, G., Peter-Katalinic, J., Eagge, H., Van Halbeek, H., Mutsaers, J.H.G.M., and Vliegenthart, J.F.G., 1987, Structure of the monosialyl oligosaccharides derived from salivary gland mucin glycoproteins of the Chinese Swiftlet (Genus *Collocalia*). *J. Biol. Chem.* 262: 6650-6657.

Wilson, M.P., Carrow, G.M., and Levitan, I.B., 1992, Modulation of growth of *Aplysia* neurons by an endogenous lectin. *J Neurobiol.* 23: 739-750.

Wu, A.M., 1988, Structure concepts of the blood group A, B, H, Le^a, Le^b, I and i active glycoproteins purified from human ovarian cyst fluid. *Adv. Exp. Med. Biol.* 228: 351-392.

Wu, A.M., and Sugii, S., 1988, Differential binding properties of GalNAc and/or Gal specific lectins. *Adv. Exp. Med. Biol.* 228: 205-263.

Wu, A.M., and Sugii, S., 1991, Coding and classification of D-galactose N-acetyl-D-galactosamine, and β-D-Galp-[1→3(4)]-β-D-GlcpNAc specificities of applied lectins. *Carbohydr. Res.* 213: 127-143.

Wu, A.M., Herp, A., Song, S.-C., Wu, J.H., and Chang, K.S.S., 1995, Interaction of native and asialo rat sublingual glycoproteins with lectins. *Life Sci.* 57: 1841-1852.

Wu, A.M., Wu, J.H., Song, S.-C.and Kabat,E,A., 1996, *Bandeiraea (Griffonia) simplicifolia* lectin-I, isolectin A4, reacting with Tn (GalNAcα 1→Ser/Thr) or galabiose (Galα 1→4Gal) containing ligands. FEBS Letters. 398, 183-186.

Wu, A.M., Song, S.-C., Sugii, S., and Herp, A., 1997, Differential binding properties of Gal/GalNAc specific lectin available for characterization of glycoreceptors. *Ind. J. Biochem. Biophys.* 34: 61-71.

Wu, A.M., Song, S.-C., Chen, Y.-Y., and Gilboa-Garber, N., 2000, Defining the carbohydrate specificities of *Aplysia* gonad lectin exhibiting a peculiar D-galacturonic acid affinity. *J. Biol. Chem.* 275: 14017-14024.

Zipris, D., and Gilboa-Garber, N., 1987, Stimulation of human peripheral lymphocytes and induction of interleukin 2 production by a lectin from the gonad of the sea hare, *Aplysia fasciata. Dev. Comp. Immunol.* 11: 501-511.

Zipris, D., Gilboa-Garber, N., and Susswein, A.J., 1986, Interaction of lectins from gonads and haemolymph of the sea hare *Aplysia* with bacteria. *Microbios* 46: 193-198.

LECTIN AND ANTI-CARBOHYDRATE ANTIBODY ASSAYS USING CHEMICALLY MODIFIED LIGANDS

Maria Duk, Albert M. Wu* and Elwira Lisowska

*Department of Immunochemistry, Ludwik Hirszfeld Institute of Immunology and Experimental Therapy, Polish Academy of Sciences, 53-114 Wroclaw, Poland; *Glycoimmunochemistry Research Laboratory, Chang-Gung Medical College, Kwei-san, Tao-yuan, Taiwan 33332*

Studies on lectins nad anti-carbohydrate antibodies include a determination of their specificity, and once this specificity is known, these reagents are useful tools for characterization of glycoconjugates of unknown structure. Despite great progress in the development of instrumental methods for oligosaccharide structure determination, lectins or antibodies are still the reagents of choice for detection of oligosaccharide structures in biological material (e.g. tissue sections, or cells), or when the amount of material is too small for instrumental analysis. In this short review, we present simple immunoenzymatic and 'lectinoenzymatic' methods, with application of chemically modified glycoprotein ligands, developped recently in our laboratories.

I. DETERMINATION OF LECTIN- OR ANTIBOBY-CARBOHYDRATE INTERACTION

Undoubtedly, a most accurate method (which does not require instrumentation) to measure quantitatively an interaction of a macromolecular glycoconjugate with a lectin or antibody is a quantitative precipitation (Kabat, 1961; Wu *et al.*, 1996 and refs. therein). However, this method is time- and material-consuming. A modern excellent method of measuring an interaction beteween macromolecules in real time is the surface plasmon resonance (SPR). This method does not need derivatization of interacting molecules, allows to determine kinetics of the reaction (association and dissociation coefficients), affinity constants, and has several other applications, but it requires an expensive BIAcore instrument (Myszka, 1999; Haseley *et al.*, 1999). Radioisotopic methods are avoided, if possible, because of the health hazard. Therefore, most common methods are enzymatic solid phase assays, first of all the microtiter plate enzyme-linked immunosorbent or lectinosorbent assays.

The estimation of antibody bound to the antigen immobilized on the plate is usually done using respective anti-Ig antibodies conjugated with an enzyme (most frequently

The Molecular Immunology of Complex Carbohydrates–2.
Edited by Albert M. Wu, Kluwer Academic/Plenum Publishers, 2001.

alkaline phosphatase or horseradish peroxidase). Such a conjugate is an universal reagent for all antibodies of a given species, independently of the antibody specificity. The same method can be applied for lectins, but it requires preparation of antibodies to each individual lectin. To overcome this difficulty, the lectins are labeled with a non-radioactive compound and detected with an universal reagent which specifically binds to a label. An example of such method is labeling lectins with digoxigenin and detection with anti-digoxigenin antibody/enzyme conjugate (Haselbeck *et al.*, 1990).

Another possibility is labeling the lectins with biotin and detection with avidin conjugated with an enzyme. Native avidin is a glycosylated protein and should be avoided in lectin studies, and nonglycosylated bacterial streptavidin has a lower affinity for biotin. The best high-affinity reagent is a recombinant nonglycosylated egg avidin, e.g ExtrAvidin (product of Sigma). Biotinylated lectins are commercially available, but the procedure of biotinylation is simple and can be easily done in a laboratory. We tested two methods of biotinylation, *via* oligosaccharide chains of glycosylated lectins, using biotin hydrazide, and *via* lectin amino groups, using biotinamidocaproate-N-hydroxysuccinimide ester (Duk *et al.*, 1994a; Lisowska *et al.*, 1996). The second method was found to be better, because it (i) can be applied to any lectin, independently of its glycosylation, (ii) has not caused a decrease of binding activity or specificity of any lectin studied so far, and (iii) gives higher sensitivity of the lectin detection than modification with biotin hydrazide. Biotinylated lectins were successfully used in our laboratories for the plate assay and Western blotting (Duk *et al.*, 1994a, 1994b, 1997 and 2000; Wu *et al.*, 1995, 1998, 1999a, 1999b and 2000).

Biotinylation can be also applied to antibodies, but requires more caution, because in some cases this modification can change the antibody properties (Hoyer-Hansen *et al.*, 2000). We found, that sensitivity of the assay with weak mouse monoclonal antibodies, which hardly detect the antigen on the blot using enzyme-conjugated anti-mouse Ig antibodies, can be enhanced over 100-times when biotinylated anti-mouse Ig antibodies and ExtrAvidin-enzyme conjugate are used. Generally, the biotinylation is a versatile method which is applied for many purposes, and in our experience is a convenient method for lectin studies.

II. APPLICATION OF SMITH DEGRADATION OF GLYCOPHORINS FOR STUDIES WITH ANTI-Tn REAGENTS

A major oligosaccharide chain of human glycophorin A (GPA) and of minor glycophorins (B, C and D) is the tetrasaccharide, disialylated Galβ1-3GalNAc. No lectin specific for this type of O-glycans is known. The sialylated glycophorin O-glycans react only with sialic acid specific lectins which, however, recognize sialic acid residues also on other oligosaccharide chains, including N-glycans. On the other hand, there are many lectins specific for the Thomsen-Friedenreich (TF) antigen (Galβ1-3GalNAcα-Ser/Thr chains) and Tn antigen (GalNAcα-Ser/Thr residues) (Peumans and Van Damme, 1998). These antigens were originally detected on eyrthocytes and identified as asialo- and asialo-agalacto-glycophorin, respectively, and later were found to be tumor-related markers (Springer, 1984;

Lisowska, 1995). The Tn (GalNAc-Ser/Thr) and sialyl-Tn (NeuAcα2-6GalNAc-Ser/Thr) determinants are present on some animal mucins (refs. in Duk *et al.*, 1994a). Persons with Tn syndrome are extremely rare. Preparation of the Tn glycophorin by enzymatic (sialidase and β-galactosidase) treatment of normal glycophorin encounters diffculties, because the galactose residues of glycophorin O-glycans are resistant to many β-galactosidases. Therefore, we obtained an 'artificial' Tn glycophorin by less expensive and more effective chemical degradations of O-glycans, i.e. desialylation by mild acid hydrolysis, and degalactosylation of asialoglycophorin by Smith degradation, i.e. periodate oxidation, $NaBH_4$ reduction and release of 'degraded' galactose residues under mild acidic conditions.

$$\text{NeuAc}\alpha\text{2-3Gal}\beta\text{1-3(NeuAc}\alpha\text{2-6)GalNAc-} \xrightarrow[\textit{Desialylation}]{} \text{Gal}\beta\text{1-3GalNAc-} \xrightarrow[\textit{Smith degradation}]{} \text{GalNAc-}$$

Both procedures can be performed either in solution, or directly on ELISA plates coated with native glycophorin, or on blots to which glycophorins or erythrocyte membrane proteins seperated by SDS-PAGE are transferred (Duk *et al.*, 1994a; Wu *et al.*, 1995) Performing the degradation *in situ* (plates, blots) has several advantages: (i) it allows to save material, by modification of exactly that amount of glycoconjugate which is used for an assay; (ii) the problems with decreased solubilty or increased aggregation of modified proteins are avoided; (iii) degradation on the blots leaves the modified glycoproteins in the same positions as native counterparts that facilitates their identification; (iv) the procedure of modification is simpler and faster, since dialyses or chromatographic separations of products are not necessary.

The 'artificial' Tn glycophorin found several applications. It was used as immunogen for obtaining murine anti-Tn monoclonal antibodies and as a reagent for identification and characterization of these antibodies (Duk *et al.*, 1992b). The Tn and TF glycophorins were applied for affinity purification and characterization of human anti-Tn and anti-TF antibodies, respectively (Duk and Lisowska, 1998), and were used for characterization of lectin specificities. A spectacular example is application of the 'artificial' Tn glycophorin (without using rare Tn erythrocytes) for identification of anti-Tn specificity of the *Moluccella laevis* lectin (MLL) (Duk *et al.*, 1992a). The MLL agglutinates erythrocytes of blood group A or N. This specificity was a puzzle, because there is no structural relation between blood group A (GalNAcα1-3[Fucα1-2]Galβ-) and N antigens (Leu_1 and Glu_5 in GPA^N, *vs* Ser_1 and Gly_5 in GPA^M). Finding the anti-Tn specificity of MLL explained this surprising serological specificity. MLL, like many other anti-Tn reagents, crossreacts weakly with blood group A antigen, and reacts with unsubstituted GalNAc residues, present in 'trace' amount in glycophorins, due to microheterogeneity of glycosylation. The anti-N specificity of MLL results from its preferential reaction with GPA^N which has a greater number of nongalactosylated GalNAc residues in the N-terminal portion than GPA^M (Duk *et al.*, 1994b; Krotkiewski *et al.*, 1997). This finding demonstrates an example of modulation of GalNAc-Ser/Thr galactosylation by adjacent amino acid residues. The desialylation and Smith degradation on blots have also been applied for characterization of unknown O-glycans of glycophorin from chicken erythrocytes (Duk *et al.*, 2000). The desialylated chicken glycophorin bands were stained by anti-TF (peanut) lectin, and when desialylation was followed by Smith degradation, they were stained with anti-Tn lectins (MLL and *Vicia villosa*) that indicated the presence of sialylated Gal-GalNAc- chains. Generally, a detection of on-blot modified glycophorins with anti-TF or anti-Tn lectins may be a method replacing a less sensitive Schiff-staining of electrophoretic gels. This method is also useful for screening blood samples for the presence of variant glycophorins which form unusual electrophoretic bands (Wu *et al.*, 1995).

III. β-ELIMINATION OF O-GLYCANS FROM GLYCOPROTEINS BLOTTED TO THE MEMBRANES

The β-elimination procedure (alkaline tratment in the presence of an excess $NaBH_4$) in solution has been frequently applied for obtaining O-linked oligosaccharide alditols for structural analysis. This method is less useful for studying de-O-glycosylated proteins, since they are usually partly degraded and/or aggregated. In order to study the role of protein O-glycosylation in binding lectins or antibodies, we developped a method of specific release of O-glycans from glycoproteins blotted to the membranes (Duk *et al.*, 1997).

The following conditions of the procedure are important. Immobilon P, but not nitrocellulose, must be used for blotting, since nitrocellulose sheets are destructed under alkaline conditions, and Immobilon P membranes are stable. Secondly, the $NaBH_4$, used to reduce and stabilize the structure of released O-glycans, is not only not necessary, but for unknown reasons it partly destroys or removes de-O-glycosylated polypeptide chains. Using untreated or desialylated erythrocyte membrane proteins as a model, the optimal results are obtained by treating the membranes with 0,055 M NaOH for 16 h at 40°C. Exhaustive de-O-glycosylation of glycophorins occurring under these conditions is demonstrated by a loss of their reactivity with lectins or monoclonal antibodies recognizing O-glycans. On the other hand, glycoproteins on NaOH-treated blots show an unchanged reactivity with N-glycan specific lectin (*Phaseolus vulgaris* erythroagglutinin) and with monoclonal antibodies specific for peptidic epitopes of glycophorins or band 3 protein.

The described procedure can be used for several purposes. It allows to examine whether antibodies have anti-peptidic specificity, or recognize epitopes dependent on O-glycosylation. This application pertains to various mucins and anti-mucin antibodies which are important in cancer research. Moreover, certain carbohydrate structures recognized by lectins and anti-cabohydrate antibodies (e.g. blood group ABH determinants or selectin ligands) may be present at the nonreducing end of O-glycans and N-glycans as well, and β-elimination on the blots enables to define a type of glycans carrying a given structure. This procedure does not require purification of studied glycoproteins and can be applied to complex mixtures, e.g. cell or tissue extracts.

SUMMARY

Microtiter plate assays and 'lectinoblotting' with the use of biotinylated lectins are sensitive and easy to perform methods that can be combined with simple procedures of chemical modifications of glycoproteins immobilized on ELISA plates or blots (desialylation by mild acid hydrolysis, Smith degradation, β-elimination). These modifications are helpful in the determination of lectin and anti-carbohydrate antibody specificities, or in the characterization of glycoconjugates by means of lectins and antibodies.

ACKNOWLEDGMENT

Part of this work was supported by Grants NO. 676 and NO.1028 from the Chang-Gung Medical Research Project, Kwei-san, Taiwan, and Grant NSC 89-2320-B-182-003 from the National Science Council, Taipei, Taiwan.

REFERENCES

Duk, M. and Lisowska, E., 1998, Purification of human anti-TF (Thomsen-Friedenreich) and anti-Tn antibodies by affinity chromatography on glycophorin A derivatives and characterization of the antibodies by microtiter plate ELISA, *Arch. Immunol. Ther. Exp.* 46: 69.

Duk, M., Mitra, D., Lisowska, E., Kabat, E.A., Sharon, N. and Lis H., 1992a, Immunochemical studies on the combining site of the A+N blood type specific *Moluccella laevis* lectin. *Carbohyd. Res.* 236: 245.

Duk, M., Steuden, I., Du□, D., Radzikowski, C. and Lisowska, E., 1992b, Application of chemically desialylated and degalactosylated human glycophorin for induction and characterization of anti-Tn monoclonal antibodies. *Glycoconjugate J.* 9: 148.

Duk, M., Lisowska, E., Wu, J.H. and Wu, A.M., 1994a, The biotin/avidin-mediated microtiter plate lectin assay with the use of chemically modified glycoprotein ligand, *Anal. Biochem.* 221: 266.

Duk, M., Wu, A.M. and Lisowska, E., 1994b, *Vicia villosa* B4 lectin is the second anti-Tn lectin shown to react better with blood group N than M. antigen, *Glycoconjugate J.* 11: 371.

Duk, M., Ugorski, M. and Lisowska, E.,1997, β-elimination of O-glycans from glycoproteins transferred to Immobilon P membranes: method and some applications, *Anal. Biochem.* 253: 98.

Duk, M., Krotkiewski, H., Stasyk, T.V., Lutsik-Kordowski, M., Syper, D. and Lisowska, E., 2000, Isolation and characterization of glycophorin from nucleated (chicken) erythrocytes, *Arch. Biochem. Biophys.* 375: 111.

Haselbeck, A., Schickaneder, E., von der Eltz, H. and Hösel, W., 1990, Structural caharacterization of glycoprotein carbohydrate chains by using digoxigenin-labeled lectins on the blots, *Anal. Biochem.* 191: 25.

Haseley, S.R., Talaga, P., Kamerling, J.P. anf Vliegenthart, J.F.G., 1999, Characterization of the carbohydrate binding specificity and kinetic parameters of lectins by using surface plasmon resonance, *Anal. Biochem.* 274: 203.

Hoyer-Hansen, G., Hamers, M.J.A.G., Pedersen, A.N., Nielsen, H.J., Brunner, N., Dano, K. and Stephens, R.W., 2000, Loss of ELISA specificity due to biotinylation of monoclonal antibodies, *J. Immunol. Meth.* 235: 91.

Kabat, E.A., 1961, Kabat and Mayer's Experimental Immunochemistry, 2nd edn., Thomas, C.C., Springfield, IL.

Krotkiewski, H., Duk, M., Syper, D., Lis, H., Sharon, N. and Lisowska, E., 1997, Blood group MN-dependent difference in degree of galactosylation od O-glycans of glycophorin A is restricted to the GalNAc residues located on amino acid residues 2-4 of the glycophorin polypeptide chain, *FEBS Lett.* 406: 296.

Lisowska, E., 1995, Tn antigens and their significance in oncology, *Acta Biochim. Polon.* 42: 11.

Lisowska, E., Duk, M. and Wu, A.M., 1996, Preparation of biotinylated lectins and application in microtiter plate assays and Western blotting, in A Laboratory Guide to Biotin Labeling in Biomolecule Analysis, Meier, T. and Fahrenholz, F., eds., Birkhäuser Verlag, Basel, p. 115.

Myszka, D.G., 1999, Survey of the 1998 optical biosensor literature, *J. Mol. Recognit.* 12: 390.
Peumans, W.J. and Van Damme, E.J.M., 1998, Plant lectins: specific tools for the identification, isolation, and characterization of O-linked glycans, *Crit. Rev. Biochem. Mol. Biol.* 33: 209.
Springer, G.F., 1984, T and Tn, general carcinoma markers, *Science* 224; 1198.
Wu, A.M., Duk, M., Lin, M., Broadberry, R.E. and Lisowska E., 1995, Identification of variant glycophorins of human red cells by lectinoblotting: application to the Mi.III variant that is relatively frequent in the Taiwanese population, *Transfusion* 35: 571.
Wu, A.M., Wu, J.H., Song, S-C. and Kabat, E.A., 1996, *Bandeiraea (Griffonia) simplicifolia* lectin-I, isolectin A4, reacting with Tn (GalNAcα1→Ser/Thr) or galabiose (Galα1→ 4Gal) containing ligands, *FEBS Lett.* 398: 183.
Wu, A.M., Wu, J.H., Watkins, W.M., Chen, C-P., Song, S-C. and Chen Y-Y., 1998, Differential binding of human blood group Sd(a+) and Sd(a-) Tamm Horsfall glycoproteins with *Dolichos biflorus* and *Vicia villosa*-B4 agglutinins, *FEBS Lett.* 429: 323.
Wu, A.M., Wu, J.H., Chen, Y-Y., Tsai, M.-S. and Herp, A., 1999a, Forsmann pentasaccharide and polyvalent Galβ1-4GlcNAc as major ligands with affinity for *Caragana aborescens* agglutinins, *FEBS Lett.* 463: 225.
Wu, A.M., Wu, J.H., Chen Y-Y., Song, S-C. and Kabat, E.A., 1999b, Further characterization of the combining sites of *Bandeiraea* (*Griffonia*) *simplicifolia* lectin-I, isolectin A4, *Glycobiology* 9: 1161.
Wu, A.M., Song, S-C., Chen, Y-Y.and Gilboa-Garber, N., 2000, Defining the carbohydrate specificities of Aplysia gonad lectin exhibiting a peculiar D-galacturonic acid affinity, *J. Biol. Chem.* 275: 14017.

INTRAMOLECULAR CARBOHYDRATE-PROTEIN INTERACTION[1]

Johannes F.G. Vliegenthart

Bijvoet Center, Department Bio-Organic Chemistry, Utrecht University, Utrecht, The Netherlands.

The widespread occurrence of protein glycosylation has stimulated the thorough investigation of the structure and function of this posttranslational protein modification [1,2]. Significant progress has been made with respect to the determination of the primary structures of glycans. Numerous structures were derived with regard to glycans as well as to glycosylation sites in glycoproteins. Although the N and O linked chains are still the most abundant types of protein decoration, a number of rare types have been discovered as well. Even today, it should be realised that the occurrence of multiple glycosylation sites in one protein, in conjunction with (micro)heterogeneity make it extremely difficult to define the complete structure of such a protein. In fact, most of the glycoproteins represent ensembles of glycoforms.

On the important question on the function of glycoprotein glycans, it is impossible to give an unambiguous answer. A large array of glycosylation effects have been described. However, it is not feasible to interpret these effects in terms of unique glycan structures being exclusively responsible for well-defined features.

In general, it is considered that glycosylation effects vary from physico-chemical to biological. Parts of the influence of glycans on the macromolecular behaviour are

[1]The author is much indebted to Prof. Dr. J.P. Kamerling, Drs. J.P.M. Lommerse, J.A. van Kuik, B.R. Leeflang, A.L.J. de Beer, C.W.E.M. Thijssen-van Zuylen, F. Casset, P.A.J. Erbel and Y. Karimi-Nejad, of the Bijvoet Center; and Drs J.Hofsteenge and J. Löffler of the Friedrich Miescher Institute, Basel, Switzerland for fruitful collaboration. The studies in the author's laboratory were supported by the Council for Chemical Sciences of the Netherlands Organisation for Scientific Research and by the European Commission.

The Molecular Immunology of Complex Carbohydrates–2.
Edited by Albert M. Wu, Kluwer Academic/Plenum Publishers, 2001.

obvious, such as molecular mass, solubility, charge, stability and resistance against proteolytic attack. However, some could be subtler and exhibit effects that represent phenomena occurring at the atomic level.

Huge is the diversity in claims for involvement of the glycans in biological processes, ranging from cell-cell, cell-pathogen, and cell-molecule interactions to signal transduction and protein folding. It is remarkable that sometimes in the same glycoprotein identical types of glycan structure are postulated to be responsible for different effects. However, sometimes the opposite is found. Different glycan structures are held responsible for similar kinds of effects. Apparently, the local molecular environment of glycans in a glycoprotein is a decisive factor in determining the functional behaviour. The presentation of glycans on the outside of glycoproteins and in membrane-bound glycoproteins on the membrane surface is crucial for recognition and interaction processes.

To gain insight into the molecular events that control the glycan presentation in space and time, it is necessary to investigate the intramolecular carbohydrate-protein interactions. Some pertinent information in this respect is available from X-ray studies on N-glycoproteins [3]. In a number of cases, details could be obtained for the structure of the core parts of the glycans. It was shown that the Asn-linked GlcNAc could occupy various positions. The α- or β-face of this residue may be oriented towards the protein or the O5 edge. Alternatively, the residue can be solvent exposed. Another variable is the local protein structure in which the Asn is located. So far turns, extended structures, β-strands, and ends of α-helices have been found as locations for N-glycosylation in the protein. These data show clearly that already at the level of the first GlcNAc the presentation of the glycan to the outside world can be completely different. It is reasonable to assume that upon extension of the carbohydrate chain beyond this first residue these differences in local structure will exert their effects until the outer parts.

It is most relevant to see how this works out in solution, because many interactions that take place under physiological conditions occur in an aqueous environment in a solution-like state. Advanced NMR techniques in conjunction with molecular dynamics calculations are the methods of choice to obtain such results. In this report, our studies on a few glycoproteins will be summarised.

I. Pineapple stem bromelain

Pineapple stem bromelain [E.C. 3.4.22.4] is a cysteine proteinase with a broad specificity for the hydrolysis of peptide bonds. It has a molecular mass of 33 kDa and contains one N-glycosylation site. Interestingly, the N-glycan exhibits no heterogeneity and consists of: Manα1-6[Xylβ1-2]Manβ1-4GlcNAcβ1-4[Fucα1-3] GlcNAcβ1-R {R represents peptide or protein}.

To investigate the influence of a folded polypeptide chain on the conformation of the glycan, studies were carried out on the carbohydrate chain in a glycotetrapeptide containing the intact glycan [4], and on the carbohydrate chain in the integral protein [5]. This should allow a comparison of the glycan conformations. Standard NMR measurements were performed providing chemical shift and NOESY data. Theoretical NOESY cross peak intensities were calculated on the basis of MD simulations, by using the CROSREL program [6]. A number of calculation methods were applied, each accounting differently for internal flexibility. The methods use generalised order parameters and/or individual rotational correlation times for every constituting monosaccharide.

For the glycopeptide, wherein R corresponds to Asn-Glu-Ser-Ser, we found that the Xylβ1-2Manβ1-4GlcNAcβ1-4GlcNAcβ part is rather rigid and has mainly one conformation. The Fucα1-3GlcNAc and Manα1-6Man linkages are relatively flexible. It is interesting to note that the characteristic times for conformational transitions of glycosidic linkages were calculated to be in the order of 1 –100 ns.

NMR measurements on the intact glycoprotein could easily be performed if, in view of the autodigestion of the protein, the enzyme was irreversibly inhibited. The NMR spectra contained two sets of signals. One corresponds to a denatured form of the protein, the other to the native form. The data of the protein in its denatured form fit the information obtained for the glycopeptide excellently. The narrow lines for the glycan in the denatured form indicate a considerable flexibility for this part of the molecule.

However, for the native intact form the situation is different from the glycopeptide with regard to the glycan conformation and flexibility. The Fucα1-3GlcNAc linkage has an altered distribution over the two minima ϕ,ψ = -90,-100 and ϕ,ψ = -140,-135. In comparison to the glycopeptide, the former conformation is more abundant in the glycoprotein. An obvious explanation for this observation is an interaction with the protein. In addition, the behaviour of the Manα1-6Man linkage is different. In the glycoprotein the Man *gt* ($\omega = 60$) rotamer is more abundant.

The mean rotation correlation time of the glycan in the intact glycoprotein is larger than that in the glycopeptide ($\tau_0 \pm 9$ ns versus $\tau_0 \pm 0.7$ns), reflecting the large difference in mass of the protein and the peptide moiety. In the glycoprotein the mobilities in the glycosidic linkages of the Xylβ1-2Manβ1-4GlcNAcβ1-4GlcNAcβ part are rather similar, but much less mobile are the linkages in the Fucα1-3GlcNAc and Manα1-6Man elements. The reduced mobilities have to be ascribed to interactions with the protein. It is noteworthy that the terminal Man and Fuc residues are located on the same side of the carbohydrate chain. The unaffected mobility of the Xyl residue indicates that it is exposed to the bulk solution.

This study has led to the important conclusion that conformational data on oligosaccharides or glycopeptides in solution cannot simply be transferred to glycans in glycoproteins.

II Ribonuclease-2

Non-secretory ribonuclease 2 (RNase2) is a heat and acid stable enzyme involved in the digestion of RNA. It is predominantly found in kidney, liver, lung, spleen and leukocytes. It has five N-glycosylation sites. The amino acid sequence, comprising 134 amino acids, has a high homology to other RNases. Surprisingly, Trp at position 7 bears an α-D-mannopyranosyl residue via a C-C bond [7,8]. This represents a novel type of glycosidic linkage in eukaryotic glycoproteins. Site directed mutagenesis has revealed that the sequence Trp-X-X-Trp is required to achieve C-mannosylation of the first Trp residue [9]. This sequence motif occurs in 336 mammalian proteins suggesting that this new form of glycosylation is not necessarily a rare feature. From a conformational point of view, the behaviour of the C-Man residue is interesting. For the glycopeptide Phe-Thr-[Manα]Trp-Ala-Gln-Trp, we concluded that, based on vicinal ^{1}H-^{1}H coupling constants and rotating frame nuclear Overhauser enhancements, the mannopyranosyl ring adopts an ensemble of conformations [8]. The usual $^4C_1$ conformation gives only a small, if any, contribution to the ensemble.

By means of NMR spectroscopy, we showed that the situation in the protein is even more complicated. In a denatured form, the Man residue behaves rather similar as in the peptide. However, in the native form the Man residue has a different orientation around the C-C linkage. Obviously, this must be the result of interaction with the protein. To gain further insight into the conformational behaviour of the Manα1-Trp linkage dynamic simulations were carried out in vacuum starting from the $^1C_4$, $^4C_1$, and $^1S_3$ conformations with various force fields. It appeared that the $^1C_4$ conformation is preferred. Simulations in water with the GROMOS programme starting with the $^1S_3$ conformation showed the transition to the $^1C_4$ conformation after about 40 ps. The NMR data exclude almost completely the $^4C_1$ conformation. The contribution of the $^1S_3$ conformation can only be small, because a NOE between H1 and H4 of Man, which is characteristic for this conformation, can hardly be detected.

For modelling experiments, the X-ray structural data of recombinant eosinophil-derived neurotoxin (EDN) were used. EDN has the same amino acid sequence as RNase 2 and is also C-mannosylated. In conjunction with NMR data of native RNase, the modelling results demonstrate that the mannose residue has interactions with loop residues 115-123, the end of β strand Met105-Arg114 and with the beginning of β strand Pro124-Ile134. By these interactions, Trp7 is kept in a specific conformation [10]. For RNase2 the carbohydrate-protein interaction is important to realise a well-defined protein conformation.

III. Human Chorionic Gonadotropin

Human chorionic gonadotropin (hCG) belongs to a family of glycoprotein hormones. It is crucial for the maintenance of the corps luteum in early pregnancy. HCG consists of a non-covalently associated α,β heterodimer. The X-ray structure of the chemically deglycosylated hCG has been determined [11,12]. The subunits can be dissociated and the bioactivity can completely be restored upon recombination. This important experiment leads to the conclusion that the dissociation is reversible. Apparently, the subunits are not irreversibly denatured.

The α-subunit is of special interest, because in addition to the hormonal action when associated with the β-subunit, it has its own bioactivity [13]. The α-subunit consists of 92 amino acids and has two N-glycosylation sites at Asn52 and Asn78, respectively. The α-subunit is a good starting compound for analysing the solution structure of the hormone. Its size and the occurrence of two N-glycosylation sites make it an attractive example of a glycoprotein and it is suited for the study of the mutual structural influences of intramolecular carbohydrate and protein chains.

Three probes are available for the determination of the solution structure of the α-subunit by NMR spectroscopy: intact αhCG, PNGase -F- deglycosylated native αhCG carrying only the glycan at Asn78 (pdαhCG), and endo-B-treated αhCG possessing only GlcNAc (fdαhCG) at Asn52 and Asn78. All NMR experiments were performed at natural abundance [14-18]. By application of various 2D NMR techniques, the ^1H signals of the amino acids and carbohydrates of the different NMR probes were assigned. Based on the NOE data, the protein structure of fdαhCG was modelled. Comparison with the crystallographic data for the heterodimer reveals some interesting differences in structure. Not surprisingly, there is a larger overall flexibility. This is in particular evident in the loop of the residues 33 – 57, which exhibit structural disorder, and also by conformational heterogeneity in the hairpin loop α70 – 74 and in the tight turn α20 – 23. The backbone conformation near Val76 and Glu77 is also altered. Further studies on pdαhCG and intact free αhCG showed that the local protein structure around Asn78 is influenced by glycosylation. This feature can be demonstrated by removal of the carbohydrate chain beyond the Asn linked GlcNAc, leading to a decreased structural order. The conformation of the glycan at Asn78 is in good accordance with the energy minima, present in rigid geometry maps of free model disaccharides. Furthermore, the protein sterically hinders the glycosidic linkages GlcNAcβ1-4GlcNAc and GlcNAcβ1-4Man. In addition to macroscopic effects like solubility and increased protease stability, the Asn78 linked GlcNAc functions as a stabiliser of the three-dimensional structure. This residue shields the protein surface from the environment through interactions with hydrophobic amino acid residues.

The glycan at Asn52 behaves quite differently and is far more flexible and solvent exposed. The first indication for this phenomenon stems from the accessibility of this glycosylation site in native αhCG for PNGase F. In sharp contrast with this observation is the resistance of this glycosylation site towards PNGase F in the not denatured heterodimer. Further indications are obtained from the NMR data. These first comprise the small number of NOE cross peaks between glycan and protein in comparison to the glycan at Asn78 and secondly the line widths.

Most of the differences in three-dimensional structure and flexibility between the free α-subunit and the α-subunit as part of the heterodimer have to be ascribed to interactions with the β-subunit. Currently, studies are carried out to define these interactions in more detail.

SUMMARY

In this report, an overview is presented of studies on the three-dimensional structure in solution of glycoproteins. The investigations were carried out by using high resolution NMR spectroscopy in conjunction with molecular dynamics calculations and molecular modelling. For pine apple stem bromelain, ribonuclease 2 and the free α-subunit of human chorionic gonadotropin it could be shown that intramolecular interactions between carbohydrate and protein exist. Glycosylation sites are rather unique structural entities in glycoproteins, because of the local spatial arrangements. This is reflected in different intramolecular interactions between carbohydrate and protein.

REFERENCES

1. Montreuil, J., Vliegenthart, J.F.G. and Schachter, H. (1995) Glycoproteins, New Comprehensive Biochem. Vol 29a, Elsevier, Amsterdam.
2. Montreuil, J., Vliegenthart, J.F.G. and Schachter, H. (1997) Glycoproteins, New Comprehensive Biochem. Vol 29b, Elsevier, Amsterdam
3. Protein Brookhaven Databank.
4. Lommerse, J.P.M., Kroon-Batenburg, L.M.J., Kroon, J., Kamerling, J.P. and Vliegenthart, J.F.G. (1995) Conformations and internal mobility of a glycopeptide derived from bromelain using molecular dynamics simulations and NOESY analysis. J. Biomol. NMR, 5:79-94.
5. Lommerse, J.P.M., Kroon-Batenburg, L.M.J., Kamerling, J.P. and Vliegenthart, J.F.G. (1995) Conformational analysis of the xylose-containing N-glycan of pineapple stem bromelain as part of the intact glycoprotein. Biochemistry, 34:8196-8206.
6. Kroon-Batenburg, L.M.J., Kroon, J., Leeflang, B.R. and Vliegenthart, J.F.G. (1993) Conformational analysis of methyl beta-cellobioside by ROESY NMR spectroscopy and MD simulations in combination with the CROSREL method. Carbohydr. Res., 245:21-42.
7. Hofsteenge, J., Müller, D.R., De Beer, T., Löffler, A., Richter W.J. and Vliegenthart, J.F.G. (1994) New type of linkage between a carbohydrate and a protein: C-glycosylation of a specific tryptophan residue in human RNase Us. Biochemistry, 33:13524-13530.
8. De Beer, T., Vliegenthart, J.F.G., Löffler, A. and Hofsteenge, J. (1995) The hexopyranosyl residue that is C-glycosidically linked to the side chain of tryptophan-

7 in human RNase Us is alpha-mannopyranose. Biochemistry, 34:11785-11789.
9. Krieg, J., Hartmann, S., Vicentini, A., Gläsner, W., Hess, D. and Hofsteenge, J. (1998) Recognition signal for C-mannosylation of Trp-7 in RNase 2 consists of sequence Trp-x-x-Trp. Mo.Biol.Cell, 9:301-309.
10. Vliegenthart, J.F.G. and Casset, F. (1998) Novel forms of protein glycosylation. Curr. Opin. Struct. Biol., 8:565-571.
11. Lapthorn, A.J., Harris, D.C., Littlejohn, Lustbader, J.W., Canfield, R.E., Machin, K.J., Morgan, F.J. and Isaacs, N.W. (1994) Crystal structure of human chorionic gonadotropin, Nature, 369:455-461.
12. Wu, H., Lustbader, J.W., Liu, Y., Canfield, R.E. and Hendrickson, W.A. (1994) Structure of human chorionic gonadotropin at 2.6 Å resolution from MAD analysis of the selenomethionyl protein, Structure, 2:545-558.
13. Blithe, D.L., Richards, R.G. and Skarulis, M.C. (1991) Free alpha human chorionic gonadotropin molecules from pregnancy stimulate secretion of prolactin from human decidual cells: a novel function or free alpha hcg in pregnancy, Endocrinology, 129:2257-2259.
14. Van Zuylen, C.W.E.M., De Beer, T., Rademakers, G.J., Haverkamp, J., Thomas-Oates, J.E., Hård, K., Kamerling, J.P. and Vliegenthart, J.F.G. (1995) Site-specific and complete enzymatic deglycosylation of the native human chorionic gonadotropin alpha-hcg subunit. Eur. J. Biochem., 231:754-760.
15. De Beer, T., van Zuylen, C.W.E.M., Leeflang, B.R., Hård, K., Boelens, R., Kaptein, R., Kamerling, J.P. and Vliegenthart, J.F.G. (1996) NMR studies of the free alpha-subunit of human chorionic gonadotropin; structural influences of N-glycosylation and the beta-subunit on the conformation of the alpha-subunit. Eur. J. Biochem., 241:229-242.
16. Van Zuylen, C.W.E.M., Kamerling, J.P. and Vliegenthart, J.F.G. (1997) Glycosylation beyond the Asn78-linked GlcNAc residue has a significant enhancing effect on the stability of the α subunit of human chorionic gonadotropin. Biochem. Biophys. Res. Commun., 232:117-120.
17. Thijssen-Van Zuylen, C.W.E.M., De Beer, T., Leeflang, B.R., Boelens, R., Kaptein, R., Kamerling, J.P. and Vliegenthart, J.F.G. (1998) Mobilities of the inner three core residues and the Man(alpha1-6) branch of the glycan at Asn78 of the alpha-subunit of human chorionic gonadotropin are restricted by the protein. Biochemistry, 37:1933-1940.
18. Erbel, P.J.A., Karimi-Nejad, Y., De Beer, T., Boelens, R., Kamerling, J.P. and Vliegenthart, J.F.G. (1999) Solution structure of the α-subunit of human chorionic gonadotropin. Eur. J. Biochem., 260:490-498.

RED BLOOD CELL ANTIGENS RESPONSIBLE FOR INHERITED TYPES OF POLYAGGLUTINATION

Elwira Lisowska and Maria Duk

Department of Immunochemistry, Ludwik Hirszfeld Institute of Immunology and Experimental Therapy, Polish Academy of Sciences, 53-114 Wroclaw, Poland

Human sera contain several 'natural' anticarbohydrate antibodies. An example are alloantibodies which recognize carbohydrate antigens of ABO(H) and P1/P2 blood group systems, and are present in circulation of only those individuals who do not express a target blood group antigen. Other 'natural' antibodies, which are present in all or most human sera, are directed against antigens not exposed on normal human tissues. Polyagglutination occurs when an unusual antigen, recognized by these antibodies, is expressed on erythrocyte surface. Some polyagglutination-related antigens are present in common oligosaccharides as 'cryptic' structures, and are transiently exposed in polyagglutinable erythrocytes. For example, the anti-Thomsen-Friedenreich (TF) antibodies recognize glycoproteins carrying the Galβ1-3GalNAc-Ser/Thr units which in normal tissues are core structures of sialylated or otherwise elongated or branched O-glycans (rev. by Hanisch and Baldus, 1997). The TF antigens of human erythrocytes are desialylated glycophorins which are easily obtained *in vitro* and may be transiently formed *in vivo* by the action of bacterial or viral sialidases in the course of infections. Another example is 'permanent mixed-field polyagglutinability' related to the acquired Tn syndrome, recently reviewed by Berger (1999). The Tn antigenic epitope is an unsubstituted α-N-acetylgalactosamine (or cluster of such residues), O-linked to Ser/Thr residues of protein. Recent data indicate that expression of the Tn-antigen is the result of persistent but reversible repression of the 3β-galactosyltransferase gene in some pluripotent stem cells, which give rise to various Tn-positive hematopoietic cell lineages, co-existing with normal cells. Interestingly, TF and Tn antigens, which are mostly 'cryptic' in normal tissues, are expressed on the surface of various cancer cells and are considered to be carcinoma-related markers.

Besides the acquired and transient types of polyagglutination, there are also known the inherited types. The inherited polyagglutinations are related to 'stable' and inherited genetic alterations, leading either to impaired biosynthesis of normal glycans (HEMPAS), or to the appearance of new structures on the erythrocyte surface (Cad, NOR).

The Molecular Immunology of Complex Carbohydrates–2.
Edited by Albert M. Wu, Kluwer Academic/Plenum Publishers, 2001.

I. HEMPAS (CDA II)

HEMPAS (hereditary erythroblastic multinuclearity with positive acid serum test), known also as CDA II (congenital dyserythropoietic anemia type II) is the most frequent disorder of the three major types of rare congenital dyserythropoietic anemia (CDAs I-III) (Marks and Mitus, 1996). HEMPAS, a condition inherited in autosomally recessive mode, is manifested by mild to moderate anemia, an inefficient erythropoiesis and multinuclearity of erythroblasts in the bone marrow, and these features are frequently accompanied by other symptoms (for revs. see Fukuda, 1990 and 1999). In this review, the attention is focused on alterations in HEMPAS erythrocytes which give rise to HEMPAS antigens recognized by human antibodies.

HEMPAS erythrocytes released to peripheral blood are short-lived, show some morphological abnormalities (e.g. so-called double membrane) and are agglutinated and lysed by complement-fixing IgM antibodies present in about 30% of normal human sera.

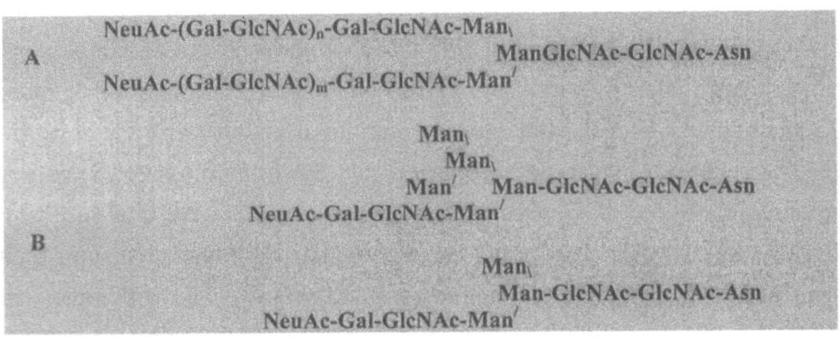

Fig. 1. Polylactosamine type oligosaccharide chains of normal erythrocytes (A) and hybrid chains accumulated in HEMPAS erythrocytes (B).

The examination of surface antigens of HEMPAS erythrocytes showed abnormalities in glycosylation of both, proteins and lipids. Glycosylation defect concerns mainly band 3 (anion transporter) and band 4.5 (glucose transporter) proteins, which in normal erythrocytes carry complex N-glycans of polylactosamine type. In HEMPAS erythrocytes the amount of polylactosamine type chains is strongly decreased and oligomannosyl and hybrid type N-glycans are accumulated (Mawby *et al.*, 1983; Fukuda *et al.*, 1990 and 1999) (Fig. 1). HEMPAS erythrocytes also show an abnormal glycolipid profile (Bohours *et al.*, 1985). The polylactosamine chains, present in low amount in glycoproteins of HEMPAS erythrocytes, accumulate as polyglycosylceramides (Fukuda *et al.*, 1986; Zdebska *et al.*, 1987). This may explain why HEMPAS erythrocytes are agglutinated by anti-i antibodies (refs. in Fukuda, 1990).

The question arose whether the observed morphological abnormalities of HEMPAS erythrocytes are related to defective glycosylation of erythrocyte proteins. Kameh *et al.*

(1998) reported that HEMPAS glycosylation defect of band 3 and band 4.5 erythrocyte proteins does not impair their transport functions, but produces subtle structural changes in the band 3 protein that result in its increased cytoskeletal interaction and self-association. On the other hand, De Franceschi *et al.* (1998) found a decreased anion exchange activity and clusterization of band 3 in HEMPAS erythrocytes. These authors suggest that the mild hemolysis and premature demise of erythrocytes observed in patients with CDA II may be ascribed to clusterization of band 3, followed by increased IgG binding and phagocytosis, and not to suggested modification of the cytoskeletal structure of the cells.

The defective glycosylation was not found in lymphoid cells of patients with HEMPAS, but accumulation of hybrid and high mannose type oligosaccharide chains was observed in some secreted glycoproteins, such as transferrin and α_1-acid glycoprotein (Fukuda *et al.*, 1992). These serum glycoproteins are synthesized in hepatocytes which indicates that glycosylation defects in HEMPAS are not limited to the erythroid lineage.

The genetic background underlying an altered glycosylation in HEMPAS is not fully understood. The structure of accumulated incomplete N-glycans suggested that disruption of their biosynthesis occurs at the level of α-mannosidase II (MII), or N-acetyl-glucosaminyltransferase II (GnT-II). The decreased expression of GnT-II, or defects in the gene encoding MII were detected in some HEMPAS cases (Fukuda *et al.*, 1987 and 1990). The incompletness of glycosylation defect in HEMPAS, despite the lack of MII, suggested an alternative pathway of N-glycan processing and led to identification of an isoenzyme of MII, α-mannosidase IIx (MIIx) (Misago *et al.*, 1995). However, later genetic studies showed that a defect in the GnT-II gene occurs in carbohydrate-deficient glycoprotein syndrome type II (CDGS-II), but not in HEMPAS (Charuk *et al.* 1995). Moreover, there are several HEMPAS cases in which the mutation of neither GnT-II gene, nor MII genes was detected. Iolascon *et al.* (1997) looked for the causative CDA II gene by linkage analysis, using microsatelite markers flanking MII, MIIx and GnT-II genes, and excluded the linkage to these three candidate genes in HEMPAS. The same group obtained the conclusive evidence for linkage of HEMPAS to microsatelite markers on the long arm of chromosome 20 (20q11.2, locus CDAN2) (Gasparini *et al.*, 1997). Recently, Iolascon *et al.* (1998) reported two families in which CDA II is not linked to the CDAN2 locus. These results demonstrate that CDA II, which is homogeneous regarding biochemical features, is genetically a heterogeneous disease.

The data on HEMPAS leave many unsolved problems. The questions remain how different genetic defects result in similar glycosylation impairment, and how a defect in gene encoding protein which is not a glycosyltransferase results in disruption of N-glycan biosynthesis. Moreover, the specificity of common human antibodies reacting with HEMPAS erythrocytes has not been yet elucidated. It is not known, whether these antibodies have an unique specificity, or are a group of antibodies with different subspecificities, and it has not been directly shown that they recognize defective N-glycans in HEMPAS erythrocytes.

II. Cad AND RELATED Sd^{a} AND CT ANTIGENS

Cad is a rare human red blood cell antigen inherited as an autosomal dominant character. The Cad erythrocytes react with antibodies present in all human sera and with several GalNAc-specific lectins (e.g. *Dolichos biflorus*, *Helix pomatia, H. aspersa, Salvia horminum* and *Leonorus cardiaca*) (refs in Watkins, 1995). Reactions with lectins suggested that N-acetylgalactosamine residue is an important component of the Cad determinant. This was confirmed by finding that O-glycans carried by glycophorins in the Cad erythrocytes have an additional GalNAc residue which forms a branched structure GalNAcβ1-4(NeuAcα2-3)Gal- (Blanchard *et al.*, 1983; Herkt *et al.*, 1985). Moreover, Cad erythrocytes accumulate an abnormal ganglioside with an oligosaccharide chain terminated with the same branched trisaccharide (Blanchard *et al.*, 1985a; Gillard *et al.*, 1988) (Fig. 2). Examination of glycophorin A from three Cad-positive donors showed that they contained different numbers of the Cad-specific pentasaccharide chain, from about 12 chains per molecule in one donor, to only 2-3 chains per molecule in the remaining two donors (Blanchard *et al.*, 1985b). This finding is in agreement with the known quantitative differences in the proportion of Cad erythrocytes in different donors.

A NeuAcα2-3Galβ1-3GalNAcα-O-Ser/Thr
NeuAcα2/^{6}

B GalNAcβ1\$_{4}$
NeuAcα2-3Galβ1-3GalNAcα-O-Ser/Thr
NeuAcα2/^{6}

C GalNAcβ1\$_{4}$
NeuAcα2-3Galβ1-4GlcNAcβ1-3Galβ1-4Glc-Cer

D GalNAcβ1\$_{4}$
NeuAcα2-3Galβ1-4GlcNAcβ1-3-R

Fig. 2. The structures of the major O-glycans of glycophorins in normal erythrocytes (A) and in Cad erythrocytes (B), ganglioside of Cad erythrocytes (C), and the structure of Sd(a+) N-glycan fragment of the Tamm-Horsfall glycoprotein (D).

The Cad antigen is structurally related to the Sd^{a} antigen of the Sid blood group system (rev. by Watkins, 1995). The Sd^{a} antigen is serologically detected in over 90% of human erythrocytes. As in the case of Cad, the expression of Sd^{a} in erythrocytes of various Sd(a+) persons shows quantitative differences and it may be sometimes difficult to define the phenotype by means of hemagglutination. The anti-Sd^{a} alloantibodies are rather weak and are detectable only in about a half or less of Sd(a-) individuals. The Sd^{a} antigen is also expressed in other human tissues and secretions, with the highest concentration occurring in urine, where the Sd^{a} determinant is carried mainly by the Tamm-Horsfall (TH) N-glycosylated protein (van Rooijen *et al.*, 1998 and 1999). The TH glycoprotein of Sd (a+)

persons contains 1-2% N-acetylgalactosamine which is located mostly in N-glycans terminated with the structure shown in Fig. 2D, closely resembling the Cad determinant (Donald *et al.*, 1983 and 1987; Donald and Feeney, 1986; van Rooijen *et al.*, 1998). The Sd^a determinant shows a higher expression on antennae which display N-acetyllactosamine elongation, suggesting a better accessibility of these sites for β-N-acetylgalactosaminyl-transferase (van Rooijen *et al.*, 1998). The TH glycoproteins from Sd(a+) and Sd(a-) donors show quantitative differences in reactivity with several lectins (Wu *et al.*, 1995b and 1996). However, the most distinct difference is that the Sd(a+) TH glycoprotein, in contrast to that from Sd(a-) donors, strongly inhibits the Sd^a-mediated hemagglutination and precipitates with *D. biflorus* lectin. These assays, or determination of N-acetylgalactosamine in the TH glycoprotein, allow to determine the Sd(a+) or Sd(a-) phenotype more precisely than hemagglutination (Watkins, 1995).

The similarity of the Cad and Sd^a structures initially suggested that Cad is a superstrong form of Sd^a. However, there are several lines of evidence that they are not identical antigens. Both, anti-Cad and anti-Sd^a antibodies require the branched trisaccharide with terminal N-acetylgalactosamine and sialic acid residues. However, hemaglutination-inhibition assays with model oligosaccharides showed that anti-Sd^a reacts strongly only with oligosaccharides in which the trisaccharide is linked to N-acetylglucosamine residue (structures C and D in Fig. 2), and not with those in which the trisaccharide is linked to a sialic acid-substituted N-acetylgalactosamine residue (structure B in Fig. 2) (Watkins, 1995). Moreover, no Cad pentasaccharide was identified in glycophorins of Sd(a+) donors (Blanchard *et al.*, 1985b). The Sd^a-active oligosaccharides of Sd(a+) erythrocytes have not been unequivocally identified, but the candidates are the ganglioside (Fig. 2C), accumulated in Cad erythrocytes but detected also in Sd(a+) red cells (Blanchard *et al.*, 1985a), and protein-linked glycans containing the structure found in the Sd(a+) TH glycoprotein (Fig. 2D) (Watkins, 1995).

The genetic background of the Cad and Sd(a+) condition and genetic relationship between both of them is not known, because the human N-acetylgalactosaminyltransferase (GalNAcT) involved in the biosynthesis of Sd^a determinants has not been cloned yet. The initial studies based on determination of activity and substrate specificity showed that a characteristic feature of this enzyme is transferring the GalNAc residue to the subterminal galactose residue substituted at C3 with sialic acid, and not to the terminal nonsubstituted galactose residue. The activity of β1,4GalNAcT, which preferably transfers GalNAc residue to sialylated N-glycans in TH (from Sd(a-) donor) and other glycoproteins has been identified in human kidney (Piller *et al.*, 1986), plasma (Takeya *et al.*, 1987), urine (Serafini-Cessi *et al.*, 1988) and large intestine (Malagolini *et al.*, 1989). The enzyme from kidney microsomes transfers N-acetylgalactosamine to both N- and O-linked oligosccharide chains of fetuin and to tryptic glycopeptides of glycophorin A, but not to native glycophorin A, while the enzymes from other sources show a low capacity to transfer N-acetylgalactosamine to glycophorin A. The use of gangliosides as substrates indicated that these enzymes are different from β-1,4-N-acetygalactosaminyltransferases involved in the biosynthesis of GalNAcβ1-4(NeuAcα1-3)Gal-terminated carbohydrate chains of GM2 or GD2 gangliosides. After a murine counterpart of Sd^a-GalNAcT had been cloned (Smith and Lowe, 1994), a cDNA segment was amplified from the total mRNA of human gastric mucosa by reverse-transcription-polymerase chain reaction (RT-PCR), using primers designed according to the sequence of a murine Sd^a-GalNAcT. Specific primers to the human RT-PCR product were found to be useful for the estimation of the Sd^a-GalNAcT mRNA expression in human tissues (Dohi *et al.*, 1996).

Characteristic changes of Sd^a/Cad-related antigen expression are observed in cancer cells. Hiraiwa *et al.* (1990) obtained two monoclonal antibodies, 2A3D2 and 2D11E2, which react with ganglioside oligosaccharide chains and protein-linked glycans terminated with GalNAcβ1-4(NeuAcα2-3)Galβ- structure, including the GM2 ganglioside and glycophorin A from Cad erythrocytes. By using these antibodies, it was shown that cultured human hepatocarcinoma cell lines and cancer tissues from patients with hepatocellular carcinoma contain Sd^a/Cad-related glycoprotein antigens and gangliosides, which are not detectable in appreciable amounts in normal liver tissues. On the other hand, the Sd^a structure is abundantly expressed on glycolipids and glycoproteins in normal human gastrointestinal mucosa, but its expression in gastrointestinal cancer tissues is drastically reduced or abolished, and may be increased when cancer cells are induced to differentiate. These conclusions are mostly based on determination of the Sd^a-GalNAcT activity or mRNA expression in normal and cancer tissues (Malagolini *et al.*, 1989; Serafini-Cessi *et al.*, 1995; Dohi *et al.*, 1996).

The Sd^a determinant is widely distributed in animal (mammalian) tissues and activity of a β-1,4GalNAcT able to synthesize the Cad and Sd^a type oligosaccharides was characterized in guinea pig kidney (Dall'Olio *et al.*, 1987) and mouse T lymphocytes (Conzelman and Kornfeld, 1984b). Interestingly, a differentiation marker of murine cytotoxic lymphocytes (CTL), called CT antigen, is structurally related to Sd^a and Cad. Conzelman and Kornfeld (1984a and 1984b) studied the cloned CTL lines, susceptible or resistant to the cytotoxic effect of *Vicia villosa* lectin. They found that the resistant CTL line, in contrast to the susceptible one, did not contain β-linked GalNAc residues at the nonreducing termini of O-linked oligosaccharide chains and was deficient in β-1,4GalNAcT involved in formation of the GalNAcβ1-4(NeuAcα2-3)Galβ- structure. Conzelman and Lefrancois (1988) presented evidence that two earlier obtained monoclonal antibodies (MAbs), CT1 and CT2, specific for CTL clones and blocking specific target cell lysis by CTLs, recognize the terminal branched trisaccharide structure, present in O-glycans of the CD45 glycoprotein of CTLs and in human Cad and Sd^a determinants. The cDNA encoding an N-acetylgalacosaminyltransferase synthesizing a CT epitope (CT-GalNAcT) was cloned from a CTL cDNA library and transiently expressed on Chinese hamster ovary (CHO) cells (Smith and Lowe, 1994). The nucleotide sequence predicts a 510-amino acid type II transmembrane protein characteristic of other mammalian glycosyltransferases. The cDNA shares 51% nucleotide sequence homology with a cDNA encoding the human GM2/GD2 synthase. As shown by Southern blot analysis, the cDNAs encoding CT-GalNAcT and GM2/GD2 synthase represent distinct loci in the murine genome. The originally CT-negative CHO cells transfected with the CT-GalNAcT cDNA express structures recognized by CT1 and CT2 MAbs and by anti-Sd^a human antibodies. Moreover, the enzyme present in the microsomal fraction of transfected CHO cells transfers the GalNAc residue to fetuin and 3'-sialyl-N-acetyllactosamine, but not to their asialo counterparts. All these data demonstrate that murine CT-GalNAcT shows properties of an Sd^a-synthase.

Examination of the CT1 epitope expression on murine thymocyte subsets shows that CT1 is a temporally regulated antigen during early thymocyte development (Reed *et al.*, 1998). CT1 is present primarily on pro-T cells ($CD44^+/CD25^+$) and early pre-T cells ($CD44^-/CD25^+$), and is missing in non-T-lineage lymphoid progenitors ($CD44^+/CD25^-$). Thymocytes in mice deficient in recombination activating gene-2 (Rag-2), which are blocked at the early pre-T stage, express a CT1 epitope on most cells. Expression of CT1 by thymocytes of $Rag\text{-}2^{-/-}$ mice is rapidly lost when mice are injected with the anti-CD3ε MAb, and the decline of CT1 induced by CD3 signaling is accompanied by a loss of mRNA for the CT-GalNAcT. The CT1 and CT2 MAbs also stain distinct synaptic sites on skeletal muscle fibers, and motoneuron-like and muscle cell lines (Martin *et al.*, 1999). The results show that pre- as well as postsynaptic membranes bear CT-related epitopes.

The data obtained so far suggest a possible role of CT determinants in cell differentiaion and perhaps in other biological processes. Monoclonal antibodies raised against human Sd^a or murine CT antigens crossreact with the Cad antigen that suggests their broader specificity, compared to human anti-Sd^a antibodies. A detailed specificity of common human anti-Cad agglutinins and their relation to human anti-Sd^a antibodies has not yet been elucidated.

III. NOR

The first case of NOR polyagglutination was reported by Harris *et al.* (1982), who found that NOR erythrocytes are agglutinated by most ABO-compatible human sera. The agglutination is enhanced after proteolytic treatment of erythrocytes and is inhibited by avian P1 substance. However, the character of an antigen responsible for the polyagglutination remained unknown. A second case of NOR has been recently identified in a two-generation Polish family (Kusnierz-Alejska *et al.*, 1999). Red blood cells of the proposita (TS), her two siblings and one of three children, showed all described features of NOR characteristics. Inhibition of the polyagglutination by P1 antigen, terminated with Galα1-4Galβ- sequence (Naiki *et al.*, 1975) suggested that the NOR antigen may be an oligosaccharide terminated with an α-galactose residue. Indeed, treatment of NOR erythrocytes with coffee bean α-galactosidase strongly decreased their agglutination by human sera. On the other hand, the agglutination was enhanced by treatment of the NOR red blood cells with sialidase.

Analysis of TS's erythrocyte membranes in SDS-PAGE revealed a normal protein pattern, but staining the gels with periodic acid-Schiff reagent and staining glycophorins on the blots by means of monoclonal antibodies and lectins (Wu *et al.*, 1995a) showed the presence of variant glycophorin bands. The variant was identified as St^a glycophorin, a known hybrid glycoprotein containing N-terminal fragment of glycophorin B and C-terminal portion of glycophorin A (Huang *et al.*, 1991), which has not been reported as a reason for polyagglutination. Examination of glycophorins in erythrocyte membranes of TS's family members showed that the St^a glycophorin and NOR antigen are inherited independently and the occurrence of two rare antigens in TS's family is coincidental.

Resistance of the NOR antigen to proteolysis and its better exposure after treatment of erythrocytes with proteases and sialidase suggested that an antigen may be a neutral glycosphingolipid. This suggestion is confirmed by examination of neutral glycolipids isolated from the TS's (phenotype: A2, ccDEE, MNss, P1) and control (blood group A1,P1, A2,P1 and B,P1) erythrocytes. The polyagglutination is strongly inhibited by neutral glycolipids from TS's red blood cells, but is not inhibited by neutral glycolipids from control erythrocytes. Although NOR polyagglutination is inhibited by P1, it does not represent overexpression of P1 glycolipid. First of all, anti-NOR agglutinins are present in sera of P1-positive individuals, which do not contain anti-P1 antibodies. Secondly, in contrast to anti-NOR, a monoclonal anti-P1 antibody was similarly (and rather weakly) inhibited by neutral glycosphingolipids from NOR and control erythrocytes. The thin-layer chromatography (TLC), followed by orcinol staining of the TLC plates, shows in glycolipids from the NOR erythrocytes the presence of an additional band, migrating in the region of pentaglycosylceramides, which is not seen in control preparations (Fig.3A). This band is more distinctly detected by overlaying the TLC plates with *Ricinus communis* agglutinin I (RCA-I), which recognizes terminal β- and α-galactose residues (Wu *et al.*, 1993). Besides staining a NOR-related glycolipid, RCA-I strongly reacts with paragloboside (*neo*lactotetraosylceramide, nLc4) in NOR and control samples as well (Fig. 3B). *Griffonia simplicifolia* (known also as *Bandeiraea simplicifolia*) lectin IB4 (GSL-IB4) does not react with any glycolipid derived from control blood group A red cells, but stains strongly two bands in glycolipids from NOR erythrocytes. One of these bands is that reacting with RCA-I (called here NOR1) and the second band migrates slower (NOR2) (Fig. 3C). GSL-IB4 is specific for Galα1-3Gal-terminated glycans (Wood *et al.*, 1979), and its strong reaction with NOR glycosphingolipids suggested that their oligosaccharide chains also are terminated with this sequence (Kusnierz-Alejska *et al.*, 1999). This possibility seemed to be more likely in view of the fact that human sera contain common 'natural' antibodies with specificity similar to that of GSL-IB4.

The anti-α-galactosyl antibodies (commonly called anti-Gal), recognize the Galα1-3Galβ1-4GlcNAc structure which is abundantly expressed in glycoproteins and glycolipids of nonprimate mammals and New World monkeys, and is not present in Old World monkeys, apes and humans (Galili, 1993). These antibodies are one of the major reasons of a hyperacute rejection of xenotransplants (Joziasse and Oriol, 1999). The galactosyltransferase involved in the synthesis of Galα1-3Gal (α1,3GalT) is different from human blood group B transferase, since the latter enzyme requires a fucosylated galactose residue as an acceptor. The murine, bovine, and porcine cDNAs encoding α1,3GalT have been cloned and sequenced (Larsen *et al.*, 1989; Joziasse *et al.*, 1989; Katayama *et al.*, 1998). The homologous gene is present in the human genome, but it is a nonfunctional pseudogene, due to multiple frameshift mutations and nonsense codons (Larsen *et al.*, 1990; Joziasse *et al.*, 1991). The suggestion that rare NOR-positive individuals are able to synthesize the Galα1-3Gal structure and that anti-NOR agglutinins are the common anti-αGal antibodies seemed attractive, but required direct evidence.

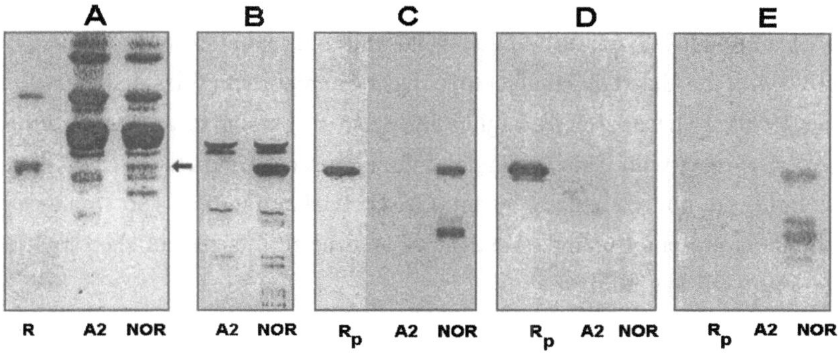

Fig. 3. Characterization of NOR and control glycosphingolipids on TLC plates by means of lectins and human antibodies. The plates were either stained with orcinol (**A**), or overlayed with RCA-I (**B**), GSL-IB4 (**C**), purified human anti-αGal antibodies (**D**), human anti-NOR antibodies isolated by adsorption/elution from glutaraldehyde-fixed NOR RBCs (**E**). The following glycolipids were analyzed: total neutral glycolipid fraction (**R**) and purified Gal-nLc4 ($\mathbf{R_p}$) from rabbit RBCs; neutral glycolipid fraction from control blood group A2 RBCs (**A2**) and from NOR RBCs (**NOR**). Arrow indicates NOR1 in the orcinol-stained plate. The procedures used were as described (Kusnierz-Alejska *et al.*, 1999)

Our further (unpublished) studies were undertaken to confirm this assumption. It was expected that NOR1 may be identical to one of the major glycosylceramides from rabbit erythrocytes, Galα1-3Galβ1-4GlcNAcβ1-3Galβ1-4Glcβ1-Cer (galactosyl-paragloboside, Gal-nLc4). Therefore, the latter glycolipid was purified, but our preliminary observations indicated that Gal-nLc4 and NOR1 are different. The NOR1 was detected on TLC plates by GSL-IB4 with higher sensitivity than Gal-nLc4, in relation to orcinol staining. Both glycolipids were digested with the coffee bean α-galactosidase, but NOR1 required a higher enzyme concentration and/or longer incubation time for completion of the reaction, defined by the disappearance of RCA-I- or GSL-IB4-stained substrate bands in TLC plates. Moreover, the digestion of Gal-nLc4 produced, as expected, a distinct band detectable with RCA-I, migrating as nLc4 (paragloboside), and no RCA-I-positive band was detectable as the α-galactosidase digestion product of NOR1. The most important finding was that NOR glycolipids and Gal-nLc4 react with different antibodies. The anti-Galα1-3Gal antibodies, purified from human serum by affinity chromatography on Sepharose-linked porcine glycophorin, and murine monoclonal antibody Gal-13, kindly offered by Dr. U. Galili (Galili *et al.*, 1987), reacted only with Gal-nLc4, and did not react with the NOR glycolipids (Fig. 3D). In turn, the serum from which anti-Galα1-3Gal antibodies were removed, or antibodies isolated by adsorption on and elution from glutaraldehyde-cross-linked NOR erythrocytes, reacted only with NOR glycolipids, and not with rabbit Gal-nLc4 (Fig. 3E). These results confirmed that human 'natural' anti-αGal antibodies are a heterogeneous population. Most of them are anti-Galα1-3Gal antibodies, which are not homogenous either, since they may crossreact to different extent with other Gal-Gal structures (Galili *et al.*, 1985; Parker *et al.*, 1999), or even with GalNAc/GlcNAcβ1-3Gal-terminated oligosaccharide chains (Teneberg *et al.*, 1996). Moreover, the existence in human sera of anti-Galα1-2Gal, anti-Galα1-4Gal and anti-Galα1-6Gal antibodies, which are not reactive with the Galα1-3Gal glycotope, has been reported (Wieslander *et al.*, 1990; Parker *et al.*, 1999). The anti-NOR agglutinins seem to belong to anti-αGal antibodies which do not cross-react with Galα1-3Gal, and the oligosaccharide chains of NOR-related glycosphingolipids are not recognized by anti-Galα1-3Gal antibodies. The structure of the NOR glycolipids is presently under investigation and the results should shed light on the specificity of anti-NOR agglutinins.

SUMMARY

The three described types on inheritable polyagglutination are related to altered carbohydrate structures in glycoproteins or/and glycolipds on the erythrocyte surface. HEMPAS, a condition causing anemia and other pathological symptoms, is characterized by impaired biosynthesis of N-glycans, mostly those carried by band 3 and band 4.5 erythrocyte membrane proteins. Cad erythrocytes have abnormal glycophorin O-glycans, structurally related to the more common human Sd^a and murine CT determinants, and accumulate an Sd^a-like ganglioside. NOR erythrocytes express recently detected abnormal α-galactose-terminated glycosphingolipids, which strongly react with *G. simplicifolia* IB4 isolectin, but do not react with human anti-Galα1-3Gal antibodies.

ACKNOWLEGEMENTS

This work was supported by Grant 4PO5A 124 18 of the Committee for Scientific Studies (KBN), Warsaw. Poland.

REFERENCES

Berger, E.G., 1999, Tn-syndrome, *Biochim. Biophys. Acta* 1455:255.

Blanchard, D., Cartron, J.-P., Fournet, B., Montreuil, J., van Halbeek, H. and Vliegenthart, J.F.G., 1983, Primary structure of the oligosaccharide determinant of blood group Cad specificity, *J. Biol. Chem.* 258: 7691.

Blanchard, D., Piller, F., Gillard, B., Marcus, D.M. and Cartron, J.-P., 1985a, Identification of a novel ganglioside on erythrocytes with blood group Cad specificity, *J. Biol. Chem.* 260: 7813.

Blanchard, D., Capon, C., Leroy, Y., Cartron, J.-P. and Fournet, B., 1985b, Comparative study of glycophorin A derived O-glycans from human Cad, Sd(a+) and Sd(a-) erythrocytes, *Biochem. J.* 232: 813.

Bouhours, J.-F., Bouhours, D.and Delaunay, J., 1985, Abnormal fatty acid composition of erythrocyte glycosphingolipids in congenital dyserythropoietic anemia type II, *J. Lipid Res.* 26: 435.

Charuk, J.H.M., Tan, J., Bernardini, M., Haddad, S., Reihtmeier, R.A.F., Jaeken, H. And Schachter, H., 1995, Carbohydrate-deficient glycoprotein sundrome type II. An autosomal recessive N-acetylglucosaminyltransferase II deficiency different from typical hereditary erythroblastic multinuclearity, with a positive acidified-serum lysis test (HEMPAS), *Eur. J. Biochem.* 230: 797.

Conzelman, A. and Lefrancois, L., 1988, Monoclonal antibodies specific for T cell-associated carbohydrate determinants react with human blood group antigens Cad and Sd^a, *J. Exp. Med.* 167: 119.

Conzelman, A. and Kornfeld, S., 1984a, β-linked N-acetylgalactosamine residues present at the nonreducing termini of O-linked oligosaccharides of a cloned murine cytotoxic T lymphocyte line are absent in a *Vicia villosa* lectin-resistant mutant cell line, *J.Biol. Chem.* 259: 12528.

Conzelman, A. and Kornfeld, S., 1984b, A murine cytotoxic T lymphocyte cell line resistant to *Vicia villosa* lectin is deficient in UDP-GalNAc:β-galactose β1,4-N-acetylgalactosaminyltransferase, *J.Biol. Chem.* 259: 12536.

Dall'Olio F., Malagolini, N., Joziasse D.H., van den Eijnden, D.R. and Serafini-Cessi, F., 1987, Purification of β-D-(1-4)-N-acetylgalactosaminyltransferase from guinea pig kidney. *Biochem Soc. Trans.* 15: 399.

De Franceschi L., Turrini, F., del Giudice, E.M., Perrotta, S., Olivieri, O., Corrocher, R., Manu, F. and Iolacson, A., 1998, Decreased band 3 anion transport activity and band 3 clusterization in congenital dyserythropoietic anemia type II, *Exp. Hematol.* 26: 869.

Dohi, T., Yuyama, Y., Natori, Y., Smith, P.L., Lowe, J.B. and Oshima, M., 1996, Detection of N-acetylgalactosaminyltransferase mRNA which determines expression of Sd^a blood group carbohydrate structure in human gastrointestinal mucosa and cancer, *Int. J. Cancer* 67: 626.

Donald, A.S.R. and Feeney, J., 1986, Oligosaccharides obtained from a blood group Sd(a+) Tamm-Horsfall glycoprotein, *Biochem. J.* 236: 821.

Donald, A.S.R., Yates, A,D., Soh, C.P.C., Morgan, W.T.J. and Watkins, W.M., 1983, A blood group Sd^a-active pentasaccharide isolated from Tamm-Horsfall urinary glycoprotein, *Biochem. Biophys. Res. Commun.* 115: 625.

Donald, A.S.R., Soh, C.P.C., Yates, A.D., Feeney, J., Morgan, W.T.J. and Watkins, W.M., 1987, Structure, biosynthesis and genetics of the Sd^a antigen, *Biochem. Soc. Trans.* 15; 606.

Fukuda, M.N., 1990, HEMPAS disease: genetic defect of glycosylation, *Glycobiology* 1: 9.

Fukuda, M.N., 1999, HEMPAS, *Biochim. Biophys. Acta* 1455: 231.

Fukuda, M.N., Bothner, B., Scartezzini, P. and Dell, A., 1986, Isolation and characterization of poly-N-acetyllactosaminyl ceramides accumulated in the erythrocytes of congenital dyserythropoietic anemia type II patients, *Chem. Phys. Lipids* 42:185.

Fukuda, M.N., Dell, A. and Scartezzini, P., 1987, Primary defect of congenital dyserythropoietic anemia type II: failure in glycosylation of erythrocyte glycosaminoglycan proteins caused by lowered N-acetylglucosaminyltransferase II, *J. Biol. Chem.* 262: 7195.

Fukuda, M.N., Masri, K.A., Dell, A., Luzzato, L. and Moremen, K.W., 1990, Incomplete biosynthesis of N-glycans in congenital dyserythropoietic anemia type II caused by a defect in the gene encoding α-mannosidase II, *Proc. Natl. Acad. Sci. USA* 87: 7443.

Fukuda, M.N., Gaetani, G.F., Izzo, P., Scartezzini, P. and Dell, A., 1992, Incompletely processed N-glycans of serum glycoproteins in congenital dyserythopoietic anemia type II (HEMPAS), *Br. J. Haematol.* 82: 745.

Galili, U., 1993, Evolution and and pathophysiology of the human natural anti-α-galactosyl IgG (anti-Gal) antibody.*Springer Semin. Immunopathol.* 15: 155.

Galili, U., Macher, B.A., Buehler, J. and Shohet, S.B., 1985, Human neutral anti-α-glactosyl IgG. II. The specific recognition of α(1-3) linked galactose residues, *J. Exp. Med.* 162: 573.

Galili, U., Basbaum, C.B., Shohet, S.B., Buehler, J. and Macher, B.A., 1987, Identification of erythrocyte Galα1-3Gal glycosphingolipids with a mouse monoclonal antibody, Gal-13, *J. Biol. Chem.* 262: 4683.

Gasparini, P., Miraglia del Giudice, E, Delaunay, J., Totaro, A., Granatiero, M., Merchionda, S., Zelante, L. and Iolascon, A., 1997, Localization of the congenital dyserythropoietic anemia II locus to chromosome 20q11.2 by genomewide search, *Am. J. Hum. Genet.* 61: 1112.

Gillard, B.K., Blanchard, D., Bouhours, J.-F., Cartron, J.-P., van Kuik, J.A., Kamerling, J.P., Vliegenthart, J.F.G. and Marcus, D.M., 1988, Structure of a ganglioside with Cad blood group antigen activity, *Biochemistry* 27:4601.

Hanisch, F.-G. and Baldus, S.E., 1997, The Thomsen-Friedenreich (TF) antigen: a critical review on the structural, biosynthetic and histochemical aspects of a pancarcinoma-associated antigen, *Histol. Histopathol.* 12: 263.

Harris, P.A., Roman, G.K., Moulds, J.J., Bird, G.W.G and Shah, N.G., 1982, An inherited RBC characteristic, NOR, resulting in erythrocyte polyagglutination, *Vox Sang.* 42: 134.

Herkt, F., Parente, J.P., Leroy, Y., Fournet, B., Blanchard, D., Cartron, J.-P., van Halbeek, H. and Vliegenthart, J.F.G., 1985, Structure determination of oligosaccharides isolated from Cad erythrocyte membranes by permethylation analysis and 500 MHz ^{1}H-NMR spectroscopy, *Eur. J. Biochem.* 146: 125.

Hiraiwa, N., Tsuyuoka, K., Li, Y-T., Tanaka, M., Seno, T., Okubo, Y., Fukuda, Y., Imura, H. and Kannagi, R., 1990, Gangliosides and sialoglycoproteins carrying a rare blood group antigen determinant, Cad, associated with human cancers as detected by specific monoclonal antibodies. *Cancer Res.* 50: 5497.

Huang, C.H. and Blumenfeld, O.O., 1991, Multiple origins of the human glycophorin St^a gene. Identification of hot spots for independent unequal recombinations. *J. Biol. Chem.* 266: 23306.

Iolascon, A., Miraglia del Giudice, E., Perrotta, S., Granatiero M., Zelante, L. and Gasparini, P., 1997, Exclusion of three candidate genes as determinants of congenital dyserythropoietic anemia type II (CDA II), *Blood* 90: 4197.

Iolascon, A., De Matia, D., Perrotta, S.., Carella, M., Gasparini, P. and Deliliers, G.L., 1998, Genetic heterogeneity of congenital dyserythropoietic anemia type II, *Blood* 92: 2593.

Joziasse, D.H. and Oriol, R., 1999, Xenotransplantation: the importance of the Galα1,3Gal epitope in hyperacute vascular rejection. *Biochim. Biophys. Acta* 1455: 403.

Joziasse, D.H., Shaper, J.H., Van den Eijnden, D.H., Van Tunen, A.H. and Shaper, N.L., 1989, Bovine α1-3 galactosyltransferase: isolation and characterization of a cDNA clone. Identification of homologous sequences in human genomic DNA. *J. Biol. Chem.* 264: 14290.

Joziasse, D.H., Shaper, J.H., Jabs, E.W. and Shaper, N.L., 1991, Characterization of an α1-3-galactosyltransferase homologue on human chromosome 12 that is organized as a processed pseudogene, *J. Biol. Chem.* 266: 6991.

Kameh, H., Landolt-Marticorena, C., Charuk, J.H.M., Schachter, H. and Reithmeier, R.A.F., 1998, Structural and functional consequences of an N-glycosylation mutation (HEMPAS) affecting human erythrocyte membrane glycoproteins, *Biochem. Cell Biol.* 76: 823.

Katayama, A., Ogawa, H., Kadomatsu, K., Kurosawa, N., Kobayashi, T., Kaneda, N., Uchimura, K., Yokoyama, I., Muramatsu T. and Takagi, H., 1998, Porcine α-1,3-galactosyltransferase: full length cDNA cloning, genomic organization, and analysis of splicing variants, *Glycoconjugate J.* 15: 583.

Kusnierz-Alejska, G., Duk, M., Storry, J.R., Reid, M.E., Wiecek, B., Seyfried, H. and Lisowska E., 1999, NOR polyagglutination and St^a glycophorin in one family: relation of NOR polyagglutination to terminal α-galactose residues and abnormal glycolipids, *Transfusion* 39: 32.

Larsen, R.D., Rajan, V.P., Ruff, M., Kukowska-Latallo, J., Cummings R.D. and Lowe, J.B., 1989, Isolation of a cDNA encoding murine UDP-galactose: β-D-galactosyl-1,4-N-acetyl-D-glucosamine α1,3-galactosyltransferase: expression cloning by gene transfer. *Proc. Natl. Acad. Sci. USA* 86: 8227.

Larsen, R.D., Rivera-Marrero, C.A., Ernst, L.K., Cummings, R.D. and Lowe, J.B., 1990, Frameshift and nonsense mutations in a human genomic sequence homologous to a murine UDP-Gal:β-D-Gal(1,4)-D-GlcNAc α(1,3)-galactosyltransferase cDNA, *J. Biol. Chem.* 265: 7055.

Malagolini, N., Dall'Olio, F., Di Stefano, G., Minni, F., Marrano, D. and Serafini-Cessi, F., 1989, Expression of UDP-GalNAc:NeuAcα2-3Galβ-R β1,4(GalNAc to Gal) N-acetylgalactosaminyl-transferase involved in the synthesis of Sd^a antigen in human large intestine and colorectal carcinoma, *Cancer Res.* 49: 6466.

Marks, E.W. and Mitus, A.J., 1996, Congenital dyserythropoietic anemias, *Amer. J. Hematol.* 51: 55.

Martin, P.T., Scott, L.J., Porter, B.E. and Sanes, J.R., 1999, Distinct structures and functions of related pre- and postsynaptic carbohydrates at the mammalian neuromuscular junction, *Mol. Cell. Neurosci.* 13: 105.

Mawby, W.J., Tanner, M.J.A., Anstee, D.J. and Clamp, J.R., 1983, Incomplete glycosylation of erythrocyte membrane proteins in congenital dyserythropoietic anaemia type II, *Brit. J. Haematol.* 55: 357.

Misago, M., Liao, Y.-F., Kudo, S., Eto, S., Mattei, M.-G., Moremen, K.W. and Fukuda, M.N., 1995, Molecular cloning and expression of cDNAs encoding human α-mannosidase II and a novel α-mannosidase IIx isozyme. *Proc. Natl. Acad. Sci. USA* 92: 11766.

Naiki, N., Fong, J., Ledeen, R. and Marcus, D.M., 1975, Structure of the human erythrocyte blood group P1 glycosphingolipid, *Biochemistry* 14: 4831.

Parker, W., Lin, S.S., Yu, P.B., Sood, A., Nakamura, Y.C., Song, A., Everett, M.L. and Platt, J.L., 1999, Naturally occurring anti-αgalactosyl antibodies: relationship to xenoreactive anti-αgalactosyl antibodies, *Glycobiology* 9: 865.

Piller, F., Blanchard, D., Huet, M. and Cartron, J.-P., 1986, Identification of a α-NeuAc-(2-3)-β-D-galactopyranosyl N-acetyl-β-D-galactopyranosyl-transferase in human kidney, *Carbohydr. Res.* 149: 171.

Reed, D.S., Olson, S. and Lefrancois, L., 1998, Glycosyltransferase regulation mediated by pre-TCR signaling in early thymocyte development, *Int. Immunol.* 10: 445.

Serafini-Cessi, F., Malagolini, N. and Dall'Olio, F., 1988, Characterization and partial purification of β-N-acetylgalactosaminyltransferase from urine of Sd(a+) individuals, *Arch. Biochem. Biophys.* 266: 573.

Serafini-Cessi, F., Malagolini, N., Guerrini, S. And Turrini, I., 1995, A soluble form of Sd^a-β1,4-N-acetylgalactosaminyltransferase is released by differentiated human colon carcinoma CaCo-2 cells, *Glycoconjugate J.* 12: 773.

Smith, P.L. and Lowe, J.B., 1994, Molecular cloning of a murine N-acetylgalactosamine transferase cDNA that determines expression of the lymphocyte-specific CT oligosaccharide differentiation antigen. *J. Biol. Chem.* 269: 15162.

Takeya, A., Hosomi, O. and Kogure, T., 1987, Identification and characterization of of UDP-GalNAc:NeuAcα2-3Galβ1-4Glc(NAc) β-1-4(GalNAc to Gal) N-acetylgalactosaminyltransferase in human plasma, *J. Biochem.* 101: 251.

Teneberg, S., Lönnroth, I., Torres Lopez, J.F., Galili, U., Halvarsson, M.O., Angström, J. and Karlsson, K.-A., 1996, Molecular mimicry in the recognition of glycosphingolipids by Galα3Galβ4GlcNAcβ-binding *Clostridium difficile* toxin A, human natural anti-α-galactosyl IgG and the monoclonal antibody Gal13: characterization of a binding-active human glycosphingolipid, non-identical with the animal receptor, *Glycobiology* 6: 599.

van Rooijen, J.J.M, Kamerling, J.P. and Vliegenthart, J.F.G., 1998, The abundance of additional N-acetyllactosamine units in N-linked tetraantennary oligosaccharides of human Tamm-Horsfall glycoprotein is a donor specific feature, *Glycobiology* 8: 1065.

van Rooijen, Voskamp, A.F., J.J.M, Kamerling, J.P. and Vliegenthart, F.F.G., 1999, Glycosylation site and site-specific glycosylation in human Tamm-Horsfall glycoprotein, *Glycobiology* 9: 21.

Watkins, W.M., 1995, Sd^a and Cad antigens, in *Blood Cell Biochemistry*, vol. 6: *Molecular Basis of Major Human Blood Group Antigens* (Cartron, J.-P. and Rouger, P., eds.), Plenum Press, New York, p. 351.

Wieslander, J., Manson, O., Kallin, E., Gabrielli, A., Nowack, H. and Timpl, R., 1990, Specificity of human antibodies against Galα1-3Gal carbohydrate epitope and distinction from natural antibodies reacting with Galα1-2Gal or Galα1-4Gal. *Glycoconjugate J.* 7: 85.

Wood, C., Kabat, E.A., Murphy A.L. and Goldstein, I.J., 1979, Immunochemical studies of the combining site of the two isolectins, A_4 and B_4, isolated from Bandeiraea simplicifolia seeds. *Arch. Biochem. Biophys.* 198: 1.

Wu, A.M., Duk, M., Lin, M.C., Broadberry, R.E. and Lisowska, E., 1995a, Identification od variant glycophorins of human erythrocytes by lectinoblotting. Application to MiIII variant which is relatively frequent in Taiwanese population. *Transfusion* 35: 571.

Wu, A.M., Watkins, W.M., Chen, C-P., Song, S-C., Chow, L-P. and Lin, J-Y., 1995b, Native and/or asialo-Tamm-Horsfall glycoproteins Sd(a+) are important receptors for *Triticum vulgaris* (wheat germ) agglutinin and for three toxic lectins (abrin-a, ricin and mistletoe toxic lectin-I), *FEBS Lett.* 371: 32.

Wu, J.H., Herp, A. and Wu, A.M., 1993, Defining carbohydrate specificity of *Ricinus communis* agglutinin as Galβ1-4GlcNAc (II) > Galβ1-3GlcNAc (I) > Galα1-3Gal (B) > Galβ1-3GalNAc (T). *Mol. Immunol.* 30: 333.

Wu, J.H., Watkins, W.M., Chen, C-P., Song, S-C. and Wu, A.M., 1996, Interaction of a human blood group Sd(a-) Tamm-Horsfall glycoprotein with applied lectins, *FEBS Lett.*, 384: 231.

Zdebska, E., Anselstetter, V.B., Pacuszka, T. and Koscielak, J., 1987, Glycolipids and glycopeptides of red cell membranes in congenital dyserythropoietic anemia type II, *Br. J. Haematol.* 66: 385.

ANTIGENIC PROPERTIES OF HUMAN GLYCOPHORINS - AN UPDATE

Elwira Lisowska

Department of Immunochemistry, Ludwik Hirszfeld Institute of Immunology and Experimental Therapy, Polish Academy of Sciences, 53-114 Wroclaw, Poland

Glycophorins are major sialoglycoproteins of human and animal erythrocytes. They are transmembrane type 1 glycoproteins of relatively low molecular size (20-30 kDa), carrying sialylated O-glycans. Some glycophorins also contain one or more N-glycans. Besides these similarities, the glycophorins of various species, or even within the same species, show structural differences (rev. by Krotkiewski, 1988).

Human erythrocytes carry four glycophorins (A, B, C, D). The structure and genetics of human glycophorins and their common (MN and Ss) and rare genetic variants are well known and have been a subject of several reviews (Huang *et al.*, 1991; Dahr *et al.*, 1992; Tippett *et al.*, 1992; Fukuda, 1993; Cartron *et al.*, 1993; Reid, 1994; Reid and Spring, 1994; Blumenfeld and Huang, 1995). A dense O-glycosylation of human glycophorins and the existence of their variant forms creates various kinds of antigenic epitopes and makes these glycoproteins interesting model antigens to study the contribution of glycosylation to protein antigenic properties. Moreover, this problem is important, because various anti-glycophorin antibodies occur in human sera, mostly as alloantibodies in persons whose erythrocytes are glycophorin-deficient or carry a variant glycophorin. The antigenic properties of normal and variant glycophorins, disclosed by means of human and rabbit polyclonal antisera and monoclonal antibodies, were reported at the first MICC Meeting held at College Station, TX, in 1985 (Lisowska, 1988). Since that time many new monoclonal anti-glycophorin antibodies were obtained, and participation in the three Workshops on „Monoclonal Antibodies against Red Blood Cell and Related Antigens" (Paris, 1987, Lund, 1990 and Nantes, 1996) allowed us to study many of them, with the use of new techniques. In this review, antigenic properties of glycophorins are summarized, with a focus on the progress accomplished in recent years with the use of monoclonal antibodies. Low incidence antigens of human erythrocytes carried by variant glycophorins have been already discussed in the cited reviews, and are not included in this article. Here, only few variant glycophorins are mentioned, mainly those which have been useful for identification and characterization of antigenic epitopes of common forms of GPA and GPB or GPC and GPD.

The Molecular Immunology of Complex Carbohydrates–2.
Edited by Albert M. Wu, Kluwer Academic/Plenum Publishers, 2001.

I. STRUCTURE OF HUMAN GLYCOPHORINS

A major sialoglycoprotein of human erythrocytes (0.5-1x10^6 molecules *per* cell) is glycophorin A (GPA) which is a product of two allelic genes, giving rise to the MN blood group system. GPA-M and GPA-N differ in two amino acid residues at positions 1 (Ser/Leu) and 5 (Gly/Glu). A minor component, glycophorin B (GPB, 2x10^5 copies *per* cell) shows a high degree of sequence homology with GPA, including the sequence of amino acid residues 1-26 which is identical in GPA-N. GPB also exists in two allelic forms differing in one amino acid residue at position 29 (Met/Thr) and representing blood group S and s antigens, respectively. Two other minor glycophorins, C (GPC, 1.3x10^5 copies *per* cell) and its shorter form D (GPD, lacking amino acid residues 1-21 of GPC, 5x10^4 copies *per* cell) have amino acid sequences different from those of GPA and GPB. Amino acid sequences of human glycophorins were determined by chemical methods and generally were are in agreement with those deduced from the sequences of respective cDNAs (Fig. 1).

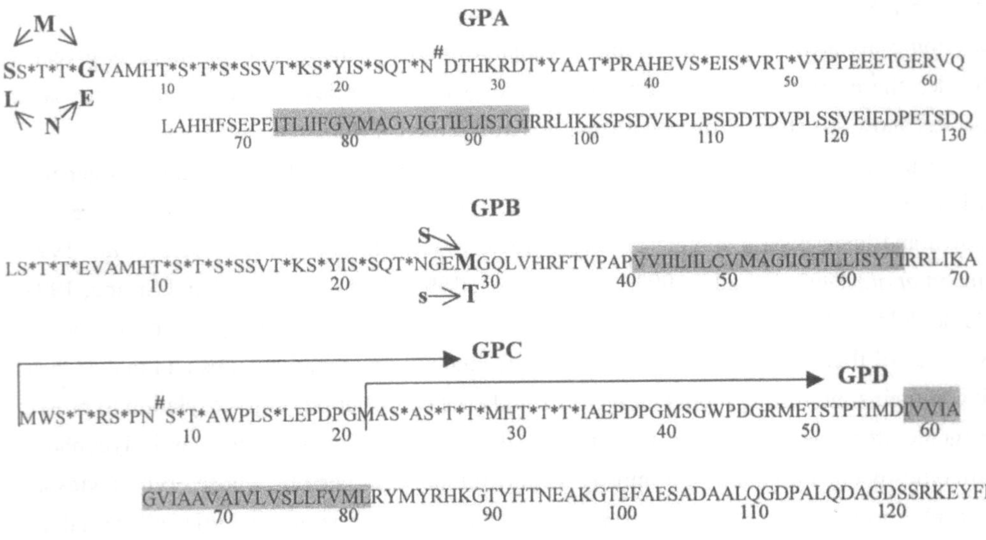

Fig. 1. Amino acid sequences of GPA (M and N), GPB (S and s) and GPC/GPD. The transmembrane fragments are shadowed. Asterisks (*) denote O-glycosylated amino acids, # shows N-glycosylated Asn residues in GPA and GPC.

Most GPA O-glycans are NeuAcα2-3Galβ1-3(NeuAcα2-6)GalNAc- chains. They are accompanied by minor amounts of other O-glycans, which are either more complex (e.g. trisialylated pentasaccharides, chains with blood group A, B or H structures), or truncated (e.g. monosialylated linear trisaccharide, sialylated or nonsubstituted GalNAc residues), compared to the major tetrasaccharide (Fukuda *et al.*, 1987; Krotkiewski *et al.*, 1993, and 1997; Podbielska and Krotkiewski, 2000). The distribution of O-glycans was established during amino acid sequence determination of GPA, GPB (Tomita *et al.*, 1978; Dahr *et al.*, 1980) and GPC (Dahr and Beyreuther, 1985) by Edman degradation methods. The O-glycosylation sites in GPA have been later re-investigated by solid phase Edman degradation (Pisano *et al.*, 1993) and these results for GPA and, *per* analogy, for GPB, are given in Fig. 1. The results generally agree with earlier studies, except that Pisano *et al.* did find no glycosylation of Ser$_{14}$, Ser$_{15}$, Ser$_{23}$ and Thr$_{28}$, residues which were formerly thought to be at least partly glycosylated. Although there are 16 O-glycosylated Ser/Thr residues in GPA, some are incompletely glycosylated, and calculation based on carbohydrate composition of GPA indicates the presence of 12 O-glycans *per* molecule (Zdebska and Koscielak, 1999). A single N-glycosidic chain is present on the GPA (Asn$_{26}$) and GPC (Asn$_{8}$) molecules. The N-glycan of GPA is a complex biantennary chain with a bisecting GlcNAc residue. The structure of oligosaccharide chains was determined only in crude (containing other glycophorins) or purified GPA preparations, and not in purified GPB or GPC. However, it is asumed that all human glycophorins contain similar glycans.

II. ANTIGENIC EPITOPES OF GLYCOPHORIN A (GPA)

Characterization of epitopes recognized by a number of anti-GPA monoclonal antibodies (MAbs) reveals that some fragments of the GPA molecule are distinctly immunogenic. The antibodies tested so far recognize blood group M- and N-related glycopeptidic epitopes, located in the N-terminal portion of GPA, and internal peptidic epitopes in the extracellular and cytoplasmic portions of GPA. For mapping and examination of peptidic epitopes, immobilized peptides synthesized on plastic pins (Pepscan analysis) have been used. This technique was introduced by Geysen *et al.* (1987) and commercial kits are now available.

1. Blood group M- and N-related epitopes

The epitopes recognized by anti-M and anti-N antibodies are interesting due to their complexity. The GPA-M and GPA-N have different amino acid residues at positions 1 and 5 which are separated by three O-glycosylated Ser/Thr residues (Fig. 2).

✱ ✱ ✱

GPA-**M**: H_2N-**Ser** - Ser - Thr - Thr - **Gly** - Val - Ala - Met - His-

✱ ✱ ✱

GPA-**N** **&** GPB: H_2N-**Leu**- Ser - Thr - Thr - **Glu** - Val - Ala - Met - His-

✱ ✱ ✱

GPA-**M^c**: H_2N-**Ser** - Ser - Thr - Thr - **Glu** - Val - Ala - Met - His-

GPA-**M^g**: H_2N-**Leu**- Ser - Thr-**Asn** - **Glu**-Val - Ala - Met - His-

✱ ✱ ✱

GPB-**He**: H_2N-**Trp** -Ser - Thr - Ser - **Gly** -Val - Ala - Met - His-

✱ - tetrasaccharide NeuAcα2-3Galβ1-3GalNAc-
6
NeuAcα2/

Fig. 2. N-terminal fragments of GPA-M, GPA-N and of low-incidence variant antigens GPA-M^c, GPA-M^g and GPB-Henshaw(He).

Anti-M and anti-N monoclonal antibodies differ in the degree of their specificity, from highly blood group M- or N-specific ones to those showing only a preferential reaction with one of the alloantigens. This specificity is based on the recognition of amino acid residues which are different in GPA-M and GPA-N. However, it has been known for a long time that sialic acid residues are also involved in blood group M and N epitopes. Examination of monoclonal anti-M and anti-N antibodies has been performed by testing their reaction with normal and variant antigens, or with chemically modified normal glycophorins, using inhibition of hemagglutination, plate ELISA and Western blotting. The results show a great diversity in their subspecificities (Lisowska *et al.*, 1987; Lisowska, 1988; Anstee and Lisowska, 1990; Jaskiewicz *et al.*, 1990a and 1990b; Duk *et al.*, 1994; Duk and Lisowska, 1997; Reid *et al.*, 1997).

Extended testing of anti-M antibodies confirmed that they do not require the presence of both blood group M-typical amino acid residues in the antigen, but recognize epitopes dependent on Ser_1 or Gly_5 (anti-M_{Ser} and anti-M_{Gly} subspecificities, respectively). This subspecificity can be defined by testing the antibodies with variant glycophorins: anti-M_{Ser} react with GPA-M^c, and anti-M_{Gly} react with GPB-He (Fig. 2). The rare variant cells can be replaced by chemically modified normal GPA-M. The anti-M_{Ser} antibodies do not react, while anti-M_{Gly} do react, with GPA-M after one step of Edman degradation which removes the N-terminal amino acid residue. Another simple test is periodate oxidation: the antigen-coated wells of ELISA plate are shortly treated with periodate and binding of the antibodies to periodate-untreated and -treated wells is compared. Although periodate oxidizes the nonreducing end of sialic acid residues in blood group MN epitopes, an effect of oxidation is not related to a role of sialic acid residues in the epitope. We have found that oxidized GPA-M reacts with all anti-M_{Gly} antibodies and does not react with anti-M_{Ser} antibodies (Duk *et al.*, 1994; Duk and Lisowska, 1997). The anti-N antibodies are more uniform, regarding the recognition of amino acid residues in the epitope. All anti-N antibodies tested by us recognized an epitope dependent on Leu_1, since they did not react with the Edman-degraded or periodate-oxidized GPA-N. Another element of the polypeptide chain playing a role in blood group MN epitopes (mostly in those dependent on the first aa residue) is the amino group of the N-terminal amino acid. N-acetylation of the target antigen abolishes or strongly decreases the reaction with most anti-N (an exception is a strong increase of activity found with one anti-N), but only with part of anti-M_{Ser} antibodies, and ususally has no distinct effect on M_{Gly} epitopes. Therefore, the effect of amino group blocking cannot be used for defining the epitope dependence on the first amino acid residue. Anti-M and anti-N antibodies show various patterns of pH-dependence, that may be mediated by the charge of the antigen's amino group or by charged groups of the antibody combining site.

The role of the carbohydrate moiety in blood group M and N epitopes is also diversified. Most anti-M and anti-N antibodies do not react with desialylated antigens. To answer the question whether there is any difference in the role of two sialic acid residues present in glycophorin O-glycans (Fig. 2), we constructed the GPA-M and GPA-N with monosialylated O-linked chains (Duk *et al.*, 1994). GPAs with linear trisaccharide chains, containing only Gal-linked sialic acid residues, were obtained by exhaustive desialylation followed by resialylation with α2,3-sialyltransferase, and those with branched trisaccharide chains, containing GalNAc-linked sialic acid residues, were obtained by selective release of Gal-linked sialic acid with sialidase from Newcastle disease virus. Reactivity of these GPA forms with anti-M and anti-N antibodies recognizing sialic acid-dependent epitopes showed that for most antibodies the presence of only one type of sialic acid residues, Gal-linked or GalNAc-linked, is sufficient. The requirement for Gal-linked or GalNAc-linked sialic acid residues by anti-M and anti-N antibodies is not correlated with their other properties.

The role of other O-glycan components (Gal, GalNAc) has been less studied, since there are not many anti-M and anti-N antibodies which react with desialylated antigens. However, the Gal-GalNAc units seem to be necessary for reactivity of sialic acid-independent epitopes, since anti-N antibodies reacting with asialoGPA-N did not react with enzymatically de-O-glycosylated antigen (Jaskiewicz *et al.*, 1990a, and unpublished). Unique type of dependence on O-glycosylation show anti-M_{Gly} antibodies cross-reacting with variant M^g antigen (see next chapter).

2. Variant M^g antigen-related epitopes

A low incidence M^g antigen has the amino acid sequence of GPA-N in which Thr^4 is replaced by Asn and due to this replacement Ser_2 and Thr_3 are not or hardly glycosylated (Fig. 2). Agglutination of M^g erythrocytes by some anti-M monoclonal antibodies was a serological puzzle (for refs. see Lisowska, 1988). We solved this problem with the use of synthetic peptides (Jaskiewicz *et al.*, 1994). The three selected anti-M antibodies (E3, E6 and 425/2B), which agglutinate M^g erythrocytes, belong to the group with anti-M_{Gly} specificity, and do not react, or react only weakly with asialoGPA-M. However, these antibodies react strongly with synthetic octapeptides corresponding to the sequence of amino acid residues 1-8 of GPA-M, GPA-N and GPA-M^g. Examination of multiple analogs of these peptides showed that the most essential amino acid residues in the epitope are Val_6 and Met_8 and that the sequence of amino acid residues 1-5 is of minor importance. The anti-M/M^g antibodies represent an interesting example of bidirectional modulation of the epitope activity by glycosylation. The Val_6/Met_8-dependent epitope is recognized in sialylated GPA, provided that Gly, and not Glu, is present at position 5. Desialylation of GPA changes the steric arrangement so that the epitope is not or hardly available for the reaction, and the nonglycosylated peptide is active, independent of which amino acid residue is present at position 5. This type of effect of O-glycans on peptidic epitopes is not limited to glycophorins, because similar properties are shown by some anti-leukosialin (CD43) monoclonal antibodies that react with the native and nonglycosylated (recombinant) antigen, but not with desialylated leukosialin (Cyster *et al.*, 1991). A simple method of identification of anti-M/M^g antibodies is electrophoresis (SDS-PAGE) of erythrocyte membranes, blotting to Immobilon P and de-O-glycosylation of glycophorins by treatment of the blots with NaOH. Anti-M/M^g antibodies react only with GPA-M on untreated blots and, in contrast to other anti-M which do not react with deglycosylated antigen, stain GPA-M, GPA-N and GPB bands on NaOH-treated blots (Duk *et al.*, 1997).

Anti-M^g monoclonal antibodies elicited by immunization with M^g erythrocytes show different specificities. Two anti-M^g antibodies tested by Pepscan analysis reacted with nonglycosylated N-terminal octapeptide of GPA-M^g and a number of its analogs, showing differences in fine specificity. One of the antibodies recognizes an epitope in which the N-terminal Leu is a most essential residue. In turn, Leu^1 can be absent or replaced in the epitope reacting with the second anti-M^g, for which the residues Ser_2-Val_6 are more important (Jaskiewicz *et al.*, 1999). Altogether, these results demonstrate diverse fine specificities of monoclonal antibodies reacting with an M^g determinant.

3. Internal peptidic epitopes of GPA

Initially, an approximate location of internal epitopes (called En^a) in the GPA molecule was based on the reaction of antibodies with protease-treated erythrocytes. Three categories of epitopes were identified, trypsin-sensitive (En^aTS), trypsin-resistant and ficin sensitive (En^aFS), and ficin-resistant (En^aFR). Competitive binding assays showed that examined monoclonal anti-GPA antibodies recognized two groups of similar or overlapping internal epitopes in the extracellular portion of GPA (Gardner *et al.*, 1989). The antigenicity algorithms predicted that GPA segments containing amino acid residues 27-34 and 52-62 would be antigenic (Rearden, 1993). The corresponding peptides were synthesized and the predictions were confirmed for the GPA 52-62 segment only, by using the specific antipeptide sera.

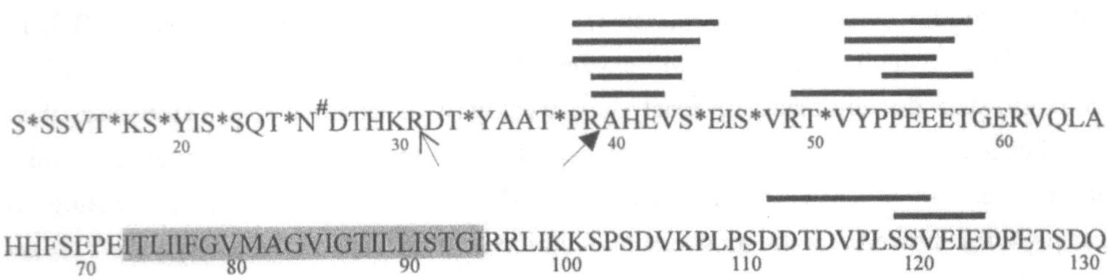

Fig. 3. Fragment of GPA molecule in which peptidic epitopes identified by Pepscan analysis of various monoclonal antibodies are marked with horizontal lines. Arrows indicate the sites which are partiallly (↑) or totally (▲) digested by trypsin on the erythrocyte surface.

More precise localization of internal peptidic epitopes of GPA was possible by using multipin peptide synthesis/Pepscan analysis. Most of the antibodies reacting with the internal (blood group MN-unrelated) portion of the exofacial GPA domain recognize two groups of overlapping peptidic epitopes located within amino acid residues 38-45 and 49-58 (Fig. 3) (Wasniowska *et al.*, 1992 and 1997; Rasamoelisolo *et al.*, 1997). These two groups of epitopes correspond to En^aTS and En^aFS antigens, because treatment of erythrocytes with trypsin destroys epitopes of the first group (38-45) and does not affect or enhances the reactivity ot the epitopes of the second group (49-58), which, in turn, are ficin-sensitive.

These peptidic epitopes show differences in dependence on glycosylation of adjacent amino acid residues that is reflected by differences in response of antibodies to GPA desialylation, from no effect to a strong enhancement of reaction. Interestingly, three monoclonal anti-GPA antibodies specific for the same epitope (^{39}RAHEV^{43}) showed unaltered (MAb OSK4-1), moderately increased (MAb GPA105), or strongly increased (MAb GPA33) reactivity after GPA desialylation. To explain these differences, the specificity of these antibodies was studied in detail, by using multiple replacement analogs of RAHEV (Wasniowska *et al.*, 1992). As shown in Fig. 4, the antibodies differ in requirement for individual amino acid residues in the epitope. The results allow to conclude that different amino acid residues essential for the reaction with each antibody are 'masked' in different degree by adjacent sialic acid residues.

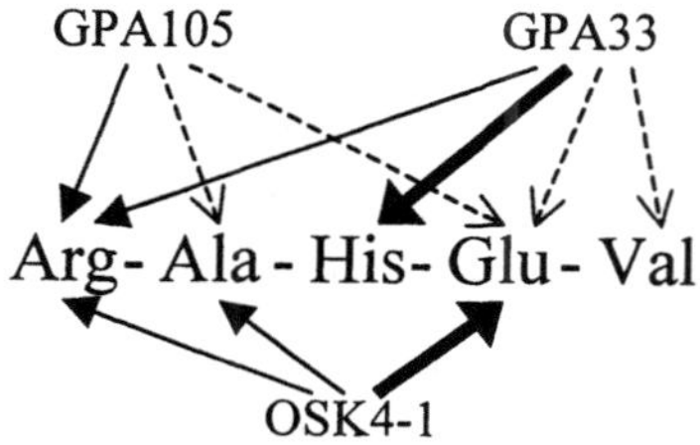

Fig. 4. Schematic presentation of the fine specificity of anti-GPA (anti-RAHEV) monoclonal antibodies, GPA105, GPA33 and OSK4-1. Thick arrows indicate amino acid residues which cannot be replaced in the epitope by other amino acid residues, thin or broken arrows indicate amino acid residues replaceable only by few, or by more than a half, of other 19 amino acid residues, respectively; no arrow – the residue can be replaced by any of other amino acid residues.

Besides the peptidic epitopes, the extracellular portion of GPA also contains blood group MN-independent glycopeptidic epitopes that loose activity after antigen desialylation (Lisowska, 1988; Gardner *et al.*, 1989; Rasamoelisolo *et al.*, 1997). These epitopes are probably located within densely glycosylated amino acid residues 6-26 of GPA, because they are usually shared with GPB which has the same amino acid sequence in this region of the molecule (Fig. 1). Moreover, some human and mouse monoclonal anti-GPA antibodies are directed against conformational epitopes (En^{a}FR) located close to the erythrocyte membrane. These epitopes are difficult to define and they may result from interaction of GPA with membrane phospholipids or with band 3 protein (see chapter on Wright antigens).

Another immunogenic site in GPA is the cytoplasmic tail (Fig. 3). Three monoclonal antibodies against epitopes comprising amino acid residues 112-121 (MAb PEP80, Duk *et al.*, 1992) or 119-124 (MAbs 4E4 and 5F4, Rasamoelisolo *et al.*, 1997) were obtained by immunization of mice with solubilized glycophorins, or with cord erythrocytes (MAb 4E4). These antibodies are useful tools for many purposes. For example, they have been used for characterization of hybrid glycophorins (Wasniowska *et al.*, 1993; Wu *et al.*, 1995; Kusnierz-Alejska *et al.*, 1999), for identification of GPA degradation products left in protease-treated erythrocytes (Bykowska *et al.*, 1993), and recently for identification of the recombinant fusion protein containing the N-terminal Duffy domain and cytoplasmic portion of GPA (Wasniowska *et al.*, 2000).

4. The Wright (Wr) antigens

Two blood group antigens, a low-incidence Wr^a and a high-incidence Wr^b, attracted attention due to their not clearly defined relation to GPA. Studies with human antibodies and later with mouse monoclonal antibodies showed that a common Wr^b antigen is not expressed either in GPA-deficient erythocytes, or in erythrocytes carrying variant glycophorins lacking amino acid residues 60-70 of GPA. This suggested that a labile (conformation- and phosholipid-dependent) Wr^b determinant is located within these residues. On the other hand, GPA in the rare Wr^a erythrocytes was found to be normal, and anti-Wr antibodies (similarly as other anti-En^aFR) do not react with solubilized GPA (rev. by Lisowska, 1988). Later studies were focused on the possible involvement of band 3 (anion transporter) erythrocyte membrane protein in the Wr epitopes (Dahr *et al.*, 1986). This concept was supported by studies showing the coprecipitation of band 3 and GPA by anti-Wr^b antibodies (Telen and Chasis, 1990; Ring *et al.*, 1994; Leddy *et al.*, 1994). Direct evidence was supplied by sequencing cDNAs encoding GPA and band 3 in a Wr(a+b-) individual. The sequence of the GPA cDNA was normal, but the band 3 cDNA showed an allelic change (A1972G) causing the substitution of Glu_{658} by Lys in the band 3 protein (Bruce *et al.*, 1995). This finding suggested that Glu_{658} of band 3 interacts with Arg_{61} of GPA to form the Wr^b determinant. However, further studies on variant human erythrocytes and red cells of nonhuman primates demonstrated that this problem is more complex. The homozygote erythrocytes carrying a SAT-glycophorin variant express a new SAT antigen, and do not express the Wr^b antigen, despite that these cells contain the band 3 with Glu_{658} and that SAT-glycophorin (composed of amino acid residues 1-70 of GPA and 40-72 of GPB) contains amino acid residues 60-70 of GPA (Huang *et al.*, 1996). Although the problem has not been fully elucidated, the Wr^b antigen is an interesting example of GPA contribution to a complex epitope formed by two proteins interacting within the cell membrane.

5. Antigenic properties of recombinant GPAs

The GPA-M, GPA-N and some rare GPA variants were stably expressed in Chinese hamster ovary (CHO) cells (Remaley *et al.*, 1991; Ugorski *et al.*, 1993; Blackall *et al.*, 1994). The expression of variant GPA forms, containing Ser_1/Glu_5 (corresponding to the M^c antigen) and Leu_1/Gly_5 (not found in humans), and examination of the binding of antibodies to recombinant GPA forms confirmed anti-M_{Ser} or anti-M_{Gly} specificity of anti-M antibodies and anti-N_{Leu} specificity of anti-N antibodies tested. Moreover, using the Leu_1/Gly_5 'artificial' GPA variant allowed to show for the first time that Glu_5 is not essential for anti-N_{Leu} antibodies (Blackall *et al.*, 1994). However, some anti-M and anti-N antibodies do not bind to the recombinant GPAs, most likely due to differences in glycosylation between CHO cell-expressed and erythrocyte GPAs. The major O-glycan in GPA expressed in CHO cells is monosialylated linear trisaccharide, while the disialylated tetrasaccharide is not detected (Pahlsson *et al.*, 1994). Therefore, the recombinant GPAs do not bind antibodies which require the GalNAc-linked sialic acid residues in the epitope. The recombinant GPAs also bind antibodies against internal peptidic epitopes. The successful expression of GPA variant forms in CHO cells opens further possibilities of replacing rare glycophorin variants with their recombinant forms and construction of new variants that could be helpful in elucidating unsolved serological problems.

III. ANTIGENIC PROPERTIES OF GLYCOPHORIN B (GPB)

An extracellular domain of GPB is identical to GPA-N up to the 26th amino acid residue (Fig. 1) and GPB reacts with all anti-N antibodies and with those antibodies that recognize epitopes located within the amino acid residues 7-26 of GPA or GPB. The short region with the sequence unique for GPB (aa 27-40) includes Met/Thr polymorphism at position 29, responsible for the S and s blood groups. The anti-S and anti-s antibodies rarely occurr in human sera, and a few murine anti-S are the only known monoclonal antibodies specific for GPB. Besides the Ss antigens, GPB expresses a common U antigen (defined by human antibodies) located in a region close to the erythrocyte membrane (aa 33-39). The U-related epitope is not reactive in the isolated GPB and it may require an interaction of GPB with another membrane protein, like the Wr^b antigen (Reid and Lomas-Francis, 1997).

IV. ANTIGENIC PROPERTIES OF GLYCOPHORIN C (GPC) AND D (GPD)

GPC and its shorter form, GPD, are structurally and antigenically distinct from GPA and GPB (Fig. 1). GPC and GPD are involved in interactions with protein 4.1 of erythrocyte membranes and contribute to the regulation of shape and mechanical properties of erythrocytes. In contrast to GPA and GPB, which are expressed in erythroid cells only, GPC and GPD are expressed in both, erythroid and nonerythroid tissues.

GPC and GPD molecules are encoded by single gene (*GYPC*) containing four exons, by the use of alternative translation initiation sites at the mRNA AUG codons at positions 1 and 22, respectively (Le Van Kim *et al.*, 1996). Antigenic sites on GPC and GPD molecules (Ge antigens which form the Gerbich blood group system) have been defined by human antibodies, mostly alloantibodies produced in persons whose erythrocytes do not express GPC (Leach phenotype) or carry variant GPC forms (refs. in: Cartron *et al.*, 1993; Reid and Spring, 1994; Reid and Lomas-Francis, 1997). These variants include Yusseff (Yus) and Gerbich (Ge) types (Fig. 5), lacking amino acid residues 17-35 (encoded by exon 2), or 36-63 (encoded by exon 3), respectively (Fig. 5). In the Leach phenotype GPC and GPD are not expressed, due to lack of exons 3 and 4, or due to other alterations resulting in premature stop codon in the *GYPC* gene. The individuals of Gerbich, Yus and Leach phenotypes produce several anti-GPC alloantibodies (elicited by transfusion or pregnancy, or naturally occuring) directed against epitopes located in GPC fragments missing in these phenotypes. The major common antigenic sites of GPC and GPD defined by serological methods are located at the amino terminal portions of GPC (Ge4) and GPD (Ge2, absent from the Gerbich and Yus phenotypes), or within amino acid residues 40-50 of GPC or 19-29 of GPD (Ge3, present in Yus and absent from the Gerbich phenotype).

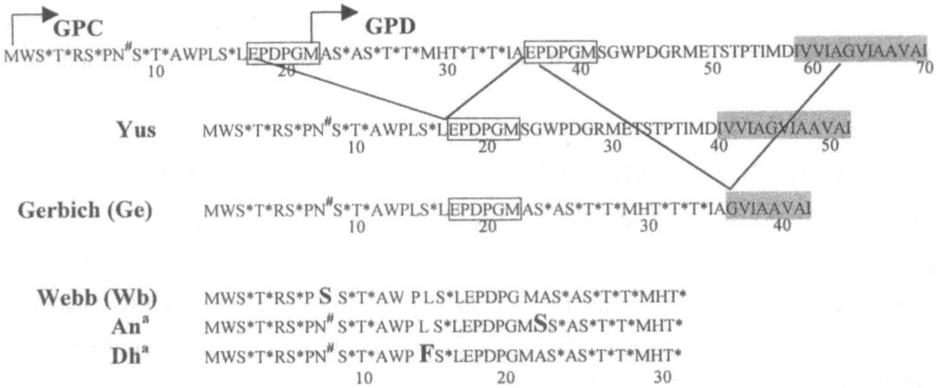

Fig. 5. The extracellular portions of GPC and its deletion variants, Yus (lacking aa 17-35 of GPC), and Gerbich (lacking aa 36-63 of GPC), and replacement variants: Wb ($Asn_8 \rightarrow Ser$), An[a] ($Ala_{23} \rightarrow Ser$ and $Ala_2 \rightarrow Ser$ in GPD) and Dh[a] ($Leu_{14} \rightarrow Phe$), in which the changed amino acid residues are in bold face.Antigenic EPDGM motifs in GPC and in Yus and Ge variants are in frames.

Other GPC variants carry unusual antigenic epitopes resulting from single amino acid replacements (Wb, Dh[a], An[a], shown in Fig. 5). In the GPC-Wb variant the replacement $Asn_8 \rightarrow Ser$ results in the absence of N-glycan. Interestingly, the An[a] mutation is structurally expressed on GPC ($Ala_{23} \rightarrow Ser$) and GPD ($Ala_2 \rightarrow Ser$), but anti-An[a] antibodies recognize this antigen only on GPD. In another rare phenotype, exon 3 in *GYPC* gene is duplicated, and elongated GPC (156 aa residues) and GPD (138 aa residues) carry a new antigen, Ls[a], formed at the junction site of two exon 3-encoded fragments (MDIVVIA/EPDPG). Recently, a combination of two genetic events occurring in the *GYPC* gene has been detected: a deletion of exon 3 in one allele, accompanied by a point mutation in exon 3 of the second allele, resulting in $Asp_{58/37} \rightarrow Val$ replacement in the encoded glycoproteins, designated GPC.RW and GPD.RW (King *et al.*, 1997)

A number of murine monoclonal anti-GPC or anti-GPC/GPD antibodies has been obtained. Using normal and variant erythrocytes these antibodies were characterized as anti-Ge4, anti-Ge3 or anti-Ge2 (Dahr *et al.*, 1989; Loirat *et al.*, 1992 and 1994; Smythe *et* al., 1994; Janvier *et al.*, 1998). A monoclonal antibody (BGRL-100) reactive with GPC and GPD was obtained by immunization of mice with the conjugate of keyhole limpet hemocyanin and a synthetic peptide corresponding to the C-terminal amino acid residues 112-128 of GPC (King *et al.*, 1995).

A panel of 13 anti-GPC antibodies was tested at the Nantes 1996 Workshop (Wasniowska *et al.*, 1997; Reid *et al.*, 1997). Six of these antibodies were specific for N-terminal epitopes of GPC. These epitopes, similarly to blood group MN-related epitopes of

GPA, are strongly or moderately dependent on sialic acid residues, N-terminal amino acid (methionine) residue and its amino group. The remaining seven antibodies, directed against internal sites, were tested by the synthetic peptide/pin technology, and we were successful in defining the epitopes for four of them. The MAb NAM57-1F6 recognizes the epitope ^{110}QGDPAL^{115} in the cytoplasmic tail of GPC and GPD. This antibody is an useful reagent, since it reacts with all normal and variant forms of GPC and GPD. Interestingly, other three antibodies, which were tested for binding to synthetic overlapping octapeptides covering the sequence of amino acid residues 11-60 of GPC, recognize epitopes related to the EPDPGM sequence. This sequence is repeated twice in the GPC molecule (residues 17-22 and 36-41) and occurs only once in GPD, and in Gerbich and Yus variants (Fig. 5). Two antibodies (NAM89-2G11 and 002-56) react with peptides containing ^{16}LEPDPGM^{22} , and not with those containing ^{35}AEPDPGM^{41}, that indicates an essential role of Leu_{16} in the epitope. In agreement with this specificity, both antibodies react with GPC, GPC-Ge and GPC-Yus, and not with GPD. Therefore, they are classified by serological methods as anti-Ge4, but have entirely different specificity from the antibodies specific for N-terminal glycosylated epitopes of GPC. The MAb NAM98-3C1 reacts with larger set of peptides corresponding to both EPDPGM-containing regions of GPC, since it requires a shorter sequence, $^{17/36}$EPDP$^{20/39}$, for the reaction. However, this antibody reacts only with GPC and GPD, both containing the ^{34}IA**EPDP**GMS^{42} sequence of GPC, and does not react with Gerbich and Yus variants which have EPDP only in the context of the ^{15}S*L**EPDP**GMA^{23} sequence. These results indicate that the MAb NAM98-3C1 recognizes an epitope located in the region around 36-39 amino acid residues of GPC, and does not react with that related to the residues 17-20, possibly due to the role of flanking amino acid residues or glycosylation in the native glycoproteins.

V. ANTI-GLYCOPHORIN ANTIBODIES CROSS-REACTING WITH OTHER ERYTHROCYTE MEMBRANE PROTEINS

Recently, two monoclonal antibodies, anti-GPA and anti-GPC, each cross-reacting with another erythocyte membrane component, have been characterized. The MAb NAM10-3C10 binds on immunoblots to GPA and to a 80 kDa protein, identified as protein 4.1 (Rasamoelisolo *et al.*, 1998). Epitope analysis with immobilized peptides indicated that the antibody recognizes the sequence ^{53}PPEEE^{58}, located within one of earlier defined internal immunogenic GPA sites (Fig. 3). A sequence with PPE motif is present in the protein 4.1:

GPA: ^{49}RTVYPPEEETGERVQLAH^{66}

Protein 4.1: ^{391}KEDEPPEQAEPEPTEAWK^{408}

Since the MAb NAM10-3C10 binds distinctly to PPEQ and PPEE tetrapeptides (70-80% of binding to PPEEE and PPEEE-containing octapeptides), it is likely that it recognizes an epitope in the shown region of protein 4.1.

The second 'bispecific' antibody NAM26-4C6 shows binding to GPC and to the band 3 protein (Loirat *et al.*, 1999). Peptide mapping allowed localization of the band 3 epitope on its 23 kDa N-terminal intracellular fragment. Binding of the antibody to immobilized synthetic peptides corresponding to sequences of GPC and N-terminal portion of the band 3 protein indicated that the antibody recognizes sequences ^{16}LEPDPGM^{22} of the GPC and ^{22}EDPDIP^{27} of band 3. These sequences are different, but contain common amino acid residues (Glu, Asp, Pro), and the replacement analysis showed that Asp and Pro residues are essential for the antibody binding. Despite the binding to the GPC 16-22 peptide, the antibody does not bind to desialylated or de-O-glycosylated GPC. Moreover, the antibody does not react (agglutination, immunoblotting) with the Gerbich and Yus variants of GPC, despite that both GPC variants contain the LEPDPGM sequence. These data suggest that the epitope NAM26-4C6 is properly exposed only in the fully glycosylated GPC molecule, and that Gerbich and Yus GPC variants may have an altered glycosylation. Interestingly, the MAb 002-56 specific for the same ^{16}LEPDPGM^{22} sequence of GPC also was reported to cross-react with band 3 protein, but this antibody reacts with GPC, GPC-Ge and GPC-Yus, and the GPC epitope is not sensitive to desialylation (Reid *et al.*, 1997). The MAbs NAM26-4C6 and 002-56 represent another example of differences in subspecificity of antibodies recognizing the same fragment of an antigen polypeptide chain.

SUMMARY

Glycophorins are complex heavily glycosylated antigens carrying peptidic and glycopeptidic epitopes. Detailed immunochemical studies showed that GPA/GPB and GPC/GPD molecules have defined sites which are particularly immunogenic. These sites include N-terminal portions of all glycophorins, internal fragments of their extracellular domains, and cytoplasmic tails. The extracellular epitopes involve directly oligosaccharide chains (e.g. blood group M- and N-related epitopes, or N-terminal epitopes of GPC) or have peptidic character, shown by the reaction of respective antibodies with synthetic peptides. Peptidic eitopes are independent of glycosylation, or are variably affected by adjacent O-glycans which may mask the epitopes or may be required for a proper exposure of an antibody binding site. Several low incidence epitopes are present on variant glycophorin molecules. Among anti-glycophorin antibodies there are the 'bispecific' ones, or antibodies recognizing an epitope formed by an interaction of two proteins (Wr^b). Alltogether, the glycophorins serve as convenient model antigens for studying Ag-Ab interaction and a role of O-glycosylation in protein antigenic properties. Moreover, well defined specificty of monoclonal anti-glycophorin antibodies makes them more precise tools in serological investigation and identification of normal and variant antigens. Last but not least, elucidation of antigenic properties of glycophorins is important for identification and characterization of human anti-glycophorin antibodies, which in some cases create medical problems at transfusion or pregnancy.

REFERENCES

Anstee, D.J. and Lisowska, E., 1990, Worhshop 2. Monoclonal antibodies against glycophorins and other glycoproteins, *J. Immunogenet.* 17: 301.

Blackall, D.P., Ugorski, M., Pahlsson, P., Shakin-Eshleman H. and Spitalnik, S.L., 1994, A molecular biologic approach to study the fine specificity of antibodies directed to the MN human blood group antigens, *J. Immunol.* 152: 2241.

Blumenfeld, O.O. and Huang, C-H., 1995, Molecular genetics of the glycophorin gene family, the antigens for MNSs blood groups: multiple gene rearrangements nad modulation of splice site usage result in extensive diversification, *Hum. Mutat.* 6: 199.

Bruce, L.J., Ring, S.M., Anstee, D.J., Reid, M., Wilkinson, S. and Tanner, M.J.A., 1995, Changes in the blood group Wright antigens wre associated with a mutation at amino acid 658 in human erythrocyte band 3: a site of interaction between band 3 and glycophorin A under certain conditions, *Blood* 85: 541.

Bykowska, K., Duk, M., Kusnierz-Alejska, G., Kopec, M. and Lisowska, E., 1993, Degradation of human erythrocyte surface components by human neutrophil elastase and cathepsin G: preferential digestion of glycophorins, *Brit. J. Haematol.* 84: 736.

Cartron, J.-P., Le Van Kim, C. and Colin, Y., 1993, Glycophorin C and related glycoproteins: structure, function and regulation, *Semin. Hematol.* 30: 152.

Cyster, J.G., Shotton, D.M. and Williams, A.F., 1991, The dimensions of the T lymphocyte glycoprotein leukosialin and identification of linear protein epitopes that can be modified by glycosylation, *EMBO J.* 10: 893.

Dahr, W., 1992, Miltenberger subsystem of the MNSs blood group system, *Vox Sang.* 62: 129.

Dahr, W. and Beyreuther, K., 1985, A revision of the N-terminal structure of sialoglycoprotein D (glycophorin C) from human erythrocyte membranes, *Biol. Chem. Hoppe-Seyler* 366: 1067.

Dahr, W., Beyreuther, K., Steinbach, H., Gielen, W. and Kruger, J. 1980, Structure of the Ss blood group antigens. II. A methionine/threonine polymorphism within the N-terminal sequence of the Ss glycoprotein. *Hoppe-Seyler's Z. Physiol. Chem.* 361: 895.

Dahr, W., Wilkinson, S., Issitt, P.D., Beyruther, K., Hummel, M. and Morel, P., 1986, High frequency antigens of human erythrocyte membrane sialoglycoproteins, III Studies on the En^aFR, Wr^a and Wr^b antigens, *Biol. Chem. Hoppe-Seyler*, 367: 1033.

Dahr, W., Blanchard, D., Kiedrowski, S., Poschmann, A., Cartron, J.P. and Moulds, J.J., 1989, High frequency antigens of human erythrocyte membrane sialoglycoproteins. VI. Monoclonal antibodies reacting with the N-terminal domain of glycophorin C, *Biol. Chem. Hoppe-Seyler* 370; 849.

Duk, M. and Lisowska, E., 1997, Subspecificities of anti-M. and anti-N antibodies tested with chemically modified antigens, *Transfus. Clin. Biol.* 1: 69.

Duk, M., Czerwinski, M. and Lisowska, E., 1992, Identification of an epitope recognized by the monoclonal antibody PEP80 in the C-terminal cytoplasmic fragment of glycophorin A, *Hybridoma*, 11: 181.

Duk, M., Sticher, U., Brossmer, R., and Lisowska, E., 1994, The differences in significance of α2,3Gal-linked and α2,6GalNAc-linked sialic acid residues in blood group M- and N-related epitopes recognized by various monoclonal antibodies, *Glycobiology* 4: 175.

Duk, M., Ugorski, M. and Lisowska, E., 1997, β-elimination of O-glycans from glycoproteins transferred to Immobilon P membranes: method and some applications, *Anal. Biochem.* 253: 98.

Gardner, B., Parsons, S.F., Merry, A.H. and Anstee, D.J., 1989, Epitopes on sialoglycoprotein α: evidence for heterogeneity in the molecule, *Immunology* 68: 283.

Geysen, H.M., Rodda, S.J., Mason, T.J., Tribbick, G. and Schoofs, P.G., 1987, Strategies for epitope analysis using peptide synthesis, *J. Immunol. Meth.* 102: 259.

Fukuda, M., 1993, Molecular genetics of the glycophorin A gene cluster, *Semin. Hematol.* 30: 138.

Fukuda, M., Lauffenburger, M., Sasaki, H., Rogers, M.E. and Dell, A., 1987, Structures of novel sialylated O-linked oligosaccharides isolated from human erythrocyte glycophorins, *J. Biol. Chem.* 262: 11952.

Huang, C-H., Johe, K.K., Seifter, S. and Blumenfeld, O.O., 1991, Biochemistry and molecular biology of MNSs blood group antigens, *Bailliere's Clin. Haematol.* 4: 821.

Huang, C-H., Reid, M.E., Xie, S-S., and Blumenfeld, O.O., 1996, Human red blood cell Wright antigens: a genetic and evolutionary perspective on glycophorin A - band 3 interaction, *Blood*, 87:3942.

Janvier, D., Veaux, S. and Benbunan, M., 1998, New murine monoclonal antibodies, directed against glycophorins C and D, have anti-Ge2 specificity, *Vox Sang.* 74: 101.

Jaskiewicz, E., Lisowska, E. and Lundblad, A., 1990a, The role of carbohydrate in the blood group N-related epitopes recognised by three new monoclonal antibodies, *Glycoconjugate J.* 7: 255.

Jaskiewicz, E., Moulds, J.J., Kraemer, K., Goldstein, A.S. and Lisowska, E., 1990b, Characterization of the epitope recognized by a monoclonal antibody highly specific for blood group M antigen, *Transfusion* 30: 230.

Jaskiewicz, E., Czerwinski, M., Syper, D. and Lisowska, E, 1994, Anti-M monoclonal antibodies crossreacting with variant M^g antigen: an example of modulation of antigenic properties of peptide by its glycosylation, *Blood* 84: 2340.

Jaskiewicz, E., Blanchard, D., Rasamoelisolo, M., Loirat, M., Moulds, J.J. and Lisowska E., 1999, Fine specificities of murine anti-M^g monoclonal antibodies, *Transfus. Med.* 9: 161.

King, M.-J., Holmes, C.H., Mushens, R.E., Mawby, W., Reid, M.E. and Scott, M.L., 1995, Reactivity with erythroid and non-erythroid tissues of a murine monoclonal antibody to a synthetic peptide having amino acid sequence common to cytoplasmic domain of human glycophorins C and D, *Brit. J. Haematol.* 89: 440.

King, M.-J., Kosanke, J., Reid, M.E., Poole, J., Banks, J., Hemming, N.J., Smythe, J. and Martin, P.G., 1997, Co-presence of a point mutation and a deletion of exon 3 in the glycophorin C gene and concomitant production of a Gerbich-related antibody, *Transfusion* 37: 1027.

Krotkiewski, H., 1988, The structure of glycophorins of animal erythrocytes, *Glyoconjugate J.* 5: 35.

Krotkiewski, H., Lisowska, E., Nilsson, G., Gronberg, G. and Nilsson, B., 1993, An improved approach to the analysis of the structure of small oligosaccharides of glycoproteins – application to the O-linked oligosaccharides from human glycophorin A, *Carbohyd. Res.* 239: 35.

Krotkiewski, H., Duk, M., Syper, D., Lis, H., Sharon, N. and Lisowska E., 1997, Blood group MN-dependent difference in degree of galactosylation of O-glycans of glycophorin A is restricted to the GalNAc residues located on the amino acid residues 2-4 of the glycophorin polypeptide chain, *FEBS Lett.* 406: 296.

Kusnierz-Alejska, G., Duk, M., Storry, J.R.,. Reid, M.E., Wiscek, B., Seyfried H. and Lisowska E., 1999, NOR polyagglutination and St^a glycophorin in one family. Relation of NOR polyagglutination to terminal α-galactose residues and abnormal glycolipids. *Transfusion* 39: 32.

Le Van Kim, C., Piller, V., Cartron, J.-P. and Colin, Y., 1996, Glycophorins C and D are generated by the use of alternative translation initiation sites, *Blood* 88: 2364.

Leddy, J.P., Wilkinson, S.L., Kissel, G.E., Passador, S.I., Falany, J.L. and Rosenfeld, S.I., 1994, Erythrocyte membrane proteins reactive with IgG (warm-reacting) anti-red blood cell autoantibodies: II. Antibodies coprecipitating band 3 and glycophorin A, *Blood* 84: 650.

Lisowska E., 1988, Antigenic properties of human erythrocyte glycophorins, in *The Molecular Immunology of Complex Carbohydrates,* (Wu, A.M., ed.), p. 265, Plenum, New York and London.

Lisowska, E., Messeter, L., Duk, M., Czerwinski, M. and Lundblad, A., 1987, A monoclonal anti-glycophorin A antibody recognizing the blood group M determinant: studies on the subspecificity, *Mol. Immunol.* 24: 605.

Loirat, M..J., Gourbil, A., Frioux, Y., Muller, J.Y. and Blanchard, D., 1992, A murine monoclonal antibody directed against the Gerbich 3 blood group antigen, *Vox Sang.* 62: 45.

Loirat, M.J., Dahr, W., Muller, J.Y. and Blanchard, D., 1994, Characterization of new murine monoclonal antibodies directed against glycophorin C and D, *Transfus. Med.* 4: 147.

Loirat, M.J., Czerwinski, M., Duk, M. and Blanchard, D., 1999, The murine monoclonal antibody NAM26-4C6 identifies a common structure on band 3 and glycophorin C, *Transfus. Med.* 9: 69.

Pahlsson, P., Blackall, D.P., Ugorski, M., Czerwinski, M. and Spitalnik, S.L., 1994, Biochemical characterization of the O-glycans on recombinant glycophorin A expressed in Chinese hamster ovary cells, *Glycoconjugate J.* 11: 43.

Pisano, A., Redmond, J.W., Williams, K.L. and Gooley, A.A., 1993, Glycosylation sites identified by solid-phase Edman degradation: O-linked glycosylation motifs on human glycophorin A, *Glycobiology* 3: 429.

Podbielska, M. and Krotkiewski, H., 2000, Identification of blood group A and B antigens in human glycophorin, *Arch. Immunol. Ther. Exp.* 48: 211.

Rasamoelisolo, M., Czerwinski, M., Bruneau, V., Lisowska, E. and Blanchard, D., 1997, Fine characterization of a series of new monoclonal antibodies directed against glycophorin A., *Vox Sang.* 72: 185.

Rasamoelisolo, M., Czerwinski, M., Willem, C. and Blanchard, D., 1998, Shared epitopes of glycoprotein A and protein 4.1 defined by antibody NAM10-3C10, *Hybridoma* 17: 283.

Rearden, A., 1993, Immunogenecity and antigenicity of defined linear determinants of the MN sialoglycoprotein glycophorin A, *Transfusion*, 33: 476.

Reid, M.E., 1994, Some concepts relating to the molecular genetic basis of certain MNSs blood group antigens, *Transfus. Med.* 4: 99.

Reid, M.E. and Lomas-Francis, C., 1997, The Blood Group Antigen Factsbook, Academic Press, San Diego and London.

Reid, M.E. and Spring, F.A., 1994, Molecular basis of glycophorin C variants and their associated blood group antigens, *Transfus. Med.* 4: 139.

Reid, M.E., Lisowska, E. and Blanchard, D., 1997, Report on Section 2B: glycophorin/band 3 and associated antigens (Third International Workshop and Symposium on Monoclonal Antibodies against Human Red Blood Cells and Related Antigens), *Transfus. Clin. Biol.* 1: 57.

Remaley, A.T., Ugorski, M., Wu, N., Litzky, L., Burger, S.R., Moore, J.S., Fukuda, M. and Spitalnik, S.L., 1991, Expression of human glycophorin A in wild type and glycosylation-deficient Chinese hamster ovary cells. Role of N-linked and O-linked glycosylation in cell surface expression, *J. Biol. Chem.* 266: 24176.

Ring, S.M., Tippett, P. and Swallow, D.M., 1994, Comparative immunochemical analysis of Wr^a and Wr^b red cell antigens, *Vox Sang.* 67: 226.

Smythe, J., Gardner, B. and Anstee, D.J., 1994, Quantitation of the number of molecules of glycophorins C and D on normal red blood cells using radioiodinated Fab fragments of monoclonal antibodies, *Blood* 83: 1668.

Telen, M.J. and Chasis, J.A., 1990, Relationship of the human erythrocyte Wr^b antigen to an interaction between glycophorin A and band 3, *Blood* 76: 842.

Tippett, P., Reid, M.E., Poole, J., Green, C.A., Daniels, G.L. and Anstee, D.J., 1992, The Miltenberger susbsystem: is it obsolescent? *Transfus. Med. Rev.* 6: 170.

Tomita, M., Furthmayr, H. and Marchesi, V.T., 1978., Primary structure of human erythrocyte glycophorin A. Isolation and characterization of peptides and complete amino acid sequence, *Biochemistry* 17: 4756.

Ugorski, M., Blackall, D.P., Pahlsson, P., Shakin-Eshleman, S.H., Moore, J.S. and Spitalnik, S.L., 1993, Recombinant Miltenberger I and II human blood group antigens: the role of glycosylation in cell surface expression and antigenicity of glycophorin A, *Blood* 82: 1913.

Wasniowska, K., Duk, M., Czerwinski, M., Steuden, I., Dus, D., Radzikowski, C., Bartosz-Bechowski, H., Konopinska, D. and Lisowska, E., 1992, Analysis of peptidic epitopes recognized by the three monoclonal antibodies specific for the same region of glycophorin A but showing different properties, *Molec. Immunol.* 29: 783.

Wasniowska, K., Czerwinski, M., Halasa, J., and Lisowska, E., 1993, An immunoblotting procedure for screening glycophorins and band 3 protein in the same blots. Identification of glycophorin and band 3 variant forms, *J. Immunol. Meth.* 160: 253.

Wasniowska, K., Jaskiewicz, E., Czerwinski, M., Syper, D. and Lisowska, E., 1997, Mapping of peptidic epitopes of glycophorins A (GPA) and C (GPC) with peptides synthesized on plastic pins (Pepscan analysis), *Transfus. Clin. Biol.* 1: 73.

Wasniowska, K., Czerwinski, M., Jachymek, W. and Lisowska E., 2000, Expression and binding properties of a soluble chimeric protein containing the N-terminal domain of the Duffy antigen. *Biochem. Biophys. Res. Commun.* 273: 705.

Wu, A.M., Duk, M., Lin, M., Broadberry, R.E. and Lisowska, E., 1995, Identification of variant glycophorins of human red cells by lectinoblotting: application to the Mi.III variant that is relatively frequent in the Taiwanese population, *Transfusion* 35: 571.

Zdebska, E. and Koscielak, J., 1999, A single-sample method for determination of carbohydrate and protein contents glycoprotein bands separated by sodium dodecylsulfate-polyacrylamide gel electrophoresis, *Anal. Biochem.* 275: 171.

KDN-CONTAINING GLYCOPROTEIN FROM LOACH SKIN MUCUS

Hiroki Nakagawa,[1] Yoichiro Hama,[1] Toshihisa Sumi,[1] Su-Chen Li,[2] and Yu-Teh Li[2]

[1]*Department of Applied Biological Sciences, Faculty of Agriculture, Saga University, Saga 840-8502, Japan*
[2]*Department of Biochemistry, Tulane University School of Medicine, New Orleans, Louisiana 70112, U.S.A.*

The mucus coat of fish skin contains a variety of secretions from epidermal goblet cells and epithelial cells. These secretions have been implicated in many important biological functions (1-3). In spite of its biological importance, very little is known about the chemical nature of fish skin mucus. It has been shown that the fish mucus secretions contain mucin type glycoproteins with sugar compositions similar to that of the mammalian epithelial secretions (4, 5). The presence of sialic acid in fish skin mucus has been widely reported (1, 4-9). While studying the sugar compositions of various fish skin mucus (10), we detected a thiobarbituric acid positive compound which was subtly different from NeuAc and NeuGc in the skin mucus of loach, *Misgurnus anguillicaudatus*. The structure of this compound was subsequently established to be 2-keto-3-deoxy-D-*glycero*-D-*galacto*-nononic acid (KDN) by NMR spectroscopy and mass spectrometry (11). KDN (Figure 1), a deaminated neuraminic acid analogue, was first isolated from the polysialoglycoproteins of rainbow trout eggs (12, 13). Subsequently, KDN-containing glycoconjugates have been isolated from the vitelline envelope (14, 15), ovarian fluid (16), and sperm (17) of rainbow trout, the jelly coat of amphibian eggs (18-20), and the capsule of the bacterium *Klebsiella ozaenae* (21). Loaches are freshwater fish that live in rivers and ponds and often bury themselves under the mud. The degenerate scales of loach are buried in the skin. They are widely distributed in Asia and Europe. In Japan, they are cultured in ponds and used for food. The presence of KDN in loach skin mucus represented the first report of the occurrence of KDN in animal tissues not related to the reproductive systems. Since we were intrigued by the finding that over 95 % of the sialic acid in the loach skin mucus was KDN, we have isolated KDN-containing glycoprotein in a highly purified form from the loach skin mucus and characterized its chemical nature and structure.

Abbreviations used: KDN, 2-keto-3-deoxy-D-*glycero*-D-*galacto*-nononic acid that is synonymous with 3-deoxy-D-*glycero*-D-*galacto*-nonulosonic acid, NeuAc N-acetyl-neuraminic acid; NeuGc, *N*-glycolyl-neuraminic acid; GalNAc-ol, *N*-acetylgalactosaminitol: GalNAc, *N*-acetylgalactosamine; GlcNAc, *N*-acetyl-glucosamine; Gal, galactose; Fuc, fucose; GLC, gas liquid chromatography; TLC, thin layer chromatography; HPLC, high performance liquid chromatography; LSIMS, liquid secondary ion mass spectrometry.

The Molecular Immunology of Complex Carbohydrates–2.
Edited by Albert M. Wu, Kluwer Academic/Plenum Publishers, 2001.

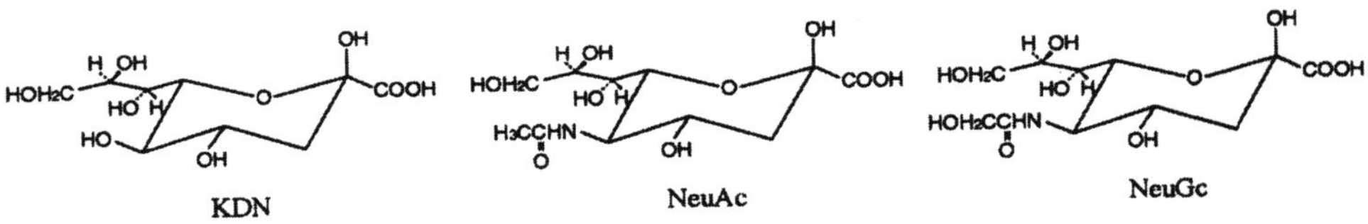

Figure 1. Structures of 2-keto-3-deoxy-D-*glycero*-D-*galacto*-nononic acid (KDN), *N*-acetylneuraminic acid (NeuAc) and *N*-glycolylneuraminic acid (NeuGc).

MATERIALS AND METHODS

Cultured loach were obtained directly from a local aquaculture farm.

Component sugars were determined as their trimethylsilyl derivatives by GLC after methanolysis and reacetylation as described previously (11,22). The release of ketosidically linked sialic acids was achieved by heating the samples with 0.1 N H_2SO_4 at 80°C for 1 h. Free sialic acids were determined by the modified periodate-thiobarbituric acid method described by Uchida *et al.* (23). Oligosaccharide alditols were also analyzed by TLC using silica gel-60 plates and n-butanol/acetic acid/ water (2:1:1, v/v/v) as the developing solvent. The plate was sprayed with diphenylamine reagent (24) and heated at 110°C for 20 min to reveal oligosaccharide alditols. Amino acid composition was determined by using a JOEL automatic amino acid analyzer. Protein samples were hydrolyzed in 6 N HCl at 110°C for 24, 48 and 72 h under vacuum. Corrections were made for decomposition of amino acids during hydrolysis.

HISTOCHEMICAL ANALYSIS OF LOACH SKIN TISSUES

Figures 2-a and 2-b show the cross sections of loach skin stained with periodate-Schiff reagent and toluidine blue, respectively. Loach skin consists of an epidermis and a dermis which are separated by basement membrane. The epidermis contains goblet and clavate cells. These mucus cells differentiate from basement membrane cells. Mature goblet cells were stained bluish red with periodate-Schiff reagent and stained reddish blue with toluidine blue, indicating that mature goblet cells (upper cells) are filled with sialic acid-containing glycoprotein (25).

PREPARATION OF LOACH SKIN MUCUS

One hundred live loaches (700 g) were first placed in plastic bag and kept frozen at -25°C overnight. Then, they were immersed in 3 liters of 50 % ethanol. This treatment caused the skin mucus to coagulate into a jelly that could be easily removed from the skin by scraping. The jelly like mucus was collected by filtration using a Buchner funnel, washed with ethanol and acetone, and dried to obtain 4 g of loach skin mucus powder.

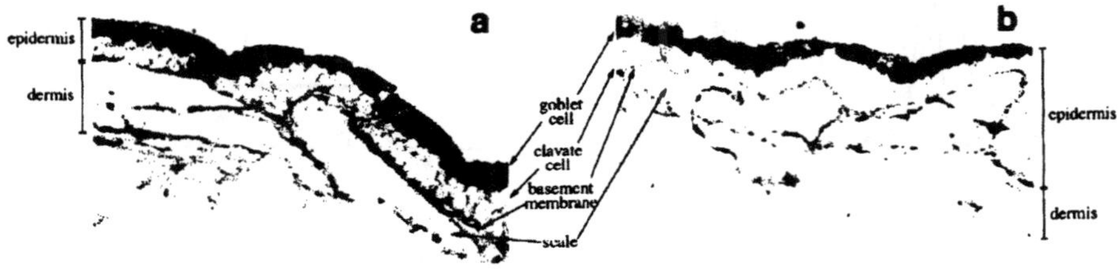

Figure 2. Histochemical observation of loach skin tissues (x 200). (a) Loach skin section stained with periodate-Schiff reagent. (b) Loach skin section stained with toluidine blue reagent.

ISOLATION OF KDN-CONTAINING GLYCOPROTEIN FROM LOACH SKIN MUCUS

The loach skin mucus powder (2 g, containing 135 mg of bound KDN) was suspended in 700 ml of 0.05 M Tris-HCl buffer, pH 8.0, and homogenized using a generator type homogenizer. The homogenate was centrifuged at 20,000 x g for 30 min. The clear supernatant which contained 122 mg of protein-bound KDN was applied to a DE52-cellulose column (3.0 x 47 cm) which had been equilibrated with 0.05 M Tris-HCl buffer, pH 8.0. After washing with this buffer, the column was eluted with a 0-0.5 M NaCl gradient in 0.05 M Tris-HCl buffer, pH 8.0 (total volume 2 liters) at 80 ml/h. The elution profile is shown in Figure 3-a. The fractions which contained KDN were pooled, concentrated, and equilibrated with 0.05 M sodium acetate buffer, pH 5.7, in an Amicon ultrafiltration stirred cell using a PM-10 membrane to give a final volume of 20 ml. This preparation was then digested with 200 units of Nuclease P1 at 37°C for 24 h in the presence of toluene. After incubation, the incubation mixture was heated at 100°C for 3 min to inactivate the enzyme, and centrifuged to remove the insoluble materials. The clear supernatant was subsequently applied to a Sepharose CL-4B column (4.0 x 80 cm) which had been equilibrated with 0.05 M Tris-HCl buffer, pH 8.0, containing 0.1 M NaCl. The column was eluted with the same buffer at 30 ml/h. The elution profile is shown in Figure 3-b. From this gel filtration, the apparent molecular weight of this glycoprotein was estimated to be 600,000 using pullulan molecular weight markers. The KDN-containing fractions were pooled, dialyzed against water, and lyophilized to obtain KDN-containing glycoprotein (210 mg).

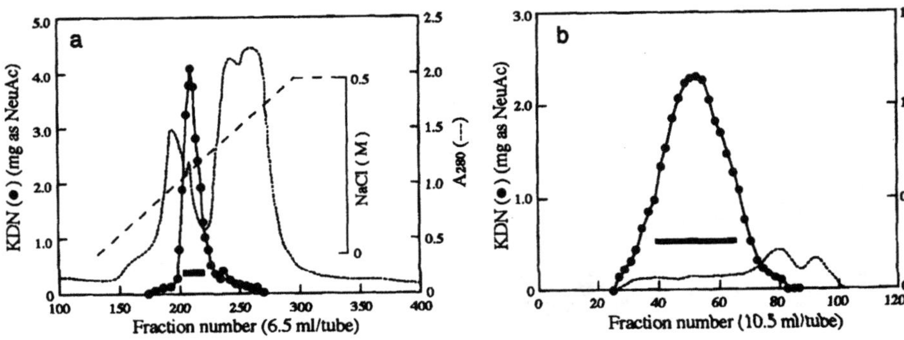

Figure 3. Isolation of KDN-containing glycoprotein from loach skin mucus. Detailed experimental conditions are described in the text.

SUGAR AND AMINO ACID COMPOSITIONS OF LOACH SKIN MUCUS GLYCOPROTEIN

Sugar Composition

The sugar composition and amino acid content of loach skin mucus glycoprotein are shown in Table I. For comparison, the compositions of the glycoproteins prepared from the skin mucus of six other species of fishes are also included in this table. Loach mucus glycoprotein contains more than 65 % of carbohydrate by weight in which KDN (38.5 %) and GalNAc (25.0 %) are the major constituents. The amino sugar in this glycoprotein consists exclusively of GalNAc. The other sugars detected are Gal (3.5 %), GlcNAc (0.5 %) and NeuAc (0.5 %). On the other hand, mucus glycoproteins of eel, rainbow trout and stingray contain exclusively NeuAc. Although KDN was first isolated from rainbow trout eggs, the skin mucus glycoprotein of rainbow trout does not contain KDN. Skin mucus glycoproteins from carp and wrasse (a marine fish) were found to contain a comparatively large amount of KDN. It should be noticed that loach and carp belong to the same family although they differ each other in their appearance. The sugar compositions of eel and rainbow trout are similar to that of ovine submaxillary glycoprotein (26). It appears that NeuAc in those glycoproteins is replaced by KDN in the loach skin mucus glycoprotein.

Amino Acid Composition before and after Mild Alkaline Treatment

Table II shows the amino acid compositions of loach mucus glycoprotein before and after mild alkaline treatment with 0.3 N NaOH at 30°C for 48 h. The protein core of this glycoprotein accounts for only 28 % of the dried weight. The glycoprotein is unusually rich in Thr (32 residues/100 residues) and 82 out of 100 residues of amino acids are Thr, Ser, Glu (or Gln), Pro, Gly and Ala. The sum of Thr and Ser that are available to link with sugar chains through GalNAc, occupy nearly half of the total amino acids. It is also noteworthy that this glycoprotein is extremely poor in Cys, Met, Tyr and Phe. Trp was not detected in

Table I
Sugar compositions and amino acid contents of mucus glycoproteins from seven species of fishes.

	Sugar compositions (%)							Amino acid	Total
	Fuc	Gal	GlcNAc	GalNAc	KDN	NeuAc	NeuGc	(%)	(%)
Loach (*Misgurnus anguillicaudatus*)	0.0	3.5	0.5	25.0	38.5	0.5	0.0	28.2	96.2
Eel (*Anguilla japonica*)	0.0	6.4	0.0	26.4	trace	30.8	0.4	25.1	89.1
Rainbow trout (*Salmo gairdneri*)	0.0	5.0	0.0	26.0	trace	30.1	trace	26.0	87.1
Carp (*Cyprinus carpio*)	1.2	9.7	8.8	23.7	14.6	2.6	0.4	27.1	88.1
Crucian carp (*Carassius auratus*)	0.1	3.8	0.0	27.8	trace	28.3	2.1	28.6	90.7
Stingray (*Dasyatis akajei*)	0.0	6.1	9.9	24.4	0.1	22.1	1.1	27.4	91.1
Wrasse (*Halichoeres tenuispinis*)	0.0	5.1	6.9	22.2	7.3	22.0	9.4	25.6	98.5

Table II
Amino acid compositions of loach mucus glycoprotein before and after alkaline treatment.

Amino acid	Before		After	
	μg/mg	res/100res	μg/mg	res/100res
Asx	6.9	2.1	7.0	3.2
Thr	93.0	32.3	16.5	8.7
Ser	28.5	11.5	10.4	6.4
Glx	37.4	10.2	37.9	15.7
Pro	31.9	11.5	30.6	16.8
Gly	14.0	8.6	13.9	13.0
Ala	12.3	6.1	12.2	9.2
Cys	2.0	0.7	2.0	1.1
Val	9.5	3.4	9.6	5.1
Met	1.6	0.4	1.6	0.7
Ile	5.4	1.7	5.3	2.5
Leu	8.7	2.7	8.4	4.0
Tyr	3.1	0.7	3.0	1.0
Phe	3.4	0.8	3.4	1.2
His	9.4	2.4	9.5	3.7
Lys	11.6	3.2	11.8	4.9
Arg	8.4	1.9	8.4	2.9
Trp	0.0	0.0	0.0	0.0
Total	287.0	100.0	191.6	100.0

the glycoprotein by using the method of Sartin *et al.* (27). The extremely low UV absorbance of the glycoprotein (Figures 3-b, 4-b) agrees with the fact that the glycoprotein is extremely poor in aromatic amino acids.

Mild alkaline treatment of this glycoprotein resulted in the destructions of 82 % of Thr and 64 % of Ser, suggesting that the majority of Thr and Ser are involved in the linkage of sugar chains to the protein core through GalNAc.

ISOLATION OF GLYCOPEPTIDE FROM LOACH MUCUS GLYCOPROTEIN AND ITS SUGAR AND AMINO ACID COMPOSITIONS

Isolation of Glycopeptide

Loach mucus glycoprotein (100 mg) dissolved in 5 ml of 0.05 M Tris-HCl buffer, pH 7.8, containing 0.02 M calcium acetate was exhaustively digested with 10 mg in total of Actinase E at 37°C for 72 h in the presence of toluene. Actinase (3-4 mg x 3 times) were added at 24-hour intervals. After digestion, the mixture was heated at 100°C for 2 min and centrifuged to remove the precipitate. The supernatant was applied to a DEAE-Sephadex A-25 column (2.2 x 19 cm) which had been equilibrated with 0.05 M Tris-HCl buffer, pH 7.8. After washing with this buffer, the column was eluted with a 0-0.5 M NaCl gradient in the same buffer. The elution profile is shown in Figure 4-a. The fractions containing KDN were pooled, concentrated and equilibrated with 0.1 M NH_4HCO_3 by ultrafiltration with a YM-3 membrane to give a final volume of 3 ml. The concentrate was applied to a Sephacryl S-200 HR column (2.2 x 90 cm) which had been equilibrated with 0.1 M NH_4HCO_3. The column was eluted with the same solution. The elution profile is shown in Figure 4-b. The fractions containing KDN were pooled and lyophilized to obtain 76 mg of glycopeptide. From this gel filtration, the apparent molecular weight of this glycopeptide was estimated to be 50,000 using pullulan molecular weight markers. This suggests that the molecular size of the glycopeptide is approximately 1/12 of the intact glycoprotein.

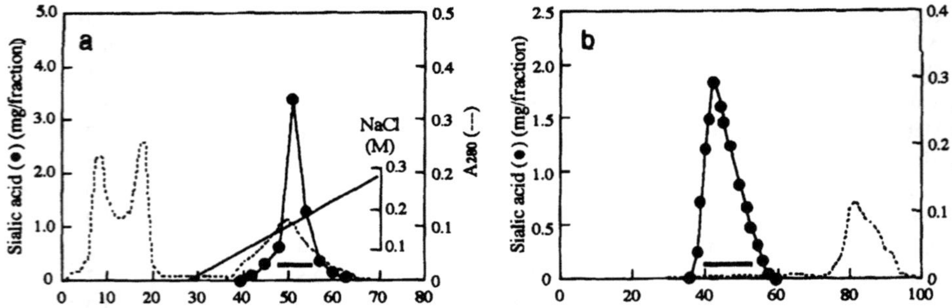

Figure 4. Isolation of glycopeptide from loach mucus glycoprotein. Detailed experimental conditions are described in the text.

Sugar and Amino Acid Compositions

Table III shows the sugar composition of the glycopeptide, together with that of the intact glycoprotein. The sugar content of the glycopeptide was considerably higher than that of the intact glycoprotein. There was no appreciable loss of sugar constituents after Actinase digestion.

Table III

Sugar compositions of loach mucus glycoprotein before and after Actinase digestion.

	Sugar composition (%)					Molecular weight[a]
	Gal	GlcNAc	GalNAc	KDN	NeuAc	
Glycoprotein	3.5	0.5	25.0	38.5	0.5	600,000
Glycopeptide	3.7	0.5	28.7	43.3	0.5	50,000

[a] Apparent molecular weights measured by gel filtration as described in the text.

Table IV

Amino acid compositions of loach mucus glycoprotein before and after Actinase digestion.

Amino acid	Glycoprotein		Glycopeptide	
	μg/mg	res/100res	μg/mg	res/100res
Asx	6.9	2.1	2.4	0.9
Thr	93.0	32.3	93.7	38.1
Ser	28.5	11.5	28.7	13.5
Glx	37.4	10.2	33.3	10.6
Pro	31.9	11.5	29.2	12.3
Gly	14.0	8.6	12.0	8.6
Ala	12.3	6.1	10.8	6.2
Cys	2.0	0.7	0.0	0.0
Val	9.5	3.4	8.5	3.5
Met	1.6	0.4	0.5	0.0
Ile	5.4	1.7	3.0	1.1
Leu	8.7	2.7	2.5	0.9
Tyr	3.1	0.7	0.0	0.0
Phe	3.4	0.8	0.0	0.0
His	9.4	2.4	6.1	1.8
Lys	11.6	3.2	5.0	1.6
Arg	8.4	1.9	3.2	0.8
Trp	0.0	0.0	0.0	0.0
Total	287.0	100.0	238.4	100.0

Table IV shows the amino acid composition of the glycopeptide, together with that of the intact glycoprotein. The total amino acid content was decreased by only 5 % in the glycopeptide. The contents of Thr and Ser in the glycopeptide became significantly higher than those of the intact glycoprotein. The glycopeptide was devoid of the minor amino acids, Cys, Met, Tyr and Phe that were found in the intact glycoprotein, suggesting that the protease hydrolyzed the polypeptide region containing these amino acids. It is evident that the glycopeptide is a polypeptide bearing a number of KDN-containing sugar chains that protected the core polypeptide against the protease digestion. From these results and molecular sizes of the glycoprotein and glycopeptide, it is estimated that the loach mucus glycoprotein contains approximately 11 highly glycosylated polypeptides (protease-resistant) which are linked in tandem through nonglycosylated peptides (protease-susceptible) to form a glycoprotein molecule. Since the glycopeptide is devoid of the minor amino acids, these nonglycosylated minor amino acids must be involved in linking the glycosylated peptides.

ISOLATION OF SUGAR CHAINS AS OLIGOSACCHARIDE ALDITOLS FROM LOACH MUCUS GLYCOPROTEIN

Loach mucus glycoprotein (100 mg) was dissolved in 50 ml of 1 M $NaBH_4$/0.05 M potassium hydroxide and incubated at 37°C for 30 h (28). The mixture was neutralized with acetic acid, diluted with ten volumes of distilled water and then passed through a Dowex 50-X8 column (H^+ form, 3.0 x 10 cm) in the cold room. The effluent was neutralized and lyophilized. The boric acid in this fraction was removed by repeated codistillation with methanol. The remaining residue was dissolved in 1 ml of 0.1 M NH_4HCO_3, applied to a Sephadex G-25 column (superfine, 1.8 x 132 cm) which had been equilibrated with 0.1 M NH_4HCO_3 and eluted with the same solution. The sugar chains (oligosaccharide alditols) liberated from the glycoprotein were resolved into three KDN-containing fractions, A, B and C by Sephadex G-25 gel filtration (Figure 5–a). Fractions A and C (the major sugar chains) gave only one spot by TLC analysis. Fraction B was found to be a mixture of four minor sugar chains when analyzed by TLC. To separate minor sugar chains by HPLC, Fraction B was dissolved in 1 ml of 30 mM potassium phosphate buffer (pH 5.2)/ acetonitrile (3:7, v/v) and aliquots (100 μl) were injected to a Shimadzu LC-9A liquid chromatograph using a μ-Bondapak NH_2 column (0.8 x 10 cm, Waters). Elution was carried out with the same solution. The eluate was monitored by adsorption at 210 nm for the acetamide group of oligosaccharide alditols. Fraction B was resolved into four sugar chains (Figure 5-b). Fractions B1, B2, B3 and B4 were separately pooled and evaporated to dryness. The remaining residues were dissolved in 0.5 ml of water and desalted by Sephadex G-15 gel filtration. Sugar chains A and C represent 55 % and 40 % (by weight), respectively , of total amounts of liberated oligosaccharide alditols, and the sum of sugar chains B1, B2, B3 and B account for only 5 %.

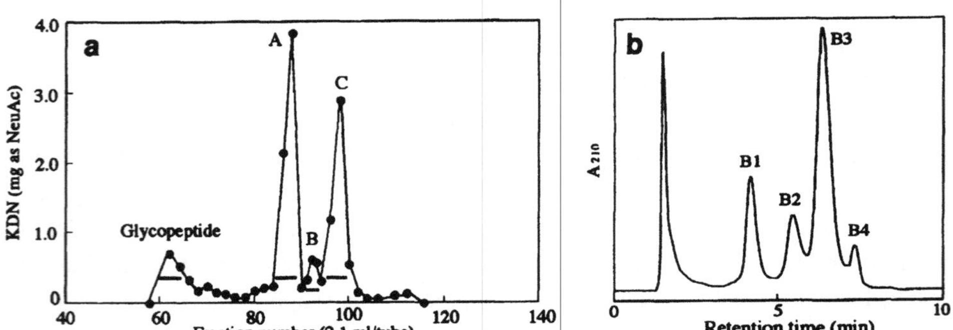

Figure 5. Separation of sugar chains (oligosaccharide alditols) liberated from loach mucus glycoprotein. (a) Separation of sugar chains by Sephadex G-25 gel filtration. (b) Separation of fraction B by HPLC. Detailed conditions are described in the text.

CHARACTERIZATION OF SUGAR CHAINS LIBERATED FROM LOACH MUCUS GLYCOPROTEIN

Major sugar chains A and C and minor sugar chains B1, B2, B3 and B4 were analyzed for sugar compositions by GLC and the results are summarized in Table V. Sugar chain C contained KDN and GalNAc-ol in a molar ratio of 1:1. By LSIMS, C gave m/z 472 for $[M-H]^-$ indicating that C is KDN-GalNAc-ol. By permethylation analysis, the structure of C was determined to be KDNα2→6GalNAc-ol. Sugar chain A contained KDN and GalNAc-ol in a molar ratio of 2:1. By LSIMS, A gave m/z 722 for $[M-H]^-$ indicating that A is $(KDN)_2$GalNAc-ol. By permethylation analysis, the structure of A was concluded to be KDNα2→3(KDNα2→6)GalNAc-ol. Sugar chains A and C were refractory to sialidases from *Arthrobacter urefaciens* and *Clostridium perfringens*, but they are hydrolyzed by KDN-sialidases from loach liver (29) and starfish entrails (30) to yield KDN, GalNAc-ol and KDN-GalNAc-ol (intermediate products from A). Sugar chain A is a novel structure that has not been reported previously. The structures of minor sugar chains B1, B2, B3 and B4 were also determined by combination of permethylation analysis, ^{1}H-NMR and glycosidase digestion to be NeuAcα2→6GalNAc-ol, KDNα2→3(GalNAcβ1→4) GalNAc-ol, KDNα2→6(GalNAcα1→3)GalNAc-ol and KDNα2→6(Galβ1→3)GalNAc-ol, respectively. Although both B2 and B3 have the same sugar composition and molecular weight, they differ in the linkages between component sugars. By ^{1}H-NMR spectroscopy, the anomeric proton signals of the GalNAc of B2 and B3 are observed at 4.6 and 5.2 ppm and the coupling constants of proton signals on C-1 and C-2 are 8.8 and 3.7 Hz, indicating that the anomeric forms of the GalNAc residues of B2 and B3 are β and α, respectively. B3 is hydrolyzed by α-*N*-acetylgalactosaminidase from skipjack liver (31) to give GalNAc and KDN-GalNAc-ol, but B2 is not hydrolyzed. Established structures of isolated sugar chains are shown in Figure 6.

Sugar chains	Structures
A	KDNα2 → 6 KDNα2 → 3 GalNAc-ol
C	KDNα2 → 6GalNAc-ol
B1	NeuAcα2 → 6GalNAc-ol
B2	KDNα2 → 3 GalNAcβ1 → 4 GalNAc-ol
B3	KDNα2 → 6 GalNAcα1 → 3 GalNAc-ol
B4	KDNα2 → 6 Galβ1 → 3 GalNAc-ol

Figure 6. Structures of sugar chains isolated from loach mucus glycoprotein.

Table V

Molar ratios of component sugars of sugar chains.

Sugar chains	KDN	NeuAc	Gal	GalNAc	GalNAc-ol
A	1.9				1.0
B1		1.0			1.0
B2	0.9			0.9	1.0
B3	1.0			1.0	1.0
B4	0.9		1.0		1.0
C	1.0				1.0

OVERALL STRUCTURE OF LOACH MUCUS GLYCOPROTEIN

The sugar chains were liberated as oligosaccharide alditols from loach mucus glycoprotein by alkaline borohydride treatment as described above. After this step, the liberated oligosaccharide alditols were analyzed for GalNAc and GalNAc-ol by GLC and 88 % of the GalNAc originally present in the glycoprotein was found as GalNAc-ol. Considering the incomplete reaction of β-elimination (32), it is estimated that more than 90 % of the GalNAc are involved in the glycosidic linkage to the protein core.

Exhaustive digestion of the glycoprotein by Actinase resulted in obtaining an extremely highly glycosylated polypeptide preparation with an apparent molecular weight of approximately 50,000 (Figure 4-b). This apparent molecular weight was 1/12 of that of the intact glycoprotein. The glycopeptide contained 28.7 % of GalNAc (Table III) and 9.4 % of Thr and 2.9 % of Ser (Table IV). Assuming that the molecular weight of this glycopeptide is 50,000, one mole of the glycopeptide is estimated to contain 65 moles of GalNAc and 53 moles of these hydroxyl amino acids. Since the majority of the GalNAc and hydroxyl amino acids are involved in the glycosidic linkage to the polypeptide, each glycopeptide molecule has more than 50 KDN-containing oligosaccharide chains that are linked to the Thr and Ser residues of the polypeptide through GalNAc. These sugar chains are spaced at an average of 2 amino acids apart on the polypeptide.

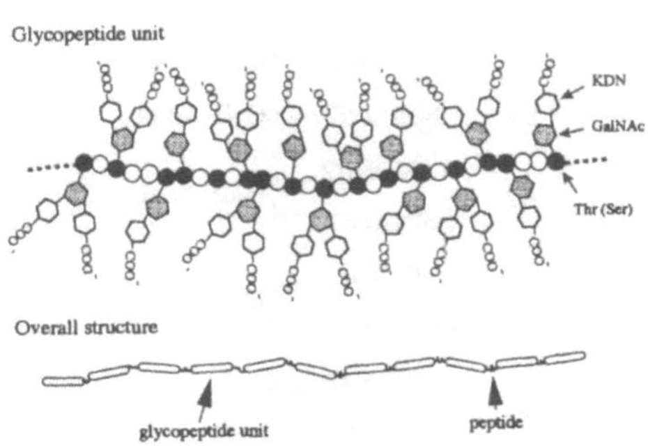

Figure 7. A model structure of loach mucus glycoprotein.

From the molecular sizes of the glycoprotein (apparent molecular weight, 600,000) and glycopeptide (50,000), a molecule of loach mucus glycoprotein is considered to consist of approximately 11 glycopeptide units that are linked in tandem through nonglycosylated peptides (Figure 7). This overall structure is essentially in agreement with mucin oligomer structures proposed for human cervical mucin (33) and porcine submaxillary mucin (34), in which several highly glycosylated polypeptides and nonglycosylated polypeptides are attached in tandem to form mucin oligomers. However, it should be pointed out that loach mucus glycoprotein bears large numbers (more than 500) of KDN-containing sugar chains, as compared with human cervical mucin and porcine submaxillary mucin.

CONCLUDING REMARKS

We have so far studied the skin mucus glycoproteins isolated from more than 10 species of fishes (10, 22, 35, 36). In the majority of cases, sialic acids of fish skin mucus glycoproteins were NeuAc (major) and NeuGc (minor). In contrast, loach skin mucus glycoprotein is exceptionally rich in KDN(38.5 %) and very poor in other sialic acids.

A molecule of loach mucus glycoprotein is considered to be composed of approximately 11 glycopeptide units. Each glycopeptide unit has approximately 50 KDN-containing sugar chains. It should be pointed out that the glycopeptide unit is extremely refractory to proteases and KDN-capped sugar chains are also resistant to bacterial sialidases. The multiple, mutually repelling and highly hydrated negative charges of the KDN-containing sugar chains would give the glycoprotein an extended rod-like conformation, which is largely responsible for the characteristic viscoelastic property in an aqueous solution. The fish skin mucus glycoproteins interact noncovalently with mucus cell secretions, cellular macromolecules, electrolytes, and remnants of epithelial cells to form a mucus gel layer on fish skin. The mucus coat of fish is in constant contact with the external aqueous environment. One of the important roles of the fish skin mucus is considered to be a biochemical barrier against microorganisms and parasites in the environmental water.

SUMMARY

It has been widely recognized that the mucus coat of fish plays a variety of important physical, chemical, and physiological functions. One of the major constituents of the mucus coat is mucus glycoprotein. We found that sialic acids in the skin mucus of the loach, *Misgurnus anguillicaudatus*, consisted predominantly of KDN. Subsequently, we isolated KDN-containing glycoprotein from loach skin mucus and characterized its chemical nature and structure.

Loach mucus glycoprotein was purified from the Tris-HCl buffer extract of loach skin mucus by DEAE-cellulose chromatography, Nuclease P1 treatment, and Sepharose CL-6B gel filtration. The purified mucus glycoprotein was found to contain 38.5 % KDN, 0.5 % NeuAc, 25.0 % GalNAc, 3.5 % Gal, 0.5% GlcNAc and 28 % amino acids. Exhaustive Actinase digestion of the glycoprotein yielded a glycopeptide with a higher sugar content and higher Thr and Ser contents. The molecular size of this glycopeptide was approximately 1/12 of the intact glycoprotein. These results suggest that approximately 11 highly glycosylated polypeptide units are linked in tandem through nonglycosylated peptides to form the glycoporotein molecule.

The oligosaccharide alditols liberated from the loach mucus glycoprotein by alkaline borohydride treatment were separated by Sephadex G-25 gel filtration and HPLC. The purified sugar chains were analyzed by permethylation analysis, ^{1}H-NMR spectroscopy, and glycosidase digestion. Their structures were established to be: KDNα2→6(KDNα2 →3)GalNAc-ol, KDNα2→6GalNAc-ol, KDNα2→3(GalNAcβ1→4)GalNAc-ol, KDNα2 →6(GalNAcα1→3)GalNAc-ol, KDNα2→6(Galβ1→3)GalNAc-ol, and NeuAcα2→6Gal NAc-ol. It is estimated that one loach mucus glycoprotein molecule contains more than 500 KDN-containing sugar chains that are linked to Thr and Ser residues of the protein core through GalNAc.

REFERENCES

1. Whitear, M., 1984, Epidermis, in: *Biology of the Integument 2, Vertebrates*, J. Bereiter-Hahn, A.G. Motoltsy, K.S. Richards, eds., Springer-Verlog, Berlin.
2. Rosen, M.W., and Cornford, N.E., 1971, Fluid friction of fish slimes, *Nature* 234: 49.
3. Oosten, J.V., 1957, The skin scales, in : *The Physiology of Fishes, Vol. 1*, pp. 207-244, M.E. Brown, ed., Academic Press, New York.
4. Faillard, H., and Schauer, R., 1972, Glycoproteins as lubricants, protective agents, carriers, structural proteins, and as participants in other functions, in: *Glycoproteins Part B*, pp. 1246-1267, A. Gottschalk, ed., Elsevier, Amsterdam.
5. Buddecke, E., 1972, Occurrence in animals and plants, in: *Glycoproteins Part A*, pp. 535-564, A. Gottschalk, ed., Elsevier, Amsterdam.
6. Pickering, A.D., 1974, The distribution of mucous cells in the epidermis of the brown trout *Salmo trutta* and the char *Salrelinus alpinus*, *J. Fish Biol.* 6: 111.
7. Wold, J.K., and Selset, R., 1977, Glycoproteins in the skin mucus of the char (*Salmo alpinus*), *Comp. Biochem. Phys.* 56B: 21.
8. Asakawa, M., 1974, Sialic acid-containing glycoprotein in the external mucus of eel, *Anguilla japonica* -II Carbohydrate and amino acid compositions, *Nippon Suisan Gakkaishi* 40: 303.
9. Asakawa, M., 1983, Structure of disaccharide from eel skin mucous glycoprotein and its linkage to protein moiety, *Nippon Suisan Gakkaishi* 49: 1601.
10. Nakagawa, H., Asakawa, M., and Enomoto, N., 1988, Diversity in the carbohydrate moieties of mucus glycoproteins of various fishes, *Nippon Suisan Gakkaishi* 54: 1653.
11. Kimura, M., Hama, Y., Sumi, T., Asakawa, M., Narasinga Rao, B.N., Horne, A.P., Li, S.-C., Li, Y.-T., and Nakagawa, H., 1994, Characterization of a deaminated neuraminic acid-containing glycoprotein from the skin mucus of the loach, *Misgurnus anguillicaudatus*, *J. Biol. Chem.* 269: 32138.

12. Nadano, D., Iwasaki, M., Endo, S., Kitajima, K., Inoue, S., and Inoue, Y., 1986, A naturally occurring deaminated neuraminic acid, 3-deoxy-D-*glycero*-D-*galacto*-nonulosonic acid (KDN), *J. Biol. Chem.* 261: 1150.
13. Iwasaki, M., Inoue, S., Nadano, D., and Inoue, Y., 1987, Isolation and structure of a novel deaminated neuraminic acid containing oligosaccharide chain present in rainbow trout egg polysialoglycoprotein, *Biochemistry* 26: 1452.
14. Inoue, S., Kanamori, A., Kitajima, K., and Inoue, Y., 1988, KDN-glycoprotein: a novel deaminated neuraminic 153: 172.
15. Kanamori, A., Inoue, S., Iwasaki, M., Kitajima, K., Kawai, G., Yokoyama, S., and Inoue, Y., 1990, Deaminated neuraminic acid-rich glycoprotein of rainbow trout egg vitelline envelope, *J. Biol. Chem.* 263: 21811.
16. Kanamori, A., Kitajima, K., Inoue, S., and Inoue, Y., 1989, Isolation and characterization of deaminated neuraminic acid-rich glycoprotein (KDN-gp) in the ovarian fluid of rainbow trout (*Salmo gairdneri*), *Biochem. Biophys. Res. Commun.* 162: 744.
17. Song, Y., Kitajima, K., Inoue, S., and Inoue, Y., 1991, Isolation and structural elucidation of a novel type of ganglioside, deaminated neuraminic acid (KDN)-containing glycosphing-lipid, from rainbow trout sperm, *J. Biol. Chem.* 266: 21929.
18. Strecker, G., Wieruszeski, J-M., Michalski, J-C., Alonso, C., Boilly, B, and Montreuil, J., 1992, Characterization of Le(x), Le(y) and A Le(y) antigen determinants in KDN-containing *O*-linked glycan chains from *Pleurodeles waltlii* jelly coat eggs, *FEBS Lett.* 298: 39.
19. Strecker, G., Wieruszeski, J-M., Michalski, J-C., and Montreuil, J., 1992, ^{1}H- and ^{13}C-n.m.r. spectroscopy of 2-oxo-3-deoxy-D-glycero-D-galactononulosonic acid-containing oligo-saccharide-alditols bearing Lewis X, Lewis Y and A-Lewis Y determinants isolated from the jelly coat of *Pleurodeles waltl* eggs, *Biochem. J.* 287: 905.
20. Strecker, G., Wieruszeski, J-M., Michalski, J-C., Alonso, C., Leroy, Y., Boilly, B, and Montreuil, J., 1992, Primary structure of neutral and acidic oligosaccharide-alditols derived from the jelly coat of the Mexican axolotl. Occurrence of oligosaccharides with fucosyl (α1-3)fucosyl(α1-4)-3-deoxy-D-*glycero*-D-*galacto*-nonulosonic acid and galactosyl(α1-4)[fucosyl(α1-2)]galactosyl(β1-4)-N-acetylglucosamine sequences, *Eur. J. Biochem.* 207: 995.
21. Knierel, Y.A., Kocharova, N.A., Shashkov, A.S., Kochetokov, N.K, Mamontova, V.A., and Solov'era, T.F., 1989, Structure of the capsular polysaccharide of *Klebsiella ozaenae* serotype K4 containing 3-deoxy-D-*glycero*-D-*galacto*-nonulosonic acid, *Carbohydr. Res.* 188:145.
22. Sumi, T., Hama, Y., Maruyama, D., Asakawa, M., and Nakagawa, H, 1997, Purification and characterization of glycoprotein from the skin mucus of the rainbow trout, *Salmo gairdneri*, *Biosci. Biotech. Biochem.* 61: 675.
23. Uchida, Y., Tsukada, Y., and Sugimori, T., 1977, Distribution of neuraminidase in *Arthrobacter* and its purification by affinity chromatography, *J. Biochem.* 82: 1425.
24. Harris, G., and MacWilliam, I.C., 1954, A dipping technique for revealing sugars on paper chromatograms, *Chem. Ind. (London)* 249.
25. Nakagawa, H., Asakawa, M., and Enomoto, N., 1987, Histochemical localization of β-*N*-acetylhexosaminidase in the epidermis of eel, *Nippon Suisan Gakkaishi* 53: 1251.
26. Gottschalk, A., Bhargava, A.S., and Murty, V.L.N., 1972, Submaxillary gland glycoproteins, in: *Glycoproteins Part B*, pp. 810-829, A. Gottschalk, ed., Elsevier, Amsterdam.
27. Sartin, J.L., Hugli, T.E., and Liao, T.-S., 1980, Reactivity of the tryptophan residues in bovine pancreatic deoxyribonuclease with *N*-bromosuccinimide, *J. Biol. Chem.* 255: 8633.
28. Slomiany, B.L., Murty, V.L.N., and Slomiany, A., 1980, Isolation and characterization of oligosaccharides from rat colonic mucus glycoprotein, *J. Biol. Chem.* 255:9719.

29. Li, Y.-T., Yuziuk, J.A., Li, S.-C., Nematalla, A., Hasegawa, A., Kimura, M., and Nakagawa, H., 1994, A novel sialidase capable of cleaving 3-deoxy-D-*glycero*-D-*galacto*-2-nonulosonic acid (KDN), *Arch. Biochem. Biophys.* 310: 243.
30. Yuziuk, J.A., Nakagawa, H., Hasegawa, A., Kiso, M., Li, S.-C., and Li, Y.-T., 1996, Two different sialidases, KDN-sialidase and regular sialidase in the starfish *Asterina pectinifera*, *Biochem. J.* 315: 1041.
31. Nakagawa, H., Asakawa, M., and Enomoto, N., 1987, Purification and characterization of α-*N*-acetylgalactosaminidase from skipjack liver, *J. Biochem.* 101: 855.
32. Downs, F., Peterson, C., Murty, V.L.N., and Pigman W., 1977, Quantitation of the β-elimination reaction as used on glycoproteins, *Int. J. Pept. Protein Res.* 10:315.
33. Sheehan, J.K., and Carlstedt, I., 1990, Electron microscopy of cervical - mucus glycoproteins and fragments therefrom - the use of colloidal gold to make visible naked protein regions, *Biochem. J.* 265: 169.
34. Eckhardt, A.E., Timpte, C.S., Abernethy, J.L., Zhao, Y., and Hill, R.L., 1991, Porcine submaxillary mucin contains a cystine-rich, carboxyl-terminated domain in addition to a highly repetitive, glycosylated domain, *J. Biol. Chem.* 266: 9678.
35. Sumi, T., Hama, Y., Maruyama, D., Asakawa, M., and Nakagawa, H., 1997, Isolation and properties of a sialoglycoprotein from the skin mucus of the stingray *Dasyatis akajei*, *Fisheries Science* 63: 453.
36. Sumi, T., Hama, Y. Maruyama, D., Asakawa, M., and Nakagawa, H., 1997, Structural characterization of the carbohydrate chains from the skin mucus sialoglycoprotein of the stingray *Dasyatis akajei*, *Fisheries Science* 63: 779.

Glycoconjugates from Parasitic Helminths : Structure Diversity and Immunobiological Implications [1]

Kay-Hooi Khoo[a] and Anne Dell[b]

[a]*Institute of Biological Chemistry, Academia Sinica, Taipei, Taiwan; and*
[b]*Department of Biochemistry, Imperial College of Science, Technology and Medicine, London SW7 2AZ, UK*

The word "helminth" refers to a large and diverse group of roundworms and flatworms which include some of the most important parasites of medical and veterinary significance. Infection with parasitic helminths are highly prevalent globally, particularly among the population of the developing world (Bundy, 1997). Despite some remarkable successes in our concerted efforts to achieve helminth control, about a third of the world's population is still estimated to harbor infection with one of the most prevalent intestinal nematodes, namely the large roundworm (*Ascaris lumbricoides*), whipworm (*Trichuris trichuria*), and hookworm species (*Ancylostoma duodenale* and *Necator americanus*); about 200 million people are infected with blood flukes (*Schistosoma mansoni, Schistosoma haematobium, Schistosoma japonicum*); and over 150 million more are afflicted by lymphatic filariasis and onchocerciasis due to infection with one of the filarial nematodes (*Wuchereria bancrofti, Brugia malayi, Onchocercia volvulus*).

Parasitic helminths, in general, persist for years but do not multiply within the infected human hosts. Their life cycles typically entail passage through the environment and a second intermediate host or vector, usually of invertebrates. The developing parasites must therefore adapt in turn to very different host environments, employing a repertoire of immunomodulatory or evasion strategies. Like the kinetoplastid protozoan parasites which have been more thoroughly examined, the parasitic helminths are equally endowed with an abundance of glycosylated entities in the form of surface coats or glycocalyx, or on the excreted-secreted (ES) antigens. Two lines of experimental approaches have persistently argued for the significance of glycosylation in helminthic infections. First, immuno-dominant glycosylated epitopes are often the major targets of natural and experimental host

---

[1] The abbreviations used are : CAA, circulating anodic antigen; CCA, circulating cathodic antigen; Cer, ceramide; Con A, con-canavalin A; ES, excretory-secretory; GSL, glycosphingolipids; HexNAc, GlcNAc or GalNAc; lacNAc, *N*-acetyl-lactosamine; lacdiNAc, *N,N'*-diacetyl-lactosediamin; Le^x, Lewis X; mAb, monoclonal antibody.

humoral responses, as demonstrated by loss of antibody recognition through deglycosylation of the parasite glycoprotein antigens or destruction of the glyco-epitopes by periodate oxidation. Second, immuno-staining by glycan-recognizing monoclonal antibodies or lectins against whole parasite or parasite extracts often show developmental stage-specific expression profiles of glycosylation. Yet, for all their dominant presence and functional implications in host-parasite interactions, too few immunobiological observations are actually founded on a firm structural basis.

Detail structural characterization of helminth glycans has always been hampered by limited availability of parasite samples compounded by considerable structural novelty. The latter factor implies that significantly larger amount of sample materials are required for complete definition of their unique features which cannot be comparatively derived from known standard reference nor probed by the usual range of glycosidase repertoire. Another problem relates to the difficulty in distinguishing between genuine parasite-synthesized glycans and those derived from host tissues, usually as contaminants during isolation and extraction. With a knowledge of the mammalian host glycosylation pattern, this is less a problem when structural studies are focused on "unusual" glycans with "unusual" substituents, glycosyl linkages, or residues, which are obviously parasite-specific. In cases where such glycans do not exist, one can perhaps only sensibly conclude that the glycosylation profile of a particular parasite under investigation resembles that of its immediate host where it resides. The subtractive approach using samples from non-infected hosts as a control is useful in identifying the possible contribution of host glycans in the overall profile.

Another approach involves the use of metabolic labeling of newly synthesized glycans by *in vitro* cultured parasites in the absence of host cells or media. Subsequent analysis can then be focused only on radiolabeled glycans which, by definition, must be synthesized *de novo* by the parasites. However, one may argue that the *in vitro* culture conditions do not genuinely reflect the complicated interplay between the host environment and the parasites which may influence or modify the actual glycosylation pattern. This approach is also obviously limited to cases where a particular developmental stage of the parasite under investigation can be successfully cultured. In fact, a severely limiting factor in defining the developmental stage specific glycosylation profile is often that all stages are not readily available for experimental purposes.

To date, most helminth glycan structures characterized were performed on those released from a total mixture of extracted glycoproteins, rather than from a highly purified glycoprotein of known function. A typical approach is to first obtain a total glycosylation profile of a particular developmental stage and then home in on a few characteristic features for detailed analysis. Although chromatographic mapping is feasible for this purpose, a mass spectrometry analysis as adopted by us and others represents a more powerful alternative since the assignment of peaks is not dependent on availability of standards. Each component within the mixture would afford at least a molecular mass based on which possible glycosyl composition with or without other non-saccharide substituent can be derived. Additional fragment ions produced either by direct fragmentation or tandem mass spectrometry analysis further provide information related to the kind of terminal epitopes presented. When the aim is to demonstrate the presence of a previously defined or immunologically implicated epitope, prior monoclonal antibody-based enrichment or sub-fractionation is particularly useful in deriving such glycans of interest from a pool of extremely heterogeneous mixtures otherwise dominated by irrelevant or less interesting glycans.

Current literature on helminth glycosylation includes many reports on potential immunological activities attributed to glycosylated molecules. We make no attempt here to provide a comprehensive review on these but choose instead to focus on one system, the schistosome, with an emphasis on the progress we and others have made in establishing a structural basis for many of the implicated immunological activities. We then end this article with a brief account on other helminth glycoconjugates that have been structurally defined, and a synthesis of the underlying principles and patterns governing helminth glycosylation.

1. PATTERN AND DIVERSITY IN SCHISTOSOMAL GLYCOSYLATION

The glycans from the trematode *Schistosoma mansoni* are among the first biochemically characterized helminth glycans (Nyame *et al.*, 1987; Nyame *et al.*, 1988a; Nyame *et al.*, 1988b; Nyame *et al.*, 1989). To date, the glycobiology of schistosomes remains the best studied system within the parasitic helminths (for recent reviews, see Cummings and Nyame, 1996; Cummings and Nyame, 1999; Dell *et al.*, 1999). Apart from one developmental stage within the intermediate mollusc host, glycosylation for all other stages have been investigated to different extents, both structurally and immunologically. The tremendous research effort devoted to schistosomes is perhaps a just reflection of its importance among the tropical infectious disease.

Morbidity inflicted by schistosomiasis is due primarily to deposition of parasite eggs which induces a delayed hypersensitivity granulomatous response, leading to the development of fibrosis, and eventually hepatic destruction, cirrhosis and liver failure in chronic infections. Although the general pathology is similar, there are notable differences in host specificity and distribution of egg deposition among the major species which infect humans. Of the two which cause human intestinal schistosomiasis, *S. japonicum* infects a broader range of hosts including domestic animals *e.g.* pigs, water buffalo, and rodents, whereas *S. mansoni* infects only human, primates, and a few rodents. Adult *S. mansoni* localizes to the portal veins draining the large intestine and deposits eggs in the wall of the colon whereas adult *S. japonicum* lives and deposits eggs in the small intestine. Those eggs which adhere and successfully penetrate the vessels enter the faeces and hatch in freshwater into miracidia that infect specific snail hosts. Human infection is initiated when the free-swimming cercariae released from the snails penetrate the mammalian host skin, followed by transformation into schistosomula and subsequent development into adult worms.

1.1 Mammalian host-like protein *N*-glycosylation

Experimentally, the cercariae obtained from infected snail could be mechanically transformed into schistosomula which could then be metabolically labeled *in vitro* with radioactive precursor sugars for up to 48 hr (Nyame *et al.*, 1988b). Likewise, eight weeks old *S. mansoni* adults obtained from the portal veins of infected mice could be metabolically radiolabeled for 24 hr (Nyame *et al.*, 1987). Using this experimental approach, Cummings's group were the first to show that the mechanically transformed, 48 hr-old schistosomula and the adult worms of *S. mansoni* both synthesize the typical range of high mannose and complex type *N*-glycans resembling those of mammalian, but notably contain no detectable sialic acid (Nyame *et al.*, 1988a; Nyame *et al.*, 1988b; Nyame *et al.*, 1989).

In their studies, radiolabeled glycopeptides were prepared by pronase digestion of total parasite glycoprotein extracts, fractionated by serial lectin-affinity column chromatography, and different classes of *N*-glycans were inferred from the specificity of lectin binding (Cummings, 1994). Analysis of the endoglycosidase H released glycans from the strong ConA binding fraction demonstrated the presence of mostly $Man_6GlcNAc_2$ to $Man_9GlcNAc_2$ among the high mannose type (Nyame *et al.*, 1988a; Nyame *et al.*, 1988b). Of the complex types examined, both bi- and triantennary-like structures were found and a portion were inferred to contain the typical fucosylated chitobiose core. Interestingly, terminal β-GalNAc appeared to dominate over terminal β-Gal among the weak ConA binding biantennary chains (Nyame *et al.*, 1989), a portion of which were subsequently identified to be terminating with the lacdiNAc or fucosylated lacdiNAc sequences, GalNAcβ1→4(±Fucα1→3)GlcNAcβ1→2Manα1-R (Srivatsan *et al.*, 1992b). On the other hand, starting from a glycopeptide fraction bound to immobilized serum immunoglobulins from schistosome-infected hamsters, the major antigenic glycans synthesized by *S. mansoni* adults were shown to carry the Lewis X (Le^x) determinant, Galβ1→4(Fucα1→3)GlcNAcβ1 →, on poly-*N*-acetyllactosamine chains of tri- and tetraantennary complex type structures (Srivatsan *et al.*, 1992a).

From these pioneering studies using metabolic radiolabeling, an early picture thus emerged showing unambiguously that the trematode, as represented by the adult *S. mansoni*, is capable of *N*-glycosylating its proteins in a mammalian-like fashion. It should be noted that the relatively high abundance of lacdiNAc and the presence of antigenic Le^x -carrying polylactosamines on the adult schistosomal *N*-glycans are unusual. Yet, these structures are neither parasite-specific nor particularly novel as compared against those subsequently defined glycan structures from the eggs and cercariae.

1.2 Stage specific expression of novel multifucosylated epitopes and Le^x

It is well known that during schistosomal infection, the level of circulating anti-schistosome antibodies increases with the onset of egg-laying. A majority of antibodies from normally infected hosts or animal vaccinated with highly irradiated cercariae actually recognize glycan epitopes which are immunologically cross-reactive to different extents with antigens from each of the developmental stages, leading to the early perception that there are common as well as developmental stage-specific carbohydrate epitopes (reviewed in Dunne, 1990; Simpson, 1990). Among those that were further investigated, two monoclonal antibodies (mAb) raised against the egg antigens, mAb E.5 and 504B1, were subsequently shown to recognize the Le^x epitope (Ko *et al.*, 1990; Köster and Strand, 1994), giving the first evidence of its presence on the egg antigens. These mAbs also bind to the surface of both schistosomula and adult worms. Apparently, mAb 504B1 did not react with the surface of cercariae but did recognize secreted material near the acetabular gland openings and in the ventral sucker (Köster and Strand, 1994), as well as whole cercarial extracts (Weiss and Strand, 1985).

On the other hand, a mAb (128C3/3) raised against *S. mansoni* cercarial glycoproteins has been shown to recognize a uniquely fucosylated epitope present in both glycolipid and glycoprotein fractions of the egg antigens (Weiss and Strand, 1985; Weiss *et al.*, 1986; Dalton *et al.*, 1987; Levery *et al.*, 1992), which was also exposed on the surface of cercariae, schistosomula, and miracidia, but not the adult worms (Köster and Strand, 1994). Immunogold labeling further indicated that this mAb 128C3/3 defined epitope is associated with the outermost glycocalyx of the cercariae which was shed into the surrounding medium after transformation (Köster and Strand, 1994).

1.3 Multifucosylated structures on egg GSLs and glycoproteins

The first possible structure of the uniquely fucosylated epitope recognized by mAb 128C3/3 was provided by Levery *et al* from their work on the glycosphingolipids (GSL) isolated from *S. mansoni* eggs (Levery *et al.*, 1992). It was shown that virtually every neutral glycolipid fractionated by silica gel chromatography and preparative thin layer chromatography was reactive against mAb 128C3/3. Based on monosaccharide, fatty acid, linkage, and mass spectrometry analyses, the glycolipids were characterized as a series of novel fucose-containing GSL with a ceramide moiety composed primarily of t20:0 sphingosine with 16:0, 2-OH 16:0, and 18:0 fatty acids (Fig 1A). Together with additional periodate oxidation data from an independent structural study by us (Khoo *et al.*, 1997a), the revised structures were shown to consist of a highly unusual, multifucosylated HexNAc backbone extending from a glucosylceramide (Fig 1B). Monosaccharide inhibition studies indicated that the fucosylated epitope recognized by mAb 128C3/3 probably resides in the novel $Fuc_2HexNAc$ moiety characterized, distinct from other conventional epitopes including Fucα1→2Gal, Fucα1→3GlcNAc, and Fucα1→4GlcNAc tested (Levery *et al.*, 1992).

The characterized ceramide core of the egg GSLs was consistent with earlier work by Makaaru *et al* who first identified a novel core structure for the GSLs from *S. mansoni* adult worm and coined the term "schisto-core", referring to GalNAcβ1→4Glc-ceramide as a possible core structure for all other GSLs from schistosomes (Makaaru *et al.*, 1992). Using metabolic radiolabeling of adult worms and structural characterization of the ceramide glycanase released glycan, it was shown that the major GSLs newly synthesized by cultured adults comprise both galactosylceramide and glucosylceramide in addition to GalNAcβ1→4Glc-ceramide, but not lactosylceramide or any other larger GSLs. The egg glycolipids from *S. mansoni* therefore share the same "schisto-core" ceramide as the adult GSLs but are multifucosylated with the mAb 128C3/3 defined epitope which is absent from the adults.

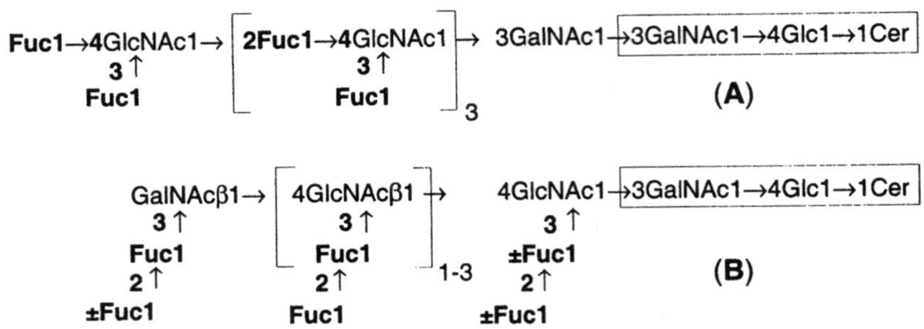

Figure 1. Structures of the major GSLs from *S. mansoni* eggs. Structures (A) and (B) differ primarily in the deduced arrangement of the internally linked Fuc residues. The boxed structural moiety represents the "schisto-core" unique to schistosomal GSLs..

The multifucosylated terminal structures as defined for the *S. mansoni* egg GSLs afforded a very characteristic series of fragment ions during FAB-mass spectrometry analysis of their permethyl derivatives, resulting from cleavage at each of the HexNAc residues (Levery *et al.*, 1992; Khoo *et al.*, 1997a) which can be assigned as $(Fuc_{1\ or\ 2}HexNAc_1)_n^+$. Likewise, terminal Le^x and fucosylated lacdiNAc containing structures, if present, also yield their respective fragment ions corresponding to $Fuc_1Hex_1HexNAc_1^+$ and $Fuc_1HexNAc_2^+$. Detection of these abundant and diagnostic fragment ions allowed us to demonstrate that both *N*-glycans and *O*-glycans released from total glycoprotein extracts of *S. mansoni* eggs carry a complement of all three differently fucosylated terminal structures, along with the non-fucosylated version, namely HexNAc, lacNAc and lacdiNAc termini (Khoo *et al.*, 1997b). It is unclear if the terminal sequence of repeating $Fuc_2HexNAc$ unit identical to that on egg GSLs was also present but ions corresponding to $Fuc_{0\text{-}2}HexNAc_1$, $Fuc_{0\text{-}4}HexNAc_2$ and $Fuc_{0\text{-}2}HexNAc_3$ could be found. The structural data obtained by us and others therefore support the immunological observation that a uniquely fucosylated epitope as defined by mAb 128C3/3 is present on both egg glycolipid and glycoprotein fractions but not those of the adults. In contrast, Le^x, fucosylated lacdiNAc, and their non fucosylated versions are common to both adults and eggs.

1.4 Core fucosylation and xylosylation on the egg *N*-glycans

Our mass spectrometry analysis of the glycans from the total glycoprotein extracts of schistosomal eggs was the first attempt to profile the full range of *N*-glycans and *O*-

glycans that are synthesized. The aim then was not to isolate and characterize each of the glycans present but to gain an insight into the glycosylation pattern and diversity of a particular parasite and to ask the question whether there is significant developmental stage and/or species specificity. The facile production and detection of the non-reducing terminal fragment ions described above afford a molecular mass mapping of the non-reducing terminal structures which often constitute the most important epitopes in the immunopathogenesis of parasitic infections. Additionally, we can address the question if the normal core structures of *N*- and *O*-glycans are conserved and/or modified since the total complement of glycans isolated invariably contain the smallest core structures based on which the detected terminal sequences are extended. Using this approach, we have identified a few notable structural features of schistosomal egg glycosylation (Fig 2) in comparison with the adult stage and the related species, *S. japonicum* (Khoo *et al.*, 1997b).

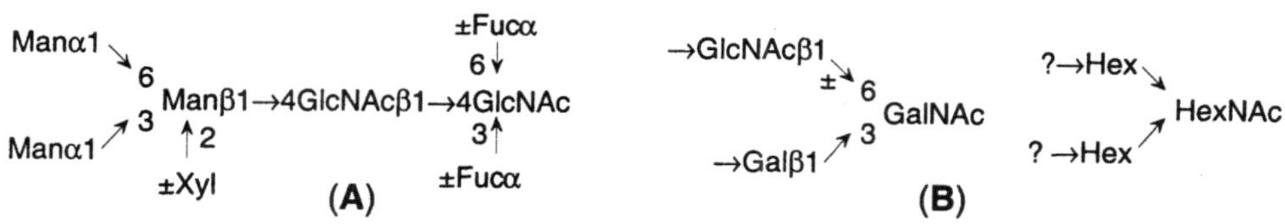

Figure 2. Core structures of the schistosomal egg *N*-glycans (A) and *O*-glycans (B). The smallest *N*-glycans may have either one or both of the α-Man also trimmed off. Typically, the cores of pauci-mannose, hybrid and complex types are at least α6-fucosylated. A significantly higher proportion of *S. japonicum* core structures is also α3-fucosylated resulting in preferential association of the additional xylosylation with difucosylated core whereas α3-fucosylation is relatively less abundant in *S. mansoni* and not found to be accompanied by xylosylation. The *O*-glycan cores are assumed to be typical mucin type cores 1 and 2 as confirmed in the analysis of CCA. The novel $Hex_2HexNAc_1$ core is also found in CCA and the cercariae.

First, in common to plants, insects, and nematodes (Dell *et al.*, 1999), a high abundance of truncated or pauci-mannose type structures ($Man_{2\text{-}4}GlcNAc_2$) were found, most of which are core fucosylated whereas the high mannose structures ($Man_{5\text{-}9}GlcNAc_2$) found are not. These pauci mannose type structures could also be found in other developmental stages including the adults and cercariae (Khoo *et al.*, 1999). Second, a portion of the *N*-glycan trimannosyl *N,N'*-diacetyl-chitobiose core is additionally 2-xylosylated at the β-Man in a manner similar to *N*-glycans from plants (Lerouge *et al.*, 1998) and molluscan hemocycanin (Kamerling and Vliegenthart, 1997), and/or α3-fucosylated to yield difucosylated cores as described for certain insect (Altmann *et al.*, 1999) and nematode (Haslam *et al.*, 1996) *N*-glycans. Interestingly, either core α3-fucosylation or xylosylation can occur on the egg *N*-glycans of both schistosomal species examined but co-existence of both core modification appears to be restricted to *S. japonicum*.

Like other more typical core structures, these modified cores (Fig 2A) can be non-extended, or extended with a single GlcNAc, lacNAc units, or lacdiNAc, and end with the terminal sequences described above, as hybrid or complex type structures. Some minor differences exist in the range of fucosylated terminal structures which decorate the *N*-glycans of the two species. Notably, the terminal $Fuc_{1,2}HexNAc$ epitope appears to be lacking in *S. japonicum* eggs which probably also relates to the absence of $Fuc_{2\text{-}4}HexNAc_2$ terminal structures. On the other hand, Le^x and $Fuc_1HexNAc_{2,3}$ are common features of both species. The same range of terminal structures were also detected in the analysis of the released *O*-glycans (Khoo *et al.*, 1997b).

1.5 Unusual range of *O*-glycosylation

Most of the egg *O*-glycans identified from both schistosomal species are consistent with them being based on typical core 1, R→Galβ1→3GalNAc, or core 2, R→Galβ1→3 (R'→GlcNAcβ1→6)GalNAc structures, where R represents further elongation or termination with a variety of sequences (Fig 2B). A definitive proof for either core is still lacking but existence of the branched core 2-like structure is supported by periodate oxidation data (Khoo *et al.*, 1997b). It is of interest to note that the identification of a stretch of up to 4 HexNAc directly linked to the core, with variable degree of fucosylation is highly unusual. This HexNAc backbone may resemble those found on the egg GSLs. In addition, a third core structure as represented by $Hex_2HexNAc$ was also detected and it is unclear at present if this core can also be further elongated.

In comparison, *O*-glycans from the *S. mansoni* adult worms and schistosomula appeared to be rather short and only simple *O*-GlcNAc, GalNAc, and Gal-GalNAc and have been identified based on characterization of metabolically radiolabeled glycans from total glycoprotein extracts of the parasite homogenates (Nyame *et al.*, 1987; Nyame *et al.*, 1988b). However, other work has demonstrated that the adult worm is capable of *O*-glycosylating its secreted protein in a highly unusual manner. The gut-associated antigens that are regularly released by the adult worms into the host circulation have long been recognized to carry highly immunoreactive glycans. Both the circulating cathodic antigen (CCA) and anodic antigen (CAA) isolated by 40% ammonium sulfate precipitation (supernatant fraction) and immunoaffinity chromatography from adult *S. mansoni* worms have been characterized in detail with respect to their *O*-glycosylation (Fig 3) using a complement of NMR, mass spectrometry and chemical analyses (Bergwerff *et al.*, 1994; van Dam *et al.*, 1994).

From CCA, the *O*-glycans released by alkaline reductive elimination were fractionated by gel-filtration into two main pools - a major polysaccharide fraction and a smaller amount of oligosaccharides. Like the egg *O*-glycans, the oligosaccharides were characterized as being based on mucin type core 1 and 2 structures elongated by lacNAc and possibly Le^x structures. Interestingly, among the incompletely defined structures was an unusual $Hex_2HexNAcitol$ branched core extended by a Hex-HexNAc unit. The polysaccharide fractions were shown to comprise poly-*N*-acetyllactosamine chains where the internal and non-reducing terminal repeating lacNAc units are respectively near 100% and 80% fucosylated to yield poly-Le^x epitopes (Fig 3A). It was hypothesized that these polysaccharide chains of an estimated 25 repeating units are conjugated to the protein backbone via core types 1 and 2, namely an extension from the characterized oligosaccharides (van Dam *et al.*, 1994). The CAA, on the other hand, was shown to contain polymeric chains of 6-linked β-GalNAc backbone with a β-glucuronic acid residue attached to its C-3 position (Fig 3B). The mode of linkage to protein core was not established (Bergwerff *et al.*, 1994). The major antigenicity of CAA was thought to arise from this unique repeating -6(GlcAβ1→3)GalNAcβ1→ epitope, estimated to have a degree of polymerization of more than 30, whereas cross-reactivity with anti-CCA mAb is probably due to the presence of a small amount of CCA-poly Le^x chains on CAA.

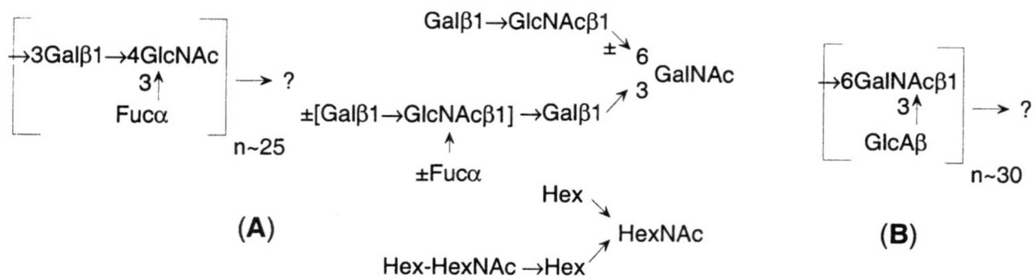

Figure 3. Structures of the unusual *O*-glycans from CCA (A) and CAA (B). In both, the polysaccharide chains were inferred to be O-linked to protein but the exact mode of attachment has not been established.

1.6 Cercarial glycocalyx and glycoconjugates

O-Glycans with composition similar to those identified for the oligosaccharide fraction of CCA including the unusual $Hex_2HexNAc$ core were similarly found in the total glycoprotein extracts of the cercariae of both schistosome species (Khoo, unpublished data) which represent the third source of some remarkably complex schistosomal *O*-glycans. The free living and infectious cercariae is covered by a dense 1-µm thick glycocalyx which is an extremely potent immunogen, with an apparent molecular mass of more than 5×10^6 Da, and an unusually high fucose content (Samuelson and Caulfield, 1985; Caulfield *et al.*, 1987). Among the mAb which would react against the *S. mansoni* cercarial glycoprotein extracts as well as binding to the cercariae surface is the mAb 128C3/3 (Dalton *et al.*, 1987; Köster and Strand, 1994) whereas mAb against Le^x only binds to the surface of transformed schistosomula after the glycocalyx has been shed off.

Experimentally, the fucose-rich glycocalyx was typically obtained from the phenol-water extracts of cercariae and purified away from the accompanying large amount of glycogen-like materials using a fucose-binding lectin (*Lotus* or *Anguilla anguilla* lectins) affinity column (Nanduri *et al.*, 1991; Xu *et al.*, 1994). Detail structural studies on this purified material from *S. mansoni* cercariae (Khoo *et al.*, 1995) revealed the presence of a very heterogeneous population of highly fucosylated *O*-glycans (Fig 4A) related to, but distinct from, the multifucosylated egg glycans (Fig 1). The main structural unit is the hexasaccharide, →3GalNAcβ1→4 (Fucα1→2 Fucα1→2Fucα1→3)GlcNAcβ1→3Galα1, which is repeatedly extended, probably until the terminal GalNAc is capped with α-Fuc instead of being further elongated with another α-Gal residue. It is still uncertain how these chains are linked to the protein backbone but most likely via conventional core 1 and 2 structures (Fig 4A). The smallest detectable *O*-glycans within this series are $Fuc_{2\text{-}4}Hex_1HexNAc_3$ while the next set up is precisely a $Fuc_3Hex_1HexNAc_2$ increment (Khoo, unpublished data). The former would likely correspond to a core 1 structure elongated with a lacdiNAc and fucosylated at both GalNAc and GlcNAc.

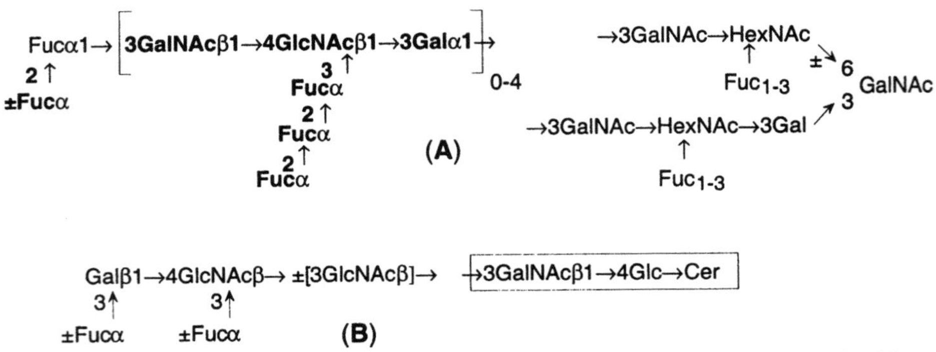

Figure 4. Fucosylated structures as found on the *S. mansoni* cercariae. (A) The highly fucosylated large *O*-glycans derived from fucose-lectin purified glycocalyx material. Heterogeneity resides mainly on i) the degree of fucosylation; ii) the number of repeating units; and iii) the incomplete branching off the 6-arm of the core GalNAc. (B) The major GSLs characterized. The fucosylated terminal may be directly linked to the "schisto core" or via an additional GlcNAc residues. Other glycocalyx *O*-glycan like terminal fucosylation structures were reported to be present as well but not further defined.

It is conceivable that the mAb 128C3/3 defined, cross reactive fucosylated epitope may reside in the non-reducing terminal $Fuc_{1,2}$GlaNAc, or the penultimate $Fuc_{2,3}$GlcNAc, or a composite of both. The polymeric nature of these cercarial glycocalyx *O*-glycans with an estimated 12 to more than 60 glycosyl residues may explain its immuno-dominance since any other underlying smaller or shorter glycoconjuagtes will be masked and rendered invisible to the host. In fact, further structural studies on the materials not bound to the *Anguilla anguilla* lectin column has led to identification of a range of smaller *O*-glycans of unusual core structures, as well as xylosylated *N*-glycans carrying an abundance of Le^x termini (Khoo, unpublished data).

The presence of the Le^x epitope among the *S. mansoni* cercariae glycoconjugates was further demonstrated by both immunostaining and structural studies on the cercarial glycolipid extracts (Wuhrer *et al.*, 2000). Apart from mass spectrometry and chemical analysis of the isolated GSLs, the glycan chains were further released from the ceramide by endoglycoceramidase, fractionated and characterized in detail (Fig 4B). In addition to confirming the already established "schisto-core" and the presence of Le^x, another novel fucosylated epitope was defined, namely Fucα1→3Gal which appeared to be associated with Le^x, in a manner similar to Le^y and was therefore coined a "pseudo-Lewis Y" structure.

1.7 From miracidia to cercariae - the missing link

To date, the vast majority of structural studies on schistosome glycoconjugates have been focused on three stages, namely the cercariae, adults, and eggs, as described above. To a certain extent, the egg antigens also comprise those actually synthesized by the developing miracidia encased within the egg shell. It could therefore be expected that the glycosylation of miracidia may be fairly similar to those of eggs and that the highly unusual glycan structures identified may well carry recognition signal or chemical identity compatible with the targeted intermediate molluscan host. Virtually nothing is known about the glycosylation of the sporocysts stage and how it may be related to that of the released cercariae.

In incompatible or resistant snail hosts, developing sporocysts are known to be encapsulated and destroyed by hemocytes whereas such cellular response is not elicited in susceptible strains following miracidial penetration. Recognition by the internal defense system of snails which does not produce immunoglobulins is widely believed to be mediated by lectins (reviewed in Fryer and Bayne, 1996; van der Knaap and Loker, 1990). It has been shown that most of the sporocyst tegumental proteins are glycosylated and that these surface glycans constitute binding determinants for the snail plasma- or hemocyte-derived lectins (Johnston and Yoshino, 1996). Lectins have also been identified in the hemolymph of *B. alexandrina* which bind specifically to fucosylated determinant on the miracidia (Mansour *et al.*, 1995) of *S. mansoni*. It is as yet unclear how these lectin-glycan interactions translate into a 'self versus non-self' recognition system. Probing with panels of plant lectins or antibodies have demonstrated common carbohydrate epitopes on the snail host and parasite tissues (Dissous *et al.*, 1986; Zelck and Becker, 1990). On the other hand, it was suggested that the predominant carbohydrate epitope displayed on the surface of the miracidia and sporocysts differ and that there are species-specific differences (Daniel *et al.*, 1992).

2. From Schistosome to Nematodes - Glycans in the Wormy World

As discussed in section 1.2, the glyco-immunonology of schistosomes with respect to interactions with mammalian host is dominated by different fucosylated epitopes, some of which are truly schistosome-specific while others can also be found on mammalian hosts and/or other lower organisms. Can what we learn from schistosome serves as a paradigm for helminthic glycosylation in general ?

The roundworms, or nematodes, are sufficiently different from the flatworms which include the trematode schistosomes, and the cestodes. Most nematodes share highly conserved developmental stages. The adult nematodes sexually produce eggs that hatch to release L1 larvae which, with intervening cuticular moults and size increases, develop through three more larval phases (L2, L3, and L4) before attaining full sexual maturity. The soil-transmitted intestinal nematodes have evolved a direct life cycle; their eggs or larvae normally leave the host via the faeces and become infective during a period of development in the soil. Infections with the insect vector-transmitted filarial species are initiated when the blood sucking vectors bite; the L3 larvae emerge from the insect mouth parts and actively penetrate the feeding wound in the skin before migrating to the final site of infection.

In addition to the cuticle which is extensively re-modeled with each moult stage, an outer envelope in the form of a loosely attached carbohydrate-rich surface coat or glycocalyx, has further been identified in many nematodes (Blaxter *et al.*, 1992; Maizels *et al.*, 1993). It was demonstrated that in nematode species which possess such a surface coat, their surface-labeled proteins could be distinguished as either integral cuticle proteins synthesized and secreted from the underlying epidermis, or those associated with the dynamic surface coats derived from specialized secretory glands, therefore sharing the same biosynthetic origins with at least some of the excretory/secretory (ES) products. The glycosylated structures identified so far for the nematodes have recently been reviewed (Dell *et al.*, 1999) and are summarized in Fig 5. From these limited examples and those of schistosomes (Figs 1-4), several general themes and principles can be developed.

2.1 Variation on a familiar core

Helminth *N*-glycans appear to follow the general route of biosynthesis, giving rise to high mannose, hybrid and complex type structures, with a typical trimmanosyl core. Pauci mannose ($Man_{2-4}GlcNAc_2$) are highly abundant and a significant portion can be core α6-fucosylated. These have been found in a diversity of nematodes including the ES antigens from ascarid larvae *Toxocara canis*, and *Ascaris suum* (Khoo *et al.*, 1993), the adult and larvae extracts of strongylids *Haemonchus contortus* (Haslam *et al.*, 1996; Haslam *et al.*, 1998), and the extracts of the adult filarial worms *Acanthocheilonema vitae*, *Onchocerca volvulus*, and *Onchocerca gibsoni* (Haslam *et al.*, 1999). Among these, the unusual core α3-fucosylation as found on the schistosomal egg glycans to give difucosylated cores have only been identified in the *N*-glycans from the L3 larvae and adult of *H. contortos*, and only the latter stage can be further modified by a unique third Fuc on the distal GlcNAc (Fig 5B) (Haslam *et al.*, 1996; Haslam *et al.*, 1998). On the other hand, β2-xylosylation has so far not been structurally identified in any nematodes. It is likely that both core α3-Fuc and β2-Xyl may have a wider occurrence among the *N*-glycans from various nematodes, as indicated by Western blotting studies using affinity purified core α3-Fuc and β2-Xyl specific antisera (van Die *et al.*, 1999). Significantly, it was demonstrated that IgE from serum from *S. mansoni* infected mice and *H. contortus* infected sheep recognizes *N*-glycans carrying the core α3-Fuc epitope. Thus, unusual modification on conventional core structures can potentially generate antigenic or allergenic epitopes which may contribute to antibody response during parasitic helminth infection, in a manner similar to allergenic responses to plant and insect glycoproteins carrying core α3-Fuc and/or β2-Xyl substituents (Garcia-Casado *et al.*, 1996; Tretter *et al.*, 1993).

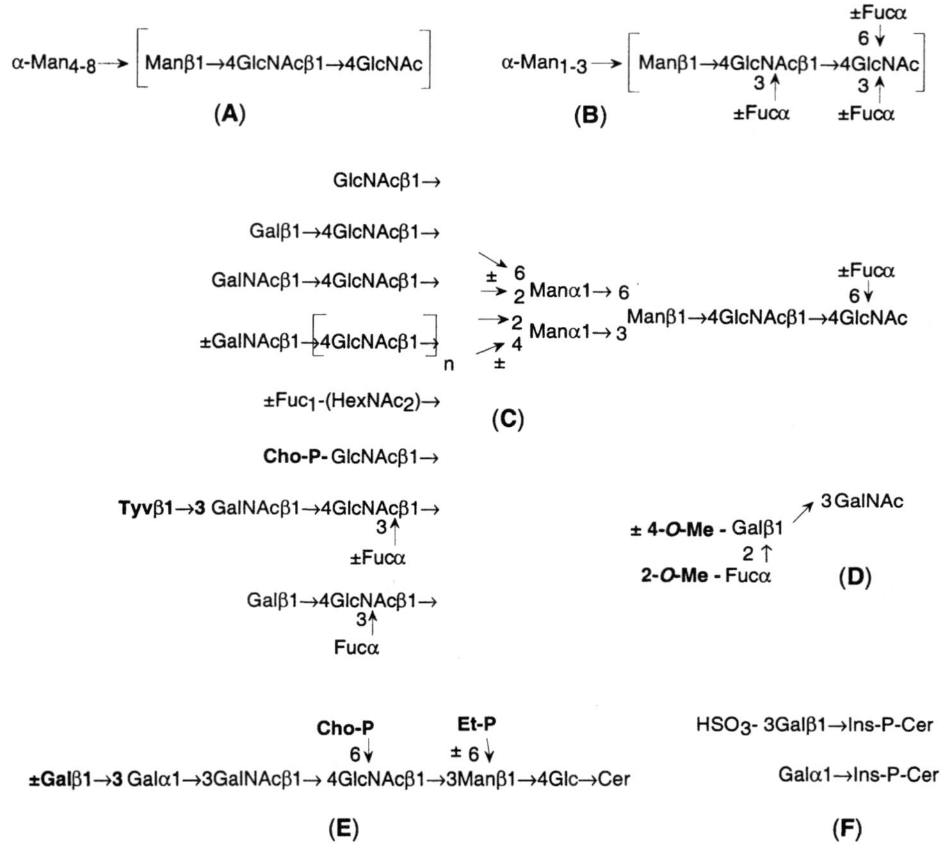

Figure 5. Glycan structures from nematodes. The *N*-glycans from all nematodes examined so far consist of high mannose (A), core fucosylated pauci-mannose (B), and hybrid/complex types (C). Core α3-fucosylation (B) has so far been identified only in *H. contortus*. The presence of tri- and tetraantennnary structures (C) was mostly inferred from the detection of 2,6-linked and 2,6-linked mannose in linkage analysis, except for *T. spiralis N*-glycans which appear to be exclusively of tri- and tetraantennary without the smaller bi- or monoantennary structures and terminate only with tyvelosylated lacdiNAc structures. Le^x has only been found in *D. viviparus*; phosphocholine substituted GlcNAcs and chito-oligomers in the filarial nematodes; other termini seem to be common especially lacdiNAc and fucosylated $HexNAc_2$. The unusually *O*-methylated *O*-glycans (D) have only been identified in *T. canis* and *T. cati*. The GSLs of arthro-series (E) have only been characterized for *A. suum* and *C. elegans* whereas the unusual acidic glycosyl inositol-P-ceramides (F) were found only in the former. Cho-P, phosphorylcholine; Et-P, phosphoethanolamine; Tyv, tyvelose.

Apart from the *O*-glycans from ascarids (Khoo *et al.*, 1991) which were found to be based on conventional mucin type core 1 structure (Fig 5D), there has been a lack of investigation into *O*-glycans from nematodes and thus generalization is probably premature. As exemplified by studies on the schistosomes, *O*-glycans can carry a potentially very complex range of structures as well as unusual cores. However, only core

1 and 2 structures have been definitively characterized to date. Unlike protein glycosylation, the glycosphingolipids as found in the schistosomes and nematodes appear to be based on ceramide cores distinct from those of mammalian. The so-called "schisto-core" characterizes schistosomal GSLs from cercariae, adults and eggs. In nematodes, both GSLs from the parasitic *Ascaris suum* (Lochnit *et al.*, 1997) and free living *Caenorhabditis elegans* (Gerdt *et al.*, 1997) were found to be based on arthro-series core ceramides (Fig 5E) similar to those found on other arthropods including insects (Dennis and Wiegandt, 1993). It differs from the "schisto-core" primarily in having a 3-linked β-Man replacing the 3-linked β-GalNAc. Another series of glycoinositol phosphoceramide based short acidic glycolipids have also been identified in *A. suum* (Lochnit *et al.*, 1998b; Sugita *et al.*, 1996) (Fig 5F). These inositol phosphate ceramide based glycolipids are typical of plants, fungi, yeast and protozoa but not animals (Lester and Dickson, 1993). Interestingly, GSLs from the other class of helminths, the cestodes, are based on yet another novel "neogala" series of GSLs, Galβ1(→6Galβ1)$_{0\text{-}3}$-Cer (Dennis and Wiegandt, 1993). Variations on this core structure are characterized by additional Fuc and α-Gal substituents, with an exception of a GSL from the plerocercoids of *Spirometra erinacei*, shown to be Galβ1→4(Fucα1→3) Glcβ1→3Galβ-Cer (Kawakami *et al.*, 1993).

As with the *O*-glycans, we probably have too few GSLs structurally defined from too few representative nematodes at present, to allow firm conclusions with respect to the kind and range of core ceramide structures synthesized. Nevertheless, it is clear that mammalian-like GSLs cores such as the lactoceramides or gangliosides are generally not found in parasitic helminths. Instead, the parasitic helminths appear to adopt protein and lipid glycosylation modes more akin to lower animals, especially insects, and plants.

2.2 Chain elongation - variants of the lacNAc pathway

Chain elongation from *N*- and *O*-glycan core structures in vertebrates are typically through either type 1 (Galβ1→3GlcNAc) or type 2 (Galβ1→4GlcNAc, lacNAc) chains, the latter may be repeated to give poly-*N*-acetyllactosamine structures. Thus far, all structures defined for parasitic helminths are based on lacNAc and variants of it (Van den Eijnden *et al.*, 1997) although the presence of the type 1 chain cannot be ruled out. The schistosomes appear to be capable of utilizing a variety of common and novel pathways. In addition to simple, omnipresent lacNAc and lacdiNAc units found in all stages examined, poly-lacNAc was identified in the adult worm extracts as well as on CCA, most likely also on egg antigens. Modified lacdiNAc pathways are exemplified by repeating α-galactosylated lacdiNAc on the cercarial glycocalyx *O*-glycans, and the stretches of HexNAc chains on the egg glycoproteins. It is as yet unclear if these represent a stretch of β4-linked GlcNAc conforming to the "chitobio pathway" or interspersed with GalNAc or differently linked GlcNAc. A likely structure is represented by elongation via the chitobio pathways but terminate with a lacdiNAc unit as found on the egg GSLs. The branched β6-linked GalNAc chain of CAA represent yet another novel type of elongation pathway.

In nematodes, lacdiNAc appears to be the preferred pathway over lacNAc as found for the *N*-glycans from *Trichinella spiralis* larvae (Reason *et al.*, 1994) and the microfilariae of *Dirofilaria immitis* (Kang *et al.*, 1993), while a series of chito-oligomers were found to be attached to the *N*-glycans from the adult extracts of the filarial worms *A. vitae*, *O. volvulus*, and *O. gibsoni* (Haslam *et al.*, 1999). Yet, in the cattle lung worm *Dictyocaulus viviparus*, lacNAc seems to be more abundant than lacdiNAc (Haslam *et al.*, 2000). Interestingly, the α3-galactosylated lacdiNAc as found on the *S. mansoni* cercarial glycocalyx *O*-glycans was identified as the main glycan chain motif conjugated to the 3Manβ1→4Glc-Cer of *A. suum* and *C. elegans* (Lochnit *et al.*, 1997; Gerdt *et al.*, 1999). In summary, all three pathways, namely lacNAc, lacdiNAc and chitobio are commonly utilized in parasitic helminths to elongate their *N*-, *O*-glycans and GSLs. In contrast, lacdiNAc is relatively rare in mammals and chito-oligomers have not been identified.

2.3 Immunogenic terminal structures - the functional implications

Arguably, the most important part of the parasite glycans attached to either protein or lipid are the non-reducing termini which can directly interact with various receptors, lectins, and are most readily recognized by the host immune system. It is perhaps of no coincidence that the most characteristic features of each of the helminth glycoconjugates also localize to the non-reducing terminal epitopes. In the case of schistosomes, apart from the CAA, all immunogenic structures appear to be related to fucose residues in various linkages and recognized by a panel of mAbs raised. The immunogenicity of the cercariae and egg antigens may be primarily due to the cross-reactive multifucosylated HexNAc epitopes but contribution of Le^x and fucosylated lacdiNAc may also be important. In fact, most of the studies investigating the possible functions of schistosomal glycocnjugates have been focused on the possible roles mediated by Le^x, especially in relation to its interactions with selectins (El Ridi *et al.*, 1996) and adhesion to endothelium (Lejoly-Boisseau *et al.*, 1999).

Le^x has also been implicated in the switch from Th1 response to Th2 in schistosomiasis which is known to occur at the onset of egg deposition (Pearce *et al.*, 1991). The Le^x epitope, as represented by lacto-N-fucopentaose III (LNFP) conjugated to human serum albumin, was shown to induce peripheral blood mononuclear cells from infected human to proliferate (Velupillai *et al.*, 2000) and the splenic B cells of infected mice to produce large amounts of IL-10 and prostaglandin E_2, which down regulate the Th1 response (Velupillai and Harn, 1994). Furthermore, mice sensitized with LNFP mounted an increased cellular response towards soluble egg antigen coupled beads implanted in the liver, resulting in the formation of larger periparticular granulomas (Jacobs *et al.*, 1999). In addition to mediating the host's cellular immunity to schistosomal infection, Le^x also induces humoral responses. Sera from infected human patients, monkeys and rodents have been shown to contain antibodies reactive to Le^x (Ko *et al.*, 1990; Nyame *et al.*, 1996; van Dam *et al.*, 1996; Nyame *et al.*, 1997; Nyame *et al.*, 1998) which can mediate specific complement-dependent cytolysis of human leukemic cells, thus implying a possible autoimmune pathological consequence manifested as neutropenia associated with chronic schistosomiasis.

Although the adult worm extracts do contain Le^x, the main contributing source probably originates from the abundant CCA and egg glycoproteins, and possibly the cercariae, all of which have been shown to carry the Le^x determinant. On the other hand, it is not known if the fucosylated lacdiNAc or other multifucosylated epitopes can mediate the same range of immunological effects as these have not received the same depth of investigation. *S. mansoni* infected mice have been shown to generate antibodies against lacdiNAc determinants (Nyame *et al.*, 1999) which may have important implications in infections by other nematodes. It was originally shown, by using mAbs against Le^x, that Le^x expression may be restricted to trematodes since it could not be detected among extracts from a panel of nematodes examined including *D. immitis*, *H. contortus*, and *C. elegans* (Nyame *et al.*, 1998). More recent structural studies have, however, shown that Le^x can be found on the *N*-glycans of *D. viviparus* (Haslam *et al.*, 2000). In such case, it is not known if some of the implicated immunobiological activities of Le^x in schistosomiasis can be extended to other parasitic helminth systems.

Unlike Le^x, the multifucosylated epitopes of schistosomes remain schisto-specific as they have yet to be detected elsewhere. Other nematodes are endowed with other equally unique terminal structures, some may turn out to be truly species-specific while others may be more wide spread. For example, the *O*-methylated Fuc and Gal epitopes are restricted to *Toxocara canis* and *T. cati O*-glycans and not found in the related ascarid, *A. suum* (Khoo *et al.*, 1991; Khoo *et al.*, 1993), and the terminal β-tyvelose (3,6-dideoxy-D-*arabino*-hexose) on fucosylated lacdiNAc is specific to *Trichinella spiralis N*-glycans (Reason *et al.*, 1994). Each was implicated in a variety of immunological response associated with infection (for review, see Dell *et al.*, 1999). On the other hand, the phosphorylcholine epitope has long been recognized to occur widely on a variety of nematodes, particularly the filarial nematodes (Maizels and Selkirk, 1988; Harnett and Parkhouse, 1995), and shown to modulate a range of host immune responses (reviewed in Harnett *et al.*, 1999; Harnett and Harnett, 1999). Structurally, it has been conclusively identified only on the *N*-glycans of the secreted protein ES62 from *A. vitae* (Haslam *et al.*, 1997), the adult extracts from *A. vitae*, *O. volvulus*, and *O. gibsoni* (Haslam *et al.*, 1999), and the GSLs from *A. suum* (Lochnit *et al.*, 1998a) and *C. elegans* (Gerdt *et al.*, 1999). For the *N*-glycans, the exact attachment site of phosphorylcholine has not been determined but known to be directly linked to GlcNAc residues attached to the trimannosyl core. In the GSLs, the phosphorylcholine was shown to be attached to the 6-position of the 4-linked GlcNAc (Fig 5E).

It is intriguing to conjecture upon the raison d'être of a host of unique terminal epitopes on the glycoconjugates from parasitic helminths. These may involve common monosaccharide residues *e.g.* Fuc in unusual linkage position like Fucα1→2Fucα1→; unusual monosaccharide residues *e.g.* tyvelose; and non-saccharide substituents such as *O*-Me and phosphorylcholine. Each is highly immunogenic and may thus be involved in the diversionary immune evasion strategy . For example, the glycocalyx of the invading schistosome cercariae is rapidly shed and the antigenic fucosylated epitopes are no longer detected on the developing schistosomula. Likewise, the host immune response may be directed towards the highly immunogenic ES antigens and be protective against re-infection while the host dwelling parasites remain unharmed. On the other hand, epitope such as Le^x could also be found on host tissue particularly on the granulocytes and may thus be utilized by the parasite as a kind of host mimicry strategy or even to exploit the host recognition system for its own good. Immunogenic structures including phosphorylcholine and Le^x may have immunomodulatory properties and thus induce a specific sets of cytokines and/or T cell response beneficial to the parasite's survival. There has been no lack of immunological studies in search of the answer, too many in fact to be covered in this article. Yet, definitive proof for a functional role for any of the structures characterized is still elusive.

SUMMARY

We have provided an account of the progress we and others have made over the last decade on the structural characterization of glycans from parasitic helminths. We hope to have illustrated a few principles and patterns governing helminth glycosylation, as well as the experimental approaches adopted and their associated strengths and limitations. Schistosomes remain the best studied systems but are still punctuated with gaps of knowledge. An important theme developed here is the regulated developmental stage-specific expression of various glycan epitopes and their interplay with immediate host environments for successful parasitism. It is anticipated that more novel or unusual structures will continuously be uncovered in the future and that despite many difficulties, current analytical techniques should be well up to meet the challenge in at least elucidating the major or key glycoconjugates from each of the diverse range of worms. The bottle neck will in fact reside in finding suitable experimental models to test their putative immunobiological functions from which the intricate host-parasite interactions can be delineated and rational vaccine design be achieved. The glycobiology of parasitic helminths is an area waiting to be more fully explored and the rewards should be sweet.

REFERENCES

Altmann, F., Staudacher, E., Wilson, I.B. and Marz, L., 1999, Insect cells as hosts for the expression of recombinant glycoproteins, *Glycoconj J*, 16: 109-123.

Bergwerff, A.A., van Dam, G.J., Rotmans, J.P., Deelder, A.M., Kamerling, J.P. and Vliegenthart, J.F., 1994, The immunologically reactive part of immunopurified circulating anodic antigen from *Schistosoma mansoni* is a threonine-linked polysaccharide consisting of → 6)-(β-D-Glc*p*A-(1→ 3))-β-D-Gal*p*NAc-(1→repeating units, *J Biol Chem*, 269: 31510-31517.

Blaxter, M.L., Page, A.P., Rudin, W. and Maizels, R.M., 1992, Nematode surface coats: Actively evading immunity, *Parasitol Today*, 8: 243-247.

Bundy, D.A.P., 1997, This wormy world - then and now, *Parasitol Today*, 13: 407-408.

Caulfield, J.P., Cianci, C.M.L., McDiarmid, S.S., Suyemitsu, T. and Schmid, K., 1987, Ultrastructure, carbohydrate, and amino acid analysis of two preparations of the cercarial glycocalyx of *Schistosoma mansoni*, *J Parasitol*, 73: 514-522.

Cummings, R.D., 1994, Use of lectins in analysis of glycoconjugates, *Methods Enzymol*, 230: 66-85.

Cummings, R.D. and Nyame, A.K., 1996, Glycobiology of schistosomiasis, *FASEB J*, 10: 838-848.

Cummings, R.D. and Nyame, A.K., 1999, Schistosome glysoconjugates, *Biochim Biophys Acta*, 1455: 363-374.

Dalton, J.P., Lewis, S.A., Aronstein, W.S. and Strand, M., 1987, *Schistosoma mansoni*: immunogenic glycoproteins of the cercarial glycocalyx, *Exp Parasitol*, 63: 215-226.

Daniel, B.E., Preston, T.M. and Southgate, V.R., 1992, The *in vitro* transformation of the miracidium to the mother sporocysts of *Schistosoma margrebowiei*; changes in the parasite surface and implications for interactions with snail plasma factors, *Parasitology*, 104: 41-49.

Dell, A., Haslam, S.M., Morris, H.R. and Khoo, K.H., 1999, Immunogenic glycoconjugates implicated in parasitic nematode diseases, *Biochim Biophys Acta*, 1455: 353-362.

Dennis, R.D. and Wiegandt, H., 1993, Glycosphingolipids of the invertebrata as exemplified by a cestode platyhelminth, *Taenia crassiceps*, and a dipteran insect, *Calliphora vicina*, *Adv Lipid Res*, 26: 321-351.

Dissous, C., Grzych, J.M. and Capron, A., 1986, *Schistosoma mansoni* shares a protective oligosaccharide epitope with freshwater and marine snails, *Nature*, 323: 443-445.

Dunne, D.W., 1990, Schistosome carbohydrates, *Parasitol Today*, 6: 45-48.

El Ridi, R., Velupillai, P. and Harn, D.A., 1996, Regulation of schistosome egg granuloma formation: host-soluble L- selectin enters tissue-trapped eggs and binds to carbohydrate antigens on surface membranes of miracidia, *Infect Immun*, 64: 4700-4705.

Fryer, S.E. and Bayne, C.J., 1996, Host-parasite interactions in molluscs, in: *Invertebrate Immunology, Progress in Molecular and Subcellular Biology series, vol. 15*, Rinkevich, B. and Müller, W.E.G. eds., Springer-Verlag, Berlin Heidelberg, pp. 131-153.

Garcia-Casado, G., Sanchezmonge, R., Chrispeels, M.J., Armentia, A., Salcedo, G. and Gomez, L., 1996, Role of complex asparagine-linked glycans in the allergenicity of plant glycoproteins., *Glycobiology*, 6: 471-477.

Gerdt, S., Lochnit, G., Dennis, R.D. and Geyer, R., 1997, Isolation and structural analysis of three neutral glycosphingolipids from a mixed population of *Caenorhabditis elegans* (Nematoda:Rhabditida), *Glycobiology*, 7: 265-275.

Gerdt, S., Dennis, R.D., Borgonie, G., Schnabel, R. and Geyer, R., 1999, Isolation, characterization and immunolocalization of phosphorylcholine- substituted glycolipids in developmental stages of *Caenorhabditis elegans*, *Eur J Biochem*, 266: 952-963.

Harnett, W. and Parkhouse, R.M.E., 1995, Nature and function of parasitic nematode surface and excretory-secretory antigens., in: *Perspectives in Nematode Physiology and Biochemistry.*, Sood, M.L. and Kapur, J. eds., Narenda Publishing House, Delhi, pp. 207-242.

Harnett, W., Deehan, M.R., Houston, K.M. and Harnett, M.M., 1999, Immunomodulatory properties of a phosphorylcholine-containing secreted filarial glycoprotein., *Parasite Immunol*, 21: 601-608.

Harnett, W. and Harnett, M.M., 1999, Phosphorylcholine: friend or foe of the immune system ?, *Immunol Today*, 20: 125-129.

Haslam, S.M., Coles, G.C., Munn, E.A., Smith, T.S., Smith, H.F., Morris, H.R. and Dell, A., 1996, *Haemonchus contortus* glycoproteins contain *N*-linked oligosaccharides with novel highly fucosylated core structures, *J Biol Chem*, 271: 30561-30570.

Haslam, S.M., Khoo, K.H., Houston, K.M., Harnett, W., Morris, H.R. and Dell, A., 1997, Characterisation of the phosphorylcholine-containing *N*-linked oligosaccharides in the excretory-secretory 62 kDa glycoprotein of *Acanthocheilonema viteae*, *Mol Biochem Parasitol*, 85: 53-66.

Haslam, S.M., Coles, G.C., Reason, A.J., Morris, H.R. and Dell, A., 1998, The novel core fucosylation of *Haemonchus contortus* N-glycans is stage specific, *Mol Biochem Parasitol*, 93: 143-147.

Haslam, S.M., Houston, K.M., Harnett, W., Reason, A.J., Morris, H.R. and Dell, A., 1999, Structural studies of *N*-glycans of filarial parasites. Conservation of phosphorylcholine-substituted glycans among species and discovery of novel chito-oligomers, *J Biol Chem*, 274: 20953-20960.

Haslam, S.M., Coles, G.C., Morris, H.R. and Dell, A., 2000, Structural characterization of the *N*-glycans of *Dictyocaulus viviparus*: discovery of the Lewis^{X} structure in a nematode, *Glycobiology*, 10: 223-229.

Jacobs, W., Deelder, A. and Van Marck, E., 1999, Schistosomal granuloma modulation. II. Specific immunogenic carbohydrates can modulate schistosome-egg-antigen-induced hepatic granuloma formation., *Parasitol Res*, 85: 14-18.

Johnston, L.A. and Yoshino, T.P., 1996, Analysis of lectin- and snail plasma-binding glycopeptides associated with the tegumental surface of the primary sporocysts of *Schistosoma mansoni.*, *Parasitology*, 112: 469-479.

Kamerling, J.P. and Vliegenthart, J.F.G., 1997, Hemocyanins, in: *Glycoproteins II*, Montreuil, J., Vliegenthart, J.F.G. and Schachter, H. eds., Elsevier, Amsterdam, pp. 123-142.

Kang, S., Cummings, R.D. and McCall, J.W., 1993, Characterization of the N-linked oligosaccharides in glycoproteins synthesized by microfilariae of *Dirofilaria immitis*, *J Parasitol*, 79: 815-828.

Kawakami, Y., Nakamura, K., Kojima, H., Suzuki, M., Inagaki, F., Suzuki, A., Sonoki, S., Uchida, A., Murata, Y. and Tamai, Y., 1993, A novel fucosylated glycosphingolipid with a Galβ1-4Glcβ1-3Gal sequence in plerocercoids of the parasite, *Spirometra erinacei*, *J Biochem (Tokyo)*, 114: 677-683.

Khoo, K.-H., Maizels, R.M., Page, A.P., Taylor, G.W., Rendell, N.B. and Dell, A., 1991, Characterization of nematode glycoproteins: the major *O*-glycans of *Toxocara* excretory-secretory antigens are *O*-methylated trisaccharides, *Glycobiology*, 1: 163-171.

Khoo, K.-H., Morris, H.R. and Dell, A., 1993, Structural characterisation of the major glycans of *Toxocara canis* ES antigens, in: *Toxocara and Toxocariasis: Clinical, Epidemiological and Molecular Perspectives*, Lewis, J.W. and Maizels, R.M. eds., Institute of Biology, pp. 133-140.

Khoo, K.-H., Sarda, S., Xu, X., Caulfield, J.P., McNeil, M.R., Homans, S.W., Morris, H.R. and Dell, A., 1995, A unique multifucosylated -3GalNAcβ1→4GlcNAcβ1→3Galα1- motif constitutes the repeating unit of the complex *O*-glycans derived from the cercarial glycocalyx of *Schistosoma mansoni*, *J Biol Chem*, 270: 17114-17123.

Khoo, K.-H., Chatterjee, D., Caulfield, J.P., Morris, H.R. and Dell, A., 1997a, Structural characterization of glycophingolipids from the eggs of *Schistosoma mansoni* and *Schistosoma japonicum*, *Glycobiology*, 7: 653-661.

Khoo, K.-H., Chatterjee, D., Caulfield, J.P., Morris, H.R. and Dell, A., 1997b, Structural mapping of the glycans from the egg glycoproteins of *Schistosoma mansoni* and *Schistosoma japonicum*: identification of novel core structures and terminal sequences, *Glycobiology*, 7: 663-677.

Khoo, K.-H., Fan, K.-C. and Lee, K.-M., 1999, Structural probing of the developmental stage-specific glycosylation profile of schistosomes, in: *Sialobiology and Other Novel Forms of Glycosylation*, Inoue, Y., Lee, Y.C. and Troy II, F.A. eds., Gakushin, Osaka, pp. 255-258.

Ko, A.I., Drager, U.C. and Harn, D.A., 1990, A *Schistosoma mansoni* epitope recognized by a protective monoclonal antibody is identical to the stage-specific embryonic antigen 1, *Proc Natl Acad Sci USA*, 87: 4159-4163.

Köster, B. and Strand, M., 1994, *Schistosoma mansoni*: immunolocalization of two different fucose-containing carbohydrate epitopes, *Parasitology*, 108: 433-446.

Lejoly-Boisseau, H., Appriou, M., Seigneur, M., Pruvost, A., Tribouley-Duret, J. and Tribouley, J., 1999, *Schistosoma mansoni*: in vitro adhesion of parasite eggs to the vascular endothelium. Subsequent inhibition by a monoclonal antibody directed to a carbohydrate epitope, *Exp Parasitol*, 91: 20-29.

Lerouge, P., Cabanes-Macheteau, M., Rayon, C., Fischette-Lainé, A.-C., Gomord, V. and Faye, L., 1998, N-Glycoprotein biosynthesis in plants: recent developments and future trends, *Plant Mol Biol*, 38: 31-48.

Lester, R.L. and Dickson, R.C., 1993, Sphingolipids with inositolphosphate-containing head groups, *Adv Lipid Res*, 26: 253-274.

Levery, S.B., Weiss, J.B., Salyan, M.E., Roberts, C.E., Hakomori, S., Magnani, J.L. and Strand, M., 1992, Characterization of a series of novel fucose-containing glycosphingolipid immunogens from eggs of *Schistosoma mansoni*, *J Biol Chem*, 267: 5542-5551.

Lochnit, G., Dennis, R.D., Ulmer, A.J. and Geyer, R., 1998a, Structural elucidation and monokine-inducing activity of two biologically active zwitterionic glycosphingolipids derived from the porcine parasitic nematode *Ascaris suum*, *J Biol Chem*, 273: 466-474.

Lochnit, G., Nispel, S., Dennis, R.D. and Geyer, R., 1998b, Structural analysis and immunohistochemical localization of two acidic glycosphingolipids from the porcine, parasitic nematode, *Ascaris suum*, *Glycobiology*, 8: 891-899.

Lochnit, G., Dennis, R.D., Zähringer, U. and Geyer, R., 1997, Structural analysis of neutral glycosphingolipids from *Ascaris suum* adults (Nematoda: Ascaridida), *Glycoconj J*, 14: 389-399.

Maizels, R.M. and Selkirk, M.E., 1988, Immunobiology of nematode antigens, in: *The Biology of Parasitism*, Englund, P.T. and Sher, F.A. eds., Alan R. Liss, New York, pp. 285-308.

Maizels, R.M., Blaxter, M.L. and Selkirk, M.E., 1993, Forms and functions of nematode surfaces, *Exp Parasitol*, 77: 380-384.

Makaaru, C.K., Damian, R.T., Smith, D.F. and Cummings, R.D., 1992, The human blood fluke *Schistosoma mansoni* synthesizes a novel type of glycosphingolipid, *J Biol Chem*, 267: 2251-2257.

Mansour, M.H., Negm, H.I., Saad, A.H. and Taalab, N.I., 1995, Characterization of *Biomphalaria alexandrina*-derived lectins recognizing a fucosyllactose-related determinant on schistosomes, *Mol Biochem Parasitol*, 69: 173-184.

Nanduri, J., Dennis, J.E., Rosenberry, T.L., Mahmoud, A.A.F. and Tartakoff, A.M., 1991, Glycocalyx of bodies *versus* tails of *Schistosoma mansoni* cercarieae, *J Biol Chem*, 266: 1341-1347.

Nyame, K., Cummings, R.D. and Damian, R.T., 1987, *Schistosoma mansoni* synthesizes glycoproteins containing terminal *O*-linked *N*-acetylglucosamine residues, *J Biol Chem*, 262: 7990-7995.

Nyame, K., Cummings, R.D. and Damian, R.T., 1988a, Characterization of the high mannose asparagine-linked oligosaccharides synthesized by Schistosoma mansoni adult male worms, *Mol Biochem Parasitol*, 28: 265-274.

Nyame, K., Cummings, R.D. and Damĭan, R.T., 1988b, Characterization of the *N*- and *O*-linked oligosaccharides in glycoproteins synthesized by *Schistosoma mansoni* schistosomula, *J Parasitol*, 74: 562-572.

Nyame, K., Smith, D.F., Damian, R.T. and Cummings, R.D., 1989, Complex-type asparagine-linked oligosaccharides in glycoproteins synthesized by *Schistosoma mansoni* adult males contain terminal β-linked *N*-acetylgalactosamine, *J Biol Chem*, 264: 3235-3243.

Nyame, A.K., Pilcher, J.B., Tsang, V.C. and Cummings, R.D., 1996, *Schistosoma mansoni* infection in humans and primates induces cytolytic antibodies to surface Le(x) determinants on myeloid cells, *Exp Parasitol*, 82: 191-200.

Nyame, A.K., Pilcher, J.B., Tsang, V.C. and Cummings, R.D., 1997, Rodents infected with *Schistosoma mansoni* produce cytolytic IgG and IgM antibodies to the Lewis x antigen, *Glycobiology*, 7: 207-215.

Nyame, A.K., Debose-Boyd, R., Long, T.D., Tsang, V.C. and Cummings, R.D., 1998, Expression of Le^x antigen in *Schistosoma japonicum* and *S. haematobium* and immune responses to Le^x in infected animals: lack of Le^x expression in other trematodes and nematodes, *Glycobiology*, 8: 615-624.

Nyame, A.K., Leppanen, A.M., DeBose-Boyd, R. and Cummings, R.D., 1999, Mice infected with *Schistosoma mansoni* generate antibodies to LacdiNAc (GalNAcβ1→4GlcNAc) determinants, *Glycobiology*, 9: 1029-1035.

Pearce, E.J., Caspar, P., Grzych, J.M., Lewis, F.A. and Sher, A., 1991, Downregulation of Thl cytokine production accompanies induction of Th2 responses by a parasitic helminth, *Schistosoma mansoni*, *J Exp Med*, 173: 159-166.

Reason, A.J., Ellis, L.A., Appleton, J.A., Wisnewski, N., Grieve, R.B., McNeil, M., Wassom, D.L., Morris, H.R. and Dell, A., 1994, Novel tyvelose-containing tri- and tetra-antennary *N*-glycans in the immunodominant antigens of the intracellular parasite *Trichinella spiralis*, *Glycobiology*, 4: 593-603.

Samuelson, J.C. and Caulfield, J.P., 1985, The cercarial glycocalyx of *Schistosoma mansoni*, *J Cell Biol*, 100: 1423-1434.

Simpson, A.J.G., 1990, Schistosome surface antigens: Developmental expression and immunological function, *Parasitol Today*, 6: 40-45.

Srivatsan, J., Smith, D.F. and Cummings, R.D., 1992a, The human blood fluke *Schistosoma mansoni* synthesizes glycoproteins containing the Lewis X antigen, *J Biol Chem*, 267: 20196-20203.

Srivatsan, J., Smith, D.F. and Cummings, R.D., 1992b, *Schistosoma mansoni* synthesizes novel biantennary Asn-linked oligosaccharides containing terminal β-linked *N*-acetylgalactosamine, *Glycobiology*, 2: 445-452.

Sugita, M., Mizunoma, T., Aoki, K., Dulaney, J.T., Inagaki, F., Suzuki, M., Suzuki, A., Ichikawa, S., Kushida, K., Ohta, S. and Kurimoto, A., 1996, Structural characterization of a novel glycoinositolphospholipid from the parasitic nematode, *Ascaris suum*, *Biochim Biophys Acta*, 1302: 185-192.

Tretter, V., Altmann, F., Kubelka, V., Marz, L. and Becker, W.M., 1993, Fucose α1,3-linked to the core region of glycoprotein N-glycans creates an important epitope for IgE from honeybee venom allergic individuals, *Int Arch Allergy Immunol*, 102: 259-266.

van Dam, G.J., Bergwerff, A.A., Thomas-Oates, J.E., Rotmans, J.P., Kamerling, J.P., Vliegenthart, J.F. and Deelder, A.M., 1994, The immunologically reactive O-linked polysaccharide chains derived from circulating cathodic antigen isolated from the human blood fluke *Schistosoma mansoni* have Lewis x as repeating unit, *Eur J Biochem*, 225: 467-482.

van Dam, G.J., Claas, F.H., Yazdanbakhsh, M., Kruize, Y.C., van Keulen, A.C., Ferreira, S.T., Rotmans, J.P. and Deelder, A.M., 1996, *Schistosoma mansoni* excretory circulating cathodic antigen shares Lewis-x epitopes with a human granulocyte surface antigen and evokes host antibodies mediating complement-dependent lysis of granulocytes., *Blood*, 88: 4246-4251.

Van den Eijnden, D.H., Bakker, H., Neeleman, A.P., van den Nieuwenhof, I.M. and van Die, I., 1997, Novel pathways in complex-type oligosaccharide synthesis: new vistas opened by studies in invertebrates, *Biochem Soc Trans*, 25: 887-893.

van der Knaap, W.P.W. and Loker, E.S., 1990, Immune mechanisms in trematode-snail interactions, *Parasitol Today*, 6: 175-182.

van Die, I., Gomord, V., Kooyman, F.N., van den Berg, T.K., Cummings, R.D. and Vervelde, L., 1999, Core α1→3-fucose is a common modification of *N*-glycans in parasitic helminths and constitutes an important epitope for IgE from *Haemonchus contortus* infected sheep, *FEBS Lett*, 463: 189-193.

Velupillai, P. and Harn, D.A., 1994, Oligosaccharide-specific induction of interleukin 10 production by B220+ cells from schistosome-infected mice: a mechanism for regulation of CD4+ T-cell subsets, *Proc Natl Acad Sci USA*, 91: 18-22.

Velupillai, P., dos Reis, E.A., dos Reis, M.G. and Harn, D.A., 2000, $Lewis^X$-containing oligosaccharide attenuates schistosome egg antigen-induced immune depression in human schistosomiasis, *Human Immunol*, 61: 225-232.

Weiss, J.B. and Strand, M., 1985, Characterization of developmentally regulated epitopes of *Schistosoma mansoni* egg glycoprotein antigens, *J Immunol*, 135: 1421-1429.
Weiss, J.B., Magnani, J.L. and Strand, M., 1986, Identification of *Schistosoma mansoni* glycolipids that share immunogenic carbohydrate epitopes with glycoproteins, *J Immunol*, 136: 4275-4282.
Wuhrer, M., Dennis, R.D., Doenhoff, M.J., Lochnit, G. and Geyer, R., 2000, *Schistosoma mansoni* cercarial glycolipids are dominated by lewis X and pseudo-lewis Y structures, *Glycobiology*, 10: 89-101.
Xu, X., Stack, R.J., Rao, N. and Caulfield, J.P., 1994, *Schistosoma mansoni*: Fractionation and characterization of the glycocalyx and glycogen-like material from cercariae, *ExpParasitol*, 79: 399-409.
Zelck, U. and Becker, W., 1990, Lectin binding to cells of *Schistosoma mansoni* sporocyts and surrounding *Biomphalaria glabrata* tissue, *J Invertebr Pathol*, 55: 93-99.

GLYCOSYL PHOSPHATIDYLINOSITOL-LINKED GLYCOCONJUGATES: STRUCTURE, BIOSYNTHESIS AND FUNCTION[1-3]

Kuo-yuan Hwa

Institute of Biological Chemistry, Academia Sinica, Taipei, 115, Taiwan

Since the discovery of glycosyl phosphatidylinositols (GPIs), there has been rapid progress in understanding their structure, biosynthesis and function. Several excellent reviews have been written on various aspects of GPIs in recent years (1-7). This review emphasizes only recent developments in this field and it focuses on the conservation and variation of different GPIs among various organisms. The review is divided into three parts: structure, biosynthesis, and functions, with a closing remark on the application of GPI molecules in biomedical research and biotechnology.

I. Structure

GPIs have a conserved core structure of Manα1-4GlcNα1-6*myo*-inositol phosphate. They either serve as protein anchors or exist as free glycolipids. Many different surface glycoproteins such as enzymes, adhesion molecules, and surface antigens, are anchored on the cell surface via GPIs (see Table II). The protein-linked GPIs have a longer conserved core structure, ethanolamine-phosphate-Manα1-2Manα1-6Manα1-4GlcNβ1–4*myo*-inositol. The C-terminus of the protein is linked via an amide bond, to the ethanolamine-phosphate of GPI. In most surface glycoproteins, lipid groups are attached to the inositol phosphate of GPI core at the other end; hence the proteins are anchored to the cell membrane. The known exception to this arrangement is the cell wall protein of yeast, in which the GPI core is linked to a glucose polymer (8), not a lipid. A variety of sidechains are also found attaching to the GPI core (see review 4). GP-63, a surface cysteine protease of *Leishmania* has the simplest structure with only the conserved core, without any sidechains. Procyclin,

---

[1] Corresponding address: 128 Academia Road, Taipei, 115, Taiwan, telephone number: +886-2-785-5696, fax number: +886-2-788-9759, e-mail:kyhwa@gate.sinica.edu.tw.

[2] The author thanks Dr. Paul T. Englund, Dr. Anthony Herp and Dr. Kay-hooi Khoo for correcting this review.

[3] The abbreviations used are: GIPL, glycoinositolphospholipid; GlcNAc, *N*-acetyl glucosamine; GlcN, glucosamine; GPI, glycosyl phosphatidylinositol; LPG, lipophosphoglycan; LPPG, lipopeptidophosphoglycan; LPG, lipophosphoglycan; Man, mannose; PI, phosphatidylinositol.

The Molecular Immunology of Complex Carbohydrates–2.
Edited by Albert M. Wu. Kluwer Academic/Plenum Publishers. 2001.

the major surface antigen of the *Trypanosoma brucei* insect stage, has the most complicated GPI anchor with a large side chain of highly branched poly-*N*-acetyl-lactosamines (9,10).

The lipid structures of the GPI anchor can be grouped into two general categories. One has a glycerol moiety, including sn-1-alkyl-2-acyl-glycerol-3-phosphate-inositol (alkyl-acyl-PI), diacyl-PI, lyso-acyl-PI, lyso-alkyl-PI. The other type has a ceramide moiety such as lipopeptidophosphoglycan (LPPG), a free GPI from *Trypanosoma cruzi*. In mammalian protein anchor GPIs, a glycerol moiety is commonly found whereas in yeast and *Dictyostelim discoideum* protein anchors, a ceramide moiety is common. As with the protein-linked GPIs, significant structural and mass differences are found in free GPIs. They can be as simple as *Leishmania* glycoinositolphospholipids (GIPLs) which are essentially free GPI core with few additional sugar residues, or as complicated as *Leishmania* lipophosphoglycans (LPG), which has a long complex carbohydrate chain linked to a GPI anchor.

II. Biosynthesis and Genetics

The trypanosome, a protozoan parasite, provided a good biochemical model for studying the biosynthesis of the protein-linked GPI anchor. This is due to the high copy number of GPI-anchored variant surface glycoproteins (VSGs) in this parasite (10 million molecules per cell). Studies of trypanosome GPI have provided us with insights into the synthesis of the conserved core structure (for reviews see ref. 2, 4, 6). Furthermore, studies with *Leishmania* GPIs provided information on the biosynthesis of lipid-linked and protein-linked GPIs (11, 12, 13). Subsequent genetic work on mammalian and yeast GPI mutants accelerated the discovery of genes involved in the synthesis of protein-linked GPIs (for review see ref. 14).

The steps for the biosynthesis of protein-linked GPIs are similar in various organisms (for review see ref. 4, 6). Initiation of GPI synthesis occurs in the ER, with the addition of GlcNAc to PI. Following the addition, the second step involves de-acetylation of GlcNAc to GlcN. Mannoses are transferred from dolichol-phosphate-mannose (Dol-P-Man) and ethanolamine-phosphate is added, transferred from phosphatidylethanoamine. Although the synthesis of all protein-linked GPIs follows this pathway, there are additional steps and differences observed among organisms. These differences can be due to the properties of enzymes involved in the same steps for the GPI core or additional structure for which different enzyme is required. For example, many GPIs have additional side chain, such as ethanoamine-phosphate in mammalian cells and α-linked Gal in trypanosome VSG GPI. Syntheses of these side chains require additional enzymes. Moreover, in mammalian cells, addition of palmitate to the inositol residue is required prior to addition of mannoses (15) whereas in the trypanosome, acylation of inositol is required for additions of phospho-ethanolamine. Furthermore, comparative studies using synthetic substrate analogs indicate that the first mannosyl-transferase of HeLa cells is different from that of *Trypanosoma brucei* although both enzymes are involved in the addition of Man with the same linkage (15).

Steps involved in the biosynthesis of non-protein-linked GPIs, such as LPG, GIPLs and LPPG, are not as well conserved as in protein-linked GPIs. Different steps (hence different enzymes) are required to synthesize various structural components of GPIs. However, the initial step of non-protein-linked GPI biosynthesis, is similar to those for protein-linked GPIs, in which GlcNAc is transferred onto PI. Starting with the addition of the second mannose, the biosynthesis of non-protein-linked GPIs can diverge from that of protein-linked GPIs. The best documented system is from *Leishmania*, in which after addition of the first mannose, an α1-3-Man is added for the LPG anchor and an α1-6 Man is added to the protein-linked GPI precursor. Furthermore, it has been proposed that after addition of second mannose, LPGs are transported to Golgi whereas precursors for GIPLs and protein-anchors GPIs are retained in a special sub-compartment of ER (12).

Genetic mutants in mammalian and yeast cells have been extremely useful in identifying the genes involved in the biosynthesis of protein-linked GPIs (summarized in Table I). Another series of genetic mutants in *Leishmania* has also been very useful in finding the genes involved in the biosynthesis of non-protein-linked GPIs (see review 16).

Paroxysmal nocturnal hemoglobinuria is the only known human disease that might be due to mutation in the GPI anchor. Mutation in the *PIG-A* gene, which encodes one of the proteins involved in the first step transferring GlcNAc to PI (see Table I).

Table I
Genes encoding proteins involved in protein-linked GPI anchor synthesis

Biosynthesis Step	Gene	Note[1]
GlcNAc transfer	Mammalian	
	PIG-A	Yeast homologue, *GPI3/SPT14/CWH6*
	GPI-1	Yeast homologue, *GPI1*
	PIG-C	Yeast homologue, *GPI2*
	PIG-H	No homologue in yeast
	Yeast	
	GPI1	
	GPI2	
	GPI3	
Deacetylation of GlcNAc-PI	Mammalian	
	PIG-L	Yeast homologue, *GPI12* (YM8021.07)
	Yeast	
	GPI-12	
Third Man transfer	Mammalian	
	PIG-B	Yeast homologue, *GPI-10*(YGL142C)
	Yeast	
	GPI-10	
EtN-P transfer	**Mammalian**	
	PIG-F	Possible yeast homologue, ORF YDR302W
GPI transfer	**Mammalian**	
	GAA1	Yeast homologue, *GAA1*
	GPI-8	Yeast homologue,*GPI-8*
	Yeast	
	GAA1	
	GPI-8	
Side chain addition	**Yeast**	
	GPI-7	Addition of ethanoamine-phosphate

[1] For a more comprehensive review for this, see ref. 13.

III. Functions of GPI-linked Glycoproteins

A short list of GPI-linked proteins in human is summarized in Table II. Various human proteins, such as enzymes, surface antigens, and adhesion molecules are anchored on the cell surface by GPIs. Other than for anchoring protein, GPI anchor can have other functional roles such as in protein sorting and in signal transduction. In the following section, two cell systems will be presented to elucidate some of the functions of protein-linked GPI-anchors.

Table II
GPI-linked Proteins in Human

Protein Name	Location	Function	Note
semaphorin K1	brain and spinal cord	ıxon guidance	similar to viral, semaphorin K1
GFR alpha and GFR beta	neuron	neuronal survival	form complex with RET
RAGS (repulsive axon guidance singnal)	human adult brain, heart, spleen and ovary	axon guidance	
transferrin receptor	epithelium	iron transport	
prion protein	brain		cause scrapie
Thy-1 (CD 90)	lymphocyte		associate with p56lck and p60fyn. GPI anchor structure is known.
contactin (F3)	granule cell	adhesion molecule	
T-cadherin	vascular smooth muscle cells	homophilic cell adhesion	
urokinase receptor(CD 87)	many	receptor for urokinase plasminogen activator	signaling via JAK/STAT1
thyrotropin receptor	thyroid	thyroid function	associate with Graves disease
TRAIL-3 receptor	blood lymphocytes and spleen		
ecto-5'-nucleotidase (CD 73)	T-cells, B-cells and apical surface of epithelium	nucleotide dephosphorylation	
decay-accelerating factor (CD 55)	peripheral blood mononuclear cells, T- and B-cells	complement activation	signaling via p56lck and p59fyn
diphtheria toxin receptor	pacrease and fibroblasts	mitogenic responses to FGF-2 and HB-EGF	cancer antigen
folate receptor	kidney and liver	binding to folate	
CAMPATH-1 (CD 52)	B- and T-cells	surface antigen	GPI structure is known
LPS receptor (CD14)	monocytes macrophage	binding to LPS/LPS binding protein	signaling via p56lck
membrane attack complex (CD 59)	leukocytes	complement regulation	signaling via p56lck and G protein
Fc-R III (CD 16)	natural killer cell and monocytes	binding to IgG	
AchE receptor		acetylcholinesterase	
Dipeptidase	Kidney and pancrease	peptidase	GPI structure is known
LFA-3 (CD 58)	leukocytes	counter receptor	signaling via p56lck
CD 24	B-cell graulocytes	adhesion molecules	
BLAST-1 (CD 48)	T- and B-cells, macrophage, natural killer cells	counter receptor for CD2	signaling via p56lck and G protein
CDw109	platelets	T-cell activation	

A. GPIs in protein sorting and localization

Studies from Madin-Darby canine kidney (MDCK) cells, polarized cells with apical and basolateral surfaces, suggest that the GPI anchor can act as an active signal for protein sorting and localization. In polarized cells (e.g. MDCK cell), most of GPI-linked proteins are sorted to the apical surface (17, 18). However, the direction of sorting is cell type dependent (for review see 19) and protein-dependent. For example, carboxypeptidase M is sorted to the basolateral surface in MDCK cells (20). In some proteins, other post-translational modifications, such as *N*- or *O*-linked glycans affect the sorting mechanism of GPI-linked glycoproteins (21, 22).As a summary, multiple factors can play a synergistic role in the sorting of GPI-anchor proteins.

B. GPI anchor in T-cell activation

T-cells have many surface proteins (or receptors) that are GPI-anchored (some listed in Table II). Cross-linking of these GPI-linked proteins by multivalent antibody results in T-cell activation, even though GPI-anchors are localized on the outer leaflet of the membrane and do not penetrate through the lipid bilayer. How are signals transmitted across the cell membrane via a GPI anchor (see review 23, 24, 5, 25)? Before answering this question, it is essential to describe the distribution of GPI-linked proteins on the cell surface. It is currently accepted that GPI-anchored proteins are localized in special micro-domains on the cell surface. Within these micro-domain, many of the protein tyrosine kinases (PTKs) are associated with GPI-linked receptors in a domain of the plasma membrane, though they are not covalently attached to GPI-linked receptors. For example, the *src*-like PTKs are at the inner leaflet of the membrane. How do GPI-linked receptors on the outside communicate with PTKs on the inside? One possibility is that when the receptors are cross-linked on the outside, the PTKs are brought into closer proximity with each other. This could induce phosphorylation of kinases and initiates the signal transduction pathway. Moreover, the clustering effect of GPI-anchored receptors and kinases is sensitive to detergents, which suggests that lipid-lipid hydrophobic interaction has a role in maintaining the effect. Interaction of the T-cell receptor with its ligands can also bring closer the micro-domains consisting of GPI-linked receptors and kinases, which enhance the signal for T-cell activation. Other mechanism of signal transduction by GPI-linked receptors might involve linker molecules, such as integrin (26, 27).

IV. Closing Remark

After a decade of progress, an overwhelming amount of information has been accumulated concerning structure, biosynthesis and function of GPIs. Moreover, the information has possible applications for chemotherapy of protozoan parasitic diseases and for presentation of surface antigens. These applications in biomedical and biotech-nology research are only the beginning of the use of GPI anchors.

A. GPI Chemotherapy in Parasitic Disease

The availability of synthetic GPI-anchor molecules allows comparison studies on the enzymatic properties of protozoan parasites and their host (28, 29, 15, 30, 31). These studies suggest that there are significant differences between parasite and host enzymes. Thus, it is possible to design anti-parasitic compounds specifically targeting GPI biosynthesis since the GPI-anchored molecules are essential for the survival of parasites in their prospective hosts. Hopefully, they will be useful in the treatment of parasitic diseases, for which no efficient chemotherapy is available (for review see 32).

B. GPI in Antigen Presentation

GPI-linked surface proteins can be transferred to or inserted into membranes readily. Thus, it is possible to insert any proteins into the cell surface, via a GPI-anchor. Using GPI anchor as an exogenous mechanism to express different proteins on the cell surface via genetic engineering of fusion proteins with GPI-anchored addition sequence at the C-terminus, has been successfully performed (for review see 33). Others have used recombinant GPI-protein in a novel anti-cancer approach (34). Besides inserting recombinant GPI-linked proteins onto cell surface for antigen presentation, GPI-anchored proteins can serve potent antigens for antibody production via a CD-1 presentation pathway by natural killer cells (35).

SUMMARY

The purpose of this review is to summarize the most recent advances on GPI research. Structural studies on GPI-linked glycoconjugates indicate that there are significant variations in different organisms, although there is a conserved core structure. Furthermore, structural studies suggest that in different cell types, there is an army of glycosyltransferases dedicated to the synthesis of GPI-linked glycoconjugates. Biochemical studies on the synthesis of these GPI-linked glycoconjugates suggest that not only many different enzymes are involved but also that enzymes from different cell types, involving in the conserved core structure can have different substrate specificity. Genetic cloning of the yeast genes involved in synthesizing the core structure suggests that many of these enzymes also have human homologues. However, paroxysmal nocturnal hemogobinuria (PNH) is the only known human disease associated with the synthesis of GPI-linked glycoconjugates. Functional studies suggest that GPI-anchor can act as a signal for protein sorting and localization. Furthermore, GPI-linked receptors play an important role in T-cell activation.

REFERENCES

1. Takeda, J. and Kinoshita, T. (1995) GPI-anchor biosynthesis. Trends Biochem Sci., 20: 367-71.
2. Menon, A. K., Baumann, N. A., van't Hof, W., and Vidugiriene, J. (1997) Glycosylphosphatidylinositols: biosynthesis and intracellular transport. Biochem Soc Trans., 25: 861-5.
3. Brodbeck, U. (1998) Signalling properties of glycosylphosphatidylinositols and their regulated release from membranes in the turnover of glycosylphosphatidylinositol-anchored proteins. Biol Chem., 379: 1041-4.
4. Ferguson, M. A. (1999) The structure, biosynthesis and functions of glycosylphosphatidylinositol anchors, and the contributions of trypanosome research. J Cell Sci., 112: 2799-809.
5. Horejsi, V., Drbal, K., Cebecauer, M., Cerny, J., Brdicka, T., Angelisova, P., and Stockinger, H. (1999) GPI-microdomains: a role in signalling via immunoreceptors. Immunol Today., 20: 356-61.
6. Tiede, A., Bastisch, I., Schubert, J., Orlean, P., and Schmidt, R. E. (1999) Bio-synthesis of glycosylphosphatidylinositols in mammals and unicellular microbes. Biol Chem., 380: 503-23.
7. Morita, Y., Acosta-Serrano, A., and Englund, P. T. (in press) The biosynthesis of GPI anchors, (Ernst, B., Sina, P. and Hart, G. eds), Weinheim, Wiley-VCH.
8. Kapteyn, J. C., Montijn, R. C., Vink, E., de la Cruz, J., Llobell, A., Douwes, J. E., Shimoi, H., Lipke, P. N., and Klis, F. M. (1996) Retention of Saccharomyces cerevisiae cell wall proteins through a phosphodiester-linked beta-1,3-/beta-1,6-glucan heteropolymer. Glycobiology., 6: 337-45.
9. Mehlert, A., Zitzmann, N., Richardson, J. M., Treumann, A., and Ferguson, M. A. (1998) The glycosylation of the variant surface glycoproteins and procyclic acidic repetitive proteins of Trypanosoma brucei. Mol Biochem Parasitol., 91: 145-52.
10. Ferguson, M. A. (1997) The surface glycoconjugates of trypanosomatid parasites. Philos Trans R Soc Lond B Biol Sci., 352: 1295-302.
11. Ralton, J. E. and McConville, M. J. (1998) Delineation of three pathways of glycosylphosphatidylinositol biosynthesis in Leishmania mexicana. Precursors from different pathways are assembled on distinct pools of phosphatidylinositol and undergo fatty acid remodeling. J Biol Chem., 273: 4245-57.
12. Vidugiriene, J., Sharma, D. K., Smith, T. K., Baumann, N. A., and Menon, A. K. (1999) Segregation of glycosylphosphatidylinositol biosynthetic reactions in a subcompartment of the endoplasmic reticulum. J Biol Chem., 274: 15203-12.
13. Hilley, J. D., Zawadzki, J. L., McConville, M. J., Coombs, G. H., and Mottram, J. C. (2000) Leishmania mexicana Mutants Lacking Glycosylphosphatidylinositol (GPI):Protein Transamidase Provide Insights into the Biosynthesis and Functions of GPI-anchored Proteins. Mol Biol Cell., 11: 1183-1195.
14. Kinoshita, T., Ohishi, K., and Takeda, J. (1997) GPI-anchor synthesis in mammalian cells: genes, their products, and a deficiency. J Biochem (Tokyo)., 122: 251-7.
15. Smith, T. K., Sharma, D. K., Crossman, A., Dix, A., Brimacombe, J. S., and Ferguson, M. A. (1997) Parasite and mammalian GPI biosynthetic pathways can be distinguished using synthetic substrate analogues. EMBO J., 16: 6667-75.
16. Descoteaux, A. and Turco, S. J. (1999) Glycoconjugates in Leishmania infectivity. Biochim Biophys Acta., 1455: 341-52.
17. Ali, S., Hall, J., Hazlewood, G. P., Hirst, B. H., and Gilbert, H. J. (1996) A protein targeting signal that functions in polarized epithelial cells *in vivo*. Biochem J., 315: 857-62.
18. Lipardi, C., Nitsch, L., and Zurzolo, C. (2000) Detergent-insoluble GPI-anchored proteins are apically sorted in fischer rat thyroid cells, but interference with cholesterol or sphingolipids differentially affects detergent insolubility and apical sorting. Mol Biol Cell., 11: 531-42.
19. Nosjean, O., Briolay, A. Roux, B. (1997) Mammalian GPI proteins: sorting, membrane residence and functions. Biochim Biophys Acta., 1331: 153-86.

20. McGwire, G. B., Becker, R. P., and Skidgel, R. A. (1999) Carboxypeptidase M, a glycosylphosphatidylinositol-anchored protein, is localized on both the apical and basolateral domains of polarized Madin-Darby canine kidney cells. J Biol Chem., 274: 31632-40.
21. Benting, J. H., Rietveld, A. G., and Simons, K. (1999) N-Glycans mediate the apical sorting of a GPI-anchored, raft-associated protein in Madin-Darby canine kidney cells. J Cell Biol., 146: 313-20.
22. Alfalah, M., Jacob, R., Preuss, U., Zimmer, K. P., Naim, H., and Naim, H. Y. (1999) O-linked glycans mediate apical sorting of human intestinal sucrase-isomaltase through association with lipid rafts. Curr Biol., 9: 593-6.
23. Seed, B. (1995) Initiation of signal transduction by receptor aggregation: role of nonreceptor tyrosine kinases. Semin Immunol., 7: 3-11.
24. Robinson, P. J. (1997) Signal transduction via GPI-anchored membrane proteins. Adv Exp Med Biol., 419: 365-70.
25. Ilangumaran, S., He, H. T. and Hoessli, D. C. (2000) Microdomains in lymphocyte signalling: beyond GPI-anchored proteins. Immunol Today., 21: 2-7.
26. Petty, H. R. and Todd, R. F. D (1996) Integrins as promiscuous signal transduction devices. Immunol Today., 17: 209-12.
27. Porter, J. C. and Hogg, N. (1998) Integrins take partners: cross-talk between integrins and other membrane receptors. Trends Cell Biol., 8: 390-6.
28. Smith, T. K., Cottaz, S., Brimacombe, J. S., and Ferguson, M. A. (1996) Substrate specificity of the dolichol phosphate mannose: glucosaminyl phosphatidylinositol alpha1-4-mannosyltransferase of the glycosylphosphatidylinositol biosynthetic pathway of African trypanosomes. J Biol Chem., 271: 6476-82.
29. Sharma, D. K., Smith, T. K., Crossman, A., Brimacombe, J. S., and Ferguson, M. A. (1997) Substrate specificity of the N-acetylglucosaminyl-phosphatidylinositol de-N-acetylase of glycosylphosphatidylinositol membrane anchor biosynthesis in African trypanosomes and human cells. Biochem J., 328: 171-7.
30. Sharma, D. K., Smith, T. K., Weller, C. T., Crossman, A., Brimacombe, J. S., and Ferguson, M. A. (1999) Differences between the trypanosomal and human GlcNAc-PI de-N-acetylases of glycosylphosphatidylinositol membrane anchor biosynthesis. Glycobiology., 9: 415-22.
31. Smith, T. K., Sharma, D. K., Crossman, A., Brimacombe, J. S., and Ferguson, M. A. (1999) Selective inhibitors of the glycosylphosphatidylinositol biosynthetic pathway of Trypanosoma brucei. ENBO J., 18: 5922-30.
32. Ferguson, M. A., Brimacombe, J. S., Brown, J. R., Crossman, A., Dix, A., Field, R. A., Guther, M. L., Milne, K. G., Sharma, D. K., and Smith, T. K. (1999) The GPI biosynthetic pathway as a therapeutic target for African sleeping sickness. Biochim Biophys Acta., 1455: 327-40.
33. Medof, M. E., Nagarajan, S., and Tykocinski, M. L. (1996) Cell-surface engineering with GPI-anchored proteins. FASEB J., 10: 574-86.
34. Rushmere, N. K., Van Den Berg, C. W., and Morgan, B. P. (2000) Production and functional characterization of a soluble recombinant form of mouse CD59. Immunology., 99: 326-32.
35. Schofield, L., McConville, M. J., Hansen, D., Campbell, A. S., Fraser-Reid, B., Grusby, M. J., and Tachado, S. D. (1999) CD1d-restricted immunoglobulin G formation to GPI-anchored antigens mediated by NKT cells. Science., 283: 225-9.

Enzymatic and Chemical Approaches for the Synthesis of Sialyl Glycoconjugates[1]

Chun-Hung Lin[2] and Chun-Cheng Lin[3]

Institutes of Biological Chemistry and Chemistry, Academia Sinica
Nan-Kang, Taipei, 11529, Taiwan

Due to their important role in the chemical and biological diversity of glycoconjugates, sialic acids mainly exist as terminal components of cell surface glycoproteins and glycolipids (1). These biomolecules are highly glycosylated by complex carbohydrate chains terminated by one or several sialic acid residues. Many significant biological events are associated with sialyl glycoconjugates. For instance, changes in either amount, type or linkage of sialic acids in tumor cell glycoconjugates can affect tumor growth and metastasis (2,3). Selectins, cell adhesion molecules implicated in the recruitment of leukocytes to lymphoid tissues and to sites of inflammation, require ligands possessing α2,3-sialic acids for proper recognition (4). The B cell-specific differentiation antigen CD_{22}, which is involved in cell activation, binds to cellular lactosamine sequences containing α2,6-sialic acids (5,6). Furthermore, striking differences have been found in the sialylation pattern of cells during development, activation, aging and oncogenesis (1). Consequently, tremendous efforts have been dedicated to understand the chemical and biological significance of sialic acid containing glycoconjugates, as well as to study their structures, metabolism and immunological activities. Since natural carbohydrates exist as diverse forms, it still remains a difficult challenge to obtain identical molecules in a reasonable amount for further investigations. Synthetic methods provide an efficient approach for the synthesis of sialylated glycoforms in terms of quantity and molecular variety.

Of all the reactions of glycosidic bond formation, addition of sialic acids is often considered to be the most laborious task due to problems of low reactivity, yield, and stereoselectivity (7,8). Here we present a review how the glycosidic linkages of sialic acids are generated by enzymatic and chemical methods.

---

[1] This work was supported by Grants from Academia Sinica, Taiwan, the National Science Council (NSC 89-2113-M-001-058 for C.-H. Lin and 89-2113-M-001-062 for C.-C. Lin), Taipei, Taiwan.
[2] Institute of Biological Chemistry, Fax: +886-2-2651-4705, E-mail: chunhung@gate.sinica.edu.tw
[3] Institute of Chemistry, Fax: +886-2-2783-1237, E-mail: cclin@chem.sinica.edu.tw

I. Enzymatic Approaches

Enzymatic methods to incorporate sialic acid residues into glycoconjugates require the utilization of cytidine-5'-monophosphate-*N*-acetylneuraminic acid (CMP-sialic acid or CMP-NeuAc) synthetase and a specific sialyltransferase (Figure 1). The former protein assists in the activation of sialic acid and catalyzes the reaction of sialic acid and cytidine-5'-triphosphate. The latter transfers the sialic acid moiety to the non-reducing end of an acceptor carbohydrate. Different types of sialyltransferases are required depending on various acceptors and glycosidic linkages of sialic acid. Therefore, how to obtain both enzymes is a prerequisite to execute the enzymatic sialylation.

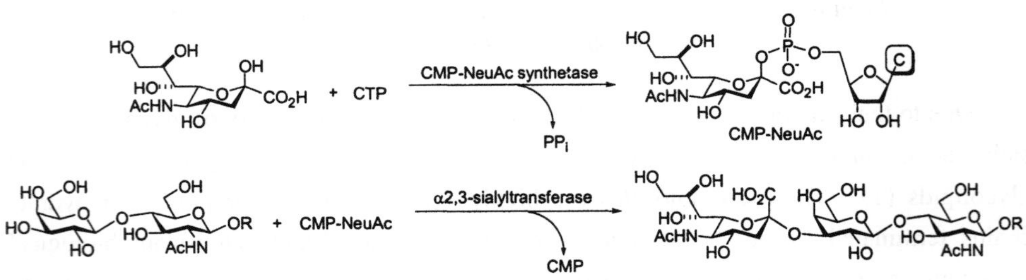

Figure 1. Reactions of CMP-NeuAc synthetase and sialyltransferase.

Although CMP-NeuAc is commercially available, its high cost (9) makes it advisable to use an efficient method of preparation. Several chemical syntheses of CMP-NeuAc have been reported (10-12). The number of reaction steps and tedious protection/deprotection procedures reflect the advantage of using CMP-NeuAc synthetase as a one-step enzymatic approach.

CMP-NeuAc synthetase, isolated from several mammalian tissues, has been used in the synthesis of CMP-NeuAc (13). The enzyme from *E. coli* was sequenced and cloned by Vann *et al.* (14-16) and further overexpressed by Shames *et al.* for large scale synthesis (17). Wong's group improved the overexpression and investigated the substrate specificity (18). The results indicated that CMP-NeuAc synthetase can tolerate the modification at the C-9 position of sialic acid without affecting the K_m value (18-21). Furthermore, sialic acid analogs, such as 9-deoxy, 7,9-dideoxy and 4,7,9-trideoxy-sialic acids were all accepted by the enzyme and converted to the corresponding CMP-NeuAc derivatives (22). However, neither the oxidation of hydroxyl group to a ketone introduced at the C-4, 7, or 8 position, nor their respective dimethylacetals were recognized as substrates by CMP-NeuAc synthetase (23).

Several α2,6- and α2,3-sialyltransferases have been used for carbohydrate synthesis (24-26). These enzymes generally transfer *N*-acetylneuraminic acid to either the 6- or 3-position of terminal Gal or GalNAc residues. The α2,8-sialyltransferase exhibits specificity for the synthesis of α2,8-linked polysialic acids (27). Nearly twenty different sialyltransferases are known to synthesize all known sialyl-oligosaccharide structures (28). Since these enzymes share the same sugar donor (CMP-NeuAc) and recognize similar acceptor substrates, it was expected that they would exhibit some sequence homology. Surprisingly, there are no significant amino acid sequence similarities among the cloned sialyltransferases, with the exception of two short consensus sequences called the sialyl motifs (29,30). Several sialyltransferases have been successfully cloned on the basis of these sialyl motifs and PCR cloning strategy (31). However, these enzymes are limited to carbohydrate synthesis because they are membrane bound proteins and mainly expressed in specific mammalian cell lines (31) in insufficient amount (in terms of synthetic purpose). Since enzymatic processes are useful and convenient to provide sialyl glycoconjugates to meet research need in glycobiology, both expressed and isolated sialyltransferases have been used to perform enzymatic sialylation. We describe some representative examples below.

The pioneering work on sialyltransferases, including isolation, characterization, cloning and synthetic application, is due to Paulson *et al.* As early as in the middle of 1980's, his group reported the enzymatic synthesis of sialylated carbohydrates by using three purified mammalian sialyltransferases which accepted type 1 (Gal-β1,3-GlcNAc), type 2 (Gal-β1,4-GlcNAc), or type 3 (Gal-β1,3-GalNAc) oligosaccharides as substrates (24). In addition, linear and branched glycopeptides with multiple sialyl-*N*-acetyl lactosamine side chains (Scheme 1) were prepared using a combined chemical and enzymatic approach (32). After solid phase synthesis to incorporate β-GlcNAc-Asn and enzymatic galactosylation to add galactose, CMP-sialic acid and α2,6-sialyltransferase were employed to form the desired α2,6-sialyl linkage, as shown in Scheme 1 (32). Calf alkaline phosphatase destroyed the inhibiting side product CMP (32).

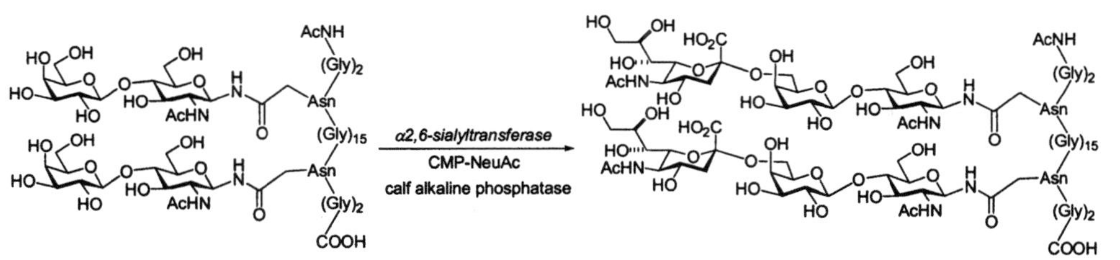

Scheme 1. Paulson's synthsis of sialyl glycopeptides using α2,6-sialytransferase

The tetrasaccharide sialyl Lewis x (SLe^x) is the carbohydrate epitope at the terminus of glycolipids displayed on the surface of neutrophils (33). SLe^x has been shown to be the ligand recognized by E-selectins, which are expressed on the surface of endothelial cells during inflammation (33, 34). The cell adhesion process of neutrophils and endothelial cells occurs through the interaction of E-selectin and SLe^x (34). The SLe^x synthesis on a large scale (kg quantities) was carried out with glycosyltransferases by Wong *et al.* and was indeed a milestone in enzymatic catalysis (35). Scheme 2 shows the consecutive glycosylation steps using galactosyl-, sialyl-, fucosyl-transferases and the corresponding sugar nucleotides.

Furthermore, multivalent SLe^x can be prepared in an efficient way based on the enzymatic remodeling of naturally existing biomolecules (36,37). For example, biantennary α2,6-sialyl *N*-acetyllactosamine has been obtained after complete delipidation and digestion of proteases (36), as shown in Scheme 3. The subsequent treatment with neuraminidase and α2,3-sialyltransferase resulted in the desired α2,3-sialyl linkage. The succeeding fucosylation with α1,3-fucosyltransferase gave the divalent SLe^x containing thirteen sugar units (36).

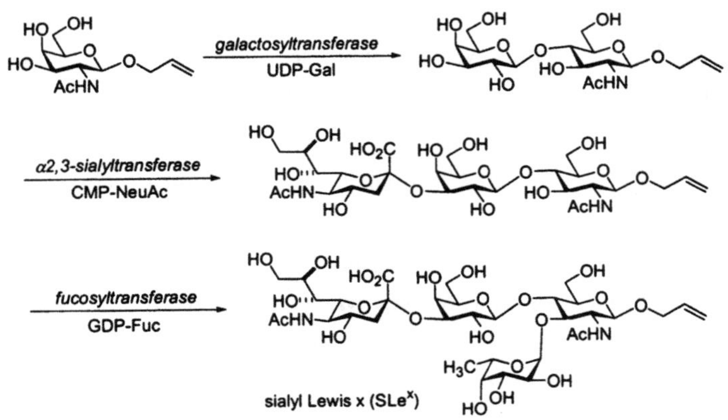

Scheme 2. Enzymatic synthesis of sialyl Lewis x developed by Wong *et al.*

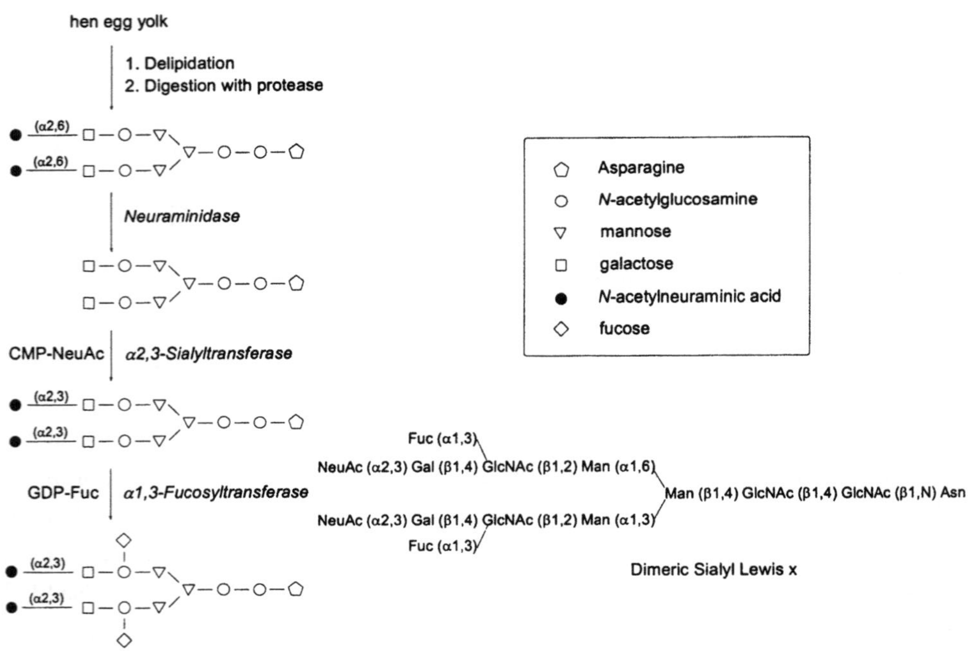

Scheme 3. Enzymatic synthsis of dimeric SLe^x glycopeptide

In addition of playing an important role in cell-cell recognition, sialic acid acts as molecular a mask or marker. The release of sialic acids from sialo-glycoproteins of the erythrocyte membrane causes exposure of the Thomsen-Friedenreich (TF) antigen (Galβ1,3-GalNAcα1-OThr)(38). Such consequence results in destruction of the red cell membrane followed by cell lysis (38). The sialylated TF antigen was recently synthesized by Thiem *et al* using a combination of β-galactosidase from bovine testes and α2,3-sialyltransferase from porcine liver (Scheme 4)(39). The former enzyme catalyzs the galactosylation in which *p*-nitrophenyl β-galactopyranoside (pNPβGal) functions as a donor. Despite the possibility of the reverse hydrolysis of the disaccharide intermediate, this reaction could be blocked by further sialylation leading to the final trisaccharide product.

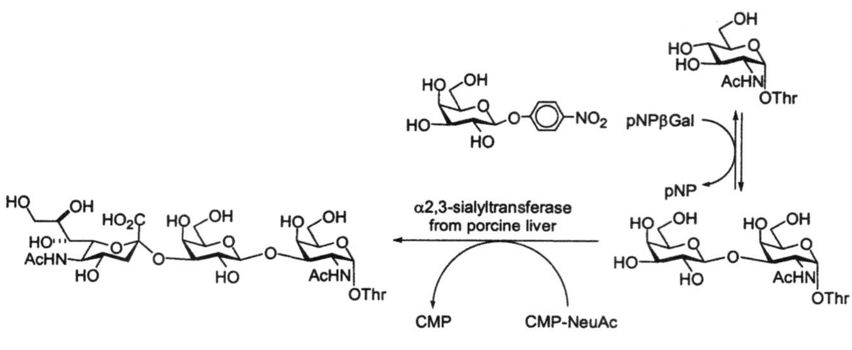

Scheme 4. Enzymatic synthesis of the Thomsen-Friedenreich antigen

The reverse reaction of glycosidases can be considered as an alternative method of enzymatic glycosylation (40) because most glycosidases, unlike glycosyltransferases, are prevalently abundant in nature. This strategy has been introduced to the enzymatic sialylation. Thiem's group utilized *p*-nitrophenyl α-*N*-acetylneuraminic acid as the sialyl donor and neuraminidase from *Vibrio cholerae* to study trans-sialylation (Scheme 5)(41). Though various unusual di- and tri-saccharides can be prepared in this way and the rate of conversion can be measured based on the release of *p*-nitrophenol, an unsatisfactory low yield (14 to 20%) remained to be improved. Moreover, a mixture of α2,3- and α2,6-isomers was obtained even in the excessive presence of sialyl acceptor (41).

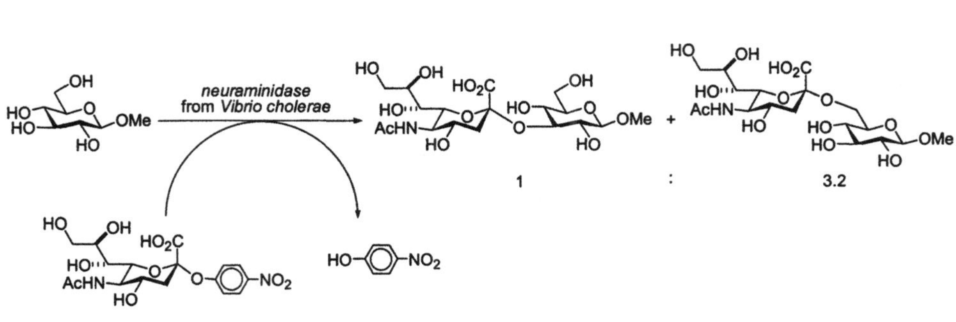

Scheme 5. Enzymatic sialylation using neuraminidase

Judging from the previous examples, silayltransferases are extremely powerful in terms of exclusive stereoselectivity and high efficiency (without protection and deprotection steps required in chemical synthesis). The enzyme reactions are, nevertheless, restricted to analytical- and small-scale synthesis due to the high cost of CMP-sialic acid (or CTP if CMP-NeuAc synthetase is used). The reactions also suffer from product inhibition caused by the released CMP. The regeneration of CMP-NeuAc *in situ* from CMP developed by Wong *et al.* is a simple solution to solve these two obstacles (42, 43). As shown in Scheme 6, nucleoside monophosphate kinase and pyruvate kinase are able to catalyze the transformation of CMP into CTP. In the regeneration system, the ultimate donor of phosphorylation comes from phospho(enol)pyruvate. As a matter of fact, Thiem's enzymatic synthesis of TF antigen was coupled with such a cofactor regeneration (39). Several advantages are offered in the *in situ* cofactor regeneration. First, the operating cost is reduced substantially since CMP-NeuAc or CTP is required in a catalytic amount. Second, product inhibition of CMP is minimized because of its low concentration. Finally, purification of the enzyme product is greatly improved without interfering CMP.

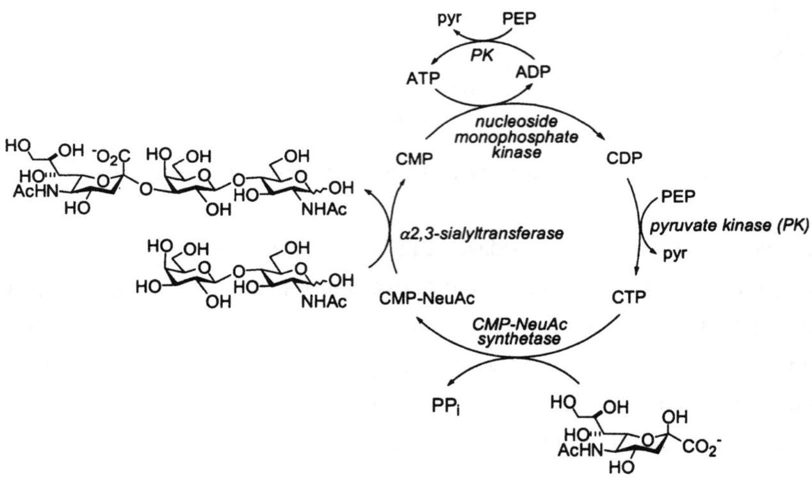

Scheme 6. *In situ* cofactor regeneration used in enzymatic sialylation

II. Chemical Approaches

Before the mid 80's, most chemical sialylations were carried out based on the Koenigs-Knorr reaction using the per-acetylated 2-halo-NeuAc **1** as the glycosyl donor. Unfortunately, for the sialylation occurring at the site of hindered alcohols (7), low α-selectivity and low yield were often observed while the elimination product **2** occurred as the major product. Recently, several new methods have been developed to obtain α-sialoglycosides as major isomers in high yields by the use of different sialyl donors, such as the 2-thioglycosides of sialic acid **3**, NeuAc phosphites **4**, S-glycosyl xanthates of

NeuAc **5**, and 3-substituted NeuAc derivatives **6** (Figure 2). In the following, we discuss the sialylation reactions on the basis of glycosyl donors.

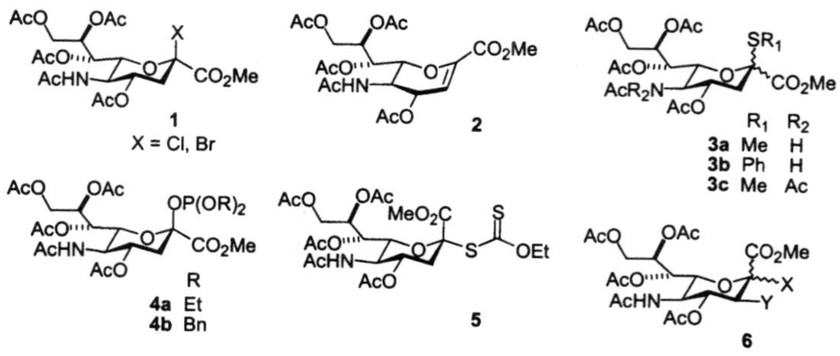

Figure 2. Various sialic acid donors

The first synthesis of thioglycosides was reported by Pivalova (44), and further investigated by Hasegawa *et al* (45-48). α-Glycosides were obtained predominantly in high yields using the anomeric mixture of either methyl (**3a**) or phenyl (**3b**) 2-thioglycoside as the sialyl donor, and DMTST (dimethyl(methylthio)sulfonium triflate) or NIS (*N*-iodosuccinimide)-T_fOH (*p*-trifluoromethanesulfonic acid) as the promoter. As shown in Figure 3 and Table 1 (45-50), various sugar acceptors were studied in acetonitrile at –40 °C. It should be noted that the secondary hydroxyls were sialylated to give exclusively the α-configuration (40-50%) in the presence of DMTST, while an anomeric mixture was obtained for the primary hydroxyls.

Table1. Glycosylation using the 2-thiolglycosides of NeuAc

Entry	Donor	Acceptor	Promoter	Yield (%) α	Yield (%) β	Reference
1	**3a**	**7**	DMTST	27	23	45
2	**3a**	**8**	DMTST	36	12	45
3	**3a**	**9**	DMTST	61	0	46
4	**3a**	**10**	DMTST	70	0	46
5	**3a**	**11**	DMTST	50	15	46
6	**3a**	**12**	NIS/T_fOH	70	0	47
7	**3a**	**13**	NIS/T_fOH	70	0	47
8	**3a**	**14**	DMTST	71	20	48
9	**3a**	**15**	DMTST	49	0	48
10	**3a**	**16**	DMTST	63	24	48
11	**3a**	**17**	DMTST	46	0	48
12	**3a**	**18**	NIS/T_fOH	59	10	46
13	**3a**	**19**	DMTST	40	0	47
14	**3a**	**20**	NIS/T_fOH	40	0	47
15	**3b**	**21**	NIS/T_fOH	65	0	49
16	**3c**	**9**	NIS/T_fOH	72	0	50
17	**3c**	**21**	NIS/T_fOH	71	0	50
18	**3c**	**22**	NIS/T_fOH	85	0	50

Moreover, under the condition of NIS/TfOH in acetonitrile, the reaction yield increased, but undesired β-glycosides were then obtained in some examples. Reducing the number of protecting groups in the acceptors improved yield and stereoselectivity. Interestingly, acceptors **10** and **11** containing different protecting groups at C-3 resulted in different α-selectivity at the C-6 glycosylation site (entries 4 and 5). There are no systematic studies about the difference in activity between sialyl donors **3a** and **3b** (49). Recently, it was discovered by Boons *et al.* that the 5-*N*-diacetyl neuraminic acid derivative **3c** is significantly more reactive and does give higher yields in sialylation than the corresponding mono-*N*-acetylated derivative **3a** (entries 16-18) (50).

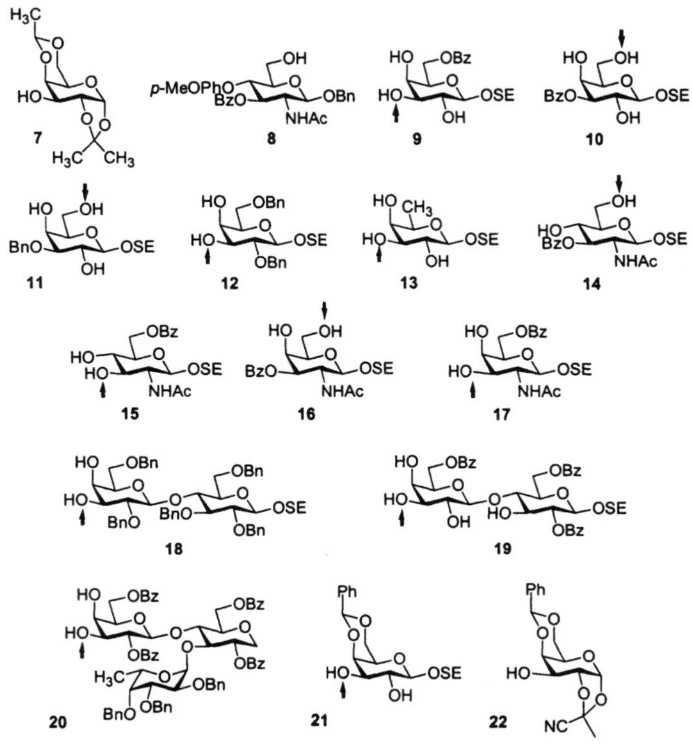

Figure 3. Structures of acceptors for glycosylation reactions using donors **3a-c**. Bold arrow indicates the glycosylation site. (SE: trimethylsilylethyl)

As shown in Figure 4, the 5-*N*-modified-2-thiosialosides (**23, 24**) and deoxy sialic acid donors (**25, 26**) also show good α-selectivity (Table 2)(51-53). In addition, the S-sialyl xanthates **28** and **29**, first reported by Sinay *et al.* (54), showed comparable reactivity to that of the thiol analogs (**23, 24**) but with less α-selectivity (Table 3 and Figure 5). The reaction conditions of these two classes of sialyl donors are similar except the promoters are different (54-61). When benzoate was used to protect the hydroxyl groups of the acceptors, sialylation yield decreased (entries 8 and 9 in Table 3). The xanthate donors, however, are not as popular as the thiol ones.

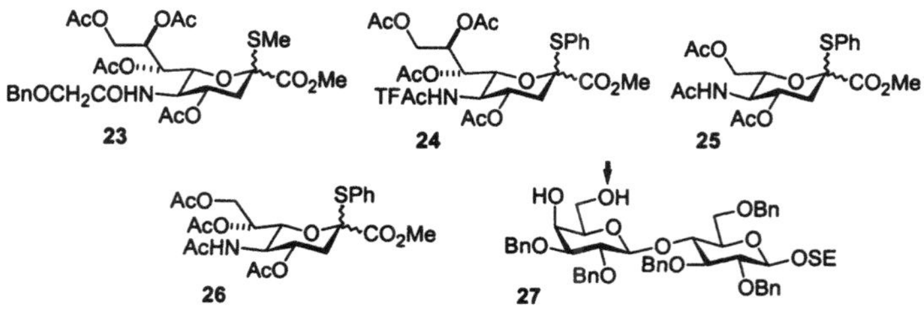

Figure 4. 2-Methyl thiol NeuAc analogues as sialic acid donors (compound **27** as the acceptor).

Table 2. Glycosylation using 2-thio NeuAc analogues

Entry	Donor	Acceptor	Promoter	Yield (%) α	Yield (%) β	Reference
1	**23**	**18**	NIS/T$_f$OH	55	0	51
2	**23**	**27**	NIS/T$_f$OH	50	13	51
3	**24**	**21**	NIS/T$_f$OH	61	0	52
4	**25**	**9**	NIS/T$_f$OH	45	0	53
5	**26**	**9**	NIS/T$_f$OH	45	0	53

Table 3. Glycosylation using the xanthates of NeuAc

Entry	Donor	Acceptor	Promoter	Yield (%) α	Yield (%) β	Reference
1	**28**	**30**	DMTST	48	16	54
2	**28**	**31**	DMTST	26	4	54
3	**28**	**32**	AgOT$_f$/MeSBr	71	4	55
4	**28**	**33**	AgOT$_f$/MeSBr	82	-	56
5	**28**	**34**	AgOT$_f$/MeSBr	56	-	57
6	**28**	**35**	AgOT$_f$/MeSBr	65	-	58
7	**28**	**36**	AgOT$_f$/PhSCl	58	-	59
8	**29**	**37**	AgOT$_f$/MeSBr	39	9	60
9	**29**	**38**	AgOT$_f$/MeSBr	15	5	60
10	**29**	**39**	AgOT$_f$/MeSBr	18	5	60
11	**29**	**40**	AgOT$_f$/MeSBr	41	18	61

Although the 2-thioglycosides of neuraminic acid have been applied for the addition of sialic acids, the requirement of at least two equivalents of thiophilic reagents is a disadvantage for this approach. Wong (12) and Schmidt (64) independently reported the utilization of sialyl phosphites as the donors of sialylations. As shown in Figure 6 and Table 4, these phosphite donors showed good α-selectivity for secondary hydroxyl acceptors. Nevertheless, as previously mentioned for most sialic acid donors, the obstacle still remains to obtain exclusive α-selectivity for sialylations occurring at primary hydroxyl groups. In order to solve this challenging problem, the anchimeric assistance by introducing an auxiliary group at C-3 has been proposed. As shown in Figures 7, 8, and Table 5 (67-73), the arylthio groups present a good directing effect with excellent α-selectivity.

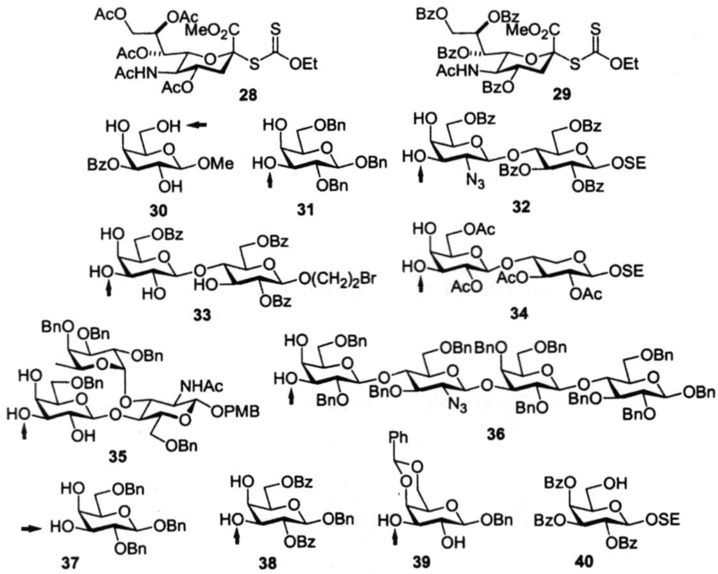

Figure 5. Structures of sialyl donors (xanthate type) and acceptors (Bold arrow indicates the glycosylation site).

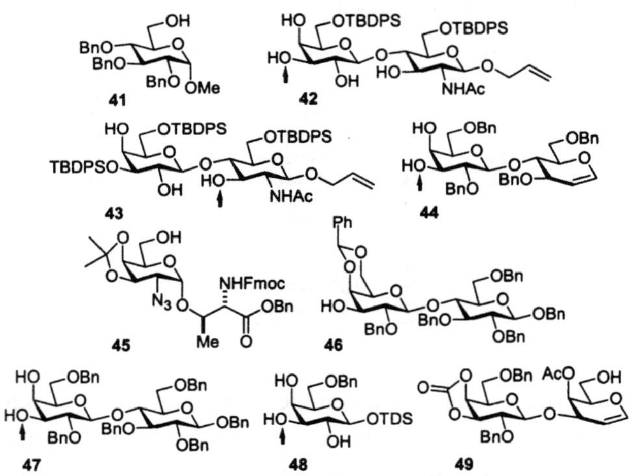

Figure 6. Structures of acceptors for the sialyl phosphite donors. (Bold arrow indicates the glycosylation site).

Table 4. Glycosylation using sialyl phosphites as donors

Entry	Donor	Acceptor	Promoter	Yield (%) α	Yield (%) β	Reference
1	**4a**	**41**	TMSOT$_f$	67	13	12
2	**4a**	**42**	TMSOT$_f$	67	11	12
3	**4a**	**43**	TMSOT$_f$	68	11	12
4	**4a**	**44**	TMSOT$_f$	75	0	62
5	**4a**	**18b**	TMSOT$_f$	77	0	62
6	**4a**	**45**	TMSOT$_f$	37	9	63
7	**4b**	**41**	TMSOT$_f$	56	14	64
8	**4b**	**46**	TMSOT$_f$	38	0	64
9	**4b**	**47**	TMSOT$_f$	55	0	65
10	**4b**	**48**	TMSOT$_f$	51	0	66
11	**4b**	**49**	TMSOT$_f$	67	17	63

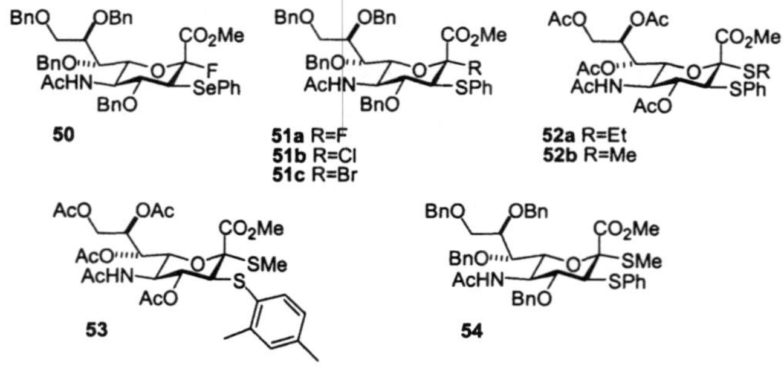

Figure 7. 3-Substituent NeuAc analogues as sialyl donors

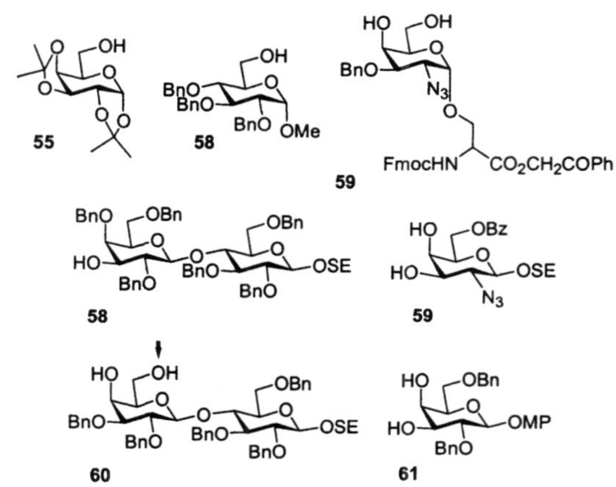

Figure 8. Structures of sialylation acceptors **55-61** for using 3-substituted NeuAc analogues (**50-54**) as sialyl donors

Table 5. Glycosylation using 3-substituted NeuAc analogues

Entry	Donor	Acceptor	Promoter	Yield (%) α	Yield (%) β	Reference
1	**50**	**55**	$AgOT_f$-$SnCl_2$	46	-	67
2	**50**	**56**	$AgOT_f$-$SnCl_2$	72	-	67
3	**50**	**47**	$AgOT_f$-$SnCl_2$	20	-	67
4	**51a**	**55**	$AgOT_f$-$SnCl_2$	68	4	68
5	**51b**	**56**	$Hg(CN)_2$-$HgBr_2$	71	0	68
6	**51b**	**47**	$Hg(CN)_2$-$HgBr_2$	64	-	68
7	**51c**	**57**	$Hg(CN)_2$-$HgBr_2$	85	-	69
8	**52a**	**58**	MSB/T_fOH	67	-	70
9	**52a**	**59**	MSB/T_fOH	77	-	70
10	**52a**	**18a**	MSB/T_fOH	71	0	70
11	**53**	**18a**	PhSCl/$AgOT_f$	78	0	71
12	**53**	**60**	PhSCl/$AgOT_f$	83	0	71
13	**52b**	**18a**	PhSCl/$AgOT_f$	83	0	72
14	**54**	**61**	NIS/T_fOH	85	0	73

The α-2,8-disialic acid, NeuAc-α(2-8)-NeuAc, is an essential component of some important glycoconjugates including gangliosides, oligo- and polysialic acids. These molecules play indispensable roles in numerous biological phenomena; e.g., tumor-associated antigens and bacterial toxins. As for the corresponding sialylations, the C-8 hydroxyl functionality as an acceptor exhibits low reactivity due to steric hindrance and intramolecular hydrogen bonding of 8-OH and 1-carboxylic ester (or 2-substituent)(74). On the other hand, attempts to prepare the α(2-8)-linkage using traditional sialyl donors, such as thioglycosides or phosphites, gave a very low yield and/or undesired β-linkage (7,45,74).

To solve such a problem, the auxiliary group at the C-3 position of sialyl donors was introduced for an efficient neighboring group participation. As shown in Table 6 and Figure 9 (74-80), dimeric sialic acid was formed as one isomer when the donor had the auxiliary group at C-3 (entries 7-9). The results appeared different from the examples without the C-3 auxiliary where α,β mixtures were obtained (entry 6). Additionally, the acceptors of 2,3-anhydroneuraminic acid derivatives, containing no hydrogen bonding as mentioned above, gave better yields of α, β mixtures when coupled with normal sialyl donors, e.g., **4b** and **3c**. Recently, Schmidt and coworkers reported that the phenoxythiocarbonyloxy group served as an excellent anchimeric group in terms of α-selectivity and yield (up to 83%) by using the phosphite as the leaving group.

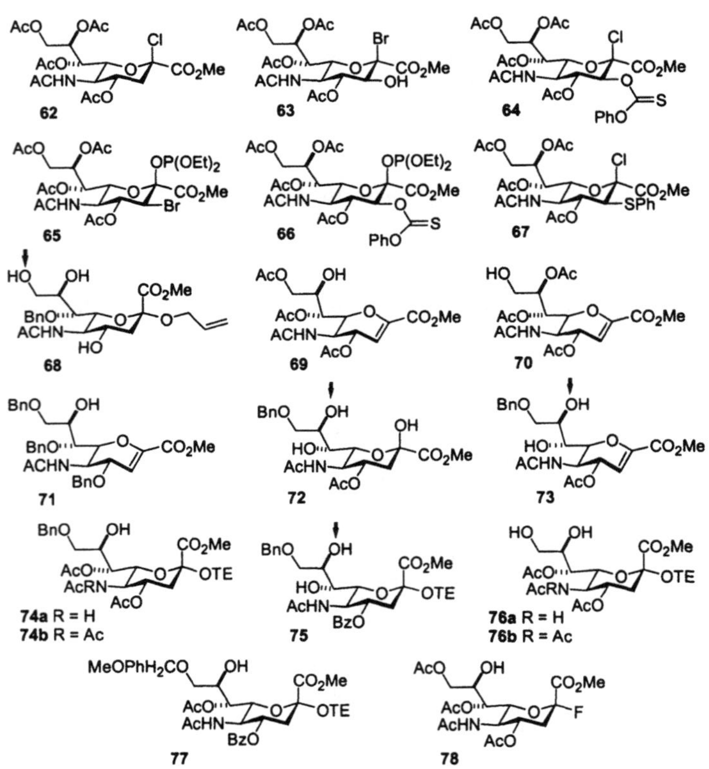

Figure 9. Sialic acid donors and acceptors for synthesis of NeuAcα(2-8)NeuAc (Bold arrow indicates the glycosylation site).

Being useful for establishing the NeuAc(2-3)Gal glycosidic linkage, the sialyl donor **3c** was also studied to prepare α-2,8 and α-2,9 sialic acid dimers. It demonstrated that the glycosyl acceptors of di-*N*-acetylated derivatives **74b** and **76b** generated significantly better results than those of mono-*N*-acetylated compounds **74a** and **76a** (entries 10, 11, 13, and 14 in Table 6). It indicated the possibility that the nucleophilicity of 8-OH could well be enhanced by removing the hydrogen bonding between 8-OH and 5-NAc.

Table 6. Synthesis of NeuAcα(2-8)NeuAc

Entry	Donor	Acceptor	Promoter	Yield (%) α	Yield (%) β	Reference
1	**62**	**68**	$Hg(CN)_2/HgBr_2$	55	28	75
2	**63**	**69**	$AgOT_f/Na_2HPO_4$	26	8	76
3	**63**	**70**	$AgOT_f/Na_2HPO_4$	42	21	76
4	**51c**	**71**	$Hg(CN)_2/HgBr_2$	64	-	77
5	**4b**	**72**	$TMSOT_f$	-	55	74
6	**4b**	**73**	$Sn(OT_f)_2$	14	54	74
7	**64**	**73**	$AgOT_f/DTBP$	68	-	74
8	**65**	**73**	$TMSOT_f$	-	58	78
9	**66**	**73**	$TMSOT_f$	83	-	78
10	**3c**	**74a**	NIS/T_fOH	8	8	79
11	**3c**	**74b**	NIS/T_fOH	32	18	79
12	**3c**	**75**	NIS/T_fOH	20	40	79
13	**3c**	**76a**	NIS/T_fOH	44	23	79
14	**3c**	**76b**	NIS/T_fOH	70	28	79
15	**52a**	**77**	$AgOT_f/MSB$	28	-	70
16	**67**	**78**	$AgOT_f/SnCl_2$	49	-	80

SUMMARY

In conclusion, either enzymatic or chemical approaches have their unique features and unavoidable disadvantages. Enzyme-catalyzed sialylations provide the desired sialo-glycosidic linkages in the two enzyme reactions (CMP-NeuAc synthetase and sialyltransferase) with exclusive stereoselectivity and high yield as long as the required sialyltransferase is available. High substrate specificity of the two enzymes is a limitation so that many unnatural glycoconjugates cannot be prepared enzymatically. As for chemical glycosylations of sialic acids, it is possible to introduce any modification in sialyl donor and acceptor, in addition to create special sugar linkages. Nevertheless, reducing the number of reaction steps (for preparing both donors and acceptors of glycosylation), and enhancing stereoselectivity, as well as reaction yield are still problems to be overcome.

REFERENCES

1. Rosenberg, A. Ed. (1995) Biology of the Sialic Acids; Plenum: New York.
2. Passaniti, A. and Hart,G. W. (1988) Cell surface sialylation and tumor metastasis. J. Biol. Chem., 263:7591-7603.
3. Santer, U. V., DeSantis, R., Hard, K. J., van Kuik, A., Vliegenthart, J. F. G., Won, B., and Glick, M. C. (1989) *N*-linked oligosaccharide changes with oncogenic transformation require sialylation of multiantennae. Eur. J. Biochem., 181:249-260.
4. Varki, A. (1994) Selectin ligands. Proc. Natl. Acad. Sci. USA, 91:7390.
5. Stamenkovic, I., Sgroi, D., Aruffo, A., Sy, M. S., and Anderson, T. (1992) The B lymphocyte adhesion molecule CD22 interacts with the leukocyte common antigen CD45RO on T cells and α2,6-sialyltransferase, CD75, on B cells. Cell, 66:1122.
6. Sgroi, D., Varki, A., Braesch-Andersen, S., and Stamenkovic, I. (1993) CD22, a B cell-specific immunoglobulin superfamily member, is a sialic acid-binding lectin. J. Biol. Chem., 268:7011.

7. Okamoto, K. and Goto, T. (1990) Glycosidation of Sialic Acid. Tetradedron, 46: 5835-5857.
8. Schmidt, R. R., Castro-Palomino, J. C., and Retz, O. (1999) New aspects of glycoside bond formation. Pure Appl. Chem. 71:729-744.
9. This enzyme is available in Calibiochem Co. (San Diego, USA). The cost is US$150/0.5 mg/1 Unit.
10. Martin, T. J. and Schmidt, R. R. (1993) Convenient chemical synthesis of CMP-*N*-acetylneuraminate (CMP-Neu-5-Ac). Tetrahedron Lett., 34:1765-68.
11. Makino, S., Ueno, Y., Ishikawa, M., Hayakawa, Y., and Hata, T. (1993) Chemical synthesis of cytidine-5'-monophospho- *N*-acetylneuraminic acid (CMP-NeuAc). Tetradedron Lett., 34:2775.
12. Kondo, H., Ichikawa, Y., and Wong, C.-H. (1992) β-Sialyl Phosphite and Phosphoramidite: Synthesis ans Application to the Chemoenzymatic Synthesis of CMP-Sialic Acid and Sialyl Oligosaccharides. J. Am. Chem. Soc., 114:8748-50.
13. (a) Simon, E. S., Bednarski, M. D., and Whitesides, G. M. (1988) Synthesis of CMP-NeuAc from *N*-acetylglucosamine: generation of CTP from CMP using adenylate kinaseJ. Am. Chem. Soc., 110:7159-63; (b) Thiem, J. and Stangier, P. (1990) Preparative-enzymatic formation of cytidine-5'-monophospho- sialate by integrated cytidine 5'-triphosphate regeneration. Liebigs Ann. Chem., 1101.
14. Vann, W. F., Silver, R. P., Abeijon, C., Chang, K., Aaronson, W., Sutton, A., Finn, C. W., Linder, W., and Kotsatos, M. (1987) Purification, properties, and genetic location of Escherichia coli cytidine-5'-monophosphate N-acetylneuraminic acid synthetase. J. Biol. Chem., 262:17556-62.
15. Silver, R. P., Finn, C. W., Vann, W. F., Aaronson, W., Schneerson, R., Kretschmer, P. J., and Garon, C. F. (1981) Molecular cloning of the K1 capsular polysaccharide genes of *E. coli.* Nature, 289:696-8.
16. Zapata, G., Vann, W. F., Aaronson, W., Lewis, M. S. and Moos, M. (1989) Sequence of the cloned Escherichia coli K1 CMP-*N*-acetylneuraminic acid synthetase gene. J. Biol. Chem., 264:14769.
17. Shames, S. L., Simon, E. S., Christopher, C. W., Schmid, W., Whitesides, G. M., and Yang, L.-L. (1991) CMP-*N*-acetylneuraminic acid synthetase of Escherichia coli: high level expression, purification and use in the enzymatic synthesis of CMP-*N*-acetyl neuraminic acid and CMP-neuraminic acid derivatives. Glycobiology, 1:187-91.
18. Liu, J. L.-J., Shen, G.-W., Ichikawa, Y., Rutan, J. F., Zapata, G., Vann, W. F., and Wong, C.-H. (1992) Overproduction of CMP-sialic acid synthetase for organic synthesis. J. Am. Chem. Soc., 114:8748.
19. Gross, H. J., Buensch, A.; Paulson, J. C., and Brossmer, R. (1987) Activation and transfer of novel synthetic 9-substituted sialic acids. Eur. J. Biochem., 168:595-602.
20. Gross, H. J.and Brossmer, R. (1988) Enzymatic introduction of a fluorescent sialic acid into oligosaccharide chains of glycoproteins. Eur. J. Biochem., 177:583-9.
21. Schauer, R., Wember, M., and do Amaral, C. F. (1972) Synthesis of CMP-glycosides of radioactive *N*-acetyl-, *N*-glycoloyl-, *N*-acetyl-7-*O*-acetyl and *N*-acetyl-8-*O*-acetylneuraminic acids by CMP-sialate synthase from bovine submaxillary glands. Hoppe-Seyler's Z. Physiol. Chem., 353:883.
22. Schreiner, E., Christian, R., and Zbiral, E. (1990) Structural variations of *N*-acetylneuraminic acid. 15. Synthesis of 9-deoxy-*N*-acetylneuraminic, 7,9-dideoxy-*N*-acetylneuraminic, and 4,7,9-trideoxy-N-acetylneuraminic acids and their behavior towards CMP-sialate synthetase. Liebigs Ann. Chem., 93-97.
23. Hartmann, M., Christian, R., and Zbiral, E. (1990) Structural variations of *N*-acetylneuraminic acid. 14. Synthesis of 4-oxo-*N*-acetylneuraminic, 7-oxo-N-acetylneuraminic, and 8-oxo-*N*-acetylneuraminic acids, and the corresponding 7,7-dimethoxy and 8,8-dimethoxy derivatives–their behavior towards CMP-sialate synthetase. Liebigs Ann. Chem., 83-91.
24. Sabesan, S. and Paulson, J. C. (1986) Combined chemical and enzymatic synthesis of sialyloligosaccharides and characterization by 500-MHz ^{1}H and ^{13}C NMR spectroscopy. J. Am. Chem. Soc., 108:2068-2080.
25. Thiem, J. and Treder, W. (1986) Synthesis of the trisaccharide NeuAc-α2,6-Gal-β1,4-GlcNAc by the use of immobilized enzymes. Angew. Chem. Int. Ed. Engl. 25:1096-7.
26. Auge, C. and Gautheron, C. (1988) An efficient synthesis of cytidine monophospho-sialic acids with 4 immobilized enzymes. Tetradedron Lett., 29:789-90.
27. (a) McCoy, R. D., Vimr, E. R. and Troy, F. A. (1985) CMP-NeuAc:poly-a-2,8-sialosyl sialyltransferase and the biosynthesis of polysialosyl units in neural cell adhesion molecules. J. Biol. Chem., 260:12695; (b) Troy, F. A., McCloskey, M. A. (1979) Role of a membranous sialyltransferase complex in the synthesis of surface polymers containing polysialic acid in Escherichia coli. Temperature-induced alteration in the assembly process. J. Biol. Chem., 254:7377-87.
28. Harduin-Lepers, A., Recchi, M.-A., and Delannoy, P. (1995) 1994, the year of sialyltransferase. Glycobiology, 5:741-758.
29. Drickamer, K (1993) A conserved disulphide bond in sialyltransferases. Glycobiology, 3: 2-3.
30. Livingston, B. D. and Paulson, J. C. (1993) Polymerase chain reaction cloning of a developmentally regulated member of the sialyltransferase gene family. J. Biol. Chem., 268:11504-11507.
31. For a representative review: Tsuji, S. (1996) Molecular cloning and functional analysis of sialyltransferases. J. Biochem., 120:1-13.
32. Unverzagt, C., Kelm, S., and Paulson, J. C. (1994) Chemical and enzymatic synthesis of multivalent sialoglycopeptides. Carbohydr. Res., 251:285-301.

33. (a) Phillips, M. L., Nudelman, E., Gaeta, F. C. A., Perez, M., Singhal, A. K., Hakomori, S., and Paulson, J. C. (1990) ELAM-1 mediates cell adhesion by recognition of a carbohydrate ligand, sialyl-Le^x. Science, 250:1130-2; (b) Graves, B. J., Crowther, R. L., Chandran, C., Rumberger, J. M., Li, S., Huang, K.-S., Presky, D. M. Familletti, P. C., Wolitzky, B. A., and Burns, D. K. (1994) Insight into E-selectin/ligand interaction from the crystal structure and mutagenesis of the lec/EGF domains. Nature, 367:532.
34. Lasky, L. A., (1992) Selectins: interpreters of cell-specific carbohydrate information during inflammation. Science, 258:964-9.
35. Ichikawa, Y., Lin, Y.-C., Dumas, D. P., Shen, G.-W., Garcia-Junceda, E., Williams, M. A., Bayer, R., Ketcham, C., Walker, L. E., Paulson, J. C., and Wong, C.-H. (1992) Chemical-enzymatc synthesis and conformational analysis of sialyl lewis x and derivatives. J. Am. Chem. Soc., 114:9283.
36. Lin, C.-H., Shimazaki, M., Wong, C.-H., Koketsu, M., Juneja, L. R., and Kim, M. (1995) Enzymatic synthesis of a sialyl lewis x dimer from egg yolk as an inhibitor of E-selectin. Bioorg. Med. Chem., 3:1625-30.
37. De Frees, S. A., Gaeta, F. C. A., Lin, Y.-C., Ichikawa, Y., and Wong, C.-H. (1993) Ligand recognition by E-selectin – Analysis of conformation and activity of synthetic monomeric and bivalent sialyl Lewis x analogs. J. Am. Chem. Soc., 115:7549-50.
38. Schauer, R., Shukla, A. K., Schröder, C., and Müller, E. (1984) The anti-recobnition of sialic acids – studies with erythrocytes and macrophages. Pure Appl. Chem. 56:907-921.
39. Thiem, J. and Gambert, U. (1999) Multienzyme system for synthesis of the sialylated Thomsen-Friedenreich antigen determinant. Eur. J. Org. Chem. 107-110.
40. Herrmann, G. F., Ichikawa, Y., Wandrey, C., Gaeta, F. C. A., Paulson, J. C., and Wong, C.-H. A new multi-enzyme system for a one pot synthesis of sialyl oligosaccharides: combined use of β-galactosidase and α(2,6)-sialyltransferase coupled with regeneration *in situ* of CMP-sialic acid. Tetrahedron Lett. 34:3091-3094.
41. Thiem, J. and Sauerbrei, B. (1991) Chemoenzymatic syntheis of sialyloligosaccharide with immobilized sialidase. Angew. Chem. Int. Ed. Engl. 30:1503-1505.
42. Ichikawa, Y., Shen, G.-W., and Wong, C.-H. (1991) Enzyme-catalyzed synthesis of sialyl oligosaccharide with in situ regeneration of CMP-sialic acid. J. Am. Chem. Soc., 113:46984700.
43. Ichikawa, Y., Liu, J. L.-J., Shen, G.-W., and Wong, C.-H. (1991) An highly efficient multienzyme system for the one step synthesis of a sialyl tetrasaccharide – in situ generation of sialic acid and *N*-acetyllactosamine coupled with regeneration of UDP-glucose, UDP-galactose, and CMP-sialic acid. J. Am. Chem. Soc., 113:6300-02.
44. Pivalova and I. M.; Kholin, A. Y. (1969) Izv. Akad. Nauk SSSR, Ser. Khim. 2785.
45. Kanie, O.; Kiso, M. and Hasegawa, A. (1988) Glycosylation using methylthioglycosides of *N*-acetylneuraminic acid and dimethyl (methylthio)sulfonium triflate. J. Carbohydr. Chem., 7:501-506.
46. Hasegawa, A.; Nagahama, T.; Ohki, H.; Hotta, K.; Ishida, H. and Kiso, M. (1991) Synthetic studies on sialoglycosylation of *N*-acetylneuraminic acid with the primary and secondary hydroxyl groups in the suitably protected galactose and lactose derivatives. J. Carbohydr. Chem., 10:493-498.
47. Murase, T.; Ishida, H.; Kiso, M. and Hasegawa, A. (1988) A facile regio- and stereo-selective synthesis of a-glycosides of *N*-acetylneuraminic acid. Carbohydr. Res., 184:C1-C4.
48. Hasegawa, A. (1996) Synthesis of sialoglycoconjugates. In *Modern Methods in Carbohydrate Synthesis*, (Khan, S. H. and O'Neil R. A., Ed.) pp. 277-300, Harwood Academic GmbH, Netherlands.
49. Komba, S.; Meldal, M.; Werdelin, O.; Jensen, T. and Bock, K. (1999) Convenient synthesis of Thr and Ser carrying the tumor associated sialyl-(2→3)-T antigen as building blocks for solid-phase glycopeptide synthesis. J. Chem. Soc., Perkin Trans. 1, 415-419.
50. Demchenko, A. V. and Boons, G.-J. (1998) A novel and versatile glycosyl donor for the preparation of glycosides of *N*-acetylneuraminic acid. Tetrahedron Lett., 39:3065-3068.
51. Yoshida, M.; Uchimura, A.; Kiso, M.; Hasegawa, A. (1993) Synthesis of chemically modified sialic acid-containing sialyl-Le^x ganglioside analogues recognized by the selectin family. Glycoconjugate J., 10:3-15.
52. Komba, S.; Galustian, C. Ishida, H. Feizi, T. annagi, R. and Kiso, M. (1999) The first total synthesis of 6-sulfo-de-*N*-acetylsialyl Lewis x ganglioside: a superior ligand for human L-selectin. Angew. Chem. Int. Ed., 38:1131-1133.
53. Sugata, T. and Higuchi, R. A (1996) Facile preparation of the methyl 2-thioglycoside of *N*-glycolylneuraminic acid, an efficient donor of NeuGc. Tetrahedron Lett., 37:2613-2614.
54. Marra, A. and Sinaÿ, P. (1990) A novel stereoselective synthesis of *N*-acetyl-α-neuraminosyl-galactose disaccharide derivatives, using anomeric *S*-glycosyl xanthates. Carbohydr. Res., 195:303-308.
55. Ray, A. K.; Nilsson, U. and Magnusson, G. (1992) Synthesis and conformational annalysis of GM_3 lactam, a hydrolytically stable analogue of GM_3 ganglioside lactone. J. Am. Chem. Soc., 114:2256-2257.
56. Lönn, H. and Stenvall, K. (1992) Exceptionally high yield in glycosylation with sialic acid: Synthesis of a GM3 glycoside. Tetrahedron Lett., 33:2613-2614.
57. Wilstermann, M. and Magnusson, G. (1997) Synthesis of XylβCer, Galβ1-4XylβCer, NeuAcα2-3Galβ1-4XylβCer and the corresponding lactone and lactam trisaccharides. J. Org. Chem., 62:7961-7971.

58. Ellervik, U. and Magnusson, G. (1998) A high yielding chemical synthesis of sialyl Lewis x tetrasccharide and Lewis x trisaccharide; examples of regio- and stereo-differentiated glycosylations. J. Org. Chem., 63:9314-9322.
59. Tietze, L. F.; Janssen, C. O. and Gewert, J.-A. (1998) Synthesis of neolacto ganglioside LM1. Eur. J. Org. Chem., 1887-1894.
60. Birberg, W. and Lönn, H. (1991) Glycosylation with sialic acid at HO-3 of three different *O*-protected D-galactosides in acetonitrile/dichloromethane at low temperature. Tetrahedron Lett., 32:7457-7458.
61. Birberg, W. and Lönn, H. (1991) α-Selectivity and glycal formation are temperature dependent in glycosylation with sialic acid. synthesis of a NeuAcα(2-6)Gal thioglycoside building block. Tetrahedron Lett., 32:7453-7456.
62. Allen, J. R. and Danishefsky, S. J. (1999) New applications of the *n*-pentenyl glycoside method in the synthesis and immunoconjugation of fucosyl GM1: A highly tumor-specific antigen associated with small cell lung carcinoma. J. Am. Chem. Soc., 121:10875-10882.
63. Schwarz, J. B.; Kuduk, S. D.; Chen, X.-T,; Sames, D.; Glunz, P. W. and Danishefsky, S. J. (1999) A broadly applicable method for the efficient synthesis of α-*O*-Linked glycopeptides and clustered sialic acid residues. J. Am. Chem. Soc., 121:2662-2673.
64. Martin, T. J. and Schmidt, R. R. (1992) Efficient sialylation with phosphite as leaving group. Tetrahedron Lett., 33:6123-6126
65. Martin, T. J.; Brescello, R. Toepfer, A. and Schmidt, R. R. (1993) Synthesis of phosphites and phosphates of neuraminic acid and their glycosyl donor properties-convenient synthesis of GM3. Glycoconjugate J., 10:16-25.
66. Gege, C.; Vogel, J.; Bendas, G.; Rothe, U. and Schmidt, R. R. (2000) Synthesis of the sialyl Lewis x epitope attached to glycolipids with different core structures and their selectin-binding characteristics in a dynamic test system. Chem. Eur. J., 6:111-122.
67. Ito, Yukishige and Ogawa, T. (1987) An efficient approach to stereoselective glycosylation of *N*-acetylneuraminic acid: Use of phenylselenyl group as a stereocontrolling auxiliary. Tetrahedron Lett., 28:6221-6224.
68. Ito, Yukishige and Ogawa, T. (1988) Highly stereoselective glycosylation of *N*-acetylneuraminic acid aided by a phenylthio substituent as a stereocontrolling auxiliary. Tetrahedron Lett., 29:3987-3990.
69. Nakahara, Y.; Iijima, H.; Shibayama, S., and Ogawa, T. (1991) Stereoselective total synthesis of glycopeptides bearing the dimeric and trimeric sialyl-Tn epitope. Carbohydr. Res., 216:211-225.
70. Ercégovic, T. and Magnusson G. (1995) Highly stereoselective α-sialylation: Synthesis of GM_3-saccharide and a bis-sialic acid unit. J. Org. Chem., 60:3378-3384.
71. Martichonok, V. and Whitesides, G. M. (1996) A practical method for the synthesis of sialyl α-glycosides. J. Am. Chem. Soc., 118:8187-8191
72. Martichonok, V. and Whitesides, G. M. (1997) Studies on a-sialylation using sialyl donors with an auxiliary 3-thiophenyl group. Carbohydr. Res., 302:123-129.
73. Singh, L.; Nakahara, Y.; Ito, Y., and Nakahara, Y. (1999) The first tatal synthesis of the core class II disialylated hexasaccharide as a building block for glycopeptide synthesis. Tetrahedron Lett., 40:3769-3772.
74. Castro-Palomino, J. C.; Tsvetkov, Y. E.; Schneider, R. and Schmidt, R. R. (1997) 8-*O*-Sialylation of neuraminic acid acceptor reactivity and anomeric stereocontrol. Tetrahedron Lett., 38:6837-6840.
75. Shimizu, C. and Achiwa, K. (1987) Synthesis of α-(2-9)-disialic acid. 166:317-323.
76. Okamoto, K., Kondo, T., and Goto, T. (1986) Synthesis of (α2-9) and (α2-8) linked neuraminylneuraminic acid derivatives. Tetrahedron Lett., 27:5229-5232.
77. Ito, Y., Numata, M., Sugimoto, M., and Ogawa, T. (1989) Highly stereoselective synthesis of ganglioside GD_3. J. Am. Chem. Soc., 111:8508-8510.
78. Castro-Palomino, J. C., Tsvetkov, Y. E., and Schmidt, R. R. (1998) 8-*O*-Sialylation of neuraminic acid. J. Am. Chem. Soc., 120:5434-5440.
79. Demchenko, A. V. and Boons, G.-J. (1999) A novel direct glycosylation approach for the synthesis of dimers of *N*-acetylneuraminic acid. Chem. Eur. J., 5:1278-1283.
80. Kondo, T.; Tomoo, T.; Abe, H.; Isobe, M.; and Goto, T. (1996) Simple construction of Neu5Ac α(2-8)Neu5Ac and total synthesis of ganglioside GD3. J. Carbohydr. Chem., 15:857-878.

THE GENETIC REGULATION OF FUCOSYLATED AND SIALYLATED ANTIGENS ON DEVELOPING MYELOID CELLS [i]

Winifred M. Watkins and Julia L. Clarke

Department of Haematology, Imperial College School of Medicine, Hammersmith Hospital, London W12 ONN, U.K.

The haemopoietic tissues provide an especially valuable model for the study of cellular differentiation, proliferation, morphogenesis, functional maturation and senescence because the blood cells exist as single free entities in the bone marrow and in the peripheral circulation. Cell-cell interactions are believed to mediate the development of blood cells and the stromal cells of the marrow are considered to provide the growth-promoting and differentiation inducing molecules[1] but, although the pathways leading to rolling, binding and extravasation of mature leukocytes have been widely explored, the processes involved in events occurring within the bone marrow are still poorly understood. Intimate contact between the haemopoietic cells and the stromal cells is required for development processes to take place and hence cell surface adhesion molecules and their ligands are assumed to be essential facilitators of these interactions. The carbohydrate moieties on glycoproteins and glycosphingolipids have been shown to be involved in many cell-cell recognition events[2] such as lymphocyte homing to peripheral lymph nodes[3], the recruitment of granulocytes to the sites of endothelial injury and inflammation[4,5,6] and the interaction between bone marrow macrophages and developing myeloid cells[7]. Therefore cell surface carbohydrate structures on myeloid cells may well have an important part to play in the differentiation and maturation processes occurring in the bone marrow and the mechanisms by which the cells are released into the circulation.

Haemopoietic cells commited to the myeloid lineage undergo final differentiation into granulocytes and monocytes. In the course of cellular maturation along the granulocytic pathway specific changes occur in the carbohydrate moieties of glycoconjugates on the cell surface[8,9,10,11] and the degree of terminal glycosylation of oligosaccharide chains with the monosaccharides fucose and sialic acid plays a major role in the changing profiles. Much valuable information on the importance of fucosylated and sialylated cell surface structures in relation to function of myeloid cells has come recently from experiments on mice rendered "null" for certain fucosyltransferase genes by gene-targeting approaches[12,13]. These experiments can, however, reveal only the end-stage effects of the absence of a gene and do not always enable the exact developmental stage to be pin-pointed at which lack of a gene may have a disadvantageous consequence. Additionally maturation changes may depend on variations in the levels of gene expression rather than the absolute presence or

[i] The experimental part of the work described in this review was supported by the Wellcome Trust.

absence of a certain gene product. In any event, since haemopoietic development in the mouse does not completely mirror the situation in humans [14,15], examination of human cells committed to iscrete lineages as they undergo specific maturation within the bone marrow may complement the functional information deduced from the "knockout" mice experiments. This article reviews the background to the regulation of the appearance of fucosylated and sialylated antigens on the surface of human myeloid cells and outlines our attempts to understand the genetic basis of the modulation of expression of the sialyl-Le^x determinant, and related antigens, that occurs in the course of cell proliferation and maturation.

I. MATURATION OF HAEMOPOIETIC CELLS

Multipotent haemopoietic stem cells can be differentiated along erythroid, myeloid or lymphoid lineages[16]. Haemopoietic progenitor cells that are destined to become granulocytes first become committed to the myeloid lineage. The early forms are recognised as *myeloblasts,* with large nuclei and little cytoplasm. The cells gradually become smaller, the nucleus occupies less space and the cytoplasm becomes more granulated as they develop into *promyelocytes.* The progeny of the promyelocytes are *myelocytes* which are the last proliferative stage of granulopoiesis. The cells nevertheless undergo further maturation in the bone marrow to *metamyelocytes* and *band* forms, until eventually the nucleus becomes fully segmented and the cells emerge into the peripheral circulation as *mature granulocytes* (**Fig 1**). The average time scale for a blast cell to mature to a polymorphnuclear cell, that is, into a mature granulocyte, is about twelve days. Proliferation of the cells ceases after about six days but maturation continues in the bone marrow until the mature cells are released into the circulation where they spend less than one day before utilisation and destruction in the tissues.

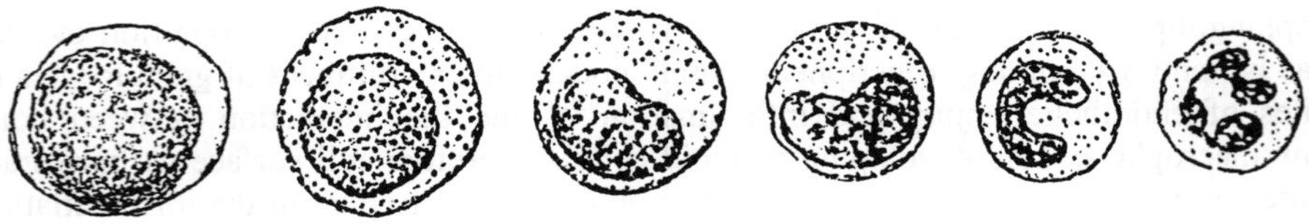

Fig 1. Diagrammatic representation of maturation stages of human myeloid cells committed to the granulocytic pathway.

II. HUMAN LEUKAEMIC CELL LINES

Human leukaemic cell lines are derived from haemopoietic cells that cease to differentiate at various stages of development due to the leukaemic process[17] but continue to proliferate and can therefore be grown to supply amounts of cells at defined differentiation stages in quantities that are not readily available from aspirates of normal human bone marrow. The KG1a cell line, which is composed of undifferentiated blast cells, represents an early myeloid cell progenitor; this cell line cannot be induced to differentiate in response to chemical inducers[18]. The parent cell line KG1 from which the subline KG1a was isolated, was derived from bone marrow cells of a patient with erythroid leukaemia and consists of cells at the myeloblast and promyelocytic stages of development[19]. These cells do not show granulocytic differentiation in response to dimethylsulfoxide (DMSO) but they do mature into macrophages in the presence of phorbol esters and respond to colony stimulating factor (CSF) in soft-gel culture. The widely used HL-60 cell line was established from a patient with acute myeloid leukaemia[20] and comprises cells which in many respects closely resemble non-leukaemic promyelocytes[16]. These cells can be induced to differentiate along the granulocytic pathway when grown in the presence of DMSO, or certain other chemical inducers, and along the monocytic pathway when cultured with phorbol esters[21]. Culturing the HL-60 cells for three to four days in DMSO yields a population that morpholgically resemble a mixture of myelocyte, metamyelocyte and early band forms of developing granulocytes[16,22]. Thus KG1a, KG1, HL-60 and differentiated HL-60 cells **(Table 1),** although they are not normal cells, have properties that enable them to be used as models to represent successive steps in the maturation pathway of granulocytes.

Table1.
Human leukaemic myeloid cell lines arrested at different maturation stages

Cell line	Maturation stage	Expression of maturation markers [a]			
		CD34	CD15	CD11b	CD16
KG1a	Myeloblast	+++	-	-	-
KG1	Myeloblast and early promyelocyte	+++	+++	-	-
HL-60	Promyelocyte	-	+++	-	-
HL-60 differentiated with DMSO	Myelocyte, metamyelocyte, Band form	-	+++	++	-
-	Mature granulocyte	-	+++	+++	+++

a. - No expression; + weak expression; ++ moderate expression; +++ strong expression

III. CELL SURFACE CHANGES ON DEVELOPING MYELOID CELLS

The possibility of studying the cell surface changes on developing myeloid cells has been greatly facilitated by the advent of the hybridoma technique that has made available a range of highly specific monoclonal antibodies for the detection of closely related carbohydrate structures[23,24]. Of equal importance for an understanding of the biochemical and genetical mechanisms underlying the changing pattern of carbohydrate cell surface structures has been the characterisation of glycosyltransferase enzymes responsible for the biosynthesis of the oligosaccharide chains[25,26,27] and the cloning of the genes encoding the fucosyl-[28] and sialyl-transferases[29,30]. Among the cell surface carbohydrate antigens that undergo modulation in the course of development of myeloid cells are the Lewis blood-group-related determinants, Le^x (also known as CD15 and SSEA-1) and sialyl- Le^x (SLe^x). Examination of normal bone marrow derived cells with monoclonal anti-Le^x has revealed that the Le^x determinant is absent from myeloblasts but is expressed on all myeloid cells from the promyelocyte stage onwards [31,32]. Terstappen *et al.* [33] undertook a detailed study of granulocytic maturation in aspirates of normal human bone marrow and, based on morphology and differential expression of the maturation markers CD15, CD11b, and CD16, identified six distinct maturational stages which they called NI to NVI. Subsequently, in order to identify membrane components involved in stage-specific interactions they studied expression of cell adhesion molecules along the granulocytic maturational pathway [34] and observed that, in contrast to Le^x , sialyl-Le^x is strongly expressed on myeloid progenitor cells but is down-modulated to a low level during intermediate stages of granulocytic maturation and up-regulated to levels similar to those in progenitor cells after termination of proliferative activity (**Fig.2**). The stage specific regulation of this determinant led these authors to suggest that sialyl-Le^x could be important for compartmentalisation of distinct phases of granulocyte maturation. The transient down- regulation of sialyl-Le^x at intermediate stages of maturation of normal bone marrow derived myeloid cells was confirmed by Le Marer *et al.* [35]

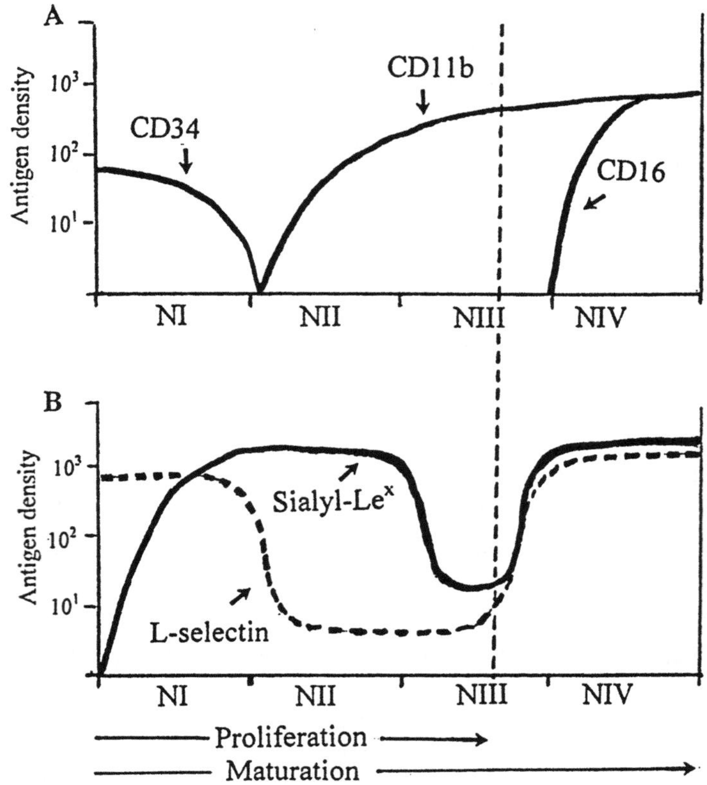

Fig. 2. Schematic representation of granulocyte maturation as defined by expression of (**A**) maturation markers CD34, CD11b and CD16 and (**B**) adhesion molecules sialyl-Le^x and L-selectin (Modified from Lund-Johansen and Terstappen [34]). NI, predominantly myeloblasts, NII, predominantly promyelocytes, NIII, early myelocytes and myelocytes, NIV, myelocytes and metamyelocytes.

IV. STRUCTURES OF FUCOSYLATED AND SIALYLATED ANTIGENIC DETERMINANTS ON MYELOID CELL SURFACES

Some of the fucosylated and sialylated epitopes that occur as terminal structures on oligosaccharide chains on myeloid cell surfaces are illustrated in **Fig.3**. They are all based on precursor Type 2 , *N*-acetyllactosamine (Galβ1-4GlcNAc) units (Structure 1, **Fig.3**). Fucose linked α1,3 to the subterminal *N*-acetylglucosamine residue gives the Le^x structure

(Structure 2, **Fig.3**) and the difucosyl compound with an α1,2 linked substituent on the terminal galactose and α1,3-linked on *N*-acetylglucosamine forms the Le^y determinant (Structure 6, **Fig.3**). Sialic acid (*N*-acetylneuraminic acid) may be linked either α2,3 or α2,6 to a terminal galactose residue to give, respectively, 3' and 6'-sialyl-*N*-acetyllactosamine (Structures 3 and 4, **Fig.3**). 3'-sialyl-*N*-acetyllactosamine substituted with α1,3-linked fucose on the *N*-acetylglucosamine residue constitutes the sialyl-Le^x determinant (Structure 5, **Fig.3**). A feature of myeloid cells is the presence of oligosaccharide chains containing repeating *N*-acetyllactosamine units that give rise to polylactosamine chains (lactosaminoglycan). Fukuda *et al.*,[36] demonstrated that that leukaemic cell lines arrested in the early stages of differentiation, such as KG1a and KG1, had a small amount of essentially unbranched lactosaminoglycan whereas a significant quantity of branched lactosaminoglycan is present on the more mature HL-60 cells. They also showed that chemically induced maturation of the promyelocyte HL-60 cells is accompanied by a great increase in the quantity and branching structure of lactosaminoglycan[36]. These polylactosamine chains function as scaffolds for internal fucosylation along the chains and for terminal substitution by sialic acid. A murine mono-

---

1. *N*-acetyllactosamine (Type 2 — Galβ1-4GlcNAc-R
2. Le^x (CD15; SSEA-1) — Galβ1-4GlcNAc-R
   3
   1
   Fucα
3. 3'-Sialyl-*N*-acetyllactosamine — NeuAcα2-3Galβ1-4GlcNAc-R
4. 6'-Sialyl-*N*-acetyllactosamine — NeuAcα2-6Galβ1-4GlcNAc-R
5. Sialyl-Le^x — NeuAcα2-3Galβ1-4GlcNAc-R
   3
   1
   Fucα
6. Le^y — Fucα1-2Galβ1-4GlcNAc-R
   3
   1
   Fucα
7. Dimeric Le^x — Galβ1-4GlcNAcβ1-3 Galβ1-4GlcNAc-R
   3 3
   1 1
   Fucα Fucα
8. VIM 2 (CDw65) — NeuAcα2-3Galβ1-4GlcNAcβ1-3 Galβ1-4GlcNAc-R
   3
   1
   Fucα
9. Sialyl-dimeric Le^x — NeuAcα2-3Galβ1-4GlcNAcβ1-3 Galβ1-4GlcNAc-R
   3 3
   1 1
   Fucα Fucα

Fig 3. Some terminal structures of oligosaccharide chains on glycoproteins and glycolipids on myeloid cells. Abbreviations: Gal, D-galactose; GlcNAc, *N*-acetyl-D-glucosamine; Fuc, L-fucose; NeuAc, *N*-acetylneuraminic acid (sialic acid); R, remainder of molecule.

clonal antibody, VIM-2, specific for an antigen on human myeloid cells[37,38], was shown to combine with a structure in which a polylactosamine chain carried a terminal sialic acid residue and also a fucosyl residue α1,3-linked to the *N*-acetylglucosamine residue of the the second *N*-acetyllactosamine unit (Structure 8, **Fig.3**); thus showing that antigenicity does not reside solely in the terminal tri- and tetra-saccharide (Le^x and sialyl-Le^x) units. The antigen detected by the monoclonal antibody VIM-2 has been given the CD classification CD_w65 but is generally also referred to as VIM-2. α1,3-Fucosylation of two adjacent lactosamine units gives the dimeric Le^x structure (Structure 7, **Fig.3**) and terminal sialylation and fucosylation of two adjacent lactosamine units gives sialyl-dimeric Le^x (Structure 9, **Fig.3**).

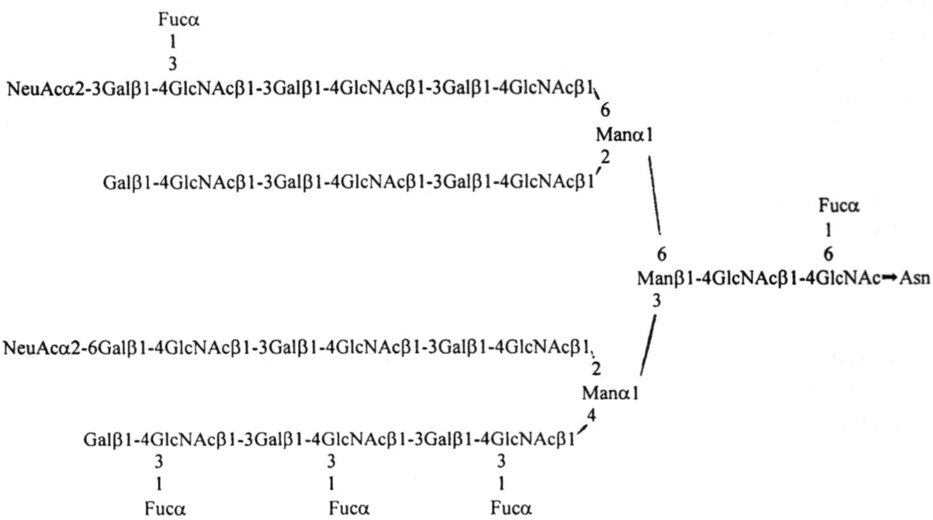

Fig. 4. Main component of *N*-linked sialylated fucosyl lactosaminoglycan isolated from human granulocytes (Fukuda *et al.*,[39]). Asn, Asparagine; other abbreviations as in **Fig.3.**

These terminal sequences may form part of both glycoproteins and glycolipids. Detailed structural analysis of sialylated fucosylated lactosaminoglycan isolated from human granulocytes[39] revealed that one of the main components had a unique structure with four linear polylactosamine side chains attached to the core portion of an *N*-linked chain (**Fig.4**). It was established that one side chain (arising from position 4 of 2,4-linked mannose) could have one, two or three α1,3-linked fucosyl residues and another (arising from position 2 of 2,4-linked mannose) terminated with sialic acid linked through an α2,6 linkage and did not carry any fucosyl units. Only one side chain (arising from position 6 of 2,6-linked mannose) carried a sialyl-Le^x determinant and the fourth side chain (arising from position 2 of 2,6-linked mannose) had a terminal unsubstituted *N*-acetyllactosamine structure and carried neither sialic acid nor fucose substitutents. Thus this one complex type of oligosaccharide chain may carry mono-, di or trimeric-Le^x, sialyl-Le^x, 6'-sialyl-*N*-acetyllactosamine and unsusbstituted *N*-acetyllactosamine determinants.

The two glycoproteins carrying *O*-glycans on human myeloid cells for which the structures of the oligosaccharide moieties have been studied in some detail are leukosialin (CD43)[40] and the P-selectin glycoprotein ligand-1 (PSGL-1)[41]. Both contain *O*-glycans with a Core-2 motif (Structure 1, **Fig.5**). PSGL-1 has many neutral Core-2 tetrasaccharides whereas CD43 has mostly disialylated Core-2 structures [42]. Leukosialin contains one to two poly-*N*- acetyllactosaminyl glycans per molecule and about 50% of these contain sialyl-Le^{x} termini [43]. Only a minority of the *O*-glycans isolated from PSGL-1 are α1,3-fucosylated and those that do occur are present as two species containing sialyl-Le^{x}: one species (Structure 2(i), **Fig.5**) is a di-sialylated monofucosylated glycan and the other (Structure 2(ii) **Fig.5**) is a monosialylated trifucosylated glycan having a polylactosamine backbone [42].

---

1) Core-2 structure

```
Galβ1-4GlcNAcβ
              1
              6
Galβ1-3GalNAc-Ser/Thr
```

2) O-glycans from PSGL-1

(i)

```
                         Fucα
                           1
                           3
NeuAcα2-3Galβ1-4GlcNAcβ
                           1
                           6
NeuAcα2-3Galβ1-3GalNAc-Ser/Thr
```

(ii)

```
                Fucα               Fucα               Fucα
                  1                  1                  1
                  3                  3                  3
NeuAcα2-3Galβ1-4GlcNAcβ1-3Galβ1-4GlcNAcβ1-3Galβ1-4GlcNAcβ
                                                          1
                                                          6
                                        Galβ1-3GalNAc-Ser/Thr
```

Fig 5. Core-2 structure in O-glycans and fucosylated O-glycans isolated from PSGL-1. Modified from Wilkins *et al*, [42]. Ser, serine: Thr, threonine; other abbreviations as in **Fig.3.**

The neutral and sialylated glycosphingolipid molecules on myeloid cells that carry the fucosylated epitopes are based on the neo-lacto-series (Galβ1-4GlcNAcβ1-3Galβ1-4Glc-ceramide) of glycolipids[44] and Nakamura *et al*,[11] showed that the neo-lacto-series of gangliosides is enriched during granulocyte maturation. In a detailed study on the monosialogangliosides isolated from HL60 cells and normal granulocytes Stroud *et al.*,[45,46] identified a large number of glycosphingolipids some of which carried single polylactosamine chains with ten or more monosaccharide units (**Fig.6**). However, in contrast to the polylactosamine chains found in the glycoproteins from myeloid cells none carried sialyl-Le^{x} determinants although some had terminal VIM-2 structures and others had fucosyl residues on more internal lactosamine units.

These examples represent only some of the macromolecules carrying terminal sialylated and fucosylated epitopes on human myeloid cell surfaces but serve to demonstrate the variety and complexity of the glycans with which these antigenic structures are associated.

1. NeuAcα2-3Galβ1-4GlcβCer

2. NeuAcα2-3Galβ1-4GlcNAcβ1-3Galβ1–4GlcβCer

3. NeuAcα2-6Galβ1-4GlcNAcβ1-3Galβ1–4GlcβCer

4. NeuAcα2-3Galβ1-4GlcNAcβ1-3Galβ1–4GlcNAcβ1-3Galβ1–4GlcβCer

5. NeuAcα2-3Galβ1-4GlcNAcβ1-3Galβ1–4GlcNAcβ1-3Galβ1–4GlcNAcβ1–3Galβ1–4GlcβCer

6. NeuAcα2-3Galβ1-4GlcNAcβ1-3Galβ1–4GlcNAcβ1-3Galβ1–4GlcNAcβ1–3Galβ1–4GlcβCer
3
1
Fucα

7. NeuAcα2-3Galβ1-4GlcNAcβ1-3Galβ1–4GlcNAcβ1-3Galβ1–4GlcNAcβ1–3Galβ1–4GlcβCer
3
1
Fucα

8. NeuAcα2-3Galβ1-4GlcNAcβ1-3Galβ1-4GlcNAcβ1-3Galβ1-4GlcNAcβ1-3Galβ1-4GlcNAcβ1-3Galβ1-4GlcβCer
3 3
1 1
Fucα Fucα

9. NeuAcα2-3Galβ1-4GlcNAcβ1-3Galβ1-4GlcNAcβ1-3Galβ1-4GlcNAcβ1-3Galβ1-4GlcNAcβ1-3Galβ1-4GlcβCer
3
1
Fucα

10. NeuAcα2-3Galβ1-4GlcNAcβ1-3Galβ1-4GlcNAcβ1-3Galβ1-4GlcNAcβ1-3Galβ1-4GlcNAcβ1-3Galβ1-4GlcβCer
3
1
Fucα

Fig 6. Major mono-sialogangliosides isolated from HL-60 cells. Modified from Stroud *et al.*,[45,46] Cer, ceramide. Other abbreviations as in **Fig.3.**

V. FUNCTIONS OF SIALYL-LE^X ON GRANULOCYTES

The sialyl-Le^x determinant on the surface of mature circulating granulocytes has been implicated in adhesion events connected with inflammatory processes. The recruitment of neutrophils into sites of tissue injury is a multi-step process mediated by a series of adhesion molecules that are required for the attachment of cells to the endothelium and their subsequent influx into the tissue. The selectin family of adhesion molecules comprises E-selectin (endothelial selectin), P-selectin (platelet selectin) and L-selectin (leukocyte selectin). Important for the adhesive properties of the selectins are their *N*-terminal carbohydrate (C-type lectin) recognition domains [47-50]. Sialyl-Le^x is believed by many to be the major ligand for E-selectin which is transiently expressed on vascular endothelial cells in response to injury. Interaction between E-selectin and granulocytes results in the initial slowing down and 'rolling' of the cells along the endothelial surface. Firm adhesion to the endothelium mediated by the β2-integrins, and extravasation of the leukocytes into the surrounding tissues, follows this initial process [51-53]. Most of the binding studies designed to identify the ligands recognised by E-selectin have been carried out with purified oligosaccharides, glycoproteins or glycolipids presented at high density in plastic wells or on TLC plates and it is difficult to be sure that these results can be extrapolated accurately to the situation pertaining when the molecules on the granulocyte cell surface interact with the selectin molecules presented on vascular endothelial cells. Although the work emanating from transfection experiments with fucosyltransferase genes (see below) and the "knock-out" mouse studies[12] appear to support the belief that sialyl-Le^x is the target epitope for E-selectin, Stroud et al.[45,46] have challenged this concept. They found that some of the mono-sialogangliosides isolated from HL-60 cells and mature granulocytes which had unbranched polylactosamine chains with 10 or more monosaccharide units showed clear E-selectin binding despite the fact that they lacked terminal sialyl-Le^x structures. The active oligosaccharides had multiple internally located fucosyl units but did not have fucose residues on the terminal sialylated *N*-acetyllactosamine unit; these authors therefore concluded that the internally fucosylated mono-sialogangliosides, which they collectively called "myeloglycan", rather than sialyl-Le^x, provide a major physiological epitope in E-selectin–dependent binding of granulocytes and HL-60 cells. These experiments were carried out only with the glycolipid molecules and it is not clear how far they could be extrapolated to the polylactosamine chains on the glycoproteins but Wagers *et al*[54] also found that a variant of HL-60 cells that was deficient in sialyl-Le^x exhibited strong adhesion to vascular selectins. These authors concluded that high levels of this epitope are not essential for interactions with vascular selectins and that other, as yet unidentified sialylated, fucosylated structures serve as physiologically relevant ligands for E- and P-selectins.

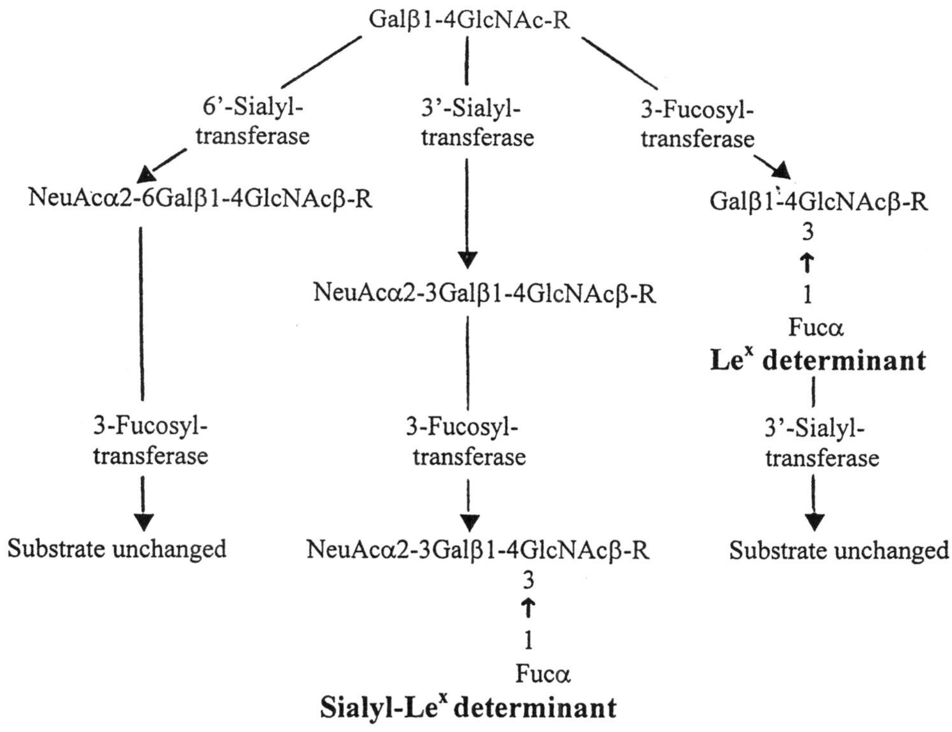

Fig 7. Biosynthetic pathways for the formation of Le^x and sialyl-Le^x. Abbreviations as in **Fig. 3**.

The P-selectin glycoprotein ligand-1 (PSGL-1) serves as a ligand for E-selectin as well as P-selectin but the two selectins are believed to interact with different parts of the ligand molecule. Adhesion of myeloid cells to P-selectin is abolished by *O*-sialylglycoprotease or chymotrypsin whereas E-selectin binding is unaffected[55] suggesting that the *O*-glycans carry the ligands recognised by the P-selectin molecule. Only a minority of the *O*-glycans in PSGL-1 are fucosylated and it is unclear whether this molecule contributes to E-selectin binding *in vivo*.

A widely expressed, cloned murine glycoprotein ESL-1 (E-selectin-ligand 1) has been described as a major ligand for E-selectin on mouse myeloid cells[56]. This molecule in the mouse carries only *N*-linked chains. The human homologue has yet to be described and doubts have been raised[48] about the cellular location of this glycoprotein because a chicken homologue of mouse ESL-1 was identified as a fibroblast growth factor[57]. Further doubts about whether this glycoprotein constitutes the cell surface E-selectin ligand in human cells has arisen because blocking the maturation of *N*-linked glycans on HL-60 cells by specific inhibitors of glycoprotein processing appeared not to affect E-selectin binding[58]. However, this aspect is still controversial since other workers reported that treatment with tunicamycin, an inhibitor of the first step in *N*-glycosylation, did inhibit the binding of HL60 cells to E-selectin[59]. Therefore, despite the vast literature that has been generated in recent years concerning E-selectin ligands on myeloid cells further studies are still required to delineate precisely which epitopes and macromolecules on the cell surface are involved in the physiological binding processes. It is evident that factors other than the simple presence of the epitopes, such as the density and clustering of sites, the presentation and accessibility of the molecules with which the structures are associated and the general topography of the cell surface may influence the recognition of the ligands. Since there is still so much to be established concerning selectin-ligand interactions on mature cells it is perhaps not surprising that even less is known about the role played by the selectins and their ligands in the course of maturation of haemopoietic cells in the bone marrow microenvironment.

L-Selectin is the only one of the three identified selectins that is expressed on myeloid cells[60] and this adhesion molecule is believed also to mediate rolling of mature granulocytes on endothelium at sites of inflammation. Fucosylation, sialylation and sulfation have been implicated as parts of ligands for L-selectin binding [61,62]. The L-selectin molecules themselves carry glycans bearing sialyl-Le^x structures and exhibit affinity for E-selectin[63] although it is not known whether interactions between L-selectin on mature granulocytes and E-selectin on endothelial cells play any part in the rolling process. L-selectin is strongly expressed in myeloid progenitor cells and may therefore be important for interactions occurring in early haemopoietic development but is sharply down-regulated at the same stage as CD34 (see **Fig. 2**) and remains at a low level until the stage where proliferative activity ceases. L-selectin is then up-regulated to the level found in progenitor cells[34]. This adhesion molecule does not therefore appear to play an active part in specific recognition events that are important for development in the intermediate stages of myeloid maturation.

In contrast to its role as an activation molecule on vascular endothelium, E-selectin is expressed constitutively on bone marrow endothelial cells and it has been suggested that this expression may be an essential initial step of the homing process of haemopoietic progenitor cells[64].

VI. FUCOSYL- AND SIALYL-TRANSFERASES INVOLVED IN BIOSYNTHESIS OF FUCOSYLATED AND SIALYLATED ANTIGENS

1. Fucosyl linkages in human glycoconjugates

Three different types of fucosyl-linkage are commonly found in terminal structures on the oligosaccharide chains of human glycoconjugates. Fucose linked α1,2 to β-galactosyl-R gives rise to blood group H structures and fucose linked α1-4 to subterminal *N*-

acetylglucosamine in Type 1 (Galβ1-3GlcNAc-R) chains gives rise to blood group Le^a and related structures. Linkage of fucose α1,3 to the *N*-acetylglucosamine residue in Type 2 (Galβ1-4GlcNAc-R) chains gives Le^x and related structures. When both α1,2- and α1,3-fucosyltransferases act on Type 1 chains the difucosyl determinant Le^b is formed and α1,2- and α1,3-fucosyltransferase acting on Type 2 chains give rise to the hybrid structure Le^y [65]. Fucose linked α1,6 to *N*-acetylglucosamine occurs only in the protein core region of *N*-linked glycans (see **Fig 4**).

Examination of haemopoietic cells for α1,2-, α1,3- and α1,4-fucosyltransferase activity revealed α1,2-fucosyltransferase and not α1,3-fucosyltransferase in platelets whereas lymphocytes, monocytes and granulocytes expressed α1,3-fucosyltransferase activity and not α1,2-fucosyltransferase[66,67] None of the haemopoietic cells expressed α1,4-fucosyl-transferase activity. Since all the haemopoietic cell lineages derive from common multipotent stem cells there must be a programme for switching on or off of the required fucosyltransferases as lineage commitment occurs. However, the precise stages in the development of committed erythroid or myeloid cells at which the α1,2- and α1,3-fucosyltransferase genes switch on, or off, is not yet established. One puzzling fact is that glycosyltransferase analyses on mature granulocytes and leukaemia cell lines arrested at the promyelocyte stage have failed to reveal convincing evidence of α1,2-fucosyltransferase activity [68] despite several independent reports of the expression of Le^y on mature granulocytes and on HL-60 promyelocyte cells [69-71].

2. Pathways of biosynthesis of Le^x and Sialyl-Le^x

The final stage in the biosynthesis of Le^x and sialyl-Le^x structures is the addition of fucose to the *O*-3 position of *N*-acetylglucosamine in a Type 2 (*N*-acetyllactosamine) structure. Despite the similarity of the two structures the unsialylated Le^x is not the precursor of sialyl-Le^x because the Le^x structure is not a substrate for any of the characterised human α2,3-sialyltransferases [72]. Hence in order to synthesise sialyl-Le^x it is necessary first to form 3'-sialyl-*N*-acetyllactosamine and then to catalyse the addition of fucose to complete the structure[73].The isomeric 6'-*N*-acetyllactosamine sequence is not a substrate for any of the human fucosyltransferases[73] and hence this terminal sequence remains unsubstituted (see, for example, *N*-glycan structure in **Fig. 4**) The biosynthetic pathways for the formation of Le^x and sialyl-Le^x are summarised in **Fig. 7**.

3. Cloned α1,3-fucosyltransferase genes and their expression in myeloid cells

It is now known that the human α1,3-fucosyltransferases do not comprise one enzyme species but a family of enzymes, with closely related, but not identical, specificities [50,74]. The first indication that more than one species might be present in myeloid cells arose from the purification of the α1,3-fucosyltransferase from normal granulocytes and from the myeloid cells of a patient with chronic myeloid leukaemia (CML)[75]. Both enzyme sources transferred fucose to unsubstituted Type 2 structures to form Le^x but the enzyme from the patient had only a very weak capacity to utilise 3'-sialyl-*N*-acetyllactosamine whereas this substrate was readily used by the enzyme(s) from the healthy individual. A polyclonal antibody raised against the α1,3-fucosyltransferase in human milk failed to react with either of the enzyme preparations from the myeloid cells [76], thus indicating the existence of yet other genes encoding enzymes with similar specificities. Subsequently six human α1,3-fucosyltransferase genes have been cloned [50,74,77] which have been designated *FUT3* to *FUT9*. The major difference in the specificity of the enzymes encoded by the human α1,3-

Table 2.
Cloned α1,3-fucosyltransferase genes; chromosomal location , specificity and expression on myeloid cells.

Gene	Chromosomal location	Acceptor specificity		Expression in myeloid cells
		Galβ1-4GlcNAc	NeuAcα2-3Galβ1-4GlcNAc	
FUT3	19	+	+	-
FUT4	11	+++	+	+++
FUT5	19	+++	+++	-
FUT6	19	+++	+++	-
FUT7	9	-	+++	+++
FUT9	-	+++	-	?

Compiled from [50,74,77].

fucosyltransferase genes lies in their utilisation of sialylated or unsialylated Type 2 acceptor substrates (**Table 2**). *FUT3, FUT5* and *FUT6* genes are clustered on chromosome 19 and are believed to have arisen by duplication and subsequent divergence [78]. The first to be cloned, *FUT3*[79], encodes an enzyme that transfers fucose primarily in α1,4-linkage to *N*-acetylglucosamine residues in Type1 (Galβ1-3GlcNAc-R) structures to form Le^a (Galβ1-3[Fucα1-4]GlcNAc) and related determinants but has a weak capacity to transfer fucose also in α1,3-linkage to Type 2 structures to form Le^x determinants [80,81]; this enzyme is therefore called an α1,3/1,4-fucosyltransferase. It is not expressed in haemopoietic tissue and the blood group Le^a and Le^b antigens, the products of this enzyme activity, are taken up onto red cells from the plasma in which they circulate[82]. The other two members of the chromosome 19 cluster, *FUT5* and *FUT6* [74,83] utilise mainly Type 2 acceptors but can transfer fucose to both sialylated and unsialylated acceptor molecules; neither of these enzymes are expressed in haemopoietic tissues. The two α1,3-fucosyltransferase genes definitely expressed in myeloid cells are *FUT4,* [84-86] sited on chromosome 11, and *FUT7* [87,88], sited on chromosome 9. These two genes encode enzymes with more restricted specificities than those encoded by the genes on chromosome 19. The *FUT4* enzyme utilises unsialylated Type 2 structures and has only very limited activity with α2,3-sialylated acceptors whereas the *FUT7* enzyme utilises 3'-sialyl-*N*-acetyllactosamine and has no activity with unsubstituted *N*-acetyllactosamine (**Table 2**). The most recently reported α1,3-fucosyltransferase gene, *FUT9* [77] shows low degrees of homology with the previously cloned α1,3-fucosyltransferases but is highly conserved between mouse and humans and is said to act primarily on Type 2 unsialylated structures and to occur in peripheral blood leukocytes as well as in human brain and stomach[89]. The level of expression of this enzyme in myeloid cells in comparison to *FUT4* and *FUT7* is not yet established. *FUT4*, and possibly also *FUT9,* are thus the genes responsible for the formation of the Le^x structures on myeloid cells whereas only the enzyme encoded by *FUT7* can complete sialyl-Le^x sequences.

A detailed study carried out on Chinese hamster ovary (CHO) cells transfected with *FUT4* and *FUT7* genes involved examination of the site specificities of the encoded enzymes acting on polylactosamine substrates[90]. The *FUT4* enzyme was found to transfer fucose effectively to all *N*-acetylglucosamine units in neutral polylactosamine chains and to the inner *N*-acetylglucosamine units of α2,3-sialylated acceptors but not to transfer fucose to the distal (outermost) 3'-sialyl-*N*-acetyllactosamine units in sialylated acceptors. In contrast the enzyme encoded by *FUT7* effectively fucosylates the distal *N*-acetyllactosamine unit in α2,3-sialylated acceptors, as previously established, but fails to transfer fucose to the inner *N*-acetyllactosamine units. Recently Nishihara *et al.*, [91] compared the specificities of the cloned α1,3-fucosyltransferases, including the enzyme encoded by *FUT9;* they confirmed the specificities of *FUT4* and *FUT7* enzymes with regard to their action on polylactosamine chains and showed that the *FUT9* enzyme, in contrast to *FUT4,* acted preferentially on the distal *N*-acetylglucosamine residue and exhibited more efficient activity for the synthesis of Le^x than the *FUT4* enzyme. If the *FUT9* encoded enzyme is expressed at levels comparable with *FUT4* in myeloid cells then it may be responsible for the formation of terminal Le^x epitopes on developing and mature myeloid cells but *FUT4* expression would be required for the synthesis of the internally fucosylated structures on the polylactosamine chains, and hence for the appearance of di- and tri-meric Le^x epitopes. However, the extent to which *FUT9* contributes to the cell surface expression of the antigens must await further assessment of the level of expression of this gene. In any event, unless there are as yet other unidentified fucosyltransferase genes expressed in the myeloid cells, the ultimate appearance of sialyl-Le^x must be dependent on expression of *FUT7,* although with so many preceding biosynthetic steps necessary to provide substrate for the enzyme, expression of the gene alone will not be sufficient to ensure cell surface appearance of the antigenic structure. The pathways for the formation of Le^x, sialyl-Le^x, VIM2 (CD_w65) and sialyl-dimeric-Le^x on polylactosmine chains are illustrated in **Fig. 8.**

4. Murine haemopoietic development in the absence of *FUT4* and/or *FUT7* α1,3-fucosyltransferase genes

In order to assess the relative roles of *FUT4* and *FUT7* genes in the control of selectin ligand expression Maly *et al.*,[12] constructed and characterised mice deficient in the *FUT7* locus. These mice exhibited leukocytosis, deficits in expression of blood leukocyte E- and P-selectin ligand activity and reduced neutrophil trafficking in inflammation. They also exhibited a deficit in expression of L-selectin ligands by high endothelial venules and severe attenuation of lymphocyte homing. These results suggested an essential role for *FUT7* in the biosynthesis of the oligosaccharide portions of the ligands for E-, P- and L-selectins and a pivotal role for this gene in leukocyte trafficking.

Subsequently Lowe *et al.*[13] reported that mice rendered deficient in *FUT 4* by the gene targetting approach exhibited qualitatively normal E-and P-ligand activities and maintained an essentially normal inflammation-dependent mobilisation phenotype; the only abnormality they showed was rolling at an increased rate *in vivo* compared to the granulocytes from normal mice. However, mice deficient in both *FUT4* and *FUT7* genes exhibited a more severe phenotype than mice lacking only *FUT7* and showed little if any rolling *in vivo* and were more deficient in leukocyte recruitment than the *FUT7* null mice. These results therefore indicated a collaborative role for *FUT4* and *FUT7* in leukocyte turnover and in ligand biosynthesis. How far these results can be extrapolated to humans is not yet clear but the phenotype exhibited by the *FUT7* null mice resembles the phenotype found in humans with leukocyte adhesion deficiency syndrome II [92] which is believed to result from a defect in the enzyme converting mannose to fucose[93], and hence from a generalised defect in fucose metabolism. The results obtained by the gene-targetting method in mice are not compatible with the E-selectin binding results which Stroud *et al.*[45,46] obtained with the mono-sialogangliosides isolated from HL-60 cells and mature granulocytes because their results indicated that expression of *FUT7* was not strictly necessary for synthesis of E-selectin-ligands on myeloid cells since the characterised E-selectin-binding molecules did not all contain sialyl-Le^x determinants. However, the fact that the combined loss of *FUT4* and *FUT7* leads to a more severe phenotype than loss of

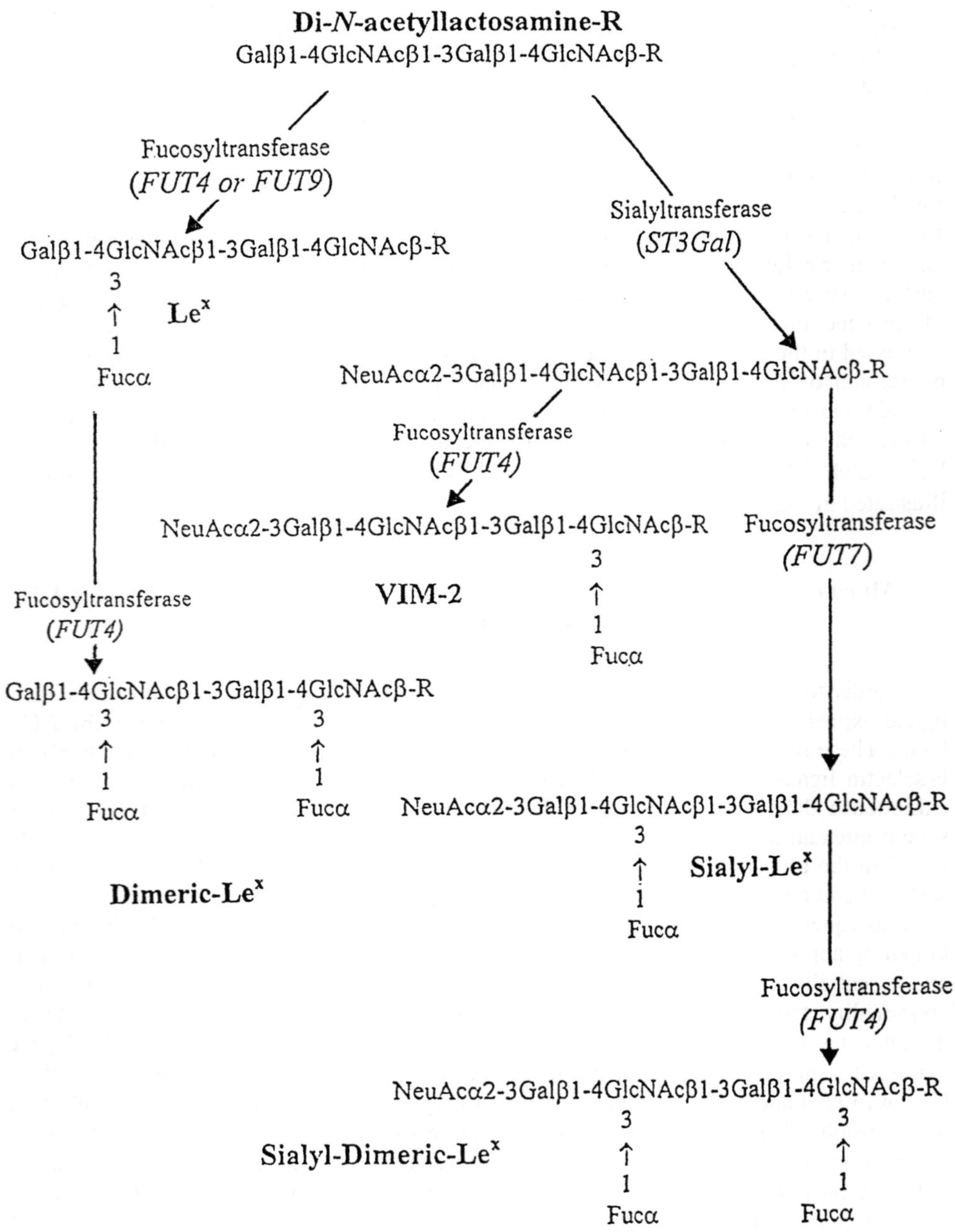

Fig 8. Pathways for the biosynthesis of Le^x, dimeric-Le^x, sialyl-Le^x, VIM2 and sialyl-dimeric-Le^x determinants on polylactosamine chains.
R = remainder of polylactosamine chain. Other abbreviations as in **Fig.3**.

FUT 7 alone would be compatible with the suggested role for multifucosylated structures in the binding to E-and P-selectins. The leukocytosis observed in the *FUT7* null mice has been attributed to an inability of these animals to effectively control bone marrow homeostasis [94] and therefore indicates a role for the fucosylated cell surface structures in these processes.

5. Cloned sialyltransferase genes

The second enzyme that is crucially important for the appearance of sialyl-Le^x is the sialyltransferase that forms the precursor substrate, 3'sialyl-*N*-acetyllactosamine, utilised by the *FUT7* encoded α1,3-fucosyltransferase. The sialyltransferases also constitute a large family of closely related genes [30,95]. Five human α2,3-sialyltransferases that transfer sialic

Table 3
Acceptor specificity of cloned human sialyltransferase genes

Acceptor substrate	*ST6Gal:* *I*	*ST3Gal:* *I*	*II*	*III*	*IV*	*VI*
Type 1 Galβ1-3GlcNAc	-	-	-	+++	-	-
Type 2 Galβ1-4GlcNAc	+++	-	-	++	+	+++
Type 3 Galβ1-3GalNAc	-	+++	+++	+	+++	-

Compiled from Kitigawa and Paulsen[96], Giodanengo *et al.,*[98] and Okajima *et al.,*[102]

acid to terminal β-galactosyl units have been cloned: *ST3Gal-1* [96] and *ST3Gal-II* [97,98] utilise predominantly Type 3 (Galβ1-3GalNAc) acceptors, *ST3Gal-III* [99] utilises Type 1, Type 2 and Type 3 acceptors but has a preference for Galβ1-3GlcNAc, *ST3Gal-IV* [100,101] utilises mainly Type 3 acceptors and *ST3Gal-VI* [102] utilises only Type 2 acceptors (**Table 3**). Therefore, either or both *ST3Gal-III* and *-VI* could be involved in the biosynthesis of 3'-sialyl-*N*-acetyl-lactosamine structures.

Only one human α2,6-sialyltransferase gene, *ST6Gal-I,* has so far been cloned [103]; the enzyme encoded by this gene transfers sialic acid exclusively to terminal β-galactosyl units in Type 2 *N*-acetyllactosamine structures and the product, 6'sialyl-*N*-acetyllactosamine, is not a substrate for any of the known human α1,3-fucosyltransferases (see **Fig.7**).

VII. ADDRESSING THE QUESTION OF THE FALL IN EXPRESSION OF SIALYL-LE^X AT INTERMEDIATE STAGES OF MYELOID DEVELOPMENT

1. Background experiments and use of cell lines

The interest of our labaoratory in the α1,3-fucosyltransferases arose out of attempts in the late 1960s to identify the proposed enzymic products of the blood-group *H* and *Lewis (Le)* genes [104]. Early work indicated that all tissues with the proposed *Le* gene encoded α1,4-fucosyltransferase activity, or with the proposed *H*-gene α1,2-fucosyltransferase activity, also exhibited α1,3-fucosyltransferase activity [105,106]. Subsequently, in an attempt to find a soluble source of *Lewis* gene encoded fucosyltransferase that could be used for genetic family studies we examined saliva and showed that, in addition to the expected transfer of fucose to the *O*-4 position of *N*-acetylglucosamine in Type 1 structures to give Le^a and Le^b determinants, salivas from those carrying an *Le* gene also transferred fucose to the *O*-3 position of glucose in lactose-based oligosaccharides[107]. Transfer of fucose to the *O*-3 position of *N*-acetylglucosamine occurred in the saliva of all individuals including those from blood group Le(a-b-) individuals, indicating the expression in saliva of an α1,3-fucosyltransferase that was independent of the *Le* gene. In the same year Prieels *et al.*,[108] purified a fucosyltransferase from human milk and demonstrated that the preparation had the capacity to transfer fucose to *N*-acetylglucosamine units in both Type 1 and Type 2 chain endings and proposed that the *Le*-gene encodes an α1,3/1/4-fucosyltransferase that can make both Le^a and Le^x antigenic structures. Although this is partially true we now know that the milk preparation was a mixture of an α1,3-fucosyltransferase with the capacity to transfer fucose only to Type 2 chains and an α1,3/1,4-fucosyltransferase that transfers fucose primarily to Type 1 chains, and to glucose in lactose-based oligosacharides, but has only a limited ability to utilise Type 2 structures [109,110]. The cloned human fucosyltransferase gene, *FUT3* [79] believed to correspond to the blood-group *Le* gene, when expressed in transfected mammalian cells was also shown to function primarily as an α1,4-fucosyltransferase [80,81].

The α1,3-fucosyltransferases purified from human milk[110], plasma[73] and liver[111] utilised 3'-, but not 6'-sialyl-*N*-acetyllactosamine, to synthesise sialyl-Le^x structures but, as described previously, purification of α1,3-fucosyltransferases from granulocytes revealed differences with regard to the utilisation of sialylated substrates by the enzymes in healthy individuals and those obtained from a patient with chronic myeloid leukaemia (CML)[75]. In order to investigate whether the observed specificity differences were leukaemia-related, we then started a programme to examine fucosyl- and sialyl-transferase levels, in normal and leukaemic cells, together with the cell surface expression of fucosylated and sialylated antigens. High levels of α1,3-fucosyltransferase activity were found in leukaemic blasts compared to the levels in normal granulocytes and this dfference was at first thought to be a manifestation of the leukaemic process[112] but the discovery that normal bone marrow cells also had high activity then suggested that the level was most probably related to the stage of maturation arrest of the cells rather than the leukaemic process [22,35]. It then appeared feasible that we could use cells arrested at different maturation stages to study the regulatory mechanisms underlying the changes that occur in cell surface expression of the fucosylated and sialylated antigens in the course of normal haemopoietic differentiation. In our first studies we examined the cell surface expression of Le^x, and sialyl-Le^x, together with α1,3-fucosyltransferase and α2,3- and α2,6 sialyl-transferase activities, in normal neutrophils, leukaemic myeloblasts and HL-60 cells induced to differentiate along the granulocytic pathway with DMSO [22]. Much higher levels of α1,3-fucosyltransferase activity were found in leukaemic blasts than in normal granulocytes when non-sialylated Type 2 substrates were used, but in agreement with earlier observations on the partially purified preparations[75] the enzyme(s) in normal granulocytes reacted with both α2,3-sialylated and non-sialylated acceptors whereas the enzyme(s) in blasts showed a marked preference for non-sialylated acceptors. α2,6-Sialyltransferase was active in blasts but was not detectable in neutrophils, whereas α2,3-sialyltransferase activity was detected in both blasts and neutrophils[22].

The most striking observation was that induction of differentiation of the HL-60 cells with DMSO led to a decrease in activity of the α1,3-fucosyltransferase with non-sialylated acceptors and a change towards the specificity pattern found in mature cells inasmuch as there was a relatively greater transfer to sialylated acceptors. This change was accompanied by a decrease in α-sialyltransferase activity and also to an increase in expression of sialyl-Le^x which was very poorly expressed on the untreated HL-60 cells[22]. With a view to pursuing these observations further we decided to use the human leukaemic cell lines KG1a, KG1, HL-60, differentiated HL-60 cells and normal neutrophils in the belief, as outlined above, that this series would provide defined models of successive stages of myeloid cell maturation. Lund-Johansen and Terstappen[34] in addition to measuring the modulation of sialyl-Le^x expression on normal bone marrow derived promyelocyte/-myelocyte cells had followed the reactions of the cells isolated from human bone marrow with other markers related to the maturation stages of the cells. CD34 reacts with the early myeloid progenitors but the activity disappears by the promyelocyte stage[113]. CD11b, a member of the family of integrins, is first expressed at the myelocyte stage[16] and CD16 first appears at the later metamyelocyte stage; both these antigens are present on the mature granulocyte (**Table 1** and **Fig.2**). When tested by flow cytometry with monoclonal antibodies directed towards these three determinants KG1a and KG1 cells both reacted with anti-CD34 in agreement with their early stage of maturation. KG1a, KG1 and HL-60 cells were negative with the antibody to CD11b but reactivity appeared on the differentiated HL-60 cells and was stronger on mature granulocytes. Neither KG1a, KG1, HL-60 nor the differentiated HL-60 cells reacted with the antibody to CD16 but normal granulocytes gave positive reactions (**Table 1** and **Fig.9**). Therefore the cell lines were behaving according to their original classification and were expressing maturation antigens that corresponded to progressive stages of differentiation along the granulocytic pathway.

2. Expression of Le^x (CD15), sialyl-Le^x (CD15s), VIM-2 (CD_w65), sialyl-dimeric Le^x and Le^y antigens measured by flow cytometry with monoclonal antibodies

The three cell lines, KG1a, KG1 and HL-60, differentiated HL-60 cells and mature granulocytes were examined by flow cytometry with mouse monoclonal antibodies directed towards Le^x, sialyl-Le^x, VIM-2, sialyl-dimeric Le^x and Le^y. Earlier workers had shown that KG1a had a lower expression of Le^x than any of the other myeloid cell lines [114]. In our experiments the KG1a cells were virtually negative (<10% positive) with antibody to Le^x but neuraminidase treatment resulted in loss of sialyl-Le^x expression and a strong development of Le^x activity showing that this determinant was masked on the cell surface by the sialic acid residues. In contrast a high proportion (>90% positive) of both KG1 and HL-60 cells reacted with the antibody to Le^x [115]. The activity of KG1 cells was considerably weaker with antibodies to sialyl-Le^x than the less mature KG1a cells and, in agreement with previous findings[22], the HL-60 cells gave only a very weak response with these reagents (**Fig.9**). Thus the increase in the expression of Le^x was matched by a decrease in expression of sialyl-Le^x as the cell matured from the myeloblast (KG1a) to the promyelocyte (HL-60) stage. The up-regulation of sialyl-Le^x expression observed on differentiation of HL-60 cells [22,115]. was maintained in an even stronger expression of this antigen on mature granulocytes. Thus examination of the cell lines arrested at successive stages of maturation appears to reflect the same pattern of a dip in expression of sialyl-Le^x at intermediate stages of maturation, followed by re-expression at a later developmental stage, as was observed for the normal bone marrow derived cells [34,35]. In agreement with earlier reports [38] very few KG1a or KG1 cells (<10%) reacted positively with antibody to VIM2 antigen (Structure 8, **Fig. 3**) whereas about 85% of HL-60 cells were positive with this reagent [68,115]. Differentiation of HL-60 cells led to a decrease in the number of positive cells but the VIM2 epitope was strongly expressed again in mature granulocytes. The antibody to sialyl-dimeric-Le^x failed to react with any of the untreated cell lines, or with differentiated HL-60 cells, but reacted weakly with mature granulocytes [68,115].

The presence of Le^y structures (Structure 6,**Fig.3**) on the surface of mature granulocytes

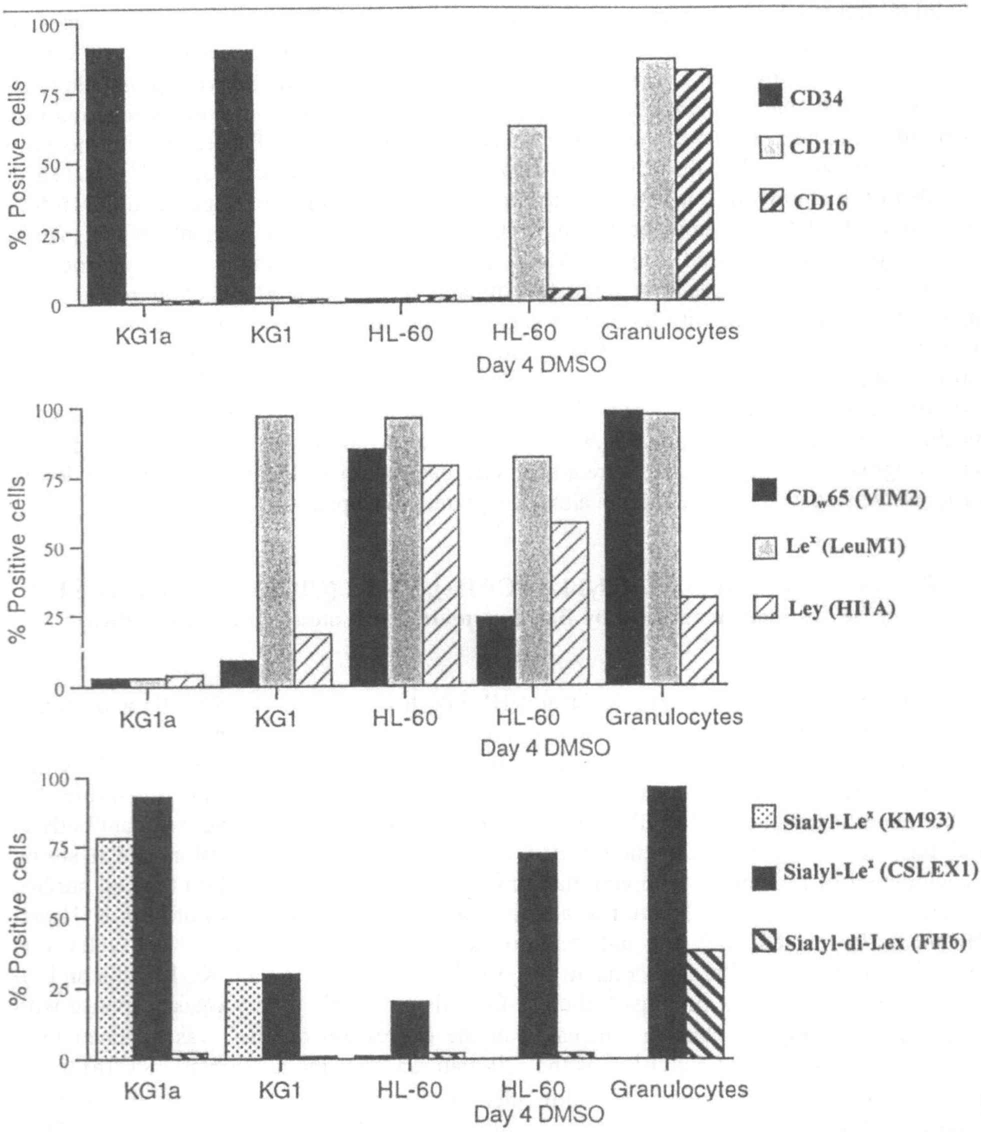

Fig.9. Flow cytometric analyses of fucosylated and sialylated antigens expressed on the surface of myeloid cell lines, KG1a, KG1, HL-60, HL-60 cells differentiated 4 days with DMSO and mature granulocytes. Monoclonal antibody clone numbers are shown in brackets. KG1a, KG1 and HL-60 cells were tested with two different monoclonal anti-sialyl-Le^x antibodies, KM-93 and CSLEX1; differentiated HL-60 cells and mature granulocytes were tested only with CSLEX1.

has been reported by several groups of workers [69-71]. Expression of Le^y antigen was very weak on KG1a cells, slightly stronger on KG1 cells and definite on HL-60 cells, with 80% giving positive signals with a monoclonal anti-Le^y reagent [68,115]. DMSO-induced differentiation of HL-60 cells led to a decrease in Le^y expression, and although mature granulocytes were still positive the activity was less than on the differentiated cells, indicating maximum activity at the promyelocytic stage [115]. The flow cytometry results are summarised in **Fig.9**.

3. Fucosyltransferase activity in cell lines and mature granulocytes

In order to compare enzyme levels in the different cell lines fucosyltransferase activities were measured in similarly prepared detergent extracts of KG1a, KG1, undifferentiated and differentiated HL-60 cells and mature granulocytes. Tests with a range of low molecular weight oligosaccharide substrates revealed that the Type 2 disaccharide , *N*-acetyllactosamine and tetrasaccharide, lacto-*N*-*neo*tetraose were the preferred acceptors for the α1,3-fucosyltransferase(s) in all three cell lines and also in mature granulocytes. Incorporation of fucose into these substrates showed a progressive rise with increasing cell maturation up to the promyelocytic HL-60 stage, followed by a drop in incorporation on differentiation of the HL-60 cells. Very low levels of incorporation were observed into these substrates by the enzymes in mature granulocytes (**Fig.10**).

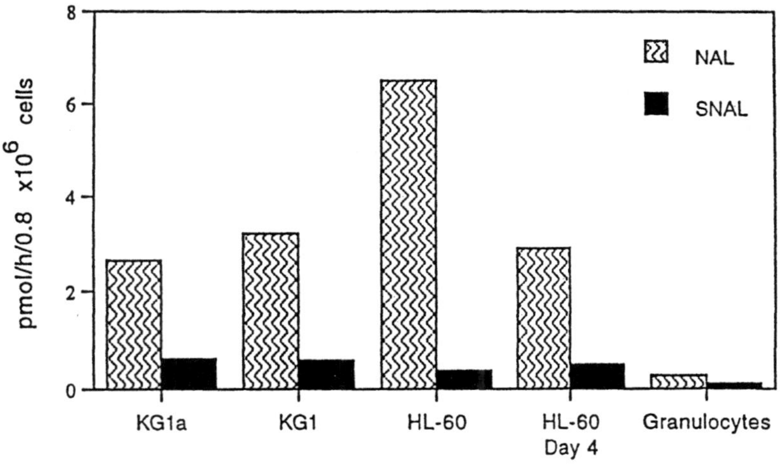

Fig.10. α1,3-Fucosyltransferase activity in KG1a, KG1, HL-60, differentiated HL-60 (HL-60 day 4 with DMSO) cells and mature granulocytes with low molecular weight substrates. NAL, *N*-acetyllactosamine: SNAL, 3'-sialyl *N*-acetyllactosamine.

Incorporation into 3'-sialyl-*N*-acetyllactosamine was much lower than that into the unsubstituted disaccharide and was similar for extracts of all three cell lines. Thus the ratio of activity with sialylated and non-sialylated acceptors changed as maturation proceeded, with a relatively greater preference for the non-sialylated compounds in the promyelocytic HL-60 cell line than in the less mature KG1 cells or the earlier KG1a myeloblasts. No activity was observed with the Type 1 disaccharide Galβ1-3GlcNAc or the tetrasaccharide, lacto-*N*-tetraose (Galβ1-3GlcNAcβ1-3Galβ1-4Glc) confirming the absence of α1,4-

fucosyltransferase activity in any of the cell lines [68-115]. DMSO-induced differentiation of HL-60 cells brought about a sharp fall in the level of activity of the fucosyltransferases with non-sialylated acceptors[22] with relatively little change in the activity with 3'-sialyl-*N*-acetyllactosamine. The ratio of activity with the two substrates fell from 17 : 1 to 6 : 1 on differentiation of the cells. The levels of α1,3-fucosyltransferase activity in mature granulocytes was very low with both substrates but the ratio in these cells was 3:1 [115]. A similar marked preference was observed at the promyelocyte HL-60 stage for glycoprotein substrates [68-115] that had been freed from sialic acid. Hence there is a progressive increase in the relative α1,3-fucosyltransferase activity with non-sialylated substrates up to the promyelocyte HL-60 stage and a progressive decrease in relative activity after that stage. The fucosyltransferase results with *N*-acetyllactosamine and 3'-sialyl-*N*-acetyllactosamine as acceptor sugars are summarised in **Fig.10**.

4. Sialyltransferase activity in cell lines and mature granulocytes

Earlier experiments had shown that, in addition to the decrease in fucosyltransferase activity, sialyltransferase activity with Type 2 substrates is reduced on DMSO-induced differentiation of HL-60 cells and the suggestion was made that competition for the common Type 2 *N*-acetyllactosamine substrate might be the reason for the down-modulation of sialyl-Le^x on the cell surface [22]. Subsequent examination of sialyltransferase activity in KG1a, KG1, HL-60 and differentiated HL-60 cells with low molecular weight substrates revealed that the preferred acceptor for all the cell lines was the Type 3 structure Galβ1-3GalNAc; however this is not a substrate for the α2,3-sialyltransferase giving rise to the 3'-sialyl-*N*-acetylactosamine structure necessary for the formation of sialyl-Le^x. The next best acceptors were the Type 2 ,Galβ1-4GlcNAc, disaccharide and the tetrasaccharide, lacto-*N-neo*-tetraose; very low activity was found with the Type 1, Galβ1-3GlcNAc, disaccharide or lacto-*N*-tetraose (**Fig.11**).

The level of activity with Galβ1-4GlcNAc increased with maturation of the cells up to the promyelocytic HL-60 stage (J.L.Clarke and W.M.Watkins unpublished observations)

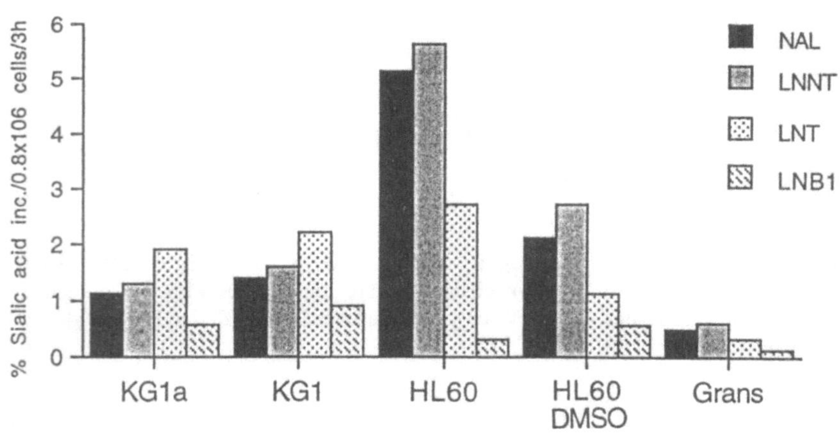

Fig 11. Sialyltransferase activity in cell lines KG1, KG1a, HL-60, HL-60 differentiated with DMSO, and mature granulocytes. NAL, *N*-acetyllactosamine, LNNT, Lacto-*N-neo*tetraose (Galβ1-4GlcNAcβ1-3Glc), LNT, lacto-*N*-tetraose (Galβ1-3GlcNAcβ1-3Glc), and LNB1, Lacto-*N*-biose 1 (Galβ1-3GlcNAc).

and then, as observed earlier[22], decreased on differentiation of the cells. Since Galβ1-4GlcNAc is an acceptor for both α2,3- and α2,6-sialyltransferases the products synthesised with this substrate were subjected to paper chromatography to separate the 3'- and 6'-sialyl-*N*-acetyllactosamine products. The amounts of both forms increased up to the promyelocyte stage and then decreased, but the decrease in the α2,6 product took place much more sharply than the decrease in the α2,3-product. Virtually no α2,6-activity was detectable in mature granulocytes while α2,3-activity persisted at a low level (**Fig.12**). Taniguchi *et al.*, [116] confirmed that on differentiation of HL-60 cells with DMSO the α2,6-sialyltransferase activity fell to levels 20-50% of that in control cells and reactivity with *Sambucus nigra*, a lectin believed to react with α2,6-linked sialic acid on cell surfaces, also decreased.

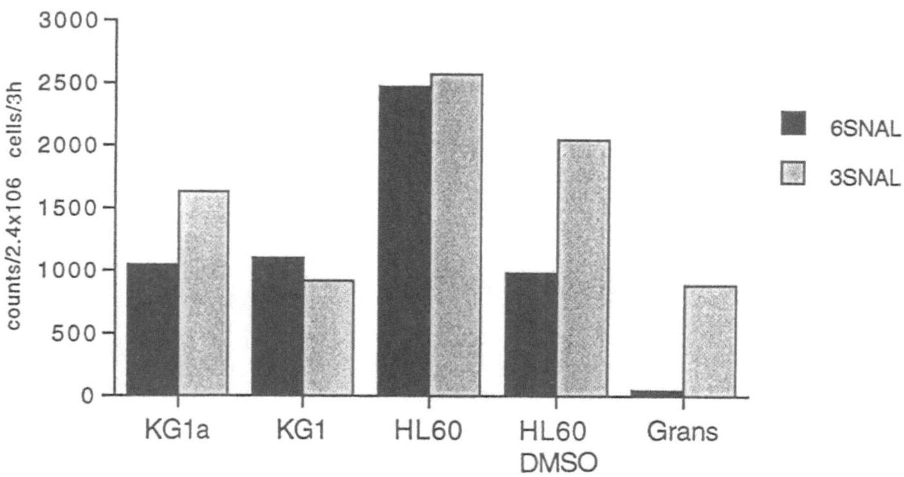

Fig.12. α2,3- and α2,6-Sialyltransferase activities in KG1a, KG1, HL-60, differentiated HL-60 cells and mature granulocytes measured with *N*-acetyllactosamine. 3SNAL, product 3'-sialyl-*N*-acetyllactosamine; 6SNAL, product 6'-sialyl-*N*-acetyllactosamine.

5. Expression levels of five cloned human fucosyltransferase genes in leukaemic cell lines and mature granulocytes

In order to determine whether the increased expression of sialyl-Le^x and the switch in specificity of the α1,3-fucosyltransferase with regard to sialylated acceptors could be correlated with loss, or change, of expression of one fucosyltransferase gene we examined [68,115] mRNA from KG1a, KG1, undifferentiated and differentiated HL-60 cells and mature granulocytes for expression of *FUT3, FUT4, FUT5, FUT6* and *FUT7* genes with probes kindly supplied by Dr John Lowe, University of Michigan. Northern blot analyses revealed the presence of *FUT4* [84-86] and *FUT7* [87,88] genes in the cell lines at each stage of maturation and the absence of expression of the three chromosome 19 located α1,3-fucosyltransferases *FUT3, FUT5* and *FUT6* [74]. Comparison of the levels of mRNA corresponding to *FUT4* and *FUT7* genes revealed definite changes as maturation proceeded [68,115]. *FUT7* mRNA was

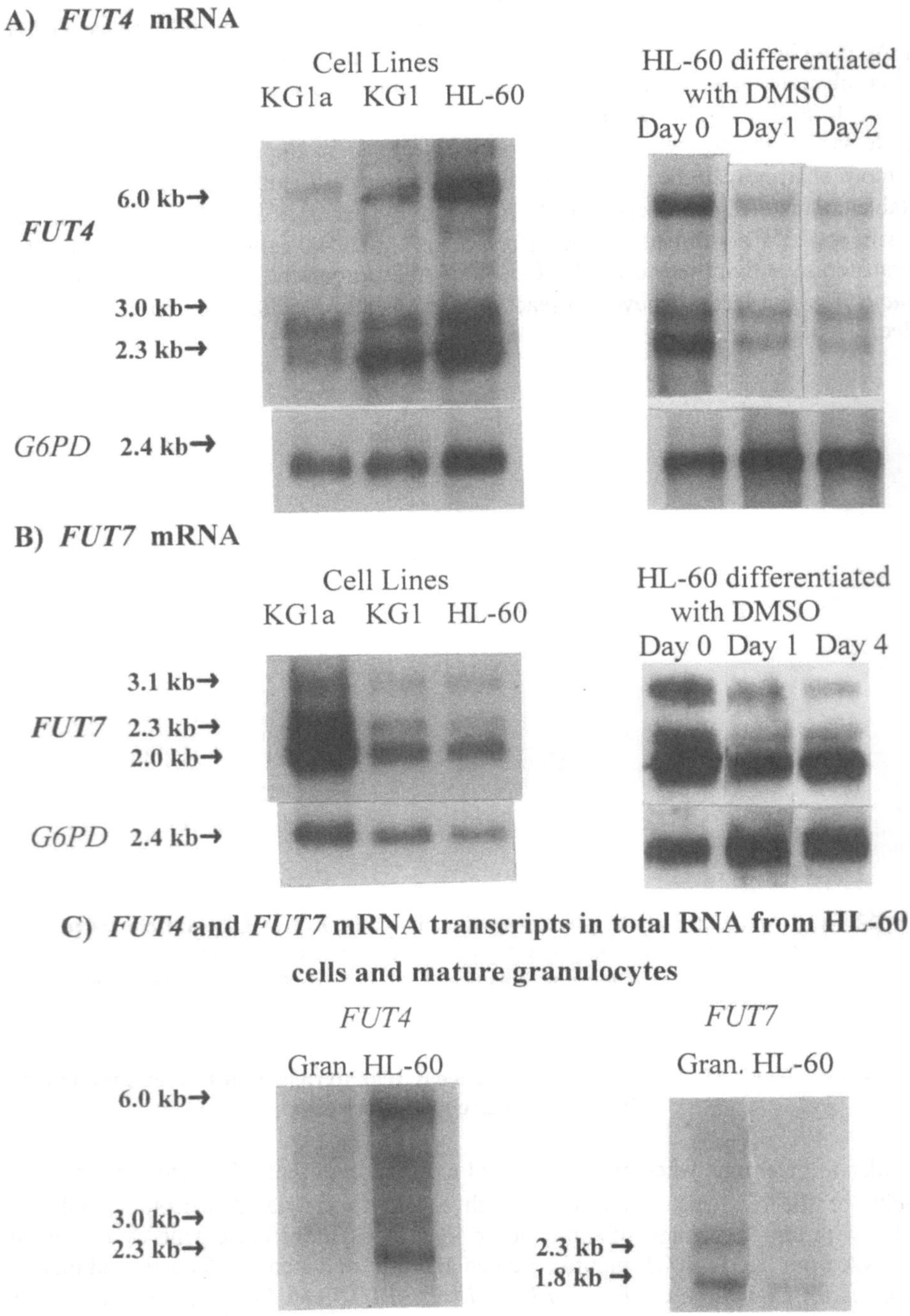

Fig. 13. Northern blot analyses. **A)** Expression of *FUT4* transcripts in mRNA from cell lines KG1a, KG1 and HL-60 mRNA and mRNA from HL-60 cells differentiated with DMSO. **B)** *FUT7* trancripts in mRNA from cell lines KG1a, KG1 and HL-60 mRNA and mRNA from HL-60 cells differentiated with DMSO. **C)** *FUT4* and *FUT7* transcripts in total RNA from mature granulocytes. Adapted from Clarke and Watkins [68,115].

most strongly expressed in the KG1a myeloblastic cell line with three distinctive transcripts 2.0 kb, 2.3kb and 3.1kb of which the 2.0 kb band was the strongest as had previously been observed for the expression of this gene in HL-60 cells [87]. The intensity of expression decreased sharply in KG1 cells and then remained at much the same level in HL-60 cells both before and after differentiation (**Fig.13**). In contrast expression of *FUT4* mRNA was low in KG1a cell mRNA, stronger in KG1 mRNA and even stronger in HL-60 cell mRNA[115]. On differentiation of HL-60 cells there was a dramatic fall in expression of *FUT4.* Three transcripts, 2.3 kb, 3.0 kb, and 6.0kb were observed in mRNA from each cell line but some differences were seen in the relative intensities of the three transcripts which may suggest that different transcription factors or promotors are used at early stages of development. In total RNA from mature granulocytes, *FUT4* expression was barely detectable whereas faint, but definite transcripts were discernible with the *FUT7* gene probe [115] (Fig. 13). Thus after the marked fall in *FUT7* expression that occurs early in maturation the next significant change is the rise in *FUT4* expression at the HL-60 promyelocyte step and the sharp fall in expression of this gene that takes place on further differentiation. In the mature granulocyte, while expression of both genes is very weak compared to the developing cells, expression of *FUT7* is stronger than expression of *FUT4*. The lack of correlation between the quantitative changes in *FUT4* and *FUT 7* mRNA levels at different maturation stages shows that these two genes are independently regulated [68]. The mRNA results taken together with the evidence from flow cytometry results suggested that, although *FUT7* is undoubtedly responsible for the final stage of the biosynthesis of sialyl-Le^x, modulation of the appearance of this antigen on the cell surface at the promyelocyte/myelocyte stage is more closely associated with the changes in expression of *FUT4* than with changes in expression of *FUT7.*

6. Expression of sialyltransferase genes in in leukaemic cell lines and mature granulocytes

Reverse transcription of mRNA and amplification (RT-PCR) of cDNA from KG1a, KG1, undifferentiated and differentiated HL-60 cells and mature granulocytes revealed expression of all five cloned human ST3Gal genes (encoding α2,3-sialyl-transferases) (**Fig.14**). in the amplified cDNAs (J.L.Clarke and W.M.Watkins unpublished observations). It was therefore not possible to clearly delineate the gene responsible for the

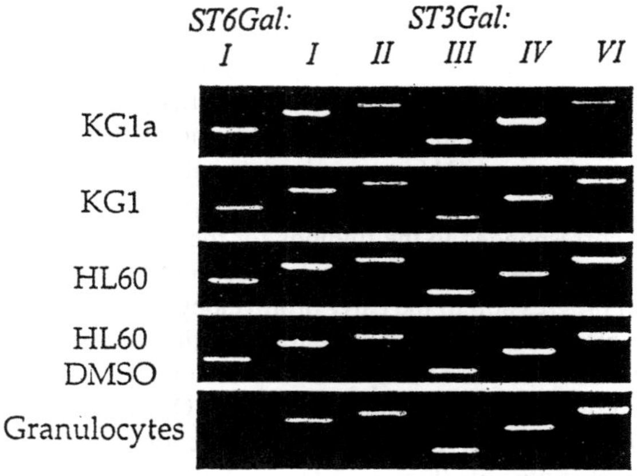

Fig.14. RT-PCR expression of sialyltransferase genes in mature granulocytes and in KG1a, KG1, HL-60 and HL-60 cells differentiated with DMSO.

formation of the precursor of sialyl-Le^x but other workers have suggested that the α2,3-sialyltransferase found in HL-60 cells represents mainly *ST3Gal IV* [117].

In agreement with the failure to detect α2,6-sialyltransferase in mature granulocytes, *ST6Gal-I* was not detected in amplified cDNA from granulocytes but was observed in RNA from all the leukaemic cell lines (**Fig.14**). Taniguchi *et al.*,[116] reported that on differentiation of HL-60 cells with DMSO the levels of *ST6Gal-I* mRNA fell dramatically after 1 days stimulation.

VIII. DISCUSSION AND CONCLUSIONS

Exploration of the biochemical, genetic and functional changes that take place in normal maturation of haemopoietic cells is restricted by the limited numbers of cells at a defined differentiation stage that can be isolated from human bone marrow aspirates. Therefore in attempts to understand the genetic basis of the changes in cell surface patterns of fucosylated and sialylated carbohydrate antigens that occur during myeloid cell development within the bone marrow microenvironment we have used the human leukaemic cell lines arrested at different maturation stages to represent steps from early myeloblasts to mature granulocytes (see **Fig.1**). Other studies have been made on the expression of fucosylated antigens in human leukaemic cell lines [118,119] and increased amounts of sialoglycoconjuagtes had been observed in granulocytes from CML patients in comparison with that in normal cells [120], but no other investigations have followed the changes in successive stages of maturation. Similarly others had found by examination of a random series of haemopoietic cell lines that *FUT4* and *FUT 7* were both expressed in myeloid cells [121] but no systematic analysis had previously been made of expression levels at progressive developmental stages.

Use of cell lines in place of normal bone marrow derived cells enables relatively unlimited amounts of cells to be grown for the experiments but a concern about using leukaemic cell lines is how far they resemble normal cells at the stage they are considered to represent. Pasqualetto *et al.* [122] examined the cell surface glycan features of HL-60 cells chemically induced to differentiate and found certain aspects of complex *N*-glycan structures that differed from those found in normal leukocytes. Nevertheless, insofar as the expression of antigens on the cell surface is concerned the cell lines we have used had the correct markers for their maturation stage (**Table 1**). The undifferentiated blast cell line, KG1a, and the myeloblast /early promyelocytic line KG1 expressed CD34, a marker of early progenitor cells that is not detectable on the promyelocytic HL-60 cells. The absence of CD11b or CD16 from HL60 cells was consistent with their promyelocytic level of maturation since these markers only appear on more highly differentiated forms (see **Fig.2**). CD11b appeared on differentiation of HL60 cells and this result was compatible with changes to myelocytic and metamyelocytic forms; CD16 was only detectable on mature granulocytes. The absence of the carbohydrate marker Le^x (CD15) on KG1a cells, together with the strong expression of sialyl-Le^x, corresponded to the findings for these antigens on early progenitor cells isolated from normal marrow [34,35] as did the development of Le^x and the decrease in expression of sialyl-Le^x on progression to the early promyelocyte KG1 cells. The HL-60 cell line that we have used, obtained from the European Collection of Animal Cell cultures , has consistently shown initial low levels of sialyl-Le^x expression which have considerably increased on DMSO-induced differentiation of the cells [22,115]. Other workers have recorded considerably stronger expression of sialyl-Le^x on the HL-60 cells they have used for their experiments [121,123] and the examples of this cell line employed in different laboratories appear to differ considerably with respect to the degree of expression of this determinant. The reason for this difference is not clear but in our laboratory continuous passage of the HL-60 cells led to a degree of spontaneous differentiation which resulted in increased activity with anti-sialyl-Le^x ; in our experiments care was therefore always taken to use cells freshly brought up from storage in liquid nitrogen and passaged only a limited number of times. Whether we have fortuitously hit upon an HL-60 sub-population with the suppressed sialyl-Le^x expression cannot be established but because the antigenic changes we have observed on the cells arrested at different stages of maturation, from the early promyelocyte KG1 cells to the differentiated HL-60 cells, appear to correspond, with respect to Le^x and sialyl-Le^x expression, to those recorded for normal bone marrow myeloid cells at intermediate stages of maturation (**Fig.2**) [34,35] we considered it reasonable to extrapolate from the changes in the glycosyltransferase enzyme and RNA levels in the cell lines to what is happening in the normal development of haemopoietic cells.

Much evidence has accumulated to indicate that, in myeloid cells, the α1.3-fucosyltransferase encoded by the *FUT7* gene is responsible for the addition of fucose to the precursor 3'-sialyl-*N*-acetyllactosamine and is therefore the gene ultimately responsible for the appearance of sialyl-Le^x [87,88,115]. Therefore the dip in expression of sialyl-Le^x at the intermediate stage of development could result from a lowering of the level of expression of the *FUT7* gene, lack of the appropriate sialylated precursor Type 2 substrate, competition of the enzyme with co-expressed glycosyltransferases for the same precursor substrate, or from changes in other regulatory mechanisms influencing the expression or functioning of this gene. Wagers *et al.*[54] isolated a variant of HL-60 cells that was completely deficient in sialyl-Le^x expression and found that the cells nevertheless expressed normal or elevated levels of *FUT7* enzyme activity; the authors concluded that the absence of cell surface sialyl-Le^x in this variant resulted from a generalised sialylation defect. A second cloned sialyl-Le^x negative variant of HL-60[124] was shown to be deficient in expression of *FUT7*; these cells had normal sialyltransferase activity but exhibited marked reduction in activity with 3'-sialylated Type 2 structures compared with the parent cell line. This variant more closely resembles the cell population that we have used as it displayed increased cell surface expression of Le^x and VIM2 epitopes, but the authors did not report whether the cells could undergo further differentiation. In our experiments the expression of the *FUT7* gene in the three cell lines, KG1a, KG1 and HL-60, revealed that changes in the level of expression of the *FUT7* mRNA did not correlate with the cell surface expression of sialyl-Le^x throughout the different maturation stages. The strong expression of sialyl-Le^x determinant on the earliest KG1a myeloblastic cell line was reflected in strong expression of *FUT7* mRNA. Similarly, the decreased expression of sialyl-Le^x in the early promyelocyte KG1 cells also correlated with a decrease in expression of mRNA corresponding to this gene (**Fig.13**). In the subsequent stages however, this correlation ceased to hold. In the promyelocyte HL-60 cells there was a sharp rise in α1,3-fucosyltranferase activity measurable with unsialylated Type 2 acceptors (**Fig.10**) which correlated with an increase in expression of *FUT4* mRNA. Despite the decrease in sialyl-Le^x activity on the cell surface of the HL-60 cells, however, the level of *FUT7 mRNA* expression was similar to that in KG1 cells. DMSO-induced differentiation of the HL-60 cells led to a mixture of cells morphologically resembling myelocyte, metamyelocyte and band forms, and resulted in a marked increase in cell surface expression of sialyl-Le^x, although the level of *FUT7* mRNA expressed in these cells did not change to any extent. In contrast, DMSO treatment resulted in a dramatic fall in the level of expression of *FUT4* mRNA and, in agreement with this change, the α1,3-fucosyltransferase activity measured with unsialylated substrates decreased without any significant change in activity with the sialylated compounds.

In addition to the changes in fucosytransferase activity, DMSO-induced differentiation of HL-60 cells leads to a fall in the level of activity of α2,6-sialyltransferase [22,] and Tanuguchi *et al.*[116] demonstrated that this fall is accompanied by a decrease in expression of mRNA corresponding to the sialyltransferase gene *ST6GalI* . Levels of both the α2,3- and α2,6-sialyltransferase enzymes fall on differentiation of HL-60 cells but the decrease in α2,6-sialyltransferase activity is much steeper than that of α2,3-fucosyltransferase and no α2,6-activity was detectable in mature granulocytes. Therefore, at the time of differentiation, sharp falls occur in the level of expression of both the *FUT4* encoded enzyme and the *ST6Gal1* encoded enzyme and these changes correlate with the increased expression of sialyl-Le^x on the cell surface despite the fact that the level of expression of *FUT7* mRNA does not appreciably alter. The enzymic product of the *FUT4* gene does not fucosylate α2,3-sialylated acceptors but to make Le^x it utilises the same Galβ1-4GlcNAc-R substrate as the α2,3- and α2,6-sialyltransferases and thus the elevated level of this enzyme in the undifferentiated HL-60 cells could cause it to compete for the terminal lactosamine structures at the expense of the α2,3-sialyltransferase and thus deprive the cell of sufficient substrate for the *FUT7* enzyme. The high level of VIM2 antigenic expression in untreated HL-60 cells is compatible with the elevated expression of *FUT4*, and the fall in the level of this determinant in the differentiated cells correlates with the decreased expression of this gene. The strongly expressed α2,6-sialyltransferase constitutes another possible competitor for the poly-lactosamine structures in the undifferentiated HL-60 cells. On differentiation the sharp fall in the level of both the *FUT4* mRNA and *ST6Gal1* mRNA could therefore make available more of the substrate required by the α2,3-sialyltransferase for the synthesis of the precursor structure acted on by the *FUT7* encoded enzyme and hence would allow the re-appearance of sialyl-Le^x on the cell surface. These arguments are, of course based on assumptions that all the other enzymes in the biosynthetic pathway needed for the formation of the terminal sialyl-Le^x structures are expressed and that the relative location of the glycosyltransferases in the Golgi apparatus, and the order in which they encounter the growing oligosaccharide chains, are compatible with the effects described. Factors such as the degree of internal fucosylation of the polylactosamine chains have been reported to influence the relative amounts of terminal α2,6- and α2-3-sialyl groups that are synthesised[117] and therefore would also have an influence the amount of substrate available for the *FUT7* encoded enzyme.

Another enzyme that has been correlated with expression of sialyl-Le^x in some cell lines is the Core-2 β1,6-*N*-acetylglucosaminyltransferase that transfers *N*-acetyl-glucosamine to GalNAc in Galβ1-3GalNac-Ser/Thr structures (see **Fig 5**) in the core region of glycoproteins. This enzyme is believed to initiate synthesis of *O*-linked poly-*N*-acetyllactosamine chains of the type found in myeloid cell glycoproteins[125] and *O*-linked structures such as those found on PSGL-1[41]. Studies on human pre-B lympoid cell lines[126] showed that expression of the Core-2 enzyme had a significant correlation with the cell surface expression of sialyl-Le^x and did not parallel expression of the *FUT7* gene, indicating that absence of *O*-linked chains was the significant factor leading to loss of substrate for the glycosyltransferases involved in the formation of the terminal antigenic structures. Mice lacking this enzyme also revealed a restricted phenotype with neutrophilia and a partial deficiency of selectin ligands[127]; hence absence of Core-2 oligosaccharide biosynthesis leads to failure of cells to react with selectins and reduces recruitment of neutrophils to sites of inflammation. These results therefore served to strengthen the suggestion that the functional ligands for E-and P-selectins are on *O*-glycans with the Core 2 structure. However, in myeloid cells the amount of polylactosamine chains increases steadily during maturation [36] and hence it does not appear that the dip in expression of sialyl-Le^x at the promyelocyte stage can be explained by a deficiency of the Core-2 β1,6-*N*-

acetylglucosaminyltransferase. We therefore conclude that the most likely explanation for loss of expression of sialyl-Le^x at the promyelocyte stage, represented in our experiments by the HL-60 cells, results from the effects of an elevated *FUT4* gene expression, and possibly elevated *ST6Gal1* gene expression, which encode enzymes that compete for the terminal *N*-acetyllactosamine structures. One way of testing the hypothesis would be to transfect HL-60 cells with anti-sense to the *FUT4* gene in order to inhibit expression of this gene; a rise in cell surface activity of sialyl-Le^x would indicate that removal of the *FUT4* enzyme had enabled. the *FUT7* encoded enzyme to function. Conversely transfection of KG1a with anti-sense to *FUT 7* could be investigated to see if this treatment led to suppression of sialyl-Le^x and expression of Le^x. Similar experiments could be carried out to test the role of the α2,6-sialyltransferase. Transfection of anti-sense cDNA of *FUT7* encoded α1,3-fucosyltransferase has been demonstrated to suppress sialyl-Le^x expression and E-selectin-mediated cell adhesion in cultured human lymphoid cells[128]. Whether or not the *FUT9* gene[89] encoded enzyme participates in the cell surface expression of the antigens on myeloid cell surfaces is not yet known. Since the enzyme encoded by this gene is said to have a strong preference for fucosylating the distal *N*-acetylglucosamine residue in polylactosamine chains, and to be without significant activity with sialylated acceptors [91], a strong expression of this enzyme at the promyelocyte stage would enhance the effect of *FUT4* in utilising substrate to produce Le^x structures. A puzzling feature of granulocyte carbohydrate antigen expression has been the persistence of strong cell surface Le^x activity in view of the apparent absence, or very weak expression, of *FUT4* mRNA [115,35]. It will therefore be interesting to discover whether the *FUT9* encoded enzyme is responsible in the final stages of granulocyte maturation for the synthesis of this epitope. The strong expression of VIM2 on mature granulocytes would nevertheless still require an explanation since the *FUT9* enzyme is said not to readily fucosylate internal *N*-acetylglucosamine residues in the polylactosamine chains [91].

The significance of the dip in expression of sialyl-Le^x at the promyelocyte/myelocyte stage is at present unknown. Our experiments have endeavoured to explain the mechanism underlying the phenomenon but have so far not produced any evidence that it has relevance to events occurring in the bone marrow. However, as it takes place just prior to the stage when the myeloid cells cease to proliferate, and enter a phase where the cells mature through several more morphological steps before release into the peripheral circulation, it is feasible to assume that it could be connected with adhesive interactions with stromal or endothelial cells within the bone marrow microenvironment that are connected with normal development. Greenberg *et al.*,[129] demonstrated a correlation between the maturity of human haemopoietic progenitor cells and their rolling efficiency on selectins and concluded that there was differential expression of selectin ligands as the cells matured. Binding experiments of myeloid cells at defined maturation stages with the isolated selectins, and with intact bone marrow endothelial and stromal cells, will be necessary in order to study whether the observed changes in carbohydrate expression can be related to events that would enhance cell-cell contact or differential migration of cells to different niches in the bone marrow microenvironment. The interaction of the enzymes encoded by the two genes *FUT4* and *FUT7* has been shown to provide a large number of related, but structurally distinct molecules, that can be distinguished by the fine specificities of monoclonal antibodies and endogenous lectins and, since a deficiency of both *FUT4* and *FUT7* genes was shown to have a more serious effect on haemopoiesis in mice than absence of either gene alone[12,13] a possible role is indicated for structures synthesised by the concerted action of the two enzymes encoded by these genes. E-selectin, known to be constitutively expressed on the bone marrow endothelial cells, may be the major adhesion molecule involved in interaction with the ligands on the surfaces of the developing cells and alterations in the level of expression of the ligands may be a driving force to release the cells and enable them to move to different compartments in the marrow for further stages of maturation to proceed. It is feasible, however, that other proteins with lectin-like properties, yet to be identified, could be expressed in the bone marrow and interact with one or more of the multiple structures that can be synthesised by the cooperative actions of these two genes. Thus the story of the part played by sialylated and fucosylated carbohydrate ligands in haemopoiesis is only just beginning to unfold but sufficient evidence has accrued to indicate that these structures have a significant role to play in normal maturation events and that changes in the patterns of expression of the surface antigens may lead to pathological conditions.

SUMMARY

The first part of this article reviews the stages of normal development of haemopoietic cells committed to the myeloid lineage, properties of leukaemic cell lines that are arrested at specific maturation stages along the granulocytic pathway, the structures of carbohydrate antigenic markers that appear on myeloid cell surfaces, with especial reference to sialyl-Le^x (NeuAcα2-3Galβ1-4[Fucα1-3]GlcNAc), and the role of this antigen on mature granulocytes as a ligand for selectin molecules. The families of fucosyl- and sialyl-transferase genes encoding enzymes responsible for the biosynthesis of sialyl-Le^x, and the pathways leading to the formation of this antigen, and more complex related structures, are described.

The second part of the article outlines the work carried out in the authors' laboratory with leukaemic cell lines in an attempt to ascertain the biochemical and genetic basis of the lowering of sialyl-Le^x expression that occurs at intermediate stages of normal haemopoietic development. Analysis of enzyme levels and mRNA expression of the fucosyl-and sialyl-transferase genes has led to the conclusion that depletion of substrate resulting from high levels of enzyme activity from co-expressed genes *FUT4* and *ST6Gal1* probably accounts for the dip in expression of sialyl-Le^x, rather than a change in the level of expression of *FUT7*, the gene in myeloid cells encoding the enzyme ultimately responsible for the synthesis of sialyl-Le^x. The possible significance of this change in relation to normal cell maturation is discussed.

REFERENCES

1. T.M.Dexter, L.H. Coutinho, E. Spooncer, C.M.Heyworth, C.P. Daniel, R Shiro, J.Chang and T.D. Allen, Molecular control of haemopoiesis, In *Ciba Foundation Symposium 148,* p.76 Wiley, Chichester (1990).
2 P.Crocker and T.Feizi, Carbohydrate recognition systems; functional triads in cell-cell interactions, *Curr.Opin.Struct.Biol.* 6:679 (1996).
3. E.Butcher Leukocyte-endothelial cell recognition; three (or more) steps to specificity and diversity, *Cell* 67:1033 (1991).
4. G.Walz, A.Aruffo, W. Kolanus, M. Bevilaqua. and B. Seed, Recognition by ELAM-1 of the sialyl-Lex determinant on myeloid and tumour cells, *Science* 250:1132 (1990).
5. R.P. McEver, K.L Moore and R.D. Cummings, Leukocyte trafficking mediated by selectin-carbohydrate interactions, *J.Biol.Chem.* 270:11025 (1995).
6. P.S.Frenette, T.N.Mayadas, H. Rayburn, R.O.Hynes and D.Wagner. Susceptibility to infection

and altered haematopoiesis in mice deficient in both P-and E-selectin, *Cell* 84:563 (1996).
7. P.Crocker, S.Freeman, S. Gordon and S. Kelm, Sailoadhesin binds preferentially to cells of the granulocytic lineage *J.Clin. Invest.* 95:635. (1995).
8. C.G.,Gahmberg, K.Nilsson and L.C.Anderson, Specific changes in the surface glycoprotein pattern of human promyelocyte leukaemic cell line HL-60 during morphologic and functional differentiation, *Proc.Natl. Acad.Sci. USA.* 76: 4087 (1979).
9. M.Fukuda, Cell surface glycoconjugates as onco-differentiation markers in hematopoietic cells. *Biochem.Biophys.Acta* 780:119 (1985).
10. H.Noriji, F. Takaku., Y. Terui., Y. Miura. and M. Saito Ganglioside GM3: An acidic membrane component that increases during macrophage-like cell differentiation can induce monocytic differentiation of human myeloid and monocytoid leukaemic cell lines HL60 and U937. *Proc.Natl.Acad.Sci.* 83:782. (1986).
11. M.Nakamura, A.Tsunoda, K.Sakoe, J.Gu, A.Nishikawa, N.Taniguchi and M.Saito, Total metabolic flow of glycosphingolipid biosynthesis is regulated by UDP-GlcNAc-lactosylceramide β–1-3N-acetylglucosaminyl-transferase and CMP-NeuAc:lactosyl-ceramide α2-3-sialyltransferase in human hematopoietic cell line HL60 during differentiation, *J.Biol.Chem.*, 267:23507 (1992).
12 P.Maly, A.D. Thall, B.Petryniak, C.E Rogers, P.L Smith, P.L Marks, R.J Kelly, K.M.Gersten, G.Cheng, T.L Saunders, S.A Camper, R.T. Camphausen, F.X .Sullivan,Y.Isogai, O. Hindsgaul, U.H. von Andrian and J.B. Lowe, The α1,3-fucosyltransferase Fuc-VII controls leukocyte trafficking through an essential role in L-, E-, and P-selectin ligand synthesis. *Cell*, 86:643 (1996).
13. J.B Lowe, A.D.Thall, P. Maly, B.Petryniak, C.E. Rogers, P.L Smith, S, Askani, R..Kelly, G.Cheng and U.H. von Adrian, *Glycoconjugate J.* 14, p760, Abstract 039, (1997).
14. A, Etzioni, C.M. Doerschuk and J.M. Harlan, Of man and mouse: Leukocyte and endothelial adhesion molecule deficiencies, *Blood*, 94:3281 (1999).
15. L.Yao, H Setiadi, L Xia, Z, Laszik, F.B Taylor and R.P McEver, Divergent inducible expression of P-selectin and E-selectin in mice and primates,Blood, 94:3820 (1999).
16. M.Lubbert, F. Herrman and H.P. Koeffler. Expression and regulation of myeloid specific genes in normal and leukaemic myeloid cells, *Blood* 77: 909 (1990).
17. H.P Koeffler, Induction of differentiation of human acute myelogenous leukaemia cells: therapeutic implications, *Blood* 62:709 (1983).
18. H.P, Koeffler, R Billing, A Lusis, R Sparkes and D.W Golde. An undifferentiated variant derived from the human acute myelogenous leukaemia cell line (KG1), *Blood* 56:265 (1980) .
19. H.P. Koeffler and D.W Golde, Acute myelogenous leukaemia: a human cell line responsive to colony-stimulating activity, *Science,* 2000:1153 (1978) .
20. S.J.Collins, R.C Gallo.and R.E Gallagher, Continuous growth and differentiation of human myeloid leukemic cells in suspension culture, *Nature,* 270:347 (1977).
21 . S.J.Collins, F.W Ruscetti, R.E. Gallagher and R.C.Gallo, Terminal differentiation of human promyelocytic leukamia cells induced by dimethylsulfoxide and other polar compounds, *Proc.Natl. Acad.Sci. USA*. 75:2458 (1978).
22. P.O.Skacel, A.J.Edwards, C.T.Harrison and W.M.Watkins, Enzymic control of the expression of the X-determinant (CD15) in human myeloid cells during maturation: The regulatory role of 6'-sialyltransferase, *Blood*, 78:1452 (1991).
23. S.Hakomori, Monoclonal antibodies directed to cell surfacecarbohydrates.in:*Monoclonal antibodies and functional cell lines,* Kennet, Bechtol and McKearn, eds., Plenum Press, New York p.67. (1984).
24. T.Feizi, Demonstration by monoclonal antobodies that carbohydrate structures of glycoproteins and glycolipids are oncodevelopmental antigens. *Nature* 314: 53 (1985).
25. W.M. Watkins, Glycosyltransferases. Early history, development and future prospects. *Carbohydr. Res.* 149:1 (1986).
26. J.C.Paulson and K.J.Colley, Glycosyltransferases: structure, localisation and control of cell type-specific glycosylation, *J.Biol.Chem.* 264:17615, (1989).
27 S. Natsuka and J.B. Lowe Enzymes involved in mammalian oligosaccharide biosynthesis, *Curr. Opin. Struct. Biol.* 4:683 (1994).

28. J.B.Lowe, Mammalian cloning, expression, and uses of mammalian glycosyltransferases, *Seminars in Cell Biology*, 2:289 (1991).
29. H.Kitagawa and J.C.Paulson, Differential expression of five sialyltransferase genes in human tissues, *J.Biol.Chem.* 269:17872 (1994).
30. A.Harduin-Lepers, M-A. Recchi and P.Delannoy, 1994, The year of sialyltransferases *Glycobiology*, 5:741 (1994).
31. C.I. Civin, J. Mirro and M.L.Banquerigo, My-1, a new myeloid specific antigen identified by a mouse monoclonal antibody, *Blood* 57:842 (1981).
32. A.Tabillio, M.C. Del Canzio, A. Henri, J.Guichard, P.Mannoni , C.I. Civin, U.Testa, H.Rochan., W. Vainchenker and J.Breton-Gorius, Expression of SSEA1 antigen (3-fucosyl-N-acetyllactosamine) on normal and leukemic haemopoietic cells: modulation by neuraminidase treatment, *Brit.J.Haematol.*, 58:697 (1984).
33. L.W.M.M.Terstappen, M.Salford and M.R.Loken, Flow cytometric analysis of human bone marrow. III. Neutrophil maturation, *Leukaemia* 4:657 (1990).
34. F.Lund-Johansson. and L.W.M.M. Terstappen, Differential surface expression of cell adhesion molecules during granulocyte maturation, *J.Leuk. Biol.* 54:47 (1993).
35. N. Le Marer, M. Palcic, J.L.Clarke, D Davies, P.O. Skacel, Developmental regulation of α1,3-fucosyltransferase expression in CD34 positive progenitors and maturing myeloid cells isolated from normal human bone marrow, *Glycobiology* 7:357 (1997).
36. M.Fukuda, H.P.Koeffler and J. Minovada, Membrane differentiation in human myeloid cells.Expression of unique profiles of cell surface glycoproteins in myeloid leukaemic cell lines blocked at different stages of differentiation and maturation. *Proc. Natl. Acad. Sci. U.S.A.*, 78:6299 (1981).
37. B.A.Macher, J.Buehler, P.Scudder, W.Knapp and T. Feizi, A novel carbohydrate differentiation antigen on fucogangliosides of human myeloid cells recognised by monclonal antibody VIM-2, *J.Biol.Chem.* 263:10186, (1988).
38. B.A. Macher. and J.H.Beckstead, Distribution of VIM-2 and SSEA1 glycoconjugate epitopes among human leukocytes and leukemia cells, *Leuk. Res.* 14:119 (1990).
39. M.Fukuda, E.Spooncer, J.E.Oates, A.Dell and J.G. Klock, Structure of sialylated fucosyl lactosaminoglycan from human granulocytes. *J.Biol.Chem.* 259:10925 (1984).
40. S.R.Carlsson, and M.Fukuda, Isolation and characterisation of leukosialin, a major sialoglycoprotein on human leukocytes. *J.Biol.Chem.* 261:12779 (1986).
41. K.L.Moore, N.L.Stults, S.Diaz, D.F. Smith., R.D. Cummings, A.Varki and R.P. McEver, Identification of a specific glycoprotein ligand for P-selectin (CD62) on myeloid cells, *J.Cell biol*.118: 445 (1992).
42. P.P.Wilkins, R.P.McEver and R.D.Cummings, Structures of the O-glycans on P-selectin glycoprotein ligand -1 from HL-60 cells, *J.Biol.Chem.* 271:18732 (1996).
43. K.Maemura and M. Fukuda, Poly-N-acetyllactosaminyl O-glycans attached to leukosialin. *J.Biol.Chem.* 267:24379 (1992).
44. B.A.Macher and J.C.Klock, Isolation and chemical characterisation of neutral glyco-sphingolipids of human neutrophils, *J.Biol.Chem.* 2092:255 (1980).
45. M.R.Stroud, K.Handa, M.E.K.Salyan, K.Ito, S.B.Levery, S Hakomori, B.B.Reinhold and V.N.Reinhold, Monosialogangliosides of human myelogenous leukaemia HL60 cells and normal human leukocytes. 1. Separation of E-selectin binding from nonbinding gangliosides, and absence of sialosyl-Le^x having tetraosyl to octaosyl core,. *Biochemistry* 35:758 (1996).
46 M.R.Stroud, K.Handa, M.E.K.Salyan, K.Ito, S.B.Levery, S Hakomori, B.B.Reinhold and V.N.Reinhold, Monosialogangliosides of huamn myelogenous leukaemia HL60 cells and normal human leukocytes. 2. Characterisation of E-selectin binding fractions, and structural requirements for physiological binding to E-selectin, *Biochemistry* 35:770 (1996).
47. T.A.Springer, Traffic signals for lymphocyte recirculationand leukocyte emigration: the multistep paradigm, *Cell* 76:301 (1994).
48. A.Varki, Selectin ligands: Will the the real ones please stand up, *J.Clin.Invest.* 99: 158 (1997).
49. G.S.Kansas, Selectins and their ligands: Current concepts and controversies, *Blood* 88:3259 (1996).
50. J.B.Lowe, Selectin ligands, leukocyte trafficking, and fucosyltransferase genes, *Kidney International* 51:1418 (1997).

51. M.P.Bevilacqua and R.M.Nelson, Selectins *J.Clin.Invest.* 91:379 (1993).
52. G.Walz, A.Aruffo, W.Kolanus, M.Bevilacqua and B.Seed, Recognition by Elam-1 of the sialyl-Le^x determinant on myeloid and tumour cells, *Science* 250-1132 (1990).
53. B.K.Brandley, S.J.Swiedler,and P.W.Robbins. Carbohydrate ligands of the LEC cell adhesion Molecules, *Cell* 63:861 (1990).
54. A.J.Wagers, L.M.Stoolman, R.Craig, R.N.Knibbs and G.S.Kansas, An Lex-deficient variant of HL-60 cells exhibits high levels of adhesion to vascular selectins: further evidence that HECA-452 and CSLEX1 monoclonal antibody epitopes are not essential for high avidity binding to vascular selectins, *J.Immunol.*160:5122 (1998).
55. K.D.Patel, K.L.Moore, M.U.Nollet and R.P.McEver, Neutrophils use both shared and distinct mechanisms to adhere to selectins under static and flow conditions. *J.Clin.Invest.* 96:1887 (1995).
56. M.Lenter, A.Levinovitz, S.Isenmann, D.Vestweber. Monospecific and common glycoprotein ligands for E- and P-selectins on myeloid cells, *J.Cell Biol.* 125:471 (1994).
57. M.A.Steegmaier, S.Levinovitz, E.Isenmann, M.Borges, H.P.Lenter, B.Kocher, B.Kleuser and D.Vestweber. The E-selectin-ligand ESL1 is a variant of a receptor for fibroblast growth Factor, *Nature (Lond.)* 373:615 (1995).
58. J.F.M. Leeuwenberg, A.Tan, T.M.A.A.Jeunhomme, H.L.Ploegh and W.A.Buurman, The ligand recognised by ELAM-1 on HL-60 cells is not carried by N-linked oligosaccharides, *Eur.J.Immunol.* 21:3057 (1991).
59. G.R.Larsen., D.Sako, T.J.Ahern, M .Shaffer., J. Erban., S.A.Sajer, R.M.Gibson, D.D.Wagner, B.C. Furie and B.Furie, P-selectin and E-selectin: distinct but overlapping leukocyte specificities, *J.Biol.Chem.* 267:11104 (1992).
60. D.M.Lewinsohn, R.F.Bargatze and E.C.Butcher, Leukocyte-endothelial cell recognition: Evidence of a common molecular mechanism shared by neutrophils, lymphocytes and other leukocytes, *J.Immunol.* 138:4313, (1997).
61. S.D.Rosen and C.R.Bertozzi, The selectins and their ligands, *Curr.Opin.Cell Biol.* 6:663 (1994).
62. T.Feizi and C.Galustian, Novel oligosaccharide ligands and ligand-processing pathways for the selectins, *TIBs* 24:369 (1999).
63. L.J.Picker, R.A.Warnock, A.R.Burns, C.M.Doerschuk, E.L.Berg and E.C.Butcher, The neutrophil selectin LECAM-1 presents carbohydrate ligands to the vascular selectins ELAM-1 and CMP-140, *Cell* 66:921 (1991).
64. K.M.Schweitzer, A.M.Drager, P.van der Valk, S.F.T.Thijsen, A.Zevenbergen, A.P.Theijsmeijer, C.E.van der Schoot and M.M.A.C. Langenhuijsen, Constitutive expression of E-selectin and vascular cell adhesion molecule-1 on endothelial cells of haemopoietic tissues, *Am.J.Pathol.* 148:165 (1996).
65. W.M.Watkins, Molecular basis of antigenic specificity in the ABO, H and Lewis blood group systems: In: *Glycoproteins*, J.Montreuil, J.F.G. Vliegenthart and H.Schachter, eds, Elsevier, Amsterdam p.313, (1995).
65. J-P Cartron, C.Mulet, B.Bauvois, C.Rahuel and C.Salmon. ABH and Lewis glycosyltransferases from human red cells, lymphocytes and platelets, *Blood Transfus. Immunohaematol.* 23:271. (1980).
67. P.O.Skacel and W.M.Watkins, Fucosyltransferase expression in human platelets and leukocytes, *Glycoconjugate J.* 4:267 (1987).
68. J.L.Clarke and W.M.Watkins, Independent regulation of Fuc-TIV and Fuc-TVII genes leading to modulation of cell surface expression in developing myeloid cells, *Glycobiology* 7:835 (1997).
69. F.W.Symington, D.L.Hedges and S.Hakomori, Glycolipid antigens of human polymorphnuclear neutrophils and the inducible HL-60 myeloid leukaemic cell line, *J.Immunol,* 134:2498 (1985).
70. R.K.M.Kong, A.Barrios, W.Knapp and B.A.Macher, Fucosylated glycosphingolipids of human myeloid cells, *Arch. Biochem.Biophys.* 300:677 (1993).
71. K.Kitamura, E.Stockert, P.Garin-Chesa, S.Welt, K.O.Lloyd, K.L.Armour, T.P.Wallace, W.J.Harris, F.J.Carr and L.J.Old. Specificity analysis of blood group Lewis-y (Le^y) antibodies generated against synthetic and natural Le^y determinants. *Proc. Natl. Acad.Sci.USA.* 91:12957 (1994).

72. T.A.Beyer, J.L.Rearick, J.C.Paulson, J-P Prieels, J.E. Sadler and R.L.Hill, Biosynthesis of mammalian glycoproteins: glycosylation pathways in the synthesis of the non-reducing terminal sequences. *J.Biol.Chem.* 254:12531 (1979).
73. P.H.Johnson and W.M.Watkins, Sialyl compounds as acceptor substrates for the human α–3- and α3/4-fucosyltransferases, *Biochem. Soc.Trans.*13:1119 (1985).
74. B.W.Weston, P.L.Smith, R.J.Kelly and J.B.Lowe, Molecular cloning of a fourth member of a human alpha (1,3) fucosyltransferase gene family, *J.Biol.Chem.* 267:24575 (1992).
75. P.H.Johnson and W.M.Watkins, Sialyl compounds as acceptor substrates for fucosyltransferases in normal and leukaemic human granulocytes, *Biochem.Soc.Trans.* 15:396 (1987).
76. W.M.Watkins, P.O.Skacel and P.H.Johnson, Human fucosyltransferases involved in the biosynthesis of X (Galβ1-4[Fucα1-3]GlcNAc) and sialyl-X (NeuAcα2-3Galβ1-4[Fucα1-3]GlcNAc) antigenic determinants, In *Carbohydrate Antigens*, P.J.Garregg and A.A.Lindberg eds, ACS Symposium Series 519 p.34, (1993).
77. T.Kudo,Y. Ikehara Y, A.Toyayachi, M. Kaneko, T.Hiraga, K.Sasaki and H.Narimatsu, Expression cloning and characterisation of a novel murine a1,3-fucosyltransferase, mFUc-TIX, that synthesises the Lewis x (CD15) epitope in brain and kidney, *J.Biol.Chem.* 273:26729 (1998).
78. M.Costache, A.Cailleau, P.Fernandez-Mateos, R.Oriol and R.Mollicone, Advances in molecular genetics of α-2- and α3/4-fucosyltransferases. *Transfus.Clin.biol.* 4:367 (1997).
79. J.F.Kukowska-Latallo, R.D.Larsen, R.P.Nair and J.B.Lowe. A cloned human cDNA determines expression of a mouse stage specific embryonic antigen and the Lewis blood group (1,3/1,4)fucosyltransferase. *Genes Dev.*4:1288 (1990).
80. T.de Vries, C.A.Srnka, M.M.Palcic, S.J.Swiedler, D.H. van den Eijnden and B.A,Macher, Acceptor specificity of different length constructs of human recombinant α1/3/4-fucosyl-Transferases, *J.Biol.Chem.* 270:8712 (1995).
81. J.Costa, E. Grabenhorst, M.Nimtz and H.S.Conrad, Stable expression of the Golgi form and secretory variants of human fucosyltransferase III from BHK-21 cells, *J.Biol.Chem.* 272:11613 (1997).
82. D.M. Marcus and LE Cass, Glycosphingolipids with Lewis blood group activity: uptake by erythrocytes, *Science* 164, 553 (1969).
83. B.W.Weston, R.P.Nair, R.D.Larsen and J.B.Lowe, Isolation of a novel human α1,3-fucosyltransferase and molecular comparison to the human Lewis blood group (α1,3/1,4)fucosyltransferase gene, *J.Biol.Chem.* 267:4252 (1992).
84. S.E.Goelz, C.Hession, D. Goff, B.Griffiths, R.Tizard, B.Newman, G.Chi-Rosso and R.Lobb, ELFT : a gene that directs the expression of an ELAM-1 ligand, *Cell*, 63:1349. (1990).
85. R.Kumar, B.Potvin, W.A.Muller and P.Stanley, Cloning of a human α(1,3)fucosyl-transferase gene that encodes ELFT but does not confer ELAM-1 recognition on Chinese hamster ovary cell transfectants, *J.Biol.Chem.* 266:21777 (1991).
86. J.B.Lowe, J.F.Kukowska-Latallo,R.P.Nair, R.D.Larsen, R.M.Marks, B.A. Macher, R.J.Kelly and Ernst L.K, Molecular cloning of a human fucosyltransferase gene that determines expression of the Lewis x and VIM-2 epitopes but not ELAM-1-dependent cell adhesion, *J.Biol.Chem.* 266:17467 (1991).
87. S.Natsuka, K.M.Gersten, K.Zenita, R.Kannagi and J.B.Lowe, Molecular cloning of a cDNA encoding a novel human leukocyte α1,3-fucosyltransferase capable of synthesising the sialyl Lewis x determinant, *J.Biol.Chem.* 269:16789 (1994).
88. K.Sasaki, K.Kurato, K.Funayama , M.Nagata, E.Watanabe, S.Ohta, N.Hanai, and T.Nishi, Expression cloning of a novel a1,3-fucosyltransferase that is involved in the biosynthesis of sialyl Lewis x carbohydrate determinants in leukocytes. *J.Biol.Chem.* 269:14730 (1994).
89. M. Kaneko, T.Kudo, H.Iwasaki, Y.Ikehar., S. Nishihara, S.Nakagawa, K.Sasaki, T.Shiina, H.Inoko, N. Saitou and H. Narismatsu, α1,3-Fucosyltransferase IX (Fuc-IX) is very highly conserved between human and mouse; molecular cloning, characterisation and tissue distribution of human Fuc-TIX, *FEBS Letters* 452:237 (1999).
90. R.Niemela, J.Natunen, M-L Majuri, H.Maaheimo, J.Helin, J.B.Lowe, O.Renkonen and

R.Renkonen, Complementary acceptor and site specificties of Fuc-TIV and Fuc-TVII allow effective biosyntheisis of sialyl-TriLex and related polylactosamines present on glycoprotein counterreceptors of selectins, *J.Biol.Chem.* 273:4021 (1998).

91. S.Nishihara, H.Iwasaki, M.Kaneko, A.Tawada, M. Ito and H.Narismatsu, α1,3-Fucosyltransferase 9 (FUT9;FucIX) preferentially fucosylates the distal GlcNAc residue of polylactosamine chain while the other four α1,3FUT members preferentially fucosylate the inner GlcNAc residue,. *FEBS Lett.* 462:289 (1999).
92. M.L.Phillips, B.R.Schwartz, A.Etzioni, R.Bayer, H.D.Ochs, J.C.Paulson and J.L.Harlan, Neutrophil adhesion in leukocyte deficiency syndrome type 2, *J.Clin. Invest.* 96:2898 (1995).
93. L.Sturla, A.Etzioni, A.Bisso, D.Zanardi, G.DeFlora, L.Silengo, A.DeFlora and M.Tonetti, Defective intracellular activity of GDP-D-mannose-4-6-dehydratase in leukocyte adhesion deficiency type II syndrome, *FEBS Lett.* 429: 274 (1998).
94. M.W. Long, The role of glycoconjugates in blood cell development, *Glycobiology*, 9:1098 Abstract 1. (1999).
95. S.Tsuji, A.K.Datta and J.C.Paulson, Systematic nomenclature for sialyltransferases H.Kitigawa and J.C. Paulson, Cloning of a novel α2,3-sialyltransferase that sialylates glycoprotein and glycolipid-carbohydrate groups. *J.Biol.Chem.* 269:1394 (1994).
96. Y-J. Kim, K-S. Kim, S-H. Kim, C.H.Kim, J.H. Ko, I-S.Choe, S. Tsuji and Y-C. Lee, Molecular cloning and expression of human Galβ1,3GalNAc α2,3-sialyltransferase (hST3Gal II). *Biochem.Biophys Res.Commun.* 228:324 (1996).
97. V.Giodanengo, S.Bannwarth, C.Laffont, V.van Miegem, A.Harduin-Lepers, P.Delannoy and J.C.Lefebre, Cloning and expression of cDNA for a human Gal(β1-3) GalNAc α2,3-sialyltransferase from the CEM T-cell line, *Eur.J.Biochem.* 274:558 (1997).
99. H.Kitigawa and J.C. Paulson, Cloning and expression of human Galβ1,3(4)GlcNAc α2,3-sialyltransferase, *Biochem.Biophys Res.Commun.* 194: 375 (1993).
100. K.Sasaki, E.Watanabe, K.Kawashima, S.Sekine, T.Dohi, M.Oshima, N.Hanmai, T.Nishi and M.Hasegawa, Expression cloning of a novel Galβ(1-3/1-4)GlcNAc sialyltransferase using lectin resistance selection,. *J.Biol.Chem.* 268:22782 (1993).
101. H.Kitigawa and J.C. Paulson, Differential expression of five sialyltransferase genes in human tissues, *J.Biol.Chem.* 269:17872 (1994).
102. T.Okajima, S.Fukumoto, H.Miyazaki, H.Ishida, M.Kiso, Ke.Furukawa, T.Urano and Ki. Furukawa, Molecular cloning of a novel α2,3-sialyltransferase (ST3GalVI) that sialylates Type II lactosamine structures on glycoproteins. *J.Biol.Chem.* 274:11479 (1999).
103. U.Grundmann, C.Nerlich, T.Rein and G.Zettlmeissl, Complete cDNA sequence encoding human β-galactoside α2,6-sialyltransferase, *Nucleic Acids Res.* 18: 667 (1990).
104. W.M.Watkins, Blood group substances: In the ABO system the genes control the arrangement of sugar residues that determines blood group specificity. *Science* 152:172 (1966).
105. M.A.Chester and W.M.Watkins, α-L-fucosyltransferases in human submaxillary glands and stomach tissues associated with the H,Le^{a} and Le^{b} blood group structures, Bio*chem.Biophys.Res.Commun.* 34:835 (1969).
106. H.Schenkel-Brunner, M.A.Chester and W.M.Watkins, α–L-fucosyltransferases in human serum from donors of different ABO, secretor and Lewis blood group phenotypes. *Eur.J.Biochem.* 30:269 (1972).
107. P.H.Johnson and W.M.Watkins, Human salivary fucosyltransferases: evidence for two distinct α–3-fucosyltransferases activities one of which is associated with the Lewis blood-group gene, *Biochem.Biophys.Res.Commun.* 100:1611 (1981).
108. J-P.Prieels, D.Monnom, M.Dolmans, T.A.Beyer and R.L.Hill, Co-purification of the Lewis blood group *N*-acetylglucosaminide α1-4 fucosyltransferase and an *N*-acetylglucosaminide α1-3 fucosyltransferase from human milk, *J.Biol.Chem.* 256:10456 (1981).
109. P.H.Johnson and W.M.Watkins, Separation of an α–3-fucosyltransferase from the blood group Le-gene specified α-3/4-fucosyltransferase in human milk, *Biochem.Soc.Trans.* 10:445 (1982).
110. P.H.Johnson, A.S.R.Donald, J.Feeney and W.M.Watkins, Reassessment of the acceptor

specificity and general properties of the Lewis blood-group gene associated α–3/4-fucosyltransferase purified from human milk, *Glycoconjugate J.* 9:251 (1992).

111. P.H.Johnson, A.S.R.Donald, J.L.Clarke and W.M.Watkins, Purification, properties and possible gene assignment of an α1,3-fucosyltransferase expressed in human liver, *Glycoconjugate J.* 12:879 (1995).

112. P.O.Skacel and W.M.Watkins, α–3-Fucosyltransferase expression in leukaemia. *Biochem.Soc.Trans.* 16:1034 (1988).

113. C.I.Civin, T.M.Trishmann, M.J.Fackler, I.D.Bernstein, H-J Buhring, L.Campos, M.F.Greaves, M.Kamoun, D.R.Katz, P.M.Lansdorp, A.T.LookB.Seed, D.R.Sutherland, R.W.Tindle and B.Uchanska-Ziegler. Summary of CD34 cluster workshop section. In: *Leukocyte Typing IV: White Cell differentiation Antigens,* W.Knapp, B.Dorken, E.P.Gilks, E.rieber, R.E.Schmidt, H.Stein and A.E.G.K.von dem Borne, eds, Oxford University Press, Oxford p.816 (1989).

114. E.D.Ball, Cluster workshop report. In Leukocyte typing V: White cell differentiation *Antigens,* S.F.Schlossman, L.Bournsell, G.Wilks, J.M.Harlan, T.Kishimoto, C.Morimoto, J.Ritz, S.Shaw, R.Silverstein, T.Springer, T.F.Tedder and R.F.Todd, eds, Oxford University Press, Oxford, p790 (1995).

115. J.L.Clarke and W.M.Watkins, α1,3-fucosyltransferase expression in developing myeloid cells: antigenic, enzymatic and mRNA analyses, *J.Biol.Chem.* 271:10317 (1996).

116. A.Taniguchi, K.Higai, Y.Hasegawa, K.Utsumi and K.Matsumoto, Differentiation elicits negative regulation of α2,6-sialyltrransferase at the mRNA level in the HL-60 cell line, *FEBS Lett.* 441:191 (1998).

117. J.Natunen, P.Parmanne, J.Helin, O.Aitio, M-L Majuri, R.Niemela, R.Renkonen and O.Renkonen, Biosynthesis of sialylated and fucosylated selectin ligands of HL-60 cells in vitro. Midchain α3-fucose units inhibit terminal α6-sialylation but not α3-sialylation of polylactosamines, *FEBS Lett.,* 452:272 (1999).

118. N.E.Robinson, T.de Vries, R.E.Davis, Stults C.L.M., S.R.Watson, D.H.van den Eijnden and B.A.Macher, Expression of fucosylated antigens in human leukaemia cell lines, *Glycobiology* 4:317 (1994).

119. K.Yago, K. Zenita , H.Ginya, M.Sawada, K.Ohmori, M.Okuma, R.Kannagi and J.B.Lowe. Expression of α1,3-fucosyltransferases which synthesise sialyl-Le^x and sialyl-Le^a, the carbohydrate ligands for E- and P-selectins, in human malignant cell lines,. *Cancer Res.* 53:5559 (1993).

120. M.A.Baker, A.Kanani, I.Brockhausen, H.Schachter, A.Hindenburg and R.N.Taub. Presence of cytidine 5'-monophospho-*N*-acetylneuraminic acid:Galβ1-3GalNAc-R α(2-3)-sialyltransferase in normal human leukocytes and increased activity of this enzyme in granulocytes from chronic myelogenous leukaemia patients. *Cancer Research* 47:2763 (1987).

121. A.Wagers, L.M.Stoolman, R.Kannagi, R.Craig and G.S.Kansas, Expression of leukocyte fucosyltransferases regulates binding to E-selectin; relationship to previously implicated carbohydrate epitopes, *J.Immunol.* 159:1917 (1997).

122. V.Pasqualetto, D.Neel, J.P.Feugeas, M.Aubery and C.Derappe, HL-60 leukaemia cells chemically induced to differentiate retain some surface glycan features of undifferentiated cells not found in normal leukocytes, *Glycobiology* 5:59 (1995).

122. N.E.Robinson, T. deVries, R.E.Davies, C.L.M.Stultz, S.R.Watson, D.H. van den Einjden and B.A.Macher, Expression of fucosylated antigens and α1,3-fucosyltransferases in human leukeamia cell lines, *Glycobiology* 4:317 (1994).

123. B.W.Weston, K.M.Hiller, J.P. Mayben, G.Manousos, C.M.Nelson, M.B.Klein and J.L.Goodman, A cloned CD15s-negative variant of HL-60 cells is deficient in expression of FUT7 and does not adhere to cytokine stimulated endothelial cells, *Eur. J.Haematol.* 63:42 (1999).

124. M.F.A.Bierhuizen, K.Maemura and M.Fukuda, Expression of a differentiation antigen and poly-*N*-acetyllactosaminyl *O*-glycans directed by a cloned Core 2 β-1,6-*N*-acetylglucosaminyltransferase, *J.Biol.Chem.* 269:4473 (1994).

125. M.Nakamura, T.Kudo, H.Narismatsu, Y.Furukawa, J.Kikuchi, S.Asakura, W.Yang, S.Iwase, K.Hatake and Y.Miura, Single glycosyltransferase, Core2 β1,6-*N*-acetyl-

glucosaminyltransferase, regulates cell surface expression in human pre-B-lymphocytic leukemia cell line KM3 treated with phorbol ester, *J.Biol.Chem.* 273:26779 (1998).

126. L.G.Ellies, S.Tsuboi, B.Petryniak, J.B.Lowe, M.Fukuda and J.D.Marth, Core 2 oligosaccharide biosynthesis distinguishes between selectin ligands essential for leukocyte homing and inflammation, *Immunity*, 9:881 (1998).

127. N.Hiraiwa, T.Dohi, N.Kawakami-Kimura, M.Yuman, K.Ohmori, M.Maeda and R.Kannagi, Suppression of sialyl lewis X expression and E-selectin-mediated cell adhesion in cultured human lymphoid cells by transfection of antisense cDNA of an α1-3 fucosyltransferase (Fuc-T VII), *J.Biol.Chem.* 272:31556 (1996).

128. A.W.Greenberg, W.G.Kerr and D.A.Hammer, Relationship between selectin-mediated rolling of hematopoietic stem and progenitor cells and progression in hematopoietic development, *Blood*, 95: 478 (2000).

TRANSCRIPTIONAL REGULATION OF EXPRESSION OF CARBOHYDRATE LIGANDS FOR CELL ADHESION MOLECULES IN THE SELECTIN FAMILY[a,b]

Reiji Kannagi

Program of Molecular Pathology, Aichi Cancer Center, Nagoya 464-8681, Japan

Selectin-mediated cell adhesion is proposed to be involved in the extravasation and hematogenous metastasis of cancer cells (1, 2). In leukemia, this cell adhesion is involved in the tissue infiltration of leukemic cells. Expression of the carbohydrate ligands for selectins, such as sialyl Lewis X and/or sialyl Lewis A, is significantly enhanced in malignant cells compared to their benign counterparts, but the exact mechanism which leads to the enhanced expression of these ligands in malignant cells has not been well-studied.

Sialyl Lewis X and tissue infiltration of adult T-cell leukemia cells

Strong and sustained expression of sialyl Lewis X is frequently detected on the leukemic cells in patients with adult T-cell leukemia, an aggressive and fatal malignancy of T-helper cells (3). This leukemia is characterized by the vigorous extravascular infiltration of leukemic cells into various tissues and organs, which sometimes leads, for instance, even to infiltration of the skin by leukemic cells, a quite unusual site for leukemic cell infiltration in leukemias other than adult T-cell leukemia. The leukemic cells infiltrating the skin strongly express sialyl Lewis X, and the degree of expression of sialyl Lewis X on leukemia cells in ATL significantly correlates with the degree of extravascular infiltration of the leukemia cells (Fig. 1) (4, 5). These lines of evidence strongly suggest that sialyl Lewis X determinant expressed on the leukemic cells is involved in the infiltration of cells from blood vessels to tissues and organs. Therefore, this leukemia provides an excellent opportunity

---

Correspondence: Reiji Kannagi, M.D., Ph.D. Program of Experimental Pathology, Research Institute, Aichi Cancer Center, 1-1 Kanokoden, Chikusaku, Nagoya 464-8681, Japan. FAX +81-52-763-5233, E-mail <rkannagi@aichi-cc.pref.aichi.jp>.

[a] This work was supported in part by grants-in-aid from the Ministry of Education, Science, Sports and Culture, Japan (11680648 and on priority areas 10178104), grants-in-aid for Cancer Research (10-27) and the Second Term Comprehensive Ten-year Strategy for Cancer Control from the Ministry of Health and Welfare, Japan, and a grant from the Princess Takamatsu Foundation for the Promotion of Cancer Research.

[b] Abbreviations used are: CRE, cyclic nucleotide responsive element; CREB protein, cyclic nucleotide responsive element-binding protein; CBP, CREB-binding protein; HTLV-1, human T-cell leukemia virus-1; LTR, long terminal repeat; sialyl Lewis X, NeuAcα2→3Galβ1→4[Fucα1→3]GlcNAcβ1→R; SRE, serum responsive element; SRF, serum responsive factor.

The Molecular Immunology of Complex Carbohydrates–2.
Edited by Albert M. Wu, Kluwer Academic/Plenum Publishers, 2001.

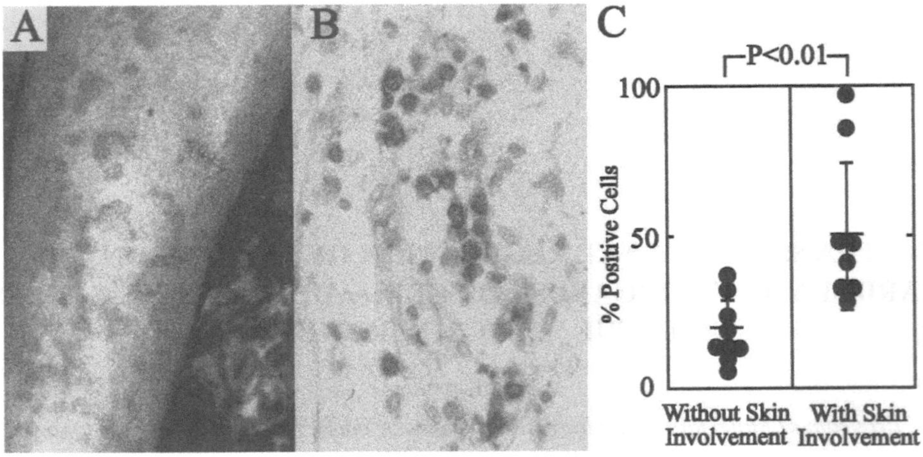

Fig. 1. Skin infiltration of leukemic cells in patients with adult T-cell leukemia. Panel A, gross appearance of skin infiltration of leukemic cells in a patient with adult T-cell leukemia. Panel B, skin biopsy specimen stained by anti-sialyl Lewis X antibody 2F3 indicating the infiltrating leukemic cells strongly express the selectin ligand. Panel C, correlation between the presence or absence of skin infiltration and expression of sialyl Lewis X on the leukemic cells in the peripheral blood of patients with adult T-cell leukemia.

for the detailed study of mechanisms involved in extravascular infiltration of malignant cells mediated by the selectins and their carbohydrate ligands.

Fucosyltransferase involved in synthesis of ligands for selectins

By the early 1990's, several $\alpha 1\rightarrow 3/4$ fucosyltransferases that are capable of transferring α-Fuc to the C-3 or -4 position of β-GlcNAc moiety were cloned, including Fuc-T III, IV, V and VI. However, the enzyme responsible for the synthesis of sialyl Lewis X in leukocytes did not seem to be included in the cloned enzymes. Myeloid cells usually contain only Fuc-T IV, which has an ability to synthesize Lewis X, but lacks the ability to synthesize sialyl Lewis X. The synthesis of sialyl Lewis X in lymphocytes was much more difficult to explain, since lymphocytes and the leukemic cell lines established therefrom usually contain a negligible amount of Fuc-T III, IV, V and VI, if any. Yet the certain subsets of lymphocytes and related lymphoid cell lines strongly express sialyl Lewis X and are significantly adherent to E- and P-selectins (4). Study of an $\alpha 1\rightarrow 3/4$ fucosyltransferase in a lymphoid cell line, YT, which strongly expresses sialyl Lewis X, but contains essentially no detectable mRNA of the fucosyltransferases known at that time even by RT-PCR method, suggested the presence of a new fucosyltransferase, which was later cloned and designated as Fuc-T VII (6, 7).

The fucosyltransferase Fuc-T VII is unique in that it can synthesize only sialyl Lewis X, and not Lewis X, Lewis A, or sialyl Lewis A. It is proposed to be responsible for the synthesis of sialyl Lewis X in granulocytes, monocytes and endothelial cells, as well as in lymphocytes. The major contribution of this enzyme to the synthesis of the selectin ligand sialyl Lewis X in leukocytes was confirmed by the transfection of anti-sense cDNA of this isozyme, which completely turned off the expression of sialyl Lewis X on human cultured lymphoid cell lines, and abrogated their binding ability to selectins (8).

Another characteristic of this isozyme is that its specific activity in leukocytes is quite low, less than one-twentieth of that known for Fuc-T III and/or VI in, for instance, colonic epithelial cells. The specific activity of Fuc-T III and/or VI in colonic epithelial cells is usually two orders higher than that of sialyltransferases, which provide synthetic precursors to fucosyltransferases in the same cells, and therefore fucosyltransferases are not regarded as a rate-limiting enzyme for the synthesis of sialyl Lewis X or sialyl Lewis A in colonic epithelial cells or colon cancer cells. On the other hand, the specific activity of Fuc-T VII in leukocytes are comparable or even less than that of sialyltransferases in the same population of cells, and may well play a role as a rate limiting enzyme in the sialyl Lewis X synthesis in leukocytes.

Transactivation of fucosyltransferase VII by the Tax protein encoded by HTLV-1

Adult T-cell leukemia is known to be caused by infection of human T-cell leukemia virus type-1 (HTLV-1) (9, 10). Among the genes in the HTLV-1 virus, the pX gene has attracted special attention since its product, the Tax protein, is regarded to be the best candidate for an oncogenic protein, having a profound transactivating effect on a wide variety of important cellular genes encoding such as IL-2, IL-2 receptor α, c-Fos and GCSF, through its binding to a set of transcription factors, NF-κB, SRF, CREM and/or CREB (Fig. 2) (11-15).

The rough correlation observed between the expression of mRNA for Tax and that for Fuc-T VII in leukemic cell lines established from patients with adult T-cell leukemia in our preliminary experiments led us to speculate that Tax may also activate transcription of

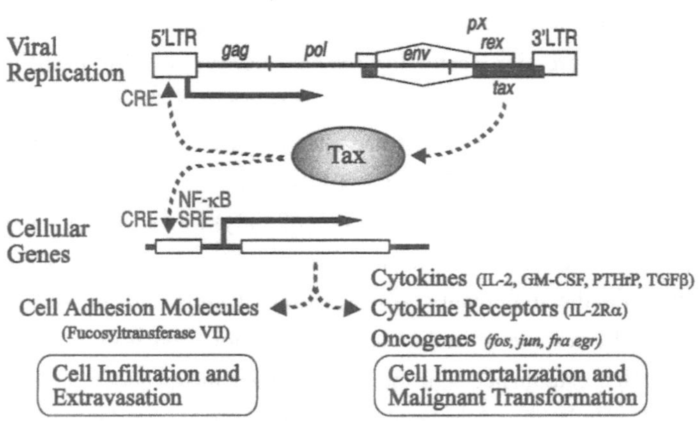

Fig. 2. Effect of Tax protein encoded by human T-cell leukemia virus-1, the putative leukemogenic virus, on viral LTR and various cellular genes involved in the regulation of cell proliferation and cell adhesion. SRE, serum responsive element.

Fuc-T VII gene (Fig. 3A). So, we investigated the effect of Tax protein on the synthesis of sialyl Lewis X using a transfectant clone (JPX-9 cells), which inducibly expresses the Tax transactivator protein through the action of murine metallothionein promoter. The induction of Tax protein resulted in an accumulation of a large amount of Fuc-T VII mRNA, and this led to the significant expression of sialyl Lewis X on the surface of the treated cells (Figs. 3B and C) (16). The amount of Fuc-T VII mRNA induced by the viral transactivator protein was comparable to that induced by the TPA-treatment of the same cells.

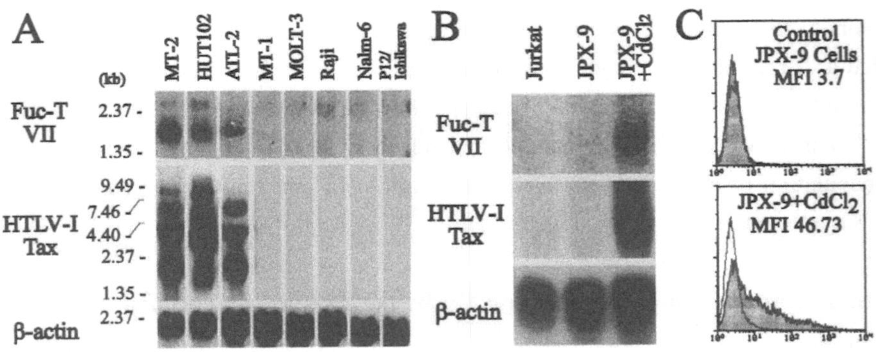

Fig. 3. Correlation of expression of mRNA of Fuc-T VII and Tax (panel A) and induction of Fuc-T VII mRNA expression in JPX-9 cells by the trans-activating factor Tax protein encoded by HTLV-1 virus (panels B and C). Panel A, Northern blot showing correlation of expression of mRNA of Fuc-T VII and Tax in cultured human leukemic cell lines. Panel B, Northern blot showing induction of Fuc-T VII mRNA expression in JPX-9 cells before and after induction of Tax with $CdCl_2$. JPX-9 is a clone of Jurkat cells that stably transfected with a metallothionein promoter-driven Tax expression plasmid, pMAXRH*neo*-1, and inducibly expressed the Tax protein by the addition of $CdCl_2$ to the culture medium (21). Panel C, flow cytometric analysis of JPX-9 cells before and 7 days after induction of Tax with $CdCl_2$. MFI, mean fluorescence intensity. Sialyl Lewis X was detected with 2F3 antibody.

Next we have cloned and sequenced the 5-prime regulatory region of the genomic Fuc-T VII gene, ligated it to the luciferase gene, and the resulting reporter construct was co-transfected with the expression vector for Tax to a human cultured T-cell line, Jurkat cells. Co-transfection of Tax with the reporter construct resulted in 30- to 40-fold activation of the luciferase transcription, and this confirmed that the Tax protein induces transcription by acting on the 5'-regulatory region of the Fuc-T VII gene.

It has long been known that expression of sialyl Lewis X is enhanced in malignant cells, and the determinant is regarded as tumor-associated antigen. However, the exact relationship between the malignant transformation of cells and augmented expression of sialyl Lewis X at the surface of malignant cells has not been fully clarified. The above-described results first established that the genetic mechanism which leads to the malignant transformation of cells is directly related to the enhancement of sialyl Lewis X expression, by showing a direct activation of Fuc-T VII transcription by the oncogenic transcriptional activator protein encoded by the etiological virus. The putative sequence of the events

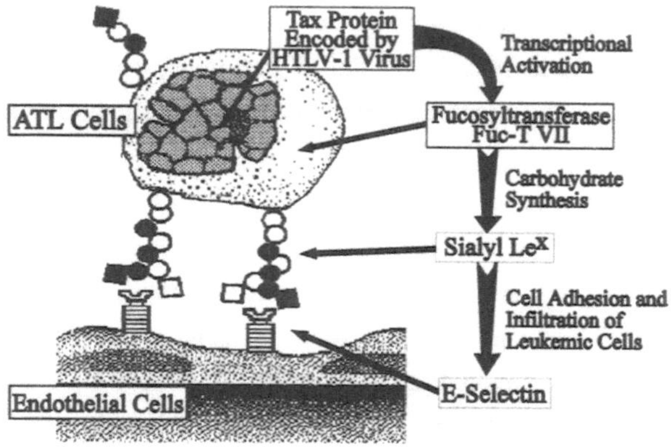

Fig. 4. Schematic illustration of mechanism leading to the augmentation of sialyl Lewis X expression in cells infected with human T-cell leukemia virus-1. ATL, adult T-cell leukemia.

which would occur in the cells infected by HTLV-1 virus is illustrated in Fig. 4. The infection of cells with HTLV-1 virus results in the production of the Tax protein in the infected cells, and while this Tax protein initiates the malignant transformation of the cells by affecting important cellular genes involved in cell-proliferation, the Tax protein also transactivates the fucosyltransferase VII gene. This in turn synthesizes sialyl Lewis X, the ligand for selectins, and this was the reason for the strong tendency for extravascular infiltration of the leukemia cells in patients with adult T-cell leukemia.

Involvement of CRE (cyclic-nucleotide responsive element) motif in transcriptional activation of Fuc-T VII

The 5'-regulatory region of the Fuc-T VII gene had no conventional TATA box upstream of the principal transcription start site, but had an initiator element juxtaposed by Spl consensus site (Fig. 5). This was compatible with a proposed notion that an Spl site

```
-852 gcggggagtctggacttcaggctggatcccttcctcttcctggagcgggtgctggcccccaacccgcttgcgtcag
-776 ggacaaaaggactccttccctttccagcctggaaagcccctctgctgcaggctggaggaagggaccctgggcccag
-700 cctatagtcagcggtgtctatgggcatggatctggacggggaaaaggacaaagcagcctccatccacagttcattc
-624 cgggaccaggcccttgcaggcacgcgctgggctcctgtgggaagacactaagggccccaggacagacctcctctcc
-548 gggcatctgggttcctagatggcagaggtggcagagtggggtgggatggcccaattgggagctttagcctccggca
-472 aagagctgagcacagtacatcttcaatgtgtaagattctcctgggagaccagggcccagctggtggtgagctgggg
-396 gaagtgggtgatactgccatgggaggagccacctggccctctggggaagtgcactcgctgtctgcagcgcccaggc
-320 ctgggtagctgggtgggggctgggggccatctgtgctcagggtgcctgcacctgggccttctctgccctgggcca
-244 agcctgcccgagcctctctgtcctctgcctgcccagctggacatctctgggcctctctggagaccagtggggtggg
-168 ctgtgggggcgtcatattgccctggcttggcatccctcttgtggctgtacccctcccagcagccccaggactagca
 -92 agtccccgagatgggggtggggacagtggttgatgccaaaggttgtgggggcaggggcggggcaggagcaggaagg
 -16 tcccctgagttccctc +1ACCTTGGGCAGAGATAAAAGGAGCACAGTTCCAGGCGGGGCTGAGCTAGGGCGTA
 +56 GCTGTGATTTCAGGGGCACCTCTGGCGGCTGCCGTGATTTGAGAATCTCGGGTCTCTTGGCTGACTGATCCT
+128 GGGAGACTGTGG ATG
```

Fig. 5. Genomic nucleotide sequence of 5'-flanking region and exon 1 of human Fuc-T VII gene. The major transcription start point is numbered +1. Exon sequences are indicated by uppercase letters, and those of untranscribed sequences by lowercase letters. The organization of Fuc-T VII exons and introns was determined by comparison with the sequence of Fuc-T VII cDNA (6, 7). The ATG codon is in *boldface* type. The CRE-like element is also underlined (core CRE motif is in *boldface* type). Another putative cis-acting motif, Spl, just 5' prime to Ets-like site (underlined with doublet), is represented by a *closed box*. A more detailed nucleotide sequence for exons and introns is registered in the DDBJ/EMBL/GenBank nucleotide sequence databases with accession number AB012668.

juxtaposed with an initiator element is critical to initiation of transcription and positioning of transcription start sites in TATA-less promoters (17, 18). The 5'-regulatory region was equipped also with multiple binding sites for important transcription activators, including a CRE-consensus motif (Fig. 5).

Subsequent analyses using deletion and/or mutated reporter constructs indicated that the Tax-induced transactivation is mediated by the CRE motif. As shown in Fig. 6, the reporter constructs p(–791) through p(–193) are activated by Tax, but this transactivation was significantly suppressed with the p(–134) and smaller reporter constructs, indicating the minimal Tax-responsive element resided between nucleotide –193 and –134, where the CRE motif is located. The involvement of the CRE motif in Tax-induced transactivation in the

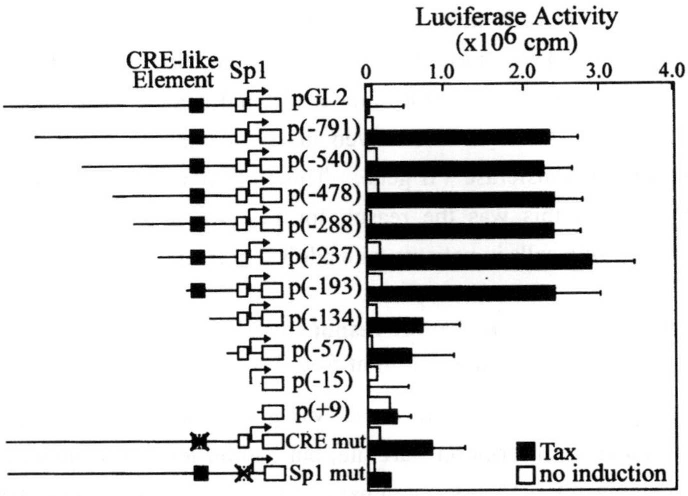

Fig. 6. Identification of Tax-responsive element in 5'-flanking region of human Fuc-T VII gene using deletion/mutation constructs. Promoter activity of 5'-deleted or mutated constructs in Jurkat cells was determined with (*black bar*) or without (*white bar*) co-transfection of the Tax expression vector, pCGTax. The Sp1 (–39 to –31) and CRE (–162 to –155) sites are represented by closed boxes and hatched boxes, respectively. Inr, an initiator sequence. The luciferase activity expressed by these constructs is shown on the right. Mean values for three independent assays are presented, and results are expressed in luciferase light units normalized to CAT units and are representative of three experiments. Mutations of either Sp1 or CRE sites as shown in this figure were performed by three-step PCR mutagenesis methods.

Fuc-T VII promoter was also confirmed by introducing a mutation in the CRE motif of the construct (Fig. 6).

Next we performed a DNA-precipitation assay using the CRE-like DNA element of Fuc-T VII so as to more directly demonstrate the interaction of Tax with other transcription factors and co-activators involved in the activation of Fuc-T VII (Fig. 7). The Fuc-T VII CRE-like element precipitated endogenous CREB-1, indicating that the CRE motif is capable of binding transcription factors of the CREB/ATF family (Fig. 7A, lane 1). The wild type CRE-like element precipitated Tax protein when the nuclear extracts of cells expressing Tax were used (Fig. 7B, lane 2), or when histidine-Tax was exogenously added

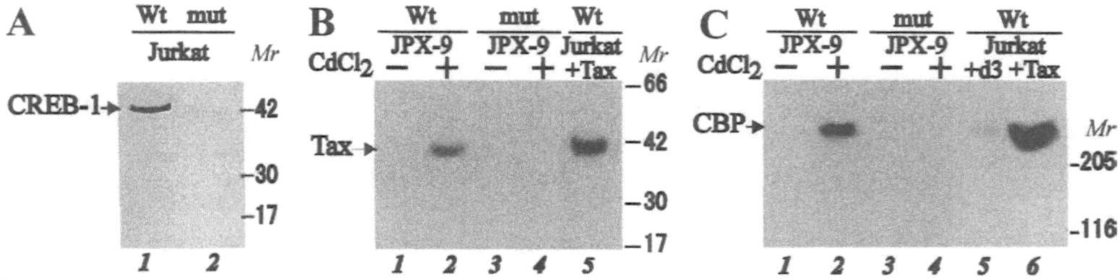

Fig. 7. Association of CREB-1 (panel A), Tax (panel B) and CBP (panel C) with the CRE motif DNA element in 5'-regulatory region of Fuc-T VII gene as ascertained by DNA-precipitation assay. Nuclear extract was prepared from Jurkat cells or $CdCl_2$-induced or control JPX-9 cells. For JPX-9 cells see legend for Fig. 2. In some experiments, nuclear extract from Jurkat cells was mixed with histidine-Tax (+Tax), or its mutant histidine-d3 (+d3). Biotinylated DNA probes consisting of the Fuc-T VII CRE-like element repeats or those having a mutation (Wt, wild type; mut, mutated) were incubated with the nuclear extracts, and the precipitated DNA-protein complexes were analyzed by immunoblotting with specific antibodies specific to CREB-1, Tax or CBP, respectively.

(lane 5). This binding was totally abrogated by introducing a mutation into the CRE motif of the DNA element (Fig. 7B, lane 4), indicating that the association of Tax with the DNA element is strictly dependent on the CRE motif. Binding of CBP /P300 was observed only in the presence of endogenous Tax or exogenously added histidine-Tax protein (Fig. 7C, lanes 3, 4 and 6), but not in the presence of a mutant Tax, d3 (lane 5), suggesting that the binding of CBP/P300 to the DNA-protein complex is mediated by the intact and functional Tax protein. CBP/P300 is an important co-activator of gene transcription having histone acetyltransferase activity. Stable recruitment of CBP/P300 to certain gene element is known to activate its transcription. As illustrated in the left panel of Fig. 8, the results described above indicate that the Tax protein first attaches to the CREB/ATF family protein bound on the CRE element in Fuc-T VII gene, and mediates the recruitment of CBP/P300, which in turn conveys the activating effect to the basal transcription machinery, thus inducing the transcription of the Fuc-T VII gene.

Deviation of transcriptional activation of Fuc-T VII in adult T-cell leukemia from normal phosphorylation-dependent regulation

The Tax protein is known to activate transcription of HTLV-1 virus by interacting with the CRE-like 21 bp motif in the viral LTR. However, as far as the Tax-mediated activation of endogenous cellular genes is concerned, most of the well-established examples in the literature are mediated by the interaction of Tax through NF-κB or SRF. Fuc-T VII can be considered to be a rare example, where Tax acts on the endogenous cellular CRE element.

The above-described mechanism for the Tax-induced activation of Fuc-T VII gene transcription is quite different from the normal regulatory mechanism for gene transcription by the usual CRE element as is typically described for cellular genes such as the somatostatin gene (Fig 8, right panel). Usually, the gene transcription mediated by the

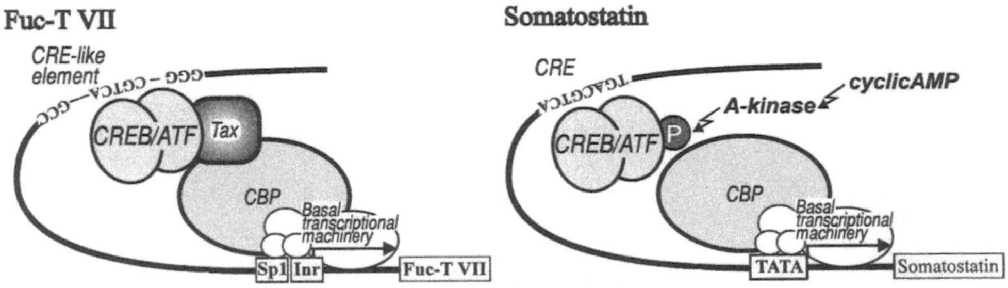

Fig. 8. Schematic illustration of mechanism involved in the Tax-mediated activation of Fuc-T VII gene transcription (left panel) and in the phosphorylation-dependent regulation of somatostatin gene transcription (right panel).

CRE element is regulated by cyclic nucleotide-dependent protein kinase (protein kinase-A). The transcription factors in the CREB/ATF family are known to require phosphorylation by protein kinase-A for recruiting CBP/P300. In contrast, the Tax protein seems to activate the Fuc-T VII CRE-site in a phosphorylation-independent manner.

To further investigate this issue, we used two functionally-impaired mutants of CREB protein, CREMα and CREMΔ(C-G), and studied the expression of Fuc-T VII reporter construct in various combinations of protein kinase A, Tax and/or these CREB/ATF family proteins. CREMα lacks the glutamine-rich domains, and CREMΔ(C-G) lacks both the glutamine-rich domains and the kinase-inducible domain (KID), but has a basic leucin zipper (bZIP) domain nearly identical to that of CREB (19, 20).

The somatostatin-luciferase reporter construct served as an example of typical intracellular genes regulated through the authentic CRE motif. In the presence of protein kinase A, transcription of the somatostatin construct was strongly activated by CREB-1, an authentic CREB/ATF family transcription factor (Fig. 9A), while the activation effect of protein kinase A was partially suppressed by CREMα (Fig. 9B), and almost completely abolished with CREMΔ(C-G) (Fig. 9C). Suppression observed with CREMα was absolved by the addition of Tax, but that observed with CREMΔ(C-G) was not. In the absence of protein kinase A, the Tax protein failed to activate transcription of the somatostatin construct with any CREB-like proteins. These results indicated that transcriptional activation mediated by the somatostatin CRE element is highly dependent on phosphorylation-dependent regulation.

Transcription of Fuc-T VII was also dependent on protein kinase A when CREB-1 or CREMα was used. However, the most remarkable difference was that the Tax-mediated significant transcriptional activation of the Fuc-T VII construct was always seen irrespective of the presence or absence of protein kinase A, even with CREMΔ(C-G). This Tax-mediated activation of the Fuc-T VII construct was only minimally affected by protein kinase A.

These results indicated that the Tax-induced transcriptional activation of Fuc-T VII is independent of the normal regulatory mechanism exerted by cyclic nucleotide-dependent

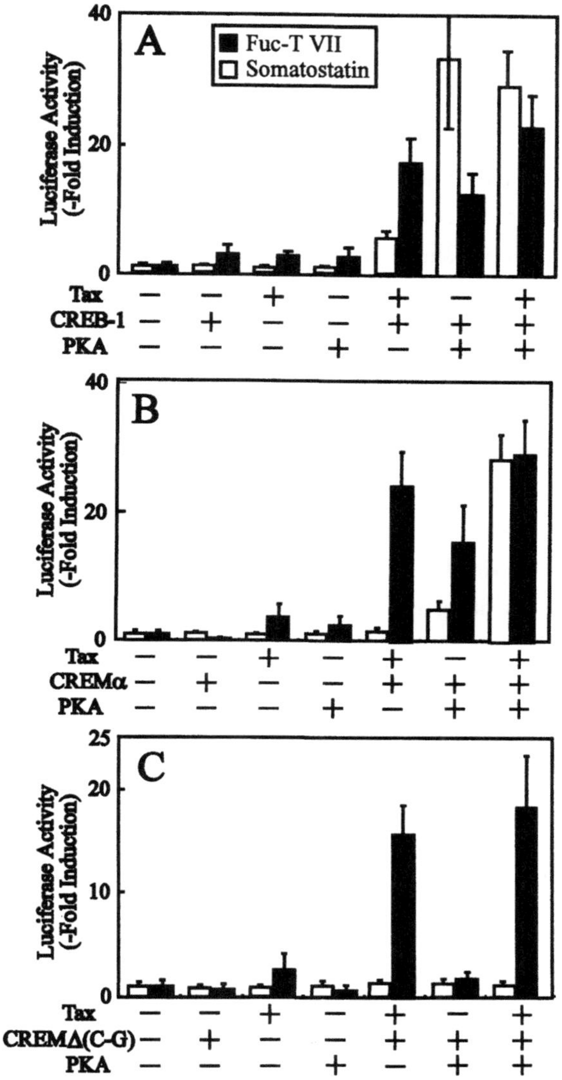

Fig. 9. Comparison of phosphorylation dependency of transcriptional activation of Fuc-T VII and somatostatin. F9 teratocarcinoma cells were transfected with 4 μg of the Fuc-T VII promoter (p(−791); *black bar*) or somatostatin promoter-luciferase construct (*white bar*), with 4 μg of RSV-CREB *(A)*, RSV-CREMα (*B*), or RSV-CREMΔ(C-G) in the presence or absence of 4 μg of RSV-PKA and/or 4 μg of pCG-Tax as indicated. F9 cells were chosen for these studies because they are known to lack functional levels of CREB and protein kinase A (22). The results are plotted as fold activation normalized for CAT activity over that in the absence of Tax. The error bars depict the standard deviations from three independent experiments.

phosphorylation. It is well known that transcriptional activation of HTLV-1 viral LTR by Tax is also phosphorylation independent (13). The mechanism of Tax-mediated transcriptional activation of Fuc-T VII gene is very similar to that of viral transcription in this aspect. For viruses, this is advantageous, since the viruses can replicate without being affected by the intracellular phosphorylation-dependent regulation. However, for the intracellular gene such as Fuc-T VII, this will cause strong and sustained activation of the transcription of the Fuc-T VII gene, which should otherwise have been regulated by phosphorylation under normal conditions. This must be the reason for the sustained strong expression of Fuc-T VII mRNA and sialyl Lewis X on adult T-cell leukemia cells. The CRE motif in the Fuc-T VII gene is distinct from other CRE-like sequences found in cellular genes in that it is non-palindromic and is preceded by a small GC-rich region at its 5'- end, which seems to facilitate the binding of Tax.

SUMMARY

Cell adhesion mediated by selectins and their carbohydrate ligands is involved in the adhesion of cancer cells to endothelial cells during the course of hematogenous metastasis of cancer. In patients with leukemia, this adhesion is involved in the extravascular infiltration of leukemic cells. Extravasation and tissue infiltration of malignant cells in patients with adult T-cell leukemia is mediated by the interaction of selectins and their carbohydrate ligand sialyl Lewis X, which is strongly and constitutively expressed on the leukemic cells. Constitutive expression of Lewis X in these cells is due to the transcriptional activation of Fuc-T VII, the rate-limiting enzyme in the sialyl Lewis X synthesis, induced by the Tax protein encoded by the human T-cell leukemia virus-1, the etiological virus for this leukemia. This transactivation is in clear contrast to the regulation of typical CRE-element found in various cellular genes in that it is independent of phosphorylation-dependent regulation. This must be the reason for the strong and constitutive expression of sialyl Lewis X, which exacerbates the tissue infiltration of leukemic cells. This is a good example corroborating the proposition that the abnormal expression of carbohydrate determinant at the surface of malignant cells is intimately associated with the genetic mechanism of malignant transformation of cells.

REFERENCES

1. Hakomori, S. (1996) Tumor malignancy defined by aberrant glycosylation and sphingo(glyco)lipid metabolism. *Cancer Res.* **56**: 5309-5318.
2. Kannagi, R. (1997) Carbohydrate-mediated cell adhesion involved in hematogenous metastasis of cancer. *Glycoconjugate J.* **14**: 577-584.
3. Takatsuki, K. (1997) Kenneth MacGredie Memorial Lectureship. Adult T-cell leukemia/lymphoma. *Leukemia* **11 Suppl 3**: 54-56.
4. Ohmori, K., Takada, A., Ohwaki, I., Takahashi, N., Furukawa, Y., Maeda, M., Kiso, M., Hasegawa, A., Kannagi, M. and Kannagi, R. (1993) A distinct type of sialyl Lewis X antigen defined by a novel monoclonal antibody is selectively expressed on helper memory T cells. *Blood* **82**: 2797-2805.
5. Furukawa, Y., Tara, M., Ohmori, K. and Kannagi, R. (1994) Variant type of sialyl Lewis X antigen expressed on adult T cell leukemia cells is associated with skin involvement. *Cancer Res.* **54**: 6533-6538.
6. Natsuka, S., Gersten, K. M., Zenita, K., Kannagi, R. and Lowe, J. B. (1994) Molecular cloning of a cDNA encoding a novel human leukocyte α-1, 3-fucosyltransferase capable of synthesizing the sialyl Lewis x determinant. *J. Biol. Chem.* **269**: 16789-16794.
7. Sasaki, K., Kurata, K., Funayama, K., Nagata, M., Watanabe, E., Ohta, S., Hanai, N. and Nishi, T. (1994) Expression cloning of a novel α1,3-fucosyltransferase that is involved in biosynthesis of the sialyl Lewis x carbohydrate determinants in leukocytes. *J. Biol. Chem.* **269**: 14730-14737.
8. Hiraiwa, N., Dohi, T., Kawakami-Kimura, N., Yumen, M., Ohmori, K., Maeda, M. and Kannagi, R. (1996) Suppression of sialyl Lewis X expression and E-selectin-mediated cell adhesion in cultured human lymphoid cells by transfection of antisense cDNA of an α1→3 fucosyltransferase (Fuc-T VII). *J. Biol. Chem.* **271**: 31556-31561.
9. Hinuma, Y., Nagata, K., Hanaoka, M., Nakai, M., Matsumoto, T., Kinoshita, K. I. and Miyoshi, I. (1981) Adult T-cell leukemia: antigen in an ATL cell line and detection of antibodies to the antigen in human sera. *Proc. Natl. Acad. Sci. USA* **78**: 6476-6480.
10. Yoshida, M., Miyoshi, I. and Hinuma, Y. (1982) Isolation and characterization of retrovirus from cell lines of human adult T-cell leukemia and its implication in the disease. *Proc. Natl. Acad. Sci. USA* **79**: 2031-2035.
11. Smith, M. R. and Greene, W. C. (1991) Molecular biology of the type I human T-cell leukemia virus (HTLV-I) and adult T-cell leukemia. *J. Clin. Invest.* **87**: 761-766.
12. Yoshida, M. (1995) HTLV-1 oncoprotein Tax deregulates transcription of cellular genes through multiple mechanisms. *J. Cancer Res. Clin. Oncol.* **121**: 521-528.
13. Yoshida, M., Suzuki, T., Fujisawa, J. and Hirai, H. (1995) HTLV-1 oncoprotein tax and cellular transcription factors. *Curr. Top. Microbiol. Immunol.* **193**: 79-89.
14. Iwanaga, Y., Tsukahara, T., Ohashi, T., Tanaka, Y., Arai, M., Nakamura, M., Ohtani, K., Koya, Y., Kannagi, M., Yamamoto, N. and Fujii, M. (1999) Human T-cell leukemia

virus type 1 tax protein abrogates interleukin-2 dependence in a mouse T-cell line. *J. Virol.* **73**: 1271-1277.

15. Arai, M., Ohashi, T., Tsukahara, T., Murakami, T., Hori, T., Uchiyama, T., Yamamoto, N., Kannagi, M. and Fujii, M. (1998) Human T-cell leukemia virus type 1 Tax protein induces the expression of lymphocyte chemoattractant SDF-1/PBSF. *Virology* **241**: 298-303.
16. Hiraiwa, N., Hiraiwa, M. and Kannagi, R. (1997) Human T-cell leukemia virus-1 encoded Tax protein transactivates $\alpha1\rightarrow3$ fucosyltransferase Fuc-T VII, which synthesizes sialyl Lewis X, a selectin ligand expressed on adult T-cell leukemia cells. *Biochem. Biophys. Res. Commun.* **231**: 183-186.
17. Smale, S. T. and Baltimore, D. (1989) The "initiator" as a transcription control element. *Cell* **57**: 103-113.
18. Smale, S. T., Schmidt, M. C., Berk, A. J. and Baltimore, D. (1990) Transcriptional activation by Sp1 as directed through TATA or initiator: specific requirement for mammalian transcription factor IID. *Proc. Natl. Acad. Sci. U. S. A.* **87**: 4509-4513.
19. Foulkes, N. S., Borrelli, E. and Sassone-Corsi, P. (1991) CREM gene: use of alternative DNA-binding domains generates multiple antagonists of cAMP-induced transcription. *Cell* **64**: 739-749.
20. Walker, W. H., Sanborn, B. M. and Habener, J. F. (1994) An isoform of transcription factor CREM expressed during spermatogenesis lacks the phosphorylation domain and represses cAMP- induced transcription. *Proc. Natl. Acad. Sci. U. S. A.* **91**: 12423-12427.
21. Ohtani, K., Nakamura, M., Saito, S., Nagata, K., Sugamura, K. and Hinuma, Y. (1989) Electroporation: application to human lymphoid cell lines for stable introduction of a transactivator gene of human T-cell leukemia virus type I. *Nucleic Acids Res.* **17**: 1589-1604.
22. Masson, N., Ellis, M., Goodbourn, S. and Lee, K. A. (1992) Cyclic AMP response element-binding protein and the catalytic subunit of protein kinase A are present in F9 embryonal carcinoma cells but are unable to activate the somatostatin promoter. *Mol. Cell Biol.* **12**: 1096-1106.

BIOSYNTHESIS AND FUNCTION OF β 1,6 BRANCHED MUCIN-TYPE GLYCANS

Paul V. Beum and Pi-Wan Cheng

Department of Biochemistry and Molecular Biology
University of Nebraska Medical Center
Omaha, NE 68198-4525 USA

PRINCIPLES OF MUCIN-GLYCAN BIOSYNTHESIS

Mucins have traditionally been defined as a family of high-molecular weight glycoproteins which are secreted from epithelial tissues of the airway, gastrointestinal tract and genitourinary tract to help form a mucus gel layer to protect the tissue from threats in the external environment. The term mucin, indeed, is simply shorthand for mucus glycoprotein. In recent years, the family has been expanded by the addition of a number of cell surface-associated glycoproteins which share the hallmark molecular features of the traditional mucins, but which perform different functions. These new additions include molecules such as MUC1, MUC4, CD43, and selectin ligands such as PSGL-1 and GlyCAM-1. Several of these molecules play key roles in biological processes such as leukocyte rolling, lymphocyte homing, and T-cell maturation and activation, greatly enlarging the functional repertoire of the mucin family.

Common to the mucins is a large peptide backbone which contains a tandem repeat region rich in serine and threonine residues, many of which are glycosylated. Mucin carbohydrate chains grow by sequential attachment of single sugar residues during transit of the mucin through the Golgi stack. The carbohydrate portion of the completed mucin product can make up the great majority of the mucin by weight, and has dramatic effects on mucin structure and function. For example, the clustering of the carbohydrate chains on the peptide causes the mucin to adopt an extended structure, and the hydrophilic character of the sugar residues enables the mucin to efficiently retain moisture. These properties, along with the ability of mucin molecules to cross-link with each other via disulfide bridges, contribute to the extreme viscosity of the mucus gel. This viscosity is critical for protection of the underlying tissue and for trapping and removal of foreign particles.

Mucin-type glycan biosynthesis begins with the attachment of *N*-acetylgalactosamine to serine or threonine in the peptide backbone:

$$\text{UDP-GalNAc + Ser/Thr-Mucin} \xrightarrow{\text{GalNAc TF}} \text{GalNAc}\alpha\text{-Ser/Thr-Mucin + UDP}$$

This step is the defining feature of mucin-type glycosylation, distinguishing it from other types of glycosylation. Numerous isoforms of polypeptide *N*-acetylgalactosaminyl-transferase (GalNAc TF), the enzyme which catalyzes this first step, have been cloned and characterized in recent years (1). The various GalNAc TF isoforms appear to possess distinct sequence preferences in the mucin peptide substrate (2-5), which may explain why a consensus structural motif has not yet been identified for initiation of mucin-type glycosylation.

Following this invariant initial glycosylation step, there is a high degree of flexibility in the construction of the remainder of the O-glycan unit. At any given step, any of several sugars can potentially be added to the acceptor, often at more than one site on the acceptor. Additionally, nonsugar moieties such as sulfate are often found as part of mucin glycan structures, and glycosidases have also been suggested to play a role in mucin glycan biosynthesis, trimming specific residues from the structure (6). These factors combine to produce a vast array of possible mucin-type glycan structures, rendering them more challenging to characterize than N-linked glycans, which show much less structural diversity.

In spite of this complexity, mucin carbohydrate biosynthesis is far from random or chaotic. The process is governed by certain rules which allow formation of some structures but not others. These rules reflect the contributions of several factors, including differential expression of glycosyltransferase genes, substrate availability, acceptor substrate structural preferences of the different glycosyltransferases, pH in the Golgi lumen, as well as the physical segregation of the different glycosyltransferases into specific Golgi subcompartments. Moreover, mucin glycan biosynthesis is tightly regulated so as to yield specific carbohydrate structures in specific cell and tissue types, as well as structures which are specific for specific stages of cellular maturation and organismal development. Intriguingly, mucin glycan structures can even differ for different types of mucins within the same cell type. For example, CD43 and PSGL-1 from HL-60 promyelocytic cells possess significantly different mucin glycan structures (7). This suggests a dependence of the final glycan structure on the particular protein being glycosylated.

The overall architecture of the mucin glycan consists of three regions, the core, backbone and periphery, arranged as follows:

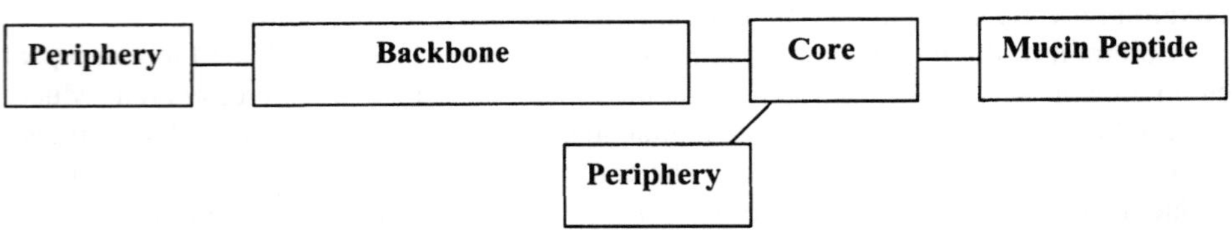

The core structure consists of the initial GalNAc residue and all sugars directly attached to GalNAc. The following eight mucin-type cores have been identified to date, with structures 1-4 being the most common:

Core 1 Galβ1,3GalNAcα-

Core 2 GlcNAcβ1,6 (branch) Galβ1,3GalNAcα-

Core 3 GlcNAcβ1,3GalNAcα-

Core 4 GlcNAcβ1,6 (branch) GlcNAcβ1,3GalNAcα-

Core 5 GalNAcα1,3GalNAcα-

Core 6 GlcNAcβ1,6GalNAcα-

Core 7 GalNAcα1,6GalNAcα-

Core 8 Galα1,3GalNAcα-

The core structure possesses several reactive hydroxyl residues where additional sugar residues can be attached, enabling elongation of the chain to form the backbone portion of the glycan. Elongation often occurs by alternating addition of galactose and *N*-acetylglucosamine residues, giving rise to commonly observed sequence motifs such as the type I structure **Galβ1,3GlcNAcβ**1,3-, the type II structure **Galβ1,4GlcNAcβ**1,3- (8), the type II-related blood group i structure **GlcNAcβ1,3Galβ1,4-** and the branched blood group I structure **GlcNAcβ1,3(GlcNAcβ1,6)Galβ1,4-**. Linear chains of repeating units of type II structure comprise a biologically important type of backbone structure known as poly-*N*-acetyllactosaminoglycan, commonly referred to simply as polylactosamine.

Attachment of certain sugar residues, such as sialic acid and fucose, prevents further elongation of the glycan chain, serving as a mechanism for "capping" the carbohydrate structure. These residues constitute the peripheral region of the mucin glycan, and, in addition to being attached to the nonreducing termini of the backbone portions of the glycan, are often found directly attached to one or more reactive sites on the core portion, as shown in the above line diagram. Peripheral structures such as sialyl Lewis x and blood group antigens attached to the ends of polylactosamine antennae are able to interact with molecules in the extracellular milieu, and are responsible for many of the biological functions of cell surface-associated mucins.

Since many excellent review articles on mucin glycan biosynthesis have been published in recent years (6, 9-13), we will not attempt an exhaustive treatment of the field here. Instead, this review will focus on the biosynthesis and functions of mucin glycans which possess the core 2, core 4 and I β1,6 branched structures. We will describe the enzymes involved in the synthesis of these three structures, including their characterization and cloning, as well as the biological roles of the structures and the changes that occur in their expression during various physiological processes. Description of the biosynthesis of β1,6 branched structures is complicated by the fact that certain branched structures can be catalyzed by more than one enzyme and by the ability of some of the β1,6 branching enzymes to form multiple types of β1,6 branched structure. This makes it difficult to pinpoint the specific glycosyltransferase responsible for a given β1,6 branched mucin glycan structure seen in nature. Compounding this problem is the fact that most of the studies in which β1,6 branching enzymes have been identified and characterized were carried out using whole cell extract rather than purified enzyme. These studies have been valuable in identifying potentially novel β1,6 branching enzymes, but the potential presence of other GlcNAc transferases makes it impossible to unequivocally assign the branching activities detected in the extract to a specific enzyme. To date, the best information regarding these issues comes from a handful of studies in which the properties of a purified or cloned β1,6 branching enzyme have been examined. However, many questions remain concerning the creation of the myriad β1,6 branched structures seen in nature.

I. MUCIN-TYPE β1,6 BRANCHING ENZYMES

I.A. ENZYME NOMENCLATURE AND STRUCTURES FORMED

The three types of β1,6 branched mucin-type glycan structure, namely core 2, core 4 and blood group I, are synthesized by activities which will henceforth be referred to as C2GnT activity, C4GnT activity, and IGnT activity, respectively. The reactions catalyzed by these activities are shown below:

1) Reaction catalyzed by C2GnT:

$$\text{UDP-GlcNAc} + \text{Gal}\beta1,3\text{GalNAc}\alpha\text{-S/T} \longrightarrow \text{Gal}\beta1,3(\text{GlcNAc}\beta1,6)\text{GalNAc}\alpha\text{-S/T} + \text{UDP}$$

GlcNAcβ1,6
\
UDP-GlcNAc + Galβ1,3GalNAcα-S/T → Galβ1,3GalNAcα-S/T + UDP

2) Reaction catalyzed by C4GnT:

GlcNAcβ1,6
\
UDP-GlcNAc + GlcNAcβ1,3GalNAcα-S/T → GlcNAcβ1,3GalNAcα-S/T + UDP

3) Reaction catalyzed by IGnT:

GlcNAcβ1,6
\
UDP-GlcNAc + R_1- GlcNAcβ1,3Galβ1,4-R_2 → R_1-GlcNAcβ1,3Galβ1,4-R_2 + UDP

where S/T represents a serine or threonine residue in the mucin peptide backbone, and R_1 and R_2 represent sugar residues. As mentioned above, several distinct enzymes containing these activities have been identified and cloned. To date, the cDNAs of four different types of β1,6 branching enzyme have been cloned, as listed in table 1 below. As shown in the table, one of these enzymes, IGnT, contains only IGnT activity (14), while two others, C2GnT-L and C2GnT-3, possess only C2GnT activity (15, 16). In contrast, the C2GnT-M enzyme is able to synthesize all three mucin-type β1,6 branched linkages (17, 18). A new nomenclature system has recently been suggested for the C2GnT isozymes (16), in which C2GnT-L would be referred to as C2GnT-1 and C2GnT-M as C2GnT-2. In this review, we shall retain the L and M nomenclature for these two isozymes.

Table 1. β1,6 mucin-type branching enzymes which have been cloned, and corresponding mucin carbohydrate structures the enzymes are able to synthesize

Cloned β1,6 Branching Enzyme	Structures Formed
C2GnT-L	core 2
C2GnT-M	core 2, core 4, blood group I
C2GnT-3	core 2
IGnT	blood group I

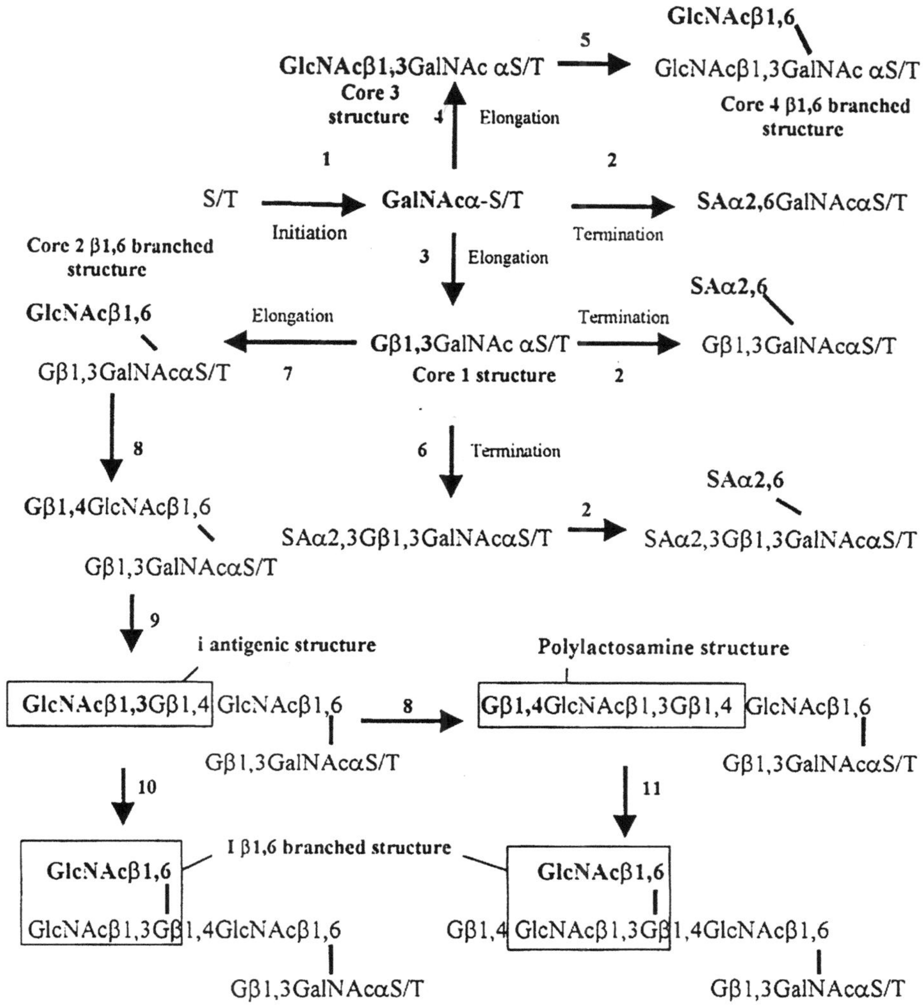

Figure 1. Biosynthetic pathways involved in the formation of branched mucin-type glycan structures. Abbreviations used are as follows: GalNAc, N-acetylgalactosamine; G, galactose; GlcNAc, N-acetylglucosamine; SA, N-acetylneuraminic acid (sialic acid); S/T, serine/threonine. Residues added during each step are highlighted. Enzymes which catalyze the reactions shown are: 1, polypeptide N-acetylgalactosaminyltransferase; 2, α2,6 sialyltransferases (ST6GalNAc I/II); 3, core 1 galactosyltransferase; 4, core 3 β1,3 N-acetylglucosaminyltransferase; 5, core 4 β1,6 N-acetylglucosaminyltransferase (C2GnT-M); 6, α2,3 sialyltransferase (ST3Gal I); 7, core 2 β1,6 N-acetylglucosaminyltransferase (C2GnT); 8, β1,4 galactosyltransferase; 9, i β1,3 N-acetylglucosaminyltransferase; 10, distally-acting I β1,6 N-acetylglucosaminyltransferase (dIGnT); 11, centrally-acting I β1,6 N-acetylglucosaminyltransferase (cIGnT).

Figure 1 shows the biosynthetic pathways which lead to formation of the core 2, core 4 and blood group I branched structures in mucin-type glycans. Following initiation by addition of GalNAc to serine or threonine, core 1 can be formed by addition of galactose in β1,3 linkage to GalNAc, enabling subsequent formation of the core 2 branched structure. Alternatively, if *N*-acetylglucosamine is added to GalNAc in β1,3 linkage, the core 3 structure is created, allowing for formation of the core 4 branched structure. This illustrates the principle of competition in mucin glycan biosynthesis, in which a single acceptor structure can potentially be acted on by any one of several enzymes. The structure which emerges from this competition then determines the subsequent synthetic pathway until another such "decision" point is reached. Once the core 2 branched structure is formed, this enables extension of the chain to form polylactosamine structure, which serves as a substrate for the addition of GlcNAc residues in β1,6 linkage to galactose residues by IGnT enzymes, forming the blood group I branched structure. As illustrated in the figure, two general types of IGnT activity have been identified, and these are distinguished from one another by their acceptor substrate preferences, as discussed later in detail. It should be pointed out that figure 1 does not represent an exhaustive description of the pathways involved in mucin-type glycosylation, or even of the formation of β1,6 branched structures. For example, while the pathways for synthesis of the core 2 and core 4 structures shown in figure 1 are the only known pathways to these structures, the blood group I structure can be formed by several other pathways in addition to that shown in figure 1.

I.B. CHARACTERIZATION OF MUCIN-TYPE β1,6 BRANCHING ENZYMES

I.B.1. INITIAL CHARACTERIZATION OF ENZYMES POSSESSING C2GnT ACTIVITY

Williams and Schachter (19, 20) were the first to identify and characterize an enzyme exhibiting C2GnT activity, demonstrating its presence in canine submaxillary glands. These investigators found that C2GnT enzyme activity requires both the galactose and GalNAc portions of the acceptor substrate, and that C2GnT action is blocked by addition of fucose in α1,2 linkage to the galactose residue of the acceptor, or of sialic acid in α2,6 linkage to the GalNAc portion. Kuhns et al (21) later determined that the 4- and 6-OH groups of the GalNAc residue, as well as the 6'-OH of the galactose residue, are required for C2GnT activity. These substituent requirements likely reflect the involvement of hydrogen bonding in binding of substrate to the active site. Information from studies such as these has been exploited in designing C2GnT inhibitors which may have therapeutic applications. For example, Hindsgaul et al (22) synthesized an analog of the acceptor for C2GnT which lacks the 6-OH on GalNAc, and found that the analog competitively inhibited C2GnT activity. Characterization of the enzymatic properties of C2GnT from dog submaxillary gland and from other mammalian tissues, including rabbit small intestine (23), bovine trachea (24), and pig trachea (25), have shown that the activity shows a pH optimum between 7.0-8.5, and is stimulated by 3-5 mM Mn^{2+} and low concentrations of a detergent such as Triton X-100 or Tween-20.

I.B.2 CHARACTERIZATION OF ENZYMES CONTAINING C4GnT ACTIVITY: IDENTIFICATION OF A TRIFUNCTIONAL β1,6 BRANCHING ENZYME

The core 4 branched structure seems to be much less common in nature than the core 2 structure, and perhaps for this reason fewer enzymes containing C4GnT activity have been identified than enzymes which contain C2GnT activity. Furthermore, no enzyme has yet been identified which possesses only C4GnT activity. Rather, C4GnT activity has invariably been detected in association with C2GnT activity in tissue extracts, and

C2GnT-M, which synthesizes all three mucin-type β1,6 branched structures (table 1), is the sole enzyme known to possess C4GnT activity to date.

Brockhausen et al performed the first detailed characterization of an enzyme containing C4GnT activity (26), detecting strong C4GnT activity in rat colon. Acceptor competition experiments suggested that the C4GnT enzyme also contained C2GnT activity. Interestingly, in spite of the apparent scarcity of core 4 structure in nature, these investigators found C4GnT activity in numerous mucin-secreting tissues in several species, including pig, human and monkey colon, pig, rat, monkey and sheep stomach, and rat and canine submaxillary gland. The authors suggested that the discrepancy between core 4 structure and C4GnT activity in tissues may result from low levels of the core 3 structure in mucin, which is the substrate for C4GnT. In a subsequent study, Brockhausen et al (27) identified an enzyme in pig gastric mucosa which appeared to contain all three mucin-type β1,6 branching activities, namely C2GnT, C4GnT and IGnT. This was the first report to suggest the existence of a trifunctional β1,6 branching glycosyltransferase. Ropp et al (28) demonstrated the existence of such an enzyme in convincing fashion by purifying from bovine tracheal epithelium a single enzyme containing all three β1,6 branching activities. The authors found that the C4GnT specific activity of the purified enzyme was roughly equivalent to its C2GnT specific activity. The enzyme utilizes a sequential binding mechanism in which UDP-GlcNAc binds first and UDP is released last. Km values of 0.36 mM and 0.14 mM were reported for UDP-GlcNAc and the core 1 acceptor Galβ1,3GalNAcα-benzyl, respectively. Kuhns et al (21) then showed that the trifunctional enzyme is expressed widely in human mucin-secreting tissues, and that a separate C2GnT enzyme which lacks C4GnT and IGnT activities is expressed in human leukemic cells. This prompted the authors to introduce the current nomenclature to distinguish the two isozymes, referring to the **L**eukemic C2GnT enzyme as C2GnT-L and the trifunctional enzyme from **M**ucin-secreting tissues as C2GnT-M.

I.B.3. CHARACTERIZATION OF ENZYMES POSSESSING IGnT ACTIVITY

The discovery of the IGnT enzyme had its roots in clinical studies from the late 1950s and 1960s on patients with a type of acquired hemolytic anemia known as cold hemagglutinin syndrome. In this disorder, the patient's blood coagulates when mixed with blood from healthy donors of the same blood type if the blood is not warmed to body temperature prior to transfusion. The adverse reaction is caused by the presence in the patient's blood of an antibody referred to as a cold autoantibody, a type of antibody which reacts with red cells at 4° but not at 37°. Wiener et al (29) identified a unique factor present in blood from healthy adult donors which was responsible for the hemolytic reaction. The authors designated this factor "I" to emphasize the highly individualistic properties of various donor blood samples tested in the study. The I factor was detected in the blood of all but 5 of 22,000 random donors tested in the study, suggesting a ubiquitous distribution of the I factor in the population. The five patients lacking the I factor were designated as type i, or I negative. Subsequently, Marsh (30) showed that all human red cells carry both the I and i antigen to some degree, and that the amounts of I and i antigen present on the cells are reciprocally related to each other. Marsh also observed that fetal and newborn human red cells express primarily the i antigen, and that this is replaced by the I antigen within the first 18 months post-partum. The I phenotype is then maintained for the remainder of the lifetime of the individual. These results indicated that the i and I antigen are related by an interconversion process, suggesting the existence of an enzyme or enzymes involved in catalyzing this interconversion. The relationship between the i and I antigens has since been examined in a large number of studies (31-39), and these have shown that these antigens cannot be represented by single carbohydrate structures, but rather by families of structures. However, the I-type structures share a common feature, the presence of the β1,6 GlcNAc branch:

GlcNAcβ1,6
\
R_1-GlcNAcβ1,3Galβ1,4-R_2

Structures recognized by anti-i antibodies, on the other hand, were found to lack this β1,6 GlcNAc branch. This suggested that a single enzyme, a β1,6 GlcNAc branching enzyme, was responsible for the developmental i-to-I conversion.

In 1983, van den Eijnden et al (40) observed that Novikoff ascites tumor extract contains an activity capable of adding GlcNAc in β1,6 linkage to the terminal galactose of asialo α1-acid glycoprotein, representing the initial discovery of a candidate IGnT enzyme. Basu and Basu (41) subsequently detected a similar enzyme in mouse T lymphoma cell extracts which catalyzed the reaction

Galβ1,4GlcNAcβ1,3Galβ1,4-Glc-Ceramide + UDPGlcNAc ⟶

UDP + GlcNAcβ1,6Galβ1,4GlcNAcβ1,3Galβ1,4-Glc-Ceramide

The activity showed a pH optimum between 7.0 and 8.0, and was stimulated by Triton CF-54 detergent and Mn^{2+}, whereas Zn^{2+} and Cu^{2+} inhibited the activity. Sialylation at the nonreducing terminus of the acceptor inhibited activity, and the truncated structure Galβ1,4Glc-Cer acted as a poor acceptor. Zielinski and Koscielak (42, 43) were the first to report a β1,6 GlcNAc transferase activity capable of synthesizing the I-type β1,6 branched structure GlcNAcβ1,3(GlcNAcβ1,6)-Galβ1,4-, detecting its presence in human serum using lactose as acceptor (43). Interestingly, serum from both I-type and i-type individuals were able to synthesize the branched structure. Piller et al (44) then identified an IGnT activity in pig gastric mucosa which catalyzed the reaction

GlcNAcβ1,6
\
GlcNAcβ1,3Galβ1,4Glc ⟶ GlcNAcβ1,3-Galβ1,4Glc

using glycolipid acceptors. The authors found that this IGnT activity could not act on the substrate Galβ1,4GlcNAcβ1,3Galβ1,4Glc, showing that a terminal galactose interferes with the enzyme activity. Brockhausen et al (27) investigated the acceptor specificity of this pig gastric IGnT further, and discovered that it could act on mucin-type acceptors in addition to the glycolipid acceptors tested by Piller et al. In particular, the enzyme was able to synthesize the core 2 and core 4 β1,6 branched structures from core 1 and core 3 acceptors, respectively, in addition to synthesizing the I β1,6 branched structure from an i-type precursor. This revealed the actual identity of the pig gastric enzyme to be C2GnT-M. The activity of the enzyme was stimulated *in vitro* by detergent, but not by the addition of Mn^{2+}. Helin et al (45) examined the ability of this enzyme to utilize the polylactosamine acceptor GlcNAcβ1,3Galβ1,4GlcNAcβ1,3Galβ1,4GlcNAc, and found that the enzyme was able to attach a β1,6 GlcNAc branch to the galactose near the nonreducing terminus, but not to the more internal galactose. The ability to act upon acceptors possessing GlcNAcβ1,3Galβ1,4- at the nonreducing terminus is also shared by the intrinsic IGnT activity of C2GnT-M isolated from bovine tracheal epithelium (28), as well as by an IGnT activity identified in Novikoff ascites tumor extract by Koenderman et al (46). The latter enzyme represents a second IGnT activity in Novikoff ascites extract, distinct from the IGnT characterized earlier from this source by van den Eijnden (40). Like the pig gastric IGnT, this second Novikoff ascites IGnT enzyme appears to be another example of C2GnT-M, since it can synthesize the core 4 branched structure in addition to the blood group I structure (46).

By the late 1980s, many examples had been reported of polylactosamine structure containing internal I-type β1,6 branches, and some even contained multiple branches (39, 47-52). Since the IGnT enzymes characterized to this point all seemed to act preferentially on the galactose residue nearest the nonreducing terminus, the existence of internally-branched polylactosamine suggested the possibility of a second type of IGnT, one which could act on internal galactose residues. Leppanen et al (53) first identified such an activity in 1991 in human serum. The enzyme added an internal β1,6 branch to a polylactosamine acceptor as shown below.

GlcNAcβ1,3Galβ1,4GlcNAcβ1,3Galβ1,4GlcNAc + UDP-GlcNAc

→ UDP + GlcNAcβ1,3-Galβ1,4GlcNAcβ1,3(GlcNAcβ1,6)Galβ1,4GlcNAc

Furthermore, this IGnT was unable to act on GlcNAcβ1,3Galβ1,4GlcNAc, in contrast to IGnT isoforms characterized previously, and the structure GlcNAcβ1,3Galβ1,4-GlcNAcβ1,3Gal served as a poor acceptor, suggesting an important role for the GlcNAc at the reducing end of the acceptor. Sakomoto et al (54) recently detected and purified a similar IGnT enzyme from hog small intestine which was capable of catalyzing the above reaction. This enzyme proved to be quite versatile, as it was also able to utilize galactose-terminated acceptors. Gu et al (55) partially purified another potentially novel IGnT activity from rat intestine and serum which could act on the highlighted residue in Galβ1,4GlcNAcβ1,3-**Galβ1,4**Glc. Like the enzyme identified by Sakomoto et al (54), this IGnT activity could act on acceptors terminated by β1,3-linked GlcNAc, such as in GlcNAcβ1,3Galβ1,4GlcNAcβ1,3Galβ1,4Glc. However, the two enzymes are probably not identical, since the enzyme characterized by Gu et al utilized the GlcNAc-terminated acceptor with much lower efficiency than did the enzyme identified by Sakomoto et al. Gu et al also observed that sialylation of the acceptor at the nonreducing terminus abolished activity, and likewise no activity was detected using an acceptor with the type I structure Galβ1,3GlcNAc- at the nonreducing terminus. Leppanen et al (56) later demonstrated the ability of this rat serum enzyme to add β1,6 GlcNAc branches at several internal galactose residues within a polylactosamine backbone, suggesting a possible mechanism for multiple branch formation in polylactosamine chains.

Table 2 below presents a summary of the different IGnT activities which have been identified, along with the tissues in which they have been found and the types of acceptors on which they act. The IGnT enzymes are classified into several categories in the table. The first type, of which the enzyme identified by Basu and Basu (41) is an example, do not appear to require β1,3 linked GlcNAc attached to the galactose in the acceptor. That is, they do not require an i-type acceptor structure, and do not actually appear to create a branched product structure. The second type of IGnT enzyme listed prefers galactose-terminated acceptors such as Galβ1,4GlcNAcβ1,3Galβ1,4GlcNAc, and can add branches to internal galactose residues within polylactosamine. They are therefore referred to as **midchain** or **centrally-acting IGnT (cIGnT)**. The third class of IGnT enzymes listed prefers β1,3-linked GlcNAc at the nonreducing terminus, such as the intrinsic IGnT activity in C2GnT-M, and are termed **distally-acting IGnT (dIGnT)**. The enzymes which show characteristics of both centrally-acting and distally-acting IGnT, such as those detected by Leppanen et al (53) and Sakomoto et al (54), are difficult to classify, but we will consider them primarily as centrally-acting IGnT enzymes since they are capable of attaching branches on internal galactose residues.

Table 2. Summary of IGnT activities which have been characterized to date. For each type of IGnT activity, the tissues in which the enzyme has been detected (by northern blotting or enzyme assay) are listed, along with the acceptor substrates on which the enzyme has been found to act. The residue within the acceptor to which the β1,6 branch becomes attached is highlighted. G = galactose, Gn = *N*-acetylglucosamine

Type of IGnT	Tissues Detected In	Preferred Acceptors	Refs
1. IGnT which do not require β1,3-linked GlcNAc	Mouse T lymphoma, Novikoff ascites	1. Asialo α1-acid glycoprotein 2. **Gβ1,4**Glc 3. **Gβ1,4**Gnβ1,3Gβ1,4Glc-Cer	40, 41
2. Centrally-acting IGnT (cIGnT)	Human teratocarcinoma, Rat brain and serum, Hog small intestine, Human serum	1. Gβ1,4Gnβ1,3**Gβ1,4**Gn 2. Gβ1,4Gnβ1,3**Gβ1,4**Gnβ1,3-Gβ1,4Gnβ1,3**Gβ1,4**Gn 3. Gβ1,4Gnβ1,3**Gβ1,4**Gnβ1,3-**Gβ1,4** Glc 4. Gβ1,4Gnβ1,3**Gβ1,4**Gnβ1,3-Gβ1,4Gn 5. Gnβ1,3Gβ1,4Gnβ1,3**Gβ1,4**Gn 6. Gα1,3Gβ1,4Gnβ1,3**Gβ1,4**Gn	53-58
3. Distally-acting IGnT (dIGnT)	Novikoff Ascites, Pig gastric mucosa, Bovine trachea, Human mucin-secreting tissues	1. Gnβ1,3**Gβ1,4**Glc 2. Gnβ1,3**Gβ1,3**GgalNAc 3. Gβ1,3**Gβ1,4**Glc 4. Gnβ1,3**Gβ1,4**Gnβ1,3Gβ1,4Gn 5. Gnβ1,3**GalNAc** 6. Gβ1,3**GalNAc** 7. Gnβ1,3**Gβ**-Me 8. Gnβ1,3**Gβ1,4**Gn	17, 18, 27, 28, 44-46

It is not yet clear exactly how distally-acting and centrally-acting IGnT enzymes work together to produce the I-type β1,6 branched structures seen in nature, particularly those with multiple branches. In principle, the branches could be added by either cIGnT or dIGnT (or both) during chain elongation each time a branch point is reached, or the branches could be added after completion of the linear chain. The latter mechanism would require that cIGnT add branches to the internal galactose residues, while dIGnT might add the branch at the nonreducing terminus (provided the chain is terminated by β1,3-linked GlcNAc). The available evidence for either of these possible mechanisms is not convincing, but generally seems to weigh in favor of cIGnT having the more significant role. The strongest evidence comes from studies on the kinetics of chain elongation from the two arms of a β1,6 branch point. Among the almost limitless variety of multiply-branched structures which could conceivably be created by the enzymes involved in polylactosamine chain synthesis, the only motif which seems to be common in nature consists of a long main chain with numerous short di- or trisaccharide β1,6 GlcNAc branches (39, 48, 49, 59-61), such as the structure depicted below.

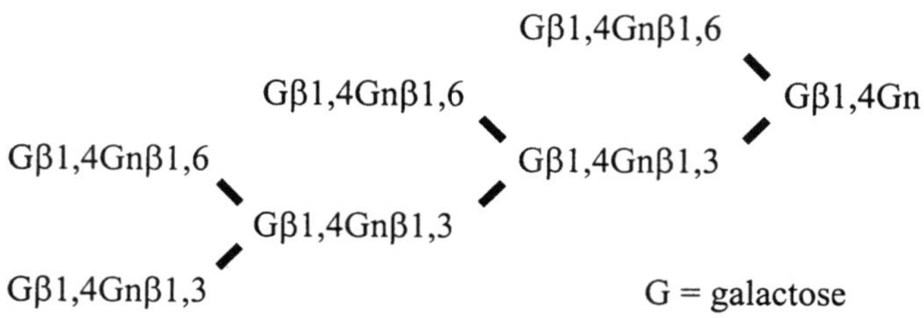

G = galactose
Gn = *N*-acetylglucosamine

In order for distally-acting IGnT to produce this type of branched polylactosamine structure during chain elongation, the kinetics of extension ought to be slower for the β1,6 arm than for the β1,3 arm of the chain. Although recent work by Ujita et al (58) suggests that this may be the case, most of the studies addressing this issue have shown that the β1,6-linked arm of an I-type branch can be extended just as efficiently as the β1,3-linked arm, if not more so (62-64). This would suggest that internal β1,6 branches arise principally by the action of cIGnT on a completed polylactosamine chain.

III. CLONING OF MUCIN-TYPE β1,6 BRANCHING ENZYMES

In 1992 Bierhuizen et al (15) reported the first cloning of a β1,6 GlcNAc branching enzyme, specifically a cDNA encoding human C2GnT-L. These investigators exploited the properties of a monoclonal antibody which recognizes the cell surface-associated glycoprotein CD43 containing core 2-based O-glycans. CHO cells were engineered to stably express the polyoma large T antigen, then were transiently transfected with an HL-60 promyelocytic cell line cDNA library contained in an expression plasmid bearing a polyoma origin of replication. This allowed for maintenance of the cDNA library as the CHO cells divided. The authors then transfected a CD43 cDNA and used the T305 monoclonal antibody to detect cells in which the expressed CD43 had been modified with the core 2 O-glycan. The substrate preferences of the cloned C2GnT showed the enzyme to be of the L type, possessing C2GnT activity, but not core 4 or blood group I β1,6 GlcNAc transferase activity. Comparison of the cDNA sequence of human C2GnT-L with that of a recently identified human C2GnT-L pseudogene (65) suggests that the pseudogene and the C2GnT-L gene arose from a common precursor gene.

Sekine and coworkers (66) have cloned two murine C2GnT cDNAs possessing high sequence similarity to human C2GnT-L. RT-PCR analysis showed that expression of one of the two transcripts was restricted to the kidney, while the other was much more widely expressed. The two cDNAs both contained three exons and shared an identical open reading frame, located in exon 3, but differed in their 5' untranslated regions, specifically within exon 1. Analysis of the genomic organization revealed the structure shown in figure 2 below, and led the authors to attribute the distinct messages and their different expression patterns to tissue-specific differences in alternative splicing and usage of alternative promoters. An interesting feature of the expressed enzyme was its ability to utilize both a mucin-type core 1 acceptor and a glycolipid-type acceptor.

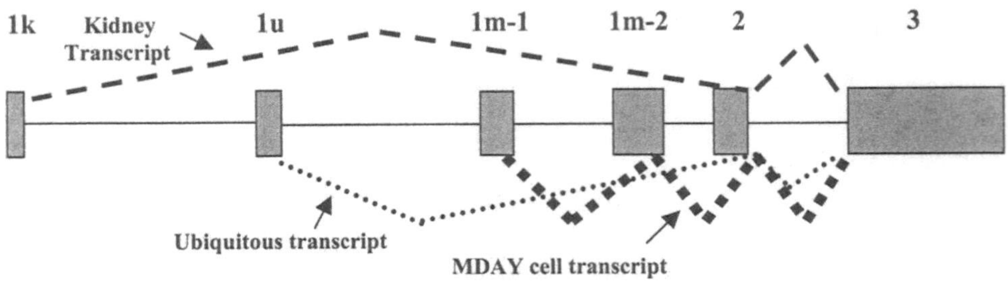

Figure 2. Genomic organization and alternative splicing pattern of murine C2GnT-L gene. Exons are represented by boxes. Numbers above boxes indicate the exon number and the tissue or cells in which the exon has been found (k = kidney, u = ubiquitous, m = MDAY cells). Exons 2 and 3 are common to all three transcripts, while exon 1 differs among the transcripts. The Genbank accession number of the MDAY C2GnT-L cDNA is MU19265. Adapted from ref 66.

In 1998 Li et al (67) reported the cloning of a bovine lung C2GnT-L cDNA. The deduced amino acid sequence showed 86% similarity to human C2GnT-L. Immunohistochemical analysis of bovine tracheal sections showed that the enzyme is present in the tracheal surface epithelium, and was especially abundant in submucosal glands (67, 68). A C2GnT-L cDNA has also recently been cloned from diabetic rat heart (69). The deduced amino acid sequence shares 83% and 92% similarity with human and murine C2GnT-L sequences, respectively.

Bierhuizen et al (14) reported the cloning of a human IGnT cDNA in 1993 from a PA-1 human teratocarcinoma cDNA library, using the same expression cloning approach they had used previously to clone the human C2GnT-L cDNA (15). Transfected cells expressing the I antigen were selected based on their production of the I antigen, detected using anti-I antibodies. Sasaki et al (70) later showed that human IGnT transcripts are restricted principally to fetal brain, fetal kidney, and adult brain, prostate, small intestine and colon. Mattila et al (57) determined that the human IGnT enzyme is a centrally-acting IGnT, since a soluble form of the enzyme expressed in Sf9 insect cells was able to attach a β1,6 GlcNAc branch to the highlighted galactose residue in Galβ1,4-GlcNAcβ1,3**Galβ1,4**GlcNAc, but could not act on GlcNAcβ1,3Galβ1,4GlcNAc. Moreover, the cloned IGnT also attached β1,6 GlcNAc branches at multiple internal galactose residues in a larger polylactosamine acceptor structure. In a follow-up study, Ujita et al (58) found that, given a choice between the two internal galactose residues highlighted in the structure Galβ1,4GlcNAcβ1,3**Galβ1,4**GlcNAcβ1,3-**Galβ1,4**GlcNAc, the human IGnT preferentially attaches a β1,6 GlcNAc branch to the residue closer to the nonreducing terminus before adding a branch at the more internal galactose.

Bierhuizen et al (71) examined the 5' upstream regions of the IGnT and C2GnT-L genes. As shown in figure 3, both contain a TATA box and a consensus site for the

hC2GnT-L

```
-270   GCTTTCTGATAGAGGTGTGGAGTGGGGTGGTGATCAAGATTGTTAGAAGAAAATGAA
              GATA-3                Sp1            CAAT
-212   TGAGCAACAAATTCAGCTTCAGGATTGTCATTAATTTGCCAGTGAGCATTGGGAATCT
                              CAAT      TATA
-153   CACAGTGGGGGAAAGGATATGTTGGATTTTCTAATCATACTGGACTGTTAAACTGTTC
        TCF-1
- 94   TTGGTATACATTAAGGGAAATGCTGCTAGAGTTTTCTTTCAAAGAACTTAACCTAAAT
                                               TCF-1
- 36   TTCAGGGTGTTTTCAAATCTTTAGCAATCCACAAAC ATAGAATTAG
                            NF-IL6
```

hIGnT

```
-170   GTCTCATATAACTCAGTGGATGTAGCTCACGACTTCTTCTTTTGACCCCAGTATTCTAG
              TATA
-110   ACAAACACCACAGCCGAAGCTGGAAGTGAGAGGAGAGGGGTTCGTTTCAGAGTCAG
                TCF-1
-53    AGCCCAAGGGGCCCCAGCAATTTCAGACCCCCTGGGACAGGAACTCCTCCTCC AGTA
          AP2                        AP2
```

Figure 3. Partial 5' untranslated regions of the genes for human C2GnT-L and IGnT . Putative transcription factor binding sites are underlined, and the major transcription start sites are indicated by the bent arrows. Sequence position numbers are relative to the major transcription start site. Adapted from ref 71.

transcription factor TCF-1, but differ in their content of other potential transcription factor binding sites. Notably, the C2GnT-L 5' upstream region contains sites for several trans-acting factors which are active in activated T lymphocytes, including NF-IL6, GATA-3 and TCF-1. This observation is consistent with the reported increase in core 2-based glycans and C2GnT activity in activated T cells (72), although direct evidence for the involvement of these factors in C2GnT-L induction in T cells has not yet been reported.

The cloning of a cDNA for the human C2GnT-M enzyme was recently achieved independently in two laboratories (17, 18). Both groups identified an expressed sequence tag having approximately 50% amino acid sequence similarity to human C2GnT-L, and the recombinant enzyme was able to synthesize core 2, core 4 and blood group I β1,6 branched oligosaccharides. Northern blotting analysis revealed the presence of C2GnT-M transcripts in mucin-secreting tissues such as colon, pancreas, trachea and small intestine, but transcripts were not detected in peripheral blood leukocytes, spleen, and thymus. In contrast, C2GnT-L mRNA was detected in many tissues, including spleen, prostate, small intestine, colon, stomach and thyroid. Taken together, the reported tissue distributions for human C2GnT-M, C2GnT-L and IGnT suggest that C2GnT-M is primarily responsible for core 2-branched and blood group I-branched mucin glycans found in mucus secretions, whereas the core 2-branched and blood group I-branched structures synthesized by C2GnT-L and IGnT probably function mainly in tissues which do not secrete mucin, although there is clearly some overlap in these assignments.

Vanderplasschen et al (73) recently reported the cloning of a C2GnT-M cDNA from bovine herpes virus type IV, the first nonmammalian β1,6 GlcNAc branching enzyme to be cloned. The deduced amino acid sequence showed 81% identity to human C2GnT-M, and the mRNA for the gene was found to be expressed during viral replication, indicating potential functional importance of the viral C2GnT-M. In keeping with this, infection of a C2GnT-negative cell line with the virus led to the appearance of core 2-based O-glycans on the surface of the cells.

Recently, a cDNA for a third type of human C2GnT, termed C2GnT-3, was cloned by Schwientek et al (16) by performing a BLAST search on human genomic sequence data. Recombinant C2GnT-3 possesses C2GnT activity, but lacks C4GnT and IGnT activity (table 1), similar to C2GnT-L. The amino acid sequence shares 42% similarity to those of human C2GnT-L and C2GnT-M and 39% similarity with human IGnT. The 5.5 kB C2GnT-3 mRNA was detected mainly in thymus, with lower levels in small intestine and peripheral blood leukocytes. The strong presence of C2GnT-3 in thymus suggests that this enzyme may play an important role in immune system function.

Comparison of the protein structure and genomic architecture of the cloned β1,6 GlcNAc transferases reveals several common features, yet significant differences can also be seen. The proteins are similar in size (ranging from 400 to 453 amino acids) and share a common domain structure containing three regions of high sequence similarity in the putative catalytic domain, as shown in figure 4A. However, the genes for the enzymes differ markedly in their genomic structure. The genes for hC2GnT-L, hC2GnT-M and hC2GnT-3 resemble one another in that the entire open reading frame is located within a single exon (16). The hIGnT gene, by contrast, is divided among three exons, with all three exons containing a portion of the open reading frame. These differences are illustrated in figure 4B, in which the genomic structures of the hC2GnT-L and hIGnT genes are compared. Additionally, the genes for the human β1,6 GlcNAc transferases show substantial differences in their chromosomal localization. The hIGnT gene has been localized to chromosome 6p24 (18), the hC2GnT-L gene resides at 9q21 (14), the hC2GnT-M gene is found at 15q22 (18), and the hC2GnT-3 gene is at 5q12 (16). These observations suggest that the present day genes for the β1,6 branching enzymes may have arisen from a common ancestral gene which underwent duplication and divergence, culminating in two distinct gene families, the family of IGnT enzymes and the family of C2GnT enzymes (16).

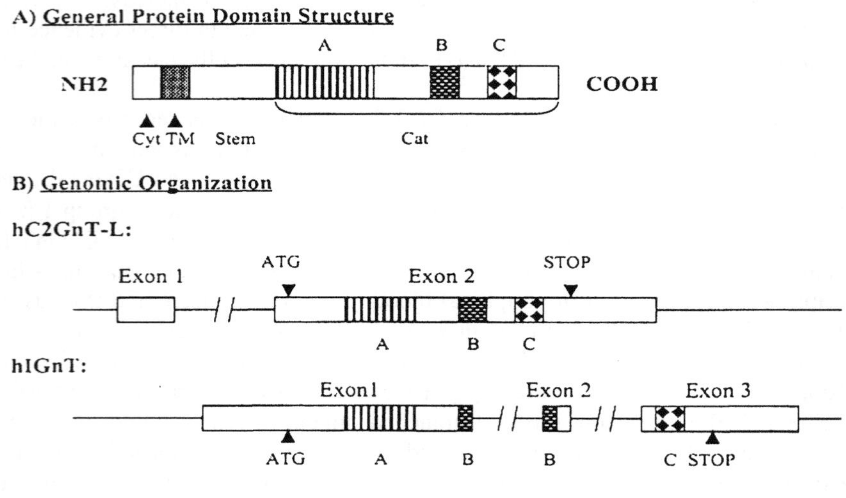

Figure 4. Protein domain structure and genomic architecture of β1,6 branching enzymes. A) Common protein domain structure for C2GnT-L, C2GnT-M, C2GnT-3 and IGnT, showing the N-terminal cytoplasmic domain (Cyt), the transmembrane domain (TM), the putative stem region (Stem) and the putative catalytic domain (Cat). The latter contains three homologous regions, A, B and C. B) Comparison of the genomic structures of the human C2GnT-L and IGnT genes. Regions corresponding to the homologous regions in the protein structure are indicated by stippling, and the approximate locations of the start and stop codons are shown to indicate the open reading frame. The genomic organization of C2GnT-M and C2GnT-3 are similar to that of C2GnT-L. Adapted from refs 14 and 71.

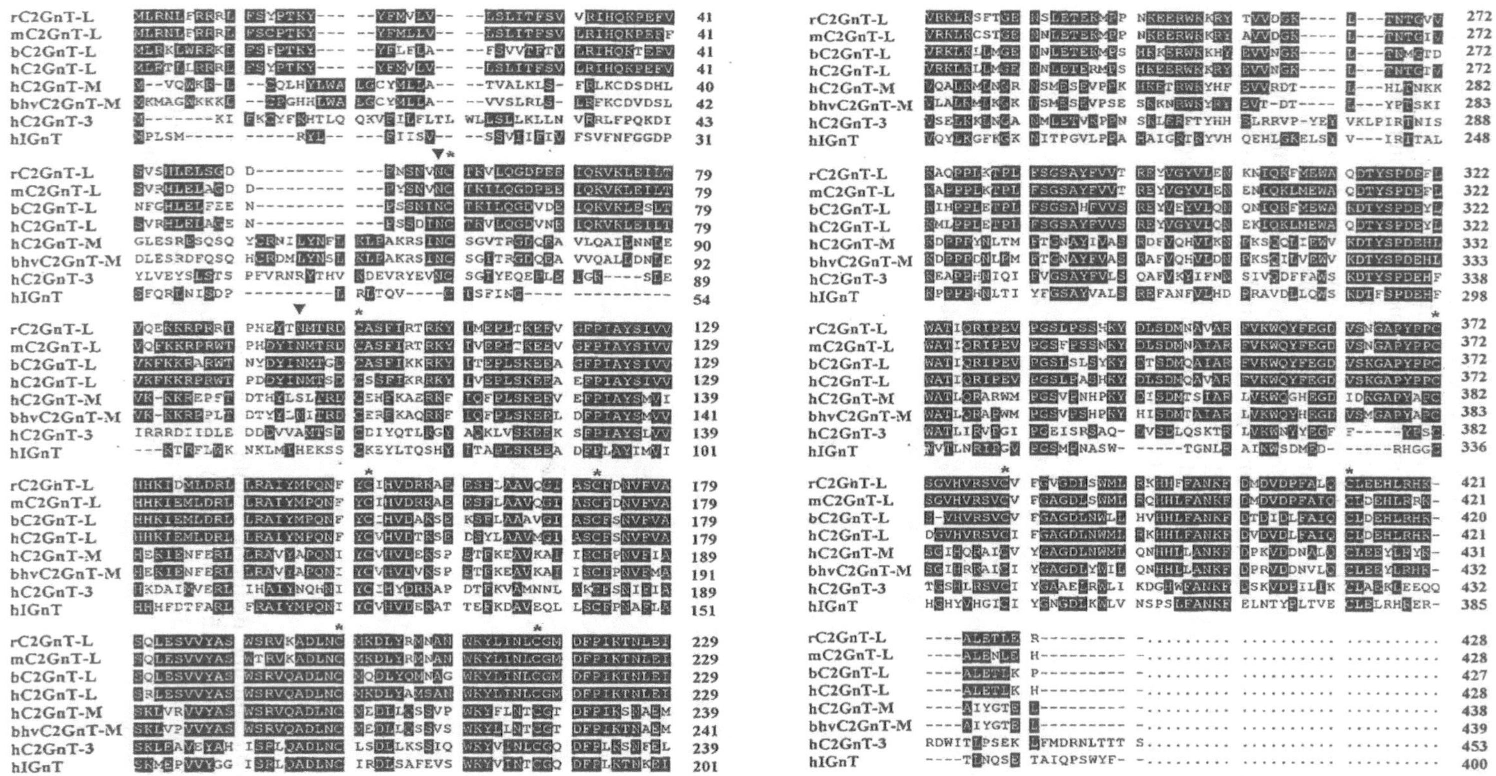

Figure 5. Alignment of deduced amino acid sequences of mucin glycan branching β1,6 GlcNAc transferases cloned from rat (r), mouse (m), bovine (b), bovine herpes virus (bhv) and human (h). Multiple sequence alignment was performed by DNASIS™ sequence analysis software. Residues common to at least one third of the sequences are boxed, and the positions of nine completely conserved cysteine residues are indicated by asterisks. Two conserved N-linked glycosylation sites at residues 58 and 95 of the C2GnT-L sequences, known to be required for activity of human C2GnT-L (74), are indicated by arrow heads. Other potential N-linked glycosylation sites include residues 69 and 289 of hC2GnT-M, residues 72, 286, 317, and 448 of hC2GnT-3, and residues 37, 212, 255, 314 and 388 of hIGnT. Genbank accession numbers for the cDNAs are as follows: rC2GnT-L, S79797; mC2GnT-L, U19265; bC2GnT-L, U41320; hC2GnT-L, M97347; hC2GnT-M, AF102542; bhvC2GnT-M, AF231105; hC2GnT-3, AF132035; hIGnT, Z19550.

Figure 5 shows an alignment of the deduced amino acid sequences of the mucin-type β1,6 GlcNAc branching enzymes cloned from several species. A high degree of similarity can be seen between the various C2GnT-L sequences, while similarity of the C2GnT-L sequences to those of human C2GnT-M, C2GnT-3 and IGnT is more limited. These sequence relationships are depicted in a more quantitative fashion in figure 6 below. In spite of the limited overall similarity among the sequences shown in figure 5, it is particularly interesting to note the existence of nine perfectly conserved cysteine residues. To date, no x-ray crystal structural information is available regarding these proteins, but the high degree of conservation of these cysteines suggests that these residues may be involved in maintaining the three-dimensional structure of the enzymes and that the overall topology of the proteins may be quite similar. Another significant feature to notice is the conservation among all four C2GnT-L sequences of two asparagine residues located at positions 58 and 95 of human C2GnT-L. These residues have been shown to be required for full activity of human C2GnT-L (74).

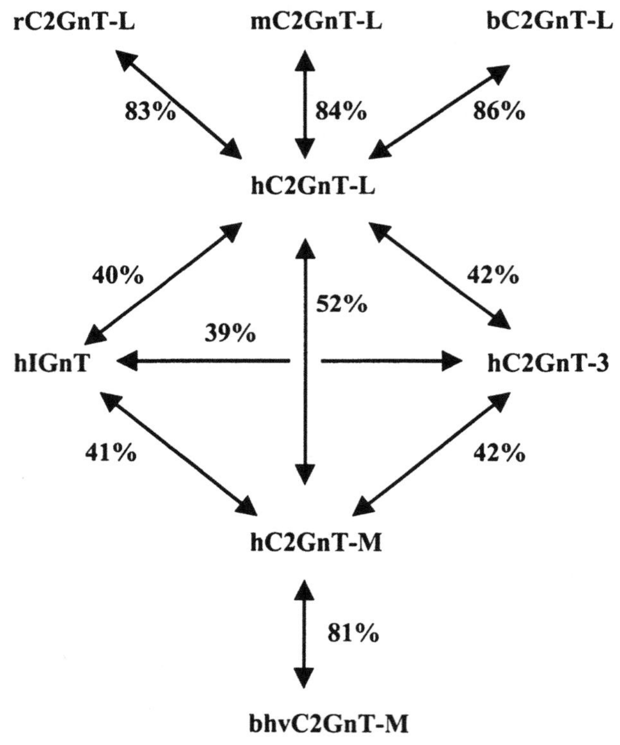

Figure 6. Similarity among deduced amino acid sequences of β1,6 branching enzymes cloned from human (h), mouse (m), bovine (b), bovine herpes virus (bhv) and rat (r). The top portion of the figure shows the percentage sequence identity between C2GnT-L enzymes from four different species, the middle portion shows the relationships among the four cloned human β1,6 branching enzymes, and the relationship between the two cloned C2GnT-M proteins is shown at the bottom. Percentage sequence identities were calculated using DNASIS™ sequence analysis software (Needleman-Wunsch algorithm).

IV. BIOLOGICAL ROLES OF MUCIN-TYPE β1,6 BRANCHED GLYCANS

IV.A. BIOLOGICAL SIGNIFICANCE OF CORE 2 STRUCTURE

IV.A.1. ROLE IN IMMUNE SYSTEM FUNCTION

A large body of experimental evidence has been amassed which suggests a significant role for C2GnT activity in generating polylactosamine chains on mucin-type glycans. Such structures are important because they serve as the framework onto which functional peripheral structures are added. An initial connection between C2GnT activity and polylactosamine came with the finding that mucin glycans from granulocytic cells (75) contain the following core 2-based polylactosamine structure:

NeuAcα2,3Galβ1,4GlcNAcβ1,3Galβ1,4GlcNAcβ1,6
\
NeuAcα2,3Galβ1,3GalNAc

Maemura and Fukuda (76) observed that the mucin glycans of CD43 in HL-60 cells, which express C2GnT activity, possessed approximately 3% polylactosamine structure. In contrast, K562 cells, which lack C2GnT activity, expressed no detectable core 2 β1,6 branched structure or polylactosamine structure. Moreover, many of the HL-60 polylactosamine chains were terminated by the sialyl Lewis x structure, suggesting a link between sialyl Lewis x structure and C2GnT activity. In a related study, Yousefi et al (77) compared glycosyltransferase activities and carbohydrate structures among several murine cell lines, and observed a general dependence of polylactosamine on C2GnT activity. Heffernan et al (78) have shown that induction of C2GnT activity in F9 teratocarcinoma cells results in increased expression of polylactosamine structure, although several other glycosyltransferases were induced to lesser degrees which could have played a role in generating the polylactosamine as well. Core 2-based polylactosamine chains have also been found as an integral component of zona pellucida oligosaccharides, which are crucial in the fertilization process (79). In a more direct demonstration of the importance of C2GnT activity in creating polylactosamine structure, CD43 transiently expressed in CHO cells (which lack detectable C2GnT activity) was found to possess both the core 2 β1,6 branched structure and polylactosamine only when the cells were cotransfected with a C2GnT-L cDNA (80). Also, a recent study of the O-linked sugar chains on the cell surface mucin-like protein CD45 revealed the presence of polylactosamine chains attached to core 2-branched glycans, but not to core 1-based glycans (81). In spite of the overwhelming evidence of a role for the core 2 branch in generating polylactosamine chains, however, synthesis of this structure independently of the core 2 structure cannot be ruled out.

Polylactosamine chains are often terminated by structures which serve as important components of the ligands for P-, E- and L-selectins (reviewed in refs 82-85). The ligand for P-selectin, P-selectin glycoprotein ligand-1 (PSGL-1), is expressed on the surface of leukocytes, as is the ligand for E-selectin, E-selectin ligand-1 (ESL-1). E-selectin is also capable of low-affinity binding to PSGL-1 (86). These selectin ligands are critical for the initial interaction between endothelial cells and leukocytes during the inflammatory response. L-selectin can interact with several ligands, including CD34, glycosylation-dependent cell adhesion molecule-1 (GlyCAM-1) and the mucosal vascular addressin cell adhesion molecule-1 (MadCAM-1). These molecules are found on the surface of peripheral lymph node high endothelial venules and activated endothelial cells, and interaction of these ligands with L-selectin has been found to be crucial for lymphocyte homing to high endothelial venules.

An essential feature common to the mucin-type glycans of the selectin ligands is the presence of the sialyl Lewis x structure, shown below:

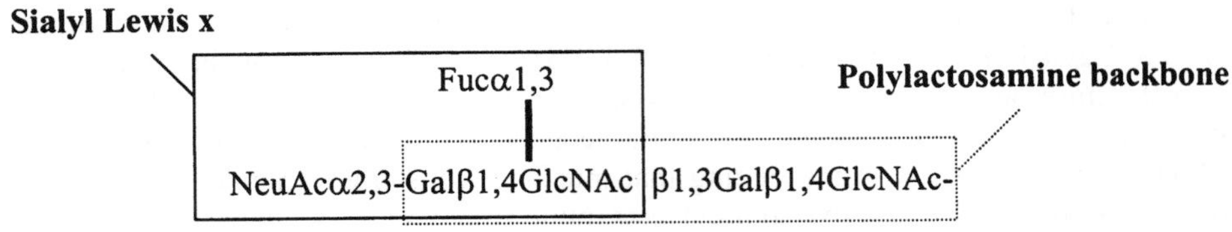

The carbohydrate structures of PSGL-1 and GlyCAM-1 have been examined extensively. In the case of PSGL-1, sialyl Lewis x is found at the nonreducing termini of a subset of its O-linked polylactosamine chains (7, 86, 87), suggesting a central role for C2GnT activity in generating the P-selectin ligand. Direct evidence for such a role came from work done independently by two research groups in 1996 (88, 89). Both groups transfected a CHO cell line with various combinations of PSGL-1, C2GnT-L, and an α1,3 fucosyltransferase involved in formation of the sialyl Lewis x structure. P-selectin and E-selectin bound to the expressed PSGL-1 only when both C2GnT and α1,3 fucosyltransferase were coexpressed in the cells. The L-selectin ligand GlyCAM-1 (90) has also been found to possess core 2-based O-glycans, but its carbohydrate chains possess functionally important sulfate moieties as well, unlike the glycans of the P- and E-selectin ligands.

Several recent reports have pointed to an *in vivo* role for C2GnT activity in generating selectin-ligands involved in inflammation and metastasis. For example, Nakamura et al (91) observed that treatment of the pre-B lymphocytic leukemia cell line KM3 with phorbol ester resulted in downregulation of both C2GnT activity and sialyl Lewis x structure. Other enzymes involved in polylactosamine and sialyl Lewis x synthesis were unaffected. Furthermore, expression of exogenous C2GnT-L in the cells prevented these changes. Using the 1D4 antibody, which recognizes CD43 carrying the core 2 branched structure (92), Mukasa et al (93) observed that T lymphocytes which bound to P and E selectins were enriched in the 1D4 antigen relative to T lymphocytes which did not bind these selectins. Significantly, T lymphocytes recruited to sites of inflammation in rheumatoid arthritis patients showed greater reactivity with 1D4 than those in normal blood, demonstrating the importance of C2GnT in the inflammatory response. Nakamura et al (94) recently observed a large increase in both C2GnT-L expression and sialyl Lewis x during activation of tonsillar B lymphocytes, and Shimodaira et al (95) have found a statistically significant association between tumor invasiveness and expression of both C2GnT and sialyl Lewis x in tissue specimens from normal and colorectal cancer patients.

In perhaps the most informative study regarding the role of C2GnT activity in selectin ligand formation and function, Ellies et al recently generated mice in which the C2GnT-L gene was deleted from the germline (96). Tissues which express high levels of C2GnT activity in wild-type mice, i.e. kidney, bone marrow and spleen, were found to be devoid of C2GnT activity in the C2GnT-L null mice. Deletion of the C2GnT-L gene had its greatest impact on the inflammatory response, resulting in a partial loss of selectin ligands and concomitant reduction in leukocyte rolling on P-, E- and L-selectins. Surprisingly, lymphocyte homing to high endothelial venules was unaffected by the C2GnT-L gene deletion. The explanation for this result may involve compensation by other C2GnT isoforms, such as C2GnT-M or C2GnT-3, or alternative mechanisms for binding of L-selectins to lymph node endothelial cells. One such alternative mechanism may involve selectin ligands which are based on the core 1 structure instead of core 2. In a striking recent finding, Fukuda et al (97) detected the presence of substantial levels of the sialylated, fucosylated and sulfated core 1-based oligosaccharide NeuAcα2,3-Galβ1,4[Sulfo6(Fucα1,3)GlcNAc]β1,3Galβ1,3GalNAcα-Ser/Thr on the L-selectin ligands in the high endothelial venules of the previously characterized C2GnT-L knockout mice. Moreover, coexpression of the cDNA for a core 1 extension enzyme and the cDNA for a specific L-selectin ligand sulfotransferase in CHO cells resulted in L-selectin dependent rolling of the CHO cells, showing functional importance of the core 1-based L-selectin ligand.

Large changes in C2GnT activity occur during T cell activation. Upon stimulation of T-lymphocytes by anti-CD3 antibodies or interleukin-2, CD43 expressed on the cell surface undergoes an electrophoretic mobility shift from 115 kD to 130 kD (98, 99). Piller et al (72) showed that this change in mobility is caused by a change in CD43 mucin glycan structure from a core 1-based structure to a core 2-based structure:

Resting T lymphocyte glycan		Activated T lymphocyte glycan
NeuAcα2,6 \ NeuAcα2,3Galβ1,3GalNAcα-S/T	→	NeuAcα2,3Galβ1,4GlcNAcβ1,6 \ NeuAcα2,3Galβ1,3GalNAcα-S/T

It was found that this transition is caused by an approximate 10-fold increase in C2GnT activity during T-cell activation. The authors also observed a substantial decrease in α2,6 sialyltransferase activity, which competes with C2GnT for carbon 6 of GalNAc, and concluded that this may play a part in the observed change in glycan structure as well.

The effects of the increase in C2GnT activity during T cell activation are felt by other lymphocyte cell surface glycoproteins in addition to CD43. Barran et al (100) transfected the EL-4 T lymphoma cell line with a C2GnT-L cDNA, and observed the anticipated upregulation of the 130 kD form of CD43 and downregulation of the 115 kD form. However, they also saw apparent molecular weight increases of 3-5 kD for each of three other lymphocyte O-glycoproteins, RPTPα, CD44 and CD45. The function of these changes is not clear at present, but it is known that CD44 and CD45, along with CD43, are involved in mediating cell-cell interactions during T cell activation (101-103).

C2GnT activity also changes during maturation of T-lymphocytes. Baum et al (104) showed that the interaction of developing thymocytes with thymic epithelial cells, which is critical for normal positive selection of thymocytes, is mediated in part by a specific protein-carbohydrate interaction between galectin-1 on epithelial cells and core 2-based glycans expressed on thymocytes and T lymphoblasts. They found that immature thymocytes bound more galectin-1 than mature thymocytes, suggesting a developmentally-related decrease in thymic core 2-based glycans. Immunohistochemical studies using the T305 antibody, which recognizes CD43 bearing core 2-based glycans, supported this conclusion. Also, a follow-up study (105) showed that the 130 kD form of CD43, which carries the core 2 branched glycan, is downregulated in thymic positive selection.

Tsuboi and Fukuda (106, 107) examined the *in vivo* role of C2GnT in immune system function by generating transgenic mice which overexpress C2GnT-L in T lymphocytes. T lymphocyte development was not impaired in the mice, in spite of the previous work showing that C2GnT activity decreases during thymocyte positive selection. However, mature T lymphocytes showed an impaired immune response as judged by their delayed-type hypersensitivity response and reduced proliferation rates. Furthermore, T lymphocytes from the transgenic mice induced B cell proliferation much less efficiently than T lymphocytes from control mice, suggesting that C2GnT overexpression weakened the interaction between the T lymphocytes and B lymphocytes.

In summary, the studies conducted to date suggest an important role for C2GnT activity in several aspects of normal immune system function. However, it is unclear which specific C2GnT enzyme (i.e. C2GnT-L, -M, or -3) is primarily responsible for this. The results of deletion of the C2GnT-L gene from the murine germline indicate that C2GnT-L does play a role, but that it is likely not the only C2GnT enzyme involved. Furthermore, Northern blotting analysis on C2GnT-L (17, 18) reveals very low expression in immune system organs such as thymus, bone marrow, lymph nodes and leukocytes, with the exception of spleen. Thus, it is probable that other C2GnT isozymes, such as C2GnT-3 or other as yet unidentified C2GnT isozymes, are important in immune function, and that selectin ligands created independently of C2GnT also play a role.

IV.A.2. ROLE OF C2GnT ACTIVITY IN DEVELOPMENT, DIFFERENTIATION AND DISEASE

Changes in C2GnT activity and the core 2 β1,6 branched structure have been examined during maturation of various cell lineages, including erythroid, myeloid and lymphoid cell types. In addition to the aforementioned decrease in core 2 structure during maturation of T lymphocytes (104, 105), Fukuda and coworkers (75) observed differences in core 2-based structure among different maturation stages of myeloid cells. Acute myelogenous leukemia cells, which represent an immature stage of myeloid development, showed a lack of core 2 β1,6 branching, whereas detectable levels of core 2 branched structure and polylactosamine were present in glycans from chronic myelogenous leukemia cells and normal granulocytes, which represent successively more mature myeloid cells.

Developmental changes in C2GnT expression have also been investigated within intact animals. Granovsky et al (108) recently examined changes in the tissue-specific expression of C2GnT-L during development of post-implantation mouse embryos using RNA *in situ* hybridization. C2GnT-L expression was widespread prior to organogenesis at embryonic day 7, then became more restricted between day 11 and day 17 to mucin-secreting tissues such as kidney proximal tubules, pancreas, the respiratory tract, the small intestine, and the isthmus of adult oviduct, as well as in cartilage. Notably, C2GnT-L transcripts were not detected in embryonic thymus. To account for the documented presence of core 2-based glycans on thymocytes (104), the authors suggested that thymic C2GnT-L may be expressed at a later, post-embryonic stage. Another possibility, however, could involve C2GnT-3 being the principal C2GnT isoform involved in thymocyte maturation, given its heavy expression in thymic tissue (16).

Several reports have shown that C2GnT activity and the degree of core 2 branching change during cellular differentiation. In general, the results of these studies indicate that differentiation leads to elevated C2GnT activity and greater complexity of mucin glycan structure. For example, as mentioned previously, Heffernan et al (78) observed a 13-fold increase in C2GnT activity upon retinoic acid-induced differentiation of murine F9 teratocarcinoma cells. Furthermore, the increased C2GnT activity was accompanied by a significant increase in polylactosamine expression. In another study, Datti and Dennis (109) witnessed a 16-fold increase in C2GnT activity in CHO cells induced to differentiate by treatment with sodium butyrate. Intriguingly, Km for the sugar nucleotide substrate was increased 9-fold by the butyrate treatment, in spite of the increased enzyme activity. Cotreatment with butyrate and protein kinase inhibitors abolished the increase in

C2GnT activity, suggesting the involvement of a protein kinase in the upregulation of C2GnT, possibly via post-translational modification of the C2GnT protein. Along these same lines, Amano and Oshima (110) recently showed that butyrate-induced differentiation of CaCo-2 colon carcinoma cells leads to elevated C2GnT activity and an increase in core 2-based glycans on glycoproteins in the cells. In subsequent work, it was found that butyrate reduced the Km of C2GnT for its disaccharide acceptor, which may account for the observed enhancement of C2GnT activity (111). These findings, combined with those of Datti and Dennis (109), suggest the existence of different mechanisms in different cell types by which butyrate can enhance C2GnT activity.

Although cellular differentiation often results in increased C2GnT activity, it should be mentioned that this is not always the case. For example, as described earlier, Nakamura et al (91) observed a 10-fold decrease in C2GnT activity upon induction of differentiation in a pre-B lymphocytic leukemia cell line using the phorbol ester TPA.

Altered C2GnT activity has been implicated in a number of disorders, ranging from certain immunodeficiency syndromes to various forms of cancer. For example, a clear role has been shown for C2GnT in the rare X-linked recessive immunodeficiency Wiskott-Aldrich Syndrome (WAS). WAS patients have greatly reduced numbers of platelets and severely compromised humoral and cellular immunity. T lymphocytes in these patients show highly abnormal morphology and function, and lymphoid tumors are common (112-115). A clue as to the molecular defect responsible for WAS came from the observation that CD43 on WAS T lymphocytes is expressed at lower levels than normal and shows aberrant electrophoretic mobility characteristics (116, 117). CD43 is involved in T lymphocyte activation and regulates lymphocyte adhesion (118, 119). Piller et al (120) showed that CD43 in resting WAS T lymphocytes carries the core 2 β1,6 branched structure, whereas CD43 from normal resting lymphocytes lacks this structure:

NeuAcα2,6
|
NeuAcα2,3Galβ1,3GalNAcα-S/T

CD43 O-Glycan Structure in Normal Resting T Lymphocytes

NeuAcα2,3Galβ1,4GlcNAcβ1,6
|
NeuAcα2,3Galβ1,3GalNAcα-S/T

CD43 O-Glycan Structure in WAS Resting T Lymphocytes

Thus, WAS T lymphocyte CD43 O-glycans constitutively display the structure normally found only on activated T lymphocytes, and this difference accounted for the aberrant electrophoretic mobility of WAS CD43. These investigators also found that C2GnT activity in WAS resting T lymphocytes was 9-fold higher than in normal resting T lymphocytes, consistent with the observed O-glycan structural differences. Higgins et al (121) then observed that activation of normal T cells was accompanied by an increase in C2GnT activity from 0.8 to 2.2 nmol/hr/mg, as expected, but that activation led to a 60% decrease in C2GnT activity in WAS T cells, going from 2.3 to 0.9 nmol/hr/mg. Thus, in terms of C2GnT activity, WAS T lymphocytes show a phenotype inversion compared to normal T cells. That is, resting WAS T lymphocytes resemble activated normal T lymphocytes, and active WAS T lymphocytes behave as resting normal T lymphocytes.

Changes in glycosyltransferase activities in platelets and B lymphocytes also may be involved in the WAS phenotype. Higgins et al (121) found that C2GnT activity was 3-fold higher in WAS platelets than normal platelets, and at least 6-fold lower in WAS B cells than normal B cells. Furthermore, activity of the α2,6 sialyltransferases ST6GalNAc

I and II, which compete with C2GnT for the same site on GalNAc, was at least 20-fold higher in WAS B cells than normal B cells. Consistent with these findings, WAS B cell CD43 possessed less polylactosamine structure than normal B cell CD43.

A possible role for C2GnT in cancer progression has been examined in several studies, both in cell lines and tissues. For example, Yousefi et al (77) measured differences in C2GnT activity between three nonmalignant murine tumor cell lines, rat 2 fibroblasts, KBL-1 lymphoma and SP1 mammary carcinoma, and several closely related metastatic cell lines. The malignant counterparts of rat 2 and SP1 showed a 70% increase in C2GnT activity, while other glycosyltransferases were not appreciably affected. This suggested a correlation between increased C2GnT activity and metastatic potential. In a similar study, Vavasseur et al (122) looked at differences in glycosytransferase activities between three colonic cell lines differing in their degree of malignancy. In contrast to the report by Yousefi et al (77), they found high C2GnT activity in all three cell lines, showing no significant correlation with metastatic phenotype. However, studies by Yang et al (123) and Shimodaira et al (95) comparing normal and cancerous colon tissues have shown a correlation between C2GnT expression and tumor invasiveness, although they also observed generally lower levels of C2GnT activity in the cancer tissue samples than in normal colon.

Changes in C2GnT activity have also been documented in cells from leukemia patients (124). In particular, leukocytes from patients with chronic myelogenous leukemia and acute myelogenous leukemia show 4-fold and 18-fold increased C2GnT activity, respectively, compared to leukocytes from normal donors. The authors proposed this as a possible contributing factor to the hypersialylation seen in leukemic leukocytes, since core 2 branching creates an additional carbohydrate chain, which is often terminated by sialic acid.

Several recent studies have examined changes in the glycosylation of mucins, particularly MUC1, in association with cancer. MUC1 is commonly overexpressed on the surface of certain cancer cells, notably breast and pancreatic cancers, and generally possesses fewer and shorter mucin glycan chains than MUC1 from normal cells (125-127). Changes in C2GnT activity appear to be at least partly responsible for this effect. For example, MUC1 from T47D breast cancer cells lacks the core 2 β1,6 branched structure, expressing primarily a nonsialylated or monosialylated core 1 structure (127), whereas MUC1 from the normal breast cell line MTSV1-7 possesses core 2-based glycans. This difference is likely a result of differences in C2GnT activity, as shown by Brockhausen et al (128). These investigators also reported 10-fold higher activity of α2,3 sialyltransferase in the T47D cells versus the normal MTSV1-7 cells, which could also contribute to the lack of core 2-based glycan structure in T47D, since this enzyme competes with C2GnT for the core 1 acceptor (figure 1). A role for α2,3 sialyltransferase in generating underglycosylated MUC1 was further substantiated by Whitehouse et al (129), who showed that overexpression of this enzyme in the MTSV1-7 cell line blocked expression of core 2-based O-glycans on MUC1. Taking a complementary approach, Beum et al (130) overexpressed a C2GnT-L cDNA in a pancreatic cancer cell line that normally expresses a tumor-associated carbohydrate antigen which lacks a core 2 β1,6 branch, detected by the CC49 antibody (131). Introduction of C2GnT-L blocked expression of this antigen, masked a MUC1 peptide-based tumor antigen, and generated sialyl Lewis x structure. These findings, along with those of Whitehouse et al (129), suggest that certain cancer cells may manipulate their expression of C2GnT to produce underglycosylated mucin and alter their antigenic profile.

Taken together, the available evidence suggests no clear relationship between C2GnT activity and cancer, since both overexpression and underexpression of C2GnT activity are seen in different cancer cell types. On the other hand, it seems clear that sialyl Lewis x plays a role in tumor cell adhesion and invasion of surrounding tissue, which suggests an important role for C2GnT activity in tumor progression because of the clear association between C2GnT activity and sialyl Lewis x. Further studies are required to elucidate the exact role of C2GnT activity and the core 2 branched structure in tumor cell metastasis and cancer progression.

Altered C2GnT activity has recently been implicated as a potential contributing factor in several other disorders, including AIDS (132), multiple sclerosis (133) and diabetes (69, 134, 135). In the case of diabetes, enhanced levels C2GnT-L mRNA were demonstrated in diabetic rat heart by a differential display method (69), and C2GnT activity was also increased in diabetic rat heart. Furthermore, transgenic mice overexpressing C2GnT-L in the heart developed cardiac hypertrophy (134), and the authors postulated that C2GnT expression may affect expression of cardiac stress markers such as c-*fos*. In keeping with the upregulation of diabetic heart C2GnT activity, Panicot et al (135) showed that serum from newly diagnosed type I diabetic patients contain an antibody directed against a core 2-based epitope present on a pancreatic enzyme, bile salt-dependent lipase.

IV.B BIOLOGICAL SIGNIFICANCE OF I-TYPE β1,6 BRANCHED STRUCTURES

IV.B.1 BIOLOGICAL ROLES OF I-TYPE β1,6 BRANCHED STRUCTURE AND POTENTIAL APPLICATIONS OF SYNTHETIC I-TYPE STRUCTURES

Polylactosamine chains bearing I-type β1,6 GlcNAc branches are widespread in nature, and many contain multiple I-type branches (39, 47-49, 59-61, 136, 137). For instance, Fukuda et al (59) have detected polylactosamine chains on embryonic carcinoma cells containing as many as nine I-type β1,6 GlcNAc branches. Based on these observations, it stands to reason that I-type β1,6 GlcNAc branching serves an important physiologic function, and several studies indicate that this may indeed be the case. For example, polylactosamine structures have been shown to be important constituents of cell surface receptors for Staphylococcal α-toxin (138), Sendai virus (139), *Mycoplasma pneumoniae* (140), and Pneumococci (141). The receptors for two of these (139, 140) are known to possess I β1,6 GlcNAc branches, and it appears that the β 1,6 branches are essential for receptor function in these cases.

As discussed earlier in this review, numerous functional structures, such as blood group antigens and the selectin ligand sialyl Lewis x, are frequently expressed at the nonreducing termini of polylactosamine chains. The ability of IGnT to synthesize multiply-branched polylactosamine suggests the potential for creation of polylactosamine chains bearing a large number of these functional structures in cells. Several studies have shown that such multivalency can greatly enhance the strength of interaction of a ligand with its cognate receptor. For example, certain antibodies, including those directed against the ABH blood group antigens (142), are able to circumvent the problem of low intrinsic binding affinity through interaction of both combining sites of the antibody with two haptens present on a single polylactosamine-based carrier molecule, potentially increasing the binding affinity by several orders of magnitude (143-145). Multivalent interactions may also play a role in the decompaction of preimplantation mouse embryos, which occurs during murine development. Decompaction roughly coincides with numerous developmental changes in carbohydrate structures on embryonic cell surfaces, including changes in Ii antigen expression, and Fenderson et al (146) showed that decompaction was facilitated greatly by a trivalent lacto-type conjugate, but not by the corresponding free monovalent lacto structure. Furthermore, the bivalent oligosaccharide

Galα1,3Galβ1,4GlcNAcβ1,3Galβ1,4GlcNAcβ1,6
\
Galα1,3Galβ1,4GlcNAcβ1,3Galβ1,4GlcNAcβ1,3Galβ1,4GlcNAc

acts as a much more effective inhibitor of mouse gamete adhesion *in vitro* than does the related unbranched monovalent structure (147).

Numerous studies using synthetic multivalent sialyl Lewis x have demonstrated the superior selectin-binding capacity of such structures relative to monovalent sialyl Lewis x. For instance, Welpy et al (148) showed that preincubation of immobilized E-selectin with multivalent sialyl Lewis x-BSA conjugates inhibits binding of HL-60 cells to the E-selectin as much as 1000-fold more efficiently than preincubation with free sialyl Lewis x. These multivalent sialyl Lewis x analogs may find application as therapeutic agents, such as in the treatment of various inflammatory conditions.

Ossi Renkonen and coworkers have focused on the development of L-selectin inhibitors, since this selectin is critical for mediating interactions between lymphocytes and endothelium during inflammation and organ rejection. These investigators have synthesized a large number of multivalent sialyl Lewis x oligosaccharides and shown them to be effective in inhibiting lymphocyte binding to lymph node high endothelium (149), as well as lymphocyte-graft endothelial cell interactions *in vitro* during kidney transplant (150, 151) and cardiac transplant (152, 153). In all cases, the inhibitory efficacy correlated with the degree of multivalency, and effective inhibition of lymphocyte-endothelium interaction was attainable using low nanomolar concentrations of tetravalent sialyl Lewis x analogs (149, 153).

IV.B.2 CHANGES IN I-TYPE β1,6 BRANCHED STRUCTURE IN DEVELOPMENT AND DISEASE

As mentioned earlier, a conversion from primarily i-type structure to I-type structure on human red cells occurs during the first 18 months of development post-birth (30). This developmental change has subsequently been confirmed in several studies (154-157). Developmental changes in I and i antigen expression have also been examined in mouse, and appear to differ from what is seen in humans. For example, Muramatsu et al (158) observed that differentiation of murine embryonic carcinoma cells, a process which mimics the early stages of murine development, is accompanied by a substantial decrease in the size and complexity of carbohydrate structure. Although not conclusive, these findings could signal a change from I-type structure in the cells to i-type structure, opposite to the transition which occurs in human development. More compelling is the work of Kapadia et al (159), who used anti-I and anti-i antibodies to show that undifferentiated mouse embryos and murine teratocarcinoma cells express primarily the I antigen, and that differentiation into primary endoderm results in upregulation of the i antigen.

Mice and humans also differ in the changes which occur in their i and I blood group antigen expression during maturation of a cell lineage at a given developmental stage of the organism. Kannagi et al (160) showed that an i-to-I conversion occurs during maturation of murine leukemic cells into a macrophage-like cell. In contrast, Fukuda et al (161) found no significant change in the amount of β1,6 GlcNAc branched polylactosamine during maturation of human erythroblasts into erythrocytes in adults.

The level of I antigen also appears to change during oncogenesis. Mucin-type glycan structure in cancer tissue is often less complex than in normal tissue, as discussed previously. Consequently, glycoproteins from tumor tissue extracts often lack the complete ABO blood group structures seen in normal tissue (162, 163), resulting in exposure of the normally cryptic I and i blood group precursors. This was initially demonstrated by Feizi et al (164), who detected the I and i antigens in colon tumor extracts. Similarly, Watanabe and Hakomori (154) used antibodies specific for the blood group H antigen or sequentially degraded derivative structures to show that, compared to normal colon tissue, colonic cancer extracts possess lower levels of complete blood group H structure and substantially higher levels of i-type and I-type precursor structures. Furthermore, Picard et al (165) and Picard and Feizi (166) have shown that gastric carcinoma extracts also express higher levels of I antigen than non-cancerous gastric mucosa, and Barnett and Clark (60) have shown a high degree of I-type β1,6 GlcNAc branching in glycolipid-associated polylactosamine chains in a pancreatic cancer cell line. In an exhaustive comparison of normal and cancer tissues, Nudelman et al (47) found that the monoclonal antibody NUH2, which specifically detects a glycan structure containing an I-type β1,6 GlcNAc branch, reacted strongly with 16 separate colonic tumor extracts, as well as with tumor extracts from breast, stomach, lung, kidney, gallbladder, and pancreas. NUH2 failed to react with normal extracts from pancreas, brain, kidney or large intestine, although it did react strongly with extracts from several other normal tissues including placenta, type A erythrocytes, and sperm.

In summary, it appears that cancer tissues show significantly higher levels of I antigen than normal tissues. This probably results primarily from increased exposure of the antigen due to incomplete synthesis of blood group structures, although there may be a contribution from increased expression of polylactosamine structure as well. In support of the latter, Holmes et al (167) showed that expression of i β1,3 N-acetylglucosaminyltransferase, which synthesizes the GlcNAcβ1,3-Gal linkage in polylactosamine chains, is greatly elevated in colon carcinoma, resulting in greater synthesis of polylactosamine structure.

SUMMARY

The contribution of carbohydrate structure to biomolecular, cellular, and organismal function is well-established, but has not yet received the attention it deserves, perhaps due to the complexity of the structures involved and to a lack of simple experimental methods for relating structure and function. In particular, β1,6 GlcNAc branching plays a key functional role in processes ranging from inflammation and immune system function to tumor cell metastasis. For instance, synthesis of the core 2 β1,6 branched structure in the mucin glycan chain by C2GnT enables the expression of functional structures at the termini of polylactosamine chains, such as blood group antigens and sialyl Lewis x. Also, IGnT can create multiple branches on the polylactosamine chain, which may serve as a mechanism for amplifying the functional potency of cell surface glycoproteins and glycolipids. The family of enzymes which creates β1,6 branched structure in mucin glycans is proving to be quite complex, since multiple isoforms appear to exist for these enzymes, and some of the enzymes are adept at forming more than one type of β1,6 branched structure, as in the case of C2GnT-M. Furthermore, the enzymes do not appear to be restricted to acting on mucin-type acceptor structures, but are able to act on glycolipid structures as well. Much remains to be learned regarding the specific biological niche filled by each of these enzymes and how their activities complement one another, as well as the manner in which the activities of these enzymes are regulated in the cell.

ACKNOWLEDGMENTS

We wish to acknowledge the NIH for funding support (RO1 HL48282) of the work on bovine core 2 branching enzymes described in this article.

REFERENCES

1. Clausen, H. and Bennett, E.P. (1996) A family of UDP-GalNAc:polypeptide *N*-acetylgalactosaminyltransferases control the initiation of mucin-type O-linked glycosylation. *Glycobiology* **6**, 635-646.
2. Elhammer, A.P., Poorman, R.A., Brown, E., Maggiora, L.L., Hoogerheide, J.G. and Kezdy, F.J. (1993) The specificity of UDP-GalNAc:polypeptide *N*-acetylgalactosaminyltransferase as inferred from a database of *in vivo* substrates and from the in vitro glycosylation of proteins and peptides. *J. Biol. Chem.* **268**, 10029-10038.
3. Wang, Y., Agrwal, N., Eckhardt, A.E., Stevens, R.D. and Hill, R.L. (1993) The acceptor specificity of porcine submaxillary UDP-GalNAc:polypeptide *N*-acetylgalactosaminyltransferase is dependent on the amino acid sequences adjacent to serine and threonine residues. *J. Biol. Chem.* **268**, 22979-22983.
4. Wandall, H.H., Hassan, H., Mirgorodskaya, E., Kristensen, A.K., Roepstorff, P., Bennett, E.P., Nielsen, P.A., Hollingsworth, M.A., Burchell, J., Taylor-Papasimitriou, J., and Clausen, H. (1997) Substrate specificities of three members of the human UDP- *N*-acetyl-α-D-galactosamine:Polypeptide *N*-acetylgalactosaminyltransferase Family, GalNAc-T1, -T2, and -T3. *J. Biol. Chem.* **272**, 23503-23514.
5. Nishimori, I., Johnson, N.R., Sanderson, S.D., Perini, F., Mountjoy, K., Cerny, R.L., Gross, M.L. and Hollingsworth, M.A. (1994) Influence of acceptor substrate primary amino acid sequence on the activity of human UDP-*N*-acetylgalactosamine: Polypeptide *N*-acetylgalactosaminyltransferase. Studies with the MUC1 tandem repeat. *J. Biol. Chem.* **269**, 16123-16130.
6. Brockhausen, I. And Kuhns, W. (1997) *Glycoproteins and Human Disease,* pp. 5-11. R.G. Landes Co., Georgetown, TX.
7. Wilkins, P.P., McEver, R.P. and Cummings, R.D. (1996) Structures of the O-glycans on P-selectin glycoprotein ligand-1 from HL-60 cells. *J. Biol. Chem.* **271**, 18732-18742.
8. Rege, V.P., Painter, T.J., Watkins, W.M. and Morgan, W.T.J. (1963) Three new trisaccharides obained from human blood group A, B, H and Le^a substances: Possible sugar sequences in the carbohydrate chains. *Nature* **200**, 532-534.
9. Brockhausen, I. and Schachter, H. (1997). Glycosyltransferases involved in N- and O-glycan biosynthesis, in *Glycosciences* (H.-J and S. Gabius, Eds.). Chapman and Hall, Weinheim, pp. 79-113.
10. Van den Steen, P., Rudd, P.M., Dwek, R.A. and Opdenakker, G. (1998). Concepts and principles of O-linked glycosylation. *Crit. Rev. Biochem. Molec. Biol.* **33**, 151-208.
11. Strous, G.J. and Dekker, J. (1992). Mucin-type glycoproteins. *Crit. Rev. Biochem. Molec. Biol.* **27**, 57-92.
12. Schachter, H. and Brockhausen, I. (1992) The biosynthesis of serine (threonine)-*N*-acetylgalactosamine-linked carbohydrate moieties in *Glycoconjugates: Composition, Structure and Function* (H. Allen and E. Kisailus, Eds.). Dekker, New York, pp. 263-332.
13. Brockhausen, I. (1997) Biosynthesis and functions of O-glycans and regulation of mucin antigen expression in cancer. *Biochem. Soc. Trans.* **25**, 871-874.
14. Bierhuizen, M.F.A, Mattei, M.-G. and Fukuda, M. (1993). Expression of the developmental I antigen by a cloned human cDNA encoding a member of a β–1,6-*N*-acetylglucosaminyltransferase gene family. *Genes and Dev.* **7**, 468-478.
15. Bierhuizen, M.F.A and Fukuda, M. (1992) Expression cloning of a cDNA encoding UDP-GlcNAc:Galβ1-3GalNAc-R(GlcNAc-to-GalNAc)β1-6 GlcNAc transferase by gene transfer into CHO cells expressing polyoma large tumor antigen. *Proc. Natl. Acad. Sci. USA* **89**, 9326-9330.
16. Schwientek, T., Yeh, J.-C., Levery, S.B., Keck, B., Merkx, G., van Kessel, A.G., Fukuda, M. and Clausen, H. (2000) Control of O-glycan branch formation: Molecular cloning and characterization of a novel thymus-associated core 2 β1,6-*N*-acetylglucosaminyltransferase. *J. Biol. Chem.* **275**, 11106-11113.
17. Schwientek, T., Nomoto, M., Levery, S.B., Merkx, G., van Kessel, A.G., Bennett, E.P., Hollingsworth, M.A. and Clausen, H. (1999) Control of O-glycan branch formation. Molecular cloning of human cDNA encoding a novel β-1,6-*N*-acetylglucosaminyltransferase forming core 2 and core 4. *J. Biol. Chem.* **274**, 4504-4512.

18. Yeh, J.-C., Ong, E. and Fukuda, M. (1999) Molecular cloning and expression of a novel β-1,6-*N*-acetylglucosaminyltransferase that forms core 2, core 4 and I branches. *J. Biol. Chem.* **274**, 3215-3221.
19. Williams, D. and Schachter, H. (1980) Mucin synthesis. I. Detection in canine submaxillary glands of an *N*-acetylglucosaminyltransferase which acts on mucin substrates. *J. Biol. Chem.* **255**, 11247-11252.
20. Williams, D., Longmore, G., Matta, K.L. and Schachter, H. (1980) Mucin synthesis. II. Substrate specificity and product identification studies on canine submaxillary glandUDP-GlcNAc:Galβ1-3GalNAc(GlcNAc-to-GalNAc)β6acetylglucosaminyltransferase. *J. Biol. Chem.* **255**, 11253-11261.
21. Kuhns, W., Rutz, V., Paulsen, H., Matta, K.L., Baker, M.A., Barner, M., Granovsky, M. and Brockhausen, I. (1993) Processing O-glycan core 1, Galβ1-3GalNAcα-R. Specificities of core 2, UDP-GlcNAc:Galβ1-3GalNAc-R(GlcNAc-to-GalNAc)β6-*N*-acetylglucosaminyltransferase and CMP -sialic acid: Galβ1-3GalNAc-Rα 2-3-sialyltransferase. *Glycoconj. J.* **10**, 381-394.
22. Hindsgaul, O., Kaur, K.J., Srivastava, G., Blaszcyzk-Thurin, M., Crawley, S.C., Heerze, L.D. and Palcic, M.M. (1991) Evaluation of deoxygenated oligosaccharide acceptor analogs as specific inhibitors of glycosyltransferases. *J. Biol. Chem.* **266**, 17858-17862.
23. Wingert, W.E. and Cheng, P.-W. (1984) Mucin biosynthesis. Characterization of rabbit small intestinal UDP-*N*-acetylglucosamine:Galactose β-3-*N*-acetylgalactosaminide (*N*-acetylglucosamine-to-*N*-acetylgalactosamine)β-6-*N*-acetylglucosaminyl-transferase. *Biochemistry* **23**, 690-697.
24. Cheng, P.-W., Wingert, W.E., Little, M.R. and Wei, R. (1985) Mucin biosynthesis. Properties of a bovine tracheal mucin β-6-*N*-acetylglucosaminyltransferase. *Biochem. J.* **227**, 405-412.
25. Sangadala, S., Sivakami, S. and Mendicino, J. (1991) UDP-GlcNAc:Galβ3GalNAc-mucin: (GlcNAc-to-GalNAc)β6-*N*-acetylglucosaminyltransferase and UDP-GlcNAc:Galβ3(GlcNAcβ6) GalNAc-mucin (GlcNAc-to-Gal)β3-*N*-acetylglucosaminyltransferase from swine trachea epithelium. *Mol. Cell. Biochem.* **101**, 125-143.
26. Brockhausen, I., Matta, K.L., Orr, J. and Schachter, H. (1985) UDP-GlcNAc:GalNAc-R β3-*N*-acetylglucosaminyltransferase and UDP-GlcNAc: GlcNAcβ1-3GalNAc-R(GlcNAc-to-GalNAc)β6-*N*-acetylglucosaminyltransferase from pig and rat colon mucosa. *Biochemistry* **24**, 1866-1874.
27. Brockhausen, I., Matta, K.L., Orr, J., Schachter, H, Koenderman, A.H.L. and van den Eijnden, D.H. (1986) Mucin synthesis. Conversion of R_1-β1-3Gal-R_2 to R_1-β1-3(GlcNAcβ1-6)Gal-R_2 and of R_1-β1-3GalNAc-R_2 to R_1-β1-3(GlcNAcβ1-6)GalNAc-R_2 by a β6-*N*-acetylglucosaminyltransferase in pig gastric mucosa. *Eur. J. Biochem.* **157**, 463-474.
28. Ropp, P.A., Little, M.R. and Cheng, P.-W. (1991) Mucin biosynthesis. Purification and characterization of a mucin β6*N*-acetylglucosaminyltransferase. *J. Biol. Chem.* **266**, 23863-23871.
29. Wiener, A.S., Unger, L.J., Cohen, L. and Feldman, J. (1956) Type-specific cold auto-antibodies as a cause of acquired hemolytic anemia and hemolytic transfusion reactions: Biologic test with bovine red cells. *Ann. Intern. Med.* **44**, 222-240.
30. Marsh, W.L. (1961) Anti-i: A cold antibody defining the Ii relationship in human red cells. *Brit. J. Hemat.* **7**, 200-209.
31. Vicari, G. and Kabat, E.A. (1970). Structures and activities of oligosaccharides produced by alkaline degradation of a blood group substance lacking A, B, H, Le^a, and Le^b specificities. *Biochemistry* **9**, 3414-3421.
32. Feizi, T., Kabat, E.A., Vicari, G., Anderson, B. and Marsh, W.L. (1971). Immunochemical studies on blood groups. XLVII. The I antigen complex: Precursors in the A, B, H, Le^a, and Le^b blood group system-hemagglutination-inhibition studies. *J. Exp. Med.* **133**, 39-52.
33. Feizi, T., Kabat, E.A., Vicari, G., Anderson, B. and Marsh, W.L. (1971). Immunochemical studies on blood groups. XLIX. The I antigen complex: specificity differences among anti-I sera revealed by quantitative precipitin studies; partial structures of the I determinant specific for one anti-I serum. *J. Immunol.* **106**, 1578-1592.
34. Watanabe, K., Laine, R.A. and Hakomori, S.-I. (1975). On neutral fucoglycolipids having long, branched carbohydrate chains: H-active and I-active glycosphingolipids of human erythrocyte membranes. *Biochemistry* **14**, 2725-2733.
35. Feizi, T., Wood, E., Auge, C., David, S. and Veyrieres, A. (1978) Blood group I activities of synthetic oligosaccharides assessed by radioimmunoassay. *Immunochem.* **15**, 733-736.
36. Niemann, H., Watanabe, K. and Hakomori, S.-I. (1978). Blood group i and I activities of "lacto-*N*-*nor*hexaosylceramide" and its analogues: The structural requirements for i-specificities. *Biochem. Biophys. Res. Comm.* **81**, 1286-1293.
37. Fukuda, M., Fukuda, M.N. and Hakamori, S.-I. (1979). Developmental change and genetic defect in the carbohydrate structure of Band 3 glycoprotein of human erythrocyte membrane. *J. Biol. Chem.* **254**, 3700-3703.

38. Okada, Y., Kannagi, R., Levery, S.B. and Hakomori, S.-I. (1984) Glycolipid antigens with blood group I and i specificities from human adult and umbilical cord erythrocytes. *J. Immunol.* **133**, 835-842.
39. Fukuda, M., Dell, A., Oates, J.E. and Fukuda, M.N. (1984) Structure of branched lactosaminoglycan, the carbohydrate moiety of Band 3 isolated from adult human erythrocytes. *J. Biol. Chem.* **259**, 8260-8273.
40. Van den Eijnden, D.H., Winterwerp, H., Smeeman, P. and Schiphorst, W.E.C.M. (1983). Novikoff ascites tumor cells contain *N*-acetyllactosaminide β1-3 and β1-6 *N*-acetylglucosaminyltransferase activity. *J. Biol. Chem.* **258**, 3435-3437.
41. Basu M. and Basu, S. (1984). Biosynthesis *in vitro* of Ii core glycosphingolipids from neolacto-tetraosylceramide by β1-3- and β1-6-*N*-acetylglucosaminyltransferases from mouse T-lymphoma. *J. Biol. Chem.* **259**, 12557-12562.
42. Zielinski, J and Koscielak, J. (1983) The occurrence of two novel *N*- acetylglucosaminyltransferase activities in human serum. *FEBS Lett.* **158**, 164-168.
43. Zielinski, J and Koscielak, J. (1983) Sera of i subjects have the capacity to synthesize the branched GlcNAcβ1,6[GlcNAcβ1,3]Gal structure. *FEBS Lett.* **163**, 114-118.
44. Piller, F., Cartron, J.-P., Maranduba, A., Veyrieres, A., Leroy, Y. and Fournet, B. (1984). Biosynthesis of blood group I antigens: Identification of a UDP-GlcNAc:GlcNAcβ1-3Gal(-R)β1-6(GlcNAc to Gal) *N*-acetylglucosaminyltransferase in hog gastric mucosa. *J. Biol. Chem.* **259**, 13385-1390.
45. Helin, J., Pentilla, L., Leppanen, A., Maaheimo, H., Lauri, S., Costello, C.E. and Renkonen, O. (1997) The β1,6-GlcNAc transferase activity present in hog gastric mucosal microsomes catalyses site-specific branch formation on a long polylactosamine backbone. *FEBS Lett.* **412**, 637-642.
46. Koenderman, A.H.L., Koppen, P.L. and van den Eijnden, D.H. (1987) Biosynthesis of polylactosaminoglycans. Novikoff ascites tumor cells contain two UDP-GlcNAc:β-galactoside β1,6-*N*-acetylglucosaminyltransferase activities. *Eur. J. Biochem.* **166**, 199-208.
47. Nudelman, E.D., Mandel, U., Levery, S.B., Kaizu, T. and Hakomori, S.-I. (1989) A series of disialogangliosides with binary 2-3 sialosyllactosamine structure, defined by monoclonal antibody NUH2, are oncodevelopmentally regulated antigens. *J. Biol. Chem.* **264**, 18719-18725.
48. Egge, H., Kordowicz, M., Peter-Katalinic, J. and Hanfland, P. (1985) Immunochemistry of I/i-active oligo- and polyglycosylceramides from rabbit erythrocyte membranes. Characterization of linear, di-, and triantennary neolactoglycosphingolipids. *J. Biol. Chem.* **260**, 4927-4935.
49. Levery, S.B., Nudelman, E.D., Salyan, M.E.K. and Hakomori, S.-I. (1989) Novel tri- and tetrasialosylpoly-*N*-acetyllactosaminyl gangliosides of human placenta: Structure determination of pentadeca- and eicosaglycosylceramides by methylation analysis, fast atom bombardment mass spectrometry, and ^{1}H NMR spectroscopy. *Biochemistry* **28**, 7772-7781.
50. Hanfland, P., Egge, H., Dabrowski, U., Kuhn, S., Roelcke, D. and Dabrowski, J. (1981) Isolation and characterization of an I-active ceramide decasaccharide from rabbit erythrocyte membranes. *Biochemistry* **20**, 5310-5319.
51. Watanabe, K., Powell, M. and Hakomori, S.-I. (1978) Isolation and characterization of a novel fucoganglioside of human erythrocyte membrane. *J. Biol. Chem.* **253**, 8962-8967.
52. Watanabe, K., Powell, M. and Hakomori, S.-I. (1978) Isolation and characterization of gangliosides with a new sialosyl linkage and core structures. II. Gangliosides of human erythrocyte membranes. *J. Biol. Chem.* **254**, 8223-8229.
53. Leppanen, A., Pentilla, L. Niemala, R., Helin, J., Seppo, A., Lusa, S. and Renkonen, O. (1991) Human serum contains a novel β1-6 *N*-acetylglucosaminyltransferase activity that is involved in midchain branching of oligo(*N*-acetyllactosaminoglycans). *Biochemistry* **30**, 9287-9296.
54. Sakamoto, Y., Taguchi, T., Tano, Y., Ogawa, T., Leppanen, A., Kinnunen, M., Aitio, O., Parmanne, P., Renkonen, O. and Taniguchi, N. (1998). Purification and characterization of UDP-GlcNAc:Galβ1-4-GlcNAcβ1-3Galβ1-4Glc(NAc)-R(GlcNAc to Gal) acting β1,6-*N*-acetylglucosaminyltransferase from hog small intestine. *J. Biol. Chem.* **273**, 27625-27632.
55. Gu, J., Nishikawa, A., Fujii, S., Gasa, S. and Taniguchi, N. (1992). Biosynthesis of blood group I and i antigens in rat tissues: Identification of a novel β1-6-*N*-acetylglucosaminyltransferase. *J. Biol. Chem.* **267**, 2994-2999.
56. Leppanen, A., Salminen, H., Zhu, Y., Maaheimo, H., Helin, J., Costello, C.E. and Renkonen, O. (1997) *In vitro* biosynthesis of a decasaccharide prototype of multiply branched polylactosaminoglycan backbones. *Biochemistry* **36**, 7026-7036.
57. Mattila, P., Salminen, H., Hirvas, L., Niittymaki, J., Salo, H., Niemela, R., Fukuda, M., Renkonen, O. and Renkonen, R. (1998). The centrally acting β–1,6-*N*-acetylglucosaminyltransferase (GlcNAc to Gal): Functional expression, purification, and acceptor specificity of a human enzyme involved in midchain branching of linear poly-*N*-acetyllactosamines. *J. Biol. Chem.* **273**, 27633-27639.

58. Ujita, M., McAuliffe, J., Suzuki, M., Hindsgaul, O., Clausen, H., Fukuda, M.N. and Fukuda, M. (1999) Regulation of I-branched poly-*N*-acetyllactosamine synthesis: Concerted actions by i-extension enzyme, I-branching enzyme, and β1,4-galactosyltransferase. *J. Biol. Chem.* **274**, 9296-9304.
59. Fukuda, M.N., Dell, A., Oates, J.E. and Fukuda, M. (1985). Embryonal lactosaminylglycan: The structure of branched lactosaminoglycans with novel disialosyl (sialyl α2-9 sialyl) terminals isolated from PA1 human embryonic carcinoma cells. *J. Biol. Chem.* **260**, 6623-6631.
60. Barnett, T. and Clark, G.F. (1992) Polyglycosylceramides with branched *N*-acetylactosamine sequences are synthesized by the human pancreatic carcinoma cell line Panc-1. *J. Biol. Chem.* **267**, 11760-11768.
61. Dabrowski, U., Hanfland, P., Egge, H., Kuhn, S. and Dabrowski, J. (1984) Immunochemistry of I/i-active oligo- and polyglycosylceramides from rabbit erythrocyte membranes. Determination of branchng patterns of a ceramide pentadecasaccharide by ^{1}H nuclear magnetic resonance. *J. Biol. Chem.* **259**, 7648-7651.
62. Vilkman, A., Niemala, R., Penttila, L., Helin, J., Leppanen, Aseppo, A., Maaheimo, H., Lusa, S. and Renkonen, O. (1992) Elongation of both branches of biantennary backbones of oligo-(*N*-acetyllactosamino)glycans by human serum 1,3-*N*-acetyl-β-D-glucosaminyltransferase. *Carbohydr. Res.* **226**, 155-174.
63. Renkonen, O., Leppanen, A., Niemala, R., Vilkman, A., Helin, J., Penttila, L., Maaheimo, H., Seppo, A. and Suopanki, J. (1992) Enzymatic *in vitro* synthesis of radiolabeled pentadecasaccharides GlcNAc β1,3(Galβ1,4GlcNAcβ1,6)Gal β1,4GlcNAc/Glc and the isomeric Galβ1,4GlcNAc β1,3(GlcNAcβ1,6)Gal β1,4GlcNAc/Glc. *Biochem. Cell Biol.* **70**, 86-89.
64. Blanken, W.M., Hooghwinkel, G.J. and van den Eijnden, D.H. (1982) Biosynthesis of blood-group I and i substances. Specificity of bovine colostrum β-*N*-acetyl-D-glucosaminide β 1,4 galactosyl-transferase. *Eur. J. Biochem.* **127**, 547-552.
65. Bierhuizen, M.F., Maemura, K. and Fukuda, M. (1995) Isolation and characterization of a pseudogene related to human core 2 β1-6 *N*-acetylglucosaminyltransferase. *Glycoconj. J.* **12**, 857-864.
66. Sekine, M., Nara, K. and Suzuki, A. (1997) Tissue-specific regulation of mouse core 2 β-1,6-*N*-acetylglucosaminyltransferase. *J. Biol. Chem.* **272**, 27246-27252.
67. Li, C.-M., Adler, K.B. and Cheng, P.-W. (1998) Mucin biosynthesis: Molecular cloning and expression of bovine lung mucin core 2 *N*-acetylglucosaminyltransferase cDNA. *Am. J. Resp. Cell Mol. Biol.* **18**, 343-352.
68. Li, C.-M., Joshee, N., Adler, K.B., and Cheng, P.-W. (1999) Development of monoclonal antibodies against bovine mucin core 2 β6 *N*-acetylglucosaminyltransferase. *Glycoconj. J.* **16**, 555-562.
69. Nishio, Y., Warren, C.E., Buczek-Thomas, J.A., Rulfs, J., Koya, D., Aiello, L.P., Feener, E.P., Miller, Jr., T.B., Dennis, J.W. and King, G.L. (1995) Identification and characterization of a gene regulating enzymatic glycosylation which is induced by diabetes and hyperglycemia specifically in rat cardiac tissues. *J. Clin. Invest.* **96**, 1759-1767.
70. Sasaki, K., Kurata-Miura, K., Ujita, M., Angata, K., Nakagawa, S., Sekine, S., Nishi, T. and Fukuda, M. (1997) Expression cloning of cDNA encoding a human β-1,3-*N*-acetylglucosaminyltransferase that is essential for poly-*N*-acetyllactosamine synthesis. *Proc. Natl. Acad. Sci. USA* **94**, 14294-14299.
71. Bierhuizen, M.F.A, Maemura, K., Kudo, S. and Fukuda, M. (1995). Genomic organization of core 2 and I branching β–1,6-*N*-acetylglucosaminyltransferases. Implications for evolution of the β-1,6-*N*-acetylglucosaminyltransferase gene family. *Glycobiology* **5**, 417-425.
72. Piller, F., Piller, V., Fox, R. and Fukuda, M. (1988) Human T-lymphocyte activation is associated with changes in O-glycan biosynthesis. *J. Biol. Chem.* **263**, 15146-15150.
73. Vanderplasschen, A., Markine-Gorlaynoff, N., Lomonte, P., Suzuki, M., Hiraoka, N., Yeh, J.-C., Bureau, F., Willems, L., Thiry, E., Fukuda, M., Pastoret, P.-P (2000) A multipotential β-1,6-*N*-acetylglucosaminyltransferase is encoded by bovine herpesvirus type 4. *Proc. Natl. Acad. Sci. USA* **97**, 5756-5761.
74. Toki, D., Sarkar, M., Yip, B., Reck, F., Joziasse, D., Fukuda, M., Schachter, H. and Brockhausen, I. (1997) Expression of stable human O-glycan core 2 β-1,6-*N*-acetylglucosaminyltransferase in Sf9 insect cells. *Biochem. J.* **325**, 63-69.
75. Fukuda, M., Carlsson, S.R., Klock, J.C. and Dell, A. (1986) Structures of O-linked oligosaccharides isolated from normal granulocytes, chronic myelogenous leukemia cells, and acute myelogenous leukemia cells. *J. Biol. Chem.* **261**, 12796-12806.
76. Maemura, K. and Fukuda, M. (1992) Poly-*N*-acetyllactosaminyl O-glycans attached to leukosialin. The presence of sialyl Le^x structures in O-glycans. *J. Biol. Chem.* **267**, 24379-24386.
77. Yousefi, S., Higgins, E., Daoling, Z., Pollex-Kruger, A., Hindsgaul, O. and Dennis, J.W. (1991) Increased UDP-GlcNAc:Galβ1-3GalNAc-R(GlcNAc-to-GalNAc)β1,6-*N*-acetylglucosaminyltrans-ferase activity in metastatic murine tumor cell lines. *J. Biol. Chem.* **266**, 1772-1782.

78. Heffernan, M., Lotan, R., Amos, B., Palcic, M., Takano, R. and Dennis, J.W. (1993) Branching β1,6-*N*-acetylglucosaminyltransferases and polylactosamine expression in mouse F9 teratocarcinoma cells and differentiated counterparts. *J. Biol. Chem.* **268**, 1242-1251.
79. Easton, R.L., Patankar, M.S., Lattanzio, F.A., Leaven, T.H., Morris, H.R., Clark, G.F. and Dell, A. (2000) Structural analysis of murine zona pellucida glycans. Evidence for the expression of core 2-type O-glycans and the Sd(a) antigen. *J. Biol. Chem.* **275**, 7731-7742.
80. Bierhuizen, M.F.A., Maemura, K., and Fukuda, M. (1994) Expression of a differentiation antigen and poly-*N*-acetyllactosaminyl O-glycans directed by a cloned core 2 β1,6-*N*-acetylglucosaminyltransferase. *J. Biol. Chem.* **269**, 4473-4479.
81. Furukawa, K., Funakoshi, Y., Autero, M., Horejsi, V., Kobata, A. and Gahmberg, C.G. (1998) Structural study of the O-linked sugar chains of human leukocyte tyrosine phosphatase CD45. *Eur. J. Biochem.* **251**, 288-294.
82. McEver, R.P., Moore, K.L. and Cummings, R.D. (1995) Leukocyte trafficking mediated by selectin-carbohydrate interactions. *J. Biol. Chem.* **270**, 11025-11028.
83. Varki, A. (1994) Selectin ligands. *Proc. Natl. Acad. Sci. USA* **91**, 7390-7397.
84. Springer, T.A. (1994) Traffic signals for lymphocyte recirculation and leukocyte emigration: The multistep paradigm. *Cell* **76**, 301-314.
85. Lasky, L.A. (1995) Selectin-carbohydrate interactions and the initiation of the inflammatory response. *Annu. Rev. Biochem.* **64**, 113-139.
86. Moore, K.L., Eaton, S.F., Lyons, D.E., Lichenstein, H.S., Cummings, R.D. and McEver, R.P. (1994) The P-selectin glycoprotein ligand from human neutrophils displays sialylated, fucosylated, O-linked poly-*N*-acetyllactosamine. *J. Biol. Chem.* **269**, 23318-23327.
87. Aeed, P.A., Geng, J.G., Asa, D., Raycroft, L., Ma, L. and Elhammer, A.P. (1998) Characterization of the O-linked oligosaccharide structures on P-selectin glycoprotein ligand-1 (PSGL-1). *Glycoconj. J.* **15**, 975-985.
88. Li, F., Wilkins, P.P., Crawley, S., Weinstein, J., Cummings, R.D. and McEver, R.P. (1996) Post-translational modifications of recombinant P-selectin glycoprotein ligand-1 required for binding to P- and E-selectin. *J. Biol. Chem.* **271**, 3255-3264.
89. Kumar, R., Camphausen, R.T., Sullivan, F.X. and Cumming, D.A. (1996) Core 2 β-1,6-*N*-acetylglucosaminyltransferase enzyme activity is critical for P-selectin glycoprotein ligand-1 binding to P-selectin. *Blood* **88**, 3872-3879.
90. Hemmerich, S. Leffler, H. and Rosen, S.D. (1995) Structure of the O-glycans in GlyCAM-1, an endothelial-derived ligand for L-selectin. *J. Biol. Chem.* **270**, 12035-12047.
91. Nakamura, M., Kudo, T., Narimatsu, H., Furukawa, Y., Kikuchi, J., Asakura, S., Yang, W., Iwase, S., Hatake, K. and Miura, Y. (1998) Single glycosyltransferase, core 2 β-1,6-*N*-acetylglucosaminyltransferase, regulates cell surface sialyl Le^x expression level in human pre-B lymphocytic leukemia cell line KM3 treated with phorbol ester. *J. Biol. Chem.* **273**, 26779-26789.
92. Mukasa, R., Homma, T., Ohtsuki, T., Hosono, O., Souta, A., Kitamura, T., Fukuda, M., Watanabe, S. and Morimoto, C. (1999) Core 2-containing O-glycans on CD43 are preferentially expressed in the memory subset of human CD4 T cells. *Int. Immunol.* **11**, 259-268.
93. Mukasa, R., Homma, T., Hosono, O., Yoshino, S., Nishioka, K., Fukuda, M., and Morimoto, C. (1999) Human T lymphocyte populations which bind to P- or E-selectin are enriched with cells expressing core 2 O-glycans. *Immunol. Lett.* **67**, 117-124.
94. Nakamura, M., Ishida, T., Kikuchi, J., Furukawa, Y. and Matsuda, M. (1999) Simultaneous core 2 β1,6-*N*-acetylglucosaminyltransferase upregulation and sialyl Le^x expression during activation of human tonsillar B lymphocytes. *FEBS Lett.* **463**, 125-128.
95. Shimodaira, K., Nakayama, J., Nakamura, N., Hasebe, O., Katsuyama, T., and Fukuda, M. (1997) Carcinoma-associated expression of core 2 β-1,6-*N*-acetylglucosaminyltransferase gene in human colorectal cancer: Role of O-glycans in tumor progression. *Cancer Res.* **57**, 5201-5206.
96. Ellies, L.G., Tsuboi, S., Petryniak, B., Lowe, J.B., Fukuda, M. and Marth, J.D. (1998) Core 2 oligosaccharide biosynthesis distinguishes between selectin ligands essential for leukocyte homing and inflammation. *Immunity* **9**, 881-890.
97. Fukuda, M., Yeh, J.-C. and Hiraoka, N. (2000) Two distinct core structures that form L-selectin ligands. The Second International Glycosyltransferases Symposium, Toronto, Ontario, Canada, abstract P9.1
98. Kimura, A.K. and Wigzell, H. (1978) Cell surface glycoproteins of murine cytotoxic T lymphocytes. T 145, a new cell surface glycoprotein selectively expressed on Ly 1-2+ cytotoxic T lymphocytes. *J. Exp. Med.* **147**, 1418-1434.
99. Andersson, L.C., Gahmberg, C.G., Kimura, A.K. and Wigzell, H. (1978) Activated human T lymphocytes display new surface glycoproteins. *Proc. Natl. Acad. Sci. USA* **75**, 3455-3458.

100. Barran, P., Fellinger, W., Warren, C.E., Dennis, J.W. and Ziltener, H.J. (1997) Modification of CD43 and other lymphocyte O-glycoproteins by core 2 β-1,6-*N*-acetylglucosaminyltransferase. *Glycobiology* **7**, 129-136.
101. Chui, D., Ong, C.J., Johnson, P., The, H.-S. and Marth, J.D. (1994) Specific CD45 isoforms differentially regulate T cell receptor signaling. *EMBO J.* **9**, 798-807.
102. Sommer, F., Huber, M., Rollinhoff, M. and Lohoff, M. (1995) CD44 plays a costimulatory role in murine T cell adhesion: Ligation of CD44 selectively costimulates IL-2 production, but not proliferation in TCR-stimulated murine Th1 cells. *Int. Immunol.* **7**, 1779-1786.
103. Sperling, A.I., Green, J.M., Mosley, R.L., Smith, P.L., DiPaolo, R.J., Klein, J.R., Bluestone, J.A. and Thompson, C.B. (1995) CD43 is a murine T cell costimulatory receptor that functions independently of CD28. *J. Exp. Med.* **182**, 139-146.
104. Baum, L.G., Pang, M., Perillo, N.L., Wu, T., Delegeane, A., Uittenbogaart, C.H., Fukuda, M. and Seilhamer, J.L. (1995) Human thymic epithelial cells express an endogenous lectin, galectin-1, which binds to core 2 O-glycans on thymocytes and T lymphoblastoid cells. *J. Exp. Med.* **181**, 877-887.
105. Ellies, L.G., Tao, W.T., Fellinger, W., The, H.-S. and Ziltener, H.J. (1996) The CD43 130-kD peripheral T-cell activation antigen is downregulated in thymic positive selection. *Blood* **88**, 1725-1732.
106. Tsuboi, S. and Fukuda, M. (1997) Branched O-linked oligosaccharides ectopically expressed in transgenic mice reduce primary T-cell immune response. *EMBO J.* **16**, 6364-6373.
107. Tsuboi, S. and Fukuda, M. (1998) Overexpression of branched O-linked oligosaccharides on T cell surface glycoproteins impairs humoral immune responses in transgenic mice. *J. Biol. Chem.* **273**, 30680-30687.
108. Granovsky, M., Fode, C., Warren, C.E., Campbell, R.M., Marth, J.D., Pierce, M., Fregrien, N. and Dennis, J.W. (1995) GlcNAc-transferase V and core 2 GlcNAc-transferase expression on the developing mouse embryo. *Glycobiology* **5**, 797-806.
109. Datti, A. and Dennis, J.W. (1993) Regulation of UDP-GlcNAc:Galβ1-3GalNAc-Rβ1,6-*N*-acetylglucosaminyltransferase (GlcNAc-to-GalNAc) in Chinese hamster ovary cells. *J. Biol. Chem.* **268**, 5409-5416.
110. Amano, J. and Oshima, M. (1999) Expression of the H type I blood group antigen during enterocytic differentiation of Caco-2 cells. *J. Biol. Chem.* **274**, 21209-21216.
111. Amano, J. and Oshima, M. (1999) Induction of glycosyltransferases on Caco-2 cells by sodium butyrate. *Glycoconj. J.* **16**, S91 (abstract 2aP#102).
112. Kenney, D.M., Cairns, L., Remold-O'donnell, E., Peterson, J., Rosen, F.S. and Parkman, R. (1986) Morphological abnormalities in the lymphocytes of patients with the Wiskott-Aldrich syndrome. *Blood* **68**, 1329-1332.
113. Perry, G.S., Spector, B.D., Schuman, L.M., Mandel, J.S., Anderson, W.E., McHugh, R.B., Hanson, M.R., Fahlstrom, S.M., Krivit, W. and Kersey, J.H. (1980) The Wiskott-Aldrich syndrome in the United States and Canada (1892-1979). *J. Pediatr.* **97**, 72-78.
114. Rosen, F.S., Cooper, M.D. and Wedgewood, R.J.P. (1983) The primary immunodeficiencies. *N. Engl. J. Med.* **311**, 300-310.
115. Ochs, H.D., Slichter, S.J., Harker, L.A., van Behrens, W.E., Clark, R.A. and Wedgewood, R.J.P. (1980) The Wiskott-Aldrich syndrome: Studies of lymphocytes, granulocytes, and platelets. *Blood* **55**, 243-252.
116. Parkman, R., Remold-O'Donnell, E., Kenney, D.M., Perine, S. and Rosen, F.S. (1981) Surface protein abnormalities in lymphocytes and platelets from patients with Wiskott-Aldrich syndrome. *Lancet* **ii**, 1387-1389.
117. Remold-O'Donnell, E., Kenney, D.M., Parkman, R., Cairns, L., Savage, B. and Rosen, F.S. (1984) Characterization of the human lymphocyte surface sialoglycoprotein that is defective in Wiskott-Aldrich syndrome. *J. Exp. Med.* **159**, 1705-1723.
118. Park, J.K., Rosenstein, Y.J., Remold-O'Donnell, E., Bierer, B.E., Rosen, F.S. and Burakoff, S.J. (1991) Enhancement of T-cell activation by the CD43 molecule whose expression is defective in Wiskott-Aldrich syndrome. *Nature* **350**, 706-709.
119. Manjunth, N., Johnson, R.S., Staunton, D.E., Pasqualini, R. and Ardman, B. (1993) Targeted disruption of CD43 gene enhances T lymphocyte adhesion. *J. Immunol.* **151**, 1528-1534.
120. Piller, F., Le Deist, F., Weinberg, K.I., Parkman, R. and Fukuda, M. (1991) Altered O-glycan synthesis in lymphocytes from patients with Wiskott-Aldrich syndrome. *J. Exp. Med.* **173**, 1501-1510.
121. Higgins, E.A., Siminovitch, K.A., Zhuang, D., Brockhausen, I. and Dennis, J.W. (1991) Aberrant O-linked oligosaccharide biosynthesis in lymphocytes and platelets from patients with the Wiskott-Aldrich syndrome. *J. Biol. Chem.* **266**, 6280-6290.
122. Vavasseur, F., Dole, K., Yang, J., Matta, K.L., Myerscough, N., Corfield, A., Paraskeva, C. and Brockhausen, I. (1994) O-glycan biosynthesis in human colorectal adenoma cells during progression to cancer. *Eur. J. Biochem.* **222**, 415-424.

123. Yang, J.-M., Byrd, J.C., Siddiki, B.B., Chung, Y.-S., Okuno, M., Sowa, M., Kim, Y.S., Matta, K.L. and Brockhausen, I. (1994) Alterations of O-glycan biosynthesis in human colon cancer tissues. *Glycobiology* **4**, 873-884.
124. Brockhausen, I., Kuhns, W., Schachter, H., Matta, K.L., Sutherland, D.R. and Baker, M.A. (1991) Biosynthesis of O-glycans in leukocytes from normal donors and from patients with leukemia: Increase in O-glycan core 2 UDP-GlcNAc:Galβ3GalNAcα-R (GlcNAc to GalNAc) β1,6-*N*-acetylglucosaminyltransferase in leukemic cells. *Cancer Res.* **51**, 1257-1263.
125. Hanisch, F.G., Uhlenbruck, G., Peter, K.J., Egge, H., Dabrowski, J. and Dabrowski, U. (1989) Structures of neutral O-linked polylactosaminoglycans on human skim milk mucins. A novel type of linearly extended poly-*N*-acetyllactosamine backbone with Galβ1-4GlcNAcβ1,6 repeating units. *J. Biol. Chem.* **264**, 872-883.
126. Hull, S.R., Bright, A., Carraway, K.L., Abe, M., Hayes, D.F. and Kufe, D.W. (1989) Oligosaccharide differences in the DF3 sialomucin antigen from normal human milk and the BT-20 human breast carcinoma cell line. *Cancer Commun.* **1**, 261-267.
127. Lloyd, K.O., Burchell, J., Kudryasov, V., Yin, B.W.T. and Taylor-Papadimitriou, J. (1996) Comparison of O-linked carbohydrate chains in MUC1 mucin from normal breast epithelial cell lines and breast carcinoma cell lines. Demonstration of simpler and fewer glycan chains in tumor cells. *J. Biol. Chem.* **271**, 33325-33334.
128. Brockhausen, I., Yang, J.-M., Burchell, J. Whitehouse, C. and Taylor-Papadimitriou, J. (1995) Mechanisms underlying aberrant glycosylation of MUC1 mucin in breast cancer cells. *Eur. J. Biochem.* **233**, 607-617.
129. Whitehouse, C., Burchell, J., Gschmeissner, S., Brockhausen, I., Lloyd, K.O. and Taylor-Papadimitriou, J. (1997) A transfected sialyltransferase that is elevated in breast cancer and localizes to the medial/*trans*-golgi apparatus inhibits the development of core 2-based O-glycans. *J. Cell. Biol.* **137**, 1229-1241.
130. Beum, P.V., Singh, J., Burdick, M., Hollingsworth, M.A. and Cheng, P.-W. (1999) Expression of core 2 β-1,6-*N*-acetylglucosaminyltransferase in a human pancreatic cancer cell line results in altered expression of MUC1 tumor-associated epitopes. *J. Biol. Chem.* **274**, 24641-24648.
131. Burdick, M.D., Harris, A., Reid, C.J., Iwamura, T. and Hollingsworth, M.A. (1997) Oligosaccharides expressed on MUC1 produced by pancreatic and colon tumor cell lines. *J. Biol. Chem.* **272**, 24198-24202.
132. Lefebvre, J.-C., Giordanengo, V., Limouse, M., Doglio, A., Cucchiarini, M., Monpoux, F., Mariani, R. and Peyron, J.-F. (1994) Altered glycosylation of leukosialin, CD43, in HIV-1-infected cells of the CEM line. *J. Exp. Med.* **180**, 1609-1617.
133. Orlacchio, A., Sarchielli, P., Gallai, V., Datti, A., Saccardi, C. and Palmerini, C.A. (1997) Activity levels of a β1,6-*N*-acetylglucosaminyltransferase in lymphomonocytes from multiple sclerosis patients. *J. Neurological Sci.* **151**, 177-183.
134. Koya, D., Dennis, J.W., Warren, C.E., Takahara, N., Schoen, F.J., Nishio, Y., Nakajima, T., Lipes, M.A. and King, G.L. (1999) Overexpression of core 2 β1,6-*N*-acetylglycosaminyltransferase enhances cytokine actons and induces hypertrophic myocardium in transgenic mice. *FASEB J.* **13**, 2329-2337.
135. Panicot, L., Mas, E., Thivolet, C. and Lombardo, D. (1999) Circulating antibodies against an exocrine pancreatic enzyme in type 1 diabetes. *Diabetes* **48**, 2316-2323.
136. Koscielak, J., Miller-Podraza, H., Krauze, R and Piasek, A. (1976) Isolation and characterization of polyglycosylceramides (megaloglycolipids) with A, H and I blood-group activities. *Eur. J. Biochem.* **71**, 9-18.
137. Jarnefelt, J., Rush, J., Li, Y.-T. and Laine, R.A. (1978) Erythroglycan, a high molecular weight glycopeptide with the repeating structure [galactosy-1,4-2-deoxy-2-acetamido-glucosyl-1,3] comprising more than one third of the protein-bound carbohydrate of human erythrocyte stroma. *J. Biol. Chem.* **253**, 8006-8009.
138. Kato, I. and Naiki, M. (1976) Ganglioside and rabbit erythrocyte membrane receptor for staphylococcal alpha-toxin. *Infection and Immunity* **13**, 289-291.
139. Suzuki, Y., Suzuki, T. and Matsumoto, M. (1983) Isolation and characterization of receptor sialoglycoprotein for hemagglutinating virus of Japan (Sendai virus) from bovine erythrocyte membrane. *J. Biochem.* **93**, 1621-1633.
140. Loomes, L.M., Uemura, K.-I., Childs, R.A., Paulson, J.A., Rogers, G.N., Scudder, P.R., Michalski, J.-C., Hounsell, E.F., Taylor-robinson, D. and Feizi, T. (1984) Erythrocyte receptors for *Mycoplasma pneumoniae* are sialylated oligosaccharides of Ii antigen type. *Nature* **307**, 560-563.
141. Andersson, B., Dahmen, J., Frejd, T., Leffler, H., Magnusson, G., Noori, G. and Eden, C.S. (1983) Identification of an active disaccharide unit of a glycoconjugate receptor for *Pneumococci* attaching to human pharyngeal epithelial cells. *J. Exp. Med.* **158**, 559-570.

142. Romans, D.G., Tilley, C.A. and Dorrington, K.J. (1980). Monogamous bivalency of IgG antibodies. I. Deficiency of branched ABHI-Active oligosaccharide chains on red cells of infants causes the weak antiglobulin reactions in hemolytic disease of the newborn due to ABO incompatibility. *J. Biol. Chem.* **124**, 2807-2811.
143. Gopalakrishnan, P.V. and Karush, F. (1974). Antibody affinity. VII. Multivalent interaction of anti-lactoside antibody. *J. Immunol.* **113**, 769-778.
144. Hornick, C.L. and Karush, F. (1972). Antibody affinity. III. The role of multivalence. *Immunochem.* **9**, 325-340.
145. Crothers, D.M. and Metzger, H. (1972). The influence of polyvalency on the binding properties of antibodies. *Immunochem.* **9**, 341-357.
146. Fenderson, B.A., Zehavi, U. and Hakomori, S.-I. (1984). A multivalent lacto-N-fucopentaose III-lysyllysine conjugate decompacts preimplantation mouse embryos, while the free oligosaccharide is ineffective. *J. Exp. Med.* **160**, 1591-1596.
147. Niemala, R., Penttila, L., Seppo, A., Helin, J., Leppanen, A., Rabina, J., Uusitalo, L., Maaheimo, H., Taskinen, J., Costello, C.E. and Renkonen, O. (1995) Enzyme-assisted synthesis of a bivalent high-affinity dodecasaccharide inhibitor of mouse gamete adhesion. The length of the chains carrying distal α1,3-bonded galactose residues is critical. *FEBS Lett.* **367**, 67-72.
148. Welpy, J.K., Abbas, S.Z., Scudder, P., Keene, J.L., Broschat, K., Casnocha, S., Gorka, C., Steininger, C., Howard, S.C., Schmuke, J.J., Graneto, M., Rotsaert, J.M., Manger, I.D. and Jacob, G.S. (1994) Multivalent sialyl Le^x: Potent inhibitors of E-selectin-mediated cell adhesion; reagent for staining activated endothelial cells. *Glycobiology* **4**, 259-265.
149. Toppila, S., Lauronen, J., Mattila, P., Turunen, J.P., Penttila, L., Paavonen, T., Renkonen, O. and Renkonen, R. (1997) L-selectin ligands in rat high endothelium: Multivalent sialyl Lewis x glycans are high-affinity inhibitors of lymphocyte adhesion. *Eur. J. Immunol.* **27**, 1360-1365.
150. Seppo, A., Turunen, J.P., Penttila, L., Keane, A., Renkonen, O. and Renkonen, R. (1996) Synthesis of a tetravalent sialyl Lewis x glycan, a high-affinity inhibitor of L-selectin-mediated lymphocyte binding to endothelium. *Glycobiology* **6**, 65-71.
151. Maaheimo, H., Renkonen, R., Turunen, J.P., Penttila, L. and Renkonen, O. (1995) Synthesis of a divalent sialyl Lewis x O-glycan, a potent inhibitor of lymphocyte-endothelium adhesion. Evidence that multivalency enhances the saccharide binding to L-selectin. *Eur. J. Biochem.* **234**, 616-625.
152. Turunen, J.P., Majuri, M.-L., Seppo, A., Tiisala, S., Paavonen, T., Miyasaka, M., Lemstrom, K., Penttila, L., Renkonen, O. and Renkonen, R. (1995) De novo expression of endothelial sialyl $Lewis^a$ and sialyl $Lewis^x$ during cardiac transplant rejection: Superior capacity of a tetravalent sialyl $Lewis^x$ oligosaccharide in inhibiting L-selectin-dependent lymphocyte adhesion. *J. Exp. Med.* **182**, 1133-1142.
153. Renkonen, O., Toppila, S., Penttila, L., Salminen, H., Helin, J., Maaheimo, H., Costello, C.E., Turunen, J.P. and Renkonen, R. (1997) Synthesis of a new nanomolar saccharide inhibitor of lymphocyte adhesion: Different polylactosamine backbones present multiple sialyl $Lewis^x$ determinants to L-selectin in high-affinity mode. *Glycobiology* **7**, 453-461.
154. Watanabe, K. and Hakomori, S.-I. (1976) Status of blood group carbohydrate chains in ontogenesis and in oncogenesis. *J. Exp. Med.* **144**, 644-653.
155. Koscielak, J., Zdebska, E., Wilczynska, Z., Miller-Podraza, H. and Dzierzkowa- Borodej, W. (1979). Immunochemistry of Ii-active glycolipids of erythrocytes. *Eur. J. Biochem.* **96**, 331-337.
156. Fukuda, M., Dell, A. and Fukuda, M.N. (1984). Structure of fetal lactosaminoglycan: the carbohydrate moiety of Band 3 isolated from human umbilical cord erythrocytes. *J. Biol. Chem.* **259**, 4782-4791.
157. Fukuda, M.N. and Levery, S.B. (1983) Glycolipids of fetal, newborn, and adult erythrocytes: Glycolipid pattern and structural study of H3-glycolipid from newborn erythrocytes. *Biochemistry* **22**, 5034-5040.
158. Muramatsu, T., Gachelin, G., Nicolas, J.F. Condamine, H., Jakob, H. and Jacob, F. (1978) Carbohydrate structure and cell differentiation: Unique properties of fucosyl-glycopeptides isolated from embryonal carcinoma cells. *Proc. Natl. Acad. Sci. USA* **75**, 2315-2319.
159. Kapadia, A., Feizi, T. and Evans, M.J. (1981). Changes in the expression and polarization of blood group I and i antigens in post-implantation embryos and teratocarcinomas of mouse associated with cell differentiation. *Exp. Cell. Res.* **131**, 185-195.
160. Kannagi, R., Levery, S.B. and Hakomori, S.-I. (1983). Sequential change of carbohydrate antigen associated with differentiation of murine leukemia cells: i-I antigenic conversion and shifting of glycolipid synthesis. *Proc. Natl. Acad. Sci. USA* **80**, 2844-2848.
161. Fukuda, M., Fukuda, M.N., Papyannopoulou, T. and Hakomori, S.-I. (1980) Membrane differentiation in human erythroid cells:Unique profiles of cell surface glycoproteins expressed in erythroblasts *in vitro* from three ontogenic stages. *Proc. Natl. Acad. Sci. USA.* **77**, 3474-3478.

162. Hakomori, S.-I., Koscielak, J., Bloch, K.J. and Jeanloz, R.W. (1967) Immunologic relationship between blood group substances and a fucose-containing glycolipid of human adenocarcinoma. *J. Immunol.* **98**, 31-38.
163. Davidsohn, I., Ni, L.Y. and Stejskal, R. (1971) Tissue isoantigens A, B, and H in carcinoma of the stomach. *Arch. Pathol.* **92**, 456-464.
164. Feizi, T., Turberville, C. and Westwood, J.H. (1975). Blood-group precursors and cancer-related antigens. *Lancet* **2**, 391.
165. Picard, J., Edward, D.W. and Feizi, T. (1978). Changes in the expression of the blood group A, B, H, Le^a, and Le^b antigens and the blood group precursor associated I (Ma) antigen in glycoprotein-rich extracts of gastric carcinomas. *J. Clin. Lab. Immunol.* **1**, 119-128.
166. Picard, J.K. and Feizi, T. (1983) Peanut lectin and anti-Ii antibodies reveal structural differences among human gastrointestinal glycoproteins. *Molec. Immunol.* **20**, 1215-1220.
167. Holmes, E.H., Hakomori, S.-I. and Ostrander, G.K. (1987) Sytnesis of type I and 2 lacto series glycolipid antigens in human colonic adenocarcinoma and derived cell lines is due to activation of a normally unexpressed β1-3*N*-acetylglucosaminyltransferase. *J. Biol. Chem.* **262**, 15649-15658.

DIFFERENTIATION-ASSOCIATED EXPRESSION OF β-*N*-ACETYLGALACTOSAMINYLATED N-LINKED OLIGOSACCHARIDES IN MAMMARY EPITHELIAL CELLS

Kiyoshi Furukawa[1], Noriaki Kitamura[1,2], Takeshi Sato[1] and Sen Hiraizumi[3]

[1]*Department of Biosignal Research, Tokyo Metropolitan Institute of Gerontology, Itabashi-ku, Tokyo 173-0015, Japan*

[2]*Faculty of Science and Technology, Science University of Tokyo, Noda, Chiba 278-0022, Japan*

[3]*Department of Surgery, Sendai City Hospital, Wakabayashi-ku, Sendai 984-0075, Japan*

A unique carbohydrate structure, GalNAcβ1→4GlcNAc (Fig. 1a), expressed on the α-mannose residues constituting the core structures of N-linked oligosaccharides was initially discovered in mammalian pituitary glycoprotein hormones, luteinizing hormone, follicle-stimulating hormone and thyroid-stimulating hormone (1-4). In this case, galactose residues in typical N-linked oligosaccharides are replaced with *N*-acetylgalactosamine residues, and most GalNAcβ1→4GlcNAc groups are sulfated (Fig. 1b). A sulfated form of this disaccharide group was shown to be recognized by the lectin present on hepatic reticuloendothelial cells, and is involved in clearance of the hormones from the circulation (5). Besides sulfation, the disaccharide group is often sialylated or fucosylated (Fig. 1c and 1d). When we analyzed the structures of N-linked oligosaccharides released by hydrazinolysis from a bovine mammary glycoprotein, CD36, about 30% of the neutral and desialylated neutral oligosaccharides were not converted to those with the core structures by digestion with jack bean β-galactosidase and then with jack bean β-*N*-acetylhexosaminidase. Subsequent analysis of the sugar composition revealed that they contain *N*-acetylgalactosamine in addition to mannose, *N*-acetylglucosamine and galactose, sugges-

The Molecular Immunology of Complex Carbohydrates–2.
Edited by Albert M. Wu, Kluwer Academic/Plenum Publishers, 2001.

ting the presence of N-linked oligosaccharides with a novel structure in the CD36. The methylation analysis of the oligo-

a GalNAcβ1→4GlcNAcβ1→R

b SO_4→4GalNAcβ1→4GlcNAcβ1→R

c Neu5Acα2→3/6GalNAcβ1→4GlcNAcβ1→R

d GalNAcβ1→4GlcNAcβ1→R
3
↑
Fucα1

Figure 1. Outer chain structures of *N*-acetylgalactosaminylated N-linked oligosaccharides found in vertebrate glycoproteins. R: sugar chains with the core structures.

saccharides before and after treatment with β-*N*-acetylgalactosaminidase established the presence of the GalNAcβ1→4GlcNAc structure in addition to the Galβ1→4GlcNAc structure in the N-linked oligosaccharides (6). In the present article, β-*N*-acetylgalactosaminylation of N-linked oligosaccharides in the mammary gland is under the regulation of cell differentiation will be reviewed.

I. DISTRIBUTION OF GLYCOPROTEINS WITH β-*N*-ACETYL-GALACTOSAMINYLATED N-LINKED OLIGOSACCHARIDES

In addition to the pituitary glycoprotein hormones, many membrane glycoproteins in bovine pituitary gland were shown to contain N-linked oligosaccharides with the GalNAcβ1→4GlcNAc structure in their outer chain moieties (7). Similarly, most glycoproteins including CD36 which are components of milk fat globule membrane derived from bovine mammary epithelial cells were also shown to contain β-*N*-acetylgalactosaminylated N-linked oligosaccharides (8). These results prompted us to investigate whether or not this unique glycosylation occurs to glycoproteins synthesized only in bovine pituitary and mammary glands. Lectin blot analysis of membrane glycoprotein samples prepared from twelve bovine organs obtained from a local slaughter house using *Wistaria floribunda* agglutinin (WFA) which interacts with oligosaccharides terminated with β-*N*-acetylgalactosamine (9), revealed that glycoproteins from kidney, brain, liver, spleen, thymus, adrenal gland, lung, muscle and stomach but not from small and large intestine are β-*N*-acetylgalactosaminylated in addition to those from pituitary and mammary glands (10). Since treatment of blots with N-glycanase prior to incubation with WFA showed no lectin binding, β-*N*-acetylgalactosaminylation occurs only to N-linked oligosaccharides. WFA-agarose column chromatography of the oligosaccharides released by hydrazinolysis from the membrane glycoprotein samples showed that only 4.4-17.9% of the total oligosaccharides depending on the tissues are β-*N*-acetylgalactosaminylated and the rest are galactosylated (8, 10). These results indicate that β-*N*-acetylgalactosaminylation occurs to glycoproteins of many bovine tissues and organs.

In addition to bovine glycoproteins, this glycosylation occurs to many glycoproteins derived from other mammals, reptiles, amphibians, insects, mollusca, nematoda and trematoda (reviewed in 11), indicating that β-*N*-acetylgalactosaminylation is also one of the ubiquitous modifications of the outer chain moieties of N-linked oligosaccharides. However, in the case of IgG molecules, those from bovine, goat and human but not from rabbit and mouse were *N*-acetylgalactosaminylated, suggesting that this glycosylation occurs in a species-specific manner (12, 13). Furthermore, several reports described the occurrence of glycoproteins containing O-linked oligosaccharides with the GalNAcβ1→4GlcNAc structure (14-17), indicating that β-*N*-acetylgalactosaminylation is not specific to N-linked oligosaccharides.

II. BIOSYNTHESIS OF THE GalNAcβ1→4GlcNAc STRUCTURE

The presence of an enzyme which transfers *N*-acetylgalactosamine from UDP-GalNAc to the *N*-acetylglucosamine residues of $GlcNAc_2Man_3GlcNAc_2$-Asn was shown in bovine pituitary gland. Quite interestingly, the efficiency of *N*-acetylgalactosamine transfer by the enzyme was enhanced when the acceptor oligosaccharide was attached to the hormone peptide (18). Later, β-*N*-acetylgalactosaminyltransferase (β-1,4-GalNAcT) specific to the hormones was shown to recognize the tripeptide sequence motifs (Pro-Xaa-Arg/Lys) located near the N-glycosylation sites (19). In contrast, we detected β-1,4-GalNAcT activities in lactating bovine mammary gland using the acceptors of asialo (As), agalacto (Ag)-human chorionic gonadotropin (hCG) which contains the tripeptide motif and AsAg-human transferrin (hTF) which does not contain the peptide motifs (20). Quite interestingly, when the same concentrations of acceptor oligosaccharides attached to the glycoproteins were incubated with the enzyme source, similar levels of the activities were obtained towards AsAg-hCG and AsAg-hTF (Fig. 2), indicating that β-1,4-GalNAcT in the mammary gland is not specific to the hormones. In sup-

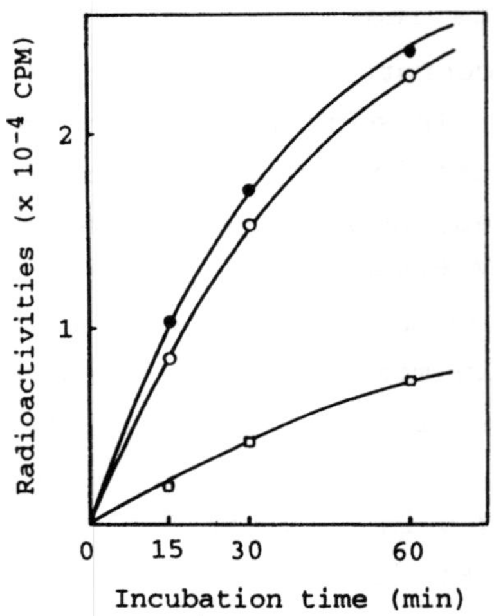

Figure 2. β-1,4-GalNAcT activities in bovine mammary gland. Six nmoles of AsAg-hCG (closed circle), AsAg-hTF (open circle) or AsAg-hTF-derived oligosaccharides (open square) in terms of oligosaccharides were incubated with the detergent-solubilized microsomal proteins from lactating bovine mammary gland in the presence of UDP-[^{3}H]GalNAc. After incubation reaction mixtures were digested with pronase and the digests were subjected to Con A-Sepharose column chromatography. Radioactivities eluted from the column with 100 mM α-methyl-mannoside were determined.

port of this finding, the presence of the hormone-nonspecific β-1,4-GalNAcTs was reported (21-23). However, lower β-1,4-GalNAcT activities were obtained towards oligosaccharides released from AsAg-hTF when compared to those towards AsAg-hTF. This probably indicates that glycoproteins are better acceptors than oligosaccharides. Since β-1,4-GalNAcT and β-1,4-galactosyltransferase (β-1,4-GalT) share the same acceptor oligosaccharides, β-1,4-GalNAcT cDNAs can be isolated by homology cloning using the conserved sequences found between β-1,4-GalT and β-1,4-*N*-acetylglucosaminyltransferase, the latter of which also shares the same acceptor *N*-acetylglucosamine and form the GlcNAcβ1→4GlcNAcβ1→Man structure (24).

On the other hand, β-1,4-GalT I was shown to be able to transfer *N*-acetylgalactosamine from UDP-GalNAc to *N*-acetylglucosamine and to the GlcNAcβ1→2Man group of AsAg-hTF, thus forming the GalNAcβ1→4GlcNAc structure, in the presence of α-lactalbumin (25, 26). α-Lactalbumin binds β-1,4-GalT I and modifies its acceptor specificity from *N*-acetylglucosamine to glucose to induce lactose synthesis in the lactating mammary gland (27). However, its limited localization indicates that β-1,4-GalT I is not an enzyme which is involved in the biosynthesis of the GalNAcβ1→4GlcNAc group found in glycoproteins of many tissues other than the mammary gland. In fact, no changes in the reactivity of WFA was observed in most brain glycoproteins from β-1,4-GalT I-knockout mouse when compared with that of the control animal (26). Interestingly, β-1,4-GalNAcTs in bovine mammary gland and snail can transfer *N*-acetylgalactosamine to glucose in the presence of α-lactalbumin (21, 28), suggesting again that β-1,4-GalNAcT is closely related to β-1,4-GalT I.

III. DIFFERENTIATION-ASSOCIATED EXPRESSION OF β-*N*-ACETYLGALACTOSAMINYLATED N-LINKED OLIGOSACCHARIDES

Among bovine tissues examined, the expression level of β-*N*-acetylgalactosaminylated N-linked oligosaccharides was highest in glycoproteins from the mammary gland (10). Since the β-*N*-acetylgalactosaminylation of bovine milk proteins appeared to increase during lactation (29), this glycosylation was considered to be associated with the functional differentiation of the mammary epithelial cells. To address this issue, the primary cultured cells were established from the lactating bovine mammary gland and cultured in the presence or absence of lactogenic hormones. When the cells were cultured on plastic dishes without the hormones, no milk proteins such as casein were secreted into media (lane a in Fig. 3), while casein was secreted when the cells were grown on collagen-coated dishes in the presence of insulin, hydrocortison and prolactin for 6 days (lane b in Fig.3). Lectin blot analysis of membrane glycoprotein samples prepared from the cells treated without and with the hormones showed that WFA-positive bands appear only in the differentiated cells (lane d in Fig. 3), and the lectin-positive bands disappeared upon treatments of blots with β-*N*-acetylgalactosaminidase or N-glycanase, indicating that the β-*N*-acetylgalactosaminylation occurs exclusively to N-linked oligosaccharides synthesized in the hormone-treated cells, most

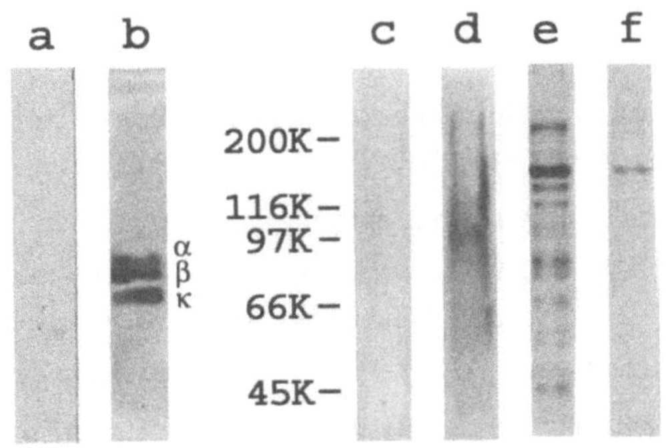

Figure 3. Lectin blot analysis of membrane glycoproteins from primary cultured bovine mammary epithelial cells and from human breast tumor. Lanes a and b contained proteins from spent media of the cells treated without and with lactogenic hormones and were incubated with anti-casein antibody. Lanes c and d contained samples from the cells without and with treatment of lactogenic hormones and were incubated with WFA. Lanes e and f contained samples from the unaffected and affected regions of human breast tumor, respectively, and were incubated with WFA.

probably mediated by the activation of β-GalNAcT (30). In support of this finding, the expression of the Neu5Acα2→6GalNAcβ1→4GlcNAc structure on N-linked oligosaccharides of prolactin/growth hormone family members produced by rat spongiotrophoblasts was shown to be regulated developmentally (31). Therefore, the expression of β-1,4-GalNAcT should be regulated at developmental and differentiation stages at least in the mammary gland and placenta. Whether or not the hormone-responsive elements are present in the promoter region of the transferase gene remains to be elucidated.

Since our study showed that functional differentiation of bovine mammary epithelial cells enhances the expression of β-*N*-acetylgalactosaminylated N-linked oligosaccharides (30), it was of interest to investigate whether this glycosylation decreases in membrane glycoproteins by malignant transformation of bovine mammary epithelial cells. However, no such tumors and cell lines are available except for the human and mouse cell lines.

Since our previous study showed that human milk proteins contain N-linked oligosaccharides with the GalNAcβ1→4GlcNAc structure (Taka, J., Fujisawa, H., and Furukawa, K., unpublished data), we initially analyzed structures of N-linked oligosaccharides from human breast tumor cell line, MRK-nu-1, and found that about 6% of complex-type oligosaccharides are β-*N*-acetylgalactosaminylated (32). These results indicated that glycoproteins produced in human mammary gland and in human breast tumors also contain β-*N*-acetylgalactosaminylated N-linked oligosaccharides. In order to elucidate differences if any in the expression levels of β-*N*-acetylgalactosaminylated N-linked oligosaccharides between human breast and its tumor, membrane glycoprotein samples were prepared from the unaffected and affected regions of the breast tumor excised by surgical treatment, which were confirmed by histological examination, and subjected to lectin blot analysis. Composition of proteins from the unaffected tissue regions and that from the tumors were almost the same among the four patients. However, the reactivity of WFA to glycoproteins from the tumor was much reduced compared to those from the unaffected region in a patient (lanes f and e, respectively, in Fig. 2 and manuscript in preparation). Comparison of the lectin reactivity to glycoprotein samples between patients at the stages-I and -II showed that it is weaker in the tumor samples from patients at the stage-II (unpublished data). The results indicated that β-*N*-acetylgalactosaminylation of N-linked oligosaccharides is greatly reduced upon malignant transformation of the mammary cells and progression of the tumor. Because of limited amounts of the materials available from the patients, further characterization of the oligosaccharides could not be performed.

The results accumulated in our laboratory indicate that β-*N*-acetylgalactosaminylation of N-linked oligosaccharides is regulated under the differentiation and dedifferentiation of mammary epithelial cells. Whether β-*N*-acetylgalactosaminylation of N-linked oligosaccharides found in tissues other than the mammary gland and placenta is also under the regulation of some physiological events remains to be clarified. Although the biological significance of this glycosylation and its regulation in the mammary gland remain to be elucidated, the lectin blot analysis of secretes squeezed from female nipples could be employed as an another diagnostic tool for human breast cancer.

SUMMARY

Not only mammalian pituitary glycoprotein hormones but also many glycoproteins from a variety of animal species have been shown to contain N-linked oligosacchardies with the GalNAcβ1→4GlcNAc structure. Two types of β-1,4-GalNAcT were found; one transfers *N*-acetylgalactosamine to acceptor oligosaccharides, which is stimulated by the hormone peptide and the other simply transfers sugar without such activation. In the case of bovine mammary membrane glycoproteins, the expression of β-*N*-acetylgalactosaminylated N-linked oligosaccharides was associated with the functional differentiation of the epithelial cells. In contrast, the expression level of such oligosaccharides was much reduced in glycoprotein samples from human breast tumors compared with those from the unaffected regions. These results strongly suggest that the β-*N*-acetylgalactosaminylation is one which is regulated under cellular differentiation and dedifferentiation of the mammary gland. Whether or not β-*N*-acetylgalactosaminylated N-linked oligosaccharides have unique functions in addition to clearance of the hormone from the circulation remains to be elucidated.

ACKNOWLEDGMENTS

This work was supported by Grants-in-Aid for Scientific Research (10680696 and 09240104) from the Ministry of Education, Science, Culture and Sports of Japan.

REFERENCES

1. Green, E.D., van Halbeek, H., Boime, I., and Baenziger, J.U. (1985) Structural elucidation of the disulfated oligosaccharide from bovine lutropin. J. Biol. Chem,. 260: 15623-15630.
2. Green, E.D., and Baenziger, J.U. (1988) Asparagine-linked oligosaccharides on leutropin, follitropin, and thyrotropin. I. Structural elucidation of the sulfated and sialylated oligosaccharides on bovine, ovine, and human

pituitary glycoprotein hormones. J. Biol. Chem., 263:25-32.
3. Green, E.D., and Baenziger, J.U. (1988) Asparagine-linked oligosaccharides on leutropin, follitropin, and thyrotropin. II. Distributions of sulfated and sialylated oligosaccharides on bovine, ovine, and human pituitary glycoprotein hormones. J. Biol. Chem., 263:36-44.
4. Smith, P.L., Bousfield, G.R, Kumar, S., Fiete, D,.and Baenziger, J.U. (1992) Equine lutropin and chorionic gonadotropin bear oligosaccharides terminating with SO_4-4-GalNAc and Siaα2,3Gal, respectively. J. Biol. Chem., 268: 795-802.
5. Fiete, D., Srivastava, V., Hinsgaul, O., and Baenziger, J.U. (1991) A hepatic reticuloendothelial cell receptor specific for SO_4-4GalNAcβ1,4GlcNAcβ1,2Manα that mediates rapid clearance of lutropin. Cell 67:1103-1110.
6. Nakata, N., Furukawa, K., Greenwalt, D.E., Sato, T., and Kobata, A. (1993) Structural study of the sugar chains of CD36 purified from bovine mammary epithelial cells - Occurrence of novel hybrid-type sugar chains containing the Neu5Acα2→6GalNAcβ1→4GlcNAc and the Manα1→2Manα1→3-Manα1→6Man groups. Biochemistry, 32:4369-4383.
7. Taka, J., Sato, T., Sakiyama, T., Fujisawa, H., and Furukawa, K. (1996) Bovine pituitary membrane glycoproteins contain β-*N*-acetylgalactosaminylated N-linked sugar chains. J. Neurochem. 66:852-859.
8. Sato, T., Furukawa, K., Greenwalt, D.E., and Kobata, A. (1993) Most bovine milk fat globule membrane glycoproteins contain asparagine-linked sugar chains with GalNAcβ1→4Glc-NAc groups. J. Biochem., 114:890-900.
9. Torres, B.V., McCrumb, D.K., and Smith, D.F. (1988) Glycolipid-lectin interactions: Reactivity of lectins from *Helix pomatia*, *Wistaria floribunda*, and *Dolichos biflorus* with glycolipids containing *N*-acetylgalactosamine. Arch. Biochem. Biophys., 262:1-11.
10. Sakiyama, T., Kabayama, M., Tomita, M., Nakamura, J., Mukai, H., Tomita,Y., and Furukawa, K. (1988) Distribution of glycoproteins with β-*N*-acetylgalactosaminylated N-linked sugar chains among bovine tissues. Biochim. Biophys. Acta 1380:268-274.
11. van den Eijnden, D.H., Bakker, H., Neeleman, A.P., van den Nieuwenhof, I.M., and van Die, I. (1997) Novel pathways in complex-type oligosaccharide synthesis: new vistas opened by studies in invertebrates. Biochem. Socie. Transact., 25:887-893.
12. Aoki, N., Furukawa, K., Iwatsuki, K., Noda, A., Sato, T., Nakamura, R., and Matsuda, T. (1995) A bovine IgG heavy chain contains N-acetylgalactosaminylated N-linked sugar chains. Biochem. Biophys. Res. Commun., 210:275-280.
13. Aoki, N., Matsuda, T., Sakiyama, T., Iwatsuki, K., and Furukawa, K. (1997) Species-specific β-*N*-acetylgalactosaminylation of serum IgG proteins. Biochim. Biophys. Acta, 1334:207-213.
14. Siciliano, R.A., Morris, H.R., Bennett, H.P., and Dell, A. (1994) O-Glycosylation mimics N-glycosylation in the 16-kDa fragment of bovine pro-opiomelanocortin. J. Biol. Chem., 269:910-920.

15. Strecker, G., Wieruszeski, J.M., Fontaine, M.D., and Plancke, Y. (1994) Structure of the major neutral oligosaccharide-alditols released from the egg jelly coats of *Axolotol maculatum*. Characterization of the carbohydrate sequence GalNAc(β1-4)[Fuc(α1-3)]GlcNAc(β1-3/6). Glycobiology 4:605-609.
16. Ohta, M., Matsuura, F., Kobayashi, Y., Shigeta, S., Ono, K., and Oka, S. (1991) Further characterization of allergenically active oligosaccharitols isolated from a sea squirt H-antigen. Arch. Biochem. Biophys. 290:474-483.
17. Khoo, K.H., Sarda, S., Xu, X.F., Caulfield, J.P., McNeil, M.R., Homans, S.W., Morris, H.R., and Dell, A. (1995) A unique multifucosylated -3GalNAcβ1-4GlcNAcβ1-3Galα1- motif constitutes the repeating unit of the complex O-glycans derived from the cercarial glycocalyx of *Schistosoma mansoni*. J. Biol. Chem., 270:17114-17123.
18. Smith, P.L., and Baenziger, J.U. (1988) A pituitary *N*-acetylgalactosamine transferase that specifically recognizes glycoprotein hormones. Science 242:930-933.
19. Smith, P.L., and Baenziger, J.U. (1992) Molecular basis of recognition by the glycoprotein hormone-specific *N*-acetylgalactosaminetransferase. Proc. Natl. Acad. Sci. USA 89: 329-333.
20. Furukawa, K., Nakata, N., Sato, T., Greenwalt, D. E., and Kobata, A. (1992) Structural study of the sugar chains of CD36 purified from bovine mammary epithelial cells. XIVth Japanese Carbohydrate Symposium, 91-92.
21. van den Nieuwenhof, I.M., Schiphorst, W.E., van Die,I., and van den Eijnden, D.H. (1999) Bovine mammry gland UDP-GalNAc: GlcNAcβ1→4R β1→4-*N*-acetylgalactosaminyltransferase is glycoprotein hormone nonspecific and shows interaction with α-lactalbumin. Glycobiology 9:115-123.
22. Dharmesh, S.M., Skelton, T.P., and Baenziger, J.U. (1993) Coordinate restricted expression of the ProXaaArg/Lys-specific GalNAc-transferase and GalNAcβ1,4GlcNAcβ1,2Manα-4-sulfotransferase. J. Biol. Chem., 268, 17096-17102.
23. Neeleman, A.P., van der Knaap, W.P., and van den Eijnden, D.H. (1994) Identification and characterization of a UDP-GalNAc:GlcNAc β1→4-*N*-acetylgalactosaminyltransferase from cercariae of the schistosome *Trichobiharzia ocellata*. Catalysis of a key step in the synthesis of N,N'-diacetyllactosediamino (lacdiNAc)-type glycans. Glycobiology 4: 641-651.
24. Bakker, H., Agterberg, M., van Tetering, A., Koeleman, C.A., van den Eijnden, D.H., and van Die, I. (1994) A *Lymnaea stagnalis* gene, with sequence similarity to taht of mammalian β1→4galactosyltransfearses, encodes a novel UDP-GlcNAc:GlcNAcβ-R β1→4*N*-acetylglucosaminyltransferase. J. Biol. Chem., 269:30326-30333.
25. Do, K.Y., Do, S.I., and Cummings, R.D. (1995) α-Lactalbumin induces bovine milk β1,4-galactosyltransferase to utilize UDP-GalNAc. J. Biol. Chem., 270:18447-18451.
26. Kido, M., Sakiyama, T., Ichinose, M., Miki, K., and Furukawa, K. (2000) β-*N*-Acetylgalactosaminylation of N-linked oligosaccharides is independent of the presence of β-1,4-

galactosyltransferase I. Res. Commun. Biochem. Cell & Molec. Biol., in press.

27. Brodbeck, U., Denton, W.L., Tanahashi, N., and Ebner, K.E. (1967) The isolation and identification of the B protein of lactose synthetase as α-lactalbumin. J. Biol. Chem. 242:1391-1397.
28. Neeleman, A.P., and van den Eijnden, D.H. (1996) α-lactalbumin affects the acceptor specificity of *Lymnaea stagnalis* albumen gland UDP-GalNAc:GlcNAcβ-R β1→4-*N*-acetylgalactosaminyltransferase: synthesis of GalNAcβ1→4-Glc. Proc. Natl. Acad. Sci. USA 93:10111-10116.
29. Ujita, M., Furukawa, K., Aoki, N., Sato, T., Noda, A., Nakamura, R., Greenwalt, D.E., and Matsuda, T. (1993) A change in soybean agglutinin binding patterns of bovine milk fat globule membrane glycoproteins during early lactation. FEBS Lett., 332:119-122.
30. Sato, T., Taka, J., Aoki, N., Matsuda, T., and Furukawa, K. (1997) Expression of β-*N*-acetylgalactosaminylated N-linked sugar chains is associated with functional differentiation of bovine mammary gland. J. Biochem., 122:1068-1073.
31. Manzella, S.M., Dharmesh, S.M., Cohick, C.B., Soares, M.J., and Baenziger, J.U. (1997) Developmental regulation of a pregnancy-specific oligosaccharide structure, Neu5Ac-α2,6GalNAcβ1,4GlcNAc, on select members of the rat placental prolactin family. J. Biol. Chem., 272:4775-4782.
32. Sato, T., Kobata, A., Greenwalt, D.E., and Furukawa, K. (1995) Differential *N*-acetylgalactosaminylation of bovine and human mammary epithelial glycoproteins. Glycoconj. J., 12:418.

METABOLISM AND ROLE OF *O*-ACETYLATED SIALIC ACIDS

Roland Schauer, Hiltrud Schmid, Jörg Pommerencke, Matthias Iwersen
and Guido Kohla
Biochemisches Institut, Christian-Albrechts-Universität
Olshausenstraße 40, D-24098 Kiel, Germany
Tel. +49-431-8802210; Fax. +49-431-8802238;
e-mail: Schauer@biochem.uni.kiel.de

In the still growing family of sialic acids with more than 40 different derivatives of neuraminic acid, the interest of the scientific community is increasingly directed towards the *O*-acetylated species. In the past two decades it evolved that *O*-acetylated sialic acids play fundamental roles in the development of organisms, in the regulation of the immune system, in cancer processes and many other biological and pathophysiological events[1-4].

Naturally occurring sialic acids can be *O*-acetylated at all their hydroxyl groups, i.e. at positions C4, C7, C8 and C9 of these carboxylated sugars as it is shown in Figure 1. They mostly are mono-*O*-acetylated, but combinations of acetyl groups at two or more positions generate oligo-*O*-acetylated derivatives. The biological effect is depending on the position relative to the 9-carbon scaffold and whether it is the *N*-acetylneuraminic acid

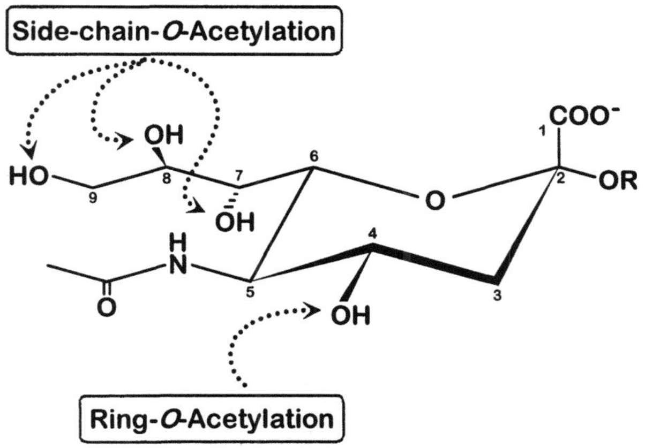

Figure 1. Naturally occurring *O*-acetylated sialic acids
Sialic acid *O*-acetylation can take place at the positions C4, C7, C8 and C9. They can be mono-*O*-acetylated or oligo-*O*-acetylated, e.g. a combination at C7 and C9 leads to the di-*O*-acetylated Neu5,7,9Ac_3.

The Molecular Immunology of Complex Carbohydrates–2.
Edited by Albert M. Wu, Kluwer Academic/Plenum Publishers, 2001.

(Neu5Ac) or the *N*-glycolylneuraminic acid (Neu5Gc). For example, these modifications prevent a fast degradation of sialylated glycoconjugates by prokaryotic and eukaryotic sialidases, trans-sialidases and sialate lyases[3]. Other biological effects will also be discussed in the paragraphs "Biological functions" of the specific sections.

Although *O*-acetylated sialic acids occur mainly in the animal kingdom, there are some exceptions. Among the prokaryotes, certain sialic acids-expressing bacteria, most of them pathogenic for mammalians, contain *O*-acetylated sialic acids[5], as will be shown later.

It is well known that eukaryotic sialic acid anabolism is a complex enzymatic event taking place in the cytosol and ending up with the activation of free Neu5Ac by CTP in the nucleus. The resulting CMP-Neu5Ac and CMP-Neu5Gc, the latter being synthesised by a cytosolic CMP-Neu5Ac hydroxylase (EC 1.14.99.18)[6], serve as donor substrates in various acceptor-substrate-specific sialyltransferase reactions in the Golgi apparatus. Finally, the

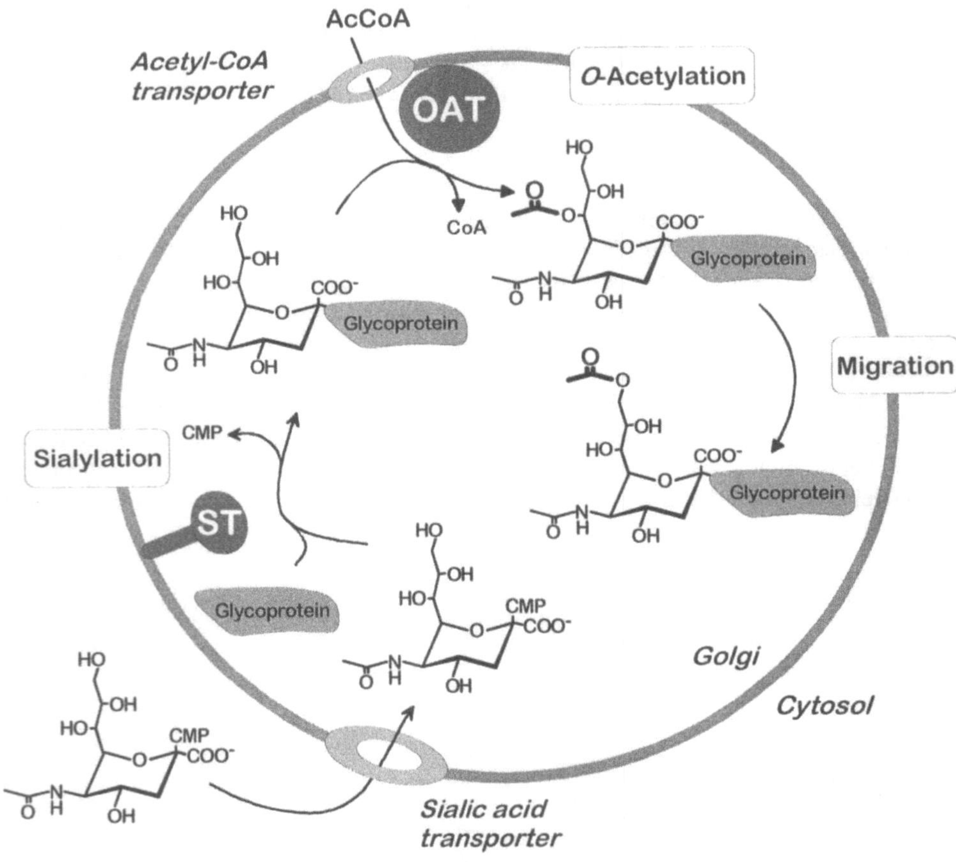

Figure 2: Model of side-chain *O*-acetylation of sialic acids
Sialic acids are transported in their CMP-activated form into the Golgi apparatus, where they serve as substrates for different sialyltransferases (ST). As an example here the transfer onto a glycoprotein is shown, although the same mechanism may function for glycolipid sialylation. AcCoA enters the Golgi via an AcCoA-transporter to provide it for the *O*-acetyltransferase (OAT), which transfers the acetyl moiety to the glycosidically bound sialic acid probably at position C7. After a migration to position C9, which could be enzymatically catalyzed, an additional *O*-acetylation reaction can take place at position C7. This model, which has not yet been firmly established in all details, is based mainly on the references [7] and [8]. From our unpublished analytical data the *O*-acetylation at the stage of CMP-Neu5Ac can not be completely excluded. Sialic acid 4-*O*-acetylation may occur in a similar way.

sialate-*O*-acetylation occurs at the α-glycosidically linked sialic acid[1].

The understanding of the metabolism of *O*-acetylated sialic acids can be divided into two parts. Firstly, the formation of acetyl esters at the different hydroxyl groups of sialic acids by the transfer of activated acetyl groups from acetyl coenzyme A (AcCoA) by diverse *O*-acetyltransferases (Figure 2). Differing in their regioselectivity, basically two groups of sialate-*O*-acetyltransferases exist: the acetyl-CoA:sialate 7(9)-*O*-acetyltransferase (EC 2.3.1.45) is responsible for the frequently found side chain *O*-acetylation which can most impressively be observed in bovine submandibular glands[9], whereas the reaction of the acetyl-CoA:sialate 4-*O*-acetyltransferase (EC 2.3.1.44) leads to the pyranose ring-*O*-acetylation at C4. As shown for the biosynthesis of 9-*O*-acetylated sialic acid in rat liver and bovine submandibular glands, the 4-*O*-acetyltransferase reaction investigated in guinea-pig liver takes place in the Golgi apparatus or trans-Golgi-network. However, up to now no structural information about a purified enzyme or its gene from either tissue is available, although several attempts to isolate the enzyme from different tissues were made by using various techniques including classical enzyme purification or expression cloning[10].

Secondly, several enzymes degrading free or glycosidically bound *O*-acetylated sialic acids are well characterized, like esterases for $Neu5,9Ac_2$ and $Neu4,5Ac_2$, respectively, which are found e.g. in influenza C virus and frequently in lysosomes of mammalian cells on the one hand or in mouse hepatitis virus and in horse liver, on the other[3,11-13].

ANALYSIS OF *O*-ACETYLATED SIALIC ACIDS

Before discussing the biochemistry and biology of sialic acid *O*-acetylation in pro- and eukaryotic systems, the very subtle process of the analysis of these sugars will be shortly addressed. A severe problem in the analysis of glycoconjugate-bound *O*-acetylated sialic acids is the initial step of glycosidic bond cleavage, since the mildest hydrolysis, which is possible by sialidases, will lead to an incomplete release of derivatives acetylated at the side chain, whereas 4-*O*-acetylated sialic acids are not released at all[2,3], with the exception of some viral sialidases[14]. Alternatively, hydrolysis with various mild acids is often applied[15], which allows the liberation of all kinds of sialic acids but is a compromise between a maximum release of sialic acids, however, slightly dependent on the type of carbohydrate linkage and the character of the glycoconjugate itself, and a minimal loss of *O*-acetylation. Alkaline treatment, of course, has to be excluded during this kind of analysis. Due to these reasons we have to keep in mind that 20-30% of *O*-acetylated derivatives are lost during careful hydrolysis and the subsequent purification procedure of the liberated monosaccharides by ion-exchange chromatography[16].

Presently, methods for a fast, sensitive and reliable sialic acid analysis, beginning with colorimetric determination, TLC, HPLC[17], Reversed Phase-HPLC of fluorescently labelled compounds[18], also in combination with Electrospray-Ionisation-MS[19], and GC-MS after pertrimethylsilylation[20], to NMR[21] have greatly improved and enable detection of these compounds in the fmol range.

In addition to these physico-chemical techniques, other methods for the analysis of *O*-acetylated sialic acids are known, among them antibody recognition, lectin staining and detection by influenza C viruses. The different methods for the use of influenza C virus as analytical tool[22,23] are based on its receptor-destroying enzyme, the sialate 9(4)-*O*-acetyl esterase (EC 3.1.1.53)[24]. Only few lectins are known to recognize specifically either 9-*O*-acetylated[25,26] or both, 4-*O*-acetylated and 9-*O*-acetylated sialic acids, as found in crustaceans[27,28]. The majority of sialic acid-specific lectins only recognise glycosidically bound Neu5Ac or Neu5Gc[1]. In contrast to the use of lectins, monoclonal antibodies against sialate side chain *O*-acetylation are well known[29-32], whereas a recognition of 4-*O*-acetyl sialic acids by antibodies has been rarely reported[33].

SIALIC ACID 4-*O*-ACETYLATION

Occurrence

Due to the lack of 4-*O*-acetylated sialic acids in man, where 9-*O*-acetylation is predominantly found, not much interest was focussed to the only possible pyranose ring *O*-acetylation of sialic acids in former sialic acid analyses, though a report of Neu4Ac5Gc-GM3 describes this ganglioside to be a tumor-associated antigen in human colon cancer[33]. However, no additional examples of sialate-4-*O*-acetylation in man have been given.

Another reason for this neglect may have been the previously mentioned poor preparative and analytical tools for sialic acid analysis. In parallel to the development of techniques protecting the *O*-acetylated sialic acids during purification, reports showed these derivatives to occur in a Japanese dace[34], a South American pit-viper[35], the Australian monotreme echidna[36], the horse, the donkey[3], and in the rabbit[37] besides the guinea-pig[38].

Metabolism

As previously mentioned in the introduction, several species- and substrate-specific *O*-acetyltransferases (OAT) occur, none of which has been purified or cloned so far. In this respect the investigation of the 4-OAT from guinea-pig liver, which has been described for the Golgi-membrane bound enzyme[39], seems to be a promising approach.

The enzyme revealed a high affinity to acetyl-CoA and CoA, expressed by an app. K_M and an apparent K_i of 0.6 and 4.2 μM, respectively; CoA inhibited the enzyme in a mixed competitive manner. Therefore, affinity chromatography by either acetyl-CoA or CoA matrices is supposed to be a good strategy to increase the yield of specific enzyme activity and to purify the enzyme to homogeneity. We were able to solubilize the 4-OAT, which is stable even after freeze- and thawing cycles. Among various substrates, including free sialic acids, sialosaccharides and sialoglycoproteins, gangliosides have been shown to be *O*-acetylated most effectively.

The Golgi-membrane-bound 9-OAT in rat liver, extensively investigated by Varki and co-workers[1,7,40], obviously shows parallels with regard to the kinetic and inhibition parameters and thus allows deduction of basic properties for this enzyme family in general. However, the existence of a transmembrane enzyme complex, which was favoured in rat liver, has never been unequivocally confirmed.

Another enzyme system under investigation, which greatly differs from the system described above, is the 4-OAT from horse submandibular glands, first mentioned by Schauer[41]. A soluble enzyme within this tissue, as observed in bovine submandibular glands, could not be verified. Due to the unusual constitution of mucin-secreting cells, an isolation of Golgi-membranes is difficult. However, the endogenous 4-OAT activity was detected in the microsomal fraction of this tissue. Work is in progress to characterize and solubilize the membrane-bound enzyme. It will be very interesting to look at its substrate specificity, since *O*-glycan-bound Neu4,5Ac$_2$ is the main *O*-acetylated sialic acid in the horse submandibular gland[42], whereas Neu4Ac5Gc-GM3 is the main *O*-acetylated component on horse erythrocyte membranes[43].

The finding of the 4,8-anhydro derivative of Neu5Ac, first described in collocalia mucin from edible birds nest substance by Pozsgay et al.[44], raises the question for its origin.

These authors suggested that this sialic acid modification originates from Neu4,5Ac$_2$ during the release of glycosidically bound sialic acids under acidic conditions. Detailed investigations by Manzi et al.[45], however, revealed a conversion of free, but not glycosidically bound Neu4,5Ac$_2$ to its anhydro compound under alkaline conditions. A small amount of 4,8-anhydro-Neu5Ac during acid hydrolysis of horse serum confirmed the assumption of Pozgay, but does not explain the relatively high amount (10 %) of this anhydro-sialic acid in collocalia mucin. In our laboratory this sialic acid was isolated only from an older preparation of this mucin, whereas two fresh samples obtained from Indonesia and Vietnam, kindly provided by A. Wu, Taipei, contained Neu5Ac exclusively.

Because eukaryotic sialidases are not able to liberate 4-*O*-acetylated sialic acids, and acylneuraminate lyases (EC 4.1.3.3), which split free sialic acids into acylmannosamines and pyruvate, do not accept 4-*O*-acetylated sialic acid as substrates, too, the existence of a further enzyme has been assumed that removes such *O*-acetyl esters prior to further degradation of the sialic acid. This led to the discovery of a sialate 4-*O*-acetylesterase (4-OAE) in horse liver which has been partially purified (Schauer *et al.*, unpublished results). It exists in isoforms and also hydrolyses sialate-9-*O*-acetyl esters, although to a lesser extent[24]. Recently, a viral 4-OAE has been reported to occur in mouse hepatitis virus (MHV)[13], as already mentioned above.

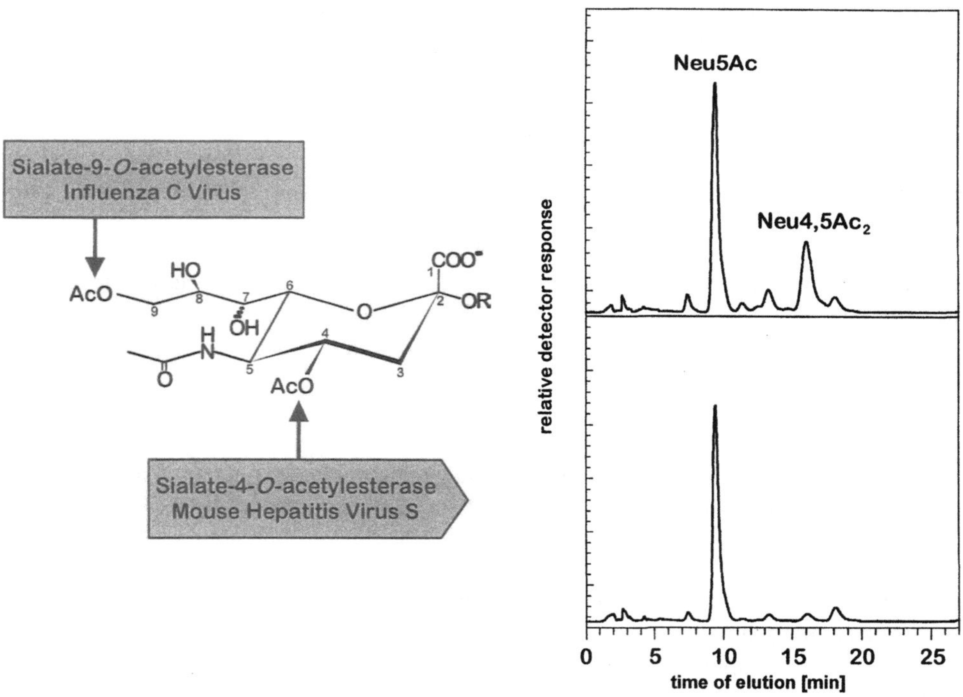

Figure 3: Viral esterases as receptor-destroying enzymes
Mouse hepatitis virus (MHV) is able to de-*O*-acetylate guinea-pig serum glycoprotein, which carry glycosidically bound Neu4,5Ac$_2$, as shown on the right site. The upper chromatogram shows the sialic acid composition of untreated serum glycoproteins, the lower panel displays the situation after treatment of the glycoproteins with MHV containing 4-OAE activity. Sialic acids were analysed after hydrolysis by fluorimetric HPLC. On the left a comparison of the different sites of OAE activities from influenza C virus and MHV are indicated.

Biological function

Still little is known about the role of 4-*O*-acetylated sialic acids compared to the wealth of information on the corresponding side chain modifications. The most obvious effect of the sialic acid 4-*O*-acetyl group is its protective function of the parent molecule towards the degrading enzymes sialidase and acylneuraminate-lyase. As a potential inhibitor of influenza virus infection, Neu4,5Ac_2 does not appear to increase the avidity of interaction with the viruses, since isolated oligosaccharides of equine α_2-macroglobulin are no stronger inhibitors of adsorption than oligosaccharides of human α_2-macroglobulin, which is a relatively poor inhibitor and contains only Neu5Ac. Since Neu4,5Ac_2 is almost resistant to cleavage by viral sialidases, it may protect the inhibitor from inactivation by these enzymes[46,47]. There is no evidence for 4-*O*-acetylated sialic acids serving as ligands for endogenous acceptors as it was investigated with 9-*O*-acetylated species, but the discovery of the viral 4-OAE may give us a hint in this direction.

Thus, the predominant role of 4-*O*-acetylated sialic acids seems to be its influence on the catabolism of respective carbohydrate chains. Therefore, the 4-OAE is involved in the initial step of degradation of such carbohydrate moieties, as was discovered in horse liver[3]. Recent investigations revealed also a 4-OAE in mouse hepatitis viruses (MHV, Figure 3). The function of this esterase is not clear at present, as there is no evidence for any sialate-4-*O*-acetylation in adult murine brain. As MHV infects the juvenile murine brain, a developmentally regulated expression of 4-*O*-acetylated sialic acids in this brain may be expected. Interestingly, lectin-binding investigations in mature male mice point to the existence of sialate-4-*O*-acetylation[48], though the frequent occurrence of sialate-9-*O*-acetylation in this species is well known[49,50].

SIALIC ACID 7(9)-*O*-ACETYLATION

Occurrence

Based on the above mentioned improved analytical techniques increased knowledge of *O*-acetylation of the glycerol side chain of sialic acid and its widespread distribution is now available. The *O*-acetylated derivatives of Neu5Ac and Neu5Gc are present in almost all deuterostomal species[2-4], in a yeast strain[51] and also in a few bacteria, as will be shown later in this article.

The expression is tissue- and development-specific; for example, the human colon mucosa shows high levels of *O*-acetylation of sialic acids, whereas during cancerogenesis the *O*-acetylation decreases already in the first stage of the adenoma-carcinoma sequence[52]. In contrast to this, melanoma[2] as well as basalioma[53] are characterized by a higher expression of *O*-acetylated gangliosides than normal skin; these glycosphingolipids can therefore be considered as tumor-associated antigens. A similar phenomenon was observed during the development of chicken erythrocytes. After hatching, the 9-*O*-acetylation of sialic acids on the surface of erythrocytes is increasing[54].

As another example, side-chain *O*-acetylated sialic acids are also expressed on human leukocytes in a range of 1 to 5 % of the total sialic acid content (G. Kohla, unpublished results,[55]), where they can be used as markers for specific subsets of T-helper cell, known as CDw60, when the 9-*O*-acetylated sialic acid is the terminal sugar of the disialoganglioside GD3[56], which is also expressed on activated B-cells[57]. Flow cytometric analysis showed that each major class of human leukocytes contains a significant fraction of cells expressing 7-*O*-acetyl GD3[30]. Large amounts of various *O*-acetylated sialic acids were also found in bovine submandibular gland[9,19], rat liver[58], healthy human colon[52] and the erythrocytes from rat, mouse and rabbit[59].

Metabolism

The enzymatic *O*-acetylation of sialic acids was first described in bovine and equine submandibular glands[41,60]. In comparison to the *O*-acetylation of the ring hydroxyl group at C4, described in the previous chapter, here the acetyl esters are formed with the hydroxyl groups of the glycerol side chain of sialic acids. All three hydroxyl groups can be *O*-acetylated, but up to now it is not known whether there are different enzymes responsible for the *O*-acetylation of each hydroxyl group. Following our favoured hypothesis[61], only one enzyme catalyses the acetylation reaction at position C7 and a subsequent migration of the *O*-acetyl groups to C8 and C9 occurs (see Figure 2). Therefore, this transferase is addressed as AcCoA:sialate 7(9)-*O*-acetyltransferase. Data from the bovine submandibular gland system, which represents a biosynthetically highly active tissue, support the latter hypothesis[8].

Only few successful attempts are known for partial purification and characterization of the enzyme. Diaz and co-workers studied the enzyme from rat liver[62]. They found that it is membrane-bound and localized in the Golgi membranes, similar to the 4-*O*-acetyltransferase from guinea pig liver[39]. At least in rat liver, it seems to be co-localized with an acetyl-CoA-transport protein, which can enhance the transferase activity efficiency. The localisation in Golgi membranes supports the momentary hypothesis that glycoconjugates are sialylated in the trans-Golgi or the trans-Golgi network and *O*-acetylation of sialic acids is an event, which modifies glycosidically bound sialo-glycoconjugates at a late step of oligosaccharide maturation.

Another tissue, used for the purification and characterisation of the enzyme, is bovine submandibular gland. As in rat liver, the enzyme activity is membrane-bound and co-localized in microsomes with sialyltransferase activity, typical for Golgi-membranes[8]. In addition, a soluble enzyme activity was found in the preparations of this tissue, and it was partially purified (R. Schauer, unpublished results). Whether this is a product of proteolysis or a splicing variant of the membrane-bound enzyme, is yet unknown.

The primary attachment site of the acetyl residue transferred by this enzyme in bovine submandibular gland seems to be at C7[8]. It is well established that the *O*-acetyl ester of sialic acids can migrate from C7 to a free hydroxyl group at C9[63]. Based on these findings we demonstrated that influenza C viruses exclusively bind to 9-*O*-acetylated ganglioside GD3, but not to the 7-*O*-acetylated derivative (Figure 4). After weak alkaline treatment, which induces migration from position C7 to C9, staining by influenza C virus appeared at the location of previous 7-*O*-acetyl-GD3.

Single *O*-acetyl groups at C8 seem to be very labile, presumably because they migrate quickly to C9. Di- or tri-*O*-acetylated sialic acids may be therefore the result of a combination of such a transfer of acetyl groups, and the additional transfer of a second and later eventually a third acetyl group again to the free hydroxyl group at C7 [3,61].

Since *O*-acetylated sialic acids more or less withstand the attack of sialidases or lyases, sialate-9-*O*-acetylesterases (9-OAE) play an important role in the catabolic pathway of these sialic acid derivatives[64]. Such esterases have been studied in detail in rat liver, bovine brain and influenza C virus[4,11,24,65,66]. So far two different OAE-systems were described in mammalian tissues, a cytosolic enzyme that acts on free sialic acids, whereas another enzyme is a membrane-bound, complex- and high-mannose-type *N*-glycans carrying protein[4], which can act on both, free and glycosidically bound *O*-acetylated sialic acids, although on the latter only to a lesser extent[4,67].

Biological functions

By introducing a more hydrophobic moiety into the sialic acid molecule the *O*-acetyl group changes parameters like size, net charge, hydrogen bonding and conformation of glycoconjugates[68]. One consequence of this is a modulation of their biological functions (see also Table 1).

For example, 7(9)-*O*-acetylated sialic acids influence the haematopoietic development[69] and they prevent the recognition of cell surface glycoconjugates by siglecs, which are sialic acid-dependent adhesion molecules of the immunoglobulin superfamily like sialoadhesin and CD22[70], supposed to function in macrophage and in lymphocyte interactions, respectively, or in the maintenance and integrity of myelinated axons in the case of the myelin-associated glycoprotein (MAG)[71]. A further "anti-adhesion effect" exerted by such *O*-acetyl groups is the hindrance of influenza A and B viruses to sialic acids[72].

For several subsets of T-lymphocytes (e.g. CDw60, 9-*O*-acetylated ganglioside GD3) *O*-acetylation of the glycerol side chain of sialic acids is known to be a differentiation marker [57,73] and is therefore of particular interest from the immuno-cytological point of view. Binding of an antibody, specific for 7-*O*-acetylated GD3, to T-cells stimulates their proliferation, indicating the involvement of this ganglioside in signal transduction processes[30].

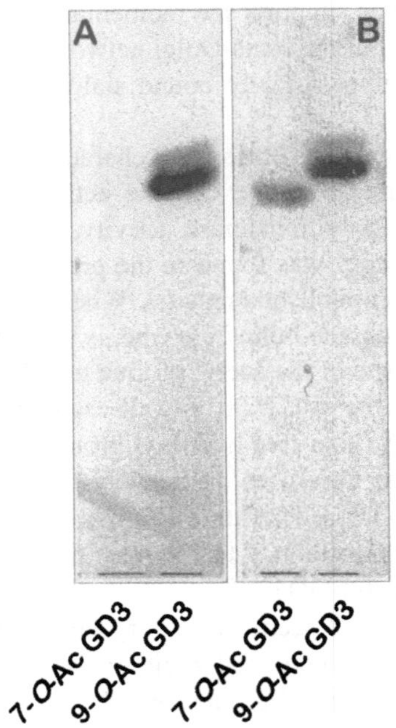

Figure 4: Binding specificity of influenza C virus
The migration of *O*-acetyl groups from position C7 to C9 is demonstrated by staining bands of 7- and 9-*O*-acetylated GD3 using influenza C virus specific for 9-*O*-acetylated sialic acids. In each lane 1µg of the respective ganglioside was applied. Staining was achieved by using the esterase activity of the virus towards the unspecific esterase substrate α-naphtyl acetate. Before alkaline treatment (A), binding only occurs to 9-*O*-AcGD3. After weak alkaline treatment at pH 10 (B) a band corresponding to the former 7-*O*-Ac-GD3 is also stained, indicating a migration to the C9 position of the sialic acid of this ganglioside. *O*-Acetylated gangliosides were a kind gift of Bernhard Kniep (Institut für Immunologie, Technische Universität Dresden, Germany).

Like its increased expression in mammary tumors[74], the presence of 9-*O*-Ac-GD3 on human melanoma[75] and basalioma[53] cells is considered as a tumor-associated antigen. Enhanced expression of melanoma-associated 9-*O*-acetyl-GD3 ganglioside stimulates cellular growth and suppresses certain differentiated phenotypes such as dendrite formation as shown for hamster melanoma cell lines[76]. The expression of 9-*O*-acetyl gangliosides on murine dorsal root ganglions correlates with periods of cell migration and axonal pathfinding[77]. Here the addition of monoclonal antibodies directed towards these gangliosides leads to a collapse of the growth cones. As in other brain regions, the pattern of expression of 9-*O*-acetyl-GD3 in the developing rat hippocampus also correlates with the migration of cells and their processes[78].

The influenza C virus enters the human organism via the respiratory tract, its natural target for binding are the 5-10 % Neu5,9Ac_2 of human nasal mucin[79], although the function of the viral hemagglutinin esterase may be, similar to the function of the sialidases from influenza A and B viruses, the detachment of the budding virus particle from its host cell[80]. By expression of the influenza C virus hemagglutinin esterase in murine neuroblastoma cell lines[81] or in transgenic mice, the significance of *O*-acetylation for the morphogenesis and development of the retina, cerebellum and adrenal gland, and for the organisation and function of neuronal tissue was clearly demonstrated[50].

Table 1: Biological effects of *O*-acetylated sialic acids

Effect	References
Modification of the physicochemical properties of glycoconjugates by increasing the hydrophobicity	82
Inhibition of the activities of sialidase, *trans*-sialidase, endoglycosidase and sialat-pyruvate lyase	2,82,83
Decreased cell degradation, e.g. of erythrocytes	82,84,85
Decreased complement pathway activation	86,87
Decreased binding of influenza A and B viruses, reovirus, *Plasmodium falciparum*, sialoadhesin, CD22 and MAG	72,88,89
Prevention the infection of cattle by *Trypanosoma brucei*	90
Increased binding of influenza C virus, Mouse hepatitis virus, Coronavirus, Encephalomyelitis virus	11,91,92
Decreased antigenicity	82,93,94
Increased bacterial pathogenicity	95
Ligands for binding antibodies and lectins	28,82,96,97
Differentiation- and tumor-associated antigens	50,54,75,82,94,98-105
Stimulation of cell growth	69,74,76,77,81,106

The previously mentioned highly *O*-acetylated sialic acids of human colonic mucins may lead to a better protection of the colon against certain microorganisms[64]. During cancerogenesis, concomitantly with the decrease of sialic acid *O*-acetylation, the expression of sialyl-Lewis(x) antigens on sialo-mucins increases, especially in metastases, because these structures were masked by the ester groups before[107].

BACTERIAL *O*-ACETYLATION

Occurrence

In the bacterial kingdom *O*-acetylated sialic acids or other *O*-acetylated carbohydrates have so far only been found in two surface components of bacteria: in the capsule or in the lipopolysaccharide (LPS). The latter is a surface glycolipid molecule found in the outer leaflet of the outer membrane in Gram-negative bacteria. It consists of a hydrophobic moiety called lipid A, a core oligosaccharide, and the O-specific side chain or O-antigen made of repeating units from different sugars[108]. In contrast to this, the capsular polysaccharides are only carbohydrate polymers comprised of repeating saccharide units.

The first detection of an *O*-acetylated sugar in the O-antigen was in 1969 by Hellerqvist and co-workers, who described an *O*-acetylated abequose residue in the repeating unit of *Salmonella typhimurium*[109]. Later other *O*-acetyl groups on various carbohydrates were found in different enteric bacteria like *Escherichia coli*[110], *Shigella dysenteriae*[111] or *Klebsiella pneumoniae*[112]. In the last years more and more *O*-acetylated sugars of the repeating unit were described in many bacterial strains. Interestingly, nearly all of them are important pathogenic organisms, for example *Proteus penneri*, which is a common cause of urinary tract infections[113], *Acinetobacter calcoaceticus*, a causative agent of various severe nocosomial infections in immuno-compromised patients[114], or the sheep and cattle pathogen *Campylobacter fetus*[115]. *O*-Acetylation was also found in the capsular polysaccharides type 5 and type 8 of the Gram-positive *Staphylococcus aureus*, a major cause of nocosomial infections[116].

O-Acetylated sialic acids were only detected as capsular components of the serotypes A, C, W-135 and Y of *Neisseria meningitidis*[117] and in the capsule or the O-antigen of the LPS in some *E. coli* strains. One of them, the serotype O104:K⁻:H12 (strain-number: 21701) is an isolate from human faeces. It has no capsule and its O-antigen includes a sialic acid residue. The *O*-acetylation pattern of this sialic acid was discussed controversial. Kogan et al.[118] described an only 9-*O*-acetylated tetrasaccharide and Gamian et al.[119] a 7- and 9-di-*O*-acetylated form as repeating unit. Now it is clear, that the O-antigen of this *E. coli* strain 21701 consists of multiple copies of a tetrasaccharide with an internal sialic acid residue that is always di-*O*-acetylated on C7 and C9, as it is shown in Figure 5.

Metabolism

As outlined above, a sialic acid-specific *O*-acetyltransferase could not yet be purified. Therefore, the strategy to obtain more information about this enzyme had to be changed. In the case of the production of *O*-acetyl-deficient bacterial mutants, it would be possible to get more information about the role of *O*-acetyl groups. But neither with direct mutagenesis using transposons nor with a random mutagenesis system using UV-light it was possible to generate such mutants. A second approach involves the finding of the gene encoding the OAT-protein and the construction of the desired mutants by knocking out this gene. The amino acid sequences of some cloned *O*-acetyltransferases specific for various carbohydrates already exist in the database, and they appear to be more or less similar. One of them occurs in *Shigella flexneri*, designated "*oac*", and is responsible for the *O*-acetylation of the O-antigen. Lysogens of these bacteria harbour the temperate bacteriophage Sf6 and the gene for the modification is surprisingly encoded by the phage and not by its host[120]. The integrated prophages can be induced with mitomycin C or UV-light, but both methods had no effect on the *E. coli* strain 21701. Thus, in this strain and in all other bacteria the so far known genes, responsible for *O*-acetylation, are encoded by the bacteria themselves. Slauch *et al.*[121] aligned a series of these proteins that define a family of membrane proteins involved in the acylation of carbohydrate moieties on extracytoplasmic molecules and they showed some homologies between them. These proteins include OfaA from *Salmonella typhimurium* (*O*-acetylation of the O-antigen), GumF from *Xanthomonas campestris pv. campestris* (*O*-acetylation of the O-antigen), Sf6 and several proteins from *Streptomyces spp.* and *Rhizobium spp.* Also the *lag-1* locus responsible for *O*-acetylation of the *Legionella pneumophila* O-antigen that was cloned and characterized by Zou et al.[122] showed homologies to this family.

The ability of this locus to produce *O*-acetylation in former non-*O*-acetylated strains leads to the suggestion that *lag-1* indeed encodes an *O*-acetyltransferase. In order to confirm this, it is necessary to overexpress this or another possible *O*-acetyltransferase and to measure transferase activity, but until now, there is no *in vitro* assay available. To establish an enzyme test, we examined with the *E. coli* strain 21701 various exogenous substrates like de-*O*-acetylated bacteria, their isolated LPS, different gangliosides (GD3, GT1b or GQ1b), sialyllactose or free Neu5Ac in the presence or absence of substances, e.g. different detergents like CHAPS and Triton X-100 as well as DFP as an esterase-inhibitor and AcCoA, but in no case radioactively labelled *O*-acetyl groups were incorporated. So far even the acetyl donor for this reaction is unclear. We assume AcCoA as the donor, although there is no direct evidence for this hypothesis in bacteria. Another open question is where and when the *O*-acetylation reaction takes place. This may occur at the cytoplasmic side of the inner membrane before the O-unit is flipped and polymerized[123], although there exists no direct evidence for this mechanism. Clarification of this situation affords a lot of future work.

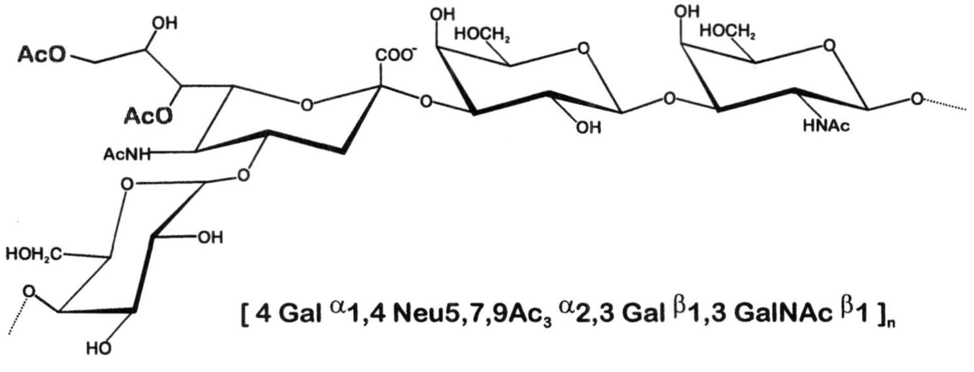

Figure 5: Oligosaccharide repeating unit of the O-antigen in theLPS from *E. coli* strain 21701, Serotype: O104:K⁻:H12
The structure of this unit was analyzed using NMR-Spectroscopy by Ulrich Zähringer (Forschungszentrum Borstel, Zentrum für Medizin und Biowissenschaften, Germany).

Biological functions

In contrast to higher organisms, there exist only little data about the function of this sialic acid modification in microorganisms. As mentioned above, nearly all bacteria with *O*-acetylated sugar residues are important pathogens, indicating that there might be a correlation between *O*-acetylation and pathogenicity. This was shown for *Salmonella typhi*, the causative agent of typhoid fever, with the capsular polysaccharide as a virulence factor that is stabilized by *O*-acetyl groups[124]. Another example for a direct relationship between *O*-acetylation and pathogenicity was demonstrated in *Neisseria meningitidis* group C, which is responsible for approximately 35% of meningococcal diseases in the USA. Most strains of this group produce a homopolymer of α2,9-linked Neu5Ac with *O*-acetyl groups at C7 and C8. About 85% of the strains causing diseases were reported to be *O*-acetyl-positive[125]. This immuno-dominance was also found for *E. coli* K1[126] and was shown for *E. coli* strain 21701, too. Mice immunized with *O*-acetylated bacteria gave a significant higher immune response than mice immunized with de-*O*-acetylated material. Since nothing more is known about the effects of *O*-acetylated sialic acids in bacteria there is still a controversial discussion about the role of these sialic acids in the literature.

CONCLUSIONS AND OUTLOOK

Sialate *O*-acetylation is a crucial cellular event especially in the animal kingdom and in man and causes a vast spectrum of biological effects. Due to their prevailing presence on glycoconjugates of the outermost leaflet on the plasma membrane, sialic acids, *O*-acetylated at their side chains, mainly as $Neu5,9Ac_2$, mediate or prevent binding of cells to each other and of viruses or bacteria to their host cells. They therefore revealed to be important in cell differentiation, neuronal development, non-immunogenic protection and immuno reactivity, or on the other hand, they are involved in cancer progression and virus invasion. *O*-Acetylated sialic acids thus are great mediators of cellular cross talk.

In contrast, our knowledge about the function of 4-*O*-acetylated sialic acids is limited and restricted to the resistance towards sialic acid degrading enzymes. However, the discovery that mouse hepatitis viruses recognize free and glycosidically bound $Neu4,5Ac_2$ stimulates the discussion that sialate-4-*O*-acetylation is more than just influencing enzyme reactivity, but may also be important for recognition phenomena similar to the side chain *O*-acetylation.

It is therefore no question that *O*-acetylation is an exciting and promising field of research, which increased rapidly during the past ten years, due to the very sensitive analytical tools that have been developed within this time on the one hand and to the great interest in this sialic acid modification on the other.

Due to the extraordinary interest of the scientific community in sialate-*O*-acetylation, different approaches to purify or clone the sialate-*O*-acetyltransferases from diverse kinds of animals were made. In the tissues and cells investigated, sialate-*O*-acetyltransferases are membrane-bound enzymes situated in the Golgi apparatus. *O*-Acetylation seems to take place preferentially on α2,6- or α2,8-linked sialic acids. The transfection of an α2,6-specific sialyltransferase into CHO-cells provides the substrates for an *O*-acetyltransferase, which was "quiet" before, when the cells expressed α2,3-linked sialic acids only[127]. Remarkably, the richest natural sources of *O*-acetylated sialic acids known are human and bovine mucins as well as 9-*O*-acetylated gangliosides.

All efforts to clone the enzymes failed so far, although various reports seem to get closer to the possibility to identify unequivocally a sialate-*O*-acetyltransferase gene. It was, however, possible to clone and sequence the gene for one of the components of the hypothetical sialate *O*-acetyltransferase complex, the AcCoA-transporter[128]. In contrast to several attempts made from the molecular biologists point of view, our present approach is a more classical way of enzyme purification to resolve the amino acid sequence. Once the miracle is solved, it will be of great scientific impact to learn more about the regulation of *O*-acetylated sialic acid expression in embryogenesis, immunology and tumor biology. Although the product of the biosynthesis is well characterized, the way it is formed is still waiting to be elucidated.

REFERENCES

1. S. Kelm and R. Schauer (1997) Sialic acids in molecular and cellular interactions. Int Rev Cytol, 175:137-240.
2. R. Schauer, S. Kelm, G. Reuter, P. Roggentin and L. Shaw (1995) Biochemistry and role of sialic acids, *in*: Biology of the Sialic Acids (A. Rosenberg, ed.) pp 7-67, Plenum Press, New York.
3. R. Schauer and J.P. Kamerling (1997) Chemisty, biochemistry and biology of sialic acids, *in*: Glycoproteins II (J. Montreuil, J.F.G. Vliegenthart and H. Schachter, eds.) pp 243-402, Elsevier, Amsterdam.
4. A. Varki (1992) Diversity in the sialic acids. Glycobiology, 2:25-40.
5. C. Traving and R. Schauer (1998) Structure, function and metabolism of sialic acids. Cell Mol.Life Sci., 54:1330-1349.
6. L. Shaw and R. Schauer (1988) The biosynthesis of *N*-glycoloylneuraminic acid occurs by hydroxylation of the CMP-glycoside of *N*-acetylneuraminic acid. Biol Chem Hoppe Seyler, 369:477-486.
7. A. Varki and S. Diaz (1985) The transport and utilization of acetyl coenzyme A by rat liver Golgi vesicles. *O*-acetylated sialic acids are a major product. J Biol Chem, 260:6600-6608.
8. V. Vandamme-Feldhaus and R. Schauer (1998) Characterization of the enzymatic 7-*O*-acetylation of sialic acids and evidence for enzymatic *O*-acetyl migration from C-7 to C-9 in bovine submandibular gland. J Biochem (Tokyo), 124:111-121.
9. R. Schauer, S. Stoll and G. Reuter (1991) Differences in the amount of *N*-acetyl- and *N*-glycoloylneuraminic acid, as well as *O*-acylated sialic acids, of fetal and adult bovine tissues. Carbohydr Res, 213:353-359.
10. W.X. Shi, R. Chammas and A. Varki (1998) Induction of sialic acid 9-*O*-acetylation by diverse gene products: implications for the expression cloning of sialic acid *O* -acetyltransferases. Glycobiology, 8:199-205.
11. G. Herrler, R. Rott, H.D. Klenk, H.P. Müller, A.K. Shukla and R. Schauer (1985) The receptor-destroying enzyme of influenza C virus is neuraminate-*O*-acetylesterase. EMBO J, 4:1503-1506.
12. H. Takematsu, S. Diaz, A. Stoddart, Y. Zhang and A. Varki (1999) Lysosomal and cytosolic sialic acid 9-*O*-acetylesterase activities can be encoded by one gene via differential usage of a signal peptide-encoding exon at the N terminus. J Biol Chem, 274:25623-25631.
13. G. Regl, A. Kaser, M. Iwersen, H. Schmid, G. Kohla, B. Strobl, U. Vilas, R. Schauer and R. Vlasak (1999) The hemagglutinin-esterase of mouse hepatitis virus strain S is a sialate-4-*O*-acetylesterase. J Virol, 73:4721-4727.
14. A.P. Corfield, M. Sander-Wewer, R.W. Veh, M. Wember and R. Schauer (1986) The action of sialidases on substrates containing *O*-acetylsialic acids. Biol Chem Hoppe Seyler, 367:433-439.
15. T.P. Mawhinney and D.L. Chance (1994) Hydrolysis of sialic acids and *O*-acetylated sialic acids with propionic acid. Anal Biochem, 223:164-167.
16. G. Reuter and R. Schauer (1994) Determination of sialic acids. Methods Enzymol, 230:168-199.
17. A.E. Manzi, S. Diaz and A. Varki (1990) High-pressure liquid chromatography of sialic acids on a pellicular resin. Anal Biochem, 188:20-32.
18. S. Hara, M. Yamaguchi, Y. Takemori, K. Furuhata, H. Ogura and M. Nakamura (1989) Determination of mono-*O*-acetylated *N*-acetylneuraminic acids in human and rat sera by fluorometric high-performance liquid chromatography. Anal Biochem, 179:162-166.
19. A. Klein, S. Diaz, I. Ferreira, G. Lamblin, P. Roussel and A.E. Manzi (1997) New sialic acids from biological sources identified by a comprehensive and sensitive approach: liquid chromatography-electrospray ionization-mass spectrometry (LC-ESI-MS) of SIA quinoxalinones. Glycobiology, 7:421-432.

20. J.P. Kamerling and J.F. Vliegenthart (1975) Identification of *O*-acetylated *N*-acylneuraminic acids by mass spectrometry. Carbohydr Res, 41:7-17.
21. J.F. Vliegenthart, L. Dorland, H. van Halbeek and J. Haverkamp (1982) NMR Spectroscopy of sialic acids, *in*: Sialic Acids (R. Schauer, ed.) pp 127-172, Springer-Verlag, Wien.
22. G. Harms, G. Reuter, A.P. Corfield and R. Schauer (1996) Binding specificity of influenza C virus to variably *O*-acetylated glycoconjugates and its use for histochemical detection of *N*-acetyl-9-*O*-acetylneuraminic acid in mammalian tissues. Glycoconj J, 13:621-630.
23. G. Zimmer, G. Reuter and R. Schauer (1992) Use of influenza C virus for detection of 9-*O*-acetylated sialic acids on immobilized glycoconjugates by esterase activity. Eur J Biochem, 204:209-215.
24. R. Schauer, G. Reuter, S. Stoll, d.R. Posadas, G. Herrler and H.D. Klenk (1988) Isolation and characterization of sialate 9(4)-*O*-acetylesterase from influenza C virus. Biol Chem Hoppe Seyler, 369:1121-1130.
25. M.M. Iglesias, G.D. Cymes and C. Wolfenstein-Todel (1996) A sialic acid-binding lectin from ovine placenta: purification, specificity and interaction with actin. Glycoconj J, 13:967-976.
26. C. Mandal and S. Basu (1987) An unique specificity of a sialic acid binding lectin AchatininH, from the hemolymph of *Achatina fulica* snail. Biochem Biophys Res Commun, 148:795-801.
27. G.A. Fragkiadakis and E.K. Stratakis (1997) The lectin from the crustacean *Liocarcinus depurator* recognizes *O*-acetyl sialic acids. Comp Biochem Physiol B Biochem Mol Biol, 117:545-552.
28. M.H. Ravindranath and J.C. Paulson (1987) *O*-acetylsialic acid-specific lectin from the crab *Cancer anntenarius*. Methods Enzymol, 138:520-527.
29. E. Cerato, S. Birkle, J. Portoukalian, A. Mezazigh, J.F. Chatal and J. Aubry (1997) Variable region gene segments of nine monoclonal antibodies specific to disialogangliosides (GD2, GD3) and their *O*-acetylated derivatives. Hybridoma, 16:307-316.
30. B. Kniep, C. Claus, J. Peterkatalinic, D.A. Monner, W. Dippold and M. Nimtz (1995) 7-*O*-acetyl-GD3 in human T-lymphocytes is detected by a specific T-cell-activating monoclonal antibody. J.Biol.Chem., 270:30173-30180.
31. M. Koga, N. Yuki, T. Ariga and K. Hirata (1999) Antibodies to GD3, GT3, and *O*-acetylated species in Guillain-Barré and Fisher's syndromes: their association with cranial nerve dysfunction. J Neurol Sci, 164:50-55.
32. G. Zhang, L. Ji, S. Kurono, S.C. Fujita, S. Furuya and Y. Hirabayashi (1997) Developmentally regulated *O*-acetylated sialoglycans in the central nervous system revealed by a new monoclonal antibody 493D4 recognizing a wide range of *O*-acetylated glycoconjugates. Glycoconj J, 14:847-857.
33. I. Miyoshi, H. Higashi, Y. Hirabayashi, S. Kato and M. Naiki (1986) Detection of 4-*O*-acetyl-*N*-glycolylneuraminyl lactosylceramide as one of tumor-associated antigens in human colon cancer tissues by specific antibody. Mol Immunol, 23:631-638.
34. S. Inoue, M. Iwasaki, K. Ishii, K. Kitajima and Y. Inoue (1989) Isolation and structures of glycoprotein-derived free sialooligosaccharides from the unfertilized eggs of *Tribolodon hakonensis*, a dace. Intracellular accumulation of a novel class of biantennary disialooligosaccharides. J Biol Chem, 264:18520-18526.
35. G. Lochnit and R. Geyer (1995) Carbohydrate structure analysis of batroxobin, a thrombin-like serine protease from *Bothrops moojeni* venom. Eur J Biochem, 228:805-816.
36. J.P. Kamerling, L. Dorland, H. van Halbeek, J.F. Vliegenthart, M. Messer and R. Schauer (1982) Structural studies of 4-*O*-acetyl-α-*N*-acetylneuraminyl-(2 goes to 3)-lactose, the main oligosaccharide in echidna milk. Carbohydr Res, 100:331-340.
37. A. Klein and P. Roussel (1998) *O*-acetylation of sialic acids. Biochimie, 80:49-57.
38. K. Hanaoka, T.J. Pritchett, S. Takasaki, N. Kochibe, S. Sabesan, J.C. Paulson and A. Kobata (1989) 4-*O*-acetyl-*N*-acetylneuraminic acid in the *N*-linked carbohydrate structures of equine and guinea pig α_2-macroglobulins, potent inhibitors of influenza virus infection. J Biol Chem, 264:9842-9849.
39. M. Iwersen, V. Vandamme-Feldhaus and R. Schauer (1998) Enzymatic 4-*O*-acetylation of *N*-acetylneuraminic acid in guinea-pig liver. Glycoconj J, 15:895-904.
40. H.H. Higa, C. Butor, S. Diaz and A. Varki (1989) *O*-acetylation and de-*O*-acetylation of sialic acids. *O*-acetylation of sialic acids in the rat liver Golgi apparatus involves an acetyl intermediate and essential histidine and lysine residues--a transmembrane reaction? J Biol Chem, 264:19427-19434.
41. R. Schauer (1970) Biosynthesis of *N*-acetyl-*O*-acetylneuraminic acids. I. Incorporation of [^{14}C] acetate into sections of the submaxillary salivary gland of ox and horse. Hoppe Seylers Z Physiol Chem, 351:595-602.
42. R. Schauer (1978) Characterization of sialic acids. Methods Enzymol, 50:64-89.
43. S. Gasa, A. Makita and Y. Kinoshita (1983) Further study of the chemical structure of the equine erythrocyte hematoside containing *O*-acetyl ester. J Biol Chem, 258:876-881.

44. V. Pozsgay, H. Jennings and D.L. Kasper (1987) 4,8-anhydro-*N*-acetylneuraminic acid. Isolation from edible bird's nest and structure determination. Eur J Biochem, 162:445-450.
45. A.E. Manzi, A. Dell, P. Azadi and A. Varki (1990) Studies of naturally occurring modifications of sialic acids by fast-atom bombardment-mass spectrometry. Analysis of positional isomers by periodate cleavage. J Biol Chem, 265:8094-8107.
46. M.N. Matrosovich, A.S. Gambaryan and M.P. Chumakov (1992) Influenza viruses differ in recognition of 4-*O*-acetyl substitution of sialic acid receptor determinant. Virology, 188:854-858.
47. T.J. Pritchett and J.C. Paulson (1989) Basis for the potent inhibition of influenza virus infection by equine and guinea pig α_2-macroglobulin. J Biol Chem, 264:9850-9858.
48. G. Menghi, L. Marchetti, A.M. Bondi and G. Materazzi (1996) Sialylation patterns of the mouse parotid secretory granules. Combined deacetylation, enzymatic degradation and lectin-gold binding. Eur J Morphol, 34:181-185.
49. A. Varki and S. Kornfeld (1980) An autosomal dominant gene regulates the extent of 9-*O*-acetylation of murine erythrocyte sialic acids. A probable explanation for the variation in capacity to activate the human alternate complement pathway. J Exp Med, 152:532-544.
50. A. Varki, F. Hooshmand, S. Diaz, N.M. Varki and S.M. Hedrick (1991) Developmental abnormalities in transgenic mice expressing a sialic acid-specific 9-*O*-acetylesterase. Cell, 65:65-74.
51. M.L. Rodrigues, S. Rozental, J.N.S.S. Couceiro, J. Angluster, C.S. Alviano and L.R. Travassos (1997) Identification of *N*-acetylneuraminic acid and its 9-*O*-acetylated derivative on the cell surface of *Cryptococcus neoformans*: Influence on fungal phagocytosis. Infection and Immunity, 65:4937-4942.
52. A.P. Corfield, N. Myerscough, B.F. Warren, P. Durdey, C. Paraskeva and R. Schauer (1999) Reduction of sialic acid *O*-acetylation in human colonic mucins in the adenoma-carcinoma sequence. Glycoconjugate J., 16:307-317.
53. U. Hubl, C. Fahr, E. Proksch and R. Schauer (1997) Nachweis *O*-acetylierter Ganglioside mit Influenza Viren in Melanomen und Basaliomen. Z.Hautkrankheiten, 72:258-264.
54. G. Herrler, G. Reuter, R. Rott, H.D. Klenk and R. Schauer (1987) *N*-acetyl-9-*O*-acetylneuraminic acid, the receptor determinant for influenza C virus, is a differentiation marker on chicken erythrocytes. Biol Chem Hoppe Seyler, 368:451-454.
55. J.P. Kamerling, J. Makovitzky, R. Schauer, J.F. Vliegenthart and M. Wember (1982) The nature of sialic acids in human lymphocytes. Biochim Biophys Acta, 714:351-355.
56. B. Kniep, W.A. Flegel, H. Northoff and E.P. Rieber (1993) CDw60 glycolipid antigens of human leukocytes: structural characterization and cellular distribution. Blood, 82:1776-1786.
57. M. Vater, B. Kniep, H.J. Gross, C. Claus, W. Dippold and R. Schwartz-Albiez (1997) The 9-*O*-acetylated disialosyl carbohydrate sequence of CDw60 is a marker on activated human B lymphocytes. Immunol Lett, 59:151-157.
58. C. Butor, S. Diaz and A. Varki (1993) High level *O*-acetylation of sialic acids on *N*-linked oligosaccharides of rat liver membranes. Differential subcellular distribution of 7- and 9-*O*-acetyl groups and of enzymes involved in their regulation. J Biol Chem, 268:10197-10206.
59. G. Reuter, S. Stoll, J.P. Kamerling, J.F. Vliegenthart and R. Schauer (1988) Sialic acids on erythrocytes and in blood plasma of mammals, *in*: Sialic Acids 1988 - Proceedings of the Japanese-German Symposium on Sialic Acids (R. Schauer and T. Yamakawa, eds.) pp 88-89, Verlag Wissenschaft und Bildung, Kiel.
60. R. Schauer (1970) Biosynthesis of *N*-acetyl-*O*-acetylneuraminic acids. II. Substrate and intracellular localization of bovine acetyl-coenzyme A: *N*-acetylneuraminate-7- and 8-*O*-acetyltransferase. Hoppe Seylers Z Physiol Chem, 351:749-758.
61. R. Schauer (1987) Sialic acids: metabolism of *O*-acetyl groups. Methods Enzymol, 138:611-626.
62. S. Diaz, H.H. Higa, B.K. Hayes and A. Varki (1989) *O*-acetylation and de-*O*-acetylation of sialic acids. 7- and 9-*O*-acetylation of α2,6-linked sialic acids on endogenous *N*-linked glycans in rat liver Golgi vesicles. J Biol Chem, 264:19416-19426.
63. J.P. Kamerling, R. Schauer, A.K. Shukla, S. Stoll, H. van Halbeek and J.F. Vliegenthart (1987) Migration of *O*-acetyl groups in *N,O*-acetylneuraminic acids. Eur J Biochem, 162:601-607.
64. A.P. Corfield (1992) Bacterial sialidases - roles in pathogenicity and nutrition. Glycobiology, 2:509-521.
65. R. Schauer, G. Reuter and S. Stoll (1988) Sialate *O*-acetylesterases: key enzymes in sialic acid catabolism. Biochimie, 70:1511-1519.
66. R. Schauer, G. Reuter, S. Stoll and A.K. Shukla (1989) Partial purification and characterization of sialate *O*-acetylesterase from bovine brain. J Biochem (Tokyo), 106:143-150.
67. H.H. Higa, A. Manzi and A. Varki (1989) *O*-acetylation and de-*O*-acetylation of sialic acids. Purification, characterization, and properties of a glycosylated rat liver esterase specific for 9-*O*-acetylated sialic acids. J Biol Chem, 264:19435-19442.

68. H.C. Siebert, C.W. Vonderlieth, X. Dong, G. Reuter, R. Schauer, H.J. Gabius and J.F.G. Vliegenthart (1996) Molecular dynamics-derived conformation and intramolecular interaction analysis of the *N*-acetyl-9-*O*-acetylneuraminic acid- containing ganglioside GD1a and NMR-based analysis of its binding to a human polyclonal immunoglobulin G fraction with selectivity for *O*-acetylated sialic acids. Glycobiology., 6:561-572.
69. W.X. Shi, R. Chammas and A. Varki (1996) Regulation of sialic acid 9-*O*-acetylation during the growth and differentiation of murine erythroleukemia cells. J Biol Chem, 271:31517-31525.
70. S. Kelm, R. Schauer, J.C. Manuguerra, H.J. Gross and P.R. Crocker (1994) Modifications of cell surface sialic acids modulate cell adhesion mediated by sialoadhesin and CD22. Glycoconj J, 11:576-585.
71. K. Strenge, R. Schauer and S. Kelm (1999) Binding partners for the myelin-associated glycoprotein of N(2)A neuroblastoma cells. FEBS Lett., 444:59-64.
72. H.H. Higa, G.N. Rogers and J.C. Paulson (1985) Influenza virus hemagglutinins differentiate between receptor determinants bearing *N*-acetyl-, *N*-glycolyl-, and *N*,*O*-diacetylneuraminic acids. Virology, 144:279-282.
73. A. Gocht, G. Rutter and B. Kniep (1998) Changed expression of 9-*O*-acetyl GD3 (CDw60) in benign and atypical proliferative lesions and carcinomas of the human breast. Histochem Cell Biol, 110:217-229.
74. G. Marquina, H. Waki, L.E. Fernandez, K. Kon, A. Carr, O. Valiente, R. Perez and S. Ando (1996) Gangliosides expressed in human breast cancer. Cancer Res., 56:5165-5171.
75. D.A. Cheresh, R.A. Reisfeld and A.P. Varki (1984) *O*-acetylation of disialoganglioside GD3 by human melanoma cells creates a unique antigenic determinant. Science, 225:844-846.
76. S. Birklé, S. Ren, A. Slominski, G. Zeng, L. Gao and R.K. Yu (1999) Down-regulation of the expression of *O*-acetyl-GD3 by the *O*-acetylesterase cDNA in hamster melanoma cells: effects on cellular proliferation, differentiation, and melanogenesis. J Neurochem, 72:954-961.
77. H. Araujo, M. Menezes and R. Mendez-Otero (1997) Blockage of 9-*O*-acetyl gangliosides induces microtubule depolymerization in growth cones and neurites. Eur J Cell Biol, 72:202-213.
78. L.E.A.M. Mello and R. Mendezotero (1996) Expression of 9-*O*-acetylated gangliosides in the rat hippocampus. Neurosci.Lett., 213:17-20.
79. G. Reuter, R. Schauer and P. Bumm (1988) Sialic acids of human nasal mucin, possible targets of the influenza C viruses, *in*: Sialic Acids (R. Schauer and T. Yamakawa, eds.) pp 258-259, Berlin.
80. T. Ito, J.N. Couceiro, S. Kelm, L.G. Baum, S. Krauss, M.R. Castrucci, I. Donatelli, H. Kida, J.C. Paulson, R.G. Webster and Y. Kawaoka (1998) Molecular basis for the generation in pigs of influenza A viruses with pandemic potential. J Virol, 72:7367-7373.
81. T. Ariga, G.M. Blaine, H. Yoshino, G. Dawson, T. Kanda, G.C. Zeng, T. Kasama, Y. Kushi and R.K. Yu (1995) Glycosphingolipid composition of murine neuroblastoma cells: *O*-acetylesterase gene downregulates the expression of *O*-acetylated GD3. Biochemistry, 34:11500-11507.
82. R. Schauer (1982) Chemistry, metabolism, and biological functions of sialic acids. Adv Carbohydr Chem Biochem, 40:131-234.
83. A.P. Corfield and R. Schauer (1992) Metabolism of sialic acids, *in*: Sialic Acids - Chemistry, Metabolism and Function (R. Schauer, ed.) pp 195-261, Springer, Wien.
84. K. Kiehne and R. Schauer (1992) The influence of α- and β-galactose residues and sialic acid *O*-acetyl groups of rat erythrocytes on the interaction with peritoneal macrophages. Biol Chem Hoppe Seyler, 373:1117-1123.
85. R. Schauer (1991) Biosynthesis and function of *N*- and *O*-substituted sialic acids. Glycobiology, 1:449-452.
86. W.X. Shi, R. Chammas, N.M. Varki, L. Powell and A. Varki (1996) Sialic acid 9-*O*-acetylation on murine erythroleukemia cells affects complement activation, binding to I-type lectins, and tissue homing. J Biol Chem, 271:31526-31532.
87. V. Koistinen (1992) Limited tryptic cleavage of complement factor H abrogates recognition of sialic acid-containing surfaces by the alternative pathway of complement. Biochem J, 283 (Pt 2):317-319.
88. F.W. Klotz, P.A. Orlandi, G. Reuter, S.J. Cohen, J.D. Haynes, R. Schauer, R.J. Howard, P. Palese and L.H. Miller (1992) Binding of Plasmodium falciparum 175-kilodalton erythrocyte binding antigen and invasion of murine erythrocytes requires *N*-acetylneuraminic acid but not its *O*-acetylated form. Mol Biochem Parasitol, 51:49-54.
89. A.F. Pacitti and J.R. Gentsch (1987) Inhibition of reovirus type 3 binding to host cells by sialylated glycoproteins is mediated through the viral attachment protein. J Virol, 61:1407-1415.
90. A. Shugaba, I. Umar, J. Omage, N.D. Ibrahim, J. Andrews, A.I. Ukoha, D.I. Saror and K.A. Esievo (1994) Biochemical differences (*O*-acetyl and glycolyl groups) in erythrocyte surface sialic acids of trypanotolerant N'dama and trypanosusceptible Zebu cattle. J Comp Pathol, 110:91-95.

91. B. Schultze, H.J. Gross, R. Brossmer, H.D. Klenk and G. Herrler (1990) Hemagglutinating encephalomyelitis virus attaches to *N*-acetyl-9-*O*-acetylneuraminic acid-containing receptors on erythrocytes: comparison with bovine coronavirus and influenza C virus. Virus Res, 16:185-194.
92. B. Schultze, H.J. Gross, R. Brossmer and G. Herrler (1991) The S protein of bovine coronavirus is a hemagglutinin recognizing 9-*O*-acetylated sialic acid as a receptor determinant. J Virol, 65:6232-6237.
93. M. Larkin, W. Knapp, M.S. Stoll, H. Mehmet and T. Feizi (1991) Monoclonal antibodies VIB-E3, IB5 and HB9 to the leucocyte/epithelial antigen CD24 resemble BA-1 in recognizing sialic acid-dependent epitope(s). Evidence that VIB-E3 recognizes NeuAc α2-6GalNAc and NeuAc α2-6Gal sequences. Clin Exp Immunol, 85:536-541.
94. R. Schauer (1988) Sialic acids as antigenic determinants of complex carbohydrates. Adv Exp Med Biol, 228:47-72.
95. H. Frasa, J. Procee, R. Torensma, A. Verbruggen, A. Algra, M. Rozenberg-Arska, K. Kraaijeveld and J. Verhoef (1993) *Escherichia coli* in bacteremia: *O*-acetylated K1 strains appear to be more virulent than non-*O*-acetylated K1 strains. J Clin Microbiol, 31:3174-3178.
96. G. Sen and C. Mandal (1995) The specificity of the binding site of AchatininH, a sialic acid-binding lectin from *Achatina fulica*. Carbohydr Res, 268:115-125.
97. F.Y. Zeng and H.J. Gabius (1992) Sialic acid-binding proteins: characterization, biological function and application. Z Naturforsch.[C.], 47:641-653.
98. S. Agawa and J.R. Jass (1990) Sialic acid histochemistry and the adenoma-carcinoma sequence in colorectum. J Clin Pathol, 43:527-532.
99. A.S. Blum and C.J. Barnstable (1987) *O*-acetylation of a cell-surface carbohydrate creates discrete molecular patterns during neural development. Proc Natl Acad Sci U S A, 84:8716-8720.
100. C.B. Huang, J.A. Xu, K.R. Wu and W.M. Wang (1988) *O*-acetyl sialomucin and differentiation of stomach cancer: a histochemical study. Anticancer Res, 8:829-832.
101. J.T. Hutchins and C.L. Reading (1988) Characterization of mono-, di-, and tri-*O*-acetylated sialic acids on human cells. J Cell Biochem, 37:37-48.
102. J.R. Jass and A.M. Roberton (1994) Colorectal mucin histochemistry in health and disease: a critical review. Pathol Int, 44:487-504.
103. A.S. Paller, S.L. Arnsmeier, J.K. Robinson and E.G. Bremer (1992) Alteration in keratinocyte ganglioside content in basal cell carcinomas. J Invest Dermatol, 98:226-232.
104. J. Reivinen, H. Holthofer and A. Miettinen (1994) *O*-acetyl GD3 ganglioside in human peripheral blood T lymphocytes. Int Immunol, 6:1409-1416.
105. G. Sen, M. Chowdhury and C. Mandal (1994) *O*-acetylated sialic acid as a distinct marker for differentiation between several leukemia erythrocytes. Mol Cell Biochem, 136:65-70.
106. M. Heidenheim, E.R. Hansen and O. Baadsgaard (1995) CDw60, which identifies the acetylated form of GD3 gangliosides, is strongly expressed in human basal cell carcinoma. Br J Dermatol, 133:392-397.
107. B. Mann, E. Klussmann, V. Vandammefeldhaus, M. Iwersen, M.L. Hanski, E.O. Riecken, H.J. Buhr, R. Schauer, Y.S. Kim and C. Hanski (1997) Low *O*-acetylation of sialyl-Le(x) contributes to its overexpression in colon carcinoma metastases. Int.J.Cancer, 72:258-264.
108. D.S. Gupta, B. Jann and K. Jann (1984) *Escherichia coli* O18ac antigen: structure of the O-specific polysaccharide moiety. Infect Immun, 45:203-209.
109. C.G. Hellerqvist, B. Lindberg, S. Svensson, T. Holme and A.A. Lindberg (1969) Structural studies on the O-specific side chains of the cell wall lipopolysaccharides from *Salmonella typhi* and *S. enteritidis*. Acta Chem Scand, 23:1588-1596.
110. B.A. Dmitriev, Y.A. Knirel, N.K. Kochetkov, B. Jann and K. Jann (1977) Cell-wall lipopolysaccharide of the 'Shigella-like' *Escherichia coli* 058. Structure of the polysaccharide chain. Eur J Biochem, 79:111-115.
111. B.A. Dmitriev, L.V. Backinowsky, Y.A. Knirel, N.K. Kochetkov and I.L. Hofmann (1977) Somatic antigens of Shigella. The strucuture of the specific polysaccharide chain of *Shigella dysenteriae* type 5 lipopolysaccharide. Eur J Biochem, 78:381-387.
112. B. Lindberg, J. Lönngren and W. Nimmich (1972) Structural studies of the Klebsiella O group 9 lipopolysaccharide. Carbohydr Res, 23:47-55.
113. N.P. Arbatsky, A.S. Shashkov, G. Widmalm, Y.A. Knirel, K. Zych and Z. Sidorczyk (1997) Structure of the O-specific polysaccharide of *Proteus penneri* strain 25 containing *N*-(L-alanyl) and multiple *O*-acetyl groups in a tetrasaccharide repeating unit. Carbohydr Res, 298:229-235.
114. E.V. Vinogradov, R. Pantophlet, S.R. Haseley, L. Brade, O. Holst and H. Brade (1997) Structural and serological characterisation of the O-specific polysaccharide from lipopolysaccharide of *Acinetobacter calcoaceticus* strain 7 (DNA group 1). Eur J Biochem, 243:167-173.

115. S.N. Senchenkova, A.S. Shashkov, Y.A. Knirel, J.J. McGovern and A.P. Moran (1997) The O-specific polysaccharide chain of *Campylobacter fetus* serotype A lipopolysaccharide is a partially *O*-acetylated 1,3-linked α-D-mannan. Eur J Biochem, 245:637-641.
116. A.I. Fattom, J. Sarwar, L. Basham, S. Ennifar and R. Naso (1998) Antigenic determinants of *Staphylococcus aureus* type 5 and type 8 capsular polysaccharide vaccines. Infect Immun, 66:4588-4592.
117. X. Lemercinier and C. Jones (1996) Full ^{1}H NMR assignment and detailed *O*-acetylation patterns of capsular polysaccharides from *Neisseria meningitidis* used in vaccine production. Carbohydr Res, 296:83-96.
118. G. Kogan, B. Jann and K. Jann (1992) Structure of the *Escherichia coli* 0104 polysaccharide and its identity with the capsular K9 polysaccharide. FEMS Microbiol Lett, 70:135-140.
119. A. Gamian, E. Romanowska, J. Ulrich and J. Defaye (1992) The structure of the sialic acid-containing *Escherichia coli* O104 O-specific polysaccharide and its linkage to the core region in lipopolysaccharide. Carbohydr Res, 236:195-208.
120. C.A. Clark, J. Beltrame and P.A. Manning (1991) The *oac* gene encoding a lipopolysaccharide O-antigen acetylase maps adjacent to the integrase-encoding gene on the genome of *Shigella flexneri* bacteriophage Sf6. Gene, 107:43-52.
121. J.M. Slauch, A.A. Lee, M.J. Mahan and J.J. Mekalanos (1996) Molecular characterization of the *oafA* locus responsible for acetylation of *Salmonella typhimurium* O-antigen: oafA is a member of a family of integral membrane trans-acylases. J Bacteriol, 178:5904-5909.
122. C.H. Zou, Y.A. Knirel, J.H. Helbig, U. Zähringer and C.S. Mintz (1999) Molecular cloning and characterization of a locus responsible for *O*-acetylation of the O-polysaccharide of *Legionella pneumophila* serogroup 1 lipopolysaccharide. J Bacteriol, 181:4137-4141.
123. G. Stevenson, B. Neal, D. Liu, M. Hobbs, N.H. Packer, M. Batley, J.W. Redmond, L. Lindquist and P. Reeves (1994) Structure of the O-antigen of *Escherichia coli* K-12 and the sequence of its *rfb* gene cluster. J Bacteriol, 176:4144-4156.
124. S. Bystricky and S.C. Szu (1994) *O*-acetylation affects the binding properties of the carboxyl groups on the Vi bacterial polysaccharide. Biophys Chem, 51:1-7.
125. G. Arakere and C.E. Frasch (1991) Specificity of antibodies to *O*-acetyl-positive and *O*-acetyl-negative group C meningococcal polysaccharides in sera from vaccinees and carriers. Infect Immun, 59:4349-4356.
126. F. Ørskov, I. Ørskov, A. Sutton, R. Schneerson, W. Lin, W. Egan, G.E. Hoff and J.B. Robbins (1979) Form variation in *Escherichia coli* K1: determined by *O*-acetylation of the capsular polysaccharide. J Exp Med, 149:669-685.
127. W.X. Shi, R. Chammas and A. Varki (1996) Linkage-specific action of endogenous sialic acid *O*-acetyltransferase in Chinese hamster ovary cells. J.Biol.Chem., 271:15130-15138.
128. A. Kanamori, J. Nakayama, M.N. Fukuda, W.B. Stallcup, K. Sasaki, M. Fukuda and Y. Hirabayashi (1997) Expression cloning and characterization of a cDNA encoding a novel membrane protein required for the formation of *O*-acetylated ganglioside: A putative acetyl-CoA transporter. Proc.Natl.Acad.Sci.USA, 94:2897-2902.

CONTRADICTORY FUNCTIONS OF THE SULFATIDE IN THE BLOOD COAGULATION SYSTEM

Mamoru Kyogashima[1]

Seikagaku Corporation Tokyo Research Institute, 1253, Tateno 3-chome Higashiyamato-shi, Tokyo207-0021, Japan

Sulfatide, sulfuric ester of galactosylceramide at the C3 of galactose residue, is the representative of sulfated glycolipids, which naturally occurs in animal cells, especially in nervous system cells, kidney cells and some tumor cells (1).

It has long been controversial whether sulfatide stimulates blood coagulation or inhibits blood coagulation. In 1961, Wago reported possible anticoagulant and antiatherosclerotic activities of sulfatide (2). He administered sulfatide orally or intravenously to rabbits, and observed not only a prolongation of the plasma clotting time but also a decrease in the plasma cholesterol level with an improvement of atheromatous involvement in the aorta. About 20 years later, sulfatide was reported for the first time to be able to activate blood coagulation factor XII (Hageman factor) *in vitro* (3). Since then, sulfatide has been believed to be one of the important factors in the initiation of the intrinsic coagulation pathway, although the physiological importance of the intrinsic pathway starting from factor XII still remains unclear (4,5).

In 1989, Taketomi *et al.* found that Watanabe hereditable hyperlipidemic (WHHL) rabbits administered sulfatide (10 mg/kg) as a bolus shot every other day for three months showed neither definite symptoms nor pathological changes related to thrombotic disorders during the process and at autopsy (6). In addition they recently found that sulfatide can prolong clotting times and bleeding time when it was administered to animals (7). These findings lead to the hypothesis that sulfatide may work as an anticoagulant rather than a coagulant. With respect to the following points, sulfatide may be similar to heparin. First, both are sulfated glycoconjugates (1,8). Heparin can prolong clotting times as well as bleeding time and it was also reported to be able to activate blood coagulation factor XII *in vitro* (9). The precise mechanism of heparin as an anticoagulant is its ability to bind to antithrombin III (ATIII) through its well-known domain of the pentasaccharide enhancing the activities of ATIII towards thrombin as well as blood coagulation factor Xa (8,10,11,12). A different mechanism has been suggested for sulfatide as an anticoagulant because sulfatide does not accelerate the activity of ATIII but it binds to fibrinogen resulting in disturbance of fibrin formation (7,13). Therefore, we examined its possible

E-mail: kyogashi@to.seikagaku.co.jp

anticoagulant activity using a rat deep vein thrombosis (DVT) model in which thrombus is experimentally induced by ligating the inferior vena cava (14). Sulfatide was administered to rats 1 minute before the vein ligation to intend to prevent thrombus formation. Unex-pectedly, however the sulfatide markedly enhanced the thrombogenesis in these rats. These observations suggest that sulfatide may exhibit contradictory functions in the blood coagulation system.

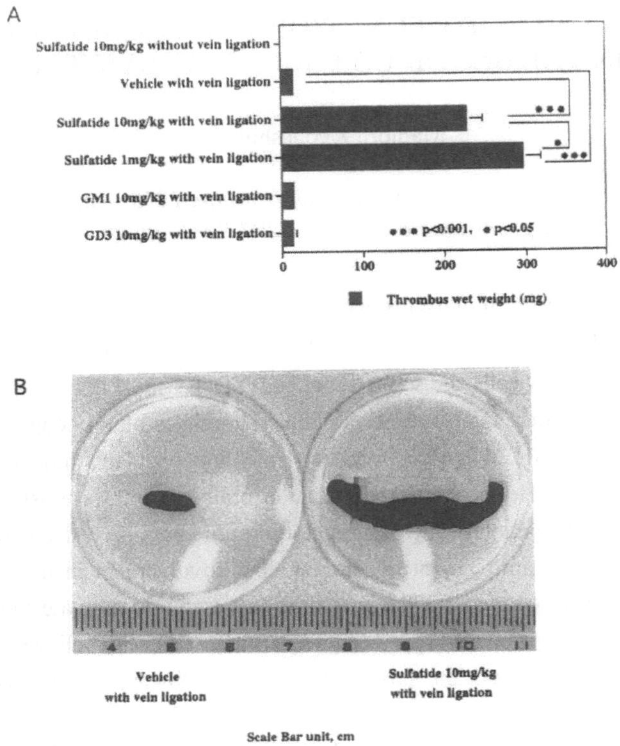

Figure 1. A. The effect of sulfatide on thrombogenesis in a rat deep vein thrombosis model. The thrombi weights when sulfatide was administered at 10 mg/kg or 1 mg/kg with vein ligation were significantly (P<.001) increased compared with those when vehicle or gangliosides was administered (n=3, mean ± SE). **B.** The left shows a thrombus from the control group administered vehicle with vein ligation, and the right is from a rat given 10 mg/kg of sulfatide with vein ligation. From reference (14) with permission.

I. Coagulant Effect

To elucidate the role of sulfatide on thrombogenesis *in vivo*, the rat thrombosis model was chosen because this model has conventionally been used to evaluate the potential of anticoagulant agents (15). After the rat was laparotomized under anesthesia, the inferior vena cava was ligated below the branch of the left renal vein with a surgical thread. Sulfatide and gangliosides, suspended homogeneously in phosphate buffered saline (PBS) were administered into rats as a bolus shot through the tail vein 1 minute prior to the ligation. Two control groups were prepared. In the first group, PBS (vehicle) was injected the same way through the tail vein 1 minute prior to ligation. In the second group, sulfatide was injected through the tail vein without vein ligation. The rats in all groups were laparotomized under ether anesthesia 3 hours after the ligation and the veins were incised to extirpate thrombi, the weights of which were measured. The results are shown in Figure 1. When 10 mg/kg of sulfatide was injected without vein ligation, there was no thrombus observed. However, 1 mg/kg or 10 mg/kg of sulfatide remarkably enhanced throm-bogenesis in the rats with vein ligation. The thrombi induced by 1 mg/kg of sulfatide were slightly but significantly ($p<0.05$) larger than those induced by 10 mg/kg. The gangliosides did not show any effect on thrombogenesis.

To evaluate the coagulant activity of plasma, kinetic turbidmetry was carried out (14, 16). Citrated plasma diluted with the same volume of PBS containing various amounts of sulfatide or other glycoconjugates were added to the wells of a 96-well microtiter plastic plate which was installed in a microplate reader. $CaCl_2$ was added to each well to start coagulation. The increase of absorbance by the increased turbidity due to fibrin formation was monitored every 15 seconds at 405 nm. The onset time of fibrin formation was defined as the first time when absorbance was continuously elevated by more than 0.02 above the base line. As shown in Figure 2a, normal human

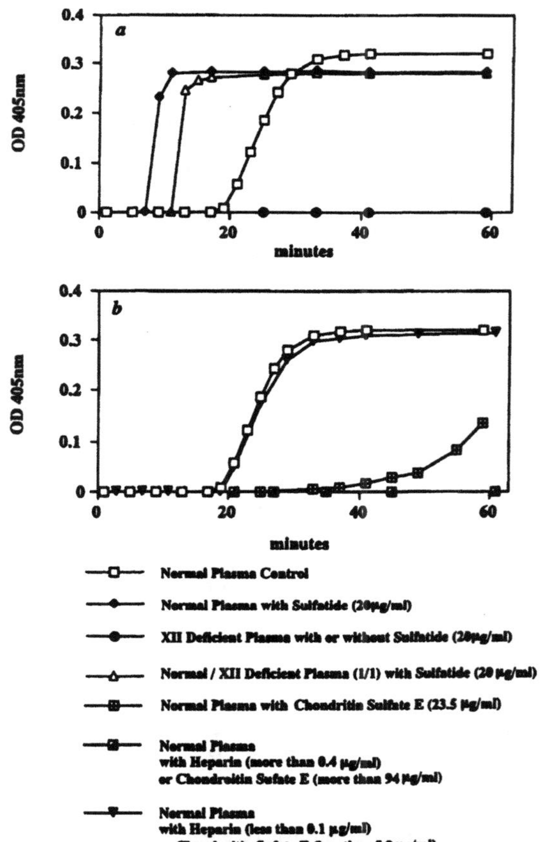

Figure 2. Effects of sulfatide (*a*) and heparin and chondroitin sulfate E (*b*) on coagulation of human plasma by kinetic assay using a 96-well microtiter plate. Ganglioside GM1 and GD3 showed no effect in this test. From reference (16) with permission.

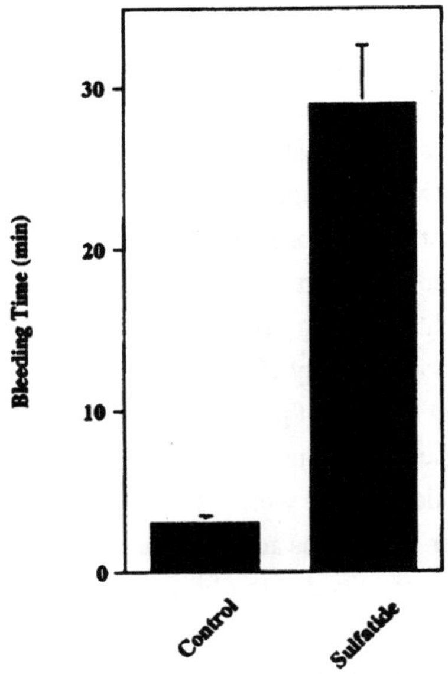

Figure 3. Effect of sulfatide on bleeding time. Sulfatide (20mg/kg) was injected to mice through tail veins. After 30 minutes, tail tips of the mice were cut and bleeding time was measured. Sulfatide significantly prolonged bleeding time (<0.0002, n=3, mean $\pm$ SE)

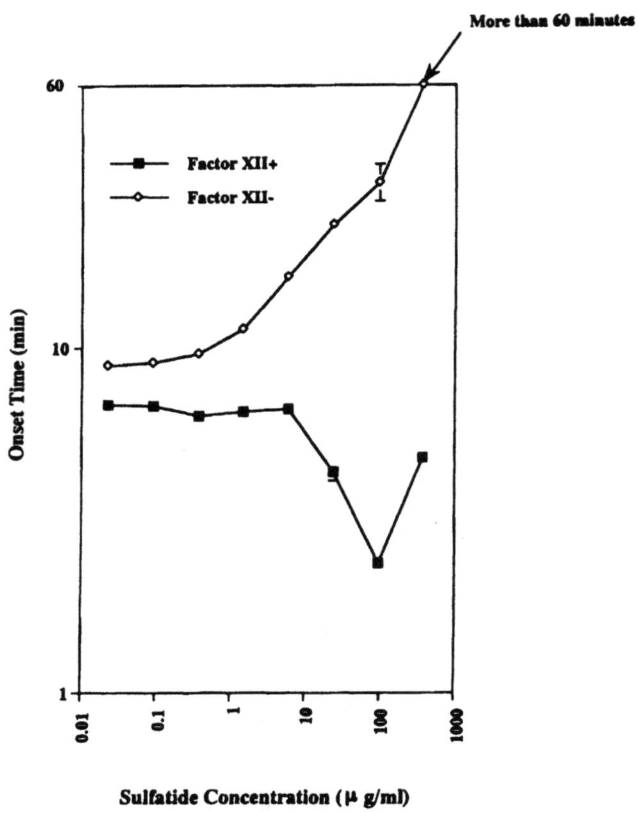

Figure 4. The relationship between sulfatide [log10(μg/ml)] and coagulation onset time [log10(min)] initiated by tissue factor plus $CaCl_2$ in human plasma. Factor XII+, normal plasma; Factor XII-, factor XII deficient plasma. From reference (14) with permission.

plasma slowly started to coagulate at about 19 minutes but sulfatide (6.25 microgram/well) apparently accelerated this process starting at about 5 minutes. On the other hand, factor XII-deficient plasma did not to start to coagulate either in the presence or the absence of sulfatide. However, when normal and the deficient plasma were mixed, the coagulation was rapid in the presence of sulfatide. As found in the *in vivo* test, ganglioside GM1 and GD3 showed no effect on this *in vitro* experiment. Both heparin and chondroitin sulfate E are highly sulfated glycosamino-glycans and were reported to activate factor XII *in vitro* (8). However, as shown in Figure 2b, neither of them did accelerate, but they rather prolonged normal plasma coagulation onset times in this study (16). Generally, the haemostatic system acting via the extrinsic pathway (starting from the tissue factor/factor VII) is believed to be physiologically important, but role of the system via the intrinsic pathway in which factor XII is involved seems to be questionable because there are no bleeding symptoms in patients lacking this factor (4,5). Also the question as to the roles of the intrinsic pathway could have been raised because, even heparin was reported to activate factor XII *in vitro* (9) but normally heparin can not stimulate the blood coagulation system. Furthermore, as everybody knows, heparin is a typical anticoagulant widely used in current clinical practice. The intrinsic pathway starting from factor XII should be re-evaluated from the perspective of sulfatide.

II. Anticoagulant Effect

Although it may be possible that sulfatide works coagulantly in some pathological conditions, the reverse possibility that sulfatide acts anticoagulantly still remains. It should be noted that sulfatide could not induce thrombus without vein ligation (14). Furthermore, sulfatide prolonged the bleeding time when it was simply injected into animals (Figure 3). To elucidate the mechanism of this discrepancy, we stimulated the blood coagulation system *in vitro* via not only the intrinsic system by adding $CaCl_2$ but also the extrinsic pathway by adding tissue factor by using kinetic turbidmetry (14). In the normal plasma, between under 6-100 microgram /ml of sulfatide, the coagulation was accelerated remarkably, although it was abated more than 100 microgram /ml of sulfatide. On the contrary, in the factor XII deficient plasma, the coagulation was prolonged dose-dependently (Figure 4). These observations suggested that sulfatide may have anticoagulant activity under the inactivation of factor XII where the animals injected with sulfatide without vein ligation may coincide. The contradictory roles of sulfatide in the blood coagulant system should be noted.

III. Possible Roles of Sulfatide Related to Cancer Associated Thrombosis

It has been reported that thrombotic diseases occasionally occur in patients with cancers (17,18). The major causes of such diseases have been attributed to procoagulants or coagulants such as factor X activator or tissue factor that are aberrantly produced by cancer cells. However, uncharacteristic factors including membrane vesicles shed from cancer cells, which may be related to sulfatide, have also been reported (18). There have been many papers that various cancer cells express or shed sulfatide (19-28). It may be possible that such cancers induce thrombotic diseases. The thrombotic diseases related to cancers have been reported as venous/arterial thrombosis, disseminated intravascular coagulation or catheter-associated thrombosis etc. (18). We have experimentally demon-strated in this paper that sulfatide could induce venous thrombosis by vein ligation, situation which seemed to be similar to cancer cells shedding sulfatide, which invade vessels and obstruct blood flow. "Catheter-associated thrombosis" seems to be a very characteristic thrombotic disease. We continuously infused sulfatide to a rat through plastic canulae, which resembled a clinical catheter, to determine whether sulfatide can induce thrombosis (16). As shown in Figure 5, continuously infused sulfatide but not gangliosides via the canulae caused a giant thrombosis at the tip of the canulae. Thus the thrombotic diseases related to cancers should be re-evaluated from the perspective of sulfatide.

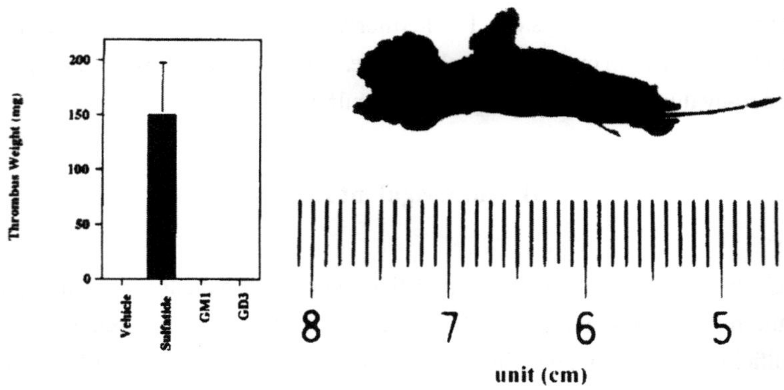

Figure 5. Thrombus induced by continuous infusion of sulfatide. Sulfatide was continuously infused through plastic canulae at a rate of 1.25 mg/kg per hour for 4 hours. From reference (16) with permission.

SUMMARY

It has long been controversial whether sulfatide stimulates blood coagulation or inhibits blood coagulation. In this paper, I demonstrated not only anticoagulant activity but also coagulant activity of sulfatide *in vivo* by using experimental animal models and presented the possibility that sulfatide may function contradictorily under physiological and/or pathological conditions. For instance, I presented possible involvement of sulfatide in thrombotic diseases associated with cancers.

REFERENCES

1. Ishizuka, I. (1997) Chemistry and functional distribution of sulfo-glycolipids. Prog. Lipid. Res. 36 : 245-319.
2. Wago, K. (1961) Antiatherosclerotic and anticoagulant activity of cerebroside sulfate. Jpn. Heart. J. 2 : 354-367.
3. Fujikawa, K., Heimark, R.L., Kurachi, K. and Davie, E.W. (1980) Activation of bovine factor XII (Hageman factor) by plasma kallikrein. Biochemistry 19 : 1322-1330.
4. Mann K.G. Jenny, R.J., and Krishnaswamy, S. (1988) Cofactor proteins in the assembly and expression of blood clotting enzyme complex. Annu. Rev. Biochem. 57 : 915-956.
5. Schmaier, A.H. (1997) Contact activation : A revision. Thromb. Haemost. 78 : 101-107.
6. Taketomi, T., Hara, A., Uemura, K., and Kyogashima, N. (1989) Effect of long-term treatment with sulfatide on hyperlipidemia and progression of atherosclerosis in WHHL rabbits. Jpn. J. Exp. Med. 59 : 221-231.
7. Hara, A., Uemura, K., and Taketomi, T. (1996) Sulfatide prolongs blood-coagulation time and bleeding time by forming a complex with fibrinogen. Glycoconjugate J. 13 : 187-194.
8. Bjork, I., Olson, S.T., and Shore, J.D. (1989) Molecular mechanisms of the accelerating effect of heparin on the reactions between antithrombin and clotting proteinases; In Heparin (Lane, D.A. and Lindahl, U. eds.) pp. 229-255, Edward Arnold, London.
9. Hojima, Y., Cochrane, C.G., Wiggins, R.C., Austen, K.F., and Stevens, R.L. (1984) In vitro activation of the contact (Hageman factor) system of plasma by heparin and chondroitin sulfate E. Blood 63 : 1453-1459.
10. Casu, B., Oreste, P., Torri, G., and Zoppetti, G. (1981) The structure of heparin oligosaccharide fragments with high anti-(factor Xa) activity containing the minimal antithrombin III-binding sequence : chemical and ^{13}C nuclear-magnetic-resonance studies. Biochem. J. 197 : 599-609.
11. Thunberg, L., Backstrom, G., and Lindahl, A.U. (1982) Further characterization of the antithrombin-binding sequence in heparin. Carbohydr. Res. 100 : 393-410.
12. Bourin, M.-C., and Lindahl, U. (1993) Glycosaminoglycans and the regulation of blood coagulation. Biochem.J. 289 : 313-330.
13. Hara, A., Kutsukake, Y., Uemura, K., and Taketomi, T. (1993) Anticoagulant activity of sulfatide and its anti-thrombotic effect in rabbit. J. Biochem. 113 : 781-785.
14. Kyogashima, M., Onaya, J., Hara, A., and Taketomi, T. (1998) Sulfatide can markedly enhance thrombogenesis in rat deep vein thrombosis model. Glycoconjugate J. 15 : 915-922.
15. Reyer, I., Mussoni, L., Donati, M.D., and Gaetano, G.D. (1980) Failure of aspirin at different doses to modify experimental thrombose in rats. Thromb. Res. 18 : 669-674.
16. Kyogashima, M., Onaya, J., Hara, A., and Taketomi, T. (1998) Contradictory functions of sulfatide in the blood coagulation system as coagulant and anticoagulant. Acta. Biochim. Pol. 45 : 493-499.
17. Edwards, R.L., and Rickles, F.R. (1993) Role of procoagulant activity in neoplastic disease. In Roles of procoagulant activity in health and disease (Levy, G.A., and Cole, E.H., eds) pp. 153-176, CRC Press Inc, Boca Raton.
18. Green, K.B., and Silverstein R.L. (1996) Hypercoagulability in cancer. Hematolo. Oncol. Clin. North. Am. 10 : 499-530.
19. Siddiqui, B., Whitehead, J., and Kim, Y.S. (1978) Glycosphingolipids in human colonic adenocarcinoma. J. Biol. Chem. 253 : 2168-2175.
20. Yoda, Y., Gasa, S., Makita, A., Fujioka, Y., Kikuchi, Y., and Hashimoto, M. (1979) Glycolipids in human lung carcinoma of histologically different types. J. Natl. Cancer Inst. 63 : 1153-1160.

21. Hattori, H., Uemura, K., and Taketomi, T. (1981) Glycolipids of gastric cancer. The presence of blood-group A-active glycolipids in cancer tissues from blood group O patients. Biochim. Biophys. Acta. 666 : 361-369.
22. Roberts, D.D. (1988) Interaction of thrombospondin with sulfated glycolipids and proteoglycans of human melanoma cells. Cancer Res. 48 : 6785-6793.
23. Sakakibara, N., Gasa, S., Kamio, K., Makita, A. and Koyanagi, T. (1989) Association of elevated sulfatides and sulfotransferase activities with human renal cell carcinoma. Cancer Res. 49 : 335-339.
24. Hiraiwa, N., Fukuda, Y., Imura, H., Tadano-Aritomi, K., Nagai, K., Ishizuka, I., and Kannagi, R. (1990) Accumulation of highly acidic sulfated glycosphingolipids in human hepatocellular carcinoma defined by a series of monoclonal antibodies. Cancer Res. 50 : 2917-2928.
25. Aruffo, A., Kolanus, W., Walz, G., Fredman, P., and Seed B. (1991) CD62/P-selectin recognition of myeloid and tumor cell sulfatides. Cell 67 : 35-44.
26. Hattori, H., Uemura, K., Ishihara, H., and Ogata, H. (1992) Glycolipid of human pancreatic cancer; the appearance of neolacto-series (type 2 chain) glycolipid and the presence of incompatible blood group antigen in tumor tissues. Biochim. Biophys. Acta. 1125 : 21-27.
27. Kiguchi, K., Takamatsu, K., Tanaka, J., Nozawa, S., Iwamori, M., and Nagai, Y. (1992) Glycosphingolipids of various human ovarian tumors : a significantly high expression of I^3SO^3GalCer and Lewis antigen in mucinous cystadenocarcinoma. Cancer Res. 52 : 416-421.
28. Kuboshiro, K., Tsukazaki, K., Tanaka, J., Takamatsu, K., Kiguchi, K., Mikami, M., Nozawa, S., Nagai, Y., and Iwamori, M. (1992) Human uterine endometrial adenocarcinoma: characteristic acquirement of synthetic potentials for II^3SO^3-LacCer and ganglio series sulfoglycosphingolipids after transfer of the cancer cells to culture. Cancer Res. 52: 803-809.

INERACTION OF GM2 ACTIVATOR PROTEIN WITH GLYCOSPHINGOLIPIDS

Su-Chen Li, Yoichiro Hama, and Yu-Teh Li

Department of Biochemistry, Tulane University, School of Medicine, New Orleans, Louisiana 70112, USA

It has been well documented that the sugar chains of glycosphingolipids in higher animals are catabolized by sequential actions of lysosomal exo-glycosidases (1). In addition to exo-glycosidases, it has also been established that the hydrolyses of several glycosphingolipids require the assistance of low molecular weight protein cofactors, named sphingolipid activator proteins (1,2). To date, five such activator proteins have been isolated, cloned, and studied in detail. Among them, four are derived from a common precursor of 65-75 kDa polypeptide and they are named saposins A, B, C and D, according to their placement from the N-terminus of the precursor protein, prosaposin (3). The gene of prosaposin was mapped to chromosome 10 (3). The fifth protein cofactor is GM2 activator, which is encoded by a separate gene on chromosome 5 (2). GM2 activator does not share sequence homology with any of the saposins. Functional aspects of these protein cofactors have also been extensively studied. Saposins A and C show overlapped

---

The abbreviations used are: GalNAc, N-acetylgalactosamine; TLC, thin layer chromatography; HPLC, high performance liquid chromatography; GM1, Galβ1→3GalNAcβ1→4(NeuAcα2→3)Gal-β1→4Glcβ1→1'Cer; GM2, GalNAcβ1→4(NeuAcα2→3)-Galβ1→4Glcβ1→1'Cer; GM3, NeuAcα2→3Galβ1→4Glcβ1→1'Cer; GM4, NeuAcα2→3Galβ→1'Cer; GalCer, Galβ1→ 1'Cer; GalNAc-GD1a, GalNAcβ1→4(NeuAcα2→3)-Galβ1→3GalNAcβ1→4(NeuAcα2→3)-Galβ1→4Glcβ1→1'Cer; GD2, GalNAcβ1→4(NeuAc-α2→8NeuAcα2→3)Galβ1→4Glcβ1→ 1'Cer; Sulfatide, HSO_3→2Galβ1→1'Cer; SM3, HSO_3→3Galβ1→4Glcβ1→1'Cer; NeuGc-GM1, Galβ1→3GalNAcβ1→4(NeuGcα2→3)Galβ1→4Glcβ1→1'Cer; KDN-GM2, KDNα2→3Galβ1→4Glcβ1→1'Cer; $IV^6KDNLnOse_4Cer$, KDN-α2→6Galβ1→4GlcNAcβ1→4Galβ1→4Glcβ1→1'Cer; $IV^6KDNLnOse_4Cer$, KDNα2→6Galβ1→

The Molecular Immunology of Complex Carbohydrates–2.
Edited by Albert M. Wu, Kluwer Academic/Plenum Publishers, 2001.

activity in stimulating the hydrolysis of glucosylceramide by β-glucosidase. Saposin D, the least studied saposins among the four, was reported to enhance the hydrolyses of sphingomyelin (3) and ceramide (4,5). These functions, however, require substantiation. Saposin B has multiple stimulatory activities, thus, is also known as a non-specific activator protein (3). GM2 activator is the only activator protein that shows a distinct and effective stimulatory activity toward the hydrolysis of the GalNAc from GM2 by β-hexosaminidase A (1,3). To understand the mechanism of action of activator proteins, it is important to know the interactions of these activator proteins with various glycosphingolipids and/or glycosidases. Because of the amphipathic nature of glycosphingolipids, the bindings between the lipids and the proteins are far more complex than that between lectins and their ligands. This chapter reviews the studies on the interactions of GM2 activator protein with glycosphingolipids.

I. GM2 Activator Protein

The history of GM2 activator protein started in 1973 from the observation by Li *et al.* that the heat-treated human liver extract was able to stimulate the conversion of GM2 into GM3 carried out by a purified β-hexosaminidase A (6). Subsequently, it took about ten years to firmly establish the existence of GM2 activator by purifying this protein from human kidney (7), liver (8), and brain (9). The physiological importance of GM2 activator has been shown by the presence of type AB Tay-Sachs disease that is caused by the deficiency or the defect of this protein cofactor (10). The properties of human GM2 activator have already been reviewed in detail (11). Briefly, the native matured GM2 activator is a hydrophobic protein with a molecular mass of 18.5 kDa and with an isoelectric point of pH 4.8. The human cDNA encoding GM2 activator has been cloned independently

---

3GlcNAcβ1→4Galβ1→4Glcβ1→1'Cer; PE→GM2, GalNAcβ1→4(NeuAcα2→3)Galβ1→4Glc-PE; PE, dipalmi-toylphosphatidylethanol-amine; GlcCer, Glcβ1→1'Cer; LacCer, Galβ1→4Glcβ1→ 1'Cer. GA1, Galβ1→3GalNAcβ1→4Galβ1→4Glcβ1→1'Cer; GA2, GalNAcβ1→4Galβ1→4Glcβ1→1'Cer;HO-GM2, GalNAcβ1→4(NeuAcolα2→3)Galβ1→4Glcβ1→1'Cer; Me-GM2, GalNAcβ1→4(NeuAcMeα2→3)Galβ1→4Glcβ1→1'Cer; DSC, differential scanning calorimetry; DMPC, dimyristoylglycerophosphocholine; DMPG, dimyristoyl-glycerophosphoglycerol.

by three laboratories (11). The recombinant GM2 activator expressed as His-tag protein in CHO cells was produced as soluble protein and was shown to have the same specific activity as that of the native GM2 activator (12). When the same cDNA with His-tag was expressed in *E. coli*, the produced GM2 activator was insoluble and was accumulated in the inclusion bodies. The activator was, then, extracted from the inclusion bodies with a denaturing buffer, purified through Ni-NTA resin, and refolded to gain the re-natured protein. In this case, the specific activity of the His-tag GM2 activator was about 52% that of the native protein (12). Wu *et al.*(13) have also expressed a cDNA in *E. coli*, which encodes the mature GM2 activator protein without His-tag, but with extra 9 amino acids belong to the T7-7 vector at the N-terminus of GM2 activator. They also harvested the recombinant GM2 activator from inclusion bodies in denatured state and then allowed the activator protein to slowly refold into the native state. The initially refolded GM2 activator was found to have a lower specific activity (about 60%) than that of the native GM2 activator. However, further purification of the refolded activator twice through a Sephadex G-75 (fine particles) column, they were able to obtain a fraction of GM2 activator that was as active as the native protein (13). Apparently, the formation of improperly folded protein molecules during the refolding process is one of the major problems in obtaining a good recombinant GM2 activator protein from *E. coli*. The cDNA encoding the mouse homologue of human GM2 activator has also been cloned (14) and expressed (15). The mouse GM2 activator protein shares 74.1% of identity in amino acid sequence with that of the human activator. The recombinant mouse protein is more soluble and stable than of the human counterpart. Both the human and the mouse GM2 activator proteins are very specific and effective in stimulating the conversion of GM2 to GM3 by β-hexosaminidase A. The hydrolysis of a number of glycosphingolipids bearing GM2 epitope, such as GalNAc-GD1a (13) and GalNAc-GM1b (16), are also stimulated by GM2 activator protein. However, between the mouse and the human activator proteins, only the mouse GM2 activator is effective in stimulating the conversion of asialo-GM2 to lactosyl ceramide. Human GM2 activator is ineffective in stimulating this reaction (15). It has been shown that GM2 activator and each of the four saposins stimulated different hydrolytic reactions to assist the removal of sugar units from glycosphingolipids. However, when both saposin B and GM2 activator were present together in a mixture containing GM2 and β-hexosaminidase A, the rate of conversion of GM2 into GM3 was greater than that in the presence of only GM2 activator. This indicated that these two activator proteins have synergistic effect in promoting the hydrolysis of GM2 (11). The lipid moiety in GM2 is essential for GM2 activator to elicit its stimulatory activity. When the ceramide moiety of GM2 was replaced by another hydrophobic component such as dipalmitoylphosphatidyl ethanolamine, the stimulatory activity of GM2 activator remained about the same. However, when ceramide moiety was removed from GM2, the resulted water-soluble tetrasaccharide derived from GM2 was not hydrolyzed by β-hexosaminidase A with or without GM2 activator (13). In contrast, other oligosaccharides derived from glycosphingolipids could be readily hydrolyzed by glycosidases without the assistance of any activator protein.

II Interaction of GM2 Activator with Glycosphingolipids

To understand whether GM2 activator interacts with glycosphingolipids or with β-hexosaminidase A was the subject of interest since the native GM2 activator became available in 1980s. Because of the yield of the purified GM2 activator from human tissues was very low, it was not possible to perform various techniques for extensive investigations of the lipid-binding property of the native human GM2 activator. The studies of the binding between GM2 activator and glycosphingolipids have been complicated by the fact that GM2 activator has hydrophobic domains and glycosphingolipids are amphipathic molecules. The association between GM2 activator and β-hexosaminidase A was believed to be weak, since GM2 activator was readily separated from β-hexosaminidase A by Sephadex gel filtration. There has been no report on the interaction between β-hexosaminidase A and GM2 activator. Neither has there been any report on the interaction among the three parties, GM2 activator, GM2, and β-hexosaminidase A, although all three components are required for the catabolism of GM2.

The early study on the interaction of GM2 activator and glycosphingolipids were performed by using the native protein in small scales of sucrose density isoelectric focusing, sucrose density ultracentrifugation, polyacrylamide gel electrophoresis, and glycosphingolipid transfer from donor to acceptor liposomes (7, 17, 18). Availability of large quantities of recombinant GM2 activator in recent years made it possible to resume the study of the glycolipid-binding property of GM2 activator using additional techniques, such as TLC overlay (19), HPLC analysis (19), fluorescence dequenching assay (12), and differential scanning calorimetry and film balance measurements (22). To date, all studies suggest that the detected associations of GM2 activator with GM2 or other glycosphingolipids might represent the combination of specific and non-specific interactions between them. The exclusive and specific interaction between GM2 activator and GM2 has never been clearly demonstrated.

Isoelectric Focusing --- This technique, also called electrofocusing, was performed in sucrose density medium using an LKB8100 column (110ml) to examine the electro mobility profile of GM2 activator in the presence or absence of a glycosphingolipid (7, 17). GM2 activator and a glycosphingolipid were electrofocused individually, or together, in a sucrose gradient containing carrier ampholyte solution of pH 3.5-10. After electrofocusing, the gradient solution was collected into fractions and they were measured for pH and assayed for the presence of GM2 activator and the glycosphingolipid. When GM2 activator and the micellar form of GM2 were electrofocused separately, the activator protein was concentrated in a single peak at its pI of pH 4.8 and the micellar form of GM2 migrated to the anode because of its negative charge. When the two components were mixed and electrofocused, GM2 activator was found to split into two peaks, each at pH 4.4 and 4.8. The peak at pH 4.4 was found also to contain GM2, and the molar ratio of the two components was determined to be 1:1 (7). It was further found in a later study (17) that the amount of GM2 activator in the pH 4.4 peak, in terms of the total percentage of the activator, decreased from 60% at 30 h to about 33% at 63 h, as the time of electrofocusing increased. This result suggested that the majority of bindings between GM2 activator and GM2 might be weak and could not withstand the extended time of electrophoresis. When the same experiment was performed using the liposome-bound GM2 in place of the micellar form of GM2, the peak at pH 4.4 was not found (17). Only when a high concentration of GM2 was incorporated into the liposomes, could a small peak at pH 4.4 was observed. This may indicate that the association of GM2 with other lipid components in liposomes is greater than that of GM2 activator with GM2. GM2 activator was also found to form the peaks at pH 4.4 with several gangliosides. GM1 formed the complex at 10% that of GM2 and both GM3 and GD1a, at about 1%.

Ultracentrifugation --- Sucrose density ultracentrifugation in a Beckman centrifuge was used to demonstrate the complex formation between GM2 activator and glycosphingolipids (17). When separately centrifuged, the micellar form of GM2 sediment much faster than GM2 activator because of the difference in their apparent particle weights. When GM2 activator was mixed with the micellar form of GM2 and centrifuged, a portion of GM2 migrated together with GM2 activator protein, presumably the formation of a complex. The molar ratio of GM2 activator and GM2 in the complex was found to vary depending on the presence or absence of foreign proteins, such as cytochrome or gelatin. This phenomenon was also found in the binding of GM2 activator with GM1, GM3 or GD1a. In the presence of a foreign protein, one molecule of GM2 activator was found to bind one molecule of GM2 or GM1 and the corresponding value for GM3 and GD1a was 0.8 and 0.7, respectively. In the absence of a foreign protein, one molecule of GM2 activator was found to associate with two to five molecules of GM2 with poor reproducibility. Furthermore, when the liposome-bound GM2 was mixed with GM2 activator and centrifuged, the putative complex contained not only GM2 activator and GM2, but also lecithin, one of the constituents of liposomes, in a ratio of 1/1.7/5.3. In the presence of a foreign protein, the binding of lecithin was completely suppressed and the ratio of GM2 activator to GM2 was 1:0.72. These results indicate that the detected binding between GM2 activator protein and glycosphingolipids may be the results of the complicated interactions where both the specific and the nonspecific bindings contributed to the association. It is conceivable that the extent of non-specific interactions would be easily affected by the change of environmental factors. Therefore, the presence of a foreign protein significantly changed the association of lecithin in the complex. Alternatively these experiments also indicated that when GM2 was mixed with GM2 activator, the status of ganglioside might undergo the process of redistribution of components in the liposomes or re-establishment of a new equilibrium in the micellar units of glycosphingolipids.

Polyacrylamide Gel Electrophoresis --- Polyacrylamide gel electrophoresis at pH 4.0 in 7.5% gel was used to analyze the complex formation between GM2 activator and GM2 (7,17). While the micellar form of GM2 did not enter the gel, some GM2 molecules migrated with GM2 activator as a band that contained 1:1 molar ratio of the two components. When the liposome-bound GM2 was used in the same experiment, the ratio of GM2 to GM2 activator was reduced to 0.4:1. Again, the molar ratio of GM2 activator and GM2 in the complex did not hold as a constant. This suggests that the observed association of GM2 activator with GM2 may not be a straightforward process and possibly the complex were formed by multi mechanisms of interactions.

Glycolipid Transfer from Donor to Acceptor Liposomes --- Conzelmann et al.(7) were the first to propose a model for explaining the function of GM2 activator in stimulating the hydrolysis of GM2 carried out by β-hexosaminidase A. GM2 activator could extract GM2 from biological membranes to form a reversible, water-soluble protein/lipid complex that served as the true substrate for β-hexosaminidase A. They tested this hypothesis by examining the ability of GM2 activator to transfer GM2 from the donor to the acceptor liposomes (7, 17). The transfer of GM2 between two types of liposomes by GM2 activator was found to be time and pH dependent. The optimal pH of the transferring reaction corresponded to that of the hydrolytic reaction (pH 4.2). The frequency of GM2 activator to transfer GM2 was calculated to be three seconds per one molecule based on a derived theoretical equation (17). However, the glycolipid transfer reaction by GM2 activator was not limited to GM2. GM1, in fact, was transferred faster than GM2. GD1a was transferred at about 40% the rate of GM2 and both GM3 and GA2 were transferred at 10% the rate of GM2 (17). GalNAc-GD1a, a glycolipid containing a heptasaccharide headgroup with a GM2-epitope at the non-reducing terminus, was also transferred at a rate one-third that of GM2 (18). As stated in the studies with ultracentrifugation, when the micellar form of GM2 or the liposome-bound GM2 was mixed with GM2 activator, a re-distribution of components or re-establishment of equilibrium may occur within micelles. How and to what extent of this effect might also influence the glycolipid transfer reaction between liposomes by GM2 activator is unknown.

HPLC Gel Filtration --- Sephacryl S-200 gel filtration column connected to Waters 600 E HPLC system was used to separate GM2 activator/lipid complex from GM2 micelles and GM2 activator protein, based on the differences in their molecular sizes (19). Sephacryl S-200 gel offered an unusual advantage to this analysis, because this gel adsorbed GM2 (or other gangliosides) at the top of the column, but not the activator/lipid complex or GM2 activator. When GM2 activator was applied alone to the column, the protein was not adsorbed by the Sephacryl-200 gel and eluted from the column at a certain retention time. When GM2 was applied to the column alone, it was retained by Sephacryl S-200 gel until regeneration of the column. When an incubation mixture containing GM2 and GM2 activator in a molar ratio of 1:1 was applied to the column, a peak presumed to be the activator/lipid complex, which was confirmed to contain both GM2 activator and GM2 in a ratio of 1:1, was obtained. When the experiment was run with a mixture of GM2 and GM2 activator in a molar ratio of 50:1, a peak containing the complex was eluted at a slightly shorter retention time and a broader peak area. This peak was confirmed to contain GM2 and GM2 activator in a ratio of 48.2:1. This indicates that the size of the complex varied depending on the initial proportion of GM2 activator and GM2 in the mixture. Similar chromatographic profiles were obtained when GM2 activator was incubated with either GM1 or GM3. No complex formation was detected when GM2 activator was incubated with the oligosaccharide derived from GM2, indicating the importance of the lipid moiety in GM2 for complex formation. As shown in Table I, when GM2 activator was incubated with an equimolar ratio of either GM1 or GM2, the molar ratio between GM2 activator and the respective ganglioside in the complex was 1:1. However, when the activator protein was incubated with 50 molar excess of either GM1 or GM2, the molar ratio between the ganglioside and GM2 activator in the complex was about 50:1 in both cases. When the bindings between GM2 activator and a 50-fold molar excess of GM3 was examined in the same manner, the ratio of GM3 to GM2 activator in the complex was found to be about 80:1. Apparently the size differences between micelles and vesicles also affected the sizes of the complexes.

Table I, Ratio between ganglioside and GM2 activator in the lipid/protein complex

Mixture (mol:mol)	Ganglioside/GM2-Act* in the complex
GM1 : GM2-Act	
1:1	1.08
50:1	48.2
GM2 : GM2-Act	
1:1	1.03
50:1	51.0
GM3 : GM2-Act	
50:1	83.9
GM2 : Saposin B	
50:1	46.6

*GM2-Act, GM2 activator protein

Saposin B, a non-specific activator protein, was also reported to bind glycosphingolipids (2,3). The association of saposin B to GM2 was also examined for comparison. The results in Table I suggested that there was no difference in the complex formation between GM2 activator and gangliosides versus that between saposin and ganglioside in aqueous medium. To examine whether GM2 activator could specifically bind only to GM2 when GM2 was together with GM1 or with GM3, GM2 was first mixed with GM1 or GM3 in chloroform:methanol (2/1, v/v), dried, and dispersed in an aqueous buffer solution. The aqueous ganglioside mixture was then incubated with GM2 activator and subjected to Sephacryl S-200 gel filtration. The activator/lipid complex fraction was assayed for the amounts of GM2, GM1, and the activator protein (or GM2, GM3, and the activator protein). As shown in Table II, GM2 activator did not appear to bind preferentially to GM2 to form the activator/GM2 complex in 1:1 ratio. Rather, GM2 activator associated with the mixture of gangliosides in the proportion similar to that in the original ganglioside mixture. For example, when GM2 activator was incubated with a mixture containing an equimolar ratio of GM1 and GM2, the molar ratio of GM2 activator, GM1, and GM2 in the complex was found to be close to 1:1:1. However, when GM2 activator was incubated with a mixture containing 25-fold excess of GM1 and GM2, the detected ratio of the activator protein to GM1 and GM2 in the complex was 1:17.3:17.0. No selective binding of GM2 by GM2 activator from the ganglioside mixture was observed.

Table II, Analysis of the lipid/protein complex formed between GM2 activator and the mixed gangliosides

	Determined ratio of ganglioside/GM2-Act* in the complex			
	GM1	GM2	GM3	GM2-Act
Mixture	(mol/mol)			
GM1:GM2:GM2-Act				
1:1:1	0.81	0.84		1.0
25:25:1	17.3	17.0		1.0
GM2:GM3:GM2-Act				
25:25:1		23.2	25.8	1.0

*GM2-Act, GM2 activator protein

Similar result was obtained from the incubation of GM2 activator with a mixture of GM2 and GM3. These results indicate that the composition of the complexes formed under the micellar form of ganglioside was determined by the pre-existing status of the ganglioside micelle (19).

TLC Overlay --- In this technique, each glycosphingolipid in chloroform:methanol (2/1, v/v) was first applied onto a polygram SIL G TLC plate and the plate was developed with a suitable solvent system to separate the glycosphingolipids. The dried plate was then immersed in a buffer solution containing 1% each of polyvinyl pyrrolidone and bovine serum albumin to block the excess binding sites on the TLC plate. The plate was then incubated with [^{14}C]-labled GM2 activator protein. After being washed three times with the buffer solution containing 0.05% of Tween-20, the plate was air-dried. The binding of glycosphingolipids and GM2 activator were detected by radioautography of the TLC plate. GM2 activator protein was found to bind avidly to

GM1, GM2, and GM3, but very weakly to LacCer, GalCer and GlcCer. Table III summarizes the binding of various glycosphingolipids to GM2 activator when examined by TLC overlay (19).

Table III, Binding of Glycosphingolipids to GM2 activator

Bound	GM1, GM2, GM3, GM4, GalNAc-GD1a, GD2, Sulfatide, SM3, NeuGcGM1, KDN-GM3, $IV^6KDNLcOse_4Cer$, $IV^6KDNLnOse_4Cer$, PE-GM2
Not bound	GA1, GA2, HO-GM2, Me-GM2, GalCer, GlcCer, LacCer

HO-GM2: The carboxylic function of NeuAc in GM2 was reduced to alcohol.

Me-GM2: The carboxylic function of NeuAc in GM2 was converted to a methyl ester.

PE-GM2: The tetrasaccharide of GM2 is linked to dipalmitoyl-phosphatidylethanolamine

These results strongly suggest that the recognition sites on the glycosphingolipids for the binding by GM2 activator are the anionic residues such as NeuAc or sulfate group. The optimal binding condition was at pH 4.0 and low ionic strength of the buffer, which was similar to that for the conversion of GM2 to GM3 carried out by β-hexosaminidase A. High ionic strength or high pH (pH 6.8) of the buffer solution greatly attenuated the binding (19).

For comparison, the interaction between saposin B and glycosphingolipids were also examined using TLC overlay method. Saposin B was found to bind not only to gangliosides and sulfatide, but also to GA2 and LacCer. Compared with GM2 activator, saposin B bound to glycosphingolipids better at the neutral pH (pH 6.8), and the general behavior of binding was not greatly affected by the acidic pH or the high ionic strength of the buffer solution. The shortcoming of TLC overlay method is that it is difficult to make precise quantitative evaluation of the extent of the bindings. This method, however, can accurately identify the species of glycosphingolipids that bind to GM2 activator.

Fluorescence Dequenching Assay --- The self-dequenching fluorescent probe, octadecylrhodamine B chloride (R18) has been used in membrane-fusion assay based on the increase of fluorescence by the dilution of R18 upon membrane fusion (20). Using this lipid probe, the recombinant human GM2 activator and the rat GM2 activator were found to dequench the R18-loaded phosphatidylcholine-containing large unilamellar vesicles (PC-LUV) (21). This assay was to assess the R18 transfer between liposomes and, hence, to estimate the hydrophobic binding activity of GM2 activator. Subsequently, this assay was applied to examine the effects of various glycosphingolipids on the fluorescence dequenching of R18 incorporated in PC-LUV by the recombinant human GM2 activator protein (12). The fluorescence dequenching of R18 by GM2 activator was dose and pH dependent. The optimal pH was found to be pH 5 that was close to the optimal pH for the hydrolysis of GM2 by β-hexosaminidase A. When the assay was carried out in the presence of GM1, the rate of fluorescence dequenching was decreased in a dose-dependent manner. Under the standard assay condition, GM2 (1.3 μM) inhibited 90% of fluorescence dequenching, and the same concentration of GT1b inhibited 62%; GM1, 25%, globoside, 24%; GM3, 17%; and GA2, 3%. Therefore, the binding affinity between GM2 activator and the glycosphingolipids tested was highest for GM2, then GT1b>>GM1~Globoside~GM3>GA2 (12). Again, no specific binding affinity between GM2 activator and GM2 was found using R18 fluorescence dequenching assay. Since all glycosphingolipids contain the ceramide moiety, the distinct carbohydrate moiety of each glycosphingolipid might contribute to their different abilities to inhibit GM2 activator to reduce the R18 dequenching. The fluorescence dequenching analysis allows the direct measurement of glycosphingolipid binding to GM2 activator.

Differential Scanning Calorimetry and Film Balance Measurements --- This technique was used to study the interaction of GM2 activator with phospholipid/ganglioside vesicle and monolayer (22). When GM2 activator was added to dimyristoyl-glycerophosphocholine (DMPC) lipid bilayer, a destabiliztion of the gel phase was observed, suggesting the non-specific perturbations in the headgroup region of the bilayer by the protein.

When GM2 activator was added to the negatively charged DMPC/DMPG (dimyristoyl glycerophosphoglycerol) vesicles, an additional electrostatic interaction of the protein with the lipid surface, besides the non-specific interaction, was observed. When GM2 activator was added to DMPC/GM1 or DMPC/GM2 vesicles, more specific electrostatic interaction was detected between the activator protein and GM1 or GM2. This specific interaction, in the case of DMPC/GM2 vesicles, might represent the specific interaction of GM2 activator protein with GM2 headgroup (GM2 epitope). The DSC measurements indicated that, although GM2 activator showed specific binding to GM2 headgroup, there was no evidence for the activator protein to extract GM2 from lipid bilayer. More likely, upon binding, the activator molecules remained on the surface of the lipid membrane.

Discussion

GM2 activator has been shown to transfer glycosphingolipids (2, 7) and a fluorescent lipid probe, R18 (21) between the donor and the acceptor liposomes *in vitro*. Since GM2 activator has to interact with GM2 molecule for stimulating the enzymatic conversion of GM2 into GM3, it is reasonable to assume that GM2 activator should be able to bind GM2 through a specific interaction. It is known that the hydrophobic binding property of GM2 activator plays an essential part for its stimulatory function, because the lipid-free tetrasaccharide derived from GM2 was not hydrolyzed by β-hexosaminidase A, even in the presence of GM2 activator (13). In addition, human GM2 activator is very effective in stimulating the hydrolysis of GM2, but not effective in promoting the hydrolysis of GA2, the desialylated derivative of GM2, by the same enzyme (15). GM2 activator was also not involved in the hydrolysis of 6'GM2, a GM2 analogue with a GalNAc1→6Gal linkage instead of a GalNAc1→4Gal linkage (23). These results suggest that GM2 activator must be able to recognize the specific oligosaccharide head group of GM2. The evidence on the specific interaction between GM2 activator and GM2 has not been clearly demonstrated. It has not been successful to isolate the true functional activator/lipid complex and to define the specific interaction(s) between the two components. In the case of interactions between lectins and their ligands, the complexes have been isolated and crystallized. Thus, their specific interactions have been shown by crystallographic analysis (24). Although, the putative GM2 activator/GM2 complex was isolated by ultracentrifugation and the ratio between the two components was determined to be 1:1 (17), evidence for the specific binding between GM2 activator and GM2 was not provided.

Various glycosphingolipids have been shown to be associated with GM2 activator through non-specific hydrophobic interactions. Conversely, many proteins also can bind to glycosphingolipids, possibly also through the hydrophobic interactions. As shown by Tomasi et al (25), bovine serum albumin interacted with GM1 and formed multiple complexes. Conzelmann et al (17) also reported that the presence of foreign proteins such as cytochrome and gelatin affected the binding of GM2 with GM2 activator and greatly reduced the ratio between GM2 and GM2 activator in the complex. Non-specific hydrophobic interactions can occur upon the contact of compounds at their hydrophobic domains. The non-specific hydrophobic interaction between glycosphingolipids and GM2 activator is the major complication that masks the specific functional interaction between GM2 activator and GM2. Crystallographic analysis of the GM2 activator co-crystallized with GM2 should provide important information on the specific binding between GM2 activator and GM2. Among the methods reported so far for the detection of GM2 activator and GM2 interactions, ultracentrifugation, gel filtration, and electrofocusing are not able to distinguish the non-specific interaction from the specific binding. It seems that by TLC overlay, one can detect the interactions contributed by the head groups of glycosphingolipids, because the hydrophobic moieties of glycosphingolipids are bound to silica gel. Fluorescence dequenching assay is to assess the hydrophobic binding function of GM2 activator. A possible mode of action of GM2 activator in causing fluorescence dequenching of R18 has been discussed (21). GM2 activator may extract the lipid probe R18 from one membrane and transfer to another. Alternatively, GM2 activator only bound to the liposome-associated R18 and the transfer of R18 occurs when acceptor liposome collides with the donor liposome. The former mechanism was also proposed by Conzelmann et al.(7) for the action of GM2 activator to stimulate the hydrolysis of GM2 through the formation of a water-soluble complex with GM2. However, the recent report by Giehl et al. (22) suggested that GM2 activator showed specific bindings to GM2 headgroup through electrostatic interactions, and that there was no evidence for GM2 activator to extract GM2 away from the lipid bilayer as proposed by Conzelmann et al. (7). Taken together, the identification of specific interaction between GM2 activator and GM2 is still an open and challenging task. Aside from crystallography, future studies along this line require further technological development to distinguish the specific binding from the non-specific interaction between GM2 activator and GM2.

SUMMARY

GM2 activator protein is a protein cofactor that stimulates the hydrolysis of the GalNAc and the NeuAc in GM2 by β-hexosaminidase A and sialidase, respectively. To understand the mechanism of action of GM2 activator, the interaction of this protein with GM2 and/or β-hexosaminidase A has been the subject of interest since the purified GM2 activator became available. Numerous techniques including ultracentrifugation, isoelectric focusing, polyacrylamide gel electrophoresis, gel filtration, thin layer chromatogram overlay, and fluorescence dequenching assay have been used to investigate the binding and the affinity of GM2 activator to various glycosphingolipids. It has been generally accepted that GM2 activator must have a very weak binding with the enzyme, because they can be easily separated from each other by gel filtration. Therefore, the interaction of GM2 and GM2 activator has been the focus for most of the study. Although preferential association of GM2 activator with GM2 was detected by some methods, GM2 activator was found also to bind other glycosphingolipids. Isolation of the specific complex that consists of only GM2 activator and GM2 from an incubation mixture containing the activator protein and mixed glycosphingolipids has not been successfully carried out. Ultracentrifugation and gel-filtration are the mildest methods for the isolation of the complexes. However, these methods do not separate the complexes formed by specific interaction from that formed by non-specific association. In fluorescence dequenching assay, the attempt to isolate the complex of R18 lipid probe with GM2 activator was also not successful. Since GM2 activator and glycosphingolipids contain hydrophobic domains in their molecules, the non-specific hydrophobic interactions between the two components can greatly interfere with the isolation of true functional complexes. Among the reported methods, thin layer chromatography overlay and the assay based on the inhibition of fluorescence dequenching by various glycosphingolipids are more informative than the others on the binding between GM2 activator and the carbohydrate head groups of glycosphingolipids.

REFERENCES

1. Li, Y.-T. and Li, S.-C. (1984) Activator proteins related to the hydrolysis of glycosphingolipids catalyzed by lysosomal glycosidase. In: *Lysosomes in Biology and Pathology,* Vol. 7, J. Dingle, R.T. Dean and W. Sly (eds.). Elsevier Science Publishers, Amsterdam, The Netherlands, pp. 99-117.
2. Fürst, W., and Sandhoff, K. (1992) Activator proteins and topology of lysosomal sphingolipid catabolism. *Biochem. Biophys Acta*, **1126**: 1-16.
3. Kishimoto, Y., Hiraiwa, M., and O'Brien, J.S. (1992) Saposins:structure, function, distribution, and molecular genetics. *J. Lipid Res.* **33**: 1255-1267.
4. Azuma, N., O'Brien, J.S., Moser, H.W., and Kishimoto, Y. (1994) Stimulation of acid ceramidase activity by saposin D. *Arch. Biochem. Biophys.* **311**, 354-357.
5. Klein, A., Henseler, M., Klein, C., Suzuki, K., Harzer, K. and Sandhoff, K. (1994) Sphingolipid activator protein D (sap-D) stimulates the lysosomal degradation of ceramide *in vivo. Biochem. Biophys. Res. Commun.* **200**, 1440-1448.
6. Li, Y.-T., Mazzotta, M.Y., Wan, C.-C., Orth, R. and Li, S.-C. (1973) Hydrolysis of Tay-Sachs ganglioside by β-hexosaminidase A of human liver and urine. *J. Biol. Chem.* **248**: 7512-7515.
7. Conzelmann, E., and Sandhoff, K. (1979) Purification and characterization of an activator protein for the degradation of glycolipids GM2 and GA2 by hexosaminidase A. Hoppe-Seylers Z. *Physiol. Chem.*, **360**: 1837-1849.
8. Li, S.-C., Hirabayashi, Y. and Li, Y.-T. (1981) A protein activator for the enzymic hydrolysis of GM2 ganglioside. *J. Biol. Chem.* **256**: 6234-6240.
9. Hirabayashi, Y., Li, Y.-T., and Li, S.-C.(1983) The protein activator specific for the enzymic hydrolysis of GM2 ganglioside in normal human brain and brains of three types of GM2-gangliosidosis. *J. Neurochem.* **40**: 168-175.
10. Gravel, R.A., Clarke, J.T.R., Kaback, M.M., Mahuran, D., Sandhoff, K., and Suzuki, K. (1995) The GM2 gangliosidoses. In the Metabolic Basis of Inherited Diseases (Scriver, C.R., Beaudt, A.L., Sly, W.S., and Valle, D., eds.) Vol II, 7th Ed. pp. 2839-2879, McGraw-Hill, New York.

11. Sandhoff, K., Harzer, K., and Fürst, W. (1995) GM2 activator protein. In the Metabolic Basis of Inherited Diseases (Scriver, C.R., Beaudet, A.L., Sly, W.S., and Valle, D., eds.) Vol II, 7th Ed. pp. 2427-2441, McGraw-Hill, New York.
12. Smiljanic-Georgijev, N., Rigat, B., Xie, B., Wang, W., and Mahuran, D.J. (1997) Characterization of the affinity of the GM2 activator protein for glycolipids by a fluorescence dequenching assay. *Biochem. Biophys. Acta*, **1339**: 192-202.
13. Wu, Y.-Y., Lockyer, J.M., Sugiyama, E., Pavlova, N.V., Li, Y.-T. and Li, S.-C. (1994) Expression and specificity of human GM2-activator proteins. *J. Biol. Chem.* **269**: 16276-16283.
14. Bellachioma, G., Stirling, J.L., Orlacchio, A., and Beccari, T. (1993) Cloning and sequence analysis of a cDNA clone coding for the mouse GM2 activator protein. *Biochem. J.* **294**: 227-230.
15. Yuziuk, J.A., Bertoni, C., Beccari, T., Orlacchio, A., Wu, Y.-Y., Li, S.-C., and Li, Y.-T. (1998) Specificity of mouse GM2 activator protein and β-N-acetylhexosaminidase A and B: Similarities and differences with their human counterparts in the catabolism of GM2. *J. Biol. Chem.* **273**: 66-72.
16. Itoh, T., Li, Y.-T., Li, S.-C. and Yu, R.K. (1981) Isolation and characterization of a novel monosialosylpentahexosyl ceramide from Tay-Sachs Brain. *J. Biol. Chem.* **256**: 165-169.
17. Conzelmann, E., Burg, J., Stephan, G., and Sandhoff, K. (1982) Complexing of glycolipids and their transfer between membranes by the activator protein for degradation of lysosomal ganglioside GM2. *Eur. J. Biochem.* **123**: 455-464.
18. Meier, E.M., Schwarzmann, G., Fürst, W., Sandhoff, K. (1991) The human GM2 activator protein, a substrate specific cofactor of β-hexosaminidase A. *J. Biol. Chem.* **266**: 1879-1887.
19. Hama, Y., Li, Y.-T., and Li, S.-C. (1997) Interaction of GM2 activator protein with glycosphingolipids. *J. Biol. Chem.* **272**: 2828-2833.
20. Mullock, B.M., and Luzio, J.P. (1992) Reconstitution of rat liver endosome-lysosome fusion in vitro. *Methods Enzymol.* **219**: 52-60.
21. Kuwana, T., Mullock, B.M., and Luzio, J.P. (1995) Identification of a lysosomal protein causing lipid transfer, using a fluorescence assay designed to monitor membrane fusion between rat liver endosomes and lysosomes. *Biochem. J.* **308**: 937-946.
22. Geihl, A., Lemm, T., Bartelsen, O., Sandhoff, K., and Blume, A. (1999) Interaction of the GM2-activator protein with phospholipid-ganglioside bilayer membranes and with monolayers at the air-water interface. *Eur. J. Biochem.*, **261**, 650-658.
23. Li, Y.-T., Li, S.-C., Hasegawa, A., Ishida, H., Kiso, M., Bernardi, A., Brocca, P., Raimondi, L., and Sonnino S. (1999) Structural basis for the resistance of Tay-Sachs ganglioside GM2 to enzymatic degradation. *J. Biol. Chem.* **274**: 10014-10018.
24. Rini, J. M. (1995) Lectin structure. Ann. Rev. Biophys. Biomol. Struc. **24**: 551-577.
25. Tomasi, M., Roda, L.G., Ausiello, C., D'Agnolo, G., Venerando, B., Ghidoni, R., Sonnino, S., and Tettamanti, G. (1980) Interaction of GM1 ganglioside with bovine serum albunin formation and isolation of multi complexes. *Eur. J. Biochem.* **111**: 315-324.

TUMOR-ASSOCIATED CARBOHYDRATE ANTIGENS DEFINING TUMOR MALIGNANCY: BASIS FOR DEVELOPMENT OF ANTI-CANCER VACCINES[1,2]

Sen-itiroh Hakomori

Pacific Northwest Research Institute and University of Washington, 720 Broadway, Seattle, WA 98122, USA

Aberrant glycosylation has been found in essentially all tumor cells so far examined. Tumor-associated carbohydrate antigens (TACAs), formed in association with this process, are defined by monoclonal antibodies. Detailed clinicopathological studies have shown that differences in survival rates of patients are related to expression of TACAs or to degree of aberrant glycosylation (1). Thus, certain TACAs are not simply tumor markers, but constitute essential machinery to induce metastasis and invasiveness.

During the past two decades, great progress has been made in tumor biology through introduction of the concept of transforming genes (oncogenes) and tumor-suppressing genes (2-6). Tumorigenesis, metastasis, invasiveness, and tumor progression have been shown to be associated closely with balanced expression of oncogenes and tumor-suppressing genes. This concept currently dominates various areas of tumor biology. However, incidence of well-known oncogenes (*e.g.*, myc, ras^k) expressed in various common human cancers is surprisingly low (5-10% without PCR). The detection rate goes up to 30% with PCR (**Fig. 1**). Exceptionally, oncogene *Rb* is expressed in 100% of retinoblastoma cases.

The majority of oncogene products, if not all, are not good immunogens. There are a few exceptions, *e.g.*, HER-2/neu, the product of its oncogene, originally termed Erb. This protein is a good immunogen, is expressed in breast, ovarian, and various other cancers, and has received much recent interest as a target of immunotherapy (7), even though its expression rate is 30% in breast cancer, 15-20% in ovarian cancer, and possibly less in other cancers.

The general concept that initiation and progression of various types of human cancer is associated with deletion or inactivation of tumor-suppressing gene products (*e.g.*, p53, p16) is increasingly popular, yet deletion of p53 or p16 occurs in ≤50% of colorectal carcinomas (3-5). In striking contrast, incidence of certain TACAs (*e.g.*, Tn, STn, extended Le^y) expressed in colorectal carcinomas is 80-100% (8, 9) (**Fig. 1**). A similar contrasting situation may be observed in other types of human cancer. The mechanism for expression of TACAs may not be related simply to oncogene expression. This linkage may be difficult to elucidate, since expression of a single TACA requires coordination of a number of glycosyltransferase genes, which is assumed to be controlled by multiple trans-acting or cis-acting factors to promoter regions of each gene.

---

[1] Studies by the author and his colleagues cited in this review were supported by Outstanding Investigator Grant CA-42505 from the National Cancer Institute (NIH).
[2] I thank Dr. Stephen Anderson for scientific editing and preparation of the manuscript.

The Molecular Immunology of Complex Carbohydrates–2.
Edited by Albert M. Wu, Kluwer Academic/Plenum Publishers, 2001.

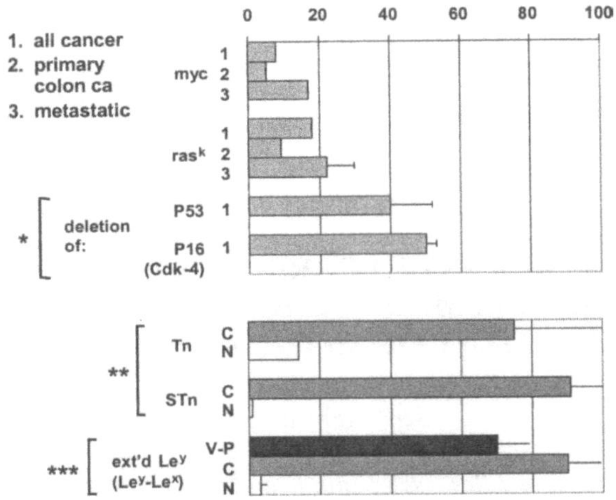

Fig. 1. Comparative expression of oncogenes, deletion (or inactivation) of anti-oncogenes (suppressor genes), and aberrant glycosylation. Expression of the most common oncogenes (myc, ras^k) in various common human cancers is <20%, even when determined by PCR. The highest expression observed for ras^k by PCR is close to 30%, for metastatic colonic cancer. *, Deletion (or inactivation) of p53 or p16 (cyclin-dependent kinase 4) is observed in various common cancers, with a maximum occurrence of 40-50% (Hollstein M, et al., *Nucleic Acid Res* 1994; Marx J, *Science* 1994; Levine AJ, *Cell* 1997, refs. 2, 3, 5 as cited in text). **, Expression of Tn (defined by two mAbs, CU1 and 1.01, and *Vicia villosa* lectin), and of STn (defined by two mAbs, TKH2 and B72.3) in colonic cancer (C; gray bar) and in normal mucosa (N; white bar). Variation in expression reflects degree of differentiation. Modified from Table 1 of Itzkowitz SH, et al., *Cancer Res* 1989 (ref 9). ***, Expression of extended Le^y (defined by mAbs CC-1, CC-2) and trifucosyl Le^y (Le^y-Le^x; defined by mAb KH1) in colonic cancer. V-P (dark gray bar), villous adenomatous polyps. C (light gray bar), cancer (from distal and proximal mucosa). N (white bar), normal distal mucosa; expression in normal proximal mucosa is up to 30%. Based on data in Figs. 1 and 5 of Kim YS, et al., *Cancer Res* 1986 (ref 8).

Not all TACAs affect tumor malignancy, and anti-cancer vaccines are effective and useful only for the TACAs that do affect tumor malignancy. A concise, excellent review on this topic was published by Muramatsu (10). In the present review, I will focus on TACAs whose relationship with malignancy of the tumor is clearly known. First, however, I will describe a few glycosylations or TACAs whose expression counteracts expression of tumor-promoting glycosylation.

I. TACA EPITOPES OR GLYCOSYLATION THAT INHIBIT HUMAN TUMOR MALIGNANCY

1. Persistent expression of histo-blood group A and B epitopes

Little attention has been paid to possible inhibitory effects of glycosylation on tumor cell malignancy, except for reduction/ deletion of histo-blood group A and B antigens, which is well established and known for many years. This phenomenon is related to expression of H/Le^y/Le^b, the precursors of A and B epitopes (see following section). Many types of human cancer from A or B patients show reduction/ deletion of A or B epitopes. This was observed initially in gastrointestinal cancer (11-13), and subsequently in many other types, *e.g.*, lung cancer (14), cervical cancer (15), oral epithelial cancer (16), and urothelial cancer (17). The degree of A/B deletion, associated with degree of metastasis or malignancy, was shown even in the early studies (18), and statistical correlation was demonstrated in more recent studies on survival rates of patients with primary non-small cell lung carcinoma (19) (**Fig. 2**). In this study, A epitope was clearly inhibitory for malignancy, whereas the effect of B epitope was variable. In an independent study, however, both A and B epitopes were found to be malignancy-inhibitory, and had equivalent effects on survival rate (20).

The biochemical rationale for the malignancy-inhibitory effect of A or B expression is still unclear. Recent studies with transfection of *A* or *B* gene to H-expressing tumor cells, and with phenotypic comparison of A^+ vs. A^- variants of a few colonic carcinomas, indicate that A^- cells have higher Matrigel-dependent (haptotactic) motility and higher proliferation than A^+ cells. Haptotactic motility of human colonic carcinoma cell lines HRT18 and MKN74 was greatly reduced when cells were transfected with histo-blood group A or B cDNA. Since motility of these cells is dominantly controlled by α6 and to a lesser extent by α3 integrin receptor, and both α6 and α3 are highly A or B glycosylated in transfected cells, reduced motility is ascribable to addition of A or B epitope (21). This trend of motility control by A determinant was also observed in two other colonic carcinoma cell lines, SW480 and HT29. Haptotactic motility of A^- clones as compared with A^+ clones was greatly enhanced for isolates of both SW480 and HT29. A^- clones showed higher cell proliferation than A^+ clones (22). The molecular mechanism causing A deletion/ reduction involves reduced activity of promoter region of A transferase, and methylation in CpG islands of promoter DNA region (23).

2. Bisecting β1→4GlcNAc counteracts synthesis of GlcNAcβ1→6Manα1→6Man side chain in N-linked structure

Increase of tri- or tetra-antennary N-linked structures resulting from addition of side chain GlcNAcβ1→6Manα1→6Man is correlated with degree of malignancy in essentially all types of cancer. This is ascribable to enhanced level of GlcNAc transferase V (GnT-V), which catalyzes transfer of GlcNAc to 6 position of αMan (24, 25). Bisecting β1→4GlcNAc, whose synthesis is catalyzed by GnT-III, inhibits GnT-V, resulting in decreased synthesis of GlcNAcβ1→6Manα1→6Man (26). Based on cloning of GnT-III and GnT-V genes, Taniguchi

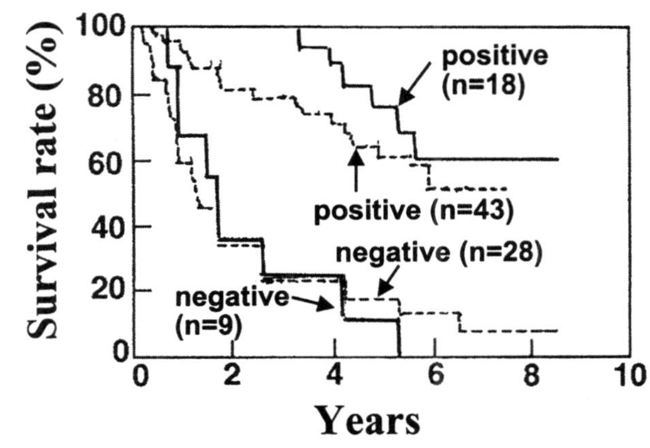

Fig. 2. Kaplan-Meier survival curve of patients with primary lung carcinoma (excluding small cell carcinoma), as related to deletion or reduction of histo-blood group A expression. Solid lines: patients with stage 1 lung carcinoma. Dashed line, overall survival of lung carcinoma patients. Numbers of cases (n) with A-positive and A-negative immunostaining are indicated. From Lee JS, et al., *N Engl J Med* 1991 (ref 19), by permission.

and colleagues demonstrated "remodeling" of tumor cell surface N-linked structures following transfection of GnT-III cDNA (27) (**Fig. 3A**). This approach, applied to mouse B16 melanoma cells, greatly inhibited lung metastasis (28) (**Fig. 3B**). Based on selective binding of PHA-L to N-linked structure having GlcNAcβ1→6Manα1→6Man side chain (29), correlation of expression of this epitope in human colorectal carcinoma with patient survival rate was demonstrated recently by immunohistological examination (30) (**Fig. 3C**). Thus, β1→4GlcNAc for bisecting structure and β1→6GlcNAc for tri- or tetra-antennary side chain competitively influence tumor malignancy. The target molecule of such glycosylation effect was suggested to be E-cadherin, integrin, or lysosomal associated membranous protein-1 (LAMP-1) (27).

3. Galactosylgloboside in seminoma

Seminoma is a rare cancer in adults, but is the major type of urogenital cancer in infants and

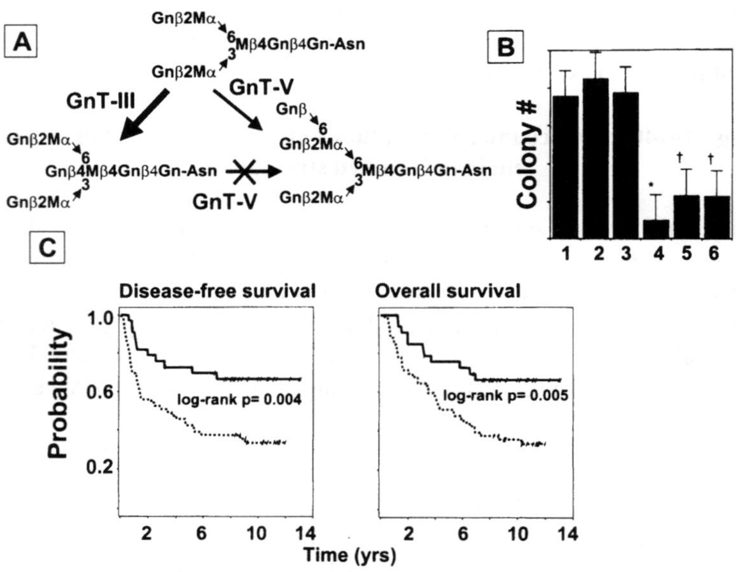

Fig. 3. Effect of N-linked glycosylation profile on tumor malignancy. **Panel A:** Relationship between bisecting GlcNAc whose synthesis is catalyzed by GlcNAc transferase III (GnT-III), and β1→6GlcNAc side chain whose synthesis is catalyzed by GnT-V. Overexpression of GnT-III and bisecting structure inhibits GnT-V and decreases β1→6GlcNAc side chain. **Panel B:** Overexpression of GnT-III in mouse melanoma B16 is associated with decreased lung metastasis. Bars 1, 2, 3: lung colony number from control B16, and B16 transfected with two types of vector alone. Bars 4, 5, 6: colony number from three clones of B16 transfected with GnT-III cDNA. Panels A and B from Yoshimura M, et al., *PNAS* 1995 (ref 28), by permission. **Panel C:** Kaplan-Meier curves of disease-free survival (left) and overall survival (right). A colorectal carcinoma with a staining index (SI) of ≤1 (solid line) is more significantly associated with longer disease-free survival and overall survival than is one with a SI of >1 (dashed line). From Seelentag WKF, et al., *Cancer Res* 1998 (ref 30), by permission.

children. Of 17 cases of seminoma, all nonmetastatic cases (n=12) contained a glycosphingolipid (GSL) identified as galactosylgloboside (Galβ1→3GalNAcβ1→3Galα1→4Galβ1→4Glcβ1→1Cer) by NMR and mass spectrometry. All metastatic cases (n=5) lacked this GSL, although the sample sizes were admittedly small. No other GSL or ganglioside examined showed any clear correlation with metastatic potential. Clinical follow-up studies showed no evidence of disease for 27-62 months following orchiectomy in all 12 cases (patients 1-12) having this GSL. Patients 13 and 14, who were Gal-globoside-negative and had Stage I seminoma before orchiectomy, showed pancreatic metastasis after 10 months and retroperitoneal lymph node metastasis after 41 months, respectively. Patients 15, 16, and 17, who were Gal-globoside-negative and had Stage IIA seminoma, showed metastasis to lymph nodes and were treated with cisplatin/ vinblastine/ peplomycin. Therefore, the presence or absence of Gal-globoside is useful for prognosis of seminoma patients, especially those at Stage I (31).

How such a single GSL can inhibit metastasis and malignancy is unknown at this time. There is a possibility that Gal-globoside may induce apoptosis, in analogy to expression of Le^x in colonic carcinoma HT29 cells (32) or Le^y in esophageal and intestinal carcinoma cells (33).

II. TACA EPITOPES THAT PROMOTE HUMAN TUMOR MALIGNANCY

1. T and Tn in various cancers

Essentially all humans have antibodies that agglutinate erythrocytes, regardless of ABO type, by an unknown transferable factor which was later identified as sialidase. The antigen involved in this pan-agglutination was called "T" or "TF" (the initials of Thomsen and Friedenreich, who first described the phenomenon) (34, 35). A patient with hemolytic anemia had antibodies showing pan-agglutination similar to that of T, but with different specificity. The antigen was called "Tn" (for "T nouvelle") (36). Subsequent studies showed that "Tn syndrome," which occurs in <1 of 100,000 individuals and is characterized by anemia, leukopenia, and thrombopenia, is based on expression of Tn antigen (37).

T (or TF) antigen was identified as desialylated core structure of common mucin-type O-linked carbohydrates, by enzymatic hydrolysis and use of various lectins by Dahr et al. (38). The structure was identical to Galβ1→3GalNAc α-linked to Ser or Thr (39). Dahr et al. also demonstrated that Tn reacts with *Helix pomatia* lectin, and assumed that Tn is GalNAc α-linked to Ser or Thr (38).

Based on a series of studies, Springer and colleagues claimed that T and Tn are expressed highly in certain human cancers, particularly breast cancer (40-42). The reagent used for detection of T was naturally-occurring pan-agglutinable anti-T antibody commonly present in all human sera. However, the reagent used for detection of Tn was serum of a patient showing Tn syndrome, adsorbed extensively with desialylated A_1 and O erythrocytes, and *Vicia graminea* lectin.[3] Their claim of T and Tn as tumor-associated antigens was not widely accepted at the time, mainly because the reagents used in their experiments were ill-defined, and the quality of reagents used was difficult for others to obtain. Pan-agglutinable antibodies in normal sera react with not only T but also many other antigens having similar structure. The antibody reagent used

---

[3] Blood group M and N antigens expressed on erythrocytes were claimed by Springer and colleagues to be converted by sialidase treatment to T and Tn, respectively. Therefore, T and Tn were presumed to be precursors of M and N. In these studies, N antigen was probed by *Vicia graminea* lectin. Tn, claimed to be present in tumors, was probed by serum from a patient with Tn syndrome, adsorbed extensively with desialylated A_1 and O erythrocytes, but was not probed by *Helix pomatia* lectin, presumably because of its cross-reactivity with many other GalNAc-terminal structures such as globoside. Because of the ill-defined nature of the probes used, the data presented were ambiguous.

for detection of Tn was ill-defined and obscure. Therefore, major questions remained concerning the idea of T and Tn as cancer antigens, until mid-1985 when monoclonal antibody (mAb) against T and Tn was introduced. Doubt was raised about T as cancer-associated antigen, whereas Tn was well-verified as cancer antigen by the establishment of mAbs NCC-LU-35 and -81 which showed highly selective reactivity with a number of human cancers. Both these mAbs were cross-reactive with blood group A antigen, and were identified as anti-Tn (43). Subsequently, non-A-crossreactive anti-Tn IgG_3 mAb CU-1 was established, showing high specificity to various types of human cancer (lung, colorectal, breast, gastric, pancreatic) and human cancer cell lines, regardless of their histo-blood group ABO status (44). Detailed analysis of Tn antigen expressed on human colonic cancer cell lines indicated that clustering of Tn is necessary for reaction with anti-Tn antibody MLS-128 (45).

There are controversial observations on the relationship between Tn expression and tumor malignancy. Tn expression in breast cancer defined by *Helix pomatia* lectin was claimed to be correlated with invasive/ metastatic potential (46), but no such correlation was found in two independent follow-up studies (47, 48). In ovarian cancer, Tn expression defined by *Vicia villosa* lectin was claimed to be correlated with patient survival, *i.e.*, 5 year survival in: Tn (-) parametrial spread (-) 95% (n=51), Tn (-) parametrial spread (+) 90% (n=10), Tn (+) parametrial spread (+) 48% (n=19) (49). In hepatic cancer, Tn expression is more frequently associated with cholangiocarcinoma with cirrhosis than are STn or T expression. Thus, Tn expression is an indicator of high malignancy of hepatocellular or cholangiocarcinoma (50).

2. Sialyl-Tn as predictor of metastasis and invasion in various types of cancer

An unknown breast cancer-associated antigen was found by Schlom and colleagues with development of mAb B72.3 (51). It was later found to be expressed widely in various types of human cancer (52). The epitope structure of this antigen was identified by us as sialyl 2→6GalNAc α-linked to Ser or Thr of mucin-type glycoprotein, since desialylation abolished antibody reactivity, and desialylated compound reacted strongly with anti-Tn mAbs NCC-LU-35 or -81, or CU-1 (53). At almost the same time, another mAb, MLS-102, was established against colonic cancer cell line LS180, and its epitope structure was identified as NeuAcα2→6GalNAc (54). Thus, the structure of this antigen is closely related to T and Tn (see **Fig. 4**). STn structure was already known to be highly enriched in ovine submaxillary mucin, but absent or scarce in other animal mucins (55); it is present in minimal quantity or virtually absent in mucins from normal human tissues (53, 54). We raised mAbs using purified ovine submaxillary mucin that specifically stained various human tumor tissues but not normal tissue (except for a few cells in crypt region of intestinal epithelia). Of these, TKH1 and TKH2 were characterized as showing highest tumor specificity (53). Interestingly, MLS-102, TKH2, and B72.3 reacted preferentially with clustered NeuAcα2→6GalNAc linked to tandemly-linked Ser or Thr residues (54, 56). Recent studies on comparative reactivity of TKH2 and B72.3 with monomeric and trimeric STn indicate that TKH2 has stronger binding affinity than B72.3, whereas B72.3 reacts more specifically with trimeric STn (56). Another factor controlling reactivity of TKH2 is 9-O-acetylation at sialic acid. De-acetylation turned ~80% of cases TKH2-positive in normal colon tissues, and increased the proportion of TKH2-positive cells in colonic cancer tissues (57). Thus, specific reactivity of STn in tumor cells with its antibody is controlled by at least two factors: (i) clusters of epitope; (ii) O-acetylation of sialic acid.

The effect of tumor STn expression on malignancy, in terms of metastasis and patient survival rate, was studied initially in colonic cancer at various Duke stages by Itzkowitz and colleagues. In TKH2-negative cases at stages A, B, or C, no patients died of cancer during a 10-year period. In contrast, in TKH2-positive cases at stage C, 45% of patients died of cancer during the same period (58) (**Fig. 5A**). Since STn antigen is released from certain types of tumor to serum,

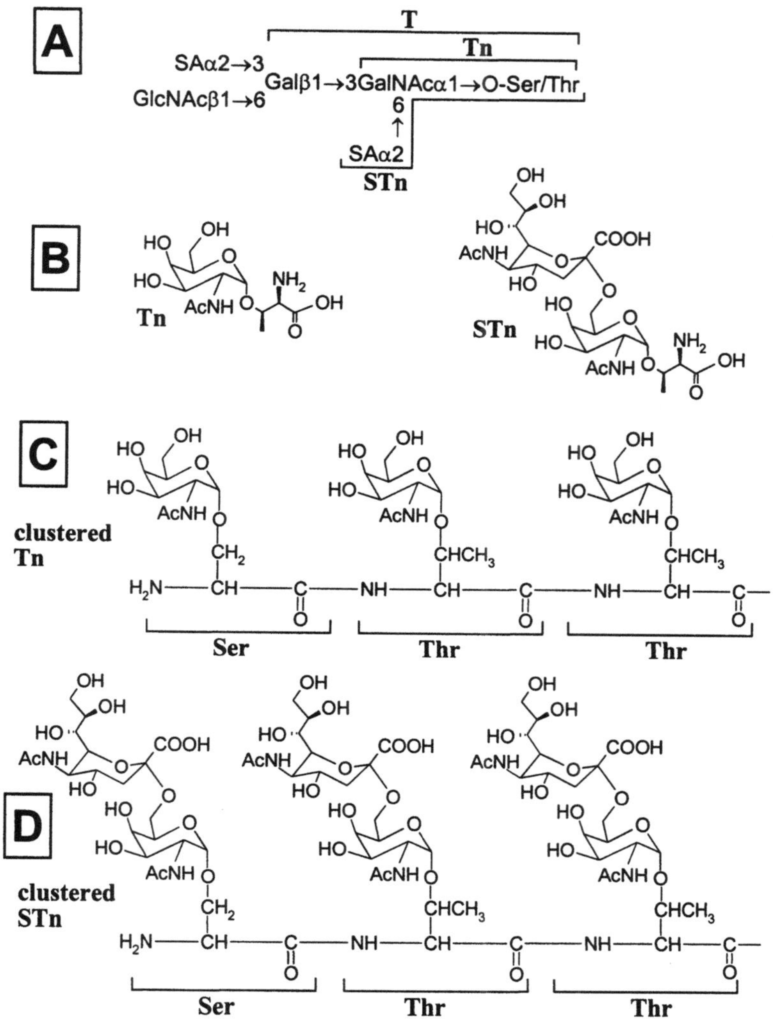

Fig. 4. **Panel A:** Basic structural unit of T, Tn, and STn.
Panel B: Conformational structure of Tn and STn.
Panel C: Conformational structure of clustered Tn.
Panel D: Conformational structure of clustered STn.

antigen level in patient sera has prognostic value. In ovarian cancer, elevation of serum STn level determined by TKH2 was correlated strongly with patient survival (59) (**Fig. 5B**). A number of subsequent clinicopathological studies of various types of cancer showed correlation between STn expression and degree of malignancy (50, 60-64). On the other hand, expression of STn is observed frequently in colorectal cancer, but its prognostic value has been questioned (65). STn expression is highly specific to tumors, and is correlated with degree of malignancy in the majority of studies. Therefore, this antigen is one of the most important targets in development of anti-cancer vaccines (see Section III-3).

The mechanism for expression for Tn and STn is basically ascribable to incomplete synthesis of O-linked mucin glycosylation, although blocking of Galβ1→3 transferase in synthesis of core structure has not been clearly demonstrated. It is possible that premature α2→6 sialylation

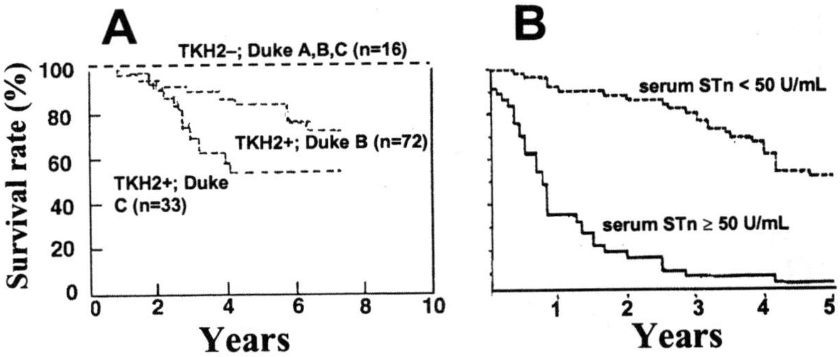

Fig. 5. Kaplan-Meier curves of patient survival as related to STn expression in primary tumor in colon carcinoma, and to serum STn level in ovarian cancer. **Panel A:** Survival rates of three groups of patients with colon carcinoma. One group shows negative immunostaining with mAb TKH2, and Duke stage A, B, or C. A second group is TKH2-positive, with Duke stage B. The third group is TKH2-positive, with Duke stage C. Note that negative-staining cases had 100% survival, regardless of Duke stage. From Itzkowitz SH, et al., *Cancer* 1990 (ref 58), by permission. **Panel B:** Solid line, progression-free survival of patients with STn serum level ≥50 U/ ml (n=37). Dashed line, survival of patients with STn serum level <50 U/ ml (n=52). From Kobayashi H, et al., *J Clin Oncol* 1992 (ref 59), by permission.

to GalNAc occurs, and blocks Galβ1→3 addition to GalNAc. If this is the case, α2→6 sialyltransferase specific to GalNAc must be highly expressed in various types of cancer, and is qualitatively different from a similar α2→6 sialyltransferase involved in sialylation of core structure in normal tissue (**Fig. 6**). Recently, an α2→6 sialyltransferase apparently specific for synthesis of STn was identified through cloning of its gene (66).

In contrast to Tn, which is expressed in many human and experimental tumor cell lines, STn is not expressed (or ceases its expression) when primary tumor cells are converted to cell lines.

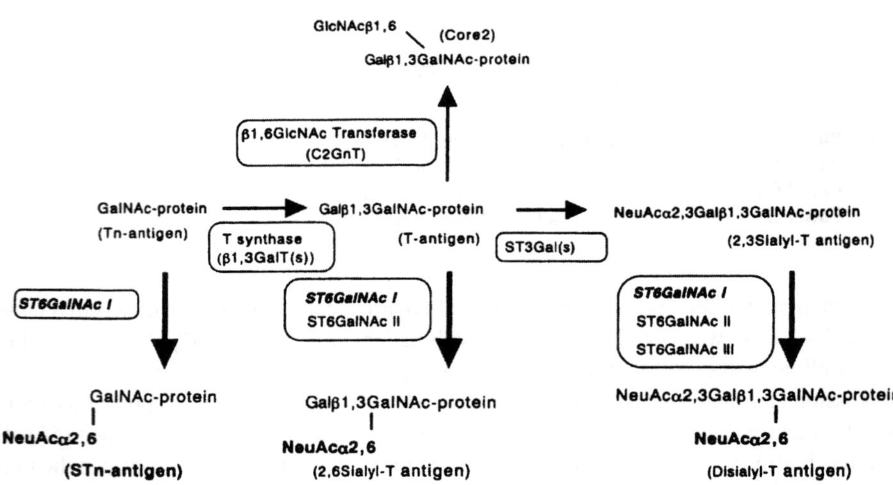

Fig. 6. Biosynthetic pathways for O-glycosylation (T, Tn, STn, and disialyl-T antigens) in mucin. STn is a direct product of sialyltransferase 6 acting on Tn (ST6GalNAc I). In contrast, 2,6 sialyl-T and disialyl-T are synthesized through cooperation of other 2,6 sialyltransferases, *i.e.*, ST6GalNAc II and III. T antigen core is converted to GlcNAcβ1→6 core 2 structure. From Ikehara Y, et al., *Glycobiology* 1999 (ref 66), by permission.

Therefore, tumor cell lines expressing STn are extremely rare; the only one known is HCT15 (66). Stable expression of this gene is apparently discontinued when primary tumor cells are converted to cell lines.

3. H/Le^y/Le^b in primary lung cancer

Le^y antigen (Fucα1→2Galβ1→4[Fucα1→3]GlcNAcβ1→R) was assumed to be present in "blood group substances" isolated from ovarian cyst mucin, because oligosaccharide having this structure was isolated and characterized from hydrolysis product of ovarian cyst mucin (67). After introduction of the mAb approach for identification of tumor-associated antigens, Le^y structure was shown to be a major antigen defined by various mAbs (68-70). Since then, this epitope has been regarded as an important target in cancer therapy, including targeting of immunotoxins (71-73).

Extended type 2 chain bearing Le^y structure (extended trifucosyl Le^y; Fucα1→2Galβ1→4[Fucα1→3]GlcNAcβ1→3Galβ1→4[Fucα1→3]GlcNAcβ1→3Galβ1→R) is expressed highly in colonic carcinoma, and specific mAbs directed to this structure were established: KH1 (74) and CC-1 and -2 (8). Reactivity of these mAbs is much more restricted in normal tissue, *i.e.*, this antigen is more tumor-specific than simple Le^y. Thus, the possibility is opened for development of anti-cancer vaccine based on this structure; in fact, "KH-1" antigen has been totally synthesized (75).

In a study on selection of motility-inhibitory mAbs directed to tumor cells, mAbs MIA-15-5 and -22-20 were established which inhibited tumor cell motility in phagokinetic as well as transmembrane migration assay. These mAbs were directed to Le^y but cross-reacted with H and Le^b, and inhibited B16 melanoma metastasis to lung *in vivo* (76). MIA-15-5 was applied for retrospective study on relationship between the antigen expression in primary lung carcinoma (excluding small cell carcinoma) and patient 10-year survival. Patients with MIA-15-5-negative tumors had much higher survival rate than those with MIA-15-5-positive tumors (**Fig. 7A**). This trend was more clear in squamous cell carcinoma than in adenocarcinoma (77). H type 2 chain and Le^y are known to be defined by *Lotus tetragonolobus* lectin (78). The histological binding pattern of this lectin to transitional bladder carcinoma sections was correlated with patient survival in the same way as MIA-15-5 reactivity (79) (**Fig. 7B**). These data indicate that high expression of H/Le^y/Le^b may promote invasive and metastatic properties of tumor cells, and thereby

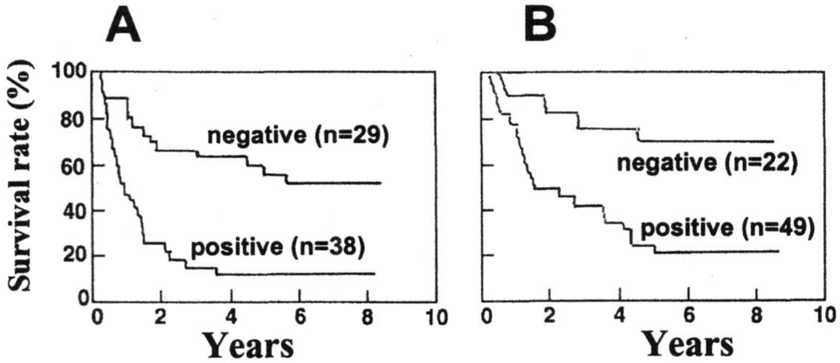

Fig. 7. Effect of fucosylated type 1 or type 2 chain (H/Le^y/Le^b) expression in two types of carcinoma on survival rate of patients. **Panel A:** Kaplan-Meier curves for patients with positive (n=38) and negative (n=29) staining of primary lung carcinoma by mAb MIA-15-5, which reacts with H/Le^y and Le^b. From Miyake M, et al., *N Engl J Med* 1992 (ref 77), by permission. **Panel B:** patients with positive (n=49) and negative (n=22) staining of bladder carcinoma by *Lotus tetragonolobus* lectin which reacts preferentially with H/Le^y. From Shirahama T, et al., *Cancer* 1993 (ref 79), by permission.

affect patient survival. Since H/Le^y/Le^b are precursors of histo-blood group A and B determinants, this phenomenon may be ascribable to deletion of A and B epitope, as discussed under Section I-1.

4. Sialyl-Le^a and sialyl-Le^x in various cancers

Classical sugar analysis of glycoconjugates, particularly by Dische, suggested that sialylation and fucosylation are competitive and do not coexist ("Dische's rule") (80). Sialyl-Le^a (SLe^a; NeuAcα2→3Galβ1→3[Fucα1→4]GlcNAcβ1→3Galβ1→R) and sialyl-Le^x (SLe^x; NeuAcα2→3Galβ1→4[Fucα1→3]GlcNAcβ1→3Galβ1→R), discovered in 1982 and 1984 respectively as tumor-associated antigens, have structures contrary to Dische's rule, *i.e.*, both sialylation and fucosylation occur at adjacent sites within the same molecule.

Discovery of SLe^a antigen, and its association with malignancy. Application of mAb approach to detect tumor-associated antigens in colorectal carcinoma led to discovery of a carbohydrate antigen expressed highly in colorectal and pancreatic cancer, whose level was elevated in patient sera (81). The antigen defined by mAb N19-9 was identified as SLe^a (82), in which α2→3 sialylation to terminal Gal occurs prior to α1→4 fucosylation to penultimate GlcNAc (83). The antigen was found in sera mainly as O-linked mucin-type macromolecule (84). Recently, its expression was found to be closely correlated with survival rate of patients with colorectal cancer in three independent studies (for review see (85) (**Fig. 8**). However, SLe^a is not found in neutrophils, monocytes, lymphocytes, or normal hematopoietic cells, and its expression is highly restricted in normal fibroblasts, epithelial cells, and other parenchymatous cells.

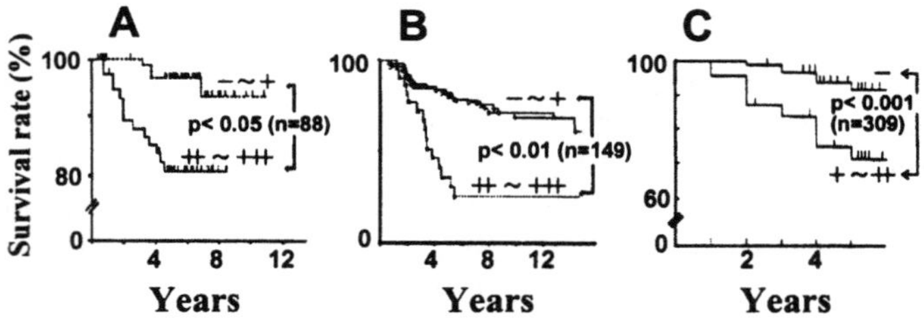

Fig. 8. Survival rates of patients with colon cancer, in relation to sialyl Le^a expression as indicated by staining with anti-SLe^a mAb. Results of three independent studies. Strongly positive cases showed significantly lower survival than negative or weakly positive cases. **Panel A:** results from Nagoya Univ. study. **Panel B:** results from Kyushu Univ. **Panel C:** results from Keio Univ. Summarized in Kannagi R, *Glycoconj J* 1997 (ref 85); by

Discovery of SLe^x and SLe^x-Le^x antigen, and their association with malignancy. During systematic analysis of gangliosides accumulated in colorectal carcinoma, a series of fucogangliosides were found having type 2 rather than type 1 chain. Some were identified chemically as SLe^x, the positional isomer of SLe^a, and others as sialylated dimeric or trimeric Le^x. mAb FH6 was established after immunization of mice with purified sialylated dimeric Le^x coated on *Salmonella minnesotae* (86). FH6 was cross-reactive with SLe^x. At almost the same time, a mAb directed to gastric cancer cell line was found to be directed to SLe^x, and was termed CSLEX (87). SLe^x is expressed highly in gastrointestinal, colorectal, lung, and breast carcinomas, but also in neutrophils, monocytes, and leukemic cell lines derived therefrom (88), and therefore cannot be regarded as a tumor-specific antigen. SLe^x expression in primary colorectal carcinoma, defined by FH6, showed a clear correlation with patient survival rate (89, 90) (**Fig. 9**).

SLe^x and SLe^a as selectin ligands that mediate adhesion of tumor cells. Soon after the "ELAM-1" (E-selectin) gene was cloned and found to contain C-type lectin domain, the carbohy-

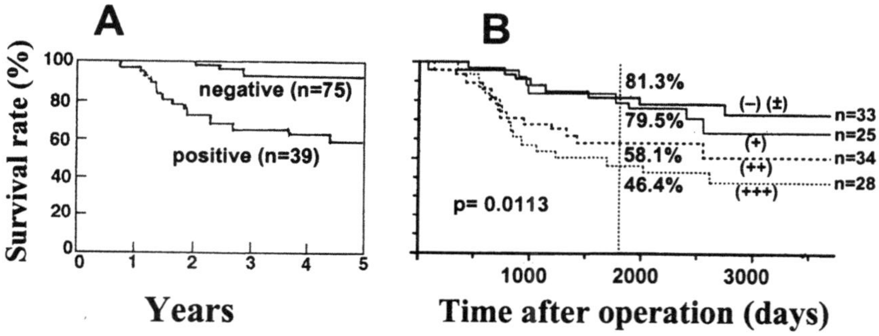

Fig. 9. Survival of patients with colon cancer, in relation to sialyl Le^x expression as indicated by immunohistological staining. **Panel A:** Negative immunostaining group with anti-SLe^x-Le^x mAb FH6 (n=75) had higher survival than positive group (n=39). From Nakamori S, et al., *Cancer Res* 1993 (ref 89), by permission. **Panel B:** Correlation of survival rate with four grades of staining with anti-SLe^x mAb CSLEX (–/±, +, ++, +++). Numbers of patients in each group are indicated. From Nakagoe T, et al., *Cancer* 1993 (ref 90), by permission.

drate ligand for E-selectin was shown to be susceptible to sialidase and its binding activity enhanced by fucosyltransferase III (91). At almost the same time, the ligand was identified as SLe^x and sialyl dimeric Le^x (SLe^x-Le^x) (92). A possibility was thus opened that tumor cells bearing SLe^x adhere preferentially to activated endothelial cells which express E-selectin, and hematogenous (blood-borne) metastasis is initiated from this point (93). Subsequently, two independent studies showed that SLe^a also binds to E-selectin, to the same or higher degree than SLe^x (94, 95). This opened a similar possibility that tumor cells bearing SLe^a have preferential binding to activated endothelial cells, and thereby promote metastasis (96). Both of these possibilities were supported by various clinicopathological data showing that expression of either SLe^a or SLe^x in primary tumors is correlated with patient survival rate (**Figs. 8, 9**).

There is a fundamental question regarding the mechanism by which tumor cells bearing SLe^a or SLe^x activate endothelial cells *in situ* prior to expression of E-selectin, since *de novo* synthesis of E-selectin requires at least 4-5 hr. Experimental conditions to elicit E-selectin expression by endothelial cells require a few μg per ml concentration of IL-1, TGF-β, or TNF-α, which are presumed to be secreted by tumor cells. In reality, many tumor cell lines tested do not secrete these factors at such high levels. However, the majority of tumor cells are capable of activating platelets to release a factor(s) that elicits E-selectin expression at endothelial cells (97). Similarly, Lewis lung carcinoma subline H59 and melanoma B16 are capable of activating E-selectin expression in liver after intrasplenic/ portal inoculation. The underlying mechanism may be interaction of tumor cells with blood cell components, which in turn activate cytokines and E-selectin mRNA synthesis (98). While the factors released from platelets and blood cells are not clearly identified, the mechanism is crucial for understanding the process of E-selectin-mediated tumor cell adhesion to endothelial cells and subsequent metastasis. A possible mechanism underlying this process is illustrated in **Fig. 10**.

Myeloglycan, rather than simple SLe^x, is the physiological major epitope for E-selectin in neutrophils and in myeloid/ monocytic leukemia. A possible pathobiological role of SLe^a and SLe^x in E-selectin-mediated binding of tumor cells to endothelial cells, the initial step in the metastatic process, was emphasized in the preceding section. However, the physiological role of selectins is to recruit neutrophils and other leukocytes to inflammatory lesions. In this process, some carbohydrate epitopes are presumed to be recognized by E- or P-selectin, followed by rolling and "tethering" of leukocytes under defined dynamic flow conditions. SLe^x was presumed in early studies to be expressed in leukocytes, and to be the physiological epitope in both E- and

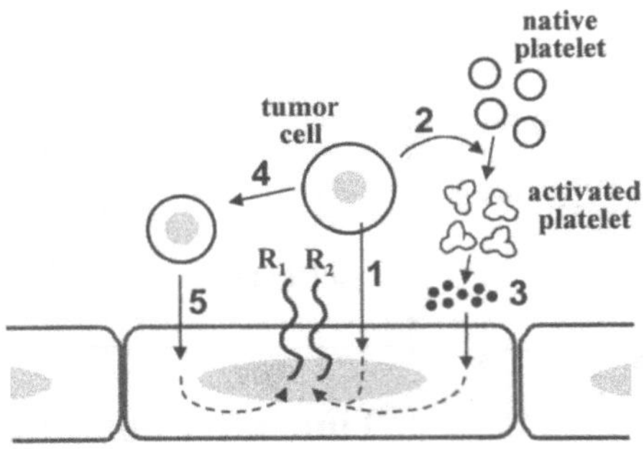

Fig. 10. Possible mechanisms for activation of endothelial cells (ECs) by tumor cells (schematic). Tumor cells may directly activate ECs through factors such as TNF-α and TGF-β (**route 1**), although tumor cell lines so far examined do not secrete levels of these growth factors sufficient to activate ECs. On the other hand, tumor cells strongly activate native platelets (**route 2**), causing release of unknown factor (**3**) which activates ECs. Alternatively, tumor cells *in vivo* may interact with neutrophils or macrophages (**route 4**), causing them to release growth factors (**5**) which activate ECs. Indirect mechanisms (routes 2/3 or 4/5) may be more practically important than direct activation. In any of these mechanisms, activation leads to *de novo* synthesis of E-selectin (R_1) or other adhesion molecules (R_2). Based on studies by S. Hakomori (*BBRC* 1994; ref 97), and by P. Brodt and associates (Khatib A-M et al., *Cancer Res* 1999; ref 98).

P-selectin-dependent rolling and tethering (92, 99); for reviews see (100, 101). However, simple SLe^x is minimally expressed in neutrophils and HL60 cells. Instead, a series of long unbranched poly-LacNAc having α2→3 sialylation at terminal Gal and α1→3 multiple fucosylation at internal (not penultimate) GlcNAc were identified as the real epitope of E-selectin, and collectively termed "myeloglycan" (102-104) (see **Table 1** for structures of typical myeloglycan). The internal α1→3 fucosylation is catalyzed by fucosyltransferase IV (FT-IV), whereas fucosylation at penultimate GlcNAc is catalyzed by FT-VII in myeloid cells. Synthesis of myeloglycan-type structure is made by overwhelming expression of FT-IV, since internal α1→3 fucosylation, once it occurs in long-chain poly-LacNAc, inhibits α1→3 fucosylation at penultimate GlcNAc, thus inhibiting SLe^x synthesis (105). Myeloglycan does not cause P-selectin-dependent adhesion, which may require the core polypeptide PSGL-1 (106). Many cell lines expressing SLe^a or SLe^x, or both, do not adhere to P-selectin unless PSGL-1 gene is transfected, even though PSGL-1 *per se* has no ability to cause P-selectin binding (107). In neutrophils, PSGL-1 having multiple O-linked myeloglycan may be the essential molecule causing P-selectin-dependent adhesion.

It should be emphasized that myeloglycan-type poly-LacNAc is also highly expressed in myelogenous and monocytic leukemia (102, 103) and in some human adenocarcinomas, including colorectal carcinoma (108), and may have an important role in tumor cell metastasis through their binding to E-selectin.

5. Extended type 1 chain in colorectal carcinoma

Many structural studies of blood cells and tissues have shown that poly-LacNAc consists of type 2 chain, either branched or unbranched (102, 103, 109-111). Type 1 chain Le^a is present at the periphery of type 2 chain (112), or Le^a is carried on Le^x (*i.e.*, Le^a-on-Le^x hybrid) (113, 114).

A novel procedure was used to establish a mAb that inhibits human tumor growth *in vitro* and *in vivo*. A human gastric cancer (St-4) xenograft was made in Balb/c nu/nu mice. During St-4 tumor growth, normal Balb/c spleen cells were injected into the mice. When tumor growth was inhibited, host spleen cells were fused with mouse myeloma cells, and hybridomas were screened based on specific reactivity with gastric cancer tissue (115). mAb NCC-ST-421 (IgG_3),

Table 1.
Structures of myeloglycan, based on unbranched poly-LacNAc with terminal sialylation and internal fucosylation (excluding penultimate fucosylation), which causes rolling and tethering under dynamic flow conditions.

```
          Galβ4GlcNAcβ3Galβ4GlcNAcβ3Galβ4GlcNAcβ3Galβ4GlcβCer
          3                  3
     NeuAcα              Fucα

          Galβ4GlcNAcβ3Galβ4GlcNAcβ3Galβ4GlcNAcβ3Galβ4GlcβCer
          3                                3
     NeuAcα                            Fucα

    Galβ4GlcNAcβ3Galβ4GlcNAcβ3Galβ4GlcNAcβ3Galβ4GlcNAcβ3Galβ4GlcβCer
    3                  3               3
 NeuAcα            Fucα            Fucα

Galβ4GlcNAcβ3Galβ4GlcNAcβ3Galβ4GlcNAcβ3Galβ4GlcNAcβ3Galβ4GlcNAcβ3Galβ4GlcβCer
3                  3
NeuAcα             Fucα

Galβ4GlcNAcβ3Galβ4GlcNAcβ3Galβ4GlcNAcβ3Galβ4GlcNAcβ3Galβ4GlcNAcβ3Galβ4GlcβCer
3                                 3
NeuAcα                            Fucα

Galβ4GlcNAcβ3Galβ4GlcNAcβ3Galβ4GlcNAcβ3Galβ4GlcNAcβ3Galβ4GlcNAcβ3Galβ4GlcβCer
3                  3              3
NeuAcα             Fucα           Fucα

Galβ4GlcNAcβ3Galβ4GlcNAcβ3Galβ4GlcNAcβ3Galβ4GlcNAcβ3Galβ4GlcNAcβ3Galβ4GlcβCer
3                  3                             3
NeuAcα             Fucα                          Fucα
```

Modified from Handa K, et al., *Biochemistry* 1997 (ref 104)

established by this procedure, strongly inhibited tumor growth *in vitro* as well as *in vivo*, based on complement-dependent cytotoxicity (CDC) and antibody-dependent cellular cytotoxicity (ADCC) (116). We found the antigen to be highly expressed in human colonic carcinoma Colo205, and it was isolated and identified as extended type 1 chain with two α1→3 fucosylations at the penultimate and internal GlcNAc, *i.e.*, Le^a-on-Le^a rather than Le^a-on-Le^x. NCC-ST-421 reacted specifically with Le^a-on-Le^a, but not with Le^a-on-Le^x or simple Le^a (117).

Another major GSL component having extended type 1 chain was found in Colo205 and the antigen was identified as Le^b-on-Le^a, based on ^{1}H-NMR, FAB-MS, and enzymatic degradation (118). Immunization of Balb/c mice with this antigen led to establishment of mAb IMH2, which reacted preferentially with Le^b-on-Le^a but not with simple Le^b, Le^a-on-Le^a, nor Le^a. IMH2 showed strong inhibition of Colo205 growth through CDC and ADCC (119). No study has been made on expression of these extended type 1 chain structures in relation to tumor malignancy, as typically studied for other antigens in terms of patient survival rate (see above). Chemical structures and antigen properties of extended type 1 chain antigens are summarized in **Table 2**.

6. Extended type 2 chain in colorectal carcinoma

Various structures collectively called poly-N-acetyllactosamine (poly-LacNAc) belong to this class, as described above. poly-LacNAc have been isolated and characterized from colorectal carcinoma as branched, disialylated antigen defined by mAb NUH2 (111), or as unbranched 10-sugar chain having α2→3 sialylation at terminal Gal and α1→3 fucosylation at internal GlcNAc, defined by mAb ACFH-18 (108). The latter antigen is a type of myeloglycan involved in tumor cell adhesion to E-selectin under dynamic flow conditions (104).

Table 2.
Extended type 1 chains as tumor-associated antigens, and monoclonal antibodies defining them.

	Antigen	mAb
1.	Galβ1→3GlcNAcβ1→3Galβ1→3GlcNAcβ1→3Galβ1→4Glcβ1→1Cer 4 4 ↑ ↑ Fucα1 Fucα1 **Le^{a}-Le^{a}**	ST421 [a]
2.	Galβ1→3GlcNAcβ1→3Galβ1→3GlcNAcβ1→3Galβ1→4Glcβ1→1Cer 2 4 4 ↑ ↑ ↑ Fucα1 Fucα1 Fucα1 **Le^{b}-Le^{a}**	IMH2 [b]
3.	Galβ1→3GlcNAcβ1→3Galβ1→4GlcNAcβ1→3Galβ1→4Glcβ1→1Cer 4 3 ↑ ↑ Fucα1 Fucα1 **Le^{a}-Le^{x}**	43/9F [c]

[a] Watanabe M, et al., Jpn J Cancer Res 1985 (ref 115); *Cancer Res* 1991 (ref 116)
[b] Ito H, et al., *Cancer Res* 1992 (ref 119)
[c] Mårtensson S, et al., *Cancer Res* 1988 (ref 113); Stroud MR, et al., *Biochemistry* 1994 (ref 114).

Type 2 chain structure having α2→6 sialylation at terminal Gal is a minor component of human blood (120). mAb 1B9 defining NeuNAcα2→6Gal was established, and found to have enhanced reactivity with colorectal carcinoma as compared to normal mucosa (121). A novel structure with extended type 2 chain having terminal α2→6 sialylation and α1→3 fucosylation at internal GlcNAc was also enhanced in colorectal carcinoma (**Table 3**). Again, no study has been made on expression of this interesting structure in relation to tumor malignancy.

7. Fucosylgalactosylgloboside (globo-H) in breast, ovarian, and prostate cancer

Globo-H was found originally as an extended globo-series structure expressed in human embryonal carcinoma 2102 cells (122). Around the same time, mAb MBr1 was established against breast carcinoma cell line MCF-7 (123). The antigen present in this cell extract was identified as globo-H GSL based on methylation analysis and mass spectrometry (124). Expression of this antigen (both as GSL and glycoprotein) in 143 cases of breast cancer was not correlated with patient survival (125). Globo-H was also found to be expressed in ovarian cancer (126) and in prostate cancer, either primary or metastatic (127) (see Section III-2).

8. Sialylgalactosylgloboside in renal cell carcinoma

Renal cell carcinoma (RCC) is a major type of kidney tumor, with high metastatic potential. Systematic analysis of GSLs in RCC tissue and a few RCC cell lines indicated the presence of a slow-migrating disialoganglioside in primary and metastatic RCC tissues, and in highly metastatic cell line TOS-1. The compound was isolated and characterized by ^{1}H-NMR, FAB-MS, and methylation analysis as disialylgalactosylgloboside (**Table 4**). mAb RM2 reacted with this ganglioside but not with monosialylgalactosylgloboside, which reacted with mAb RM1 or SSEA-4 (Table 4). Another closely-related structure is GD1α, which has the same α2→3 and α2→6 disialyl linkages to Galβ1→3GalNAc terminus. Interestingly, GD1α did not react with RM2, RM1, or SSEA-4 (128). RCC is characterized by preferential metastasis to lung tissue rather than to kidney, liver, or other organs. By Stamper-Woodruff assay, TOS-1 showed strong adhesion to the perialveolar region of lung tissue sections, but minimal or no adhesion to sections of kidney, liver, or lymph nodes. Adhesion to lung tissue sections was observed under both static and dynamic flow conditions, and was inhibited by sialidase treatment of TOS-1 cells (129).

Table 3.
Extended type 2 chains as tumor-associated antigens, and monoclonal antibodies defining them

	Antigen	mAb
1.	Galβ1→4GlcNAcβ1→3Galβ1→4GlcNAcβ1→3Galβ1→4Glcβ1→1Cer 3 3 ↑ ↑ Fucα1 Fucα1 **Le^x-Le^x (dimeric Le^x)**	FH4 [a]
2.	Galβ1→4GlcNAcβ1→3Galβ1→3GlcNAcβ1→3Galβ1→4Glcβ1→1Cer 2 3 ↑ ↑ Fucα1 Fucα1 **extended Le^y**	CC1, CC2 [b]
3.	Galβ1→4GlcNAcβ1→3Galβ1→3GlcNAcβ1→3Galβ1→4Glcβ1→1Cer 2 3 3 ↑ ↑ ↑ Fucα1 Fucα1 Fucα1 **Le^y-Le^x (trifucosyl Le^y)**	KH1 [c]
4.	Galβ1→4GlcNAcβ1→3Galβ1→3GlcNAcβ1→3Galβ1→4Glcβ1→1Cer 3 3 3 ↑ ↑ ↑ NeuAcα2 Fucα1 Fucα1 **sialyl dimeric Le^x**	FH6 [d]
5.	Galβ1→4GlcNAcβ1→3Galβ1→3GlcNAcβ1→3Galβ1→4Glcβ1→1Cer 6 3 ↑ ↑ NeuAcα2 Fucα1 **sialyl 2→6 VIM2**	1B9 [e]
6.	NeuAcα2→3Galβ1→4GlcNAcβ1→3 Galβ1→4GlcNAcβ1→3Galβ1→4Glcβ1→1Cer NeuAcα2→3Galβ1→4GlcNAcβ1→6	NUH2 [f]
7.	Galβ4GlcNAcβ3Galβ4GlcNAcβ3Galβ4GlcNAcβ3Galβ4GlcNAcβ3Galβ4GlcβCer 3 3 ↑ ↑ NeuAcα2 Fucα1	ACFH-18 [g]

a Fukushi Y, *et al.*, *J Biol Chem* 259: 4681 (1984); b Kim YS, *et al.*, *Cancer Res* 1986 (ref 8); c Kaizu T, et al., *J Biol Chem* 1986 (ref 74); d Fukushi Y, *et al.*, *J Biol Chem* 1984 (ref 86); e Hakomori S, *et al.*, *Biochem Biophys Res Comm* 113: 791 (1983). 1B9 defines only the terminal structure (underlined). There is no mAb defining the entire structure.; f Nudelman ED, *et al.*, *J Biol Chem* 1989 (ref 111); g Nudelman ED, *et al.*, *J Biol Chem* 1988 (ref 108).

These observations suggest that disialylgalactosylgloboside is the epitope structure recognized by an unknown receptor expressed on lung tissue, and that this recognition process mediates metastasis. This possibility was supported by extensive clinicopathological studies in which RCC cases (55 primary, 8 metastatic deposits) were studied for expression of mono- and disialylgalactosylgloboside as determined by RM1 and RM2 reactivity. During the follow-up period, metastasis developed in none of the RM1/RM2-negative cases which had not shown metastasis at time of initial diagnosis. RM1/RM2 reactivity was much higher at metastatic deposits than at primary tumors (130).

9. Fucosyl-GM1 ganglioside in small cell lung carcinoma (SCLC)

Chemical analysis by Nilsson et al. showed accumulation of fucosyl-GM1 in SCLC (131). This was confirmed using mAb with apparent specificity to fucosyl-GM1 in various SCLC cases. Fucosyl-GM1 accumulated in SCLC was characterized by presence of α-hydroxy fatty acid, which was absent in the trace quantity of fucosyl-GM1 detected in normal tissue (132). It appears that fucosyl-GM1 is released from SCLC cells, and antigen level is elevated in sera of patients with SCLC (133). Because essentially all SCLC cases are highly malignant, and express high

Table 4.
Monosialylgalactosylgloboside (MSGG) and disialylgalactosylgloboside (DSGG) and related structures, and reactivity to their specific antibodies.

Name	Structure	Antibody reactivity RM1	RM2	QSH2
MSGG	Galβ1→3GalNAcβ1→3Galα1→4Galβ1→4Glcβ1→1Cer 3 ↑ NeuAcα2	+	–	–
DSGG	NeuAcα2 ↓ 6 Galβ1→3GalNAcβ1→3Galα1→4Galβ1→4Glcβ1→1Cer 3 ↑ NeuAcα2	–	+	–
GD1α	NeuAcα2 ↓ 6 Galβ1→3GalNAcβ1→4Galβ1→4Glcβ1→1Cer 3 ↑ NeuAcα2	–	–	–
disialyl-T	NeuAcα2 ↓ 6 Galβ1→3GalNAcα1-O-Ser/Thr 3 ↑ NeuAcα2	–	–	+

level of fucosyl-GM1, it is not possible to perform clinicopathological studies on expression of fucosyl-GM1 as related to malignancy.

10. GM2, GD3, and GD2 gangliosides in melanoma and neuroblastoma

Accumulation of GD3 ganglioside in melanoma was initially observed through chemical analysis by Portoukalian and colleagues (134). Subsequently, this ganglioside was identified as melanoma-associated antigen through successful establishment of a few mAbs that reacted specifically or preferentially with human melanoma, *e.g.*, IgM mAb 4.2 (135, 136), and IgG_3 mAb R24 (137). R24 showed strong affinity to GD3, displayed tumor cell killing due to ADCC, and was used for a Phase I clinical trial. Some melanoma patients had remarkable response, while others did not, for some unknown reason, since all cases showed similar degree of GD3 expression (138).

Two human mAbs against oncofetal antigens (OFA-1-1, OFA-1-2) were established when lymphocytes of melanoma patients were transformed by Epstein-Barr virus. OFA-1-1 was identified as GM2 ganglioside (139, 140), and OFA-1-2 as GD2 ganglioside (141). Gangliosides GD2, GM2, and GD3 are highly expressed in human melanoma and neuroblastoma, and cell lines derived therefrom. Their levels are elevated in sera of patients, as initially observed for GD2 in melanoma patients (142). GM3 is a common ganglioside present in all types of neuroectodermal tumors, and is the major antigen of mouse melanoma B16 (143, 144). Human melanoma usually expresses GD2, GM2, and GD3 as well as GM3. Melanoma showing only GM3 expression is extremely rare in Caucasian populations, but is exemplified by acral lentiginous melanoma (ALM), the most common clinical and histopathological type of melanoma in Japanese. There is no difference in ganglioside pattern between original and metastatic ALM, nor between original and metastatic cases of other types of melanoma (145). No systematic study has been conducted on malignancy of melanomas of Caucasian populations as correlated with ganglioside pattern. There are four groups of melanoma showing different ganglioside composition, as shown in melanoma cell lines (**Table 5**).

Table 5.

Major gangliosides expressed in neuroectodermal tumors, classification of tumors according to ganglioside composition, and representative cell lines.

GM3 NeuAcα2→3Galβ1→4Glcβ1→1Cer

GD3 NeuAcα2→8NeuAcα2→3Galβ1→4Glcβ1→1Cer

GD2 NeuAcα2→8NeuAcα2→3
Galβ1→4Glcβ1→1Cer
GalNAcβ1→4

Classification	Ganglioside composition	Representative cell lines
Group 1	GM3 ≈ GD3	SK-MEL-28 (M)[a], MeWo (M)
Group 2	GM3 ≈ GD3 > GM2 > GD2	SK-MEL-31 (M), SK-MEL-37, LAN-1 (N)[b]
Group 3	GD3 << GM2 ≈ GD2	SMS-SAN (N), IMR-32 (N), BE2-C (N)
Group 4	GM2 ≈ GD1a	SK-N-MC (NE)[c], SH-EP1 (N), SH-SY5Y (N)

[a] M, melanoma cell line; [b] N, neuroblastoma cell line; [c] NE, neuroepithelioma cell line

III. SUCCESSFUL DEVELOPMENT OF ANTI-CANCER VACCINE OR REAGENTS BASED ON TACA EXPRESSION

1. GM2, GD3, and GD2 gangliosides in neuroectodermal tumors

Immediately after establishment of mAbs directed to melanoma-associated antigens GD3 and GD2, phase I clinical trials for suppression of melanoma or neuroblastoma growth were conducted (138, 146, 147). In these studies, a clear growth-suppressive effect was seen in some cases. No effect was seen in other cases, although antigen expression was comparable. There was no clear explanation for the susceptibility of tumor growth to mAb in these clinical cases. In contrast, 250 μg anti-GD2 mAb 3F8 completely inhibited micrometastasis of GD2-expressing syngeneic lymphoma EL4 in mice, even 2-4 days after tumor inoculation (148). Successful inhibition in this case may also be ascribable to lymphoma rather than adherent tumor cells. In a previous study, growth of mouse T cell lymphoma L5878 expressing Gg3Cer was successfully inhibited by anti-Gg3 IgG_3 but not IgM antibody of the same epitope (149). Thus, the effect of mAbs to suppress tumor growth is clear in experimental lymphoma cases, although high concentrations of antibody are necessary to produce the effect, and such high doses of humanized antibody are impractical for clinical cases.

Level of natural serum IgM antibodies directed to GM2 ganglioside is correlated with survival of patients with Stage III melanoma, as observed initially by a UCLA group (150). A similar phenomenon was observed by a Sloan-Kettering group in Stage III melanoma patients, particularly after treatment with tumor vaccine (151) (**Fig. 11**). These results provided a rationale for the fact that increased level of antibody (regardless of IgM or IgG) directed to GM2

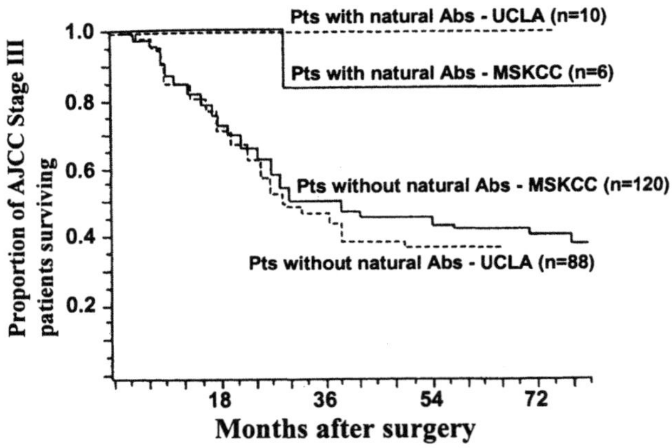

Fig. 11. Correlation between the presence of natural serum anti-GM2 antibodies and survival in AJCC (American Joint Commission on Cancer) stage III melanoma patients subsequently treated with tumor vaccines at Memorial Sloan-Kettering Cancer Center (MSKCC) (Livingston PO, et al., *J Clin Oncol* 1994; ref 151) or UCLA School of Medicine (UCLA) (Jones PC, et al., *J Natl Cancer Inst* 1981; ref 150). Note that patients who produce anti-GM2 antibody had higher survival in both studies. Modified from Livingston PO, et al., *Cancer Immunol Immunother* 1997 (ref 153), with permission.

ganglioside suppresses melanoma growth in patients, and strong motivation for development of anti-melanoma vaccine using purified GM2 with appropriate adjuvant, mainly by P.O. Livingston and colleagues (for reviews see (152, 153). Initially, they found that GM2 ganglioside coated on lipid A-exposed *Salmonella minnesotae* (R595), an effective carrier for glycolipid immunization in mice (154), was less effective than GM2 coated on Bacillus Calmette-Guérin (BCG) (155). Subsequently, they found that GM2 coated on BCG or conjugated to keyhole limpet hemocyanin (KLH), with QS-21 as adjuvant, provides an effective standard method (156). Significant improvement of clinical status and survival rate was observed for both Stage II and III melanoma patients (151, 152).

GD3 ganglioside, another major melanoma-associated antigen, was naturally considered as a target of anti-melanoma vaccine development. However, GD3 is much less immunogenic than GM2 in man, although it is strongly immunogenic in mice. After trials of various methods, including those based on lactones of sialic acid, GD3 was covalently coupled to KLH after ozonolysis of double bond of sphingosine (157). Results using such GD3-KLH conjugate as immunogen in clinical trials are not yet available.

GD2 ganglioside, a major ganglio-series antigen expressed highly in neuroectodermal tumors, is found in both melanoma and neuroblastoma. Specific anti-GD2 antibody 126 detects GD2 expressed in neuroblastoma or released in sera of patients with neuroblastoma (158). Fusion proteins of anti-ganglioside antibodies with interleukin-2 (IL2) have been shown to effectively inhibit solid tumor growth *in vivo*, and to activate CD8-positive T cells. Anti-GD2 antibody-IL2 fusion protein eradicated human hepatic neuroblastoma metastasis in mice, despite the fact that metastasis was established in severe combined immunodeficiency (SCID) mice (159). A major problem in immunotherapy of human cancer is the heterogeneity of antigen expression in tumor cells. Anti-GD3 antibody-IL2 fusion protein administered in mice overcame such heterogeneity, and suppressed melanoma metastasis (160). Clinical trials using the approach with recombinant IL2 fusion protein remain to be conducted.

2. Fucosyl galactosylgloboside (globo-H) in prostate cancer

Globo-H (type 4 chain H) is unique in its immunogenicity, as compared to type 1, 2, and 3 chain H (161), and is expressed highly in certain types of human cancer (breast, ovarian, prostate) (see Section II-7). Recently, globo-H was synthesized chemically, covalently linked to KLH, and used as immunogen. Strong immune response was observed (IgM titer 1: >10,000; IgG titer 1: >20,000 at 3 wk), declined later, but lasted for ~9 wk. Increment of prostate-specific antigen (PSA) concentration, usually observed in control patients, was not observed in subjects immunized with the antigen. Since PSA increment is used as a criterion of prostate cancer progression, immunization was claimed to be clinically effective. Interestingly, no side-effect (*e.g.*, kidney or gastrointestinal dysfunction) was observed in immunized patients, even though globo-H is also expressed highly in various normal tissues, particularly secretory borders of kidney and gastrointestinal epithelia (162). This raises an important point: tumor-associated antigens expressed in normal epithelial tissue may not be efficient targets of antibodies, and only those tumor cells moving and circulating to cause metastasis may be major targets of antibodies, either IgM or IgG.

3. Tn and STn as basic vaccines, particularly STn for breast cancer

Expression of Tn and STn in tumors is unique in terms of: (i) high specificity; (ii) wide occurrence; (iii) structural simplicity; (iv) close correlation with degree of malignancy, particularly for STn (see Sections II-1, 2). Two Tn epitopes linked to chemically well-defined synthetic lipopeptide, with Intralipid as adjuvant were used as immunogen. A clear IgG and IgM response was observed in mice. No T cell response was observed in this case (163, 164). The structure of synthetic Tn lipopeptide and immune response are shown in **Fig. 12**. In another study, Tn antigen

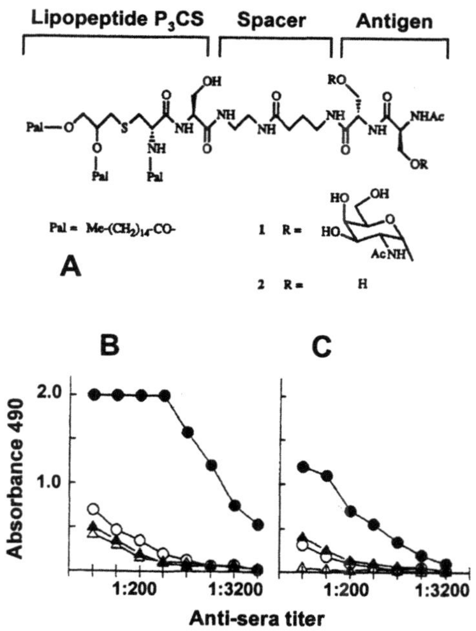

Fig. 12. Effect of synthetic lipoglycopeptide Tn antigen on IgM and IgG production. **Panel A:** Ser-Ser residue bearing α-GalNAc (R) was linked to tripalmitoyl lipopeptide as indicated. The antigen caused strong IgM (**Panel B**) and IgG (**Panel C**) response. 1, complete antigen. ∇, Intralipid adjuvant alone. Other symbols, irrelevant antigens without GalNAc. From Toyokuni T et al., *J Am Chem Soc* 1994 (ref 163); *Bioorg Med Chem* 1994 (ref 164).

was assembled with specific poliovirus "T-epitope" sequence (KLFAVWKITYKDT), and used to immunize Balb/c, SJL/J, or DBA/2 mice, followed by observation of TA3/Ha tumor growth. IgM and IgG response and T cell-dependent IL2 response were observed (165). Significant helper T cell response was originally observed for mucin-type Tn antigen (desialylated ovine submaxillary mucin) in a mouse model, to completely suppress TA3/Ha tumor growth (166). In striking contrast, mice are incapable of responding to immunization with STn antigen. Furthermore, tumor cell line expressing STn was not available. It was therefore impossible to study the effect of STn immunization on tumor growth, on any experimental basis.

Synthetic STn linked to KLH through oxidized crotyl group was used directly to immunize patients with breast cancer (167, 168). They showed appreciable antibody response (both IgG and IgM) as well as T cell response, although it was weak, as expected. An exciting finding is the effect of antibody, either IgG or IgM, following immunization with STn epitope randomly linked to KLH through oxidized crotyl group in a series of clinical trials. In one example of such study, 3-yr survival rate of patients with breast cancer who generated anti-Tn antibody greater than median titer, was 80-90%. In contrast, 100% of patients who generated less than median antibody titer were dead within 25 mo (167) **(Fig. 13)**. However, the natural form of Tn or STn is a clustered structure linked to consecutive Ser-Thr-Thr, Thr-Thr-Thr, etc. in a mucin-type carrier. The quality of antibody produced from randomly-linked STn, as described above, may be different from that of antibodies produced after immunization with clustered antigens similar to naturally-occurring form. Synthesis of clustered STn using a 4-(4-*N*-maleimidomethyl) cyclohexane-1-carboxyl hydrazide (MMCCH) linker arm was recently performed (169). In this case, amino acid carrier (Ser-Thr-Thr) was not used; therefore, specificity of antibody produced may not be identical to that of antibodies directed to naturally-occurring antigen. An alternative scheme for synthesis of clustered STn should be considered.

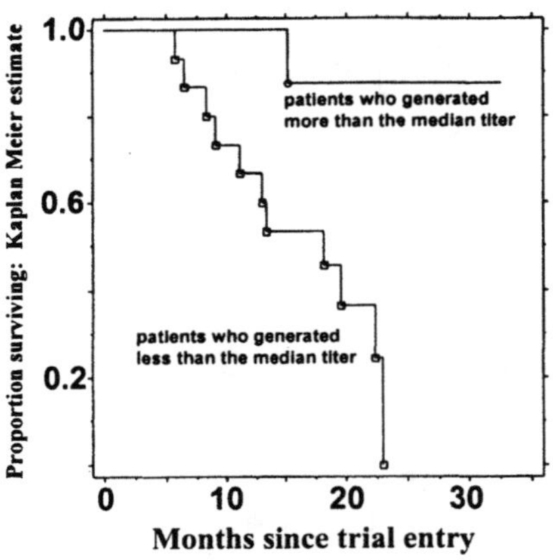

Fig. 13. Effect of active specific immunization with sialyl-Tn on patients with breast cancer. STn bound randomly to KLH through oxidized crotyl group was used as immunogen. Kaplan-Meier survival of patients showing anti-STn IgG titer (determined using ovine submaxillary mucin) greater than median value (n=9) was much higher than survival of those showing anti-STn titer less than median value (n=15). From MacLean et al., J Immunother 1996 (ref 167), by permission.

4. Anti-idiotype antibodies and "mimotopes" for cancer therapy

Immune response against carbohydrate antigens is essentially antibody response; minimal T cell response is observed unless specific methods are applied, such as adjuvant or type of carrier molecule. In contrast, anti-idiotype antibody, *i.e.*, antibody 2 directed to anti-carbohydrate antibody 1, specifically recognizes internal image of antibody 1, and mimics the surface structure of the original antigen. For example, anti-idiotype GD3 has the same surface profile as GD3 ganglioside. As discussed in Section III-1, GD3 is a poor antigen to elicit antibody response for suppression of melanoma growth. However, anti-idiotype mAb BEC2 mimics GD3 and can induce anti-GD3 IgG_3 response in melanoma patients. In this case, mixture of mAb BEC2 with BCG and adjuvant QS21 was necessary. Patients who developed high-titer IgG_3 antibodies against GD3 showed better clinical status compared to patients without immune response; 71% of patients immunized with BEC2-BCG remained alive at 2.4-yr follow-up, and 64% were free of disease (170). In further study, intradermal followed by intravenous administration of anti-idiotype GD3 antibody linked to KLH and mixed with BCG, to melanoma patients, elicited significant production of anti-GD3 antibodies. All patients (n=13) who developed such antibodies following immunization were alive and free of melanoma at 28-month follow-up (171). Thus, poor immunogenicity of GD3, the major melanoma-associated antigen, was overcome by use of anti-idiotype antibodies. The rationale for this approach was provided by crystallographic study of VH idiotope structure that recognizes anti-GD3 mAb R24 (172).

The surface profile of "VH idiotope" should be similar to that of peptide mimetics of carbohydrate antigens. Such peptide mimetics can be conveniently selected from phage display random peptide library, as applied originally for peptides (173-175), and more recently for many type-specific bacterial polysaccharides (176-178), and specific tumor-associated carbohydrate antigens such as Le^y (179) and GD1α (180, 181). These peptide mimetics, if properly assembled with carrier polypeptide, can function as immunogen, as shown by anti-idiotype carbohydrate antigens, and are termed "mimotopes" (182, 183). Antibody response to capsular lipopolysaccharides was induced when peptide mimetics selected from phage display library were used as immunogen (177). It is expected that these mimotopes may induce T cell response which is difficult to obtain by immunization with carbohydrates.

IV. PATHOBIOLOGICAL FUNCTION OF TACAs

TACAs define malignancy under these situations: (i) TACA *per se* is adhesion molecule, promoting adhesion of tumor cells to endothelial cells, platelets, or other target cells, the process regarded as the initial step of metastasis; (ii) aberrant glycosylation modifies function of transmembrane adhesion receptor, leading to increased tumor cell motility and invasiveness; (iii) tumor-associated GSLs are clustered and organized with signal transducer molecules to form GSL signaling domain ("GSD"). Therefore, GSL-dependent adhesion of tumor cells through GSD is coupled with activation of transducers (c-Src, Src family kinases, small G-proteins) to enhance signal transduction leading to increased tumor cell motility and invasiveness.

Process i is observable in three categories of mechanism: (a) TACA is SLe^x, SLe^a, or their analogue, which are capable of interacting with E- or P-selectin expressed on endothelial cells or platelets. As a result, tumor cells aggregate or extravasate from microvasculature, thus initiating blood-borne metastasis (1, 85, 93). (b) TACA is recognized by carbohydrate-binding proteins other than selectin, *e.g.*, galectin, siglec, or yet-unidentified molecule. The role of specific galectin or siglec in defining metastasis or invasiveness has been considered, but not clearly identified. Along this line, Taki and colleagues conducted an interesting series of studies on the mechanism of metastasis, using high-metastatic clone RAW117-H10 of mouse lymphosarcoma cells expressing high level of ganglioside GD1α. Their results indicate that mouse hepatic receptor expressed at sinusoidal epithelia recognizes GD1α, and such interaction initiates and promotes liver metastasis of RAW117-H10 cells (184). Peptide mimetics of GD1α (sequence WHWRHRIPLQLAAGR) bound to liver endothelial cells and inhibited such metastasis (181). (c) TACA interacts directly with glycoconjugates expressed on target cells, including endothelial cells, by carbohydrate-carbohydrate interaction. An example is interaction of GM3, highly expressed on mouse melanoma B16 cells, with LacCer, Gg3Cer, or Gb4 expressed on mouse endothelial cells, occurring prior to integrin/ICAM1-dependent adhesion (185, 186). In fact liposomes containing GM3 or Gg3, which interfere with GM3-dependent adhesion of B16 cells with endothelial cells, inhibit B16 metastasis (187). It is possible that H- or Le^y-expressing tumor cells promote metastasis through interaction of these TACAs with H-antigen expressed highly on human endothelial cells (188). SLe^x, SLe^a, or other TACAs may inhibit metastasis occurring by process i.

Process ii: A number of studies indicate that function of cell surface receptors that define cell adhesion and motility is strongly affected by N- or O-glycosylation of receptor *per se*, as well as surrounding gangliosides. Typical examples are integrins, CD44, and E-cadherin (for reviews see (1, 189). However, our knowledge on effects of specific carbohydrate antigens on function of these receptors is highly limited. A few lines of study indicate that histo-blood group A or B glycosylation strongly affects cell motility through integrin receptor function (see Section I-1).

Process iii: Some tumor-associated GSL antigens have been recovered as detergent-insoluble, low-density membrane fractions organized closely with various transducer molecules such as c-Src, Ras, Rho, and FAK. For example, >90% of GM3 in B16 melanoma is recovered in low-density fraction after Dounce homogenization followed by sucrose density centrifugation. This same fraction contains >90% of c-Src, >90% of Ras, ~50% of Rho, and ~25% of FAK present in the whole cell, even though it contains <0.5% of total protein. c-Src, Rho, and FAK were coimmunoprecipitated with GM3 by anti-GM3 antibody, indicating a close association of these transducer molecules with GM3 (190, 191). The function of GM3 in B16 melanoma as an adhesion molecule to initiate metastasis is well established. FAK activity in terms of tyrosine phosphate level, and GTP binding to Ras and Rho, are enhanced greatly when B16 cells are adhered to Gg3-coated plate. This indicates that tumor-associated antigen GM3 initiates adhesion as well as signal transduction (191).

Similarly, in human embryonal carcinoma 2102 cells, Gb4, GalGb4, and nLc_4 are concentrated in detergent-insoluble, low-density membrane fraction and organized with c-Src, Ras, Rho, and FAK (192). Adhesion of 2102 cells to Gb4-coated plate initiates signal transduction leading to activation of transcription factors AP1 and CREB (193). Note that many of the transducer molecules organized with GSL antigens and involved in signal transduction are oncogene or proto-oncogene products.

It is premature to say that all tumor-associated GSLs are organized in the same way and play essential roles in tumor cell adhesion coupled with signal transduction, but this is an intriguing possibility. GSL organization with transducer molecules may be very different in normal cells compared to tumor cells. It will be of great interest to see results of further studies along this line, using different tumor and normal cell lines.

SUMMARY

Tumors expressing a high level of certain types of tumor-associated carbohydrate antigens (TACAs) exhibit greater metastasis and progression than those expressing low level of TACAs, as reflected in decreased patient survival rate. Well-documented examples of such TACAs are: (i) H/Le^y/Le^a in primary non-small cell lung carcinoma; (ii) sialyl-Le^x (SLe^x) and sialyl-Le^a (SLe^a) in various types of cancer; (iii) Tn and sialyl-Tn in colorectal, lung, breast, and many other cancers; (iv) GM2, GD2, and GD3 gangliosides in neuroectodermal tumors (melanoma and neuroblastoma); (v) globo-H in breast, ovarian, and prostate cancer; (vi) disialylgalactosylgloboside in renal cell carcinoma.

Some glycosylations and TACAs suppress invasiveness and metastatic potential. Well-documented examples are: (i) blood group A antigen in primary lung carcinoma; (ii) bisecting β1→4GlcNAc of N-linked structure in melanoma and other cancers; (iii) galactosylgloboside (GalGb4) in seminoma.

The biochemical mechanisms by which the above glycosylation changes promote or suppress tumor metastasis and invasion are mostly unknown. A few exceptional cases in which we have some knowledge are: (i) SLe^x and SLe^a function as E-selectin epitopes promoting tumor cell interaction with endothelial cells; (ii) some tumor cells interact through binding of TACA to specific proteins other than selectin, or to specific carbohydrate expressed on endothelial cells or other target cells (carbohydrate-carbohydrate interaction); (iii) functional modification of adhesive receptor (integrin, cadherin, CD44) by glycosylation.

So far, a few successful cases of anti-cancer vaccine in clinical trials have been reported, employing TACAs whose expression enhances malignancy. Examples are STn for suppression of breast cancer, GM2 and GD3 for melanoma, and globo-H for prostate cancer. Vaccine development can be extended using other TACAs, with the following criteria for success: (i) the antigen is expressed highly on tumor cells; (ii) high antibody production depending on two factors: (a) clustering of antigen used in vaccine; (b) choice of appropriate carrier protein or lipid; (iii) high T cell response depending on choice of appropriate carrier protein or lipid; (iv) expression of the same antigen in normal epithelial tissues (*e.g.*, renal, intestinal, colorectal) may not pose a major obstacle, *i.e.*, these tissues are not damaged during immune response.

Idiotypic anti-carbohydrate antibodies that mimic the surface profile of carbohydrate antigens, when administered to patients, elicit anti-carbohydrate antibody response, thus providing an effect similar to that of TACAs for suppression of tumor progression. An extension of this idea is the use of peptide mimetics of TACAs, based on phage display random peptide library. Although examples are so far highly limited, use of such "mimotopes" as immunogens may overcome the weak immunogenicity of TACAs in general.

REFERENCES

1. Hakomori, S., Tumor malignancy defined by aberrant glycosylation and sphingo(glyco)lipid metabolism, *Cancer Res., 56*, 5309 (1996).
2. Hollstein, M., Rice, K., Greenblatt, M. S., Soussi, T., Fuchs, R., Sorlie, T., Hovig, E., Smith-Sorensen, B., Montesano, R., and Harris, C. C., Database of p53 gene somatic mutations in human tumors and cell lines, *Nucleic Acid Res., 22*, 3551 (1994).
3. Marx, J., New tumor suppressor may rival p53, *Science, 264*, 344 (1994).
4. Ko, L. J., and Prives, C., p53: Puzzle and paradigm, *Genes & Development, 10*, 1054 (1996).
5. Levine, A. J., p53, the cellular gatekeeper for growth and division, *Cell, 88*, 323 (1997).
6. Kinzler, K. W., and Vogelstein, B., Landscaping the cancer terrain, *Science, 280*, 1036 (1998).

7. Disis, M. L., and Cheever, M. A., HER-2/neu protein: A target for antigen-specific immunotherapy of human cancer, *Adv.Cancer Res., 71*, 343 (1997).
8. Kim, Y. S., Yuan, M., Itzkowitz, S. H., Sun, Q., Kaizu, T., Palekar, A., Trump, B. F., and Hakomori, S., Expression of Ley and extended Ley blood group-related antigens in human malignant, premalignant, and nonmalignant colonic tissues, *Cancer Res., 46*, 5985 (1986).
9. Itzkowitz, S. H., Yuan, M., Montgomery, C. K., Kjeldsen, T., Takahashi, H. K., Bigbee, W. L., and Kim, Y. S., Expression of Tn, sialosyl-Tn, and T antigens in human colon cancer, *Cancer Res., 49*, 197 (1989).
10. Muramatsu, T., Carbohydrate signals in metastasis and prognosis of human carcinomas, *Glycobiology, 3*, 291 (1993).
11. Oh-Uti, K., Polysaccharides and a glycidamin in gastric cancer tissue, *Tohoku J.Exp.Med., 51*, 297 (1949).
12. Masamune, H., Kawasaki, H., Abe, S., Oyama, K., and Yamaguchi, Y., Molisch-positive mucopolysaccharides of gastric cancers as compared with the corresponding components of gastric mucosae. First report: Fractionation procedure of gastric cancer and gastric mucosa, *Tohoku J.Exp.Med., 68*, 81 (1958).
13. Davidsohn, I., Kovarik, S., and Lee, C. L., A, B, and O substances in gastrointestinal carcinoma, *Arch.Pathol., 81*, 381 (1966).
14. Davidsohn, I., and Ni, L. Y., Loss of isoantigens A, B, and H in carcinoma of the lung, *Am.J.Pathol., 57*, 307 (1969).
15. Davidsohn, I., Kovarik, S., and Ni, Y., Isoantigens A, B, and H in benign and malignant lesions of the cervix, *Arch.Pathol., 87*, 306 (1969).
16. Dabelsteen, E., Roed-Petersen, B., and Pindborg, J. J., Loss of epithelial blood group antigens A and B in oral premalignant lesions, *Acta Pathol.Microbiol.Scand., 83*, 292 (1975).
17. Ørntoft, T. F., Wolf, H., Clausen, H., Dabelsteen, E., and Hakomori, S., Blood group ABH-related antigens in normal and malignant bladder urothelium: Possible structural basis for the deletion of type-2 chain ABH antigens in invasive carcinomas, *Int.J.Cancer, 43*, 774 (1989).
18. Davidsohn, I., and Stejskal, R., Tissue antigens A, B and H in health and disease, *Haematologia, 6*, 177 (1972).
19. Lee, J. S., Ro, J. Y., Sahin, A. A., Hong, W. K., Brown, B. W., Mountain, C. F., and Hittelman, W. N., Expression of blood-group antigen A: A favorable prognostic factor in non-small-cell lung cancer, *N.Engl.J.Med., 324*, 1084 (1991).
20. Matsumoto, H., Muramatsu, H., Shimotakahara, T., Yanagi, M., Nishijima, H., Mitani, N., Baba, K., Muramatsu, T., and Shimazu, H., Correlation of expression of ABH blood group carbohydrate antigens with metastatic potential in human lung carcinomas, *Cancer, 72*, 75 (1993).
21. Ichikawa, D., Handa, K., Withers, D. A., and Hakomori, S., Histo-blood group A/B *versus* H status of human carcinoma cells as correlated with haptotactic cell motility: Approach with *A* and *B* gene transfection, *Cancer Res., 57*, 3092 (1997).
22. Ichikawa, D., Handa, K., and Hakomori, S., Histo-blood group A/B antigen deletion/ reduction *vs.* continuous expression in human tumor cells as correlated with their malignancy, *Int.J.Cancer, 76*, 284 (1998).
23. Iwamoto, S., Withers, D. A., Handa, K., and Hakomori, S., Deletion of A-antigen in human cancer cell lines is associated with reduced promoter activity of CBF/NF-Y binding region, and possibly with enhanced DNA methylation of A transferase promoter, *Glycoconj.J., in press* (1999).
24. Fernandes, B., Sagman, U., Auger, M., Demetrio, M., and Dennis, J. W., β1-6 branched oligosaccharides as a marker of tumor progression in human breast and colon neoplasia, *Cancer Res., 51*, 718 (1991).
25. Dennis, J. W., and Laferté, S., Oncodevelopmental expression of -GlcNAcβ1-6Manα1-6Manβ1-6

branching of Asn-linked oligosaccharides in human breast carcinomas, *Cancer Res., 49*, 945 (1989).
26. Schachter, H., Biosynthetic controls that determine the branching and microheterogeneity of protein-bound oligosaccharides, *Biochem.Cell Biol., 64*, 163 (1985).
27. Taniguchi, N., Yoshimura, M., Miyoshi, E., Ihara, Y., Nishikawa, A., and Fujii, S., Remodeling of cell surface glycoproteins by N-acetylglucosaminyltransferase III gene transfection: Modulation of metastatic potentials and down regulation of hepatitis B virus replications, *Glycobiology, 6*, 691 (1996).
28. Yoshimura, M., Nishikawa, A., Ihara, Y., Taniguchi, S., and Taniguchi, N., Suppression of lung metastasis of B16 mouse melanoma by *N*-acetylglucosaminyltransferase III gene transfection, *Proc.Natl.Acad.Sci.USA, 92*, 8754 (1995).
29. Cummings, R. D., and Kornfeld, S., Characterization of structural determinants required for the high-affinity interaction of asparagine-linked oligosaccharides with immobilized *Phaseolus vulgaris* leukoagglutinating and erythroagglutinating lectins, *J.Biol.Chem., 257*, 11230 (1982).
30. Seelentag, W. K. F., Li, W.-P., Schmitz, S.-F. H., Metzger, U., Aeberhard, P., Heitz, P. U., and Roth, J., Prognostic value of β1,6-branched oligosaccharides in human colorectal carcinoma, *Cancer Res., 58*, 5559 (1998).
31. Ohyama, C., Orikasa, S., Kawamura, S., Satoh, M., Saito, S., Fukushi, Y., Levery, S. B., and Hakomori, S., Galactosylgloboside expression in seminoma: Inverse correlation with metastatic potential, *Cancer, 76*, 1043 (1995).
32. Akamatsu, S., Yazawa, S., Zenita, K., Matsumoto, H., Tachikawa, T., and Kannagi, R., Elevation of an α(1,3)fucosyltransferase activity correlated with apoptosis in the human colon adenocarcinoma cell line, HT-29, *Glycoconj.J., 13*, 1021 (1996).
33. Hiraishi, K., Suzuki, K., Hakomori, S., and Adachi, M., Ley antigen expression is correlated with apoptosis (programmed cell death), *Glycobiology, 3*, 381 (1993).
34. Thomsen, O., Ein vermehrungsfähiges Agens als Veränderer des isoagglutinatorischen Verhaltens der roten Blutkörperchen, eine bisher unbekannte Quelle der Fehlbestimmungen, *Z.Immunit.forsch., 52*, 85 (1927).
35. Friedenreich, V., Untersuchungen über das von O. Thomsen beschriebene vermehrungsfähige Agens als Veränderer des isoagglutinatorischen Verhaltens der roten Blutkörperchen, *Z.Immunit.forsch., 55*, 84 (1928).
36. Moreau, R., Dausset, J., Bernard, J., and Moullec, J., Anemie hemolytique acquise avec polyagglutinabilite des hematies par un nouveau factor present dans le serum humain normal (anti-Tn), *Bull.Soc.Med.Hop.Paris, 73*, 569 (1957).
37. Dausset, J., Moullec, J., and Bernard, J., Acquired hemolytic anemia with polyagglutinability of red blood cells due to a new factor present in normal human serum (anti-Tn), *Blood, 14*, 1079 (1959).
38. Dahr, W., Uhlenbruck, G., and Bird, G. W. G., Cryptic A-like receptor sites in human erythrocyte glycoproteins: Proposed nature of Tn-antigen, *Vox Sang., 27*, 29 (1974).
39. Thomas, D. B., and Winzler, R. J., Structural studies on human erythrocyte glycoproteins: Alkali-labile oligosaccharides, *J.Biol.Chem., 244*, 5943 (1969).
40. Springer, G. F., and Desai, P. R., Interaction of blood-group MN-like cancer antigen and human cytotoxin, *Naturwissenschaften, 61*, 38 (1974).
41. Springer, G. F., and Desai, P. R., Human blood-group MN and precursor specificities: Structural and biological aspects, *Carbohydr.Res., 40*, 183 (1975).
42. Springer, G. F., T and Tn, general carcinoma autoantigens, *Science, 224*, 1198 (1984).
43. Hirohashi, S., Clausen, H., Yamada, T., Shimosato, Y., and Hakomori, S., Blood group A cross-reacting epitope defined by monoclonal antibodies NCC-LU-35 and -81 expressed in cancer of blood group O or B individuals: Its identification as Tn antigen, *Proc.Natl.Acad.Sci.USA, 82*, 7039 (1985).

44. Takahashi, H. K., Metoki, R., and Hakomori, S., Immunoglobulin G3 monoclonal antibody directed to Tn antigen (tumor-associated N-acetylgalactosaminyl epitope) that does not cross-react with blood group A antigen, *Cancer Res., 48*, 4361 (1988).
45. Nakada, H., Numata, Y., Inoue, M., Tanaka, N., Kitagawa, H., Funakoshi, I., Fukui, S., and Yamashina, I., Elucidation of an essential structure recognized by an anti-GalNAcα-Ser(Thr) monoclonal antibody (MLS 128), *J.Biol.Chem., 266*, 12402 (1991).
46. Brooks, S. A., and Leathem, A. J. C., Prediction of lymph node involvement in breast cancer by detection of altered glycosylation in the primary tumour, *Lancet, 338*, 71 (1991).
47. Galea, M. H., Ellis, I. O., Bell, J., Elston, C. W., Blamey, R. W., and Baum, M., Prediction of lymph node involvement in breast cancer, *Lancet, 338*, 392 (1991).
48. Taylor, C. W., Anbazhagan, R., Jayatilake, H., Adams, A., Gusterson, B. A., Price, K., Gelber, R. D., and Goldhirsch, A., *Helix pomatia* in breast cancer, *Lancet, 338*, 580 (1991).
49. Hirao, T., Sakamoto, Y., Kamada, M., Hamada, S., and Aono, T., Tn antigen, a marker of potential for metastasis of uterine cervix cancer cells, *Cancer, 72*, 154 (1993).
50. Sasaki, M., Yamato, T., and Nakanuma, Y., Expression of sialyl-Tn, Tn and T antigens in primary liver cancer, *Pathol.Int., 49*, 325 (1999).
51. Nuti, M., Teramoto, Y. A., Mariani-Constantini, R., Hand, P. H., Colcher, D., and Schlom, J., A monoclonal antibody (B72.3) defines patterns of a novel tumor-associated antigen in human mammary carcinoma cell populations, *J.Inst.Cancer, 29*, 539 (1982).
52. Thor, A., Ohuchi, W., Szpak, C. A., Johnston, W. W., and Schlom, J., Distribution of oncofetal antigen tumor-associated glycoprotein-72 defined by monoclonal antibody B72.3, *Cancer Res., 46*, 3118 (1986).
53. Kjeldsen, T. B., Clausen, H., Hirohashi, S., Ogawa, T., Iijima, H., and Hakomori, S., Preparation and characterization of monoclonal antibodies directed to the tumor-associated O-linked associated O-linked sialosyl-2→6 α-N-acetylgalactosaminyl (sialosyl-Tn) epitope, *Cancer Res., 48*, 2214 (1988).
54. Kurosaka, A., Kitagawa, H., Fukui, S., Numata, Y., Nakada, H., Funakoshi, I., Kawasaki, T., Ogawa, T., Iijima, H., and Yamashina, I., A monoclonal antibody that recognizes a cluster of a disaccharide, NeuAcα2→6GalNAc, in mucin-type glycoproteins, *J.Biol.Chem., 263*, 8724 (1988).
55. Tettamanti, G., and Pigman, W., Purification and characterization of bovine and ovine submaxillary mucins, *Arch.Biochem.Biophys., 124*, 41 (1968).
56. Ogata, S., Koganty, R., Reddish, M., Longenecker, B. M., Chen, A., Perez, C., and Itzkowitz, S. H., Different modes of sialyl-Tn expression during malignant transformation of human colonic carcinoma, *Glycoconj.J., 15*, 29 (1998).
57. Ogata, S., Ho, I., Chen, A., Dubois, D., Maklansky, J., Singhal, A. K., Hakomori, S., and Itzkowitz, S. H., Tumor-associated sialylated antigens are constitutively expressed in normal human colonic mucosa, *Cancer Res., 55*, 1869 (1995).
58. Itzkowitz, S. H., Bloom, E. J., Kokal, W. A., Modin, G., Hakomori, S., and Kim, Y. S., Sialosyl-Tn: A novel mucin antigen associated with prognosis in colorectal cancer patients, *Cancer, 66*, 1960 (1990).
59. Kobayashi, H., Terao, T., and Kawashima, Y., Serum sialyl Tn as an independent predictor of poor prognosis in patients with epithelial ovarian cancer, *J.Clin.Oncol., 10*, 95 (1992).
60. Giuffre, G., Vitarelli, E., Tuccari, G., and Barresi, G., Immunohistochemical studyof Tn, sialosyl-Tn and T antigens in human meningiomas, *Eur.J.Histochem., 42*, 197 (1998).
61. Terashima, S., Takano, Y., Ohori, T., Kanno, T., Kimura, T., Motoki, R., and Kawaguchi, T., Sialyl-Tn antigen as a useful predictor of poor prognosis in patients with advanced stomach cancer, *Surg.Today, 28*, 682 (1998).
62. Karlen, P., Young, E., Brostrom, O., Lofberg, R., Tribukait, B., Ost, K., Bodian, C., and Itzkowitz, S.

H., Sialyl-Tn antigen as a marker of colon cancer risk in ulcerative colitis: Relation to dysplasia and DNA aneuploidy, *Gastroenterology, 115*, 1395 (1998).

63. Jensen, P., Clausen, O. P., and Bryne, M., Differences in sialyl-Tn antigen expression between keratoacanthomas and cutaneous squamous cell carcinomas, *J.Cutan.Pathol., 26*, 183 (1999).
64. Sato, T., Nishimura, G., Nonomura, A., Miwa, K., and Miyazaki, I., Serological studies on CEA, CA19-9, STn, and SLX in colorectal cancer, *Hepatogastroenterology, 46*, 914 (1999).
65. Lundin, M., Nordling, S., Roberts, P. J., Lundin, J., Carpelan-Holmstrom, M., von Boguslawsky, K., and Haglund, C., Sialyl Tn is a frequently expressed antigen in colorectal cancer: No correlation with patient prognosis, *Oncology, 57*, 70 (1999).
66. Ikehara, Y., Kojima, N., Kurosawa, N., Kudo, T., Kono, M., Nishihara, S., Issiki, S., Morozumi, K., Itzkowitz, S., Tsuda, T., Nishimura, S., Tsuji, S., and Narimatsu, H., Cloning and expression of a human gene encoding an *N*-acetylgalactosamine-α2,6-sialyltransferase (ST6GalNAc I): A candidate for synthesis of cancer-associated sialyl-Tn antigens, *Glycobiology, 9*, 1213 (1999).
67. Lloyd, K. O., Kabat, E. A., Layug, E. J., and Gruezo, F., Immunochemical studies on blood groups: XXXIV. Structures of some oligosaccharides produced by alkaline degradation of blood group A, B, and H substances, *Biochemistry, 5*, 1489 (1966).
68. Abe, K., McKibbin, J. M., and Hakomori, S., The monoclonal antibody directed to difucosylated type 2 chain (Fucα1→2Galβ1→4[Fucα1→3]GlcNAc; Y determinant), *J.Biol.Chem., 258*, 11793 (1983).
69. Brown, A., Feizi, T., Gooi, H. C., Embleton, M. J., Picard, J. K., and Baldwin, R. W., A monoclonal antibody against human colonic adenoma recognizes difucosylated type 2 blood group chains, *Biosci.Reports, 3*, 163 (1983).
70. Lloyd, K. O., Larson, G., Strömberg, N., Thurin, J., and Karlsson, K.-A., Mouse monoclonal antibody F-3 recognizes the difucosyl Type 2 blood group structure, *Immunogenetics, 17*, 537 (1983).
71. Pastan, I., Lovelace, E. T., Gallo, M. G., Rutherford, A. V., Magnani, J. L., and Willingham, M. C., Characterization of monoclonal antibodies B1 and B3 that react with mucinous adenocarcinomas, *Cancer Res., 51*, 3781 (1991).
72. Scherf, U., Benhar, I., Webber, K. O., Pastan, I., and Brinkmann, U., Cytotoxic and antitumor activity of a recombinant tumor necrosis factor-B1(Fv) fusion protein on Ley antigen-expressing human cancer cells, *Clin.Cancer.Res., 2*, 1523 (1996).
73. Bera, T. K., and Pastan, I., Comparison of recombinant immunotoxins against LeY antigen expressing tumor cells: Influence of affinity, size, and stability, *Bioconjug.Chem., 9*, 736 (1998).
74. Kaizu, T., Levery, S. B., Nudelman, E. D., Stenkamp, R. E., and Hakomori, S., Novel fucolipids of human adenocarcinoma: Monoclonal antibody specific for trifucosyl Ley (III3FucV3FucVI2FucnLc6), and a possible three-dimensional epitope structure, *J.Biol.Chem., 261*, 11254 (1986).
75. Deshpande, P. P., and Danishefsky, S. J., Total synthesis of the potential anticancer vaccine KH-1 adenocarcinoma antigen, *Nature, 387*, 164 (1997).
76. Miyake, M., and Hakomori, S., A specific cell surface glycoconjugate controlling cell motility: Evidence by functional monoclonal antibodies that inhibit cell motility and tumor cell metastasis, *Biochemistry, 30*, 3328 (1991).
77. Miyake, M., Taki, T., Hitomi, S., and Hakomori, S., Correlation of expression of H/Ley/Leb antigens with survival in patients with carcinoma of the lung, *N.Engl.J.Med., 327*, 14 (1992).
78. Goldstein, I. J., and Hayes, C. E., The lectins: Carbohydrate-binding proteins of plants and animals, *Adv.Carbohydr.Chem.Biochem., 35*, 127 (1978).
79. Shirahama, T., Ikoma, M., Muramatsu, T., Kayajima, T., Ohi, Y., Tsushima, T., Akebi, N., Ohmori, H., Hirao, Y., and Okajima, E., The binding site for fucose-binding proteins of *Lotus tetragonolobus* is a prognostic marker for transitional cell carcinoma of the human urinary bladder, *Cancer, 72*, 1329 (1993).

80. Dische, Z., Fucose and sialic acid in glycoproteins of the mucus of the digestive tract, *Fed.Proc., 19*, 904 (1960).
81. Koprowski, H., Herlyn, M., Steplewski, Z., and Sears, H. F., Specific antigen in serum of patients with colon carcinoma, *Science, 212*, 53 (1981).
82. Magnani, J. L., Nilsson, B., Brockhaus, M., Zopf, D., Steplewski, Z., Koprowski, H., and Ginsburg, V., A monoclonal antibody-defined antigen associated with gastrointestinal cancer is a ganglioside containing sialylated lacto-*N*-fucopentaose II, *J.Biol.Chem., 257*, 14365 (1982).
83. Hansson, G. C., and Zopf, D., Biosynthesis of the cancer-associated sialyl-Lea antigen, *J.Biol.Chem., 260*, 9388 (1985).
84. Magnani, J. L., Steplewski, Z., Koprowski, H., and Ginsburg, V., Identification of the gastrointestinal and pancreatic cancer-associated antigen detected by monoclonal antibody 19-9 in the sera of patients as mucin, *Cancer Res., 43*, 5489 (1983).
85. Kannagi, R., Carbohydrate-mediated cell adhesion involved in hematogenous metastasis of cancer, *Glycoconj.J., 14*, 577 (1997).
86. Fukushi, Y., Nudelman, E. D., Levery, S. B., Rauvala, H., and Hakomori, S., Novel fucolipids accumulating in human cancer: III. A hybridoma antibody (FH6) defining a human cancer-associated difucoganglioside (VI3NeuAcV3III3Fuc2nLc6), *J.Biol.Chem., 259*, 10511 (1984).
87. Fukushima, K., Hirota, M., Terasaki, P. I., Wakisaka, A., Togashi, H., Chia, D., Suyama, N., Fukushi, Y., Nudelman, E. D., and Hakomori, S., Characterization of sialosylated Lewisx as a new tumor-associated antigen, *Cancer Res., 44*, 5279 (1984).
88. Symington, F. W., Hedges, D. L., and Hakomori, S., Glycolipid antigens of human polymorphonuclear neutrophils and the inducible HL-60 myeloid leukemia line, *J.Immunol., 134*, 2498 (1985).
89. Nakamori, S., Kameyama, M., Imaoka, S., Furukawa, H., Ishikawa, O., Sasaki, Y., Kabuto, T., Iwanaga, T., Matsushita, Y., and Irimura, T., Increased expression of sialyl Lewisx antigen correlates with poor survival in patients with colorectal carcinoma: Clinicopathological and immunohistochemical study, *Cancer Res., 53*, 3632 (1993).
90. Nakagoe, T., Fukushima, K., Hirota, M., Kusano, H., Ayabe, H., Tomita, M., and Kamihira, S., Immunohistochemical expression of sialyl Lex antigen in relation to survival of patients with colorectal carcinoma, *Cancer, 72*, 2323 (1993).
91. Lowe, J. B., Stoolman, L. M., Nair, R. P., Larsen, R. D., Berhend, T. L., and Marks, R. M., ELAM-1-dependent cell adhesion to vascular endothelium determined by a transfected human fucosyltransferase cDNA, *Cell, 63*, 475 (1990).
92. Phillips, M. L., Nudelman, E. D., Gaeta, F. C. A., Perez, M., Singhal, A. K., Hakomori, S., and Paulson, J. C., ELAM-1 mediates cell adhesion by recognition of a carbohydrate ligand, sialyl-Lex, *Science, 250*, 1130 (1990).
93. Hakomori, S., Possible new directions in cancer therapy based on aberrant expression of glycosphingolipids: Anti-adhesion and ortho-signaling therapy, *Cancer Cells, 3*, 461 (1991).
94. Berg, E. L., Robinson, M. K., Mansson, O., Butcher, E. C., and Magnani, J. L., A carbohydrate domain common to both sialyl Lea and sialyl Lex is recognized by the endothelial cell leukocyte adhesion molecule ELAM-1, *J.Biol.Chem., 266*, 14869 (1991).
95. Takada, A., Ohmori, K., Takahashi, N., Tsuyuoka, K., Yago, A., Zenita, K., Hasegawa, A., and Kannagi, R., Adhesion of human cancer cells to vascular endothelium mediated by a carbohydrate antigen, sialyl Lewis A, *Biochem.Biophys.Res.Commun., 179*, 713 (1991).
96. Takada, A., Ohmori, K., Yoneda, T., Tsuyuoka, K., Hasegawa, A., Kiso, M., and Kannagi, R., Contribution of carbohydrate antigens sialyl Lewis A and sialyl Lewis X to adhesion of human cancer cells to vascular endothelium, *Cancer Res., 53*, 354 (1993).

97. Hakomori, S., Novel endothelial cell activation factor(s) released from activated platelets which induce E-selectin expression and tumor cell adhesion to endothelial cells: A preliminary note, *Biochem. Biophys.Res.Commun., 203*, 1605 (1994).
98. Khatib, A.-M., Kontogiannea, M., Fallavollita, L., Jamison, B., Meterissian, S., and Brodt, P., Rapid induction of cytokine and E-selectin expression in the liver in response to metastatic tumor cells, *Cancer Res., 59*, 1356 (1999).
99. Polley, M. J., Phillips, M. L., Wayner, E. A., Nudelman, E. D., Singhal, A. K., Hakomori, S., and Paulson, J. C., CD62 and endothelial cell-leukocyte adhesion molecule 1 (ELAM-1) recognize the same carbohydrate ligand, sialyl-Lewis x, *Proc.Natl.Acad.Sci.USA, 88*, 6224 (1991).
100. Varki, A., Selectin ligands, *Proc.Natl.Acad.Sci.USA, 91*, 7390 (1994).
101. Lasky, L. A., Selectin-carbohydrate interactions and the initiation of the inflammatory response, *Annu.Rev.Biochem., 64*, 113 (1995).
102. Stroud, M. R., Handa, K., Salyan, M. E. K., Ito, K., Levery, S. B., Hakomori, S., Reinhold, B. B., and Reinhold, V. N., Monosialogangliosides of human myelogenous leukemia HL60 cells and normal human leukocytes. 1. Separation of E-selectin binding from nonbinding gangliosides, and absence of sialosyl-Lex having tetraosyl to octaosyl core, *Biochemistry, 35*, 758 (1996).
103. Stroud, M. R., Handa, K., Salyan, M. E. K., Ito, K., Levery, S. B., Hakomori, S., Reinhold, B. B., and Reinhold, V. N., Monosialogangliosides of human myelogenous leukemia HL60 cells and normal human leukocytes. 2. Characterization of E-selectin binding fractions, and structural requirements for physiological binding to E-selectin, *Biochemistry, 35*, 770 (1996).
104. Handa, K., Stroud, M. R., and Hakomori, S., Sialosyl-fucosyl poly-LacNAc without the sialosyl-Lex epitope as the physiological myeloid cell ligand in E-selectin-dependent adhesion: Studies under static and dynamic flow conditions, *Biochemistry, 36*, 12412 (1997).
105. Handa, K., Withers, D. A., and Hakomori, S., The $\alpha 1 \rightarrow 3$ fucosylation at the penultimate GlcNAc catalyzed by fucosyltransferase VII is blocked by internally fucosylated residue in sialosyl long-chain poly-LacNAc: Enzymatic basis for expression of physiological E-selectin epitope, *Biochem.Biophys. Res.Commun., 243*, 199 (1998).
106. Sako, D., Chang, X.-J., Barone, K. M., Vachino, G., White, H. M., Shaw, G., Veldman, G. M., Bean, K. M., Ahern, T. J., Furie, B., Cumming, D. A., and Larsen, G. R., Expression cloning of a functional glycoprotein ligand for P-selectin, *Cell, 75*, 1179 (1993).
107. Handa, K., White, T., Ito, K., Fang, H., Wang, S., and Hakomori, S., P-selectin-dependent adhesion of human cancer cells: Requirement for co-expression of a "PSGL-1-like" core protein and the glycosylation process for sialosyl-Lex or sialosyl-Lea, *Int.J.Oncol., 6*, 773 (1995).
108. Nudelman, E. D., Levery, S. B., Stroud, M. R., Salyan, M. E. K., Abe, K., and Hakomori, S., A novel tumor-associated, developmentally regulated glycolipid antigen defined by monoclonal antibody ACFH-18, *J.Biol.Chem., 263*, 13942 (1988).
109. Järnefelt, J., Rush, J., Li, Y.-T., and Laine, R. A., Erythroglycan, a high molecular weight glycopeptide with the repeating structure [galactosyl-(1→4)-2-deoxy-2-acetamido-glucosyl(1→3)] comprising more than one-third of the protein-bound carbohydrate of human erythrocyte stroma, *J.Biol.Chem., 253*, 8006 (1978).
110. Levery, S. B., Nudelman, E. D., Salyan, M. E. K., and Hakomori, S., Novel tri- and tetrasialylpoly-*N*-acetyllactosaminyl gangliosides of human placenta: Structure determination of pentadeca- and eicosaglycosylceramides by methylation analysis, fast atom bombardment mass spectrometry, and 1H-NMR spectroscopy, *Biochemistry, 28*, 7772 (1989).
111. Nudelman, E. D., Mandel, U., Levery, S. B., Kaizu, T., and Hakomori, S., A series of disialogangliosides with binary 2-3 sialosyllactosamine structure, defined by monoclonal antibody NUH2, are oncodevelopmentally regulated antigens, *J.Biol.Chem., 264*, 18719 (1989).
112. Kannagi, R., Levery, S. B., and Hakomori, S., Lea-active heptaglycosylceramide, a hybrid of type 1

and type 2 chain, and the pattern of glycolipids with Lea, Leb, X (Lex) and Y (Ley) determinants in human blood cell membranes (ghosts): Evidence that type 2 chain can elongate repetitively but type 1 chain cannot, *J.Biol.Chem., 260*, 6410 (1985).

113. Mårtensson, S., Due, C., Påhlsson, P., Nilsson, B., Eriksson, H., Zopf, D., Olsson, L., and Lundblad, A., A carbohydrate epitope associated with human squamous lung cancer, *Cancer Res., 48*, 2125 (1988).

114. Stroud, M. R., Levery, S. B., Mårtensson, S., Salyan, M. E. K., Clausen, H., and Hakomori, S., Human tumor-associated Lea-Lex hybrid carbohydrate antigen IV3(Galβ1→3[Fucα1→4]GlcNAc)-III3FucnLc4 defined by monoclonal antibody 43-9F: Enzymatic synthesis, structural characterization, and comparative reactivity with various antibodies, *Biochemistry, 33*, 10672 (1994).

115. Watanabe, M., Hirohashi, S., Shimosato, Y., Ino, Y., Yamada, T., Teshima, S., Sekine, T., and Abe, O., Carbohydrate antigen defined by a monoclonal antibody raised against a gastric cancer xenograft, *Jpn.J.Cancer Res.(Gann), 76*, 43 (1985).

116. Watanabe, M., Ohishi, T., Kuzuoka, M., Nudelman, E. D., Stroud, M. R., Kubota, T., Kodaira, S., Abe, O., Hirohashi, S., Shimosato, Y., and Hakomori, S., *In vitro* and *in vivo* antitumor effects of murine monoclonal antibody NCC-ST-421 reacting with dimeric Lea (Lea/Lea) epitope, *Cancer Res., 51*, 2199 (1991).

117. Stroud, M. R., Levery, S. B., Nudelman, E. D., Salyan, M. E. K., Towell, J. A., Roberts, C. E., Watanabe, M., and Hakomori, S., Extended type 1 chain glycosphingolipids: Dimeric Lea (III4V4Fuc2Lc6) as human tumor-associated antigen, *J.Biol.Chem., 266*, 8439 (1991).

118. Stroud, M. R., Levery, S. B., Salyan, M. E. K., Roberts, C. E., and Hakomori, S., Extended type-1 chain glycosphingolipid antigens: Isolation and characterization of trifucosyl-Leb antigen (III4V4VI2Fuc3Lc6), *Eur.J.Biochem., 203*, 577 (1992).

119. Ito, H., Tashiro, K., Stroud, M. R., Ørntoft, T. F., Meldgaard, P., Singhal, A. K., and Hakomori, S., Specificity and immunobiological properties of monoclonal antibody IMH2, established after immunization with Leb/Lea glycosphingolipid, a novel extended type 1 chain antigen, *Cancer Res., 52*, 3739 (1992).

120. Watanabe, K., Powell, M. E., and Hakomori, S., Isolation and characterization of gangliosides with a new sialosyl linkage and core structure: Gangliosides of human erythrocyte membranes II, *J.Biol. Chem., 254*, 8223 (1979).

121. Hakomori, S., Patterson, C. M., Nudelman, E. D., and Sekiguchi, K., A monoclonal antibody directed to N-acetylneuraminosyl α2→6galactosyl residue in gangliosides and glycoproteins, *J.Biol.Chem., 258*, 11819 (1983).

122. Kannagi, R., Levery, S. B., Ishigami, F., Hakomori, S., Shevinsky, L. H., Knowles, B. B., and Solter, D., New globoseries glycosphingolipids in human teratocarcinoma reactive with the monoclonal antibody directed to a developmentally regulated antigen, stage-specific embryonic antigen 3, *J.Biol.Chem., 258*, 8934 (1983).

123. Mènard, S., Tagliabue, E., Canevari, S., Fossati, G., and Colnaghi, M. I., Generation of monoclonal antibodies reacting with normal and cancer cells of human breast, *Cancer Res., 43*, 1295 (1983).

124. Bremer, E. G., Levery, S. B., Sonnino, S., Ghidoni, R., Canevari, S., Kannagi, R., and Hakomori, S., Characterization of a glycosphingolipid antigen defined by the monoclonal antibody MBr1 expressed in normal and neoplastic epithelial cells of human mammary gland, *J.Biol.Chem., 259*, 14773 (1984).

125. Perrone, F., Ménard, S., Canevari, S., Calabrese, M., Boracchi, P., Bufalino, R., Testori, S., Baldini, M., and Colnaghi, M. I., Prognostic significance of the CaMBr1 antigen on breast carcinoma: Relevance of the type of recognised glycoconjugate, *Eur.J.Cancer, 29A*, 2113 (1993).

126. Mariani-Costantini, R., Colnaghi, M. I., Leoni, F., Mènard, S., Cerasoli, S., and Rilke, F., Immunohistochemical reactivity of a monoclonal antibody prepared against human breast carcinoma, *Virchows Arch.A Pathol.Anat.Histopathol., 402*, 389 (1984).

127. Zhang, S., Zhang, H. S., Reuter, V. E., Slovin, S. F., Scher, H. I., and Livingston, P. O., Expression of potential target antigens for immunotherapy on primary and metastatic prostate cancers, *Clin.Cancer Res., 4*, 295 (1998).
128. Saito, S., Levery, S. B., Salyan, M. E. K., Goldberg, R. I., and Hakomori, S., Common tetrasaccharide epitope NeuAcα2→3Galβ1→3(NeuAcα2→6)GalNAc, presented by different carrier glycosylceramides or *O*-linked peptides, is recognized by different antibodies and ligands having distinct specificities, *J.Biol.Chem., 269*, 5644 (1994).
129. Satoh, M., Handa, K., Saito, S., Tokuyama, S., Ito, A., Miyao, N., Orikasa, S., and Hakomori, S., Disialosyl galactosylgloboside as an adhesion molecule expressed on renal cell carcinoma and its relationship to metastatic potential, *Cancer Res., 56*, 1932 (1996).
130. Saito, S., Orikasa, S., Satoh, M., Ohyama, C., Ito, A., and Takahashi, T., Expression of globo-series gangliosides in human renal cell carcinoma, *Jpn.J.Cancer Res.(Gann), 88*, 652 (1997).
131. Nilsson, O., Mansson, J.-E., Brezicka, T., Holmgren, J., Lindholm, L., Sorenson, S., Yngvason, F., and Svennerholm, L., Fucosyl GM1 — a ganglioside associated with small cell lung carcinomas, *Glycoconj. J., 1*, 43 (1984).
132. Nilsson, O., Brezicka, F. T., Holmgren, J., Sorenson, S., Svennerholm, L., Yngvason, F., and Lindholm, L., Detection of a ganglioside antigen associated with small cell lung carcinomas using monoclonal antibodies directed against fucosyl-GM1, *Cancer Res., 46*, 1403 (1986).
133. Vangsted, A. J., Clausen, H., Kjeldsen, T. B., White, T., Sweeney, B., Hakomori, S., Drivsholm, L., and Zeuthen, J., Immunochemical detection of a small cell lung cancer-associated ganglioside (FucGM1) in serum, *Cancer Res., 51*, 2879 (1991).
134. Portoukalian, J., Zwingelstein, G., and Dore, J., Lipid composition of human malignant melanoma tumors at various levels of malignant growth, *Eur.J.Biochem., 94*, 19 (1979).
135. Yeh, M.-Y., Hellström, I., Abe, K., Hakomori, S., and Hellström, K.-E., A cell-surface antigen which is present in the ganglioside fraction and shared by human melanomas, *Int.J.Cancer, 29*, 269 (1982).
136. Nudelman, E. D., Hakomori, S., Kannagi, R., Levery, S. B., Yeh, M.-Y., Hellström, K.-E., and Hellström, I., Characterization of a human melanoma-associated ganglioside antigen defined by a monoclonal antibody, 4.2, *J.Biol.Chem., 257*, 12752 (1982).
137. Pukel, C. S., Lloyd, K. O., Travassos, L. R., Dippold, W. G., Oettgen, H. F., and Old, L. J., GD3, a prominent ganglioside of human melanoma: Detection and characterization by mouse monoclonal antibody, *J.Exp.Med., 155*, 1133 (1982).
138. Houghton, A. N., Mintzer, D., Cordon-Cardo, C., Welt, S. W., Fliegel, B., Vadhan, S., Carswell, E., Melamed, M. R., Oettgen, H. F., and Old, L. J., Mouse monoclonal IgG3 antibody detecting GD3 ganglioside: A phase I trial in patients with malignant melanoma, *Proc.Natl.Acad.Sci.USA, 82*, 1242 (1985).
139. Irie, R. F., Sze, L. L., and Saxton, R. E., Human antibody to OFA-1, a tumor antigen, produced *in vitro* by Epstein-Barr virus-transformed human β-lymphoid cell lines, *Proc.Natl.Acad.Sci.USA, 79*, 5666 (1982).
140. Tai, T., Paulson, J. C., Cahan, L. D., and Irie, R. F., Ganglioside GM2 as a human tumor antigen (OFA-I-1), *Proc.Natl.Acad.Sci.USA, 80*, 5392 (1983).
141. Cahan, L. D., Irie, R. F., Singh, R., Cassidenti, A., and Paulson, J. C., Identification of human neuroectodermal tumor antigen (OFA-I-2) as ganglioside GD2, *Proc.Natl.Acad.Sci.USA, 79*, 7629 (1982).
142. Watanabe, T., Pukel, C. S., Takeyama, H., Lloyd, K. O., Shiku, H., Li, L. T. C., Trabassos, L. R., Oettgen, H. F., and Old, L. J., Human melanoma antigen AH is an autoantigen ganglioside related to GD2, *J.Exp.Med., 156*, 1884 (1982).

143. Hirabayashi, Y., Hanaoka, A., Matsumoto, M., Matsubara, T., Tagawa, M., Wakabayashi, S., and Taniguchi, M., Syngeneic monoclonal antibody against melanoma antigen with interspecies cross-reactivity recognizes GM3, a prominent ganglioside of B16 melanoma, *J.Biol.Chem., 260*, 13328 (1985).
144. Nores, G. A., Dohi, T., Taniguchi, M., and Hakomori, S., Density-dependent recognition of cell surface GM3 by a certain anti-melanoma antibody, and GM3 lactone as a possible immunogen: Requirements for tumor-associated antigen and immunogen, *J.Immunol., 139*, 3171 (1987).
145. Nakano, J., Muto, M., Shimizu, T., Hirota, T., Ichimiya, M., and Asagami, C., Ganglioside expression in melanomas from Japanese individuals: Unusual pattern in two patients with metastatic lesions of acral lentiginous melanomas, *Pigment Cell Res., 10*, 201 (1997).
146. Cheung, N.-K., Lazarus, H., Miraldi, F. D., Abramowsky, C. R., Kallick, S., Saarinen, U. M., Spitzer, T., Strandjord, S. E., Coccia, P. F., and Berger, N. A., Ganglioside GD2 specific monoclonal antibody 3F8: A phase I study in patients with neuroblastoma and malignant melanoma, *J.Clin.Oncol., 5*, 1430 (1987).
147. Saleh, M. N., Khazaeli, M. B., Wheeler, R. H., Dropcho, E., Liu, T., Urist, M., Miller, D. M., Lawson, S., Dixon, P., Russell, C. H., and LoBuglio, A. F., Phase I trial of the murine monoclonal anti-GD2 antibody 14G9a in metastatic melanoma, *Cancer Res., 52*, 4342 (1992).
148. Zhang, H., Zhang, S., Cheung, N. K., Ragupathi, G., and Livingston, P. O., Antibodies against GD2 ganglioside can eradicate syngeneic cancer micrometastases, *Cancer Res., 58*, 2844 (1998).
149. Young, W. W. J., and Hakomori, S., Therapy of mouse lymphoma with monoclonal antibodies to glycolipid: Selection of low antigenic variants *in vivo*, *Science, 211*, 487 (1981).
150. Jones, P. C., Sze, L. L., Liu, P. Y., Morton, D. L., and Irie, R. F., Prolonged survival for melanoma patients with elevated IgM antibody to oncofetal antigen, *J.Natl.Cancer Inst., 66*, 249 (1981).
151. Livingston, P. O., Wong, G. Y., Adluri, S., Tao, Y., Padavan, M., Parente, R., Hanlon, C., Calves, M. J., Helling, F., and Ritter, G., Improved survival in stage III melanoma patients with GM2 antibodies: A randomized trial of adjuvant vaccination with GM2 ganglioside, *J.Clin.Oncol., 12*, 1036 (1994).
152. Livingston, P. O., Approaches to augmenting the immunogenicity of melanoma gangliosides: From whole melanoma cells to ganglioside-KLH conjugate vaccines, *Immunol.Rev., 145*, 147 (1995).
153. Livingston, P. O., Carbohydrate vaccines that induce antibodies against cancer. 1. Rationale, *Cancer Immunol.Immunother., 45*, 1 (1997).
154. Young, W. W. J., MacDonald, E. M. S., Nowinski, R. C., and Hakomori, S., Production of monoclonal antibodies specific for distinct portions of the glycolipid asialo GM2 (gangliotriosylceramide), *J.Exp. Med., 150*, 1008 (1979).
155. Livingston, P. O., Natoli, E. J., Calves, M. J., Stockert, E., Oettgen, H. F., and Old, L. J., Vaccines containing purified GM2 ganglioside elicit GM2 antibodies in melanoma patients, *Proc.Natl.Acad.Sci. USA, 84*, 2911 (1987).
156. Kitamura, K., Livingston, P. O., Fortunato, S. R., Stockert, E., Helling, F., Ritter, G., Oettgen, H. F., and Old, L. J., Serological response patterns of melanoma patients immunized with a GM2 ganglioside conjugate vaccine, *Proc.Natl.Acad.Sci.USA, 92*, 2805 (1995).
157. Helling, F., Shang, A., Calves, M. J., Zhang, S., Ren, S., Yu, R. K., Oettgen, H. F., and Livingston, P. O., GD3 vaccines for melanoma: Superior immunogenicity of keyhole limpet hemocyanin conjugate vaccines, *Cancer Res., 54*, 197 (1994).
158. Schulz, G., Cheresh, D. A., Varki, N. M., Yu, A., Staffileno, L. K., and Reisfeld, R. A., Detection of ganglioside GD2 in tumor tissues and sera of neuroblastoma patients, *Cancer Res., 44*, 5914 (1984).
159. Pancook, J. D., Becker, J. C., Gillies, S. D., and Reisfeld, R. A., Eradication of established hepatic human neuroblastoma metastases in mice with severe combined immunodeficiency by antibody-targeted interleukin-2, *Cancer Immunol.Immunother., 42*, 88 (1996).

160. Becker, J. C., Varki, N., Gillies, S. D., Furukawa, K., and Reisfeld, R. A., An antibody-interleukin 2 fusion protein overcomes tumor heterogeneity by induction of a cellular immune response, *Proc.Natl. Acad.Sci. USA, 93*, 7826 (1996).
161. Clausen, H., and Hakomori, S., ABH and related histo-blood group antigens: Immunochemical differences in carrier isotypes and their distribution, *Vox Sang., 56*, 1 (1989).
162. Slovin, S. F., Ragupathi, G., Adluri, S., Ungers, G., Terry, K., Kim, S., Spassova, M., Bornmann, W. G., Fazzari, M., Dantis, L., Olkiewicz, K., Lloyd, K. O., Livingston, P. O., Danishefsky, S. J., and Scher, H. I., Carbohydrate vaccines in cancer: Immunogenicity of a fully synthetic globo H hexasaccharide conjugate in man, *Proc.Natl.Acad.Sci. USA, 96*, 5710 (1999).
163. Toyokuni, T., Dean, B., Cai, S., Boivin, D., Hakomori, S., and Singhal, A. K., Synthetic vaccines: Synthesis of a dimeric Tn antigen-lipopeptide conjugate that elicits immune responses against Tn-expressing glycoproteins, *J.Am.Chem.Soc., 116*, 395 (1994).
164. Toyokuni, T., Hakomori, S., and Singhal, A. K., Synthetic carbohydrate vaccines: Synthesis and immunogenicity of Tn antigen conjugates, *Bioorg.Med.Chem., 2*, 1119 (1994).
165. Lo-Man, R., Bay, S., Vichier-Guerre, S., Deriaud, E., Cantacuzene, D., and Leclerc, C., A fully synthetic immunogen carrying a carcinoma-associated carbohydrate for active specific immunotherapy, *Cancer Res., 59*, 1520 (1999).
166. Singhal, A. K., Fohn, M., and Hakomori, S., Induction of Tn (α-N-acetylgalactosamine-O-serine/threonine) antigen-mediated cellular immune response for active immunotherapy in mice, *Cancer Res., 51*, 1406 (1991).
167. MacLean, G. D., Reddish, M. A., Koganty, R. R., and Longenecker, B. M., Antibodies against mucin-associated sialyl-Tn epitopes correlate with survival of metastatic adenocarcinoma patients undergoing active specific immunotherapy with synthetic STn vaccine, *J.Immunother., 19*, 59 (1996).
168. MacLean, G. D., Miles, D. W., Rubens, R. D., Reddish, M. A., and Longenecker, B. M., Enhancing the effect of THERATOPE STn-KLH cancer vaccine in patients with metastatic breast cancer by pretreatment with low-dose intravenous cyclophosphamide, *J.Immunother., 19*, 309 (1996).
169. Ragupathi, G., Koganty, R. R., Qiu, D., Lloyd, K. O., and Livingston, P. O., A novel and efficient method for synthetic carbohydrate conjugate vaccine preparation: Synthesis of sialyl Tn-KLH conjugate using a 4-(4-*N*-maleimidomethyl) cyclohexane-1-carboxyl hydrazide (MMCCH) linker arm, *Glycoconj.J., 15*, 217 (1998).
170. McCaffery, M., Yao, T.-J., Williams, L., Livingston, P. O., Houghton, A. N., and Chapman, P. B., Immunization of melanoma patients with BEC2 anti-idiotypic monoclonal antibody that mimics GD3 ganglioside: Enhanced immunogenicity when combined with adjuvant, *Clin.Cancer Res., 2*, 679 (1996).
171. Yao, T.-J., Meyers, M., Livingston, P. O., Houghton, A. N., and Chapman, P. B., Immunization of melanoma patients with BEC2-keyhole limpet hemocyanin plus BCG intradermally followed by intravenous booster immunizations with BEC2 to induce anti-GD3 ganglioside antibodies, *Clin.Cancer Res., 5*, 77 (1999).
172. Yan, X., Evans, S. V., Kaminki, M. J., Gillies, S. D., Reisfeld, R. A., Houghton, A. N., and Chapman, P. B., Characterization of an Ig VH idiotope that results in specific homophilic binding and increased avidity for antigen, *J.Immunol., 157*, 1582 (1996).
173. Scott, J. K., and Smith, G. P., Searching for peptide ligands with an epitope library, *Science, 249*, 386 (1990).
174. Devlin, J. J., Panganiban, L. C., and Devlin, P. E., Random peptide libraries: A source of specific protein binding molecules, *Science, 249*, 404 (1990).
175. Cwirla, S. E., Peters, E. A., Barrett, R. W., and Dower, W. J., Peptides on phage: A vast library of peptides for identifying ligands, *Proc.Natl.Acad.Sci. USA, 87*, 6378 (1990).
176. Scott, J. K., Loganathan, D., Easley, R. B., Gong, X., and Goldstein, I. J., A family of concanavalin

A-binding peptides from a hexapeptide epitope library, *Proc.Natl.Acad.Sci.USA, 89*, 5398 (1992).
177. Phalipon, A., Folgori, A., Arondel, J., Sgaramella, G., Fortugno, P., Cortese, R., Sansonetti, P. J., and Felici, F., Induction of anti-carbohydrate antibodies by phage library-selected peptide mimics, *Eur.J. Immunol., 27*, 2620 (1997).
178. Pincus, S. H., Smith, M. J., Jennings, H. J., Burritt, J. B., and Glee, P. M., Peptides that mimic the group B streptococcal type III capsular polysaccharide antigen, *J.Immunol., 160*, 293 (1998).
179. Lou, Q., and Pastan, I., A Lewisy epitope mimicking peptide induces anti-Lewisy immune responses in rabbits and mice, *J.Peptide Res., 53*, 252 (1999).
180. Taki, T., Ishikawa, D., Hamasaki, H., and Handa, S., Preparation of peptides which mimic glycosphingolipids by using phage peptide library and their modulation on β-galactosidase activity, *FEBS Lett., 418*, 219 (1997).
181. Ishikawa, D., Kikkawa, H., Ogino, K., Hirabayashi, Y., Oku, N., and Taki, T., GD1α-replica peptides functionally mimic GD1α, an adhesion molecule of metastatic tumor cells, and suppress the tumor metastasis, *FEBS Lett., 441*, 20 (1998).
182. Kieber-Emmons, T., Luo, P., Qiu, J., Chang, T. Y., O, I., Blaszczyk-Thurin, M., and Steplewski, Z., Vaccination with carbohydrate peptide mimotopes promotes anti-tumor responses, *Nature Biotechnol., 17*, 660 (1999).
183. Qiu, J., Luo, P., Wasmund, K., Steplewski, Z., and Kieber-Emmons, T., Towards the development of peptide mimotopes of carbohydrate antigens as cancer vaccines, *Hybridoma, 18*, 103 (1999).
184. Taki, T., Ishikawa, D., Ogura, M., Nakajima, M., and Handa, S., Ganglioside GD1α functions in the adhesion of metastatic tumor cells to endothelial cells of the target tissue, *Cancer Res., 57*, 1882 (1997).
185. Kojima, N., and Hakomori, S., Cell adhesion, spreading, and motility of GM3-expressing cells based on glycolipid-glycolipid interaction, *J.Biol.Chem., 266*, 17552 (1991).
186. Kojima, N., Shiota, M., Sadahira, Y., Handa, K., and Hakomori, S., Cell adhesion in a dynamic flow system as compared to static system: Glycosphingolipid-glycosphingolipid interaction in the dynamic system predominates over lectin- or integrin-based mechanisms in adhesion of B16 melanoma cells to non-activated endothelial cells, *J.Biol.Chem., 267*, 17264 (1992).
187. Otsuji, E., Park, Y. S., Tashiro, K., Kojima, N., Toyokuni, T., and Hakomori, S., Inhibition of B16 melanoma metastasis by administration of GM3- or Gg3-liposomes: Blocking adhesion of melanoma cells to endothelial cells (anti-adhesion therapy) via inhibition of GM3-Gg3Cer or GM3-LacCer interaction, *Int.J.Oncol., 6*, 319 (1995).
188. Hakomori, S., Role of gangliosides in tumor progression, in *Biological function of gangliosides (Progress in Brain Research, Vol. 101)*, Svennerholm, L., Asbury, A. K., Reisfeld, R. A., Sandhoff, K., Suzuki, K., Tettamanti, G., and Toffano, G., Eds., Elsevier Science BV, Amsterdam, pp. 241 (1994).
189. Herrlich, P., Pals, S., and Ponta, H., CD44 in colon cancer, *Eur.J.Cancer, 31A*, 1110 (1995).
190. Yamamura, S., Handa, K., and Hakomori, S., A close association of GM3 with c-Src and Rho in GM3-enriched microdomains at the B16 melanoma cell surface membrane: A preliminary note, *Biochem. Biophys.Res.Commun., 236*, 218 (1997).
191. Iwabuchi, K., Yamamura, S., Prinetti, A., Handa, K., and Hakomori, S., GM3-enriched microdomain involved in cell adhesion and signal transduction through carbohydrate-carbohydrate interaction in mouse melanoma B16 cells, *J.Biol.Chem., 273*, 9130 (1998).
192. Yu, S., Withers, D. A., Yamamura, S., Handa, K., and Hakomori, S., Cell recognition and subsequent signal transduction through glycosphingolipid-glycosphingolipid interaction [Abstract], *Glycoconj.J., 14 (Suppl.)*, S53 (1997).
193. Yu, S., Withers, D. A., and Hakomori, S., Globoside-dependent adhesion of human embryonal carcinoma cells, based on carbohydrate-carbohydrate interaction, initiates signal transduction and induces enhanced activity of transcription factors AP1 and CREB, *J.Biol.Chem., 273*, 2517 (1998).

CARBOHYDRATE-MEDIATED ADHESION OF HUMAN COLON CARCINOMA CELLS TO HUMAN LIVER SECTIONS [1]

Tatsuro Irimura,[2] Masayuki Ota, Yutaka Kawamura, and Yoko Nemoto-Sasaki

Laboratory of Cancer Biology and Molecular Immunology, Graduate School of Pharmaceutical Sciences, The University of Tokyo, Tokyo 113-0033, Japan

Glycans containing terminal sialyl Lewis X epitopes are known to function as ligands for selectins and to regulate cellular trafficking (Lowe et al., 1990; Phillips et al., 1990; Walz et al., 1990). Similar structures containing fucose and sialic acid are postulated to influence the pathogenesis of cancer for a number of reasons. For example, the expression of a carbohydrate epitope recognized by the monoclonal antibody (mAb)[3] FH6 (Fukushi et al., 1984) inversely correlated with the survival of colorectal carcinoma patients, strongly suggesting that this epitope plays a biological role in the malignant behavior, such as metastasis, of human colon carcinoma cells (Hoff et al., 1989; Matsushita et al., 1990; Nakamori et al., 1993). However, whether selectins are involved in this process is not known (Krause and Turner, 1999). Counter-receptors containing sialyl Lewis X carbohydrate chains for L- and P-selectins have been identified on the surface of endothelial cells and granulocytes, respectively. However, E-selectin counter receptors are still unknown (Lasky et al., 1992; Moore et al., 1994), even though such molecules are the most likely candidates for sialyl Lewis X counter-receptors on carcinoma cells involved in metastasis. Mucins containing similar carbohydrate chains, often described as being carcinoma markers, also seem to function as selectin ligands due to their multivalent configurations (Hanski et al., 1995; Kim et al., 1999). The issue of involvement of sialyl Lewis X-selectin interactions in the establishment of metastasis became further complicated by the discovery of structural variations of extended lactosamine-type selectin ligands which did not contain the sialyl Lewis X tetrasaccharide (Stroud et al., 1996; Handa et al.,

[1]This work was supported by grants-in-aid from the Ministry of Education, Science, Sports and Culture of Japan (07407063, 09254101, 11557180, and 11672162) and the Program for the Promotion of Basic Research Activities for Innovative Biosciences. We thank Ms. Chizu Hiraiwa for her assistance in preparing this manuscript.

[2]Correspondence to: Tatsuro Irimura, Ph.D., Laboratory of Cancer Biology and Molecular Immunology, Graduate School of Pharmaceutical Sciences, The University of Tokyo, Hongo 7-3-1, Bunkyo-ku, Tokyo 113-0033, Japan; Phone: +81-3-5841-4870; Fax: +81-3-5841-4879; E-mail: irimura@mol.f.u-tokyo.ac.jp

[3]The abbreviations used are: BCECF-AM, 3'-O-acetyl-2',7'-bis(carboxy ethyl)-4 or 5-carboxyfluorescein diacetoxymethyl ester; BSA, bovine serum albumin; Bzl-GalNAc, benzyl-α-N-acetyl-D-galactosaminide; Bzl-GlcNAc, benzyl-α-N-acetyl-D-glucosaminide; DME/F12, 1:1 mixture of Dulbecco's minimum essential medium and Ham's F12 medium; DPBS, Dulbecco's phosphate buffered saline; EDTA, ethylenediaminetetraacetic acid; FCS, fetal calf serum; HEPES, 4-(2-hydoxyethyl)-1-piperazineethane-sulfonic acid; mAb, monoclonal antibody.

The Molecular Immunology of Complex Carbohydrates–2.
Edited by Albert M. Wu, Kluwer Academic/Plenum Publishers, 2001.

1997). Similar variations are apparently present in specific monoclonal antibodies such as FH6, originally described as being specific for sialyl dimeric Lewis X (Fukushi et al., 1984). We have previously examined two monoclonal antibodies with similar specificities, mAb FH6 and mAb KM93, for their ability to bind to 16 human colon carcinoma cell lines (Nemoto et al., 1998). The binding profiles of these two monoclonal antibodies did not correspond, three cell lines reactive with mAb FH6 but not with mAb KM93 did not adhere to E-selectin (Nemoto et al., 1998). Thus, we concluded that a subtype of the sialyl Lewis X carbohydrate epitopes recognized by mAb FH6 do not always function as ligands for E-selectin. We have established variant cells from a human colon carcinoma cell line to study the biological function of mAb FH6-binding sites, because this mAb displayed the clinical correlation mentioned above (Matsushita et al., 1991). KM12-HX cells (high expression of mAb FH6 binding sites) and KM12-LX cells (low expression of mAb FH6 binding sites) were selected from the KM12C human colon cancer cell line. KM12-HX and KM12-LX cells exhibited similar growth rates *in vitro*. KM12-HX cells were shown to possess a greater ability to adhere to activated human umbilical vein endothelial cells (HUVEC) than KM12-LX cells. Thus, these cells are a useful *in vitro* model of human colon carcinoma cells expressing varying levels of mAb FH6 and mAb KM93 binding sites. We investigated the involvement of mAb FH6-reactive and mAb KM93-reactive sialyl Le^X in adhesion to liver tissues, using KM12-HX and KM12-LX cells and the Stamper-Woodruff assay (Stamper and Woodruff, 1976). Adhesion was shown to be dependent on endo-β-galactosidase-sensitive, mAb FH6-reactive surface glycans on KM12-HX cells.

MATERIALS AND METHODS

Monoclonal Antibodies

mAb FH6 was provided by Dr. S. Hakomori (Pacific Northwest Research Foundation, Seattle, WA). mAb KM93 (mouse IgM) was from Kyowa Hakko Kogyo (Machida, Japan). mAb CL37, specific for human E-selectin, was a gift from Dr. C. W. Smith (Baylor College of Medicine, Houston, TX).

Cell Lines

Human colon cancer cells showing different levels of mAb FH6 binding were selected from the parental moderately differentiated colon cancer cell line KM12C using a fluorescence-activated cell sorter. After three or four selection cycles, two cell lines stably expressing different levels of mAb FH6 binding sites were isolated and designated KM12-HX (high FH6 binding) and KM12-LX (low FH6 binding) (Matsushita et al., 1991). These KM12 variant cells were morphologically similar to the parental cells, and their growth rates *in vitro* were also very similar. Cell lines were grown in a 1:1 mixture of Dulbecco's minimum essential medium and Ham's F12 medium (DME/F12) containing 10% fetal calf serum (FCS) in a humidified 5% CO_2 atmosphere at 37°C.

Flowcytometric Analysis

Cells were detached from culture plates and washed three times with Dulbecco's phosphate buffered saline (DPBS), then adjusted to a concentration of 1 × 10^6 cells/ml in DPBS containing 0.1% NaN_3 and 0.1% bovine serum albumin (BSA). These cell suspensions were mixed with 111 µl of mAb KM93 (50 µg/ml) or mAb FH6 (hybridoma culture supernatant) and incubated at room temperature for 30 min. The cells were washed three times with DPBS containing 0.1% NaN_3, mixed with 5% biotin-conjugated goat anti-mouse IgM (Bio-Rad, Hercules, CA) diluted in the same solution, and incubated for 30 min. After washing the cells three times, 1.5% FITC-avidin (Vector, Burlingame, CA) in PBS containing 0.1% NaN_3 and 0.1% BSA was added and the mixture was incubated for 15 min. Flowcytometric analysis was performed using Epics XL (Coulter, Miami, FL).

Fluorescence Labeling of the Cells

KM12-HX and KM12-LX cells were rinsed with DPBS three times and treated with a mixture of 0.02% ethylenediaminetetraacetic acid (EDTA) and 0.05% Trypsin. After incubation for 5 min, detached cells were collected and adjusted to a concentration of 1 × 10^6 cells/ml in DME/F12 containing 10% FCS. To the cell suspension was added 3'-O-acetyl-2',7'-bis(carboxy ethyl)-4 or 5-carboxyfluorescein diacetoxymethyl ester (BCECF-AM) (Wako Pure Chemical, Tokyo, Japan) at a final concentration of 3 µM. The cell suspensions were incubated at 37°C for 30 min and then rinsed three times with DPBS to remove excess BCECF-AM. The labeled cells were suspended in DME/F12 containing 1% BSA {adjusted to pH 7.4 with dropwise addition of 1 M 4-(2-hydoxyethyl)-1-piperazine-ethane-sulfonic acid (HEPES)-NaOH} at a concentration of 1 × 10^6 cells/ml.

Adhesion Assays with Frozen Liver Sections

The frozen section adhesion assay was adapted from the procedure of Stamper and Woodruff (1976) . Normal mouse (male, specific pathogen-free BALB/c) liver tissues and human liver specimens obtained at surgery for operation of hepatic metastasis were rapidly frozen in liquid N_2. Cryostat sections (10 µm thick) were mounted on glass slides, and stored at -20°C until use. Cultured colon carcinoma cells were detached from culture dishes by treatment with a mixture of 0.02% EDTA and 0.05% trypsin and suspended in DME/F12 containing 10% fetal calf serum at a concentration of 10^7 cells/ml. BCECF-AM was added to the cell suspension at a final concentration of 3 µM. The cell suspensions were incubated at 37°C for 30 min and then rinsed to remove excess BCECF-AM. Labeled KM12-HX or KM12-LX cells were suspended in assay medium (DME/F12 containing 0.1% BSA, adjusted to pH 7.4 using HEPES) at a concentration of 1 × 10^6 cells/ml and maintained at 4°C. The assay was initiated by layering-100 µl of this BCECF-AM labeled colon carcinoma cell suspension onto the prepared slides. Slides were rotated at 60 rpm at 4°C for 30 min, and then unbound cells were decanted and the slides were rinsed by repeated dipping the slides repeatedly in cold DPBS. Sections were then fixed in 3% glutaraldehyde diluted in DPBS. Cells bound to the liver sections were counted under a fluorescent microscope (× 100, six fields/section). All assays were performed in triplicate and data is shown as mean ± SE.

Pretreatment with Benzyl-α-N-acetyl-D-galactosaminide (Bzl-GalNAc), Endo-β-galactosidase, or Other Agents to Modify Cell Surface Glycans

To test the effect of modification of cell surface carbohydrate chains, cells were treated with endo-β-galactosidase from *Escherichia freundii* (Seikagaku Kogyo, Tokyo, Japan) at a final concentration of 83 mU/ml for 4h at 37°C prior to BCECF-AM labeling. In some experiments, monoclonal anti-mouse E-selectin antibody was preincubated with mouse liver sections or Chinese Hamster Ovary cells transfected with mouse E-selectin cDNA cultured confluently on chamber slides at 50 μg/ml for 30 min at 4°C.

Frozen liver sections were placed on a glass slide, circled with a wax pen (diameter approx. 17 mm) and placed on ice. Two hundred microliters of cell suspension was applied to the sections at 4°C. The slides were fixed on a rotating table at 80 rpm to apply shearing force. After 10, 20, or 30 min, nonadherent cells were removed with absorbent paper, and the glass slides were immersed in 3% glutaraldehyde diluted in DPBS for 25 min, rinsed with water, and air-dried. Adherent cells were counted in 20 randomly selected fields (× 400) in each section under a fluorescence microscope (Nikon Optiphoto, Nikon, Tokyo, Japan). Fields containing portal triads or hepatic veins were excluded. Experiments were carried out in triplicate or quadruplicate.

Pretreatment of KM12-HX Cells or Liver Sections

To test the effect of cell surface carbohydrate-specific antibodies, KM12-HX cells were pretreated with mAb KM93 (100 μg/ml) or FH6 (hybridoma culture supernatant). Cells were incubated with these antibodies at 37°C for 30 min then rinsed twice with DPBS. In sialidase digestion experiments, unlabeled or fluorescence-labeled cells were incubated with sialidase (*Arthrobacter ureafaciens*, Nacalai Tesque, Kyoto, Japan) at a concentration of 100 mU/ml in DME/F12 containing 1% FCS at 37°C for 30 min. The cells were then washed twice with DPBS. Pretreatment with arylglycosides was carried out by incubating the cells in DME/F12 containing 10% FCS and either Bzl-GalNAc or benzyl-α-N-acetyl-D-glucosaminide (Bzl-GlcNAc) at a concentration of 20 mM at 37°C for 72 h. Endo-β-galactosidase treatment of unlabeled or fluorescence-labeled cells was carried out by incubating the cells with 0.05 U of endo-β-galactosidase in DME/F12 containing 0.1% BSA at 37°C for 4 h.

Pretreatment of liver sections with lactose was carried out by incubating frozen human liver sections with 200 μl of lactose solution (100 mM in DPBS) at 4°C for 30 min. The solution was then removed with absorbent paper and the slides were air-dried. Liver sections were pretreated with anti-E-selectin antibody by incubation with 200 μl of purified mAb (50 μg/ml in DPBS) at 4°C for 30 min.

RESULTS

Adhesion of Variant Human Colon Carcinoma Cell Lines to Human Liver Sections

Human colon carcinoma cell lines, selected for differential binding to mAb FH6, were examined for their ability to adhere to sections of human liver placed on glass slides. Incubation was carried out in HEPES-buffered DME/F12 medium containing 1% BSA at 4°C for 30 min under a shearing force provided by rotation at 80 rpm. Figure 1 shows typical fields of human liver section incubated with KM12-HX and KM12-LX cells observed under a fluorescence microscope. Carcinoma cells were easily distinguished from liver and other cells. Even in those cases where autofluoresence of liver tissues was also observed, carcinoma cells were easily identifiable due to their high fluorescence intensity and round shape, and the position of the carcinoma cells above the plane of the sections was detectable under a fluorescence microscope. Although adhesion of carcinoma cells was occasionally observed around portal triads, we focused our attention on liver parenchymal tissue. As shown in Figure 1, more KM12-HX cells than KM12-LX cells adhered to the liver sections. It was not possible to identify which cells in the liver sections bound the colon carcinoma cells.

Adherent cell numbers were determined by counting under a fluorescence microscope. Adhesion was time-dependent, and an increase in the number of adherent cells was observed 60 min after the beginning of the incubation (Figure 2a). KM12-HX cells did not adhere to the liver sections when 50 mM EDTA was added to the incubation medium, suggesting that the process was mediated by a calcium-dependent mechanism (Figure 2b).

Effect of Antibodies Specific for Cell Surface Carbohydrates on Colon Carcinoma Cell Adhesion to Liver Sections

Pretreatment of liver sections with anti-E-selectin mAb CL37 did not influence the adhesion of KM12-HX cells to the sections. Furthermore, immunohistochemical studies indicated that mAb CL37 did not bind to human liver specimens, whereas this antibody has been shown to bind to tumor necrosis factor (TNF)-α-activated HUVEC under similar conditions.

The Effects of Modification of Cell Surface Carbohydrate Chains

When KM12-HX cells were pretreated with sialidase, the cell surface binding of mAb FH6 was significantly reduced (Figure 3a). At the same time, the number of cells adhering to liver sections significantly increased (Figure 4a). This result was not consistent with the previous observation that adhesion of these cells to TNF-α-activated HUVEC was blocked by sialidase-treatment. The adhesion of sialidase-treated cells to liver sections may be mediated by interactions between galactosyl or N-acetylgalactosaminyl groups on carcinoma cell surfaces and galactose/N-acetylgalactosamine-specific C-type lectins present on hepatocytes. This hypothesis is supported by the observation that adhesion was blocked by the addition of 100 mM lactose to the incubation medium (data not shown). Reduction of adhesion by lactose was not observed in untreated control cells.

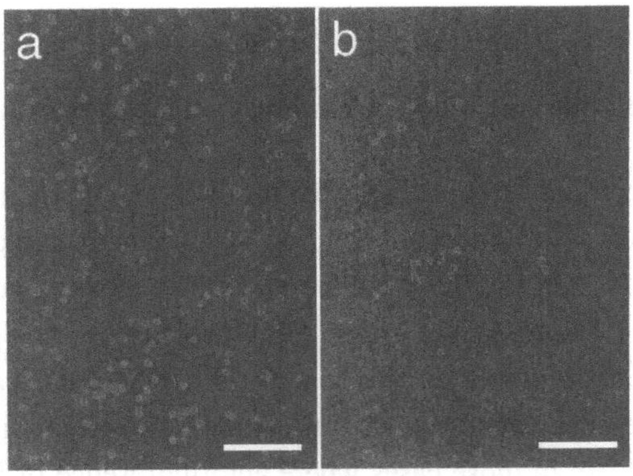

Figure 1. Fluorescence microscopic photographs of liver sections incubated with BCECF-AM-labeled KM12-HX (a) and KM12-LX (b) cells. Bars indicate 100 μm. Frozen liver sections were placed on a glass slide, circled with a wax pen (diameter approx. 17 mm) and placed on ice. Two hundred microliters of cell suspension was applied to the sections and the glass slides were fixed on a rotating table at 80 rpm to apply shearing force. After 30 min at 4°C, nonadherent cells were removed with absorbent paper, and the slides were immersed in 3% glutaraldehyde in DPBS for 25 min, rinsed with water, then air-dried.

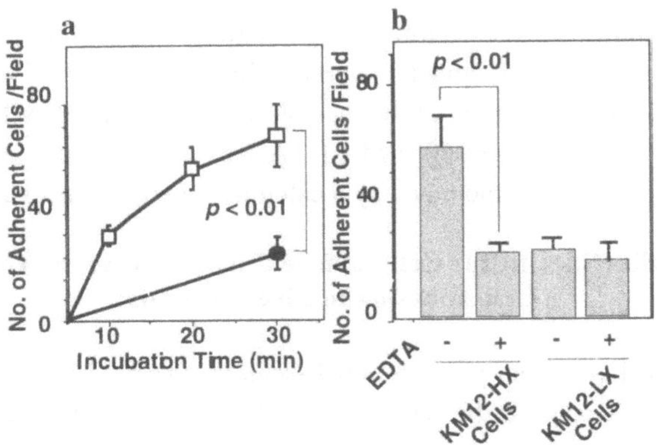

Figure 2. Conditions required for the adhesion of variant human colon carcinoma cells to liver sections. (a) Time course of adhesion of KM12-HX cells to frozen human liver sections (open squares). The number of cells which adhered to the sections increased in a time-dependent manner at least up to 30 min. The number of KM12-LX cells adherent to adjacent sections after 30 min incubation is also shown (closed circles). (b) Effect of EDTA on adhesion. 50 mM EDTA was added to the incubation medium and the KM12-HX and -LX cells adhering to the liver sections were counted 30 min later. Experiments were carried out in triplicate and data is shown as mean ± SD.

A similar phenomenon was observed when KM12-HX cells were pretreated with the arylglycoside Bzl-GalNAc, which is known to interfere with the extension of *O*-linked carbohydrate chains. A marked reduction in cell surface mAb FH6 binding sites (Figure 3b) and a significant increase in the number of cells adhering to liver sections (Figure 4b) was observed in response to incubation with Bzl-GalNAc. This was not consistent with our previous observation that similar treatment of KM12-HX cells abrogated their adhesion to activated HUVEC (Izumi et al., 1995). Incubation with Bzl-GlcNAc did not affect the

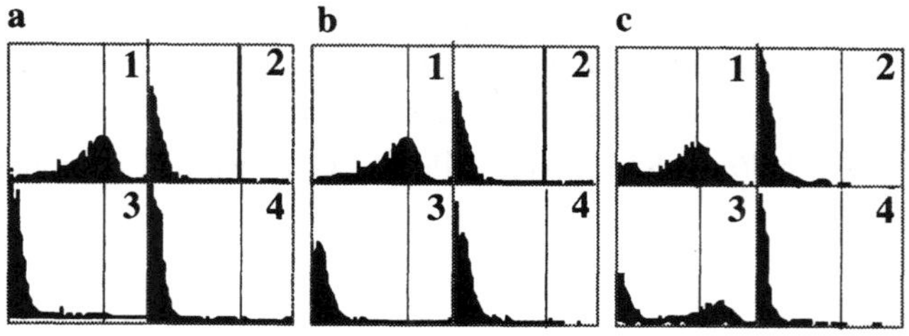

Figure 3. Flow cytometric analysis of cell surface expression of mAb FH6 binding sites on KM12-HX cells following treatment with sialidase, Bzl-GalNAc, or endo-β-galactosidase. (a) cells incubated with 100 mU/ml sialidase (*Arthrobacter ureafaciens*) at 37°C for 30 min; (b) cells incubated in DME/F12 containing 10% FCS with 2 mM Bzl-GalNAc for 72 h; (c) cells digested with 0.05U/ml endo-β-galactosidase at 37°C for 4 h. Untreated or treated cells were incubated with mAb FH6 at room temperature for 30 min then rinsed twice with DPBS. Panel 1, untreated KM12-HX cells. Panel 2, untreated KM12-HX cells processed without first antibody (negative control). Panel 3, KM12-HX cells treated with sialidase, Bzl-GalNAc, or endo-β-galactosidase. Panel 4, KM12-HX cells treated with sialidase, Bzl-GalNAc, or endo-β-galactosidase and processed without first antibody (negative control).

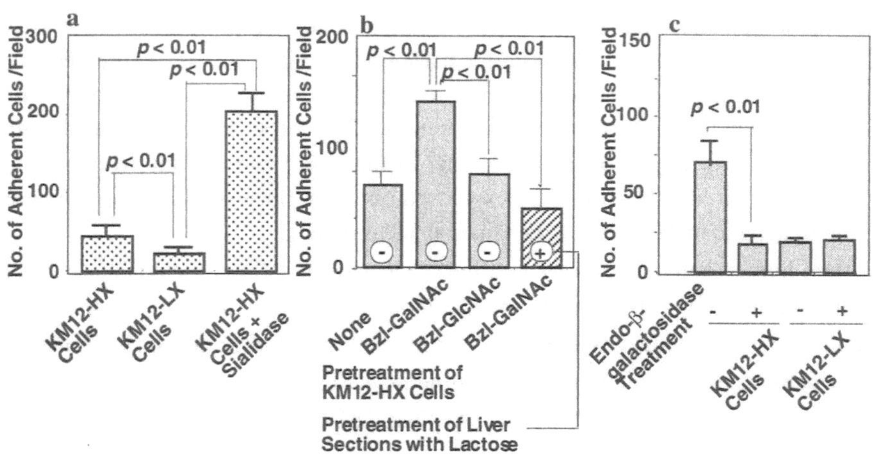

Figure 4. Effects of modification of cell surface carbohydrate chains on the adhesion of human colon carcinoma cells to human liver sections. Experiments were carried out in quadruplicate and data are shown as mean ± SD. (a) Effect of sialidase treatment of KM12-HX cells on their adhesion to human liver sections. (b) Adhesion of KM12-HX cells pretreated with Bzl-GalNAc to human liver sections. KM12-HX cells were incubated with 20 mM Bzl-GalNAc or Bzl-GlcNAc for 72 h and adhesion assays were carried out with or without pretreatment with 100 mM lactose for 30 min. (c) Effect of endo-β-galactosidase treatment of KM12-HX cells on their adhesion to human liver sections.

number of cells adhering to liver sections. The observed increase in adhesion to liver sections was also likely to be due to carcinoma cell adhesion to hepatocytes through C-type hepatic lectins. This hypothesis was supported by the observation that Bzl-GalNAc-treated KM12-HX cell adhesion was completely inhibited by the addition of 100 mM lactose to the incubation medium (Figure 4b).

Endo-β-galactosidase-treatment markedly reduced both mAb FH6 binding to the surface of KM12-HX cells (figure 3c) and KM12-LX cell adhesion to liver sections (Figure 4c). These results suggest that an endo-β-galactosidase-sensitive structure, containing the sites reactive with mAb FH6 on KM12-HX cells, is necessary for the adhesion of these cells to liver sections.

DISCUSSION

Our previous studies revealed that high molecular weight glycoproteins, containing mAb FH6 binding sites and likely to be mucins, were present in colon carcinoma tissues to a greater extent than in normal colonic mucosa. The level of mAb FH6 binding was higher in tumor tissues taken from patients at advanced stages with metastatic involvement than in tissues taken at early stages (Hoff et al., 1989; Matsushita et al., 1991; Nakamori et al., 1993). The post-surgical survival of colon carcinoma patients with low levels of mAb FH6 binding was longer than that of patients with tumors showing high levels of mAb FH6 binding (Nakamori et al., 1993). Similar carbohydrate epitopes are expressed on the surface of neutrophils and monocytes, and function as portions of ligands for E- and P-selectins. Although E- and P-selectins have been found on the surface of activated vascular endothelial cells, expression of these proteins in the vasculature of the liver is limited. Thus, whether mAb FH6 binding sites expressed on the surface of colon carcinoma cells play a role in hepatic metastases has remained unclear to date.

Variant human colon carcinoma cell lines selected for different levels of cell surface mAb FH6 binding offer a useful experimental model with which to study the involvement of this epitope in the pathogenesis of colon carcinomas and its biosynthesis in human colon carcinoma cells. In the original *in vivo* experiments, KM12-HX cells were shown to have a greater potential to develop hepatic metastasis than KM12-LX cells when injected into the spleen of nude mice (Irimura et al., 1993; Izumi et al., 1995). This difference in metastatic properties was not apparent after prolonged passage of the cell lines. Other properties of KM12-HX cells, such as high cell surface expression of mAb FH6 binding sites, did not change with long-term passage. KM12-HX cells have a greater ability than KM12-LX to adhere to activated HUVEC (Irimura et al., 1993; Izumi et al., 1995). In the present study, these cells were used to assess whether cell surface mAb FH6 binding sites were involved in the adhesion of colon carcinoma cells to molecules in the liver; KM12-HX cells were found to have a greater ability to adhere to human liver sections than KM12-LX cells. E-selectin was not likely to be involved in this adhesion, since E-selectin was not detected in the liver sections and pretreatment of liver sections with anti-E-selectin antibody did not interfere with the adhesion. Pretreatment of KM12-HX cells with either sialidase or Bzl-GalNAc resulted in increased cell adhesion to liver sections, although cell surface mAb FH6 binding was significantly reduced. Interestingly, when the sections were pretreated with lactose, adhesion of Bzl-GalNAc-treated KM12-HX cells was abolished, indicating that the adhesion was due to interaction between the galactose/N-acetylgalactosamine-specific hepatic lectin and N-acetylgalactosamine residues which appeared on the cell surfaces as a result of the Bzl-GalNAc treatment. The increased adhesion of sialidase-treated cells can also be explained by a similar mechanism. Lactose did not have an inhibitory effect on the adhesion of KM12-HX cells in the absence of sialidase or Bzl-GalNAc treatment. It therefore appears unlikely that hepatic lectin is involved in the adhesion of untreated KM12-HX cells to liver sections.

Pretreatment of KM12-HX cells with endo-β-galactosidase decreased their adhesion to liver sections, with concomitant loss of cell surface mAb FH6 binding sites. This observation indicates that endo-β-galactosidase-sensitive carbohydrate structures at the periphery of carbohydrate chains play a role in the adhesion between tumor cells and liver sections. It is interesting to note that, whereas levels of cell surface epitopes recognized by mAb FH6 were reduced after endo-β-galactosidase treatment, the level of binding to KM12-HX cells of another anti-sialyl Le^X mAb, KM93, did not change significantly in response to endo-β-galactosidase treatment. This observation seems to be consistent with the known specificity of these mAbs, in that mAb FH6 preferentially reacts with sialyl Le^X on extended backbones that are sensitive to endo-β-galactosidase (Fukushi et al., 1984).

In summary, KM12-HX cells, which express high levels of mAb FH6 binding sites, showed a greater ability to adhere to human liver sections than KM12-LX cells, which express a lower level of FH6 binding sites. Treatment of KM12-HX cells with endo-β-galactosidase decreased both cellular adhesion to liver sections and mAb FH6 binding to the cells.

SUMMARY

Clinicopathological observations have revealed that the expression level of a carbohydrate antigen recognized by the monoclonal antibody (mAb) FH6 in colorectal carcinomas is higher at advanced stages than at early stages. The present study aimed to elucidate whether human colon carcinoma cell surface glycans recognized by mAb FH6 determine the ability of these cells to adhere to sections of human liver. Variant human colon carcinoma cell lines selected for high and low binding of mAb FH6 were compared with respect to their adhesive capacity. The cells expressing the higher level of FH6 binding also showed a greater ability to adhere to liver sections. This adhesion was not blocked by anti-E-selectin monoclonal antibodies, but pretreatment of the carcinoma cells with endo-β-galactosidase significantly reduced both cell surface binding of mAb FH6 and the ability of the cells to adhere to liver sections. Our observations suggest that endo-β-galactosidase-sensitive carbohydrate chains containing an epitope recognized by mAb FH6 play an important role in the adhesion of human colon carcinoma cells to human liver sections. Whether these interactions have any relationship to the mechanism(s) of liver metastasis remains to be elucidated.

REFERENCES

Fukushi, Y., E. Nudelman, S. B. Levery, S. Hakomori and H. Rauvala, 1984, Novel fucolipids accumulating in human adenocarcinoma. III. A hybridoma antibody (FH6) defining a human cancer-associated difucoganglioside (VI3NeuAcV3III3Fuc2nLc6), *J. Biol. Chem.* 259(16):10511.

Handa, K., M. R. Stroud and S. Hakomori, 1997, Sialosyl-fucosyl Poly-LacNAc without the sialosyl-Le^X epitope as the physiological myeloid cell ligand in E-selectin-dependent adhesion: studies under static and dynamic flow conditions, *Biochemistry* 36(41):12412.

Hanski, C., M. L. Hanski, T. Zimmer, D. Ogorek, P. Devine and E. O. Riecken, 1995, Characterization of the major sialyl-Le^X-positive mucins present in colon, colon carcinoma, and sera of patients with colorectal cancer, *Cancer Res.* 55(4):928.

Hoff, S. D., Y. Matsushita, D. M. Ota, K. R. Cleary, T. Yamori, S. Hakomori and T. Irimura, 1989, Increased expression of sialyl-dimeric Le^X antigen in liver metastases of human colorectal carcinoma, *Cancer Res.* 49(24 Pt 1):6883.

Irimura, T., S. Nakamori, Y. Matsushita, Y. Taniuchi, N. Todoroki, T. Tsuji, Y. Izumi, Y. Kawamura, S. D. Hoff, K. R. Cleary and D. M. Ota, 1993, Colorectal cancer metastasis determined by carbohydrate-mediated cell adhesion: role of sialyl-Le^X antigens, *Semin. Cancer Biol.* 4:319.

Izumi, Y., Y. Taniuchi, T. Tsuji, C. W. Smith, S. Nakamori, I. J. Fidler and T. Irimura, 1995, Characterization of human colon carcinoma variant cells selected for sialyl Le^X carbohydrate antigen: liver colonization and adhesion to vascular endothelial cells, *Exp. Cell. Res.* 216(1):215.

Kim, Y. J., L. Borsig, H. L. Han, N. M. Varki and A. Varki, 1999, Distinct selectin ligands on colon carcinoma mucins can mediate pathological interactions among platelets, leukocytes, and endothelium, *Am. J. Pathol.* 155(2):461.

Krause, T. and G. A. Turner, 1999, Are selectins involved in metastasis?, *Clin. Exp. Metastasis* 17(3):183.

Lasky, L. A., M. S. Singer, D. Dowbenko, Y. Imai, W. J. Henzel, C. Grimley, C. Fennie, N. Gillett, S. R. Watson and S. D. Rosen, 1992, An endothelial ligand for L-selectin is a novel mucin-like molecule, *Cell* 69(6):927.

Lowe, J. B., L. M. Stoolman, R. P. Nair, R. D. Larsen, T. L. Berhend and R. M. Marks, 1990, ELAM-1--dependent cell adhesion to vascular endothelium determined by a transfected human fucosyltransferase cDNA, *Cell* 63(3):475.

Matsushita, Y., K. R. Cleary, D. M. Ota, S. D. Hoff and T. Irimura, 1990, Sialyl-dimeric Lewis-X antigen expressed on mucin-like glycoproteins in colorectal cancer metastases, *Lab. Invest.* 63(6):780.

Matsushita, Y., S. Nakamori, E. A. Seftor, M. J. Hendrix and T. Irimura, 1991, Human colon carcinoma cells with increased invasive capacity obtained by selection for sialyl-dimeric Le^X antigen, *Exp. Cell. Res.* 196(1):20.

Moore, K. L., S. F. Eaton, D. E. Lyons, H. S. Lichenstein, R. D. Cummings and R. P. McEver, 1994, The P-selectin glycoprotein ligand from human neutrophils displays sialylated, fucosylated, O-linked poly-N-acetyllactosamine, *J. Biol. Chem.* 269(37):23318.

Nakamori, S., M. Kameyama, S. Imaoka, H. Furukawa, O. Ishikawa, Y. Sasaki, T. Kabuto, T. Iwanaga, Y. Matsushita and T. Irimura, 1993, Increased expression of sialyl Lewis X antigen correlates with poor survival in patients with colorectal carcinoma: clinicopathological and immunohistochemical study, *Cancer Res.* 53(15):3632.

Nemoto, Y., Y. Izumi, K. Tezuka, T. Tamatani and T. Irimura, 1998, Comparison of 16 human colon carcinoma cell lines for their expression of sialyl Le^X antigens and their E-selectin-dependent adhesion, *Clin. Exp. Metastasis* 16(6):569.

Phillips, M. L., E. Nudelman, F. C. Gaeta, M. Perez, A. K. Singhal, S. Hakomori and J. C. Paulson, 1990, ELAM-1 mediates cell adhesion by recognition of a carbohydrate ligand, sialyl-Le^X, *Science* 250(4984):1130.

Stamper, H. B., Jr. and J. J. Woodruff, 1976, Lymphocyte homing into lymph nodes: in vitro demonstration of the selective affinity of recirculating lymphocytes for high-endothelial venules, *J. Exp. Med.* 144(3):828.

Stroud, M. R., K. Handa, M. E. Salyan, K. Ito, S. B. Levery, S. Hakomori, B. B. Reinhold and W. N. Reinhold, 1996, Monosialogangliosides of human myelogenous leukemia HL60 cells and normal human leukocytes. 1. Separation of E-selectin binding from nonbinding gangliosides, and absence of sialosyl-Le(x) having tetraosyl to octaosyl core, *Biochemistry* 35(3):758.

Walz, G., A. Aruffo, W. Kolanus, M. Bevilacqua and B. Seed, 1990, Recognition by ELAM-1 of the sialyl-Le^X determinant on myeloid and tumor cells, *Science* 250(4984):1132.

O-LINKED *N*-ACETYLGLUCOSAMINE AND CANCER: MESSAGES FROM THE GLYCOSYLATION OF C-MYC

Teh-Ying Chou and Gerald W. Hart*

Department of Pathology & Institute of Biochemistry, National Yang-Ming University, Taipei, Taiwan
**Department of Biological Chemistry, The Johns Hopkins University School of Medicine, Baltimore, Maryland, U.S.A.*

Altered glycosylation has been known to associate with tumorigenesis for decades (1), but its role remains enigmatic. Systemic studies on the structures and functions of cell surface carbohydrates began early this century and provided insights into the significance of carbohydrates in cancer. Specific changes in saccharide structures on cell surface glycoconjugates are frequently found in transformed cells and correlate with the potential for invasion and metastasis (2). For example, the metastatic variant of murine adrenal carcinoma cells contains about 6-fold more penultimate Gal-beta1->4GlcNAc sialylation sites than a nonmetastatic variant (3). In a study using lectins as probes, the number of experimental liver metastases and the binding of the *N*-acetyl-D-galactosamine-specific lectin, soybean agglutinin (SBA), exhibited a significant correlation (4).

Since the first oncogenes were cloned in 1980's, it has been established that sequential alterations of oncogenes and tumor suppressor genes are critical in tumor formation and progression (5). The correlation between cell surface carbohydrate alteration and oncogene/tumor suppressor gene activity has been extensively studied. Cells transformed by altered oncogene or tumor suppressor gene expression often also show altered glycosylation of membrane glycoconjugates. For instance, an increase in tri- and tetra-antennary oligosaccharides, together with a decrease in bi-antennary oligosaccharides, was observed on the cell surface of c-myc transfected cells (6).

These observations, together with numerous others (7, 8), successfully demonstrated the intriguing roles of cell surface carbohydrate in the development and behavior of cancer.

Compared to cell surface glycosylation, the studies of intracellular glycoconjugates are just in their rudimentary stage. In recent years, many oncogene and tumor suppressor gene products, such as c-Myc (9, 10) and p53 (11), were shown to be modified by *O*-linked *N*-acetylglucosamine (*O*-GlcNAc) (12), a major form of cytoplasmic and nuclear glycosylation. The dynamic glycosylation of oncoproteins and tumor suppressor proteins by *O*-GlcNAc and the localization of the saccharide at functionally key regions of these molecules suggest that intracellular glycosylation plays a critical role in the etiology of cancer.

I. *O*-linked *N*-acetylglucosamine (*O*-GlcNAc): a dynamic intracellular glycosylation

O-GlcNAc is composed of a single monosaccharide, GlcNAc, linked to the side-chain hydroxyl of serine or threonine. *O*-GlcNAc has been found almost exclusively in the nucleus and cytoplasm of eukaryotic cells (13). The known *O*-GlcNAc-bearing proteins, which include cytoskeletal proteins, transcription factors, oncoproteins, and tumor suppressor proteins, share two common features: all of them are phosphoproteins and all form reversible multimeric complexes. Although experiments to date suggest that *O*-GlcNAc modification is dynamic and appears to have a reciprocal relationship with protein phosphorylation, the definitive roles of *O*-GlcNAc are still unknown (14).

II. *O*-GlcNAc on oncoproteins and tumor suppressor proteins: c-Myc glycosylation

Early attempts to detect carbohydrate on oncoproteins were pursued on SV40 large T antigen (15, 16) and v-erb-A protein (17) in 1980's. Using lectin affinity chromatography, Privalsky showed an interaction between v-erb-A protein and wheat germ agglutinin and provided the first indirect evidence of *O*-GlcNAc on an oncoprotein.

The levels of oncoproteins and tumor suppressor proteins in either normal or transformed cells are very low. For example, there are approximately 750 molecules of c-Myc per cell in serum-starved fibroblasts and, after serum stimulation, the number increases to 6,300 per cell. HeLa cells, which are transformed cells, contain about 97,000 molecules of c-Myc per cell. The low abundance of proteins render the direct biochemical analyses of the *O*-GlcNAc modification on oncogene or tumor suppressor gene protein products extremely difficult.

The c-myc proto-oncogene encodes a helix-loop-helix-leucine zipper protein (c-Myc) that heterodimerizes with Max and regulates gene transcription during cell proliferation, cell differentiation, and programmed cell death (18). Mutations of c-myc are associated with different types of tumors in human and other species. Two regions of the c-Myc protein are required for its biological activities. A basic/helix-loop-helix/leucine zipper (bHLHZip) motif in the C-terminal sequence mediates heterodimerization with Max and DNA binding to a specific E-box sequence, CAC(A/G)TG or EMS (E-box myc site). The *N*-terminal transcription activation domain (TAD) (amino acids 1-143) is required for neoplastic transformation, inhibition of cellular differentiation, and induction of apoptosis mediated by c-Myc. In addition to Max, an increasing number of proteins have been suggested to interact with c-Myc, such as the retinoblastoma susceptibility protein Rb, the Rb-like p107 protein, the transcriptional regulators Yin-Yang-1 and AP-2, the TATA-binding protein, TFII-I, and alpha-tubulin (19).

In 1995, using proteins over-expressed in Chinese hamster ovary cells and sf9 insect cells, c-Myc was found to be modified by *O*-GlcNAc (9) and its major glycosylation site is at threonine 58 (Thr-58) in the transcription activation domain (10). Thr-58 of c-Myc, the first known *O*-GlcNAc site of oncoproteins, is a phosphorylation site in vivo (20) and can be phosphorylated in vitro by glycogen synthase kinase 3 (21). The mutations of Thr-58 in v-Myc enhance the transforming activity of c-Myc protein (22), suggesting that Thr-58 has a role in transducing a negative growth signal of c-Myc through its post-translational modifications. This working hypothesis is supported by the observations that mutations of c-Myc at or near Thr-58 are frequently found in Burkitt or AIDS-related lymphomas and Thr-58 is the most frequently mutated amino acid of c-Myc in these tumors (23).

The discovery of *O*-GlcNAc glycosylation at a known phosphorylation site of c-Myc added more complexity to deciphering the regulation of oncoprotein functions by post-translational modification. Based on our current knowledge, it appears that *O*-glycosylation and *O*-phosphorylation are mutually exclusive at the same amino acid; that is, the modification target site can only be either unmodified, glycosylated, or phosphorylated. Therefore, numerous previous experiments in which "the phosphorylation site was knocked-out" were probably not only observing the effect of loss of phosphorylation, but were also observing the effect of loss of glycosylation. On the other hand, the ability of the cell to combine all possible glycosylation/phosphorylation permutations in nearby or distant amino acids gives enormously diversified means of fine-tuning the regulation of the protein. Interestingly, Medina et al recently confirmed the presence of *O*-GlcNAc as the sole glycosylation form of SV40 large T antigen and identified a site of glycosylation also to a known phosphorylation site in vivo (24), a status similar to c-Myc.

The regulation of protein phosphorylation can be mediated by glycosylation and vice versa, through either direct competition or indirect interference. Taking c-Myc as an example, Hayes and Hart analyzed the possible glycosylation/phosphorylation interaction between Thr-58 and Ser-62 of c-Myc (25). It is known that phosphorylation of Thr-58 and/or Ser-62 regulates the transactivation activity of c-Myc. Glycosylation at Thr-58 will either suppress the basal activity or block the transcriptional activation resulting from phosphorylation at Thr-58. It is also known that phosphorylation of Thr-58 is facilitated by phosphorylation of Ser-62. Therefore, glycosylation of Thr-58 may induce the phosphorylation of adjacent Ser-62 and, once the phosphorylation of Ser-62 occurs, the phosphorylated Ser-62 may trigger the removal of *O*-GlcNAc by *O*-GlcNAcase and subsequent addition of phosphate at Thr-58. These complicated mechanisms regulating c-Myc tumorigenesis are currently being dissected by experiments using synthetic glyco-, phospho-, and glycophospho-peptides and purified kinases, phosphatases, *O*-GlcNAc transferase, and *O*-GlcNAcase.

The tumor suppressor gene p53 is the most frequently mutated tumor-associated gene in human cancer. In 1996, p53 was reported to have *O*-GlcNAc modification (11). It was shown that *O*-GlcNAcylation increases DNA binding and transcriptional activity of p53 by masking a basic region in the carboxyl terminus that represses DNA-binding. The demonstration of *O*-GlcNAc modification in tumor suppressor proteins, in addition to oncoproteins, extends the scope of cancer glycobiology and emphasizes the critical roles of glycosylation in cancer.

III. *O*-GlcNAc regulation: enzymatic addition and removal

The enzyme which catalyzes *O*-GlcNAc addition, uridine diphospho-*N*-acetylglucosamine: polypeptide beta-*N*-acetylglucosaminyl transferase (*O*-GlcNAc transferase, OGT), has been cloned and characterized from rat (26) and C. elegans and human (27). The OGT enzymes from different species are highly conserved and contain multiple tetratricopeptide repeats (TPR) and nuclear localization sequences. The rat and human pancreas have been shown to have the highest expression level of OGT by Northern analysis. By using in situ labeling techniques, OGT was shown to have abundant expression in pancreatic islet cells (28, 29) and islet cell tumors (Chou *et al*, manuscript in preparation). Kreppel and Hart further demonstrated that the TPR, which can mediate protein-protein interactions, may interact with other protein molecules to form homo- or hetero-trimers and alter the binding affinities of OGT for UDP-GlcNAc (30). Recently, a functional analysis was performed to evaluate the domain structure and substrate specificity of human OGT expressed in *E. Coli* (31).

O-GlcNAcase, the enzyme responsible for removing *O*-GlcNAc from proteins, has also been purified and characterized from rat spleen cytosol (32). Specific inhibitors of *O*-GlcNAcase activity have been extensively used to study the dynamics of *O*-GlcNAc in cells. It will be important to examine the expression of OGT and *O*-GlcNAcase in cancer cells since the levels of *O*-GlcNAc may regulate functions of the intracellular proteins and play a key role in tumorigenesis.

SUMMARY

Altered protein glycosylation is known to correlate with tumorigenesis, but its role remains enigmatic. Cells transformed by altered oncogene or tumor suppressor gene expression often also show changes of carbohydrate on cell surface glycoconjugates which correlate with the potential for tumor invasion and metastasis. In recent years, many oncogene and tumor suppressor gene products, such as c-Myc, SV40 large T antigen, and p53, were shown to be modified by *O*-GlcNAc. *O*-GlcNAc is a form of protein glycosylation found almost exclusively in the nucleus and cytoplasm of eukaryotic cells. The known *O*-GlcNAc-bearing proteins are phosphoproteins and form reversible multimeric complexes. *O*-GlcNAc modification is dynamic and appears to have a reciprocal relationship with protein phosphorylation.

The enzymes which catalyze *O*-GlcNAc addition and removal have been characterized and used as effective tools in *O*-GlcNAc studies. It is of great interest in the future to investigate the alteration of *O*-GlcNAc in different cancers since addition/removal of *O*-GlcNAc on oncoproteins, tumor suppressor proteins, and other tumor-related proteins very likely plays a key role in the pathogenesis of tumors.

REFERENCES

1. Axford, J. Glycobiology and medicine: an introduction, Journal of the Royal Society of Medicine. 90: 260-264, 1997.
2. Hebert, E. and Monsigny, M. Oncogenes and expression of endogenous lectins and glycoconjugates, Biol. Cell. 79: 97-109, 1993.
3. Jiang, M. S., Passaniti, A., Penno, M. B., and Hart, G. W. Adrenal carcinoma tumor progression and penultimate cell surface oligosaccharides, Cancer Research. 52: 2222-2227, 1992.
4. Reese, M. R. and Chow, D. A. Tumor progression in vivo: Increased soybean agglutinin lectin binding, *N*-acetylgalactosamine-specific lectin expression, and liver metastasis potential, Cancer Research. 52: 5235-5243, 1992.
5. Bishop, J. M. Molecular Themes In Oncogenesis, Cell. 64: 235-248, 1991.
6. Hiraizumi, S., Takakasaki, S., Shiroki, K., and Kobata, A. Altered protein glycosylation of rat 3Y1 cells induced by activated c-myc gene, International Jounal of Cancer. 48: 305-310, 1991.
7. Dennis, J., Granovsky, M., and Warren, C. Glycoprotein glycosylation and cancer progression, Biochimica et Biophysica Acta - General Subjects. 1473: 21-34, 1999.
8. Muramatsu, T. Essential roles of carbohydrate signals in development, immune response and tissue functions, as revealed by gene targeting, Journal of Biochemistry. 127: 171-176, 2000.
9. Chou, T.-Y., Dang, C. V., and Hart, G. W. Glycosylation of the c-Myc transactivation domain, Proc. Natl. Acad. Sci. USA. 92: 4417-4421, 1995.
10. Chou, T. Y., Hart, G. W., and Dang, C. V. c-Myc Is Glycosylated at Threonine 58, a Known Phosphorylation Site and a Mutational Hot Spot in Lymphomas, J. Biol. Chem. 270: 18961-18965, 1995.
11. Shaw, P., Freeman, J., Bovey, R., and Iggo, R. Regulation of specific DNA binding by p53: evidence for a role for *O*-glycosylation and charged residues at the carboxy-terminus, Oncogene. 12: 921-930, 1996.
12. Hart, G. W., Kreppel, L. K., Comer, F. I., Arnold, C. S., Snow, D. M., Ye, Z., Cheng, X., DellaManna, D., Caine, D. S., Earles, B. J., Akimoto, Y., Cole, R. N., and Hayes, B. K. *O*-GlcNAcylation of key nuclear and cytoskeletal proteins: reciprocity with *O*-phosphorylation and putative roles in protein multimerization, Glycobiology. 6: 711-716, 1996.
13. Hart, G. W. Dynamic *O*-linked glycosylation of nuclear and cytoskeletal proteins, Annual Review of Biochemistry. 66: 315-335, 1997.
14. Snow, D. and Hart, G. Nuclear and cytoplasmic glycosylation, International Review of Cytology. 181: 43-74, 1998.
15. Schmidt-Ullrich, R., Thompson, W., and Wallach, D. Biochemical and immunochemical characterization of two simian virus 40 (SV40)-specific glycoproteins in nuclear and surface membranes of SV40-transformed cells, Biochem. Biophys. Res. Commun. 88: 887-894, 1979.
16. Schmidt-Ullrich, R., Thompson, W., Kahn, S., Monroe, M., and Wallach, D. Simian virus 40 (SV40)-specific isoelectric point-4.7-94,000-Mr membrane glycoprotein: membrane-associated 94,000-Mr SV40 T-antigen in hamsters, J. Natl. Cancer Inst. 69: 839-849, 1982.
17. Privalsky, M. A subpopulation of the avian erythroblastosis virus v-erbA protein, a member of the nuclear hormone receptor family, is glycosylated, Journal of Virology. 64: 463-466, 1990.
18. Ryan, K. M. and Birnie, G. D. Myc oncogenes: the enigmatic family, Biochemical Journal. 314: 713-721, 1996.

19. Henriksson, M. and Luscher, B. Proteins of the Myc network: essential regulators of cell growth and differentiation, Advances in Cancer Research. 68: 109-182, 1996.
20. Lutterbach, B. and Hann, S. R. Hierarchical phosphorylation at *N*-terminal transformation-sensitive sites in c-Myc protein is regulated by mitogens and in mitosis, Mol. Cell. Biol. 14: 5510-5522, 1994.
21. Henriksson, M., Bakardjiev, A., Klein, G., and Luscher, B. Phosphorylation sites mapping in the *N*-terminal domain of c-Myc modulate its transforming potential, Oncogene. 8: 3199-3209, 1993.
22. Frykberg, L., Graf, T., and Vennstrom, B. The transforming activity of the chicken c-myc gene can be potentiated by mutations, Oncogene. 1: 415-421, 1987.
23. Raffeld, M., Yano, T., Hoang, A. T., Lewis, B., Clark, H. M., Otsuki, T., and Dang, C. V. Clustered Mutations in the Transcriptional Activation Domain of Myc in 8q24 Translocated lymphomas and Their Functional Consequences, Curr. Top. Microbiol. Immunol. 194: 265-272, 1995.
24. Medina, L., Grove, K., and Haltiwanger, R. SV40 large T antigen is modified with *O*-linked *N*-acetylglucosamine but not with other forms of glycosylation, Glycobiology. 8: 383-391, 1998.
25. Hayes, B. and Hart, G. Protein *O*-GlcNAcylation: potential mechanisms for the regulation of protein function. In: Axford (ed.) Glycoimmunology 2, pp. 85-94. New York: Plenum Press, 1998.
26. Kreppel, L. K., Blomberg, M. A., and Hert, G. W. Dynamic glycosylation of nuclear and cytosolic proteins. Cloning and characterization of a unique *O*-GlcNAc transferase with multiple tetratricopepdite repeats, The Journal of Biological Chemistry. 272: 9308-9315, 1997.
27. Lubas, W. A., Frank, D. W., Krause, M., and Hanover, J. A. *O*-linked GlcNAc transferase is a conserved nucleocytoplasmic protein containing tetratricopeptide repeats, The Journal of Biologiacl Chemistry. 272: 9316-9324, 1997.
28. Hanover, J., Lai, Z., Lee, G., Lubas, W., and Sato, S. Elevated *O*-linked *N*-acetylglucosamine metabolism in pancreatic beta-cells, Archives of Biochemistry and Biophysics. 362: 38-45, 1998.
29. Akimoto, Y., Kreppel, L., Hirano, H., and Hart, G. Localization of the *O*-linked *N*-acetylglucosamine transferase in rat pancreas, Diabetes. 48: 2407-2413, 1999.
30. Kreppel, L. and Hart, G. Regulation of a cytosolic and nuclear *O*-GlcNAc transferase: role of the tetratricopeptide repeats, The Journal of Biological Chemistry. 274: 32015-32022, 1999.
31. Lubas, W. and Hanover, J. Functional expression of *O*-linked GlcNAc transferase - Domain structure and substrate specificity, The Journal of Biological Chemistry. 275: 10983-10988, 2000.
32. Dong, D. L.-Y. and Hart, G. W. Purification and characterization of an *O*-GlcNAc selective *N*-Acetyl-B-D-glucosaminidase from rat spleen cytosol, The Journal of Biological Chemistry. 269: 19321-19330, 1994.

ANALYSIS OF A MURINE ANTI-GANGLIOSIDE GD2 MONOCLONAL ANTIBODY EXPRESSING BOTH IgG2a AND IgG3 ISOTYPES: MONOCLONALITY, APOPTOSIS TRIGGERING, AND ACTIVATION OF CELLULAR CYTOTOXICITY ON HUMAN MELANOMA CELLS

Chien-Cheng Lin,[1,2] Yung-Chi Shen,[1] Cheng-Keng Chuang,[1] and Shuen-Kuei Liao[1,2,*]

[1]*Graduate Institute of Clinical Medicine and* [2]*Graduate Institute of Basic Medical Sciences, College of Medicine, Chang Gung University, Taoyuan 333, Taiwan*

Gangliosides have been shown to be expressed on cell membranes by a variety of human tissues (1,2). They usually undergo changes in composition during differentiation and malignant transformation (3,4). Except in the brain tissue, certain gangliosides are expressed only trace amounts on normal cells. On the other hand, they are expressed strongly on most neuroectodermally derived tumors (3,5-9). It has been proposed that accumulation of specific ganglioside species may be characteristic of certain tumors (3), such as melanomas, neuroblastomas, brain tumors, small cell lung carcinomas and some sarcomas expressed high amounts of the disialoganglioside GD2 (10-14). Using monoclonal antibodies (MAbs) against GD2 (6,7,11,15) as immunologic probes, it has been shown that this glycolipid molecule could be served as an effective target for both complement-dependent cytolysis (CDC) and antibody-dependent cell-mediated cytotoxicity (ADCC) (16-19). Some of these MAbs and mouse-human chimeric antibodies derived therefrom have been used in clinical trials for patients with neuroblastoma, melanoma and osteosarcoma, with encouraging results obtained (20-22).

By the hybridoma approach, we generated two MAbs, namely Mel-11 and Mel-22, against human melanoma cells, specificities of which were shown to be gangliosides, GD2 and GD3, respectively (Fig. 1). Both MAbs belonged to the IgG3 subclass as determined initially by an enzyme-linked immunosorbent assay (ELISA) using rabbit anti-murine isotyping sera. However, we noted that the isotype of Mel-11 changed from IgG3 to IgG3 and IgG2a after the hybridoma had been subcultured in vitro for over 57 passages, whereas

---

*To whom correspondance should be addressed. FAX: 886-3-328-0170; E-mail: liaosk@mail.cgu.edu.tw
The authors wish to thank Miss Yu-Ping Perng for technical assistance, Dr. Robert K. Yu (Virginia Commonwealth University, Richmond, VA) for the confirmation of target structures recognized by the MAbs Mel-11 and Mel-22, Dr. Alice L. Yu (University of California at San Diego, San Diego, CA) for providing MAb 14.G2a to GD2 for a comparative study, and Dr William E. Grizzle (Southern Division of Cooperative Human Tissue Network, Birmingham, LA) for providing human tissues for this study. This work was supported in part by the National Science Council of the Republic of China (NSC-87-2314-B182-047) and the Chang Gung Medical Research Fund (CMRP-363).

the light chain and the target structure recognized remained unchanged. In this communication we focused on Mel-11 and provided evidence to show a single species of IgG secreted from the newly cloned hybridomas, which again showed the two immunoglobulin (Ig) subclasses with a λ light chain. In other words, our data showed that the dual isotypic specificity was not due to the presence of two or more different hybridoma populations, or to a contamination with other hybridoma(s). In addition, we demonstrated that this antibody could induce GD2-positive melanoma cells to undergo apoptosis and activate ADCC.

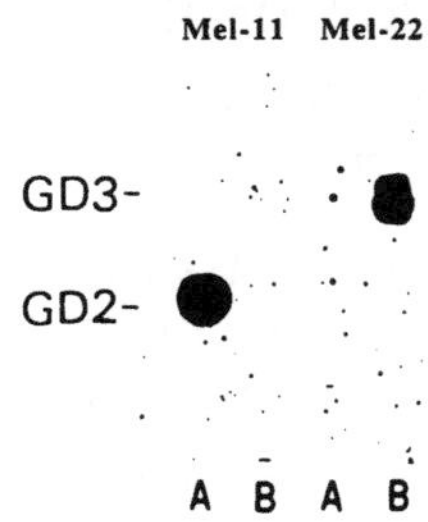

Figure 1. Target structures analyzed by TLC and immunostaining. Lane A and B contained brain gangliosides GD2 (2 μg) and GD3 (2 μg), respectively, which were immunostained with two different MAbs Mel-11 and Mel-22 indicated on the top of the figure. After washing with PBS, the plates were incubated with peroxidase-conjugated anti-mouse IgG antibodies. The plates were then incubated with cyclic hydrazine solution for 1 min and exposed to an X-ray film.

RECLONING OF THE MEL-11 HYBRIDOMA

In order to ascertain the monoclonality of the Mel-11 hybridoma and its secreted antibody, we recloned this hybridoma by limiting dilution. Out of 960 microtiter-wells in 10 microtiter-plates, 30 were shown to have good growth and to be derived from single cells as monitored daily under an inverted microscope. These clones were successfully expanded to flask cultures, and their supernatants were repeatedly tested with a panel of rabbit isotyping antisera to murine IgG1, IgG2a, IgG2b, IgG3, IgM, IgA, light chains κ and λ (Pierce, Rockford, IL) in an optimized ELISA; 27/30 (90%) clones secreted mouse IgG, all belonging to IgG2a and IgG3 with a λ-light chain specificity, and the remaining three clones had no heavy or light chains of murine Igs in their spent media detected. Examples of such reactivity profiles of positive clones are depicted in Figure 2. Both IgG2a and IgG3 at the comparable levels were detected in the supernatants of two representative clones, 9C4 and 5F5, as well as in the supernatant of the late passages of parental hybridoma Mel-11. Note that the positive controls of 140.240 and BH8.23 respectively representing IgG2a and IgG3 MAbs (23) were clearly demonstrated, with a background optical density (O.D.) value equal to or less than 0.1, close to that of the negative control, phosphate-buffered saline (PBS). Therefore, it appeared that both clones 9C4 and 5F5, like their parental Mel-11 hybridoma, were all monoclonal, so were antibodies secreted therefrom.

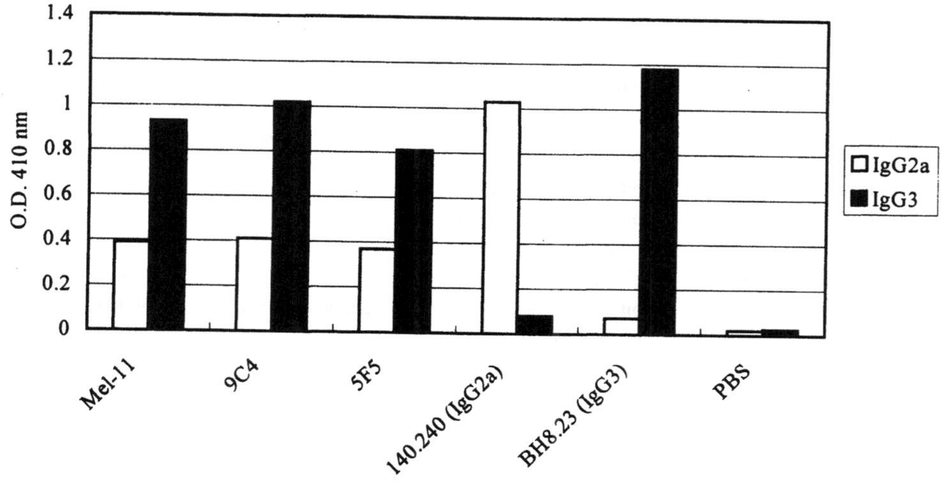

Figure 2. IgG subclass determination of MAb Mel-11 in ELISA using rabbit typing anti- murine Ig antisera

SEQUENTIAL IMMUNOPRECIPITATION STUDIES

To confirm that MAb 9C4 was a single immunoglobulin molecule, we conducted sequential immunoprecipitation experiments (24), in which 9C4 hybridoma cells were first metabolically labeled with ^{35}S-methionine, and the radiolabeled IgG secreted by these hybridoma cells were collected from the supernatants and immunoprecipitated with an excess amount of rabbit anti-mouse IgG3 antiserum absorbed on protein A-sepharose beads (Table 1). The immunodepleted material was then subjected to immunoprecipitation with rabbit anti-mouse IgG2a, and this second immunoprecipitated material was then analyzed on SDS-polyacryamide gel electrophoresis (PAGE) under reducing and non-reducing conditions. There were no specific bands detected on SDS-PAGE. It was also the case in a reciprocal manner, i.e., no bands could be visualized when rabbit anti-IgG3 was used to immunoprecipitate the supernatant material which had been immunodepleted before hand with an excess amount of rabbit anti-IgG2a antiserum absorbed on protein A-Sepharose beads. As controls, irrelevant rabbit antibodies or normal rabbit serum were used as immunodepletion antibodies or no immunodepletion step was performed by omitting the first immunoprecipitation step, two clear bands of heavy (~53 kDa) and light (~22.5 kDa) chains of radiolabeled IgG could clearly be visualized on SDS-PAGE under reducing conditions. When the very same material was analyzed on SDS-PAGE performed under non-reducing conditions, a dense band corresponding to an intact IgG molecular mass of approximately 150 kDa was detected. Similar results were also obtained with the hybridoma/MAb 5F5 when sequential immunoprecipitation experiments were conducted. Taken together, these results confirmed that the 9C4 and 5F5, each secreted a monoclonal antibody having both IgG3 and IgG2a isotypes.

Table 1. Sequential immunoprecipitation experiments

Immunodepletion with	Immunoprecipitation with	Visualization with SDS-PAGE	
		Reducing	Non-reducing
Rabbit anti-mouse IgG3	Rabbit anti-mouse IgG3	-	-
Rabbit anti-mouse IgG3	Rabbit anti-mouse IgG2a	-	-
Rabbit anti-mouse IgG2a	Rabbit anti-mouse IgG2a	-	-
Rabbit anti-mouse IgG2a	Rabbit anti-mouse IgG3	-	-
-	Rabbit anti-mouse IgG	53 kDa + 22.5 kDa	150 kDa
-	Rabbit anti-mouse IgG3	53 kDa + 22.5 kDa	150 kDa
-	Rabbit anti-mouse IgG2a	53 kDa + 22.5 kDa	150 kDa

CELL AND TISSUE BINDING STUDIES

Binding with the two MAbs, Mel-11 and 9C4, of seven human tumor lines of neuroectodermal origin was next tested by immunofluorescence/flow cytometric analysis with 5 μg/ml antibody tested. Table 2 shows the percentage of positively stained the surface tumor cells and mean fluorescence intensity in a defined population. Similar reactivity spectra were obtained with the two antibodies; 4/5 melanoma lines and 1/2 neuroblastoma lines were positive. The flow cytometric histograms of 9C4 and Mel-11 reactivities against any given target lines were very similar (results not shown). The binding of MAb Mel-11 and 9C4 to tissues representing normal skin and different stages of melanoma progression was also determined using ABC staining procedures on frozen tissues. In general, diffuse to focal staining patterns were detected in positive cells with both antibodies in 50% of nevi (specimen number: 4), 60% of primary tumors (5) and in about 86% of metastatic tumors (21), while melanocytes and keratinocytes of normal skins (2) were nonreactive with either MAb.

Table 2. Immunoreactivity pattern of MAbs, Mel-11 and 9C4, to human neuroectodermally derived tumor cell lines as determined by indirect immunofluorescence/flow cytometry

Cell line	Mean % positive cells (MCF)[a]	
	MAb Mel-11	MAb 9C4
CaCL 73-36 (melanoma)	97 (132)	100 (143)
CaCL 78-4 (melanoma)	80 (135)	84 (139)
BUR (melanoma)	3	5
SK-MEL-23 (melanoma)	96 (128)	98 (141)
Mel-21 (melanoma)	95 (112)	94 (108)
Nero5 (neuroblastoma)	73 (79)	78 (86)
SH-SY5Y (neuroblastoma)	8	7

[a]Three determinations were made for each test using 5 μg/ml MAb; MCF stands for mean channel fluorescence, representing mean fluorescence intensity. Results with < 10% positive cells were considered to be negative.

To determine the titration profile of MAb 9C4 against melanoma cells, different concentrations of purified MAb was tested against CaCL 73-36 melanoma cells in immunofluorescence/flow cytometry (Fig. 3). This titration curve resembled that of MAb Mel-11 (results not shown). In both instances, the use of rabbit anti-mouse IgG3 followed by fluorescein isothiocyanate (FITC)-conjugated goat anti-rabbit IgG antibodies as a detection system gave a comparably strong binding profile as the reactivity of primary mouse MAb detected by FITC-conjugated goat anti-mouse IgG antibodies, while the use of rabbit anti-mouse IgG2a followed by goat anti-rabbit antibodies as a detection system gave much less binding reactivity.

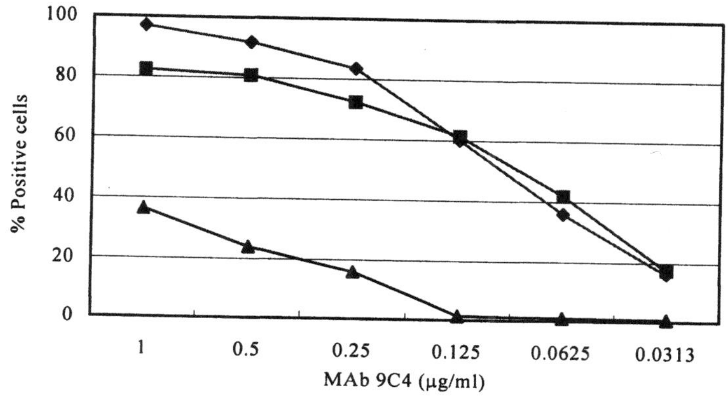

Figure 3. Binding pattern against the human melanoma CaCL 73-36 cells of MAb 9C4 secreted by one of the recloned sublines from the Mel-11 hybridoma as determined by immunofluorescence/flow cytometric analysis. Note that the tracing polyclonal antibodies used were different: ----◆----, FITC-conjugated goat anti-mouse IgG; ----■----, rabbit anti-mouse IgG3 followed by FITC-conjugated goat anti-rabbit antibodies; and ----▲----, rabbit anti-mouse IgG2a followed by FITC-conjugated goat anti-rabbit IgG antibodies.

LARGE SCALE PRODUCTION OF ANTIBODY

The Opticell Model 5200 automated bioreactor (Charles River Laboratories, Wilmington, MA) based on the ceramic core- and closed loop continuous perfusion was used to scale up the production of antibody. Note that the hybridoma 9C4 cells yielded about ~2 μg/ml in the flask culture. The hybridoma cells (1 x 10^8 per chamber) were inoculated into the bioreactor. Cell growth was monitored by oxygen consumption rate (OCR) as shown in Figure 4. Once and after the OCR reached its plateau, circulating spent medium was collected and fresh medium containing reduced fetal bovine serum content (gradually from 5, 2.5, 2, 1 and 0.5%) was re-fed into the system. Spent medium was collected and fresh medium introduced into the system every 24-40 h in a volume from 500 to 1800 ml to replace with the collected spent medium volume each time. Results of a typical run of one chamber are shown in Table 3, in which a total yield of ~0.5 g of Ig (estimated in ELISA) was harvested in about 17 days. Since two chambers in the same bioreactor unit could be operated simultaneously, the yield could be scaled up in gram quantities per run. Antibody was purified by protein A-Sepharose affinity chromatography to a high degree of homogeneity by showing a single sharp peak, as assessed by high-performance liquid chromatography. These results again point to the monoclonality of the antibody.

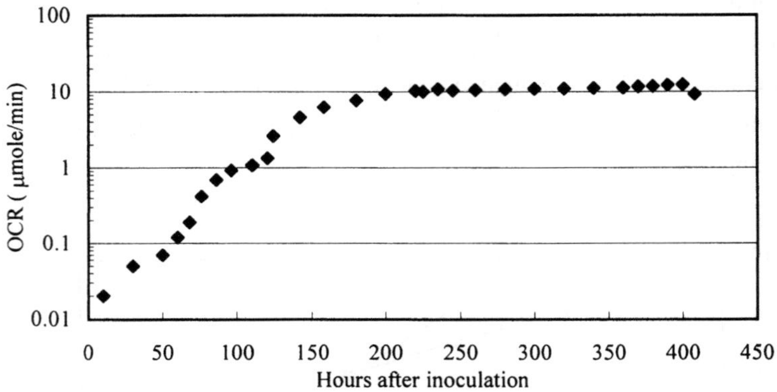

Figure 4. The oxygen consumption rate for hybridoma 9C4 cells mass produced in the bioreactor Opticell 5200. OCR stands for oxygen consumption rate.

Table 3. Mass production of immunoglobulin by culturing 9C4 hybridoma cells in a bioreactor Opticell 5200

Sample No.	Time (h)	Ig concentration (μg/ml)	Volume (ml)	Ig amount (mg)
1	70	11.6	500	5.8
2	113	24.8	500	12.4
3	167	52.3	1200	62.8
4	193	64.4	1000	64.6
5	219	59.6	600	35.8
6	256	63.5	1200	76.2
7	280	35.8	1200	43.0
8	328	40.4	1700	68.7
9	352	57.0	1000	57.0
10	399	51.3	600	30.8
11	408	37.0	1800	66.6
Total	408 h (accumulated) (17 days)		11300 (11.3 liters)	527.3 (~0.53 g)

MAb 9C4 INDUCED-CELLULAR DIFFERENTIATION AND APOPTOSIS

To determine whether MAb 9C4 could induce cellular differentiation and/or apoptosis, we incubated human melanoma CaCL 73-36 cells in the absence and presence of 5 μg/ml MAb 9C4 in vitro for 72 h. At the end of incubation period, cells were examined under an inverted microscope, about one third of the cell number was found in the treated group attached on the bottom of flask with altered cellular morphology, i.e. many elongated dendritic cells (Fig. 5A), which was consistent with a highly differentiated phenotype, as compared with the untreated cells mostly exhibiting triangular dendritic morphology with larger cellular bodies (Fig. 5B). When the treated cells were collected by scrapping, pooled with cells in the spent medium, pelleted by centrifugation and then subjected to flow cytometric DNA analysis, 10-15% of the cells were found in the pre-G1 phase, signifying that these cells underwent apoptosis. By contrast, only 2-4% of untreated melanoma cells showed the pre-G1 peak. In fact, in subsequent experiments we found that apoptotic cells were mostly in the non-adherent fraction, which could be recognized being undergoing apoptosis at the ultrastructural level by nuclear condensation and vacuolar formation in the cytoplasm (Fig. 6B,C). Of note is that bleb formation was seen with cells of in an earlier stage of MAb treatment (results not shown). On the other hand, the untreated cells rarely exhibited such appearances (Fig. 6A).

Biochemical experiments also supported the apoptotic event occurring in a fraction of MAb 9C4-treated melanoma cells (data not shown). Antitumor effects of another murine MAb (NCC-ST421) recognizing a dimeric Le^a antigen on human gastric cancer

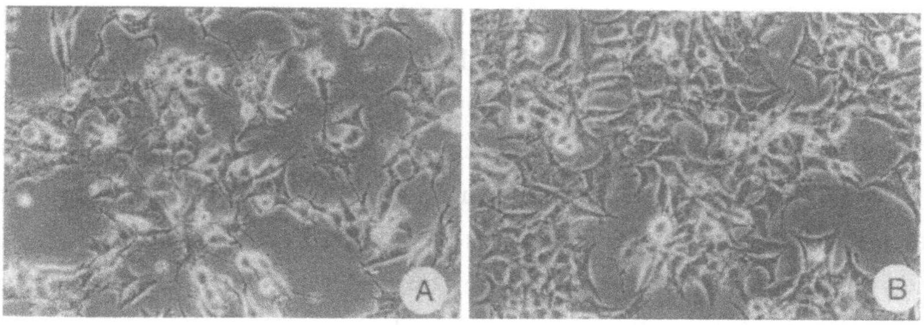

Figure 5. Morphological changes of CaCL 73-36 melanoma cells as a result of exposure to MAb 9C4 for 72 h in vitro. Note that MAb-treated group (A) had more sparse cells, most of which displayed elongated dendritic cell shape, as compared with the untreated counterpart (B).

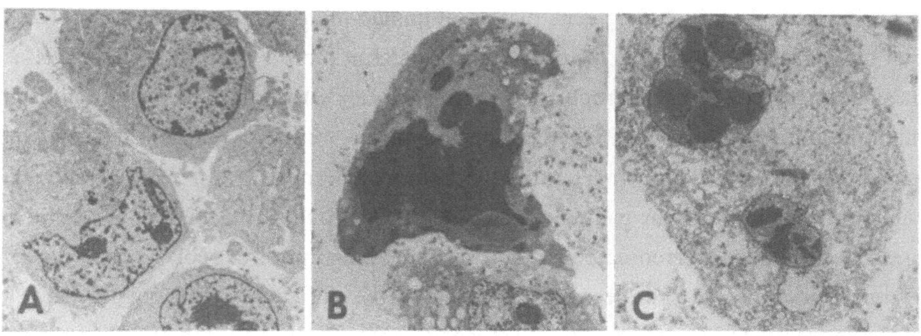

Figure 6. Effect of MAb 9C4 on electron microscopic appearance of melanoma CaCL 73-36 cells. Morphology of melanoma cells cultured in the absence (A) and presence (B, C) of MAb 9C4 (10 µg/ml) for 72 h. Note degenerating cytoplasms with vacuolar formation and apoptotic body formation shown in B and C. No apparent apoptosis was noted in A.

cells by inducing apoptosis has recently been demonstrated (25). The exact mechanism involved in the 9C4-induced apoptosis pathway in GD2-positive cells remains to be elucidated. It is interesting that there is the requirement for GD3 in Fas (APO-1, CD95)/Fas ligand- and ceramide-induced apoptosis reported recently for human cutaneous T cell lymphoma cells (26). In this context, it should be kept in mind that an inverse relationship exists between the expression of GD2 and GD3, since there is the consumption of precursor GD3 for GD2 production in the well-established common ganglioside biosynthetic pathway (27).

ANTIBODY-DEPENDENT CELL-MEDIATD CYTOTOXICITY

To determine if MAb 9C4 could activate ADCC against melanoma CaCL 73-36 cells through the use of lymphokine-activated killer (LAK) cells as effectors, we performed a 4-h ^{51}Cr release cytotoxicity assay using peripheral blood mononuclear cells (PBMCs) and LAK cells as previously described (28). Results of one such experiment showed that little or no lysis of target melanoma cells was observed with PBMCs (NK cells). However, when the PBMCs were incubated with the target cells in the presence of 9C4 as little as 1.5 μg/ml, 35 and 30% of cytotoxicity was noted at effector (E): target (T) ratios of 50:1 and 25:1, respectively. Treatment of the target cells to MAb 9C4 and LAK cells, which had been prepared by incubation of PBMCs with 1000 U/ml of IL-2 in vitro for 3 days, resulted in 55-220% increase in specific target cell lysis over LAK cells alone at the same two E:T ratios. Figure 7 shows one such experiment of ADCC/LAK killing using the M21 melanoma cell line as target cells. The degree of increased cytolysis in ADCC/LAK over LAK alone varied from case to case, depending on which target cell line was used and on individual from whom LAK cells were prepared. The magnitude of effectiveness of MAb 9C4 to exert ADCC with LAK as effector cells was comparable to that of another anti-GD2 MAb known as 14.G2a (a gift from Dr. Alice L. Yu, University of California at San Diego, CA) when tested on the M21 target cells (data not shown). We believe that there should be rooms for further improvement of 9C4 in selective and efficient tumor killing by the following approaches, including construction of human-mouse chimeric antibodies (22,29), use of chimeric antibody- IL-2 fusion proteins to recruit more Fc receptor bearing effector cells such as NK, LAK or neutrophils to the tumor site (30), and use of immunoliposomes containing anti-GD2 MAb with the entrapped synthetic retinoid fenretinide as a potent targeted apoptosis inducer for tumors of neuroectodermal origin (31). It should be pointed out that among four gangliosides, GM3, GD3, GD2 and GM2 that are commonly expressed on human melanoma cells, the first three have no direct correlation with LAK sensitivity, but only GM2 has been shown to be a marker for LAK sensitivity (32).

Aside from CDC and ADCC as the means of antibody-based killing mechanisms, it is worthwhile to note a recent finding that dendritic cells acquire and process antigen from apoptotic cells, resulting in induction of potent MHC class I-restricted cytotoxicity (33). As a consequence of apoptosis, antigens, such as MAGE-1 and tyrosinase within melanoma cells which lack costimulatory function for T cells may gain access to the dendritic cell system, thereby eliciting strong T-cell response (34). Such an apoptosis-dependent pathway has the potential to be manipulated to induce cytotoxic T cell response in melanoma patients.

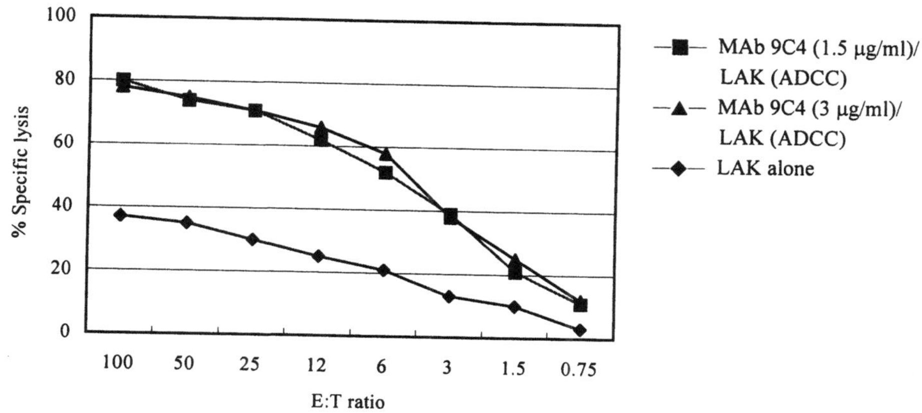

Figure 7. ADCC activity against melanoma M21 cells of allogeneic LAK obtained from a healthy donor. Three cytotoxicity curves conducted under the similar conditions in the presence and absence of MAb 9C4 in 4 hr ^{51}Cr release cytotoxicity assays. E: effector cells; T: tumor cells.

SUMMARY

In this study we have documented a hybridoma secreting an unusual MAb, which expresses both IgG3 and IgG2a subclasses with a λ-light chain. How this dual expression of isotypes was exactly brought about is not clear. To resolve this problem, it will have to wait the complete sequence analysis the heavy chain gene of MAb 9C4. Although the expression of IgG2a was about 50% that of IgG3, antibody titration studies showed the major binding affinity of MAb 9C4 to GD3-positive cells being mostly contributed by the IgG3 rather than IgG2a part of the antibody. This antibody could induce apoptosis in melanoma cells in 10 - 15% of cells in vitro, but the generality of this phenomenon is yet to be confirmed by the use of different cell targets and different anti-GD2 MAbs other than 9C4. Aside from the demonstrated indirect killing mechanisms of many anti-GD2 MAbs through CDC and ADCC, MAb 9C4 induction of apoptosis represents an alternative mechanism of tumor cell killing, by which direct killing of anti-GD2 antibody takes its effect. This apoptotic effect was demonstrated for the first time with an anti-ganglioside monoclonal antibody. From the therapeutic point of view, the cytolytic activity of MAb 9C4-targeted ADCC/LAK killing against GD2-positive tumor cells to be more effective than that of LAK alone and a possibility for dendritic cells to effectively acquire antigen through pulsing with MAb-induced apoptotic cells are both of great clinical importance. Further studies are warranted aiming at elucidating the molecular basis of bi-isotypic specificity and aberrant isotype switching, molecular pathway of anti-GD2 antibody-induced apoptosis, and ways to improve clinical utility of this unusual hybridoma/MAb 9C4.

REFERENCES

1. Hakomori, S. (1981) Glycosphingolipids in cellular interactions, differentiation and oncogenesis. Ann. Rev. Biochem., 33: 733.
2. Leeden, R.W. and Yu, R.K. (1982) Ganglioside structure, isolation and analysis, in: Methods in Enzymology, Vol 83, part D, V. Ginsburg, ed., pp139-191, Academic Press, New York.
3. Hakomori, S. and Kanagi, G. (1983) Glycosphingolipids as tumor-associated and differentiation markers. J. Natl. Cancer Instit., 71: 231.
4. Werkmeister, J.A., Triglia, T, Mackay, I.R., Dowling, J.P., Varigos, G.A., Morstyn, G., Burns, G.F. (1987) Fluctuations in expression of glycolipid antigen associated with differentiation of melanoma cells monitored by monoclonal antibody, Leo Mel 3. Cancer Res., 47: 225.
5. Portoukalian, J., Zwingelstein, G., Dore, J.F. (1979) Lipid composition of human malignant melanoma tumors at various levels of malignant growth. Eur. J. Biochem., 94: 19.
6. Cahan, L.D., Irie, R.F., Singh, F., Cassidneti, A., Paulson, J.C. (1982) Identification of a human neuroblastoma tumor antigen (OFA-1-2) as ganglioside GD2. Proc. Natl. Acad. Sci. USA, 79: 7629.
7. Cheresh, D.A., Harper, J.R., Schulz, G., Reisfeld, R.A. (1984) Localization of the gangliosides G.D2 and GD3 in adhesion plaques and on the surface of human melanoma cells, Proc. Natl. Acad. Sci. USA, 81: 5767.
8. Ladisch, S., Wu, Z.L., Feig, S., Ulsh, L., Schwartz, E., Floutsis, G., Wiley, F., Lenarsky, C., Seeger, R. (1987) Shedding of GD2 ganglioside by human neuroblastoma. Int. J. Cancer, 39: 73.
9. Chapman, P.B., Lonberg, M., Houghton, A.N. (1990) Light chain of an IgG3 anti-GD3 monoclonal antibody and the relationship among avidity, effector functions, tumor targeting and antitumor activity. Cancer Res., 50: 1503.
10. Schulz, G., Cheresh, D.A., Varki, N.M., Yu, A., Staffileno, L.K., Reisfeld, R.A. (1984) Detection of ganglioside GD2 in tumor tissues and sera of neuroblastoma patients. Cancer Res., 44: 5914.
11. Cheung, N.K., Sarinen, V.M., Neely, J., Landmeier, B., Donavan, D., Coccia, P.E. (1985) Monoclonal antibodies to glycolipid antigen on human neuroblastoma cells. Cancer Res., 45: 2642
12. Wu, Z.L., Schwartz, E., Seeger, R., Ladisch, S. (1986) Expression of GD2 ganglioside by untreated primary human neuroblastomas. Cancer Res., 46: 440.
13. Lipinski, M., Braham, K., Philip, I., Wiels, J., Philip, T., Goridis, C., Lenoir, G.M., Tursz, T. (1987) Neuroectoderm-associated antigens on Ewing's sarcoma cell lines. Cancer Res., 47: 183.
14. Liao, S-K., Perng, Y-P., Lee, L-A., Chang, K.S.S., Lai, G-M., Wong, E., Ho, Y-S. (1996) Newly established MST-1 tumor cell line and tumor-infiltrating lymphocyte culture from a patient with soft tissue melanoma (clear cell sarcoma) and their potential applications to patient immunotherapy. Eur. J. Cancer, 32A: 346.
15. Kawashima, I., Tada, N., Ikegami, S., Nakamura, S., Ueda, R., Tai, T. (1988) Mouse monoclonal antibodies detecting disialogangliosides on mouse and human T lymphomas. Int. J. Cancer, 41: 267.
16. Honsik, C.J., Jung, G., Reisfeld, R.A. (1986) Lymphokine-activated killer cells targeted by monoclonal antibodies to disialo-GD2 and GD3 specifically lyse human tumor cells of neuroectodermal origin. Proc. Natl. Acad. Sci. USA, 83: 7893.
17. Mujoo, K., Kipps, T.J., Yang, H.M., Cheresh, D.A., Wargalla, U., Sander, D.J., Reisfeld, R.A. (1989) Functional properties and effect on growth suppression of human neuroblastoma tumors by isotype switch variants of monoclonal anti-ganglioside GD2 antibody 14.18. Cancer Res., 49: 2857.
18. Mayer, P., Handgretinger, R., Bruchelt, G., Schaber, B., Rassner, G., Fieribeck, G.(1994) Activation of cellular cytotoxicity and complement-mediated lysis of melanoma and neuroblastoma cells in vitro by murine anti-ganglioside antibodies MB3.6 and 14.G2a. Melanoma Res., 4: 101.
19. Hank, J.A., Robinson, R.R., Surfus, J., Mueller, B.M., Reisfeld, R.A., Cheung, N.K., Sondel, P.M. (1990) Augmentation of antibody dependent cell-mediated cytotoxicity follwing in vivo therapy with recombinant interleukin 2. Cancer Res., 50: 5234.
20. Chueng, N.K., Lazarus, H., Miraldi, F.D., Abramowsky, C.R., Kallick, S., Sarinen, U.M., Spitzer, T., Strandjord, S.E., Coccia, P.E., Berger, N.A. (1987) Ganglioside GD2 specific monoclonal antibody 3F8: a

phase I study on patients with malignant melanoma. 1. J. Clin. Oncol., 5:1430.
21. Frost, J.D., Hank, J.A., Reaman, G.H., Frierdich, S., Seeger, R.C., Gan, J., Anderson, P.M., Ettinger, L.J., Cairo, M.S., Blazar, B.R., Krailo, M.D., Matthay, K.K., ReisfeldR.A., Sondel, P.M. (1997) A phase I/IB trial murine monoclonal anti-GD2 antibody 14.G2a plus interleukin-2 in children with refractory neuroblastoma: a report of the Children's cancer Group. Cancer, 80:317.
22. Yu, A.L., Uttenreuther-Fischer, M.M., Huang, C.S., Tsai, C.C., Gillies, S.D., Reisfeld, R.A., Kung, F.H. (1998) Phase I trial of a human-mouse chimeric anti-disialoganglioside monoclonal antibody ch14.18 in patients with refractory neuroblastoma and osteosarcoma. J. Clin. Oncol., 16: 21169.
23. Liao, S-K., Meranda, C., Avner, B.P., Romano, T., Hussein, S., Kimbro, B., Oldham, R.K. (1989) Immunohistochemical phenotyping of human solid tumors with monoclonal antibodies in devising biotherapeutic strategies. Cancer Immunol. Immunother., 28: 77.
24. Liao, S-K., Khosravi, M.J., Brown, J.P., Kwong, P.C. (1987) Difference in cell binding patterns of two monoclonal antibodies recognizing two distinct epitopes on human melanoma-associated oncofetal antigen. Mol. Immunol., 24: 1.
25. Ishizuka, H., Watanabe, M., Kubota, T., Matsuzaki, S.W., Otani, Y., Kitajima, M. (1998) Antitumor activity of murine monoclonal antibody NCC-ST-421 on human cancer cells by inducing apoptosis. Anticancer Res., 18: 2513.
26. DeMaria, R., Lenti, L., Malisan, F., d'Agostino, F., Tomassini, B., Euner, A., Rippo, M.R., Testi, R. (1997) Requirement for GD3 ganglioside in CD95- and ceramide-induced apoptosis. Science, 277: 1652.
27. Hersey, P., Jamal, O., Henderson, C., Zardarrl, D., D'Alessandro, G. (1988) Expression of the gangliosides GM3, GD3 and GD2 in tissue sections of normal skin, nevi, primary and metastatic melanoma. Int. J. Cancer, 41: 338.
28. Perng, Y-P., Lin, C-C., Perng, I-M., Shen, Y-C., Chuang, C-K., Liao, S-K. (1997) Culture medium induced morphological changes of melanoma cells associated with change in sensitivity to lysis by lymphokine-activated killer cells. Cancer Biother. Radiopharmaceut., 12: 317.
29. Saleh, M.N., Khazaeli, M.B., Wheler, R.H., Allen, L., Tilden, A.B., Grizzle, W., Reisfeld, R.A., Yu, A.L., Gillies, S.D., LoBuglio, A.F. (1992) Phase I trial of the chimeric anti-GD2 monoclonal antibody ch 14.18 in patients with malignant melanoma. Hum. Antibodies Hybridomas, 3: 19
30. Becker, J.C., Pancook, J.D., Gillies, S.D., Mendelsohn, J., Reisfeld, R.A. (1996) Eradication of human hepatic and pulmonary melanoma metastases in SCID mice by antibody-interleukin 2 fusion proteins. Proc. Natl. Acad. Sci. USA, 93: 2702.
31. Pagnan, G., Montaldo, P.G., Pastorino, F., Raffaghello, L., Kirchmeier, M., Allen, T.M., Ponzoni, M. (1999) GD2-mediated melanoma cell targeting and cytotoxicity of liposome-entrapped fenretinide. Int. J. Cancer, 81: 268.
32. Hoon, D.S., Ando, I., Sviland, G., Tsuchida, T., Okun, E., Morton, D.L., Irie, R.F. (1989) Ganglioside GM2 expression on human melanoma cells correlates with sensitivity to lymphokine-activated killer cells. Int. J. Cancer, 43: 857.
33. Albert, M.L., Sauter, B., Bhardwaj, N. (1998) Dendritic cells acquire antigen from apoptotic cells and induce class I-restricted CTLs. Nature (London), 392:86.
34. Nestle, F.O., Burg, G., Dummer, R. (1999) New perspective on immunobiology and immunotherapy of melanoma. Immunol. Today, 20:5.

PATHOGEN-HOST PROTEIN-CARBOHYDRATE INTERACTIONS AS THE BASIS OF IMPORTANT INFECTIONS[1]

Karl-Anders Karlsson

Institute of Medical Biochemistry, Göteborg University, P.O. Box 440, SE-405 30 Göteborg, Sweden

Microbe attachment to host cells has for a long time been considered as an essential initial event to establish colonization and infection (1-3). Most interactions documented with animal cells are based on specific protein-carbohydrate recognition. However, in spite of so many specificities detected (see 2), surprisingly few convincing tests in infection model systems have been performed to prove the importance for pathogenesis of such interactions (4). Only in recent time a limited number of industrial projects have been initiated concerning medical applications of protein-carbohydrate interactions (4,5). Most exciting, and of potential importance to stimulate additional efforts within glycoscience, are the current clinical trials to treat or prevent influenza with sialic acid analogues (6). Urgent efforts being made concern improvements in glycotechnology linked to a rapidly growing access to complete microbial genome sequences. As discussed below this will rationalize and facilitate the identification of microbial proteins and adhesins for vaccination purposes and drug design. Behind the corner are also mammalian genome sequences including man and mouse, important for the design of infection models to test the biological relevance of identified protein-carbohydrate interactions.

I. HOST GLYCOCONJUGATES

The complexity and diversity of host glycoconjugates are well documented. This creates a basis for the tissue and cell tropism of a large number of infections, and this is linked to distinct differences in glycosylation between animal species, individuals, tissues and cells, and also changes with development, transformation and other parameters. Some attachment sites may be

[1] The work caried out in the author's laboratory was supported by grants from the Swedish Medical Research Council, the Swedish Research Council for Engineering Sciences, the programs "Glycoconjugates in Biological Systems" and "Infection and Vaccinology" of the Swedish Foundation for Strategic Research, the Göteborg Medical Faculty, Knut and Alice Wallenberg Foundation, and IngaBritt and Arne Lundberg Foundation.

very abundant on host cells, like some sialylated glycoconjugates. However, others are only minor, like the GM1 receptor for cholera toxin in human small intestine, a level that may create experimental difficulties in identification.

Some epitopes may be expressed only in some glycoconjugates, as found for Galα4Gal in human tissues, which sequence is important for several infection systems, see below. Use of specific antibodies and microbial ligands documented that this antigen of the blood group P system is exclusively expressed in glycolipids, and therefore absent from true secretions (7). This may give an advantage for a microbial ligand, since it attaches to a strictly membrane-bound site and thus avoids elution by soluble epitopes, e.g. in secreted glycoproteins. In contrast, mammalian sequences containing Man, recognized by a large number of microbes (3), exist only in glycoproteins, both in secreted and membrane-bound form.

Differences in glycosylation between animal species should be clearly appreciated. As an example, the receptor for toxin A of *Clostridium difficile*, of importance for serious human colitis, was thought to be Galα3Galß4GlcNAc based on studies in several animals including rabbit. However, with the advent of cloned glycosyltransferases it was made likely that this sequence could not be synthetized in human tissues. Consequently, a human receptor-active sequence of red cells was shown to be different and end with the trisaccharide GalNAcß3Galß4GlcNAc (8). The potential receptor in the target epithelial cells of human large intestine, has, however, not yet been identified.

Species-specific glycosylation may cause problems in the development of infection models in animals for human infections. This aspect will be discussed below concerning *Helicobacter pylori* and transgenic mouse models.

Limited access to human tissues for ethical reasons is a serious obstacle for structural identification of relevant receptor glycoconjugates. Also, the complexity may add to the problem, illustrated by the composition of Tamm-Horsfall glycoprotein of urine, collected from one single human individual (9,10). This preparation contained more than 150 different saccharide chains, several om them earlier undescribed.

In spite of the existing species specificity, a preliminary screening in animals for realistic amounts of receptor material for working up and identification may be very helpful. However, a cross examination at a lower substance level in the human tissue for potential similarity has to be carefully performed (see toxin A above). Thus we use a panel of glycolipid mixtures of various human and animal origins and TLC overlay assaying for initial tests of receptor activity (4).

II. THE NECESSITY OF ATTACHMENT

The importance of the binding to host cell surface carbohydrate is well documented for several bacterial toxins and for viruses. In case of the more complex bacteria, biological relevance tests *in vivo* of the attachment have been performed only in a few cases.

The lethal toxigenic diarrhoea, which is inducible in newborn calves by *E. coli* K99, was successfully prevented by receptor-active glycopeptides added to the drinking water when symptoms of disease appeared (11). The adhesion specificity is based on NeuGc, which is normally not present in humans (12). Pyelonephritis in monkeys appearing after infection with a human Pap *E. coli* strain was no longer possible to induce after a selective inactivation of the pilus tip-associated PapG adhesin, which is specific for Galα4Gal (13). These model infections of clinical relevance provide strong evidence for the importance for pathogenesis of parasite-host protein-carbohydrate interactions.

III. CONSEQUENCES OF ATTACHMENT

There are additional consequences of the infection *per se* of the interaction of a microbe with the host cell. Upon contact with the eukaryotic cell, *Yersinia* species may increase the rate of transcription of virulence genes (14) and inject effector proteins into the host cytoplasm (15). The PapG-mediated adherence to the host cell glycolipid receptor may induce gene expression in *E. coli* (16), and this interaction may also stimulate cytokine responses mediated by the ceramide-signalling pathway (17). One may add lectinophagocytosis, a process where microbial carbohydrate-binding proteins mediate bacterial uptake by phagocytes (18).

IV. THE GALα 4GAL SPECIFICITY

This epitope is apparently used by several microbial ligands. Also, this is one of the most characterized of binding specificities. Note the fact discussed above that this sequence in humans is lipid-linked and thus strictly membrane-bound. The disaccharide is recognized by uropathogenic *E. coli*, *Streptococcus suis* (19), Shiga and Vero toxins (20,21), and parvoviruses (22).

1. Fine dissection of binding

S. suis is an important pig pathogen but may also cause meningitis in humans. A complete set of monodeoxy analogues of Galα4Galß4Glc was recently used in bacterial hemagglutination inhibition tests, and also as lipid-linked saccharides in the TLC overlay assay (19), to test the importance of hydroxyl groups. The hydroxyls of Glc were not critical for binding, but 4′, 6′, 2 and 3 hydroxyls of Gal-Gal were involved in hydrogen bonding. As a comparison, uropathogenic *E. coli*, which was studied earlier, was concluded to interact with 3′, 4′, 6′, and 6 of Gal-Gal. Also, by using fluorine substitution, the direction of hydrogen bonding was made likely. Thus, this elegant investigation was able to reveal substantial details of receptor-protein interaction without any knowledge of the adhesin binding site.

2. The importance of host phenotype glycosylation for the outcome of infection

As commented above the importance of Pap *E. coli* adhesion to Gal-Gal sites in the urinary tract was proven by adhesin knock-out studies (13). A large number of isoreceptor glycolipids were early identified, all with the common Gal-Gal disaccharide in terminal or internal position (23). Recently, an interesting study has made likely a relation between the human non-secretor phenotype and a unique isoreceptor glycolipid in the uroepithelium (24,25). This is of great interest since non-secretors have been shown to have a 3-4 times higher frequency of urinary infections than secretors. The glycolipid, sialylglobopentaosylceramide (NeuAcα3Galß3Gal-NAc-ß3Galα4Galß4GlcCer), which earlier had been identified in human teratocarcinoma, was shown to bind the bacterium considerably better than did other isoreceptors. It is very likely that the selective presence of this glycolipid in non-secretors improves bacterial adhesion and colonization. Its existence is consistent with the idea that an α3-sialyltransferase preferentially sialylates the precursor galactosyl globoside in the absence of the α2-fucosyltransferase encoded by the *Se* gene, which would otherwise produce the fucosylated product, which is a weaker binder of the bacterium. This example is the only known case of a molecular explanation of individual differences in glycosylation and susceptibility for an infection.

Concerning *Helicobacter pylori* and the finding of Lewis b recognition (26) it was anticipated that Lewis b positive individuals would be more prone to gastric disease. However, such a simple relation apparently does not exist (27).

V. COMPLEXITY OF INTERACTIONS

There may exist more than one receptor specificity even for simple microbial ligands like toxins and viruses (see 4). Cholera toxin (CT) and *E. coli* heat-labile enterotoxin (LT) show 80% sequence homology and both bind GM1 ganglioside with high affinity. LT, but not CT, has in addition an apparently conserved low-affinity binding to non-sialylated N-acetyllactosamine sequences (4). There is still no good explanation for this doubled specificity, but it is mediated by a single binding site and may possibly reflect different needs for the two bacteria to deliver their toxins efficiently to the GM1 site for uptake through the membrane. Hypothetically, *E. coli* cells may adhere more distant to the bilayer than *V. cholerae* cells, requiring a first-step low-affinity "picking up" binding to glycoproteins before reaching the high-affinity site at the bilayer. A difference in potency between the two toxins was recently explained as an increased stability of CT than LT holotoxin during uptake, residing in the A2-fragment (28).

VI. AN EXPERIMENTAL CHALLENGE BY COMPLEXITY OF BINDING: THE GASTRIC PATHOGEN *HELICOBACTER PYLORI*

Multiple carbohydrate-binding specificities are particularly intriguing in case of the human-specific gastric pathogen *Helicobacter pylori*, the recently discovered Gram-negative bacterium which has dramatically changed gastroenterology (27,29). *H. pylori* colonization has been shown to be the basis for chronic gastritis and peptic ulcer and is tightly linked to gastric cancer, which may develop from chronic atrophic gastritis. This bacterium is unique among microbes concerning complexity of binding. In the following I will discuss some of these specificities and speculate on their potential importance.

1. More than ten specificities detected

The first specificity described is based on sialic acid (30), probably with the sequence NeuAcα3Galß4GlcNAc being minimally required (31,32). A second binding dependent on sialic acid appears to be restricted to polyglycosylceramides of human origin, since animal polyglycosylceramides or traditional gangliosides or sialylated glycoproteins do not bind under the conditions used for assay (33-35). Apparently, a third sialic acid-dependent specificity is shown for the soluble neutrophil-activating protein, HpNAP (36). Other bindings have been reported based on sulphate lipids (37), heparan sulphate (38), fucose or Lewis b (26), ganglio glycolipids (39), lactosylceramide (40), and some as yet unpublished or unidentified sequences (S. Teneberg *et al.*, in preparation).

2. Why this complexity?

One may wonder why *Helicobacter* is in need of this set of binding properties. One hypothesis may be that the bacterium is capable of a regulated expression of surface adhesins for the targeting to various microniches of the stomach, a property required to adapt to different ecological conditions. In support of this, we have observed variability in the expression of binding for some of the specificities when tested in our technical assays under standardized conditions. For example, bacteria that apparently appear normal after growth based on various test parameters may occasionally fail to bind to particular glycolipids without any obvious reason. This is in contrast to Pap *E. coli*, or *E. coli* K99, which consistently bind to their respective glycolipids when handled properly. This labile expression may reflect a still uncharacterized swith-on and switch-off mechanism for expression of adhesins. Furthermore, the binding to sialic acid epitopes depends distinctly on cultivation conditions (33,34). Thus, the binding to the first detected epitope (30) only appears after growth on agar, while binding to the second epitope, specific for polyglycosylceramides (33-35), is present also after growth in broth.

3. Interdependence of expression of separate binding specificities

A recent finding is provocative in this respect. The only identified adhesin so far of *H. pylori* is the Lewis b-binding adhesin, BabA (41). During knock-out studies to test effects on binding, a selective inactivation of BabA resulted in the loss of binding to Lewis b (42). The particular strain used (CCGU 17875) is typically a non-binder for sialyl conjugates (32-35). However, after knock-out of BabA the sialic acid-dependent binding was clearly expressed and showed an identical binding pattern to that of wild-type strains which are positive binders (e.g. CCGU 17874). Thus the switch-off of one adhesin apparently resulted in expression of a second adhesin. This interesting result may reflect a sophisticated interdependence of the various binding properties to direct the microbe-host interaction in the varying microecology of the human stomach.

4. Separate receptor epitopes on different target cells

Most bacterial cells are found in the mucus, where they divide. Some are attached to epithelial cells and some are found adhered to inflammatory cells. Could different targets that *H. pylori* interacts with express different binding epitopes, as a potential for selective targeting? In this respect, the sialic acid recognition is of interest. We have found that sialylated glycoconjugates that are recognized by the bacterium exist in extremely low amounts in human stomach epithelium (S. Teneberg *et al.*, unpublished results). In contrast, human neutrophils are rich in receptor-active sialylated glycoconjugates (43), both gangliosides, polyglycosylceramides and glycoproteins. Therefore, the recognition of sialic acid may be for the targeting to neutrophils (including the soluble HpNAP, see above) rather than to epithelial cells (43).

Of interest for the sialic acid-dependent interaction is also the fact that inflamed epithelium (42) or transformed epithelium and stomach cancer (44,45) express receptor-active sialylated glycoconjugates which the normal epithelium (with some minor exception) does not (42). Therefore, a switch-off of the Lewis b adhesin and switch-on of a presumed sialic acid adhesin may be a way for colonization survival on an epithelium with modified glycosylation (42). A recent paper describes a transgenic mouse lacking parietal cells and which shows amplification of stem cells and daughter cells which express sialylated glycoconjugates (46). The conclusion was that this may reflect the mechanism of tumorigenesis in patients with chronic gastritis.

5. A suicide action?

The active bacterial binding to neutrophils may be considered a suicide action, since normally the inflammatory cells are parts of a host defence. It has been documented that interaction of *H. pylori* cells with neutrophils results in a rapid inflammatory burst followed by a slower phagocytosis (see 47). However, the bacteria are not necessarily damaged by this treatment. Instead, they apparently use degradation products of the inflammation to satisfy nutritional needs (48,49). Furthermore, these bacteria actively recruit neutrophils to the site of inflammation by secreting the soluble neutrophil-activating protein, HpNAP (see above), which binds to relatively unique neutrophil gangliosides (36). Thus, as discussed in detail (29), *H. pylori* has undergone its evolution in close association with the human for a long time and adapted properties for an optimal survival in a complex stomach ecology.

6. *H. Pylori* in the human stomach: an important model for the study of microbe adaptation in the host

H. pylori is an exciting object of study. Firstly, it escaped detection and was convincingly documented as a human stomach colonizer first about 15 years ago (27). Secondly, this microbe is colonizing a majority of the global population with a characteristic strain relation to population and family units (29). That only a fraction of colonizers provide symptoms of illness may indicate strain and host differences or additional environmental causative factors. In fact, some investigators consider *H. pylori* as part of the normal flora with beneficial effects on the quality of life (27,29,49,50). Therefore, eradication of the bacterium may have to be restricted to serious medical illness and not applied globally for a prevention of e.g. cancer. In support of a long and successful adaptation to the host is not only the apparent nutritional use of the inflammation induced (see above), but the fact that there is no efficient immune response from the host to get rid of the microbe (51). These parameters and the complexity of the interaction with host glycoconjugates and other properties may make *H. pylori* a suitable model to study for microbe-host relationships in general, although a model of high complexity.

VII. RECENT IMPROVEMENTS IN TECHNOLOGY

Considering the role of microbe-host protein-carbohydrate binding specificities for the outcome of an infection, it was remarked above that in only a few cases the importance for pathogenicity has been proven. However, before reaching this test level, which is one of the important prerequisites for medical treatments, one has to define both the assumed protein on the microbe and the glycoconjugate on the host, as well as their regulation and dynamics.

In spite of improved techniques for the isolation and structural analysis of glycoconjugates, including also very complex molecules (see 52), the limited access to human target tissues is a major problem. Screening for receptor epitopes in animals is rational for identification purposes (see discussion of TLC overlay in 4), but the result has to be carefully tested also for the human case (see example above for toxin A). Once a limited receptor sequence has been identified, the use of specific antibodies to analyse the expression at a cellular level in human is most valuable (see 42).

1. Rationalized identification of adhesins

When the knowledge of the host receptor saccharide is available, the corresponding adhesin on the microbe is accessible for identification by a proteomics approach, provided that the genome sequence of the microbe is known. We are applying this for *H. pylori* (unpublished results), for which the genome sequence has been published for several strains (53,54). However, one critical problem is the low abundance of bacterial adhesins which may exist at a level of only 500 copies per bacterial cell (see 41,42). An affinity probe was designed, similar to that used for traditional cloning (41), composed of albumin to which the receptor saccharide and a cross-linker tagged with biotin is coupled. This probe is incubated with the living bacteria, the cross-linker activated, followed by reduction to cleave off the albumin glycoconjugate, which may be washed away. Remaining is the adhesin tagged with a small cross-linker arm with biotin. After extraction with SDS and enrichment with streptavidin-beads the eluate is separated by electrophoresis, biotin-tagged spots detected for in-gel trypsin digestion, and the peptides analyzed by MALDI-TOF mass spectrometry. Precise masses for 5-10 peptides in the interval 1000-3000 D are then fed into a program on the Net for search against the genome sequence, and a full adhesin sequence is reliably retrieved. We will report several successful examples of this highly rationalized affinity cross-linking approach in the near future. To functionally prove the adhesin identity of the sequence obtained, the corresponding gene has to be inactivated, and the loss of binding to the receptor saccharide convincingly documented. The identified adhesin may then be expressed for vaccine studies and drug design (see below).

2. Transgenic mouse models

Before initiating an expensive development of infection prevention or therapy one has to test the relevance for pathogenicity of the adhesin-saccharide interaction in suitable animal models. However, as discussed above the characteristic differences between human and animal species may create problems in designing the glycosylation required. Transgenic approaches are inviting, and in an elegant recent study the mouse was used to create a human model for *H. pylori* colonization (55). The Lewis gene, coding for Lewis a and b antigens, is restricted to primates, which excludes the use of common rodent models to test the relevance for disease of the Lewis b binding specificity documented for *H. pylori* (26). For the actual study, the human fucosyltransferase gene was transfected to mouse and targeted to express Lewis b in mouse stomach pit cells. *H. pylori* cells bound *in vitro* to these cells of the transgenic mouse but not to cells of the normal littermates. In a following paper (56), a persistent infection *in vivo* by a clinical isolate of *H. pylori* occurred at comparable microbial densities in the two mice. However, only in the transgenic mouse the adhesion resulted in production of autoantibodies to Lewis x epitopes shared by bacteria and acid-secreting parietal cells, and in chronic gastritis, and parietal cell loss. This demonstrates that the mode of colonization is important for pathogenesis, and that the presence of the adhesin can alter disease outcome without affecting the colonization levels and density of bacteria.

Another transgenic mouse lacking parietal cells is a potential model for development of gastric cancer (46), see above. No doubt, transgenic model systems will be critical for our understanding of infection mechanisms, which is the important basis for development of therapy and prevention.

VIII. ANTI-INFECTION THERAPY AND PREVENTION

If the parasite-host protein-carbohydrate interaction is a conserved microbial property, then this property should be a logical target for the development of therapy and prevention. The protein may be used for vaccination studies, but also for structural analysis and design of synthetic saccharide analogues. Of great interest and conceptual importance are the current studies on influenza and sialic acid analogues.

Most likely, influenza is the best characterized of human infections (6), including 3D structures of the two viral surface glycoproteins, the hemagglutinin and the neuraminidase, also in complex with various natural saccharides and synthetic analogues (57). An important study (58) revealed a sialic acid-binding pocket of the hemagglutinin that is too narrow to be recognized by an antibody. Thus the conserved receptor-binding sites escape immune recognition and neutralization, while the exposed peptide surfaces of the two surface proteins continuously mutate to make long-term vaccination impossible. This finding probably triggered the efforts to design sialic acid analogues as an alternative for prevention and therapy of influenza. Two products are in current clinical tests, one as nose drops (59) and the other as a pill (60). The outcome of these tests will probably influence the will to invest in glycoscience related to treatment of infections.

The conserved adhesin protein of a bacterium is a potential vaccine component and also, after analysis of the 3D-structure, an object for structural design of antiadhesive saccharide analogues. An important recent paper described the first crystal structure of a pilus-associated Man-binding adhesin complex from uropathogenic *E. coli* (61). Within a short time, complete genome sequences will be accessible for most of the important pathogens and this makes possible the identification of adhesins by affinity proteomics approaches and use of receptor-containing probes (see above).

Access to the adhesin also invites to screening efforts to find natural receptor-active glycoconjugates of various sources that may be added to the diet to prevent infections of the gastrointestinal and upper respiratory tract. This approach may prove to be a realistic alternative in more poor parts of the world where infections are a major threat.

IX. ASPECTS OF THE FUTURE

Recent developments in methodological approaches combined with access to complete microbial genome sequences provide a promising future concerning new ways of treating and preventing important infections. Interestingly, evolution of animal glycosylation may depend on infection pressure, and recently a stimulating hypothesis concerning evolution of animal glycan diversity was presented (62). According to this, exogeneous selection pressure mediated by microbial recognition of host glycans plays a dominant role and favors intra- and interspecies diversity. Thus, species and individual differences in host cell surface glycosylation not only explain the tropism of infections but favors survival of the species from lethal infections, if groups within a population are resistant because they lack the essential saccharide receptor. New research fields may be foreseen, including exploration of the postulated evolutionary diversification of glycosylation and its potential relation to species survival from infections. In this way we may learn more about host susceptibility to infections and thus arrive at a better medical insight into this central group of global diseases where rational treatments are still lacking in most cases.

SUMMARY

Microbe-host protein-carbohydrate interactions are most likely the essential first step to produce an infection, although this has been strictly proven only in a few cases. Improved glycotechnology will help identification of new carbohydrate receptors and this knowledge may be used to identify microbial carbohydrate-binding proteins by affinity proteomics approaches. In some cases such conserved proteins may prove to be successful vaccine components, in other cases, like influenza, saccharide analogues may be the only rational alternative. The prognosis may be, based on these improvements, that infection medicine will make considerable progress in the near future.

REFERENCES

1. Beachey, E.H. (1980) "Bacterial Adherence", Receptors and Recognition, Series B, Vol. 6, Chapman & Hall, London.
2. Mirelman, D., ed. (1986) "Microbial Lectins and Agglutinins. Properties and Biological Activity", John Wiley & Sons, New York.
3. Ofek, I., and Doyle, R.J. (1994) "Bacterial Adhesion to Cells and Tissues", Chapman & Hall, London.
4. Karlsson, K.-A. (1998) Meaning and therapeutic potential of microbial recognition of host glycoconjugates. *Mol. Microbiol.*, 29:1-11.
5. McAuliffe, J.C., and Hindsgaul, O. (1997) Carbohydrate drugs- an ongoing challenge. Chemistry and Industry, No vol.:170-174.
6. Laver, W.G., Bischofberger, N., and Webster, R.G. (1999) Disarming flu viruses. *Sci. Amer.*, 280:57-65.
7. Yang, Z., Bergström, J., and Karlsson, K.-A. (1994) Glycoproteins with Galα4Gal are absent from human erythrocyte membranes, indicating that glycolipids are the sole carriers of blood group P activities. *J. Biol. Chem.*, 269: 14620-14624.
8. Teneberg, S., Lönnroth, I., Torres Lopez, J.F., Galili, U., Ölwegård Halvarsson, M., Ångström, J., and Karlsson, K.-A. (1996) Molecular mimicry in the recognition of glycosphingolipids by Galα3Galß4GlcNAc-binding *Clostridium difficile* toxin A, human natural anti α-galactosyl IgG and the monoclonal antibody Gal-13: characterization of a binding-active human glycosphingolipid, non-identical with the animal receptor. *Glycobiology*, 6:599-609.
9. H d, K., van Zadelhoff, G., Moononen, P., Kamerling, J.P., and Vliegenthart, J.F.G. (1992) The Asn-linked carbohydrate chains of human Tamm-Horsfall glycoprotein of one male. Novel sulfated and novel N-acetylgalactosamine-containing N-linked carbohydrate chains. *Eur. J. Biochem.*, 209:895-915.
10. van Rooijen, J.J.M., Voskamp, A.F., Kamerling, J.P., and Vliegenthart, J.F.G. (1999) Glycosylation sites and site-specific glycosylation in human Tamm-Horsfall glycoproetin. *Glycobiology*, 9:21-30.
11. Mouricout, M., Petit, J.M., Carias, J.R., and Julien, R. (1990) Glycoprotein glycans that inhibit adhesion of *Eschericha coli* mediated by K99 fimbriae: treatment of experimental colibacillosis. *Infect. Immun.*, 58:98-106.
12. Teneberg, S., Willemsen, P.T.J., de Graaf, F.K., Stenhagen, G., Pimlott, W., Jovall, P.-Å., Ångström, J., and Karlsson, K.-A. (1994) Characterization of gangliosides of calf small intestine with special reference to receptor-active sequences for enteropathogenic *Escherichia coli* K99. *J. Biochem.*, 116:560-574.
13. Roberts, J.A., Marklund, B.-I., Ilver, D., Haslam, D., Kaack, M.D., Baskin, G., Louis, M., Möllby, R., Winberg, J., and Normark, S. (1994) The Galα4Gal-specific tip adhesin of *Escherichia coli* P fimbriae is needed for pyelonephritis to occur in the normal urinary tract. Proc. Natl. *Acad. Sci. USA*, 91:11889-11893.
14. Petterson, J., Nordfelth, R., Dubinina, E., Bergman, T., Gustafsson, M., Magnusson, K.-E., and

Wolf-Watz, H. (1996) Modulation of virulence factor expression by pathogen target cell contact. *Science*, 273:1231-1233.

15. Cornelis, G.R., and Wolf-Watz, H. (1997) The *Yersinia* Yop virulon: a bacterial system for subverting eukaryotic cells. *Mol. Microbiol.*, 23:861-867.
16. Zang, J.P., and Normark, S. (1996) Induction of gene expression in *Escherichia coli* after pilus-mediated adherence. *Science*, 273:1234-1236.
17. Connel, H., Agace, W., Hedlund, M., Klemm, P., Shembri, M., and Svanborg, C. (1996) Fimbriae-mediated adherence induces mucosal inflammation and bacterial clearance. Consequences for antiadhesion therapy. *Adv. Exp. Biol. Med.*, 408:73-80.
18. Ofek, I., Goldhar, J., Keisari, Y., and Sharon, N. (1995) Nonopsonic phagocytosis of microorganisms. *Annu. Rev. Microbiol.*, 49:239-276.
19. Haataja, S., Zhang, Z., Tikkanen, K., Magnusson, G., and Finne, J. (1999) Determination of the cell adhesion specificity of *Streptococcus suis* with the complete set of monodeoxy analogues of globotriose. *Glycoconj. J.*, 16:67-71.
20. Lindberg, A.A., Brown, J.E., Strömberg, N., Westling-Ryd, M., and Karlsson, K.-A. (1987) Identification of the carbohydrate receptor for Shiga toxin produced by *Shigella dysenteriae* type 1. *J. Biol. Chem.*, 262:1779-1785.
21. Lacy, D.B., and Stevens, R.C. (1998) Unravelling the structures and mode of action of bacterial toxins. *Curr. Op. Struct. Biol.*, 8:778-784.
22. Brown, K.E., Anderson, S.M., and Young, N.S. (1993) Erythrocyte P antigen: cellular receptor for B19 parvovirus. *Science*, 262:114-117.
23. Bock, K., Breimer, M.E., Brignole, A., Hansson, G.C., Karlsson, K.-A., Larson, G., Leffler, H., Samuelsson, B.E., Str berg, N., Svanborg Edén, C., and Thurin, J. (1985) Specificity of binding of a strain of uropathogenic *Escherichia coli* to Galα4Gal-containing glycosphingolipids. *J. Biol. Chem.*, 260:8545-8551.
24. Stapleton, A.E., Stroud, M.R., Hakomori, S.-i., and Stamm, W.E. (1998) The globoseries glycosphingolipid sialosyl galactosyl globoside is found in urinary tract tissues and is a preferred binding receptor in vitro for uropathogenic *E. coli* expressing *pap*-encoded adhesins. *Infect. Immun.*, 66:3856-3861.
25. Stroud, M.R., Stapleton, A.E., and Levery, S.B. (1998) The P histo-blood group-related glycosphingolipid sialosyl galactosyl globoside as a preferred binding receptor for uropathogenic *Escherichia coli*: Isolation and structural characterization from human kidney. *Biochemistry*, 37:17420-17428.
26. Borén, T., Falk, P., Roth, K.A., Larson, G., and Normark, S. (1993) Attachment of *Helicobacter pylori* to human gastric epithelium mediated by blood group antigens. *Science*, 262:1892-1895.
27. Dunn, B.E., Cohen, H., and Blaser, M.J.(1997) *Helicobacter pylori*. *Clin. Microbiol. Rev.*, 10:720-741.
28. Rodighiero, C., Aman, A.T., Kenny, M.J., Moss, J., Lencer, W.I., and Hirst, T.R. (1999) Structural basis for the differential toxicity of cholera toxin and *Escherichia coli* heat-labile enterotoxin. Construction of hybrid toxins identifies the A2-domain as the determinant of differential toxicity. *J. Biol. Chem.*, 274:3962-3969.
29. Covacci, A., Telford, J.L., Del Giudice, G., Parsonnet, J., and Rappuoli, R. (1999) *Helicobacter pylori* virulence and genetic geography. *Science*, 284:1328-1333.
30. Evans, D.G., Evans Jr, D.J., and Graham, D.Y. (1988) N-acetylneuraminyllactose-binding fibrillar hemagglutinin of *Campylobacter pylori*: a putative colonization factor antigen. *Infect. Immun.*, 56:2896-2906.
31. Hirmo, S., Kelm, S., Schauer, R., Nilsson, B., and Wadström, T. (1996) Adhesion of *Helicobacter pylori* strains to α2,3-linked sialic acids. *Glycoconj. J.*, 13:1-7.
32. Miller-Podraza, H., Larsson, T., Nilsson, J., Teneberg, S., Matrosovich, M., and Johansson, L. (1998) Epitope dissection of of receptor-active gangliosides with affinity for *Helicobacter pylori* and influenza virus. *Acta Biochim. Polon.*, 45:439-449.

33. Miller-Podraza, H., Abul Milh, M., Bergström, J., and Karlsson, K.-A. (1996) Recognition of glycoconjugates by *Helicobacter pylori*: an apparently high-affinity binding of human polyglycosylceramides, a second sialic acid-based specificity. *Glycoconj. J.*, 13:453-460.
34. Miller-Podraza, H., Bergström, J., Abul Milh, M., and Karlsson, K.-A. (1997) Recognition of glycoconjugates by *Helicobacter pylori*: comparison of two sialic acid-dependent specificities based on hemagglutination and binding to human erythrocyte glycoconjugates. *Glycoconj. J.*, 14:467-471.
35. Miller-Podraza, H., Stenhagen, G., Larsson, T., Andersson, C., and Karlsson, K.-A. (1997) Screening for the presence of polyglycosylceramides in various tissues: partial characterization of blood group-active complex glycosphingolipids of rabbit and dog small intestines. *Glycoconj. J.*, 14:231-239.
36. Teneberg, S., Miller-Podraza, H., Lampert, H.C., Evans Jr, D.J., Evans, D.G., Danielsson, D., and Karlsson, K.-A. (1997) Carbohydrate binding specificity of the neutrophil-activating protein of *Helicobacter pylori*. *J. Biol. Chem.*, 272:19067-19071.
37. Saitoh, T., Natomi, H., Zhao, W., Okuzumi, K., Sugano, K., Iwamori, M., and Nagai, Y. (1991) Identification of glycolipid receptors for *Helicobacter pylori* by TLC immunostaining. FEBS Lett., 282:385-387.
38. Ascencio, F., Fransson, L.-Å., and Wadström, T. (1993) Affinity of the gastric pathogen *Helicobacter pylori* for the N-sulphated glycosaminoglycan heparan sulphate. *J. Med. Microbiol.*, 38:240-244.
39. Gold, B., Huesca, M., Sherman, P., and Lingwood, C.A. (1993) *Helicobacter mustelae* and *Helicobacter pylori* bind to common receptors in vitro. *Infect. Immun.*, 61:2632-2638.
40. Ångström, J., Teneberg, S., Abul Milh, M., Larsson, T., Leonardsson, I., Olsson, B.-M., Ölwegård Halvarsson, M., Danielsson, D., Näslund, I., Ljung, Å., Wadström, T., and Karlsson, K.-A. (1998) The lactosylceramide binding specificity of *Helicobacter pylori*. *Glycobiology*, 8:297-309.
41. Ilver, D., Arnqvist, A., Ögren, J., Frick, I.-M., Kersulyte, D., Incecik, E.T., Berg, D.E., Covacci, A., Engstrand, L., and Borén, T. (1998) *Helicobacter pylori* adhesin binding fucosylated histo-blood group antigens revealed by retagging. *Science*, 279:373-377.
42. Mahdavi, J., Sondén, B., Forsberg, L., Olfat, F.O., Sjöström, R., Roche, N., Ångström, J., Teneberg, S., Karlsson, K.-A., Kersulyte, D., Berg, D.E., El-Zimaity, H.M.T., Graham, D.Y., Dubois, A., Lindh, F., Lundskog, B.B., Hammarström, L., and Borén, T. (1999) *Helicobacter pylori* infection elicits expression of sialylated Lewis antigens that confer bacterial adherence to the inflamed gastric epithelial lining. Submitted.
43. Miller-Podraza, H., Bergstr , J., Teneberg, S., Abul Milh, M., Longard, M., Olsson, B.-M., Uggla, L., and Karlsson, K.-A. (1999) *Helicobacter pylori* and neutrophils: sialic acid-dependent binding to various isolated glycoconjugates. *Infect. Immun.*, in press.
44. Sakamoto, S., Watanabe, T., Tokumaro, T., Takagi, H., Nakazato, H., and Lloyd, K.O. (1989) Expression of Lewis a, Lewis b, Lewis x, Lewis y, sialyl-Lewis a, and sialyl-Lewis x blood group antigens in human gastric carcinoma and in normal gastric tissue. *Cancer Res.*, 49:745-752.
45. Hakomori, S.-i. (1996) Tumor malignancy defined by aberrant glycosylation and sphingo(glyco)lipid metabolism. *Cancer Res.*, 56:5309-5318.
46. Snyder, A.J., Guruge, J.L., Li, Q., Hu, Y., Oleksiewicz, C.M., Lorenz, R.G., Karam, S.M., Falk, P.G., and Gordon, J.I. (1999) *Helicobacter pylori* attaches to NeuAcα2,3Galß1,4 glycoconjugates produced in the stomach of transgenic mice lacking parietal cells. *Mol. Cell.*, 3:263-274.
47. Rautelin, H., Blomberg, B., Fredlund, H., Järnerot, G., and Danielsson, D. (1993) Incidence of *Helicobacter pylori* strains activating neutrophils in patients with peptic ulcer disease. *Gut*, 34;599-603.
48. Kirshner, D.E., and Blaser, M.J. (1995) The dynamics of *Helicobacter pylori* infection in the human stomach. *J. Theoret. Biol.*, 176:281-290.
49. Blaser, M.J. (1996) The bacteria behind ulcers. *Sci. Amer.*, 274:92-97.

50. Blaser, M.J. (1997) Not all *Helicobacter* strains are created equal: should all be eliminated? *Lancet*, 349:1020-.
51. Telford, J.L., Covacci, A., Rappuoli, R., and Ghiara, P. (1997) Immunobiology of *Helicobacter pylori* infection. *Curr. Opin. Immunol.*, 9:498-503.
52. Karlsson, H., Johansson, L., Miller-Podraza, H., and Karlsson, K.-A. (1999) Fingerprinting of large oligosaccharides linked to ceramide by matrix-assisted laser desorption/ionization time-of-flight mass spectrometry: highly heterogeneous polyglycosylceramides of human erythrocytes with receptor activity for *Helicobacter pylori*. *Glycobiology*, 9:765-778.
53. Tomb, J.-F., White, O., Kerlavage, A.R., Clayton, R.A., Sutton, G.G., Fleishmann, R.D., Ketchum, K.A., Klenk, H.P., Gill, S., Dougherty, B.A., Nelson, K., Quackenbush, J., Zhou, L., Kirkness, E.F., Peterson, S., Loftus, B., Richardson, D., Dodson, R., Khalak,H.G., Glodek, A., McKenney, K., Fitzgerald, L.M., Lee, N., Adams, M.D., Hickey, E.K., Berg, D.E., Gocayne, J.D., Utterback, T.R., Peterson, J.D., Kelley, J.M., Cotton, M.D., Weidman, J.W., Fuji, C., Bowman, C., Whattey, L., Wallin, E., Hayes, W.S., Borodowski, M., Karp, P.D., Smith, H.O., Fraser, C.M., and Venter, J.C. (1997) The complete genome sequence of the gastric pathogen *Helicobacter pylori*. *Nature*, 388:539-547.
54. Alm, R.A., Ling, L.S.-L., Moir, D.T., King, B.L., Brown, E.D., Doig, P.C., Smith, D.R., Noonan, B., Guild, B.C., deJonge, B.L., Carmel, G., Tummino, P.J., Caruso, A., Uria-Nickelsen, M., Mills, D.M., Ives, C., Gibson, R., Merberg, D., Mills, S.D., Jiang, Q., Taylor, D.E., Vovis, G.F., and T.J. Trust, T.J. (1999) Genomic-sequence comparison of two unrelated isolates of the human pathogen *Helicobacter pylori*. *Nature*, 397:176-180.
55. Falk, P.G., Bry, L., Holgersson, J., and Gordon, J.I. (1995) Expression of a human α-1,3/4-fucosyltransferase in the pit cell lineage of FVB/N mouse stomach results in the production of Le^b-containing glycoconjugates: a potential transgenic mouse model for studying *Helicobacter pylori* infection. *Proc. Natl. Acad. Sci. USA*, 92:1515-1519.
56. Guruge, J.L., Falk, P.G., Lorenz, R.G., Dans, M., Wirth, H.-P., Blaser, M.J., Berg, D.E., and J.I. Gordon, J.I. (1998) Epithelial attachment alters the outcome of *Helicobacter pylori* infection. *Proc. Natl. Acad. Sci. USA*, 95:3925-3930.
57. Karlsson, K.-A. (1995) Microbial recognition of target-cell glycoconjugates. *Curr. Opin. Struct. Biol.*, 5:622-635.
58. Weis, W., Brown, J.H., Cusack, S., Paulson, J.C., Skehel, J.J., and D.C. Wiley, D.C. (1988) Structure of the influenza virus hemagglutinin complexed with its receptor, sialic acid. *Nature*, 333:426-431.
59. von Itzstein, M., Wu, W.-Y., Kok, G.B., Pegg, M.S., Dyason, J.C., Jin, B., Van Phan, T., Smythe, M.L., White, H.F., Oliver, S.W., Colman, P.M., Varghese, J.N., Ryan, D.M., Woods, J.M., Bethell, R.C., Hotman, V.J., Cameron, J.M., and Penn, C.R. (1993) Rational design of potent sialidase-based inhibitors of influenza virus replication. *Nature*, 363:418-423.
60. Kim, C.U., Lew, W., Williams, M.A., Liu, H., Zhang, L., Swaminathan, S., Bischofberger, N., Chen, M.S., Mendel, D.B., Tai, C.Y., Laver, W.G., and Stevens, R.C. (1997) Influenza neuraminidase inhibitors possessing a novel hydrophobic interaction in the enzyme active site: design, synthesis, and structural analysis of carbocyclic sialic acid analogues with potent anti-influenza activity. *J. Am. Chem. Soc.*, 119:681-690.
61. Choudbury, D., Thompson, A., Stojanoff, V., Langermann, S., Pinkner, J., Hultgren, S.J., and Knight, S.D. (1999) X-ray structure of the FimC-FimH chaperone-adhesin complex from uropathogenic *Escherichia coli*. *Science*, 285:1061-1066.
62. Gagneux, P., and Varki, A. (1999) Evolutionary considerations in relating oligosaccharide diversity to biological function. *Glycobiology*, 9:747-755.

HOST MEDIATED VARIATION AND RECEPTOR BINDING SPECIFICITY OF INFLUENZA VIRUSES[1]

Yasuo Suzuki

Department of Biochemistry, University of Shizuoka School of Pharmaceutical Sciences 52-1 Yada, Shizuoka, 422-8526, Japan[2]

Influenza A and B viruses have 2 kinds of glycoprotein spikes, hemagglutinin and sialidase. Hemagglutinin homotrimer is responsible for the binding to the sialyl sugar chain receptors on the host cell membranes, on the other hand, the sialidase tetramer is responsible for the receptor destroying enzyme activities. The virus genome consists 8 segmented single strand nagative RNAs which encode 10 different viral proteins. Variation of the influenza A viruses occurrs by 2 major machanisms. First mechanism is gene reassortment which produces reassortant virus formed by segmented gene mixing between human and animal viruses, especially aquatic avian influenza viruses, and Second mechanism, point mutation of the RNA gene. In this case, viral RNAs are replicated by viral RNA polymerases. The lack of the proofreading among viral RNA polymarases contributes to replication errors of the order of 1 in 10^4 bases. This contrasts with the much higher replication fidelity found among DNA polymerases, their errors culculated in the order of only 1 in 10^4 bases, it means influenza virus RNA polymerases make errors more than several 100 x 10^3 fold more than those of host cell DNA polymerases. Therefore, each round of influenza virus replication results in a mixed population with many variants, most of which are not viable, but some of which have potentially advantageous mutation that can become dominant under the right selective conditions(1).

In this paper, the molecular evolution of the hemagglutinin spike and also the possible mechanism on the host mediated selection of influenza A viruses are described.

---

1. This study is collaborated with Y. Kawaoka (Wisconsin Univ., Wisconsin), R.G. Webster (St. Jude children's Research Hospiral, Memphis), T. Ito (Tottori University, Tottori, Japan), M. Kiso (Gifu University, Gifu, Japan), T. Suzuki (University of Shizuoka, Shizuoka, Japan), H. Masuda, (University of Shizuoka, Shizuoka, Japan), D. Miyamoto (University of Shizuoka, Shizuoka, Japan), K.I.-P.J. Hidari(University of Shizuoka, Shizuoka, Japan), and colleagues listed in the reference of this paper.
2. e-mail:suzukiy@ys7.u-shizuoka-ken.ac.jp and Fax: 81-54-264-5721

Neu5Ac α 2-3Gal β 1-4(3)GlcNAc β 1-
Neu5Ac α 2-6Gal β 1-4(3)GlcNAc β 1-

Sialyl-lacto Series Type I and II chains

Fig. 1. Essential Receptor sialo-sugar chains for influenza A, B Virus Binding

I The host of Influenza viruses and Influenza pandemics

Influenza A viruses have been isolated from a variety of animals, including human, pigs, horses, sea mammals, wild waterfowl, including ducks and poultry. Among these animals, wild waterfowl is the major reservoir of influenza viruses and all the influenza viruses in other animals species are thought to be originally derived from these birds. In fact, the causative viruses of the 1957 asian and 1968 Hong Kong pandemic influenza A strains are reassortant viruses between human and avian influenza viruses, indicating introduction of avian virus hemagglutinin genes into the human population.

In this century, there were at least four pandemics. First one is Spanish flu, it emerged 1918, Although human influenza virus has been isolated firstly in 1933, the origin of Spanish flu has been identidied by gene analyses as H1N1 type influenza virus (2). Second pandemic is followed by Asian flu, H2N2 in 1957, Hong Kong flu, H3N2, 1968, and most recently, H5N1 virus emerged in Hong Kong in 1997 which is suspected to be directly introduced from chicken to human population in southern china (In this case, 18 people were infected in Hong Kong and 6 persons were dead). Therefore, the molecular evolutions of the hemagglutinin and sialidase of influenza A viruses are still active in nature. All the subtypes of the hemagglutinin, H1 to H15 and sialidase N1 to N9 are maintained in the avian world. Hemaglutinin subtypes of pig influenza viruses are similar to those of humans.

II Receptor binding specificity of human and animal influenza viruses

In 1981, Wiley and Skehel's group (3) reported the three dimensional structure of the bromelain-released HA from Hong Kong H3 Viruses. There are at least 5 antigenic sites, A-E, and also receptor binding pocket in the globular region. Antigenic sites are very variable, frequent point mutation in amino acid level is induced by the antibody pressure in these regions. We found that the amino acid substitution in the receptor binding pocket is also occurred by the selective pressure caused by the type of receptor sialyl linkages and the molecular species of sialic acids such as *N*-acetyl- (Neu5Ac) and N-glycolylneuraminic acid (Neu5Gc) in the receptor molecules.

We have been interested in the molecular mechanism of the viral transmission between different hosts, especially on the basis of the receptor binding specificities (Fig. 1). The most effective, common, receptor sialyl sugar chain structures as the influenza A and B virus receptors which are expressed in sialylglycoproteins and also gangliosides contain monosialo-lactosamine type I or type II structures(4, 5). One terminal sialic acid is necessary and two sequential sialic acid and also sialic acid linked to the internal galactose in the sugar chain, such as GM1 gangliosides are not recognized at all. These are determined by using more than 80 kinds of native and synthetic sialyl sugar chains.

We determined the receptor binding specificity of human, swine, avian and equine influenza A viruses, and we also determined the sialyl linkage of the cells of the host target organs, trachea epitherium for human, pig, horse and intestinal mucosa for birds, intestinal mucosa is a major replication site of influenza viruses in bird such as duck. For this experiment, we use some lectins which recognize sialyllinkages, 2-3 or 2-6. For 2-3, Maackia amurensis (MA, this lectin binds complex *N*-glycoside type glycoprotein containing 3 or 4 antenally sugar chains. Reactivity for glycoproteins carrying O-glycosidic linkage is reported to be low), for 2-6, Sambucs nigra (SN, this lectin has wide reactivity for SA2-6Gal (GalNAc) sequences). Human influenza viruses isolated in MDCK cells binds Neu5Ac2-6Gal sequences, however, equine and avian viruses binds 2-3 predominantly. It is interesting that the swine virus binds both 2-6 and 2-3, almost equally or a little bit predominant in 2-6. If we look at the sialyl linkage of the target organ, it is very interesting that human trachea express significantly sialyl 2-6Gal, this results have already reported by Paulson's group in 1990(6), when we use the same lectin, however, horse trachea and duck intestinal mucosa express 2-3, on the other hand, pig trachea expressed both 2-3 and 2-6 linkages. These results indicates that pig has the receptor sialyl linkages which are acceptable both of avian and human viruses. These results may explain in the aspect of molecular basis that pig is an intermediate vessels or hosts between bird and human. This results also showed that the availability of the receptors such as 2-3 and 2-6 moieties, in host animals correlates with the receptor specificity of influenza viruses from the host animals and suggest that receptor specificity is one of the important determinants of the host range restriction of influenza viruses, although it is highly likely that this phenomenon is controlled by multiple host and viral genes.

We found that early H3 human viruses which transmitted from bird binds both 2-3 and 2-6 sialyl sugar chains. However, after 1977, receptor binding specificity has been changed to 2-6 (7) (Table 1). This means that early H3 express the avian receptor biding specificity (2-3 predominant), but while these viruses were maintained in the human population, their receptor specificity become 2-6, this properties may due to the selection by the host receptor sialyl sugar chains, which contain 2-6 sequence in trachea of human beings. In this experiments, we also found that when the human influenza virus cultured in different host, in this case, embryonated chicken eggs, the receptor binding properties were dramatically changed(Table1). This virus isolated in 1994 in Tottori Japan, when it isolated in MDCK cells from the clinical speciments, binding specificity was 2-6, however, when the MDCK-isolated virus passaged in embryonated chicken aminiotic membrane 2 times, and then cholio allantoic membrane 3 times, specificity changed to 2-3 predominant. This 2-3 binding property is very similar to those avian and equine viruses. So, It may be able to say that human influenza virus become avian or equine type.

Table 1.

Change of receptor binding specificity of human H3 influenza A viruses

Virus (H3N2)	Neu5Ac α 2,3 Gal	Neu5Ac α 2,6 Gal	Neu5Gc α 2,3 Gal
A/Aichi/2/68	40	60	0
A/Udorn/307/72	21	77	2
A/Victoria/3/75	11	89	0
A/Texas/1/77	0	100	0
A/Shanghai/31/80	0	100	0
A/Christ's Hospital/231/82	0	100	0
A/Los Angeles/2/87	0	100	0
A/Tottori/872/94	0	100	0
A/Tottori/872AM2AL3/94	83	17	0

III Molecular mechanism of the host range variation of influenza A viruses: Recognition of sialyllinkage, Neu5Ac2-3Gal and Neu5Ac2-6Gal

In order to eludidate the mechanism of this phenomena, we examined the sialyl chains in chicken embryonated eggs and MDCK cells. MDCK cells are now widely used for the isolation of the influenza viruses from the patiants. It was found that MDCK cells contain both 2-3 and 2-6 linkages (8). These linkages are also found in embryonated egg allantoic membranes which is also used for the first isolation from clinical samples of the virus from the patients. The presence of 2-6sialyl chains in

MDCK cells and egg amniotic membranes, makes more efficient to isolate the patient virus which binds specifically to 2-6 linkage. On the other hands, we found that cholio allantoic membranes in embryonated chicken eggs contain predominantly 2-3 linkages, and 2-6 linkage was not detected by the lectin NS. We, therefore, cultured human influenza viruses isolates, these are 2-6 specific, in several ways such as in MDCK cells, chicken eggs and cholio allantoic membranes.

The human isolate which cultured in MDCK cells which contains 2-3 and 2-6 sialyl sugar chain receptors has 2-6 specificity. In this case, no amino acid change in HA molecule observed, whereas, the virus isolates in egg cholio alantoic membranes which express 2-3 sialyl linkage showed different binding specificity. This virus shows no change of the binding specificity, and in this case, amino acid Ser186 changed to Ile. This amino acid change has no effect of the receptor binding specificity. But, other virus showed dramatical change of the receptor binding specificity, from 2-6 to 2-3. In this case, set of three amino acid or two amino acid has been changed in the hemagglutinin molecule. Both case, it was found that change of Leu 226 to Gln is in common.

This result indicates that the amino acid 226 is one of the clitical amino acid for the recognition of the 2-3 and 2-6 sialyl linkages. When the amino acid 226 is leu, HA bind to 2-6, however, 226 changed from Leu to Gln, receptor binding specificity changed to 2-3 predominant. This single amino acid alteration can be occurred frequently. Amino acid 226 is located in the left edge of the receptor binding pocket, which is near to the position of glycosidic linkage of the receptor sialyl linkage.

Recently, we also determined the sialic acid species in the target organs for influenza viruses (9). Human tissue has no Neu5Gc species in the tissue and cells, however, pig and horse trachea epitherial cells, which are target for influenza A viruses, contains significant amount of Neu5Gc. We determined the receptor binding specificity of the viruses isolated from pig and horse, and found that they bind to Neu5Gc2-3Gal-containing ganglioside GM3 and it binds also to the Neu5Ac2-3Gal GM3, whereas human isolates, PR/8/34 did not bind to Neu5Gc GM3 at all. These results indicate that the influenza viruses isolated from Neu5Gc-containing host bind to Neu5Gc containing receptor sugar chains (9).

We also found that only some strains isolated from human population agglutinate pig erythrocytes which contain significant amount of NeuGc2-6Gal sugar chains(10). We examined the receptor binding specificities using sialyl type 2 chain gangliosides analogs, which contain Neu5Ac2-3, 2-6 and Neu5Gc 2-3, 2-6. Some human isolates which agglutinate pig erythrocytes bind Neu5Gc2-6 linkage and also Neu5Ac2-6 linkage(11). So we determined amino acid sequence of the hemagglutinin from the human viruses which bind to Neu5Gc and then compared to the other human virus that bind only Neu5Ac. We found that these viruses has single amino acid substitution at 155 and/or 158. When amino acid 155 Thr and/or 158 Glu, the virus binds to Neu5Ac2-6Gal, however, after alteration of 155 Thr to Tyr and/or 188 Glu to Gly, binding specificity was changed to

Neu5Gc2-6Gal in addition of the Neu5Ac2-6Gal. Amino acid 155 and 158 are both located in very close position of the *N*-acyl (*N*-acetyl, *N*-glycolyl) residue of the sialic acid in the receptor binding pocket.

SUMMARY

There are at least two possible mechanisms for the host range variation of influenza A viruses. First mechanism is the pressure of the antibody, this results has already reported by us, in 1989(12). In this case, Ser 205 which is located in antigenic site D in the hemagglutinin trimer. When Ser 205 is substituted to Tyr, receptor binding specificity become 2-6 predominant rather than 2-3 predominant. Ser205 is located far away from the receptor binding pocket in the subunit itself, however, very close to the pocket in the next subunit. When the Ser 205 is substituted to bulky amino acid Tyr, The three dimensional structure of the receptor binding pocket in next subunit may be influenced to become 2-6 binding property rather than 2-3 binding(12).

Second mechanism is the selective pressure for the appearance of host cell variant with altered receptor binding specificities(8). In this case, amino acid 226 located in receptor binding pocket is clitical for the recognition of 2-6 and 2-3 linkages. Only single amino acid alteration of Leu 226 to Gln made remarkable change of receptor binding specificity 2-6 to 2-3.

In this study, we also identified amino acid 155 Thr, 158 Glu, and 228 Ser of the hemagglutinin which may be clitical for the recognition of the molecular species of sialic acid, such as Neu5Ac and Neu5Gc (11,15,16). Very recently, we reported the amino acid residues contributing to the substrate specificity of the influenza A virus sialidase (13, 14).

If these amino acid alteration occurres in the host animal bodies in nature, the virus aquires the new receptor binding specificity which allows the infection to the different animal species, such as bird to human population. This process may be one of the important mechanism for the host mediated variation, and the influenza virus transmission between different hosts, and also emerging of new subtypes of influenza viruses in human population.

REFERENCES

1) Yasuo Suzuki: Gangliosides as influenza virus receptor -Variation of influenza virus and their recognition of the receptor sialo-sugar chains-. Prograss in Lipid Res., 33(4), 429-457, 1994.
2) J.K. Taubenberger: Seeking the 1918 Spanish influenza virus. Gene sequences of the virus from the 1918 influenza pandemic are yielding insights into its origin but little about virulence. ASM News, 65(7) 473-478, 1999.
3) D. C. Wiley, J.J.Skehel: The structure and function of the hemagglutinin membrane glycoprotein of influenza virus. Annu. Rev. Biochem., 56, 365-394, 1987.

4) Yasuo Suzuki, Toru Nakao, Takeshi Ito, Norifumi Watanabe, Yoshihisa Toda, Guiyun Xu, Takashi Suzuki, T Kobayashi, Y Kimura, Akira Yamada, K Sugawara, H Nishimura, F Kitame, Kiyoto Nakamura, E Deya, Makoto Kiso, Akira Hasegawa: Structural determination of gangliosides that bind to influenza A, B, and C viruses by an improved sialo-sugar chains. Virology, 189, 121-131, 1992.
5) Guiyun Xu, Takashi Suzuki, Harunobu Tahara, Makoto Kiso, Akira Hasegawa:Specificity of sialylsugar chain mediated recognition by the hemagglutinin of human influenza B virus isolates. J. Biochem., 115, 202-207, 1994.
6) J.N. Couceiro, J.C. Paulson, L.G. Baum: Influenza virus strains selectively recognize sialyloligosaccharides on human respiratory epithelium: the role of the host cell selection of hemagglutinin receptor specificity. Virus Res., 29, 155-165, 1993.
7) Kathleen Pyan-Poirier, Yasuo Suzuki, William J. Bean, Darwyn Kobasa, Ayako Takada, Toshihiro Ito, Yoshihiro Kawaoka: Changes in H3 influenza A virus receptor specificity during replication in humans. Virus Res., 56, 169-176, 1998.
8) Toshihiro Ito, Yasuo Suzuki, Ayato Takada, Ayumi Kawamoto, Koichi Otsuki, Hiroyuki Masuda, Takashi Suzuki, Hiroshi Kida, and Yoshihiro Kawaoka:Differences of sialic acid-galactose linkages in the chicken egg amnion and allantois influence human influenza virus receptor specificity and variant selection. J. Virol., 71(4), 3357-3362, 1997.
9) Takashi Suzuki, Goh Horiike, Yasuhiro Yamazaki, Kaoru Kawabe, Hiroyuki Masuda, Daisei Miyamoto, Masao Matsuda, Shin-Ichiro Nishimura, Tatsuya Yamagata, Toshihiro Ito, Hiroshi Kida, Yoshihiro Kawaoka and Yasuo Suzuki: A Swine Influenza virus strains recognize sialylsugar chains containing the molecular species of sialic acid predominantly present in the swine tracheal epitherium. FEBS Lett., 404, 192-196, 1997.
10) Toshihiro Ito, Yasuo Suzuki, Lyndon Mitnaul, Angela Vines, Hiroshi Kida and Yoshihiro Kawaoka: Receptor specificity of influenza A viruses correlates with the hemagglutination of erythrocytes from different animal species. Virology, 227, 493-499, 1997.
11) Hiroyuki Masuda, Takashi Suzuki, Yoshiyuki Sugiyama, Goh Horiike, Kohji Murakami, Daisei Miyamoto, Kazuya I.-P. Jwa Hidari, Toshihiro Ito, Hiroshi Kida, Makoto Kiso, Kyoko Fukunaga, Masanobu Ohuchi, Tetsuya Toyoda, Akira Ishihama, Yoshihiro Kawaoka, Yasuo Suzuki: Substitution of amino acid residue in influenza A virus hemagglutinin affects recognition of sialyl-oligosaccharides containing N-glycolylneuraminic acid. FEBS Lett., 464, 71-74, 1999.
12) Yasuo Suzuki, Hideshige Kato, C.W. Naeve, Robert G. Webster: Single amino acid substitution in an antigenic site of influenza virus hemagglutinin can alter the specificity of binding to cell membrane associated gangliosides. J. Virol., 63, 4298-4302, 1989.
13) Darwyn Kobasa, Shantha Kodihalli, Ming Luo, Maria R. Castrucci, Isabella Donatell, Yasuo Suzuki, Takashi Suzuki, Yoshihiro Kawaoka: Amino acid residues containing to the substrate specificity of the influenza A virus neuraminidase. J. Virol., 73, 6743-6751, 1999.
14) Katsuhiko Sato, Goro Hanagata, Makoto Kiso, Akira Hasegawa, Yasuo Suzuki: Specificity of N1 and N2 sialidase subtypes of human influenza A virus for natural and synthetic gangliosides. Glycobiology, 8 , 527-532, 1998.
15) Toshihiro Ito, Yasuo Suzuki, Takashi Suzuki, Ayato Takada, Taisuke Horimoto, Krisna Wells, Hiroshi Kida, Koichi Otsuki, Makoto Kiso, Hideharu Ishida, Yoshihiro Kawaoka: Recognition of *N*-glycolyl neuraminic acid linked to galactose by the α2,3 linkage is associated with the intestinal replication of influenza A virus in ducks. J. Virol., 74, 9300-9305, 2000.
16) Yasuo Suzuki, Toshihiro Ito, Takashi Suzuki, Robert E. Holland Jr,Thomas M. Chambers, Makoto Kiso, Hideharu Ishida, and Yoshihiro Kawaoka: Sialic acid species as a determinant of the host range of influenza A viruses. J. Virol., 74, 11825-11831, 2000.

BACTERIAL POLYSACCHARIDES AS VACCINES – IMMUNITY AND CHEMICAL CHARACTERIZATION

Chi-Jen Lee[1], Lucia H. Lee[1], Cheng-shiung Lu[2], and Albert Wu[3]

Center for Biologics Evaluation and Research, FDA, Bethesda MD, U.S.A.[1]
Center for Disease Control, Taipei, Taiwan[2]; Institute of Molecular and Cellular Biology, Chang-Gung Medical College, Tao-Yuan, 333 Taiwan[3]

Studies on protective immunity and biochemical characterization of bacterial capsular polysaccharides (PSs) have led to many significant contributions to understanding the mechanisms of infectious diseases, host defense, and development of effective vaccines. The discovery of a specific soluble substance produced by pneumococci as a carbohydrate, provided evidence regarding the specificity of PS antigens for antibody production. It was the first non-protein antigen found to induce antibody response in humans (1). The molecular basis of immunogenicity to many PS antigens has been elucidated. Antibody responses to PS immunogens can be greatly affected by their physical and chemical properties including molecular size, specific determinants, and conformation (2).

Encapsulated bacterial pathogens still cause most prevalent and serious bacterial infections in the United States and many other areas of the world. The mortality and morbidity of bacteremia, meningitis, and pneumonia caused by these organisms are particularly high in children, the elderly, and immunodeficient patients (3-7). Several factors, including the high mortality and morbidity, endemic and epidemic diseases, and expanding incidence of antibiotic-resistance strains have emphasized the importance of vaccine development for prevention of bacterial infections.

The development of bacterial PS vaccines has demonstrated the full cycle of scientific discovery from laboratory on chemical structure of virulence factors to the clinical diagnosis, treatment, and eventual prevention of infectious diseases. During each stage, significant knowledge has brought forth further understanding on basic biology of bacterial infections, and importance of immunization with PS vaccines for control of infectious diseases.

*Part of this work was supported by Grant 676 from the Chang-Gung Medical Research Project, Kwei-san, Taiwan, and Grant NSC 89-2320-B-182-003 from the National Science Council, Taipei, Taiwan.

I. ISOLATION AND BIOCHEMICAL CHARACTERIZATION OF BACTERIAL PSs.

Capsular polysaccharides are cell surface polymers consisting of oligosaccharide repeating units. These PSs are the main virulence factor responsible for the pathogenesis of encapsulated bacteria. Invasiveness depends more on the composition than on the amount of capsular PS produced. For example, pneumococcal type 3 and type 37 are both heavily encapsulated, but type 3, which is composed of a polymer of glucose and glucuronic acid, is highly invasive, whereas type 37, with a homopolymer of glucose, is not pathogenic (8). Within the same serotype, however, the size of the capsule may correlate directly with the virulence to cause disease.

Most bacterial PSs are negatively charged and bind to quaternary ammonium detergent (Cetavlon), allowing their isolation from the cells by differential ethanol precipitation and purification by organic solvent extraction or column chromatography and ultracentrifugation. A diagram of isolation and purification of bacterial polysaccharides, e.g. pneumococcal PS, is shown in Figure 1 (9, 10).

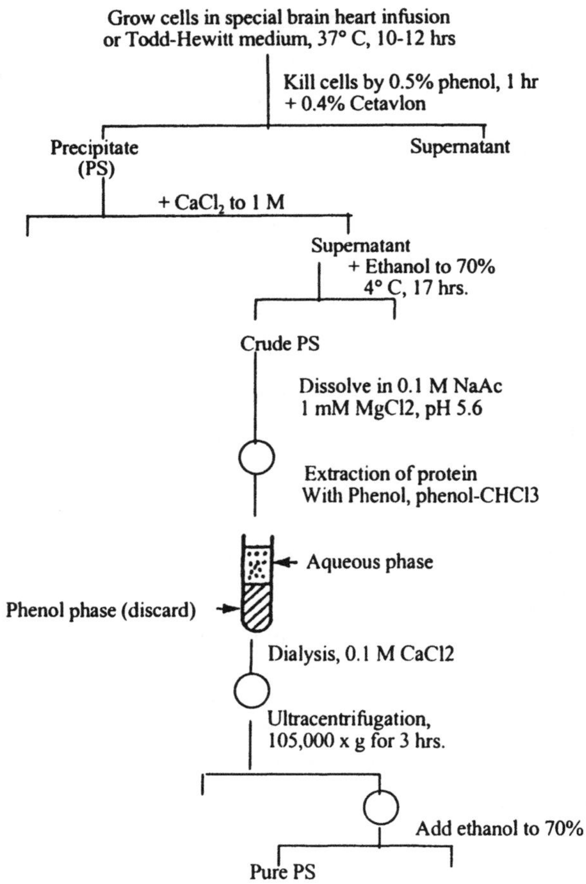

Fig. 1. Isolation and purification of bacterial PSs.

Each serotype or serogroup can be identified by its chemical structure and its reaction with specific antisera. The Quellung reaction has been used widely for serotyping test. When bacteria are reacted with homologous antiserum, a distinct capsular immuno-precipitation or swelling can be observed under the microscope after adding methylene blue dye. In addition, coagglutination and latex agglutination tests have been used for serotyping that provides equivalent response as Quellung test (11). The simplicity of the agglutination tests may permit wider use of serotyping in epidemiological studies.

Most adult animal sera contain antibodies against PS antigens of various pathogenic bacteria. For example, antibodies to pathogens, e.g. meningococcal group A, B, C, *Haemophilus* type b, and pneumococcal type 3 were detected in animals as they grew or in human adults who had never contacted with these organisms. The possible antigenic sources of these serum protective antibodies have been identified and found to be derived from cross-reactive antigens distributed in intestinal and pharyngeal bacteria (12,13). Several non-pathogenic bacteria produce PSs with identical or similar chemical structures to those of pathogenic organisms (Table 1). These cross-reactive PSs are closely related to the natural immunity developed during the early life.

Table I.
Immunologic cross-reaction between PSs of pathogenic of bacteria and other bacteria

Pathogenic bacteria	Structure of PS repeating unit	Cross-reacting bacteria
Meningococcus		
Group A	->6)α-D-ManNAc(1-PO_4-> | OAc	*E. coli* K93
Group B	->8)α-D-NeupNAc(2->	*E. coli* K1
Group C	->9)α-D-NeupNAc(2-> 7/8 | OAc	
Pneumococcus		
Type 19F	->4)β-D-ManpNAc(1->4)α-D-Glc(1->2)α-L-Rhap(1-PO_4->	*S. suis* type 8
Type 14	-4>)Glc-GlcpNAc-Gal(1-> Gal	Gr B Str III
Haemophilus b	->3)β-D-ribosef(1->1)D-ribitol(5-PO_4->	*E. coli* K100

Studies on the structure and immunological properties of various bacterial PSs have provided the knowledge for the chemical basis of immunogenicity to PSs. Immunologic cross-reactivity has been frequently observed among pneumococcal PSs. Pneumococcal group 9 includes types 9N, 9A, 9L, and 9V. Types 9N and 9V comprise about 90% of the group 9 disease isolates. In addition, among pneumococcal group 19, types 19F and 19A account for almost all disease isolates in humans from this group, whereas types 19B and 19C are not pathogenic in humans and animals. Studies on antibody response to groups 9 and 19 PSs in humans led to a recommendation to include both 9N and 9V, as well as 19F and 19A PSs in the 23-valent PS vaccine. The structures of group B streptococcal (GBS) PSs exhibit close structural similarities, but it is rare to see different types within a group. Moreover, all these structures exhibit extensive structural homology with oligosaccharide components of human glycoproteins (14). This structural similarity may act as a virulence factor by enabling the bacteria to evade the host defense mechanism. The presence of terminal sialic acid in GBS type III PS can suppress activation of the alternative complement pathway (15).

II. PROTECTIVE IMMUNITY AGAINST ENCAPSULATED PATHOGENS

Capsular PSs are involved in interactions with the immune system to produce protective immunity. These PSs are highly polar and hydrophilic and interfere with cell-to-cell interactions with phagocytes. The ability of host phagocytes to ingest and kill invading organisms requires coating of the bacteria by specific antibodies and complement. This process, called opsonization, results in phagocytosis, which destroys invading bacteria. Antibodies bound to capsules may act as bacterial cell-to-phagocytic cell ligands or as complement activator. Anticapsular antibodies increase the rate of clearance of pneumococci, primarily through the activation of the classical complement pathway (16,17). In contrast, these antibodies destroy *Haemophilus* type b through the lysis of bacteria.

The antibody response to capsular PSs is independent of control by the thymus (T-independent, TI), whereas the response to protein antigens is thymus-dependent (TD), requiring T lymphocytes for both its induction and regulation. The antibody response is regulated by T cells that have amplifier and suppressor functions. Both amplifier and helper T cells are CD4+, CD8-, and both positively enhance the antibody response, which in mice is mainly IgG3. Amplifier T cells expand an already ongoing antibody response. In contrast, suppressor T cells, exhibiting CD8+, CD4-, inhibit the B cell proliferation after antigenic binding. The TD and TI antigens are shown in Table 2. Amplifier and suppressor T cells respond to idiotypic determinants of cell-associated, type-specific antibody on the surface of immune B cells. Activated suppressor T cells are influenced by lymphokines produced by helper T cells. Interleukins IL-2, IL-4, IL-5 and interferon-r appear to be required for the activation or clonal expansion of suppressor T cells (18).

In newborn mice, suppressor T-cell activity develops within 2 weeks, whereas amplifier T cells are not fully mature until 8 to 10 weeks (19). The dominance of suppressor T cells in early life induces poor antibody response of neonate mice to TI-2 antigens. Antibody response of newborns could be improved through enhancing the function of amplifier T cells, eliminating the suppressor T cell activity, or immunization with PS-protein conjugate.

The immune system of infants differs from that of adults in both its lack of previous antigenic exposure and its functional immaturity. Infants obtain IgG almost exclusively through transplacental passage, and endogenous immunoglobulin is largely restricted to IgM antibody (20,21). In contrast, rodents acquire maternal antibodies both before and after birth (22).

Table II.
Characterization of T-dependent and T-independent antigens

Characteristics	T-dependent	T-independent (Type 2)
Th-cell activation	++	-
IgM -> IgG switch	+ IgG1	- IgM
Booster response	++	-
Affinity maturation	+	-
Immune response in children	High +	Low –
Requirement for Lyb5+ Mature B-cells	-	+
Development of Ab response	At birth	3-18 mo/humans 3-6 wk/mice
Example:	Protein Ags PS-protein Conjugate	TI-1; *B abortus* TI-2; Bacterial PSs

Immune response to most bacterial PS antigens is poor in children less than 2 years of age. This lack of responsiveness to PS antigens makes the period between 6 to 24 months of age susceptible to high incidence rate of pneumococcal, meningococcal, and *Haemophilus* type b infections (23,24).

Individuals at high risk for serious infection of encapsulated pathogens are frequently unable to respond to PS vaccine. Human immune globulin has been successfully used for both prophylaxis and treatment of pneumococcal bacteremia and otitis media. The problems of increasing antimicrobial resistance, the poor response to active immunization among immunodeficient patients, and advance in technology to produce safer, more specific human immunoglobulins have led to interests in using immunoglobulin preparations for prevention of bacterial infections. Administration of two doses of BPIG, a hyperimmune globulin obtained from adults immunized with 14-valent pneumococcal, meningococcal A, C, and Hib polysaccharide vaccines to young children with previous acute otitis media experienced significantly fewer episodes of pneumococcal otitis media (25).

III. BACTERIAL PS AND GLYCOCONJUGATE VACCINES

A. *Neisseria meningitidis*

Meningococcus, a Gram-negative organism, is divided into 12 serogroups (A, B, C, H, I, K, L, X, Y, W135, Z, and 29E) based on the chemical and immunologic specificity of capsular PSs. Approximately 90% of meningococcal diseases in various areas of the world is caused by groups A, B, and C. Among these, group A is predominant in Asia and Africa as a major cause of bacterial meningitis. Groups B and C are prevalent in Europe and Americas in causing meningococcal infections. During 1992-1995, serogroup Y accounted for an increasing proportion of meningococcal disease in the United States, while the proportion of serogroup W135 meningococcal disease remained steady (26,27). The other groups rarely cause disease.

Meningococcal groups A and C PS vaccines were licensed in 1976 in the United States whereas tetravalent A, C, Y, W135 PS vaccine was licensed in 1982 (28,29). The structure of the repeating units of these meningococcal PSs are shown in Table 3. The antibody response to meningococcal PS is age-dependent. Meningococcal group A PS is immunogenic in children aged 6 months or more (30). Children between 3 and 17 months who received a second injection of group A PS had a booster response (31,32). Booster responses are not usually observed with T-cell independent antigens. Group C PS induces significant increases in antibodies in infants aged 3 months, but antibody levels fall rapidly. In contrast to group A PS, no booster response is seen to group C PS. Immunization of adults with group Y and group W135 PSs has induced over 90% of fourfold or greater increases in bactericidal antibody levels.

Table III.
Structure of meningococcal capsular polysaccharides in vaccine

Group	Structure of repeating unit
A	-6)α-D-ManNAc(1-PO_4-> | 3 OAc
C	->9)α-D-NeupNAc(2-> 7/8 OAc
Y	->6)α-D-Glcp(1->4)α-D-NeuAc(2->(+OAc)-
W135	->6)α-D-Galp(1->4)α-D-NeuAc(2->

While the current PS vaccines are very effective in controlling outbreaks and epidemics, they do not induce antibody response in infants and long-term protective immunity in older children. These problems can be overcome by development of group A and C PS-protein conjugate vaccines. The *Haemophilus* type b PS vaccine was not effective in children under 18-24 months of age, whereas Hib glycoconjugate vaccines have been shown to be highly effective in infants 2 months of age (33,34). Meningococcal groups A, B, and C PSs were conjugated directly with tetanus toxoid by reductive amination. Furthermore, oligosaccharides from group C PS were linked through a spacer molecule to bovine serum albumin.

The group A and C conjugates were highly immunogenic in mice but the group B conjugate was not (35). Oligosaccharides derived from group A and group C PSs were covalently coupled to a non-toxic mutant of diphtheria toxin, CRM$_{197}$ using adipic acid spacer and evaluated in clinical trials recently (36).

Serogroup B PS is essentially non-immunogenic in humans. It is postulated that oligosaccharide of α(2->8)-linked sialic acid units are present on gangliosides of animal and human cell surfaces, so that the immune system will not produce antibody against an antigen having similarity to self-structure (37,38). Another possible explanation is that group B PS is easily degraded by neuraminidase and acidic pH and formed internal esters with resulting loss of antigenicity (39,40).

Attempts have been made to overcome the intrinsically poor immunogenicity of the group B PSs by chemical modification of its structure. Substitution of N-propionyl groups for the N-acetyl groups of group B PS and conjugation of this modified PS to tetanus toxoid produced a high level of IgG antibody cross-reactive with group B PS. The mouse antiserum was bactericidal for group B strains. When the modified group B PS was covalently bound to an affinity column, the column-bound IgG_{2a} and IgG_{2b} antibodies were shown to be bactericidal. In contrast, the bound IgG_1 induced from native group B PS was non-bactericidal. Furthermore, at least 10 sialic acid residues were required to form a conformational epitope capable of binding bactericidal antibodies (41). Group B PS forms a helical structure although at any point in time most of the molecule exists as a random coil (42-45). Only the helical structures appear to be responsible for antibody binding. The N-propionylated PS adopts the same extended helical structure (43).

Polysialyl linkages are present on a neural cell adhesion molecule (N-CAM) in human and animal fetal brain tissue. Only sialylated N-CAM specifically binds group B PS antibodies. This finding has led concern over possible induction of IgG antibodies to the group B PS through immunization of conjugates (43). Although there is no evidence to substantiate these concerns, the non-possibility of inducing autoimmunity due to reaction with polysialyl glycopeptides in human tissues should be established before such group B conjugates are developed.

B. *Streptococcus pneumoniae*

There are almost 90 serologically defined pneumococcal types. However, approximately 30 types account for most pneumococcal disease isolates. In order to reduce their incidences, a 14-valent PS vaccine was developed and licensed in the United States in 1977. In addition, a 23-valent PS vaccine was licensed in 1983. The disease types isolated and included in the vaccine accounted for 88.2% in the United States and Europe. These types were considerably fewer in Asia; the proportion of the vaccine types in disease isolates was 62.9% in Taiwan, 72.9% in Japan, and 79.9% in China (2,46-48). A large number of disease isolate types found in Asia, such as 15A, 23A, 24F, 27, and 34 are not present in the current pneumococcal vaccine. It may be necessary to develop a supplementary vaccine to include additional types of pneumococcal PSs to be used for the Asia population.

Another population that is not fully protected by the PS vaccine includes young children less than 2 years old. Their immune systems have not reached maturity to produce sufficient protective immunity (49,50). Several pneumococcal PSs have been conjugated covalently to proteins to improve their immunogenicity. Antigen-presenting cells take up the glyco-conjugate molecules and present the peptides in association with MHC class II molecules on their surface. This induces helper T cells to stimulate PS-specific B cells that have taken up the conjugate molecules. These B cells then produce specific antibodies and mature into memory cells (51,52).

Although it is preferable to include many different PSs in a conjugate vaccine, the total amount of carrier protein in the final vaccine should be limited. Too much carrier protein may impair the antibody response to the PS antigen through antigenic competition or carrier-mediated epitope suppression. The carrier proteins used in pneumococcal conjugate vaccines include tetanus (T) and diphtheria (D) toxoids, a non-toxic mutant of diphtheria toxin (CRM_{197}), and the outer membrane protein complex (OMPC).

Several investigators have considered the possibility of using pneumococcal protein antigens such as pneumococcal surface protein A (PspA), pneumolysin (*Ply*), or hemin-binding protein as an alternative protein vaccine (53-56). These protein antigens can provide broader, non-specific immunity. They can also be used as carrier for the conjugate to provide additional antigen or cross-protection for prevention of pneumococcal infection.

In clinical studies, the pneumococcal conjugate vaccines including Pnc-CRM, Pnc-D, and Pnc-T have shown to be immunogenic and to induce immunological memory (57,58). Compared with PS vaccine, these 3 conjugate vaccines greatly increase the serotype-specific opsonophagocytic activity of post-immune serum (59). Furthermore, individuals who do not respond to the PS vaccine may show IgG response to conjugate vaccine (60). Clinical trials conducted in pneumococcal PS-protein conjugate vaccines are shown in Table 4 (61-68).

Table IV.
Clinical trials conducted in pneumococcal PS-protein conjugate vaccine (61-68)

Conjugate vaccine (Serotype)	(Carrier)	Clinical trial I	II	III	Remark
Merck Sharp & Dohme:					
4,6B,9V,14,18C,19F,23F	MOMP	+	+	+	Immunized to infants at 2,4,6 months of age for prevention of otitis media.
Merieux-Connaught:					
1,4,5,7F,9V,19F,23F	TT				Immunized to children for prevention of
3,6B,14,18C	DT	+	+	+	otitis media.
Wyeth-Lederle:					
4,6B,9V,14,18C,19F,23F*	CRM_{197}	+	+	+	Immunized to infants at 2,4,6 mos of age.
6B,14,18C,19F	CRM_{197}				
1,5,6B,14,18C,19F	"	+	+	+	Immunized to infants and 60-85 yr of age.
SmithKline Beechem:					
1,3,4,5,6B,7F,9V	r-PD				
14,18C,19F,23F	"	+	+	+	Immunized to infants at 2,4,6 mos of age.

Note: MOMP, Meningococcal outer membrane protein; TT, Tetanus toxoid; DT, Diphtheria toxoid; CRM_{197}; Inactivated diphteria toxin; r-PD, Recombinant protein D; *, Licensed, February 2000; +, Studies and trials conducted.

At present, infants are the main target population for pneumococcal conjugate vaccine. In general, all pneumococcal conjugates studied have been immunogenic in infants who received conjugate vaccine between 2 to 6 months of age. They induced 1.4 to 14.1-fold increases in serum levels of anti-pneumococcal PS antibodies (61). In addition, the immune response of infants to the conjugate vaccination may be enhanced to maternal immunization with the same vaccine (69). Clinical studies from various areas including Israel, Gambia, Iceland, and South Africa have reported that conjugate immunization is associated with a reduction in naso-pharyngeal carriage of vaccine-type pneumococci. This could be important for preventing the spread of antimicrobial resistant pneumococcal strains in the community.

C. *Haemophilus influenzae* type b

The gram-negative bacillus *Haemophilus influenzae* can be classified serologically to types a, b, c, d, e, and f. Among these 6 types only type b (Hib) causes serious diseases. Before Hib vaccine was developed the rates of invasive disease in the United States ranged from 50-100 cases per 100,000 children below 5 years of age, with rates of meningitis between 30-60 per 100,000 (70-72). In contrast, the incidence of Hib disease in Asia (e.g. Japan) has been suggested to be substantially lower than in the United States and Europe. In 1985, Hib PS (also called polyribosyl-ribitol-phosphate, PRP) vaccine was licensed in the United States for use in children after the age of 18 to 24 months. Furthermore, Hib conjugate vaccine was licensed in 1987 for children at 18 months of age and older. The Hib conjugate vaccine produced by two manufacturers was approved for immunization of infants beginning at 2 months of age in 1990. Praxis conjugated PRP oligosaccharide to a toxoid from the mutant diphtheria toxin gene, CRM_{197} (HbOC, Lederle-Praxis) whereas Merck, Sharp and Dohme linked outer membrane protein from group B meningococci with PRP (PRP-OMP, MSD). Tetanus toxoid has also been used as carrier for the conjugate (PRP-T, Pasteur Merieux-Connaught). Each of these vaccines differed not only in the protein carrier but also in the size of the saccharide chain and "spacer" between the protein and the saccharide unit used (73).

The immunogenicity of these vaccines differs considerably (74). HbOC and PRP-T induced similar pattern of antibody response, each requiring 2 or 3 doses to reach high antibody levels. In contrast, PRP-OMP showed a markedly different pattern after giving one dose of this conjugate vaccine. Most recipients reach serum antibody levels of 1.0 ug/ml. The antibody levels increased little and reached the plateau after second and third dose. It is interesting to note that the geometric mean titer reached by PRP-OMP immunization is frequently lower than that induced by HbOC or PRP-T conjugate. The antibody levels of these conjugates decline gradually after the third dose. In the United States, a booster dose at 12-15 months of age has been recommended after a primary series of immunization (at 2, 4, 6 months of age).

In 1986, a total of 114,000 infants were involved in efficacy study of PRP conjugate vaccine (PRP-D, diphtheria toxoid), in which 4 doses were given at 3, 4, 6, and 14-18 months of age. Overall protective efficacy was 94% with 90% after 3 doses and 100% efficacy after the booster (fourth) dose (75). In addition, PRP conjugate vaccines demonstrated reduction of nasopharyngeal carriage of Hib among immunized children (76). Various surveillances were performed by the Centers for Disease Control (CDC) for reporting incidence of bacterial meningitis in the United States. Although rates of disease caused by the meningococcus and pneumococcus changed very little, the rate of Hib meningitis was reduced by over 80% during the period between 1988 and 1991 (77).

The remarkable success of Hib conjugate vaccines, with the extraordinary effect of herd immunity, led to optimistic attitude for rapid elimination of Hib diseases among children less than 5 years of age. There are 300 to 400 cases of invasive Hib disease still occurring each year in the United States (78). It is hoped that higher rates of immunization with effective conjugate vaccines will soon result in elimination of Hib diseases in the U.S.

Major difficulties in the way for achievement of elimination of Hib infection in various areas of the world include that the Hib conjugate vaccines are expensive and financially unacceptable for immunization programs in developing countries. Progress is needed in reducing costs for manufacture, distribution, and administration of Hib conjugate vaccines.

D. *Salmonella typhi*

Salmonella typhi is the only salmonella species that has a capsular PS, Vi antigen. Rarely, strains of *S. paratyphi* C also contain this antigen. It is composed of a linear homopolymer of α(1->4)-D-NAcGalpA-(variably O-acetylated to C2 and C3 (79). Purified Vi PS is of high molecular weight (approximately 4000 kDa) and immunogenic for mice and humans. Similar to capsular PSs of respiratory pathogens, Vi PS elicits protective levels of antibodies in older children and adults, and reinjection does not result in a booster response (80).

Two clinical trials were conducted in areas with a high incidence of typhoid fever to assess the safety and efficacy of the Vi PS vaccine. In Nepal, a 25 ug dose of Vi PS vaccine was immunized to 3457 individuals. There was no serious adverse reaction. Vaccine efficacy was 72% at 17 months (81). Similar results were obtained in a field trial in South Africa in which 25 ug dose of Vi PS vaccine conferred 64% protection against typhoid fever for at least 21 months (82). Later, Vi PS vaccine was licensed to Pasteur Merieux-Connaught (by FDA) in 1994. Recently, similar results were obtained in Hunan Province, China. Vi PS vaccine is now licensed in about 40 countries and the WHO has published minimum requirements for this vaccine (83).

The Vi PS conjugated to tetanus toxoid, diphtheria toxoid, and cholera toxin elicited higher levels of serum antibodies than purified Vi PS alone in mice and rhesus monkeys (84, 85). In addition, using the heterobifunctional linker SPDP, conjugates have been prepared and shown to be significantly more immunogenic than Vi PS alone (86). Clinical trials of Vi conjugate vaccines are currently ongoing.

E. Other encapsulated pathogens

1. Group B *Streptococcus*

The type-specific capsular PSs of group B *Streptococcus* (GBS) include types Ia, Ib, II, III, IV, V, VI, VII and VIII . Until 1985, the strains most commonly isolated from clinical cases belonged to types Ia, Ib, II and III, with type III being predominantly responsible for infection. Type III strains account for 66% of blood and cerebrospinal fluid isolates from all infant patients and up to 90% of isolates associated with meningitis (87,88). Recently, however, type V has been reported as a new major contributor to disease (89). These findings indicate that group B streptococci can change in the prevalence of serotypes associated with disease in different time and geographic locations.

The structure of the GBS PSs is unique in that it does not exhibit a group of capsular PSs. Moreover, all these structures exhibit extensive structural homology with oligoacharide components of human glycoprotein (14). This structure similarity may act as a virulence factor by enabling the bacteria to evade the immune defense mechanism of the human host. All PSs of the GBS types have a common trisaccharide core β-D-GlcpNAc (1->3)β-D-Galp (1->4)β-D-Glcp in their structure. The type III PS when its terminal sialic acid residue is removed, has a structure identical to that of the pneumococcal type 14 PS. It was suggested that antibodies to the type 14 pneumococcal PS would protect against type III GBS organisms. However, it was found that the individuals failed to induce significant antibody to GBS type III after immunization with pneumococcal PS vaccine (90,91).

Despite structure similarities between the type-specific GBS PSs, they are highly immunospecific. They are also immunogenic in humans (87). It has been observed that the protective epitope of types Ia, II, and III is length-dependant and conformational in nature, controlled by the presence of the terminal sialic acid residues (14).

Since GBS strains cause diseases in the newborn, the PS vaccine can not be used. Immunization of childbearing women before or during pregnancy with a GBS PS or PS-protein conjugate vaccine is a promising strategy for the prevention of GBS infection in infants. Eighty percent of the infants born to immunized mothers with type III GBS PS exhibited protective immunity at 1 month of age and 64% had protective levels at 3 months (92). Immunization of healthy women with a type III PS or III conjugate vaccine elicited a 50-fold increase in type III PS-specific antibody titers, while immunization with type III PS vaccine elicited a 7-fold increase. Even at lower doses, 90% of recipients reacted to the type III conjugate vaccine with a 4-fold increase in type III PS-specific antibody levels, in contrast to only 50% of recipients of the type III PS vaccine (93).

2. *Escherichia coli*

E. coli O157:H7 strains containing either Shiga-like toxins I and/or II cause a serious hemorrhagic enteritis, which is a major cause of hospital admissions of children in the United States (94,95). The highest incidence of *E. coli* O157:H7 enteritis is in children and in older individuals. Approximately 5% of patients with severe enteritis also develop hemolytic uremic syndrome as a complication (96,97). Other sources of this pathogen include cattle, pigs and fowl (98,99).

The O-specific PS is a linear homopolymer composed of a tetrasaccharide (100). Conjugates of the O-specific PS of *E.coli* O157 with bovine albumin, *Closstridium welchii* toxoid, or recombinant *P. aeruginosa* exprotein A (rEPA) induced serum antibodies in mice with bactericidal activity (101). A clinical trial demonstrated that O157-rEPA conjugates were safe and highly immunogenic in adults. All 75 volunteers produced high levels of IgG anti-LPS with bactericidal activity (102).

3. *Klebsiella and Pseudomonas*

During the past three decades there has been a dramatic increase in the incidences caused by gram-negative bacilli, particularly among hospitalized patients (103, 104). *Klebsiella pneumoniae* is one of the most frequently isolated pathogens causing pneumonia, bacteremia, and urinary tract infection, particularly predominant among immunodeficient patients and hospital staff (105,106). *Klebsiella* PSs are involved in the virulence factor. They are acidic PSs composed of repeating units consisting of galacturonic or glucuronic acid linked to hexoses and 6-deoxyhexoses (107) and additional moieties such as acetyl, formyl, or pyruvate groups. These PSs extensively cross-react with pneumococcal capsular PSs (108, 109).

At present, a 24-valent PS vaccine has been developed for Phase I and II clinical trials. It includes types 2, 3, 5, 9, 10, 15, 16, 17, 18, 21, 22, 25, 28, 30, 35, 43, 52, 53, 55, 60, 61, 62, 63, and 64, which account for approximately 70% of bacteremic strains. This polyvalent vaccine was administered at doses of 25 and 50 ug/PS to 73 healthy volunteers at Fort Lewis, Washington. No adverse reactions were reported by 44 and 60%, respectively, of volunteers. The vaccine elicited IgG antibody by ELISA to at least 19 of 24 antigens in more than 50% of volunteers 4 weeks after immunization, and to 10 PS serotypes not included in the vaccine. Furthermore, the pools of sera increased in opsonic activity for 21 of the 24 serotypes (110-112).

Pseudomonas aeruginosa is an opportunistic pathogen that causes fatal nosocomial infections. *P. aeruginosa* infects people whose immune systems have been weakened by severe burns, cystic fibrosis, immunosuppressive or cancer chemotherapy. Intravenous immune globulins (IVIGs) have been used with limited success for the prevention of *P. aeruginosa* infections. A IVIG has been prepared by simultaneously immunizing plasma donors with an octavalent *P. aeruginosa* O-polysaccharide-toxin A conjugate vaccine and a 24-valent *Klebsiella* PS vaccine (113,114). Both pharmacokinetic and Phase III efficacy trial have been conducted in healthy volunteers (115, 116).

The combined *Klebsiella* and *Pseudomonas* hyper-immune IVIG was examined in 1392 patients admitted to intensive care units of hospitals. The IVIG at 100 mg/kg was administered to patients and then followed for the development of type-specific infection over a 6 week period. While there was a trend towards a decrease in severe *Klebsiella* infection, the incidence of *Pseudomonas* infection was not reduced (117).

Several attempts have been made to develop safe and effective vaccines against *P. aeruginosa*. A vaccine of LPS-derived PS conjugated with toxin A or tetanus toxoid has shown to elicit antibody response which is opsonic and preventive from fatal burn wound sepsis whereas mucoid exopolysaccharides induced opsonic antibodies in humans (118,119).

Klebsiella and *Pseudomonas* vaccines were administered to patients with acute blunt and penetrating trauma wounds and the antibody responses were measured at 14 and 28 days after immunization. All of the patients responded with a fourfold rise in IgG antibody level for most of the vaccine antigens (120). No significant side effects were caused by these vaccines.

IV. CONCLUSIONS

Studies on immunity and chemical characterization of encapsulated bacterial polysaccharides have led to many significant contributions to understanding the defense mechanisms of infectious diseases and development of effective vaccines. In young children the antibody response to most capsular PSs is generally low. Enhanced antibody response to PS antigens can be achieved through the use of PS-protein conjugate vaccine, more effective adjuvant for antigen delivery system, and immunization during critical periods of neonate development.

The improvement of chemical technology to prepare PS-protein conjugate having strong and stable covalent linkage, and better understanding of the chemical characterization of and mechanisms involved in the immune response to glycoconjugate vaccines will stimulate the rational design to develop and produce new vaccines. The continued design of more effective and safe vaccines for the prevention of bacterial infections requires intensive basic research and steady support from academia, industry, and government.

SUMMARY

Studies on protective immunity and biochemical characterization of bacterial capsular polysaccharides have led to significant contributions to understanding of the mechanisms of infectious diseases and development of effective vaccines. Immunity to encapsulated bacteria is related to antibody response to polysaccharide (PS) antigen, interactions with T- and B-lymphocytes, and host defense mechanisms.

Meningococcal, pneumococcal and *Salmonella* vi PSs and *Haemophilus* type b PS-protein conjugate vaccines have been licensed and provided effective immunity for prevention of these bacterial infections. Capsular PSs are cell-surface polymers consisting of oligosaccharide repeating units. Many PSs are highly polar and hydrophilic and interfere with cell-to-cell interactions with phagocytes. Most pneumococcal PSs are negatively charged and possess acidic components such as D-glucuronic acid and phosphate in phosphodiester bonds. Extensive immunologic cross-reactivity has been observed among bacterial capsular PSs.

In infants the antibody responses to most capsular PSs are generally poor. Enhanced immunogenicity of PS antigens can be achieved through PS-protein conjugate vaccines, immunization during a critical period of perinatal development and effective antigen delivery system.

REFERENCES

1. Dochez, A.R. and O.T. Avery. The elaboration of specific soluble substance by pneumococcus during growth. J. Exp. Med., 26:477-493, 1917.

2. Lee, C.J. Bacterial capsular polysaccharides: biochemistry, immunity, and vaccine. Mol. Immunol., 24: 1005-1019, 1987.
3. Butler, J.C., R.F. Breiman, H.B. Lipman *et al.* Serotype distribution of *Streptococcus pneumoniae* infections among preschool children in the United States, 1978-1994: Implications for development of a conjugate vaccine. J. Infect. Dis., 171:885-889, 1995.
4. Siber, G.R. Pneumococcal disease: Prospects for a new generation of vaccines. Science 265:1385-1387, 1994.
5. Broome, C.V. Epidemiology of *Haemophilus influenzae* type b infections in the United States. Pediatr. Infect. Dis. J., 6: 779-782, 1987.
6. Immunization Practices Advisory Committee (ACIP). Haemophilus b conjugate vaccines for prevention of *Haemophilus influenzae* type b disease among infants and children two months of age and older. Morb. Mort. Wk. Rep., 40 RR-1:1-7, 1991.
7. Frasch, C.E. Meningococcal vaccines: Past, present and future. In *Meningococcal Disease*, Cartwright K. ed., p. 145-283, 1995, J. Wiley, Chichester.
8. Bruyn, G.A., Zegers, B.J. and van Furth, R. Mechanisms of host defense against infection with *Streptococcus pneumoniae*. Clin. Infect. Dis., 14:251-262, 1992.
9. Lee, C.J. Bacterial capsular polysaccharides: immunogenicity and vaccines. In *Polysaccharides in Medical Applications*, S. Dunitriu, ed., p.411-442, 1996, Marcel Dekker, Inc., New York, N.Y.
10. Frasch, C.E. Production and control of *Neisseria meningitidis* vaccines. In *Advances in Biotechnological Process*, A. Mizrahi, ed., Vol. 13, p. 123-145, 1990, Wiley-Liss, New York.
11. Lalitha, M.K., R. Pai, T.J. John, *et al.* Serotyping of *Streptococcus pneumoniae* by agglutination assays: A cost-effective technique for developing countries. Bull. World Health Organ., 74:387-390, 1996.
12. Robbins, J.B., R. Schneerson, M.P. Glode, *et al.* Cross-reactive antigens and immunity of disease caused by encapsulated bacteria. J. Allergy Clin. Immunol., 56:141-151, 1975.
13. Heidelberger, M. Cross-reactions of polysaccharides of staphylococci and streptococci in antipneumococcal and other antisera. Mol. Immunol., 21:1011-1014, 1984.
14. Jennings, H.L., E. Katzenellenbogen, C. Lugowski, *et al.* Structure, conformation and immunology of sialic acid containing polysaccharides of human pathogenic bacteria. Pure Appl. Chem., 56:893-905, 1984.
15. Edwards, M.S., D.L. Kasper, H.J. Jennings, *et al.* Capsular sialic acids prevents activation of the alternative complement pathway by type III group B *Streptococcus*. J. Immunol., 128:1278-1283, 1982.
16. Frank, M.M. Current concepts: complement in the pathophysiology of human disease. N. Eng. J. Med., 316:1525- 1530, 1987.
17. Bruyn, G.A., B.J.M. Zegers, and R. van Furth. Mechanisms of host defense against infection with *Streptococcus pneumoniae*. Clin. Infect. Dis., 14: 251-262, 1992.
18. Taylor, C.E., M.B. Fauntleroy, P.W. Stashak *et al.* Antigen-specific suppressor T cells respond to recombinant interleukin-2 and other lymphokines. Infect. Immun., 59:575-579, 1991.
19. Baker, P.J. Regulation of magnitude of antibody response to bacterial polysaccharide antigen by thymus-derived lymphocytes. Infect. Immun., 58:3465-3468, 1990.
20. Van Furth, R., Schuit, and R.C. Williams, Jr. The immunological development of the human fetus. J. Exp. Med., 122:1173, 1965.
21. Morito, T., A.D. Bankhurst, and R.C. Williams, Jr. Studies of human cord blood and adult lymphocyte interactions with in vitro immunoglobulin production. J. Clin. Invest., 64: 990-995,1979.
22. Schlamowitz, M. Membrane receptors in the specific transfer of immunoglobulins from mother to young. Immunol. Commun., 5: 481-500, 1976.

23. Rijkers, G., W. Kuis, E. de Graeff-Meeder, *et al.* Impaired immune response to polysaccharides. N. Eng. J. Med., 317:837-838, 1987.
24. Ambrosino, D.M., Siber, G.R., B.A. Chilmonczyk, *et al.* An immuno-deficiency characterized by impaired antibody response to polysaccharides. N. Eng. J. Med., 316:790-793, 1987.
25. Shurin, P.A., J.M. Rehmus, C.E. Johnson, *et al.* Bacterial polysaccharide immune globulin for prophylaxis of acute otitis media in high-risk children. J. Pediatr., 123:801-810, 1993.
26. Serogroup Y meningococcal disease – Illinois, Connecticut, and selected areas, United States, 1989-96. Mort. Morb. Wk. Rep., 45:1010-1013, 1996.
27. Racossin, J.A., C.Q. Whitney, C.S. Conover *et al.* Serogroup Y meningococcal disease in Chicago, 1991-97. JAMA 24:2094-2098, 1998.
28. Immunization Practice Advisory Committee (ACIP). Meningococcal polysaccharide vaccines, Mort. Morb. Wk. Rep., 27:327-329, 1978.
29. Immunization Practice Advisory Committee (ACIP). Meningococcal vaccines. Mort. Morb. Wk. Rep., 34:255-259, 1985.
30. Gold, R. Immunogenicity of meningococcal polysaccharide in man. In *Immunology of Bacterial Polysaccharides*, Rudbach, J., Baker, P., eds., p. 121-151, 1979, Elsevier, New York.
31. Goldschneider, I., M.L. Lepow, E.C. Gotschlich, et. al. Immunogenicity of serogroup A and serogroup C meningococcal polysaccharides in human infants. J Infect. Dis., 128:769-779, 1973.
32. Makela, P.H., H. Peltola, H. Kayhty, *et al.* Polysaccharide vaccines of serogroup A *Neisseria meningitidis* and *Haemophilus influenzae* type b: a field trial in Finland. J. Infect. Dis., 136 (Suppl):S43-S50, 1977.
33. Black S.B., H.R. Shinefield, Kaiser Permanente Ped Vaccine Study Gr. Immunization with oligosaccharide conjugate *Haemophilus influenzae* type b (HbOC) vaccine on a large health maintenance organization population: Extended follow-up and impact on *Haemophilus influenzae* disease epidemiology. Pediatr. Infect. Dis. J., 11:610-613, 1992.
34. Adams, W.G., K.A. Deaver, S.L. Cochi, *et al.* Decline of childhood *Haemophilus influenzae* type b (Hib) disease in the Hib vaccine era. JAMA 269:221-226, 1993.
35. Jennings, H.J. and C. Lugowski. Immunochemistry of serogroup A, B, and C meningococcal polysaccharide-tetanus toxoid conjugates. J. Immunol., 127:1011-1018, 1981.
36. Costantino, P., S. Viti, A. Podda, *et al.* Development and phase I clinical testing of a conjugate vaccine against meningococcus A and C. Vaccine 10:691-698, 1992.
37. Wyle, F.A., M.S. Artenstein, B.L. Brandt, *et al.* Immunogenic response of man to Group B meningococcal polysaccharide vaccines. J. Infect. Dis., 126:514-522, 1972.
38. Kasper, D.L., J.L. Windelhake, W.D. Zollinger, *et al.* Immunological similarity between polysaccharide antigens of *Escherichia coli* 07:K1:(NM) and Group B meningococci. J. Immunol., 110:262-268, 1973.
39. 39. Lifely, M.R., A.S. Gilbert, and C. Moreno. Sialic acid polysaccharide antigens of *Neisseria meningitidis* and *Escherichia coli*: esterification between adjacent residues. Carbohydr. Res., 94:193-203, 1981.
40. Maloney, P.C., H. Schneider, and B.L. Brandt. Production and degradation of serogroup B *Neisseria meningitidis* polysaccharide. Infect. Immun., 6:657-661, 1972.
41. Zollinger, W.D. and R.E. Mandrell. Importance of complement source in bactericidal activity of human antibody and murine monoclonal antibody to meningococcal gerogroup B polysaccharide. Infect. Immun., 40:257-264, 1983.
42. Jennings, H.J., A. Gamian, F. Michon, *et al.* Unique intermolecular bactericidal epitope involving the homosialopolysaccharide capsule on the cell surface of serogroup B *Neisseria meningitidis* and *Echerichia coli* K1. J. Immunol., 142:3585-3591, 1989.
43. Brisson, J.R., H. Baumann, A. Imberty *et al.* Helical epitope of the serogroup B meningococcal α(2-8)-linked sialic acid polysaccharide. Biochemistry 31:4996-5004, 1992.

44. Jennings, H.I., R. Roy, F. Michon. Determinant specificities of the serogroups B and C polysaccharide of *Neisseria meningitidis*. J.Immunol., 134:2651-2657, 1985.
45. Yamasaki, R. and B. Bacon. Three-dimentional structural analysis of the serogroup B polysaccharide of *Neisseria meningitidis* 6275 by two-dimentional NMR: the polysaccharide is suggested to exist in helical conformations in solution. Biochemistry, 30:851-857, 1991.
46. Fukumi, H. Studies on clinical application of pneumococcal vaccine: distribution of *Streptococcus pneumoniae* in Japan. Jpn. Infect. Dis., 58:39-53, 1984.
47. Shen, S. and the Cooperation Group on Pneumococcal Serotyping. Serotypes and epidemiological study of pneumococcal infection in 18 provinces (cities) in China. Jiangxi Inst. Med. Sci., 10:133-137, 1989.
48. Robbins, J.B., R. Austrian, C.J. Lee *et al.* Considerations for formulating the second-generation pneumococcal capsular polysaccharide vaccine with emphasis on the cross-reactive types. J. Infect. Dis., 148:1136-1159, 1983.
49. Lee, C.J. and T.R. Wang. Pneumococcal infection and immunization in children. Crit.Rev. Microbiol., 20(1):1-12, 1994.
50. Briles, D.E., and J. Carroll. Natural selection and the response to polysaccharides. In *Evolution and Vertebrate Immunity*, G. Kelsoe and D.H. Schulze, eds., p. 117-133, 1987, University of Texas Press, Austin, TX.
51. Van den Dobbelsteen, G.P., H. Kroes, van Rees, E.P. Characteristics of immune responses to native and protein conjugated pneumococcal polysaccharide type 14. Scand. J. Immunol., 41:273-280, 1995.
52. Alonso de Velasco, E., D. Merkus, S. Anderson, *et al.* Synthetic peptides representing T-cell epitopes act as carriers in pneumococcal polysaccharide conjugate vaccines. Infect. Immun., 63:961-968, 1995.
53. Ogunniiyi, A.D., D.E. Briles, and J.C. Paton. Development of a pneumococcal common protein vaccine bases a combination of pneumolysin, PspA and PsaA. Pneumococcal Vaccines for the World 1998 conference, p. 29, 1998.
54. Lock, R.A., J.C. Paton, D. Hausman. Comparative efficacy of pneumococcal neuraminidase and pneumolysin as immunogens protective against *Streptococccus pneumoniae.* Microb. Pathog., 5:461-467, 1988.
55. Tart, R.C., L.S. McDaniel, B.A. Ralph, *et al.* Truncated *Streptococcus pneumoniae* PspA molecules elicit cross-protective immunity against pneumococcal challenge in mice. J. Infect. Dis., 173:380-386, 1996.
56. Tai, S.S., T.R. Wang, and C.J. Lee. Characterization of hemininding activity of *Streptococcus pneumoniae*. Infect. Immun., 65:1083-1087, 1997.
57. Steihoff, M.C., K. Edwards, H. Keyserling, *et al.* A randomized comparison of three bivalent *Streptococcus pneumoniae* glycoprotein conjugate vaccines in young children: Effect of polysaccharide size and linkage characteristics. Pediatr. Infect. Dis. J., 13:368-372, 1994.
58. Pichichero, M.E., M.A. Shelly, and J.J. Treanor. Evaluation of a pentavalent conjugated pneumococcal vaccine in toddlers. Pediatr. Infect. Dis. J., 16:72-74, 1997.
59. Antilla, M., J. Eskola, and H. Kayhty. Opsonic activity and concentration of antibodies to *Streptococcus pneumoniae* type 6B polysaccharide. Abstracts of the 35th Annual Meeting of the Infectious Disease Society of America, p. 183, 1997, San Francisco, CA.
60. Musher, D.M., J.E. Groover, M.C. Rodriguez-Barradas, *et al.* IgG responses to protein-conjugated pneumococcal capsular polysaccharides in persons who are genetically incapable of responding to unconjugated polysaccharides. Clin. Infect. Dis., 111:222-233, 1998.
61. Kayhty, H., H. Ahman, P.R. Ronnberg, *et al.* Pneumococcal polysaccharide-meningococcal outer membrane protein complex conjugate vaccine is immunogenic in infants and children. J. Infect. Dis., 172:1273-1278, 1995.

62. Leach, A., S. Ceesay, W. Banya, *et al.* Pilot trial of a pentavalent pneumococcal polysaccharide-protein conjugate vaccine in Gambian infants. Pediatr. Infect. Dis. J., 15:333-339, 1996.
63. Anderson, E., D. Kennedy, K. Geldmacher, *et al.* Immunogenicity of heptavalent pneumococcal conjugate vaccine in infants. J. Pediatr., 128:649-653, 1996.
64. Rennels, M.b., K.M. Edwards, H.L. Keyserling, *et al.* Safety and immunogenicity of heptavalent pneumococcal vaccine conjugated to CRM_{197} in United States infants. Pediatrics, 101:604-611, 1998.
65. Sorensen, R.U., L.E. Leviva, P.A. Giangrosso, *et al.* Response to a heptavalent conjugate *Streptococcus pneumoniae* vaccine in children with recurrent infections who are unresponsive to the polysaccharide vaccine. Pediatr. Infect. Dis. J., 17(8):685-691, 1998.
66. Daum, R.S. Pneumococcal vaccines for children: An update. Pediatr. Infect. Dis. J., 17(9):823-824, 1998.
67. Obaro, S.K., Z.M. Huo, W.A.S. Banya, *et al.* A glycoprotein pneumococcal conjugate vaccine primes for antibody responses to a pneumococcal polysaccharide vaccine in Gambian children. Pediatr. Infect. Dis. J., 16(12):1135-1140, 1997.
68. Klein, D.L., R.W. Elis. Conjugate vaccine against *Streptococcus pneumoniae*. In *New Generation Vaccines*, M.M. Levine, g.C. Woodrow, J.B. Kaper, G.S. Cobon (eds.), p. 503-525, 1997, Marcel Dekker, Inc., New York.
69. Lu, C.H., C.J. Lee, and P. Kind. Immune responses of young mice to pneumococcal type 9V polysaccharide-tetanus toxoid conjugate. Infect. Immun., 62:2754-2760, 1994.
70. Wenger, J.D., A.W. Hightower, R.R. Facklam, *et al.* Bacterial meningitis in the United States, 1986: report of a multistate surveillance study. J. Infect. Dis., 162:1316-1323, 1990.
71. Claesson, B.A. Epidemiology of invasive *Haemophilus influenzae* type disease in Scandinavia. Vaccine 11:530-533, 1993.
72. Booy, R., S.A. Hodgson, M.P.E. Slack, *et al.* Invasive *Haemophilus influenzae* type b disease in the Oxford region (1985-91). Arch. Dis. Child, 69:225-228, 1993.
73. Centers for Disease Control. Recommendations for use of Haemophilus b conjugate vaccines and a combined diphtheria, tetanus, pertussis, and Haemophilus b vaccine. Mort. Morb. Wk. Rep., 42(RR13):1-15, 1993.
74. Decker, M.D., K.M. Edwards, R. Bradley, *et al.* Comparative trial in infants of four conjugate *Haemophilus influenzae* type b vaccines. J. Pediatr., 120:184-189, 1992.
75. Eskola, J., H. Kayhty, A.K. Takala, *et al.* A randomized, prospective field trial of a conjugate vaccine in the protection of infants and young children against invasive *Haemophilus influenzae* type b disease. N. Eng. J. Med., 323:1381-1387, 1990.
76. Takala, A.K., J. Eskola, M. Leinonen, *et al.* Reduction of oropharyngeal carriage of *Haemophilus influenzae* type b (Hib) in children immunized with an Hib conjugate vaccine. J. Infect. Dis., 164:982-986, 1991.
77. Adams, W.G., K.A. Deaver, S.L. Cochi, *et al.* Decline of childhood *Haemophilus influenzae* type b (Hib) disease in the Hib vaccine era. JAMA, 269:221-226, 1993.
78. Centers for Disease Control. Progress toward elimination of *Haemophilus influenzae* type b disease among infants and children -- United States, 1993-1994. Mort. Morb. Wk. Rop., 44:545-550, 1995.
79. Robbins, J.D. and J.B. Robbins. Reexamination of the protective role of the capsular polysaccharide (Vi antigen) of *Salmonella typhi*. J. Infect. Dis., 150:436, 1984.
80. Keitel W.A., N.L. Bond, J.M. Zahradnik, *et al.* Clinical and serological responses following primary and booster immunization with *Salmonella typhi* Vi capsular polysaccharide vaccines. Vaccine 12:195-199, 1994.
81. Acharya, I., C. Lowe, R. Thapa, *et al.* Prevention of typhoid fever in Nepal with the Vi capsular polysaccharide of *Salmonella typhi*. N. Eng. J. Med., 317: 1101-1104, 1987.

82. Klugman, K., H. Koornhof, and R. Schneerson. Protective ability of Vi capsular polysaccharide vaccine against typhoid fever. Lancet, 2:1165-1169, 1987.
83. World Health Organization Expert Committee on Biologic Standardization. Requirements on Vi polysaccharide for typhoid. Technical Report Series 840, p. 14-32, 1993.
84. Szu, S.C., X. Li, A.L. Stone, *et al.* Relation between the structure and immunologic properties of the Vi capsular polysaccharide. Infect. Immun., 59:4555-4561, 1991.
85. Szu, S.C., A.L. Stone, J.D. Robbins, *et al.* Vi capsular polysaccharide-protein conjugates for prevention of typhoid fever. J. Exp. Med., 166:1510-1524, 1987.
86. Szu, S.C., D.N. Taylor, A.C. Trofa, *et al.* Laboratory and preliminary clinical characterization of Vi capsular polysaccharide-protein conjugate vaccines. Infect. Immun., 62:4440-4444, 1994.
87. Baker, C.J. and D.L. Kasper. Group B streptococcal vaccines. Rev. Infect. Dis., 7:458-467, 1985.
88. Dillon, H.C., Jr., S. Khare, and B.M. Gray. Group B streptococcal carriage and disease: a 6-year prospective study. J. Pediatr., 110:31-36, 1987.
89. Wessels, M.R., J.L. DiFabio, V.J. Benedi, *et al.* Structural determination and immunochemical characterization of the type V group B *Streptococcus* capsular polysaccharide. Biol. Chem., 266:6714-6719, 1991.
90. Jennings, H.J. Capsular polysaccharides as vaccine candidates. Curr. Top. Microbiol. Immunol., 150: 97-127, 1990.
91. Kasper, D.L., C.J. Baker, M.S. Edwards *et al.* The type III Group B streptococcal capsular polysaccharide: structure, immunospecificity, immunogenicity and relationship to virulence. In *Seminars in Infectious Diseases*, Vol. 6, L. Weinstein and B.M. Fields, eds., pp. 275-278, 1982, Thieme-Stratton, New York.
92. Baker, C.J., M.A. Rench, M.S. Edwards, *et al.* Immunization of pregnant women with a polysaccharide vaccine of Group B *streptococcus*. N. Engl. J. Med., 319:1180-1185, 1988.
93. Kasper, D.L., L.C. Paoletti, M.R. Wessels, *et al.* Immune response to type III group B streptococcal polysaccharide-tetanus toxoid conjugate vaccine. J. Clin. Invest., 98:2308-2314, 1996.
94. Griffin, P.M. and R.V. Tauxe. The epidemiology of infections caused by *E. coli* 0157:H7, other enterohemorrhagic *E. coli*, and the associated hemolytic uremic syndeome. Epidemiol. Rev., 13:60-98, 1991.
95. Neill, M.A., P.L. Tarr, C.R. Clausen, et. al. *Escherichia coli* 0157:H7 and the predominant pathogen associated with hemolytic uremic syndrome: A prospective study in the Pacific Northwest. Pediatrics, 80:37-40, 1987.
96. Osstroff, S.M., P.I. Tarr, M.A. Neill, *et al.* Toxin genotypes and plasmid profiles as determinants of systemic sequelae in *E. coli* O157:H7 infections. J. Infect. Dis., 160:994-998, 1989.
97. Aleksic, S., H. Karch, J. Bockemuhl. A biotyping scheme for Shiga-like (Vero) toxin-producing *Escherichia coli* O157 and a list of serological cross-reactions between O157 and other gram-negative bacteria. Int. J. Med. Microbiol., 276:221-230, 1992.
98. Orskov, F., I. Orskov, J.A. Villar. Cattle as reservoir of verotoxin-producing *Escherichia coli* O157:H7. Lancet, 2:276, 1987.
99. Keene, W.E., J.M. McAnulty, F.G. Hoesly, *et al.* A swimming-associated outbreak of hemorrhagic colitis caused by *Escherichia coli* 0157:H7 and *Shigella sonnei*. N. Engl. J. Med., 331:579-584, 1994.
100. Perry, M.B., L. MacLean, D.W. Griffith. Structure of the O-chain polysaccharide of the phenol-phase soluble lipopolysaccharide of *Escherichia coli* 0157:H7. Biochem. Cell Biol., 64:21-28, 1986.

101. Konadu, E., J.B. Robbins, J. Shiloach, *et al.* Preparation, characterization, and immunological properties in mice of *Escherichia coli* 0157 O-specific polysaccharide-protein conjugate vaccines. Infect. Immun., 62:5048-5054, 1994.
102. Parke, J.C., Jr., E. Konadu, V. Pozsgay, *et al.* Clinical evaluation of *Escherichia coli* 0157 O-specific polysaccharide conjugates in adults and synthesis and evaluation of Shigella toxin B subunit conjugates in mice. In preparation.
103. Finland, M., W.F. Jones, and M.W. Barnes. Occurrence of serious bacterial infections since the introduction of antibacterial agents. JAMA, 170:2188-2197, 1959.
104. Rogers, D.E. The changing pattern of life-threatening microbial disease. N. Engl. J. Med., 261:677-683, 1959.
105. Bartlett, J.G., P. O'Keefe, F.P. Tally, *et al.* Bacteriology of hospital-acquired pneumonia. Arch. Intern. Med., 146:868-871, 1986.
106. Jarvis, W.R., V.P. Munn, A.K. Highsmith, *et al.* The epidemiology of nosocomial infections caused by *Klebsiella pneumoniae*. Infect. Control., 6:68-74, 1985.
107. Kenne, L. and B. Lindberg. Bacterial polysaccharides. In *The Polysaccharides*, Vol. 2, G.O. Aspinall, ed., pp. 287-363, 1983, Academic Press, New York.
108. Lee, C.J. and K. Koizumi. Immunochemical relations between pneumococcal group 19 and *Klebsiella* capsular polysaccharides. J. Immunol., 127: 1619-1623, 1981.
109. Heidelberger, M. and W. Nimmich. Immunochemical relationship between bacteria belonging to two separate families: pneumococci and *Klebsiella*. Immunochemistry, 13:67-80, 1976.
110. Cryz, S.J., P. Mortimer, A.S. Cross, *et al.* Safety and immunogenicity of polyvalent *Klebsiella* capsular polysaccharide vaccine in humans. Vaccine, 4:15-20, 1986.
111. Cryz, S.J., A.S. Cross, E. Furer, *et al.* Activity of intravenous immune globulins against *Klebsiella*. J. Lab. Clin. Med., 108:182-189, 1986.
112. Siber, G.R., P.H. Schur, A.C. Eisenberg, *et al.* Correlation between serum IgG-2 concentrations and the antibody response to bacterial polysaccharide antigens. N. Engl. J. Med., 303:178-182, 1980.
113. Cryz, S.J., Jr., J.C. Sadoff, A.S. Cross, *et al.* Safety and immunogenicity of a polyvalent *Pseudomonas aeruginosa* O-polysaccharide-toxin A vaccine in humans. Antibio. Chemother., 42:177-183, 1989.
114. Granstrom, M., B. Wretlind, B. Markham, *et al.* Enzyme-linked immunosobent assay to evaluate the immunogenicity of a polyvalent *Klebsiella* capsular polysaccharide vaccine in humans. J. Clin. Microbiol., 26:2257-2261, 1988.
115. Donta, S.T., A.S. Cross, J.C. Sadoff, *et al.* Immunoprophylaxis against *Klebsiella* and *Pseudomonas aeruginosa* infections. Clin Infect, Dis., 117:564, 1993.
116. Drabick, J.J., K. Ramsey, J.B. McClain, *et al.* The effect of passive immunotherapy with hyperimmune and non-hyperimmune gammaglobulin on circulating antigen-specific B cells in human volunteers. Clin. Res., 42:229A, 1994.
117. Campbell, W.N., E. Hendrix, S.J. Cryz, Jr., *et al.* Immunogenicity of 24-valent *Klebsiella* capsular polysaccharide vaccine and an 8-valent *Pseudomonas* O-polysaccharide conjugate vaccine administered to acute trauma victims. Clin. Infect. Dis., 23:179-181, 1996.
118. Cryz, S.J., E. furer, A.S. Cross, *et al.* Safety and immunogenicity of *Pseudomonas aeruginosa* O-polysaccharide toxin A conjugte vaccine in humans. J. clin. Invest., 80:51-56, 1987.
119. Pier, G.B., D. DesJardin, M. Grout, *et al.* Human response to *Pseudomonas aeruginosa* mucoid exopolysaccharide (alginate) vaccine. Infect. Immum., 62:3972-3979, 1994.
120. Cryz, S.J. Jr. Progress in immunization against *Pseudomonas aeruginosa* and *Klebsiella* spp. Path. Immunopath. Res., 6:147-152, 1987.

THE CONFORMATIONAL EPITOPE OF TYPE III GROUP B *STREPTOCOCCUS* CAPSULAR POLYSACCHARIDE

Wei Zou and Harold J. Jennings

Institute for Biological Sciences, National Research Council of Canada, Ottawa, Ontario, Canada K1A 0R6

Structural mimicry between complex carbohydrates of both bacterial and mammalian cells is a well-documented phenomenon, and when present on bacteria contribute to their virulence [1-3]. This mimicry occurs extensively among group B *Streptococcus*, a neonatal pathogen, which can be classified into eight serotypes based on different capsular polysaccharide structures [4-10]. In addition to homology with human tissue antigens all the group B streptococcal polysaccharides of different serotypes also share extensive structural homology. However, despite the above structural homology all the different polysaccharides are both type-specific and immunogenic. This is possible because the immune system selects length-dependent epitopes unique to the polysaccharides with which to induce protective antibodies. Because the self-antigens exist only as short chain oligosaccharides, this avoids the problem of inducing antibodies that cross-react with self-antigens [1,11]. Thus, the length-dependency of the protective epitopes of the group B streptococcal polysaccharides is an important parameter in the design of polysaccharide-based conjugate vaccines [2,3,12-14].

This review focuses on defining the protective epitope of type III group B streptococcal polysaccharide (GBSPIII[1]) although other serotypes are shown to have similar length-dependent epitopes. The key to carrying out these studies was to obtain well-defined fragments and chemically modified fragments of the GBSPIII with which to probe GBSPIII-specific antibody binding sites. We review here the synthesis of these fragment oligosaccharides and their use in defining the nature of the length-dependent epitope.

[1]Abbreviations used in this paper: GBSPIII, Type III group B *Streptococcus* polysaccharide; PN14PS, Type 14 *S. Pneumoniae* polysaccharide; HSA, human serum albumin; TT, Tetanus toxoid; MPL, Monophosphoryl lipid A; RU, complete repeating units in GBSPIII fragments derived by combined enzymatic and degradative procedures; RU^S, synthetically-derived GBSPIII repeating unit; SLe, Sialyl Lewis antigen; NMR, Nuclear magnetic resonance; SPR, surface plasmon resonance; RABA, Radio-labeled antigen binding assay; ELISA, Enzyme-linked immunosorbent assay.

The Molecular Immunology of Complex Carbohydrates–2.
Edited by Albert M. Wu, Kluwer Academic/Plenum Publishers, 2001.

SYNTHESIS OF OLIGOSACCHARIDES

Two approaches have been used in our laboratory in recent years to carry out the synthesis of oligosaccharides. One is based on a conventional building block synthesis using either chemical or chemo-enzymatic methods, and the other is executed by a tailor-assembly method that takes advantage of structural similarities among bacterial polysaccharides and the required oligosaccharide antigens. In this latter method polysaccharides were selectively depolymerized, followed by various enzymatic glycosylations on the smaller fragments obtained, or by a reverse sequence in which glycosylation was performed prior to depolymerization as illustrated in Figure 1.

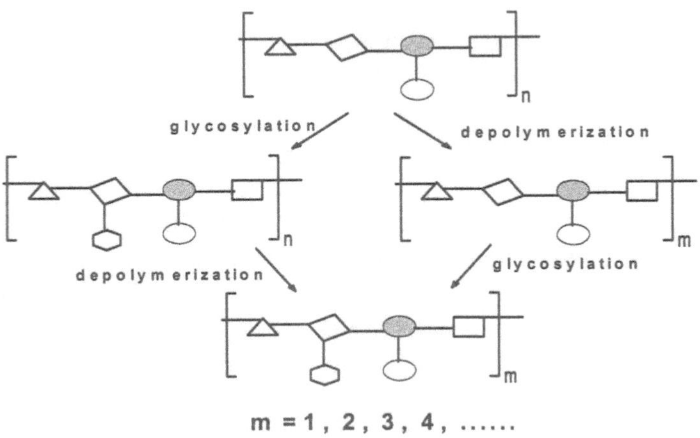

Fig. 1. Tailor-assembly method for the synthesis of oligosaccharides from polysaccharides.

Particular success with the tailor-assembly approach has been demonstrated in this laboratory by the synthesis of GBSIII oligosaccharides from PN14PS [15]. This method has the potential to become a more cost-effective way to synthesize many biologically important carbohydrates with the increasing availability of various glycosyltransferases.

THE BUILDING BLOCK SYNTHESIS OF OLIGOSACCHARIDE PROBES

The synthesis of $2RU^{S}$ and $2RU^{SM}$, oligosaccharides with a double repeating unit, was accomplished either chemically or chemo-enzymatically by building block procedures. The retrosyntheses of $2RU^{S}$ and $2RU^{SM}$ are illustrated in Figures 2 and 3, respectively. In the synthesis of $2RU^{S}$ a core octasaccharide was first constructed following a pattern of [3+2+3]. The enzymatic sialylation of the core octasaccharide afforded $2RU^{S}$. $2RU^{S}$-NeuAc, in which only one of either galactose residues was sialylated at 3-*O*-position was formed when shorter reaction times were employed. The procedures for the synthesis of $2RU^{SM}$ were similar except that methyl lactate was first introduced at 3-*O*-position of galactose. The experimental details of these syntheses have been reported [16-18].

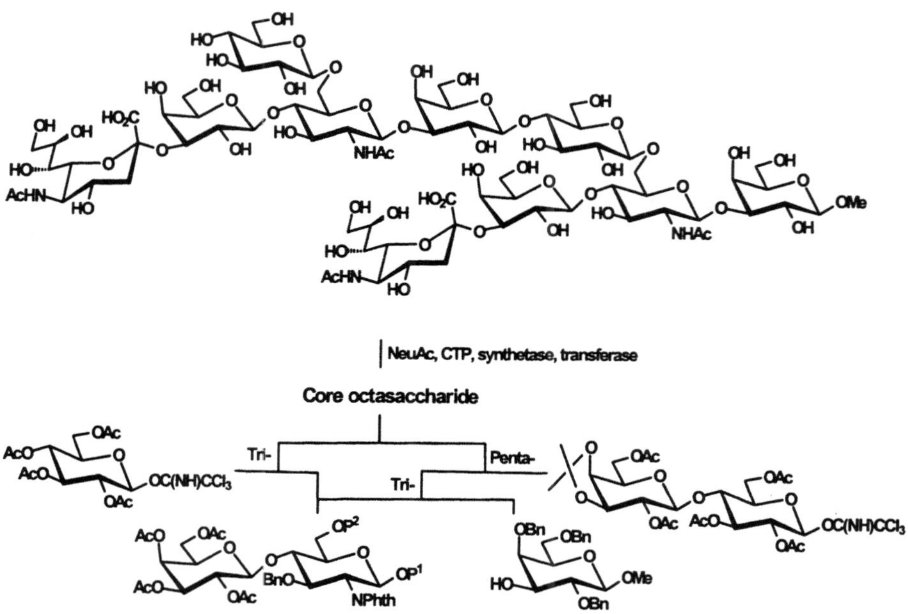

Fig. 2. Retrosynthesis of a double repeating unit of GBSPIII (2RU^{S}). Schmidt glycosylations from four building blocks were used to construct a core octasaccharide. Following enzymatic sialylation afforded 2RU^{S}, and monosialylated 2RU^{S}-NeuAc when shorter reaction times were used.

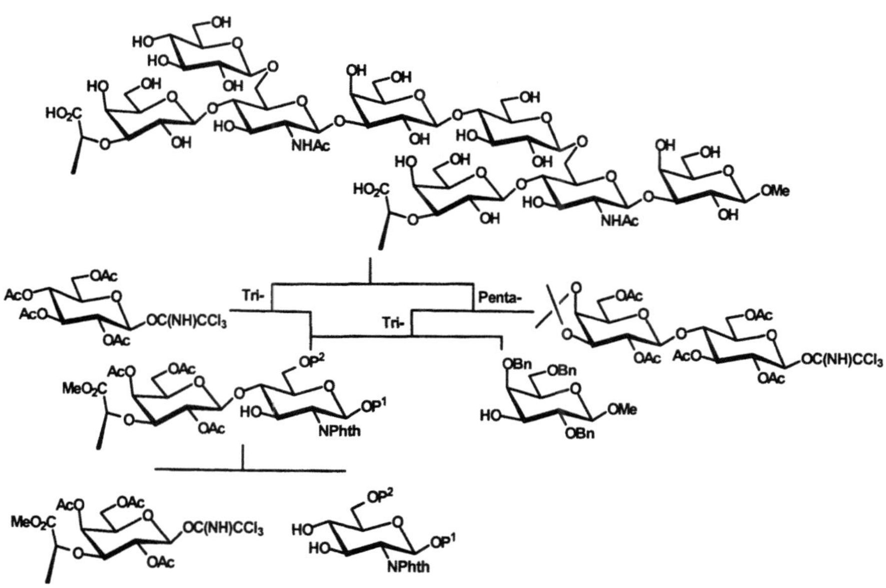

Fig. 3. Retrosynthesis of a double repeating unit mimic of GBSPIII (2RU^{SM}). (S)-Lactate was introduced to replace sialic acid at 3-*O*-positions of galactose residues to test whether the carboxylate group alone could fulfill the role of sialic acid. Using an anion group as mimicry of sialic acid has been demonstrated successfully in the design of inhibitors to the binding of sialyl-Le^{a} and -Le^{x} to E-selectin since only negative charge is required in the binding site.

To obtain higher oligomers of GBSPIII, such as 3RU up to 7RU, a tailor-assembly approach was used starting from readily available PN14PS, which has the same structure as the desialylated GBSPIII [15]. Partially de-*N*-acetylated PN14PS was enzymatically sialylated to afford partially de-*N*-acetylated GBSPIII, which was then selectively depolymerized by nitrous acid deamination into oligosaccharides containing increasing numbers of repeating units as shown in Figure 4. The separation of the oligomers was performed on a Superdex 30 column. The added advantage of this method is that the depolymerization procedure spontaneously introduces a terminal aldehyde which could then be directly conjugated to proteins, such as HSA and TT, by reductive amination to form glycoconjugates [15,19]. These conjugates were used either as probes for the characterization of antibody-antigen binding or as vaccine candidates for assaying their immunogenicity.

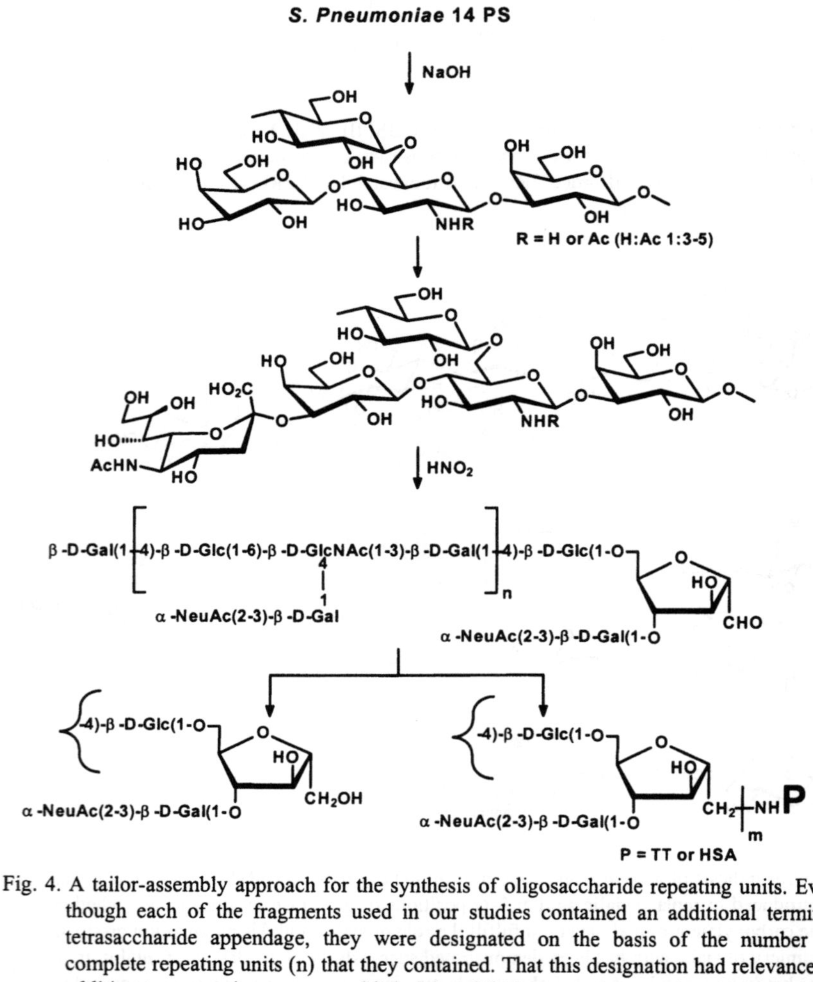

Fig. 4. A tailor-assembly approach for the synthesis of oligosaccharide repeating units. Even though each of the fragments used in our studies contained an additional terminal tetrasaccharide appendage, they were designated on the basis of the number of complete repeating units (n) that they contained. That this designation had relevance in addition to convenience was established in inhibition experiments, the result of which demonstrated that the appendage made no contribution to the inhibitory properties of the fragments.

The immunodominant epitope of GBSPIII, the repeating unit structure of which is shown in Figure 5, is length-dependent and it was established that two repeating units (decasaccharide) were required for even suboptimal binding to homologous antibodies [20], and that the affinity of binding increased with increasing numbers of repeating units. It was determined that binding also depends on the presence of terminal sialic acid residues on the side chain, particularly its carboxylate group [1,11]. The exocyclic chain of sialic acid (C-7 NeuAc) was not involved because its removal by oxidation with sodium periodate followed by sodium borohydride reduction does not abrogate binding (see Figure 5). Change of *N*-acyl substituent on sialic acid from acetyl to propionyl group also showed no effect on the antibody binding. Thus, it was concluded that although the entire sialic acid is not required for binding to occur at least its carboxylate group is essential for the formation of the conformational epitope [1,11].

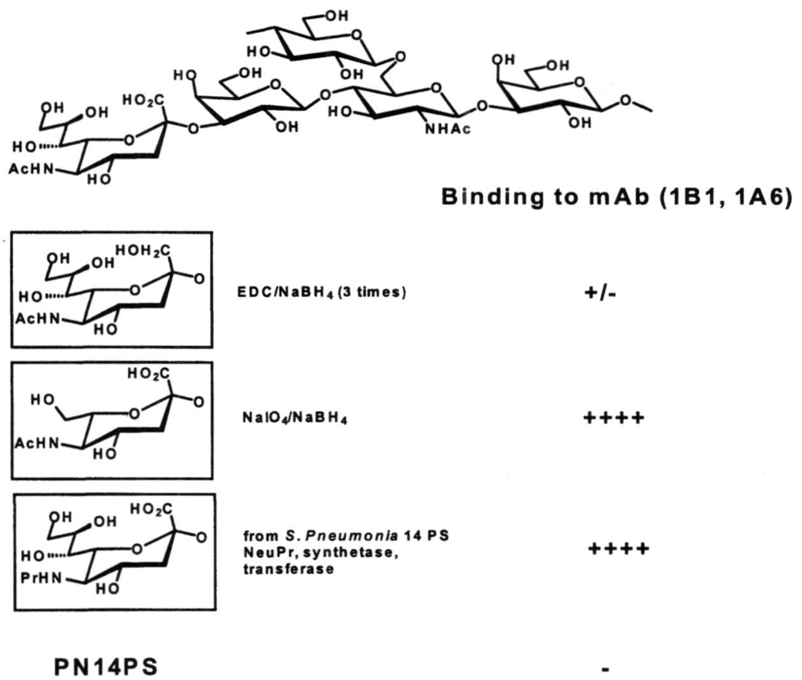

Fig. 5. The structures of the repeating unit of GBSPIII and its analogues in which the sialic acid residues were structurally modified. The reduction of carboxylate of sialic acid apparently abrogated the conformational epitope, while the changes on the exo-cyclic chain and *N*-acyl group had no effect to the antigenicity.

The characteristics of the GBSPIII conformational epitope are similar to those associ-ated with the immune responses to the capsular polysaccharides of group B *N. meningitidis* [1,21,22] PN14PS [23,24]. We have hypothesized that the epitopes of GBSPIII are located on extended helical domains of the polysaccharides rather than on its random coil form [25]. Because of lack of symmetry the random coil form is structurally related to short oligosaccharide self-antigens. Although we have no direct evidence so far for the existence of extended helical epitopes on GBSPIII or PN14PS, NMR and molecular dynamics studies indicate that both the above polysaccharides and the group B meningococcal polysaccharide are capable of forming extended helices. Additionally, there is a precedent for this type of epitope because there is strong evidence, based on X-ray crystallographic studies, that the extended epitope of the group B meningococcal polysaccharide is helical [26].

However, basic questions need to be answered to define the protective GBSPIII conformational epitope. What is its minimum antigenic structure? At what length of saccharide does the epitope optimize both antigenically and immunogenically?

MINIMUM ANTIGENIC STRUCTURE OF GBSPIII EPITOPE

Although the carboxylate groups of the sialic acid residues in GBSPIII are essential for maintaining the conformational epitope, the poor binding activity of $2RU^{SM}$ to GBSPIII-specific mAbs as shown in Figure 6 indicated that the carboxylate groups alone cannot fulfil the role of sialic acid and suggested that its pyranosyl-ring structure is also responsible. The pyranosyl ring of the sialic acid may behave as a bulky entity to keep the molecular topology or provide hydrogen bonding through its 3-OH and/or 7-OH groups. It is also possible that the lactate group of $2RU^{SM}$ is too small and too flexible to form stable hydrogen bonding with the backbone residue in order to maintain the orientation of carboxylate group as sialic acid does. The ^{1}H and ^{13}C NMR spectra of $2RU^{SM}$ were very different from $2RU^{S}$ and more like that of PN14PS, which is identical to the desialylated GBSPIII, and has no cross reactivity with anti-GBSPIII mAbs.

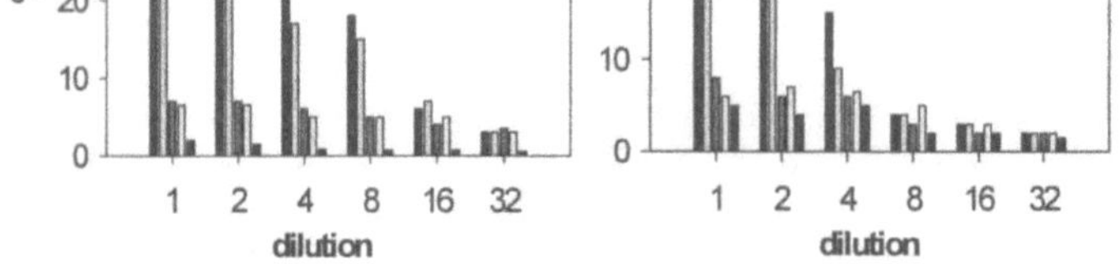

Fig. 6. The interaction of antibodies and synthetic oligosaccharides ($2RU^{S}$, $2RU^{SM}$, and $2RU^{S}$-NeuAc) were studied using RABA. Oligosaccharides, GBSPIII, and PN14PS were labeled with *N*-[^{3}H]-acetyl. The resultant saccharides showed the radioactivity within a range of 2000-3000 CPM with 25 μl of samples.

The failure of $2RU^{SM}$ to bind to GBSPIII-specific mAbs is also indicative of the fact that the carboxylate group of sialic acid in GBSPIII may not be directly involved in the binding site as in the case of SLe^{x} [27], but rather function in the control of the extended conformational epitope. The lack of binding for $2RU^{S}$-NeuAc demonstrates that the conformational epitope requires a minimum of at least two complete repeating units, because the loss of either sialic acid in a double repeating unit will abrogate binding. This also explains why any loss of sialic acid in GBSPIII results in a dramatic decline in the production of protective antibodies (Jennings et al, unpublished results).

The former conclusions were fully supported by ELISA binding experiments. The 1RU-HSA conjugate when used as a coating antigen showed no binding with anti-GBSPIII mAbs, whereas, the longer oligosaccharide-conjugates bound to mAbs with increasing affinity. These results reconfirmed that the conformational epitope required at least a complete 2RU structure for binding.

LENGTH-DEPENDENCY ON OPTIMIZATION AND FREQUENCY OF GBSPIII PROTECTIVE EPITOPE

Having established 2RU as the minimum binding structure and the critical role of sialic acid in the binding, further experiments were designed to study in more detail factors behind the increasing affinity of GBSPIII with increasing length. Using mAbs 1B1 and 1A6 and GBSPIII fragments and their HSA-conjugates, we attempted to establish the correlation between saccharide fragment length and optimization and frequency of the conformational epitope. ELISA inhibition of the binding of GBSPIII to mAbs 1B1 and 1A6 using GBSPIII fragments demonstrated that the initial formation of the conformational epitope occurs at 2RU (see Figure 7). A small increase in inhibition occurs from 2RU to 3RU with little subsequent increase in inhibition from 3RU to 7RU. The inhibition results using mAb 1A6 were very similar. These data are consistent with the presence of only one conformational epitope from 2RU to 7RU with little further optimization of epitope in this region. The above data were confirmed by studies on 1B1 binding to GBSPIII-HSA by SPA. Further increases in inhibition and binding with fragments larger than 7RU were initially attributed to epitope multivalency, but SPR data obtained using 1B1 Fab clearly demonstrated that optimization of epitope was still a factor even at this length of saccharide (see Figure 8).

Because IgG binding is bivalent both the SPR and inhibition studies could not distinguish between the contribution of epitope multivalency and epitope optimization with increasing length of GBSPIII. However, valency effects were eliminated by using a 1B1 Fab to bind to saccharide-HSA conjugates and the results using up to 7RU were again in excellent agreement with the original inhibition data [19]. The SPR data showed that oligosaccharides in the 2RU to 7RU ranges were monovalent with respect to epitope presentation and bound to the Fab with a K_D of approximately 4 μM, which was similar to the IC_{50} values for these monovalent antigens. However, the SPR data also demonstrated that affinity still increased three-fold from 7RU to 20RU, with little increase in affinity beyond 20RU. Thus, both epitope optimization and multivalency occur between 7RU and 20RU, which accounts for the increasing affinity of 1B1 IgG with increasing GBSPIII length [19].

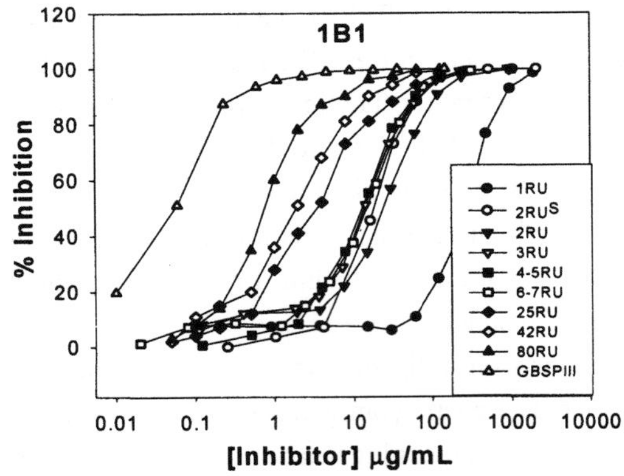

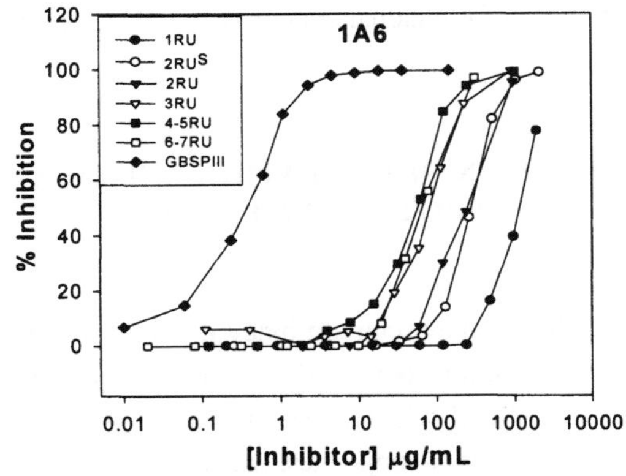

Fig. 7. Inhibition curves derived from ELISA inhibition data on the binding of GBSPIII-HSA to mAbs 1B1 and 1A6, using GBSIII oligosaccharides and larger fragments as inhibitors. Inhibitory activity is presented as percentage calculated as [OD (no inhibitor) - OD (with inhibitor)]/OD (no inhibitor) x 100%. Inhibitions vs. log concentration curves were plotted for each inhibitor and the concentrations required for 50% inhibition (IC_{50}) were determined from extrapolated curves.

Interestingly, both the length-dependency of GBSPIII epitope and its distribution are similar to those found for PN14PS and the Group B meningococcal polysaccharide. Studies on the immune responses to PN14PS conjugates in rabbits are consistent with the length-dependent epitope forming at 4RU and persisting as a single epitope up to approximately 20RU [23]. Similarly, the α2,8-polysialic acid helical epitope of group B *N. meningitidis* requires ≅10 sialic acid residues to form and the helix is not duplicated until there are approximately 40 linear sialic acid residues present [22,26].

A model of antibody binding to GBSPIII in which the binding of the first antibody induces a second epitope in the same chain was proposed previously [20], which by means of a process of epitope propagation results in the accumulation of antibodies on previously bound chains rather than on naïve chains. The SPR data do not support this model of antibody binding to GBSPIII. Such a model would not be a good fit of the 1:1 interaction (see Figure 8), since the epitope density would remain constant throughout the association at non-saturating concentrations of Fab. Our model of binding requires that helical epitopes are intermittently located between random coil segments of GBSPIII. This pattern of epitope distribution is not conducive to an induction process, which would require the formation of

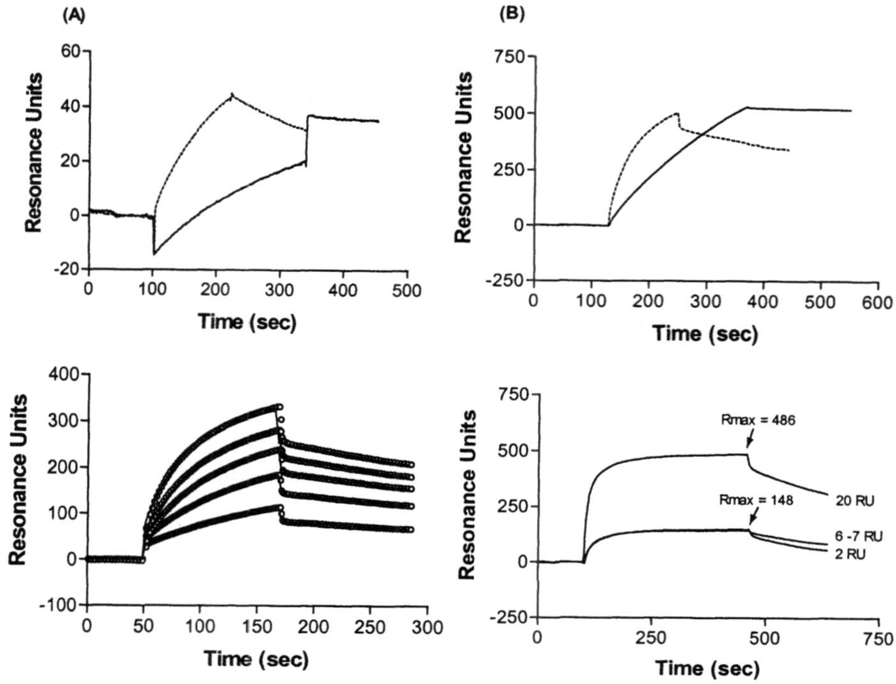

Fig. 8. The binding kinetics of saccharide-HSA conjugates (2RU at top left and 80RU at top right) to IgG (dash line) and Fab (solid line). Simultaneous fitting to a 1:1 interaction model of the association and dissociation phases of sensorgrams for 1B1 Fab binding to the 20RU conjugate (bottom left). The Fab concentrations were 5, 7.5, 10, 12.5 and 15 μM and the surface density was 200 resonance units. Sensorgrams comparing the antibody binding capacity of 2RU, 6-7RU and 20RU surfaces (bottom right). The sensorgrams show the binding of 50 μM 1B1 Fab to surfaces containing approximately 200 resonance units of immobilized conjugate in each instance.

contiguous helices. In our model, epitopes form spontaneously, and the number of epitopes per polysaccharide chain would be more restricted. This unusual antigen characteristic may be reflected in the association rate constants for 1B1 Fab binding to antigen, which are, to our knowledge, the lowest reported for antigen-antibody interactions. Values in the range of 10^4-10^5 $M^{-1}s^{-1}$ are typical for these interactions whereas those reported here for 1B1 Fab are 2 x 10^3 $M^{-1}s^{-1}$ for the longer oligosaccharides and slightly lower for the shorter chain antigens. The reason for these unusually slow rates may lie in the dynamic nature of the antigen. The extended helical conformation represents a minor proportion of the total polysaccharide with helices forming intermittently and reverting to random coil. Despite the slow association rate constant for the interaction, a slow dissociation constant contributes to an overall affinity that is unusually high for an anti-carbohydrate antibody.

LENGTH DEPENDENCY ON IMMUNOGENICITY OF GBSPIII

Antigenicity of GBSPIII fragments in terms of binding affinity to anti-GBSPIII mAb is length-dependant and the conformational epitope forms at 2RU. However, it is essential to further establish the relationship between the immunogenicity and GBSPIII chain length. All TT-conjugates of oligomers with increasing length as shown in Figure 4 elicited immune responses in rabbits using MPL as adjuvant, and ELISA detected high titers of anti-GBSPIII antibodies (Zou and Jennings, unpublished results). However, the protective antibodies measured by means of oposonophagocitic activity as illustrated in Figure 9 also revealed a length-dependency. Compared to length-dependent antigenic antibody binding a shift to even longer saccharide chain length was observed in order to produce protective antibodies. Although a 2RU oligosaccharide is able to bind to anti-GBSPIII antibodies and is immunogenic, the antisera induced by its conjugate failed to provide significant killing of GBSIII. The antisera raised by 3RU-TT conjugate were better but only those conjugates from 4RU-TT gave a comparable killing of GBSIII to antisera raised by native GBSPIII-TT. In fact, the requirement of at least 4RU to effectively raise good protective antibodies is in agreement to our observations that an optimized conformational epitope forms at 4RU as indicated by binding and inhibition experiments. The results are also consistent with our other observations on PN14PS and the group B meningococcal PS.

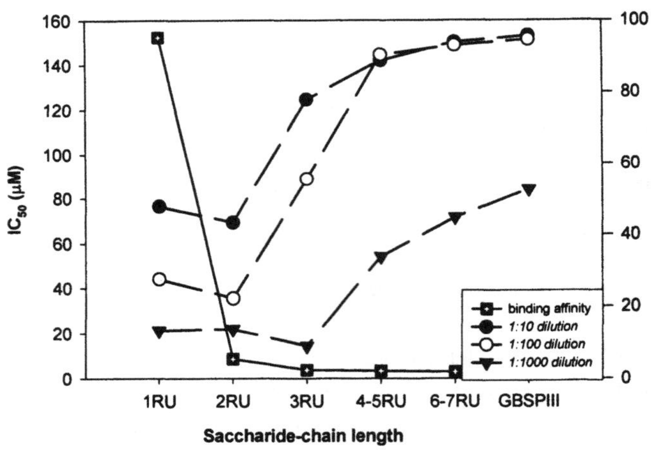

Fig. 9. The length-dependency on immunogenicity of GBSPIII.

The difference in the length dependency of the antigenicity and immunogenicity of GBSIII oligosaccharides is significant to understanding the nature of the conformational epitope. All the data indicate that antigen-antibody binding is an induced-fit process in which antibody binds to a saccharide chain with less-than-perfect conformational epitope, whereas a more stable or optimized conformational epitope is required to elicit specific immune responses for the production of high avidity protective antibodies. Therefore, the conformational epitope of GBSPIII requires twelve sugar residues in the backbone, which make up 4RU, to spontaneously form an optimized helical epitope.

SUMMARY

In order to further characterize the conformational epitope of GBSPIII, we synthesized various oligosaccharides with the GBSPIII-related structures by a tailor-assembly synthetic scheme and a more traditional block-wise chemo-enzymatic approach. The oligosaccharides were used to probe the conformational epitope of GBSPIII using number of complementary techniques. The protective epitope of GPSPIII was further defined as length-dependent and conformational. The results of the studies confirmed that two repeating units (2RU) is the minimum binding unit and the epitope optimization mainly takes place between chain length 2RU to 7RU. Epitope optimization and multivalency were observed between 7RU and 20RU. The data support our hypothesis that the conformational epitope is an extended helical segment of the GBSPIII. GBSPIII exists mainly in the random coil form, which structurally mimics short oligosaccharide self-antigens, but it can infrequently and spontaneously form extended helices. Although not prevalent in GBSPIII the immune system preferentially selects these helical epitopes because they are unique to the polysaccharide.

REFERENCES

1. H.J. Jennings, E. Katzenellenbogen, C. Lugowski, F. Michon, R. Roy, and D.L. Kasper, Structure, conformation, and immunology of sialic acid-containing polysaccharides of human pathogenic bacteria, *Pure Appl.Chem.* 56:893-905 (1984).
2. H.J. Jennings, Polysaccharides and conjugated polysaccharides as human vaccines, in: "New Developments in Industrial Polysaccharides", V. Crescenzi et al.., eds., Gordon Breach Science Publishers, New York, pp. 323-344 (1985).
3. H.J. Jennings, Capsular polysaccharides as vaccine candidates, *Curr. Top. Microbiol. Immunol.* 150:97-127 (1990).
4. H.J. Jennings, K.G. Rosell, E. Katzenellenbogen, and D.L. Kasper, Structural determination of the capsular polysaccharide antigen of type II group B *Streptococcus*, *J.Biol.Chem.* 258:1793-1798 (1983).
5. H.J. Jennings, E. Katzenellenbogen, C. Lugowski, and D.L. Kasper, Structure of native polysaccharide antigens of type Ia and type Ib group B *Streptococcus*, *Biochem.* 22:1258-1264 (1983).
6. J.L. DiFabio, F. Michon, J.R. Brisson, V.J. Benedi, M.R. Wessels, D.L. Kasper, and H.J. Jennings, Structure of the capsular polysaccharide antigen of Type IV Group B Streptococcus, *Can.J.Chem.* 67:877-882 (1989).
7. M.R. Wessels, J.L. DiFabio, V.J. Benedi, D.L. Kasper, F. Michon, J.R. Brisson, J. Jelinkova, and H.J. Jennings, Structural determination and immunochemical characterization of the type V group B Streptococcus capsular polysaccharide, *J.Biol.Chem.* 266:6714-6719 (1991).
8. G. Kogan, D. Uhrin, J.R. Brisson, L.C. Paoletti, K.L. Kasper, C. von Hunolstein, G. Orefici, and H.J. Jennings, Structure of the type VI group B Streptococcus capsular polysaccharide determined by high resolution NMR spectroscopy, *J.Carbohydr.Chem.* 13:1071-1078 (1994).
9. G. Kogan, J.R. Brisson, D.L. Kasper, C. von Hunolstein, G. Orefici, and H.J. Jennings, Structural elucidation of the novel type VII group B *Streptococcus* capsular polysaccharide by high resolution NMR spectroscopy, *Carbohydr.Res.* 277:1-9 (1995).
10. G. Kogan, D. Uhrin, J.R. Brisson, L.C. Paoletti, A.E. Blodgett, K.L. Kasper, and H.J. Jennings, Structure and immunochemical characterization of the type VIII group B *Streptococcus* capsular polysaccharide, *J.Biol.Chem.* 271:8786-8790 (1996).
11. H.J. Jennings, C. Lugowski, and D.L. Kasper, Conformational aspects critical to the

immuno-specificity of the type III Group B streptococcal polysaccharide, *Biochem.* 20:4511-4511 (1981).

12. L.C. Paoletti, D.L. Kasper, F. Michon, J. DiFabio, K. Holme, H.J. Jennings, and M.R. Wessels, An oligosaccharide-tetanus toxoid conjugate vaccine against type III group B Streptococcus, *J.Biol.Chem.* 265:18278-18283 (1990).
13. L.C. Paoletti, M.R. Wessels, F. Michon, H.J. Jennings, and D.L. Kasper, Group B *Streptococcus* type III glycoconjugate vaccines, *Trends in Glycoscience and Glycotechnology* 4:269-278 (1992).
14. L.C. Paoletti, D.L. Kasper, F. Michon, J. DiFabio, H.J. Jennings, T.D. Tosteson, and M.R. Wessels, Effects of chain length on the immunogenicity in rabbits of group B Streptococcus type III oligosaccharide-tetanus toxoid conjugates, *J.Clin.Invest.* 89:203-209 (1992).
15. W. Zou, C.A. Laferrière, and H.J. Jennings, Oligosaccharide fragments of the type III group B Streptococcal polysaccharide derived from *S. pneumoniae* type 14 capsular polysaccharide by a chemoenzymatic method, *Carbohydr.Res.* 309:297-301 (1998).
16. W. Zou and H.J. Jennings, Mimics of the structural elements of type III group B *Streptococcus* capsular polysaccharide. Part I. Synthesis of a carboxylate-containing pentasaccharide, *J.Carbohydr.Chem.* 15:257-278 (1996).
17. W. Zou and H.J. Jennings, Mimics of the structural elements of type III group B *streptococcus* capsular polysaccharide. Part III: Two repeating units (octasaccharide) with (*s*)-1-carboxyethyl groups replacing sialic acids, *Bioorg.Med.Chem.Lett.* 7:647-650 (1997).
18. W. Zou, J.R. Brisson, Q.L. Yang, M. van der Zwan, and H.J. Jennings, Synthesis and NMR assignments of two repeating units (decasaccharide) of type III group B *Streptococcus* capsular polysaccharide and its ^{13}C-labeled and *N*-propionyl subsituted sialic acid analogues, *Carbohydr.Res.* 295:209-228 (1996).
19. W. Zou, C.R. MacKenzie, L. Therien, T. Hirama, Q.L. Yang, M.A. Gidney, and H. Jennings, Conformational epitope of the type III group B *Streptococcus* capsular polysaccharide, *J.Immunol.* 163:820-825 (1999).
20. M.R. Wessels, A. Munoz, and D.L. Kasper, A model of high affinity antibody binding to type III group B *Streptococcus* capsular polysaccharide, *Proc.Natl.Acad.Sci.USA* 84:9170-9174 (1987).
21. H.J. Jennings, R. Roy, and F. Michon, Determinant specificities of the groups B and C polysaccharides of *Neisseria meningitidis*, *J.Immunol.* 134:2651-2657 (1985).
22. J.R. Brisson, H. Baumann, A. Imberty, S. Pérez, and H.J. Jennings, Helical Epitope of the Group B Meningococcal α(2-8)-Linked Sialic Acid Polysaccharide, *Biochem.* 31:4996-5004 (1992).
23. C.A. Laferrière, R.K. Sood, J.M. de Muys, F. Michon, and H.J. Jennings, *Streptococcus pneumoniae* type 14 polysaccharide-conjugate vaccines: Length stablization of opsonophagocytic conformational polysaccharide epitopes, *Infect.Immun.* 66:2441-2446 (1998).
24. M.R. Wessels and D.L. Kasper, Antibody recognition of the type 14 pneumococcal capsule, *J.Exp.Med.* 169:2121-2131 (1989).
25. J.R. Brisson, S. Uhrinova, R.J. Woods, M. van der Zwan, H.C. Jarrell, L.C. Paoletti, D.L. Kasper, and H.J. Jennings, NMR and Molecular Dynamics Studies of the Conformational Epitope of the Type III Group B *Streptococcus* Capsular Polysaccharide and Derivatives, *Biochem.* 36:3278-3292 (1997).
26. S.V. Evans, B.W. Sigurskjold, H.J. Jennings, J.R. Brisson, R. To, W.C. Tse, E. Altman, M. Frosch, C. Weisgerber, H.D. Kratzin, S. Klebert, M. Vaesen, D. Bitter-Suermann, D.R. Rose, N.M. Young, and D.R. Bundle, Evidence for the extended helical nature of polysaccharide epitopes. The 2.8 A resolution structure and thermodynamics of ligand binding of an antigen binding fragment specific of α(2-8)- polysialic acid, *Biochem.* 34:6737-6744 (1995).
27. C.-H. Wong, R.L. Halcomb, Y. Ichikawa, and T. Kajimoto, Enzymes in organic synthesis: Application to the problems of carbohydrate recognition (Part 2), *Angew.Chem.Int.Ed.Engl.* 34:521-546 (1995).

T-INDEPENDENT IgA RESPONSES TO MICROBIAL POLYSACCHARIDES[1]

Denong Wang

Immunology Laboratory, Columbia Genome Center, College of Physicians & Surgeons, Columbia University, New York, NY 10032 USA

Mucosal surfaces are protected against infectious microorganisms by the mucosal immune system, of which secretory IgA is the major humoral defense factor(1,2). Secretory IgA is polymeric and is structurally suitable for the trapping and clearance of microorganisms in mucosal secretions. A receptor-mediated mechanism effectively transports s-IgA across the epithelium to secretions(3). In the human, the amount of s-IgA exported onto mucosal surfaces can be more than 3 gm per day, far exceeding the amounts of IgG produced for circulation(1,3,4). IgA frequently reacts with microorganisms in the local mucosal microenvironment, regulating the equilibrium between the host and its resident microorganisms(5,6). In certain circumstances, such as patients with Celiac Disease(7,8), IgA antibodies specific for the food antigen Gliadin(9) and for autoantigens such as tissue transglutaminase (tTG)(10), may be detected in circulation and in intestinal juice. Thus, there is an association of hypersensitivity of IgA response to the environmental antigen and a disorder in the self- and non-self recognition with the human disease.

Antibodies of the IgA isotype can be induced by either T-dependent (TD) or T-independent (TI) antigens. Previous studies focused primarily on the T-dependent (TD) route of B-cell activation that leads to Ig-class switching of IgM to IgA. It has generally been believed that IgA response is highly or strictly T cell dependent(11-16). The IgA response to TI antigens is poorly understood. In this article, we attempt to review current evidence that suggests the presence *in vivo* of both a TI route of IgA induction and the conventional TD pathway of IgA responses. Involvement of B-1 and conventional B cells in the TI-route of IgA response; the regulatory roles of T cells, non-T cell and their soluble factors; the nature of the stimulating antigens and the necessary supporting microenvironment for the IgA response will be discussed.

---

[1] We are grateful to Professor Elvin A. Kabat for his long-term support of this project and to Ms Emily Chenette for proof-reading of this manuscript. This work is supported by NIH grant AI45326 to D.W.

The Molecular Immunology of Complex Carbohydrates–2.
Edited by Albert M. Wu, Kluwer Academic/Plenum Publishers, 2001.

I. TD AND TI ANTIGENS AND THEIR IGA INDUCTION

The structure of an antigen significantly influences the nature and characteristics of an immune response(17,18). Protein antigens fail to elicit antibody responses in athymic mice, including neonatally thymectomized(19,20) and *nude* (-/-) mice(21). In contrast, polysaccharides and other macromolecules with repetitive antigenic determinants can induce unimpaired antibody responses in these mice(19-21). These observations establish two broad categories of antigens, T-dependent (TD) and T-independent (TI). TI antigens are further divided into type I (TI-I) and type II (TI-II), based on the ability to elicit antibodies in the CBA/N mouse strain, with an X-chromosome linked immunodeficiency (*xid*)(22). Lipopolysaccharides (LPS) of gram-negative bacteria which induce antibodies in such strains are TI-I antigens; capsular polysaccharides of gram positive bacteria and exopolysaccharides, which induce no response in *xid* strains, are TI-II antigens.

TD and TI antigens activate B cells via different mechanisms(18,23,24). B cell activation via TD antigens is initiated by antigen-specific, MHC-restricted, T cell-B cell interactions. As a result of the first initiation, additional molecular interactions, such as CD40-CD40L(25,26) and CD28/CTLA4-Ig with B7-1/B7-2(27), can proceed. These molecular events further activate the interacting T-B cells. The CD40-CD40L interaction is particularly important in the TD antibody response. Blocking this molecular interaction also blocks the TD antibody response(28-30). TI antigens, such as polysaccharides, activate B cells directly by cross-linking their membrane-bound immunoglobulins, the B cell antigen receptors (BCRs)(23, 31). Cross-linking BCRs is sufficient to induce B cell proliferation, but not sufficient to induce class switching and antibody production(23). Secondary signaling is required for induction of class switching and for the terminal differentiation of B cells into antibody secreting cells (ASCs).

TD antigen-activated B cells undergo distinct pathways of activation and differentiation compared to TI antigen-activated B cells(18, 32). TD antigens induce both germinal center (GC) and antibody-secreting cells. Many critical molecular and cellular events take place in GCs: clonal selection and differentiation of B memory cells, IgH-class switching, somatic hypermutation and affinity maturation, and the production of precursors of antibody-secreting cells(32-34). By contrast, most TI antigens induce an antibody-secreting cell response without GC induction(21). As a result, the typical TI-antigen response fails to induce memory cells, somatic mutation, and class switching (IgM to IgG). Our recent studies illustrated, however, one exemption. A classical TI polysaccharide antigen, $\alpha(1\rightarrow6)$dextran (N279), elicits antigen-specifc GC response(35,36) and, to some extent, somatic mutations in IgV regions during the primary response(37).

Both TD and TI antigens can elicit an IgA response. In the mucosal microenvironment, TD antigens can induce class switching of IgM to the predominant IgA isotype(15,38). This TD pathway of IgA response has been extensively studied. There are at lest two necessary steps, i.e., B cell class switching to IgA and the terminal differentiation of surface IgA-positive (sIgA^{+}) B cells to antibody secreting plasma cells. Class switching takes place in the so-called organized lymphoid tissue, such as the Peyer's patches (PP) of the intestine. A large proportion of B cells in the GCs of PP is found to be (sIgA^{+})(39,40). These IgA-committed

B cells do not differentiate further to IgA-secreting plasma cells at the inductive sites. Instead, they leave the PP and migrate through the systemic immune system, homing back to mucosal effector sites, such as the lamina propria of the GI tract(41,42). Their terminal differentiation occurs at the effector sites, resulting in IgA secretion into the mucosal environment(15). There is a large body of evidence in the literature indicating that IgA responses are highly dependent on the type 2 T helper cell (Th-2). T cells act directly via T-B cognate interactions in the first phase of the IgA response(11-14). T cells subsequently provide the cytokines necessary for terminal differentiation of IgA-switched B cells. IL-5 and IL-6 secreted by Th-2 cells are critical for terminal differentiation of IgA^+ B-cells(15). "The strict T cell depen-dence of IgA-B cell activation and differentiation" is also illustrated by *in vitro* culture experiments(16).

It has been recognized for some time that some thymus-independent antigens, such as microbial polysaccharide $\alpha(1\rightarrow6)$dextran, can induce antigen specific IgM and IgA antibodies in normal and *nude* (-/-) mice(17,35,37,43). Thus, T-dependent antigenic stimulation is not the only way to elicit IgA antibodies *in vivo*. TI antigens of distinct structure may, however, elicit antibody responses of different Ig-isotype profiles(44-49). For example, antibody responses of both Balb/c(47) and C57BL(48) to $\alpha(1\rightarrow3)\alpha(1\rightarrow6)$dextran have been shown to be relatively restricted in IgM isotype. By contrast, both mice strains produce IgM and IgA isotypes in response to $\alpha(1\rightarrow6)$dextrans(17,35,37,49). These observations emphasize the complexity of B cell responses to polysaccharide antigens. Multiple factors, including the structural characteristics of TI antigens, lineage origins and/or subgroups of responding B cells, co-activation of macrophages or other non-T cells by a TI antigen, and the local microenvironment, may influence the induction of an antibody response. These factors will be discussed further in subsequent sections.

II. B CELL LINEAGES OR SUBGROUPS AND THEIR TI RESPONSES

B cell populations are substantially complex and heterogenic with respect to their lineage origins, differential and developmental stage, cell surface marker expression, anatomical localization and responsiveness to antigens(50-52). There are at least two stable distinct population of B lymphocytes, i.e., B-1 B cells and conventional B cells(51,53). Applying multiparameter FACS analysis, Herzenberg et al. (53) found that a cell surface glycoprotein, Ly-1 (CD5), which previously had been thought to be unique in T cells, was expressed by some murine B cell tumors and in the majority of B cells in the peritoneal cavity. Adoptive transfer experiments further showed that the Ly-1 (CD5) B cells are derived from progenitors which differed from those for conventional B cells. These studies provide supporting evidence that the Ly-1 positive B cells belong to a distinct B cell lineage, named B-1 cells. According to detailed phenotyping, the B-1 cells were further subtyped as B-1a (CD5 positive B cells) and B-1b ("sister" population, CD5 negative)(51,52). B-1a cells are derived from progenitors that are present in fetal omentum and fetal liver but are largely absent from adult bone marrow; B-1b cells are derived from progenitors that are present in fetal omentum, fetal liver and also in adult bone marrow. The progenitors of conventional B cells are absent from fetal omentum but are found in fetal liver and adult bone marrow.

The bone marrow derived conventional B cells can be further divided into subsets according to cellular compartment localization(50,54). In the un-stimulated adult spleen, conventional B cells predominantly consist of follicular B cells plus two smaller subsets, namely, marginal zone (MZ) and immature B (IB) cells. These subsets can be also identified at single cell level by advanced multiparameter FACS analysis(50,55). MZ B cells are $CD23^{-}IgM^{bright}IgD^{dull}$ and can be easily distinguished from follicular B cells, which are $CD23^{+}IgM^{interm}IgD^{bright}$. The $CD23^{-}IgM^{bright}IgD^{dull}$ phenotype, however, overlaps with splenic B-1 cells and immature B cells. Sandra et al.(56) introduced additional cellular marker, CD43/S7, to identify B-1 cells from these population. Splenic B-1 cells were identified as $CD43/S7^{+}CD23^{-}IgM^{bright}IgD^{dull}$; MZ and IB B cells as $CD43/S7^{-}CD23^{-}IgM^{bright}IgD^{dull}$. MZ B cells consist of memory B cells derived from primary immune res-ponses to T-dependent antigenic stimulation. These cells express IL-2R and complement receptors and are sessile unless reactivated by antigen.

Herzenberg et al.(52,53) suggested that B-1 and conventional B-cells are functionally distinct. B-1 populations have a repertoire which seems to be skewed toward the production of antibodies reactive with self antigens such as bromelain-treated mouse red blood cells (BrMRBC), DNA, and to many TI antigens, particularly those associated with micro-organisms such as lipopolysaccharide, phosphocholine (PC), dextran B1355S, and other bacterial carbohydrate antigens [See references(57,58) for reviews]. The majority of natural antibodies of IgM isotype in circulation is produced by B-1 cells, although B-1 cells consist only a small proportion (5%) of total B cell population (53). Herzenbergs postulated that evolution has created a "layered" B cell compartment in which the B-1 cells are specifically adapted for, and play a major role in, responses to microbial antigens, forming a first line of defense against bacterial infection(53,59).

There is, however, evidence that conventional B cells are also responsible for anti-polysaccharide responses. Studies of the ontogeny of anti-polysaccharide responses have found that the time at which the human immune system acquires full-responsiveness to polysaccharide antigens (at approximately two years of age) coincides with maturation of the MZ B cells(60,61). By contrast, splenic germinal center and mantle zone B cells reach an adult phenotype by five months, long before the maturation of their anti-polysaccharide responsiveness. Consistent to the human studies, Lane et al. found that the immune response of recovering X-irradiated mice to TI-2 antigens correlates with the rate at which their B cells reappeared in the marginal zone(62). Thus, the involvement of MZ B cells in anti-polysaccharide responses, which are bone marrow derived conventional B cells, is strongly implicated.

The delayed maturation is a common characteristic of TI antibody responses. The time required to develop a full-responsiveness to TI stimulations differs significantly among antigenic systems studied. In mice (63), antibody responses to TI-I antigens (LPS and other) and protein antigens reach adult levels within 1-2 weeks; antibodies to TI-II antigens can be detected only at 2-3 weeks. For *pneumococcal* polysaccharides SSS-III and dextran B1355S, full development of the antibody response is not reached until 4 weeks; with levan and $\alpha(1\rightarrow6)$dextran, 7 and 13 weeks respectively are needed.

B-1 and conventional B cell lineages have different kinetics of B cell development and functional maturation. It is interesting to see whether the delayed maturation of antibody responses to TI antigens coincide with B-1 or B-2 cell maturation. B-1 cell lineages (B-1a and B-1b) developed earlier than did conventional B cells(52). By four weeks of age, the mouse B-1 cell population approaches adult size and is able to mount normal levels of antibody responses. Conventional B cells reach maximal levels at 12-14 weeks of age. By comparing the kinetics of B cell development and the timing of maturation of anti-TI antigen responses, it is clear that the delayed ontogeny of antibody response to some but not all TI antigens may be attributed to the time required for B-1 cell maturation.

We have demonstrated that a specific TI-2 antigen, $\alpha(1\rightarrow6)$dextran (N279), can activate both B-1 and B-2 cells(64). In previous studies, we and others showed that two native dextrans, $\alpha(1\rightarrow6)$dextran (N279) and $\alpha(1\rightarrow3)\alpha(1\rightarrow6)$dextran (B1355S) elicit distinct patterns of B cell responses with respect to repertoires of antibodies(37,47,48,65) and the pathways of splenic B cell activation(36,44). Förster et al.(66) studied the B cell lineage association of the antibody responses to $\alpha(1\rightarrow3)\alpha(1\rightarrow6)$dextran (B1355S). Applying adopt transfer animal models, it was demonstrated that B-1 cells transferred from adult peritoneum to newborn mice could respond to $\alpha(1\rightarrow3)\alpha(1\rightarrow6)$dextran (B1355S). We studied the relative contribution of B-1 and conventional B cells to the primary responses to $\alpha(1\rightarrow6)$dextran (64). We isolated and phenotyped the antigen-specific antibody secreting cells using FACS and ELISA-spot assays. Not only are $\alpha(1\rightarrow6)$dextran-specific cells present in both the B-1 and conventional B cell populations in non-immunized mice, but the T-independent response to the antigen is also mediated by both B-1 and conventional B cells. Thus, the two microbial polysaccharides elicited B cell populations of different lineage origins.

Both B-1 and B-2 cells are capable of switching from IgM to IgA isotypes. In the small intestines, there are many IgA-producing plasma cells in the lamina propria. It was traditionally believed that these IgA producers are derived from precursor B cells that are triggered in PP and thereafter migrate to the lamina propria. Recent studies demonstrated that the B cells in the PP can be repopulated by the bone marrow B cells but not by peritoneal B cells. Peyer's patch B cells are thus considered to be part of conventional B cell lineage. There is, however, evidence that, in addition to PP conventional B cells (B-2 cells), peritoneal B-1 cells contribute significantly to the pool of IgA containing cells in the intestinal lamina propria of the mouse. Bos et al.(5,6,67) transferred peritoneal cells from BALB/c (Igh-a) mice into C.B17 (congenic strain of BALB/c, Igh-b) scid/scid mice. Six to eight months after injection, many donor-derived IgA-producing cells (Igh-a) were found in the lamina propria of the intestine. The mesenteric lymph nodes of these mice were found to be a major site of proliferation and generation of IgA plasmablasts. Eight IgA-producing hybridomas were established from the mesenteric lymph nodes of such mice. They are reactive to fecal bacterial populations, such as *Peptostreptococcus productus*, *Peptostreptococcus anaerobius* and *Clostridium perfringens* (6).

Kroese et al. (personal communication) further addressed how B-1 and conventional B cells contribute to mucosal IgA. They treated neonatal BALB/c (Igh-a) mice or congenic BAB/25 (Igh-b) mice with allotype specific anti-IgM antibodies, which resulted in a complete depletion of all B cells. During this treatment, animals were injected with B-1 cells (sorted or unsorted) from an Ig allotype congenic donor. Fecal samples were collected when animals were 3-7 months old, and the bacteria were analyzed for *in vivo* coating with IgA of both allotypes using flow-cytometry. It is demonstrated that the majority of the IgA on the fecal bacteria has the allotype of the B-1 cell donor. Nearly all IgA coated bacteria have B-1 cell derived IgA and a minority population of them carries B-2 cell derived IgA as well.

If the B-1 derived IgA+ B cells in the lamina propria of intestine are not derived from the GCs of Peyer's patches, where do they switch from IgM to IgA isotype? What is the role of T cells in regulating B-1 cell-mediated IgA responses? Investigation of these specific questions will lead to a better understanding of the biology of IgA responses.

III. T CELL REGULATION OF THE T-INDEPENDENT RESPONSES

TI-antigens can stimulate antibody production in the absence of MHC class II-restricted T cell help. T cells can, however, influence a TI-response via at least four different mechanism(23). These include 1) the T cell regulation of the idiotypic repertoire of responding B cells; 2) direct activation of T cells by TI-antigens; 3) non-classical T-B cognate interaction; and 4) T cell derived cytokines.

Baker and co-workers(68-71) demonstrated that T cells regulate anti-polysaccharide responses both positively and negatively. They applied the capsular polysaccharide of type III Streptococcus pneumoniae (SSS-III) and $\alpha(1\rightarrow3)\alpha(1\rightarrow6)$dextran (B1355S) as models and found that antigen specific T suppressor cells (Ts, CD4-CD8+) and amplifier T cells (Ta, CD4+CD8-) were involved in the regulation. When donor cell suspensions containing both Ts and Ta were transferred to athymic mice immunized with the capsular polysaccharide SSS-III, there was no change in magnitude of the resulting antibody response. If Ts activity was eliminated before transfer using the Ts specific anti-serum, antibody response increased substantially; suppression occurs after the prior removal of Ta activity. Thus, Ts and Ta may act competitively on B cells to control the magnitude of the antibody response to the antigen. T cell influence on the magnitude of antibody responses may not be unique to this carbohydrate antigen. Similar phenomena have been observed in the antibody responses to other bacterial polysaccharides that do not cross-react with the polysaccharide of type III Streptococcus pneumoniae, such as Neisseria meningitidis groups A and C capsular polysaccharides, Streptococcus mutans polysaccharide, Pseudomonas aeruginosa LPS, Serratia marcescens LPS, and E. coli. O55 LPS(68).

It was proposed by Baker et al.(71) that the antigen-specific T regulation of TI response is mediated by T recognition of the idiotypic determinant derived from the antigen-specific antibodies. Their supporting evidence is that a small number of B cells from SSS-III immunized mice can induce the T-cell regulation in non-immunized animals. Thus, the idiotypic determinant of SSS-III specific antibody on the surface of immune B cells may signal the activation of T cells of distinct functions, including Ts and Ta. By analyzing both normal mice and the *nude* strain, it was also shown that the increased magnitude of antibody response to 2,4,6-trinitrophenyl (TNP)-ficoll in *nude* mice is due to absence of the normal regulatory T cell dependent auto-anti-idiotypic response(72). Anti-levan responses, however, lack such regulation. A cyclic (oscillatory) antibody response to the polysaccharide occurs in both euthymic and athymic BALB/c mice, although auto-anti-idiotypic antibodies were not detected in athymic mice(73). The role of T cells in influencing both the magnitude and the clonal composition of the anti-phosphorylcholine (PC) response was also examined(74). It was found that T cells regulate the idiotypic repertoire of an immune response to the TI antigen phosphocholine.

Mongini et al.(75-77) suggested that T cells exert regulatory effects on the TI antibody response via a distinct population of Ig-specific regulatory cells, which is independent of carrier specificity. Using TNP-Ficoll and TNP-levan as antigens to immunize *nu/+* and *nu/nu* mice, they showed that T cell-bearing *nu/+* mice produced substantially higher levels of IgG2 anti-TNP antibodies than their athymic counterparts *(nu/nu)*. The lower IgG2 response can not be attributed to the nature of responding B cells of nu/nu mice, since a small population of transferred T cells of normal mice is able to recover the IgG2 anti-TNP responses of *nu/nu* mice.

Claasen and colleagues(78) reported that some TI-2 antigens (TNP-Ficoll and $\alpha(1\rightarrow3)\alpha(1\rightarrow6)$dextran (B1355S) stimulate T cells directly *in vivo*, although the molecular mechanism is unknown. The activated T cells were found in the periarteriolar lymphoid sheaths (PALS), where the antigen-specific ASCs are located. These T cells expressed the CD40 ligand on their surface and produced IFN-γ to the local environment. There is, however, concern that such observed effects may be mediated by other indirect mechanisms, such as contamination of LPS in the microbial antigen preparations. LPS can induce expression of certain cytokines, such as IFN-γ, by non-T cell types. The cytokine is proven to be a strong effector of TI responses. There is also a possibility that activated B cells stimulate T cells. Secreted Ig could bind to T cells expressing Fc receptors for IgG, IgM, or IgA and activate T cells. B cells may secrete soluble CD40, regulating the expression of CD40L by T cells. Activated B cells can also produce cytokines to influence T cell activation (See Ref.(23) for a recent review).

Strober and colleagues' earlier study(13,14) illustrated the importance of T-B direct interaction in regulating class switching of IgM to IgA. They established T cell clones from PP T cells. Addition of such T cells into LPS-stimulated splenic B cell culture increased the frequency with which sIgA+ B cells were observed. These T cells were termed as "switch T cells." A cognate interaction of switch T and B cells appeared to be necessary, since the isotype switching to IgA could not be induced by the culture medium of these T cell clones. The switch T cells were not, however, able to induce IgA secretion, indicating that a different type of assistance is required for the further differentiation of these sIgA+ B cells to ASCs. The molecular basis of a non-classical T-B cognate that involves T-independent responses is unknown. A well-documented example of non-classical T-B cognate interaction is the binding of non-peptide antigen by T cell receptor γδ. Whether and how such interaction influences the TI antibody response, especially the mucosal IgA responses, is a current topic of investigation.

Kiyono et al.(79) showed a different type of T help for IgA responses. They established Th clones that are specific for sheep red blood cells (SRBC) from the PP of mice fed SRBC. These cells supported IgA production by the cultures containing surface-IgA^+ B cells but not the cultures having only sIgA B cells. The action of these Th cells could be replaced by cell-free supernatants. We now know that these soluble factors may include IL-4, IL-5 and IL-6, which can be produced by Th-2 cells, enhancing the terminal differentiation of sIgA+ B cells. Cytokine regulation will be further discussed in the next section.

Recent availability of a strain of T cell deficient mice, the γδ/αβ TCR double knock-out mice (TCR-KO)(44,80,81) has made it possible to critically evaluate the roles of T cells and non-T cell type in the IgA response. The TCR-KO mice were generated by crossing TCR-β mutant mice with TCR-δ mutant mice(80). Unlike *nude* mice, whose gut-associated T-cell lineage remains unimpaired(82,83), this strain lacks mature T-cells. Interestingly, immuno-histology reveals the presence of IgA producing B cells in lymphoid tissue in TCR-KO mice (80). When this strain of T cell free mice (in C57BLX129 background) was challenged with the murine pathogen Rotavirus, a reduced yet significant number of IgA-secreting B cells was elicited, contributing to an effective protection from the infectious microorganism(81). These IgAs were found to bind exclusively to a viral structural protein, **VP6**, which forms a repetitive structure in the rotavirus particle. Although T-cells play important roles in regulating mucosal IgA responses to protein antigens, they are not absolutely necessary for the induction of a mucosal IgA response to some antigens, such as **VP6** of Rotavirus. This suggests the involvement of an alternative route of IgA induction that is completely independent of T lymphocytes.

To further investigate the T-independent IgA response, we applied two structurally well-defined TI-2 antigens(17), α(1→6)dextran N279 and α(1→3)α(1→6)dextran B1355S, for immunization. When α(1→6)dextran was injected into the strain of TCR-KO mice of C57BL background, a large number of α(1→6)dextran-specific B-cells, predominately of the IgA isotype, was elicited. The number of antigen specific IgA-ASCs in the spleen was about 30 fold higher than those in C57BL/6J mice at day 5 post-immunization. At day 12 post-immunization, a large number of antigen-specific ASCs of IgM and IgA isotypes remained in the spleen of the TCR-KO mice. In non-immunized mice, the number of α(1→6)dextran-specific ASCs was less than 10 per 10^6 splenocytes in both C57BL/6J and the TCR-KO mice. Induction of IgA in the absence of T cells is, however, not a common characteristic of all TI-antigens. α(1→3)α(1→6)dextran failed to elicit significant IgA responses in both the TCR-KO mice and normal C57BL/6J (Data not shown, Wang, D. et al.). Molecular and cellular mechanisms underlying the distinct abilities of the two dextrans in eliciting IgA responses are currently being investigated. These findings have shed light on an alternative route of IgA induction, in which the induction of IgA class switching and the terminal differentiation to IgA-secreting B cells are completely independent of T cells.

IV. CYTOKINES AND T-INDEPENDENT IGA RESPONSES

T cells with helper functions can be classified into Th1 and Th2 based on the types of cytokines they produce(84, 85). Th1 secrete IL-2, IL-3, TNF-α and IFN-γ; Th2 secrete IL-3, IL-4, IL-5, IL-6, IL-9, IL-10 and some TNF-α. In addition, Th1 and Th2 cells exhibit the preferential expression of some activation markers. For example, human Th2-like cells preferentially express CD30, CCR4, CCR8, CXCR4, and CRTH2, whereas LAG-3 and CCR5 preferentially associate with Th1 like cells(85). T cells and their cytokines have been shown to play a critical role in various stages of IgA response(1,15,86), including induction of Ig class switching of IgM to IgA (switch T cells) and of terminal differentiation of IgA-committed B cells (helper T cells). Many researchers believe that IgA responses are highly Th-2 dependent(16,75, 87,88). Evidence supporting this dominant view of IgA response includes at least the following topics: 1) Induction of a mucosal IgA response is coincident with predominant activation of $CD4^+$ Th2 cells in PP; 2) Direct demonstration of the role of Th2 cytokines, such as IL-4, IL-5 and IL-6, in inducing terminal differentiation of IgA-committed B cells; 3) The Th1 cytokine, IFN-γ, antagonizes IL-4 in a variety of systems.

Xu-Amano et al.(89) used SRBC to immunize mice either orally or systemically. Oral immunization of mice selectively induced a Th2 cell response in the PP of intestine; the systemic route produced a mixed Th1 and Th2 cell response. Using tetanus toxoid (TT) for oral immunization with cholera toxin (CT) as an adjuvant, this group was able to elicit both systemic and mucosal B cell responses with significant amounts of anti-TT IgA in both fecal extracts and serum. When PP and splenic CD4+ T cells were isolated at the peak of the mucosal IgA response and restimulated *in vitro* with TT, high numbers of IL-4 and IL-5 producing cells (Th2) were detected(90,91). By contrast, IFN-γ and IL-2 producing cells were the same as background levels. Thus, the mucosal IgA response elicited is associated with the activation of Th2 cells but not Th1 cells. Infection of mice with a pathogen *Heligmosomoides polygyrus* can elicit an increased Th2 type response, resulting in protective IgA antibodies. Evidence supporting the Th2 association of IgA response is also seen in recent studies applying IL-4 knock out mice(92). This strain is defective in Th2 cytokine responses and fails to respond to oral immunization using protein antigens, including the mix immunization with CT as an adjuvant.

The polarized Th-1 or Th-2 responses can be induced by microbial infection or by vaccination(93-96). In certain conditions, a polarized T cell response can also be responsible for immunopathological reactions [see Ref.(85,97) for recent reviews]. For examples, Th-1 dominated responses are involved in the pathogenesis of multiple sclerosis(98), Crohn's disease(99), Celiac disease(100), Helicobacter pylori-induced peptic ulcer(101,102), and some unexplained recurrent abortions(103); Th2-type responses predominate in Omenn's syndrome (85), chronic graft-versus-host disease(104), idiopathic pulmonary fibrosis and progressive systemic sclerosis(104,105). In HIV infection, activation of a Th2 response by the infection itself or by other immunological agents may trigger HIV replication(97,106, 107), resulting in more rapid evolution towards the full-blown disease. It is understood that HIV replication requires cellular factors that are expressed in activated CD4 T cell and other cells. Development of an alternative vaccination strategy that elicits IgA but not Th2 responses represents a current challenge to the field of HIV vaccination.

There is literary evidence that several critical factors necessary for IgA response can be provided by non-T cells. A notable example is Transforming Growth Factor beta (TGF-β) that can be produced by both lymphoid and non-lymphoid tissues(108). TGF-β is a pleiotropic cytokine having suppressive effects on growth, and both inhibitory and stimulatory effects on the differentiation of a variety of cell types. It acts on IgA^- B cells and directs switching to IgA from $sIgM^+sIgD^+$ B cells. TGF-β directs class switch recombination to IgA, instead of selecting committed cells. It induces expression of germline transcripts from unrearranged Ig Cα gene prior to switch recombination in both mouse and human B cells. Such germline transcripts are necessary to direct switch recombination. Molecular analysis, such as performing nuclear run-on assays, showed that the TGF-β mediated regulation acts at transcriptional level. It increases transcription of Iα-Sα-Cα but does not influence the stability of the transcripts. Th2-like cells, such as those isolated from the mesenteric lymph nodes of mice fed myelin basic protein (MBP) and those derived from PP of the intestines, may produce TGF-β. It was suggested that these T cells suppress autoimmune responses elicited by MBP or allergic responses to food antigens by producing TGF-β. T cells are not, however, the only TGF-β producers. Both normal and transformed mouse and human B cells synthesize and secrete TGF-β in the latent and/or active forms. Furthermore, activation of B cells increases the level of its mRNA expression and production of active form of TGF-β protein.

IFN-γ is another important factor for Ig class switching. IFN-γ is a characteristic Th-1 cytokine, but non-T-cells can also produce it(109). This factor plays a major role in cell-mediated immune responses. Its roles in humoral immunity are more complex. IFN-γ has been shown to influence IgA responses both positively and negatively. In an early study, Kagnoff and Murray(87) demonstrated that recombinant IFN-γ and IL-2 support the induction of IgA response to $\alpha(1\rightarrow3)\alpha(1\rightarrow6)$dextran (B1355S) in a splenic B cell culture. Recently, McIntyre et al. applied an anti-IgD-$\alpha(1\rightarrow6)$dextran (αδ-Dex) conjugate as a primary B cell cross-linking activator to study how a combination of cytokines influences the induction of class switching to IgA(110). In their *in vitro* system, IFN-γ strongly suppressed the induction of mIgA+ cells. IFN-γ enhances IgA synthesis in the existing mIgA+ cells induced by TGF-β,(111). These well-designed *in vitro* experiments clearly illustrate a combined and, in some cases, synergistic effect of multiple factors on a specific stage or step of IgA responses. There is, however, always concern about how well an *in vitro* model mimics a physiological condition *in vivo*.

As summarized above, we illustrated that a strain of TCR-KO mice were able to mount an IgA response to α(1→6)dextran but not to α(1→3)α(1→6)dextran. The induction of antigen specific IgA in the absence of functional T cells is associated with an *in vivo* induction of IL-9 and IFN-γ (Wang, D. et al, manuscript in preparation). What cell type is responsible for their production and how α(1→6)dextran induces these cytokines is currently unknown. Since these mice lack functional T cells, NK cells could be candidates for IFN-γ production. NK cells can be activated to produce high levels of IFN-γ as a consequence of direct interaction of NK cells with B cells(23). Such a direct NK-B cell interaction has been reported in a number of murine and human systems(23), including the T-independent antibody responses elicited by anti-IgD-α(1→6)dextran conjugates(112). IL-9 is believed to be produced by IL-2 activated Th2 cells(113). Detection of IL-9 in the TCR-KO mice suggests that non-T cell types can also produce IL-9.

V. MICROENVIRONMENTS AND IGA RESPONSES

It has been recognized for some time that the administration of antigen by the respiratory mucosal or gastrointestinal routes results in a greater IgA response than if administered parenterally. In PP germinal centers of intestines, it has been estimated that up to 60-70% of B cells are surface IgA positive. By contrast, in most *in vitro* studies, the frequency of switching to IgA is only of the order of 2-3% and to the best of 15-20% as recently reported by McIntyre et al.(110). Thus, the microenvironment of mucosal lymphoid tissue tends to drive B cells to an apparent IgA commitment. What are critical factors that determine the nature of B cell activation in responding to a mucosal challenge? The following factors have been discussed in the literature. 1) Environmental antigens; 2) Chronic stimulation of B cell by environmental antigens; 3) The nature of responding B cells; 4) Th2 activity in PP and 5) The specialized architecture of mucosal lymphoid tissue.

In Cebra's early work(114), the relationship between the historic nature of a stimulating antigen and the corresponding mucosal IgA response was investigated. Using the splenic fragment assay, they compared the isotype distribution of B cells stimulated by deliberately immunized foreign antigens (DNP) versus those triggered by antigens that are naturally present in the gut, such as bacterial antigen PC. They found that the PC-conjugate induced a predominance of IgA-B cell responses in the PP, which was not seen when DNP-conjugate was used for immunization. They suggested that the high proportion of induction of IgA-committed cells in PPs was linked to chronic stimulation. In the mucosal microenvironment, bacterial antigens may persist for a long time, triggering an antigen-driven clonal expansion.

This hypothesis is supported by a recent finding that class switching is cell division dependent. Hodgkin's group(115-117) developed a method, CFSE [5-(6-)carboxyfluorescein diacetate succinimidyl ester] cell labeling, to measure the frequency of isotype switching in cell population of various divisions. LPS was used to induce switching of B cells to IgG3 and, in the presence of TGF-β, to IgG2b and IgA. They found that switching to all three isotypes increased with division number according to a consistent relationship that was independent of time in culture. CD40 ligand increased switching to IgA and decreased switching to IgG2b and IgG3 when measured per division. In mouse, the IgH gene segments are arranged in the order of 5' $\mu-\delta-\gamma3-\gamma1-\gamma2\beta-\gamma2\alpha-\varepsilon-\alpha$ 3', with the $C\alpha$ region as the last segment. According to the Honjo and Kataoka deletion model of isotype expression(118), IgH class switching proceeds irreversibly along the order of the CH genes, until the last gene $C\alpha$ is reached. Thus, an antigen-driven clonal expansion triggered by chronic antigenic stimulation in the mucosal environment may lead to enriched population of IgA-B cells.

If the nature of responding B cells is taken into account, one may, however, argue Cebra's original suggestion. The anti-PC response (T15Id) is mediated by B-1 cells, whereas the anti-DNP response is a conventional B cell response. Thus, the observed IgA-dominant response to PC may be attributed to the nature of B-1 cells that are responsible for the anti-PC response. Perhaps, the self-replenishing characteristics of B-1 cells and their repertoire of antibodies selected by microbial antigen have made them ready to contribute to the mucosal IgA response.

The nature of microbial polysaccharides, including their ability to deliver prolonged and persistent signaling to B cells and their potential to stimulate non-T cells, may contribute further to their ability to elicit the T-independent IgA responses. Polysaccharides with distinct physiochemical properties may be handled differently by host cells. In the spleen, macrophages located in various cellular compartments were shown to retain polysaccharides selectively(119-121). Acidic polysaccharides predominantly localized in red pulp macrophages. Neutral polysaccharides were detectable exclusively or very intensely in macrophages of the marginal zone of the white pulp. Some polysaccharides can be trapped by follicular dendritic cells. 125 I-labeled levan appeared in germinal centers as early as 24 hours after injection and remained there for at least 10 days(120). Levan is uniquely able to activate complement C3 via the alternative complement pathway. Its follicular localization can be blocked by C3 depletion with cobra venom factor. However, since follicular dendritic cells have both C3 and Ig-Fc receptors, other polysaccharides may be localized in the germinal center as immune-complexes(35,120).

Two native dextrans, $\alpha(1\rightarrow6)$dextran and $\alpha(1\rightarrow3)\alpha(1\rightarrow6)$dextran, have served as model systems to study possible relationship of the structural characteristics of TI antigens and the B cell responses. $\alpha(1\rightarrow6)$dextran (N279) is a near-linear macromolecule of glucose with 90% $\alpha(1\rightarrow6)$, 5% $\alpha(1\rightarrow)$(terminal) and 5% $\alpha(1\rightarrow3)$ linkages(122), representing the simplest polysaccharide(35). B1355S is the prototype of dextran having significant proportions of non-$\alpha(1\rightarrow6)$linkages(122, 123). Therefore, the tertiary structure of B1355S differs strikingly from that of N279. The predominant antigenic determinant of B1355S is the backbone of alternating $\alpha(1\rightarrow3)\alpha(1\rightarrow6)$linkages(17, 123). This backbone structure is much more rigid than the $\alpha(1\rightarrow6)$glycosidic internal linear chain in dextran N279 [see Ref.(17) for a review].

The two TI-2 antigens have distinct localization patterns in the spleen(35,120). After *i.v.* injection, $\alpha(1\rightarrow6)$dextran localizes in both follicles and marginal zones (MZs);

$\alpha(1\rightarrow3)\alpha(1\rightarrow6)$dextran localizes in MZs only. Follicular dendritic cells (FDCs) are specialized to display intact antigens for recognition by B cells(124,125). They can trap and retain immune complexes by means of the C3b and Fc receptors on their membrane. In the splenic MZ, the MZ macrophages have been shown to be able to take up neutral polysaccharides (119).

$\alpha(1\rightarrow6)$dextran and $\alpha(1\rightarrow3)\alpha(1\rightarrow6)$dextran elicit distinct patterns of B cell responses with respect to repertoires of antibodies(37,47,48,65) and the pathways of splenic B cell activation(35,44). $\alpha(1\rightarrow3)\alpha(1\rightarrow6)$dextran elicits a highly restricted VHJ558/λ1 dominant responses(47,48), while the $\alpha(1\rightarrow6)$dextran responses are highly diverse(37). Moreover, unlike $\alpha(1\rightarrow3)\alpha(1\rightarrow6)$dextran, which induces only an antibody secreting cell (ASC) response, primary immunization with $\alpha(1\rightarrow6)$dextran elicits both an ASC response as well as the GC reaction(35,36). In addition, $\alpha(1\rightarrow6)$dextran but not $\alpha(1\rightarrow3)\alpha(1\rightarrow6)$dextran is able to elicit an antigen-specific IgA response in the TCR-KO mice(44). How and whether the differences in cellular compartment localization of the two dextrans contribute to their induction of distinct pattern of splenic B cell responses is, however, still unknown.

SUMMARY

There is accumulating evidence indicating the presence *in vivo* of T-independent routes of IgA response in addition to the conventional T-dependent IgA response. Factors influencing these alternative pathways of IgA responses may include the structural characteristics of a stimulating antigen, the nature of responding B cells, and the microenvironment. The structural complexity of polysaccharide antigens has made it difficult to summarize a general scheme for the antibody responses they induce. Instead, one may expect that each individual polysaccharide may be able to create a unique microenvironment by activation of specific cell populations in the repertoires of non-T cell types. A specific pattern of B cell response may thus be elicited by TI stimulation. Recognition of such a unique property of a TI antigens is necessary for us to better understand the T-independent IgA response. Information obtained may have an impact on the development of vaccination strategies directed at the mucosal immunity mediated by IgA antibodies.

REFERENCES

1. Mestecky, J., McGhee, J. R. (1987). Immunoglobulin A (IgA): molecular and cellular interactions involved in IgA biosynthesis and immune response. [Review]. *Advances in Immunology* 40: 153-245.
2. Kraehenbuhl, J. P., Neutra, M. R. (1992). Molecular and cellular basis of immune protection of musosal surfaces. *Physiological Rev.* 72: 853-879.

3. Mestecky, J., Lue, C., Russell, M. W. (1991). Selective transport of IgA: cellular and molecular aspects. *Gastroenterol. Clin. North. Am.* 20: 441-471.
4. Seilles, E., Vuitton, D., Sava, P., Claude, P., Panouse, J., Roche, A., Delacroix, D. L. (1985). IgA and its different molecular forms in the mesenteric, portal and peripheral venous blood in man. *Gastroenterol. Clin. Biol.* 9: 607-613.
5. Bos, N. A., Bun, J. C., Popma, S. H., Cebra, E. R., Deenen, G. J., van der Cammen, M. J., Kroese, F. G., Cebra, J. J. (1996). Monoclonal immunoglobulin A derived from peritoneal B cells is encoded by both germ line and somatically mutated VH genes and is reactive with commensal bacteria. *Infection & Immunity* 64: 616-23.
6. Bos, N. A., Bun, J. C., Bijma, H., Cebra, E. R., Cebra, J. J., Deenen, G. J., van der Cammen, M. J., Kroese, F. G. (1994). Analysis of IgA-producing hybridomas derived from peritoneal B1 cells. *Advances in Experimental Medicine & Biology* 355: 265-9.
7. Corazza, G. R., Gasbarrini, G. (1995). Coeliac disease in adults. *Baillieres Clinical Gastroenterology* 9: 329-50.
8. Littlewood, J. M. (1995). Coeliac disease in childhood. *Baillieres Clinical Gastroenterology* 9: 295-327.
9. Troncone, R., Ferguson, A. (1991). Anti-gliadin antibodies. *Journal of Pediatric Gastroenterology & Nutrition* 12: 150-8.
10. Dieterich, W., Ehnis, T., Bauer, M., Donner, P., Volta, U., Riecken, E. O., Schuppan, D. (1997). Identification of tissue transglutaminase as the autoantigen of celiac disease [see comments]. *Nature Medicine* 3: 797-801.
11. Kawanishi, H., Saltzman, L. E., Strober, W. (1982). Characteristics and regulatory function of murine Con A-induced, cloned T cells obtained from Peyer's patches and spleen: mechanisms regulating isotype-specific immunoglobulin production by Peyer's patch B cells. *Journal of Immunology* 129: 475-83.
12. Kawanishi, H., Strober, W. (1983). T cell regulation of IgA immunoglobulin production in gut-associated lymphoid tissues. *Molecular Immunology* 20: 917-30.
13. Kawanishi, H., Saltzman, L., Strober, W. (1983). Mechanisms regulating IgA class-specific immunoglobulin production in murine gut-associated lymphoid tissues. II. Terminal differentiation of postswitch sIgA-bearing Peyer's patch B cells. *Journal of Experimental Medicine* 158: 649-69.
14. Kawanishi, H., Strober, W. (1983). Regulatory T-cells in murine Peyer's patches directing IgA-specific isotype switching. *Annals of the New York Academy of Sciences* 409: 243-57.
15. Beagley, K. W., Black, C. A., Dunkley, M. I., McGhee, J. R. (1996). Regulation of mucosal Immunity. In *Regulation of mucosal Immunity.*, ed. C. M. Snapper, pp. 392-408. Chichester, New York, Brisbane,Toronto, Singapore: John Wiley & Sons Ltd
16. Ehrhardt, R. O., Harriman, G. R., Inman, J. K., Lycke, N., Gray, B., Strober, W. (1996). Differential activation requirements of isotype-switched B cells. *Eur. J. Immunol.* 26: 1926-1934.
17. Wang, D., Kabat, E. A. (1996). Carbohydrate Antigens(Polysaccharides). In *Carbohydrate Antigens(Polysaccharides)*, ed. M. H. V. V. Regenmortal, pp. 247-276. Boca Raton New York London Tokyo, 1996.: CRC Press.
18. Laman, J. D., Claassen, E. (1996). T-cell-independent and T-cell-dependent humoral immunity. In *T-cell-independent and T-cell-dependent humoral immunity*, ed. C. M. Snapper, pp. 24-72. Chichester, New York, Brisbane,Toronto, Singapore: John Wiley & Sons Ltd
19. Basten, A., Howard, J. G. (1973). Thymus independence. In *Thymus independence*, ed. A. J. S. Davies, pp. 265. New York: Plenum.

20. Humphrey, J. H., Parrott, D. M. V., East, J. (1964). Studies on globulin and antibody production in mice thymectomised at birth. *Immunology* 7: 419-439.
21. Weissman, I. L., Gutman, G. A., Friedberg, S. H., Jerabek, L. (1976). Lymphoid tissue architecture. III. Germinal centers, T cells, and thymus-dependent vs thymus-independent antigens. *Adv. Exp. Med. Biol.* 66: 229-237.
22. Scher, I. (1982). The CBA/N mouse strain: an experimental model illustrating the influence of the X-chromosome on immunity. *Adv. Immunol.* 33: 1-71.
23. Mond, J. J., Lees, A., Snapper, C. M. (1995). T cell-independent antigens type 2. In *T cell-independent antigens type 2*, pp. 655-92.
24. Cambier, J. C., Pleiman, C. M., Clark, M. R. (1994). Signal transduction by the B cell antigen receptor and its coreceptors. *Annu. Rev. Immunol.* 12: 457-486.
25. Durie, F. H., Foy, T. M., Masters, S. R., Laman, J. D., Noelle, R. J. (1994). The role of CD40 in the regulation of humoral and cell-mediated immunity. *Immunol Today* 15: 406-11.
26. Banchereau, J., Bazan, F., Blanchard, D., Briere, F., Galizzi, J. P., van Kooten, C., Liu, Y. J., Rousset, F., Saeland, S. (1994). The CD40 antigen and its ligand. *Annual Review of Immunology* 12: 881-922.
27. Linsley, P. S., Ledbetter, J. A. (1993). The role of the CD28 receptor during T cell responses to antigen. *Adv. Immunol* 11: 191-212.
28. Foy, T. M., Laman, J. D., Ledbetter, J. A., Aruffo, A., Claassen, E., Noelle, R. J. (1994). gp39-CD40 interactions are essential for germinal center formation and the development of B cell memory. *J Exp Med* 180: 157-63.
29. Foy, T. M., Durie, F. H., Noelle, R. J. (1994). The expansive role of CD40 and its ligand, gp39, in immunity. *Semin Immunol* 6: 259-66.
30. Foy, T. M., Shepherd, D. M., Durie, F. H., Aruffo, A., Ledbetter, J. A., Noelle, R. J. (1993). In vivo CD40-gp39 interactions are essential for thymus-dependent humoral immunity. II. Prolonged suppression of the humoral immune response by an antibody to the ligand for CD40, gp39. *J Exp Med* 178: 1567-75.
31. Dintzis, H. M., Dintzis, R. Z., Vogelstein, B. (1976). Molecular determinants of immunogenicity: The immunon model of immune response. *Proc. Natl. Acad. Sci. USA* 73: 3671-3674.
32. Kroese, F. G., Timens, W., Nieuwenhuis, P. (1990). Germinal center reaction and B lymphocytes: morphology and function. [Review]. *Current Topics in Pathology* 84: 103-48.
33. Berek, C., Ziegner, M. (1993). The maturation of the immune response. *Immunol.Today.* 14: 400-404.
34. Kraal, G., Weissman, I. L., Butcher, E. C. (1982). Germinal center B cells: antigen specificity and changes in heavy chain class expression. *Nature* 298: 377-379.
35. Wang, D., Wells, S. M., Stall, A. M., Kabat, E. A. (1994). Reaction of germinal centers in the T-independent response to the bacterial polysaccharide α(1->6)dextran. *Proc. Natl. Acad. Sci. USA* 91: 2504-2506.
36. Sverremark, E., Fernandez, C. (1998). Germinal center formation following immunization with the polysaccharide dextran B512 is substantially increased by cholera toxin. *Int Immunol* 10: 851-9.
37. Wang, D., Liao, J., Mitra, D., Akolkar, P. N., Gruezo, F., Kabat, E. A. (1991). The repertoire of antibodies to a single antigenic determinant. *Mol. Immunol.* 28: 1387-1397.
38. Cebra, J. J., Bos, N. A., Cebra, E. R., Kramer, D. R., Kroese, F. G., Schrader, C. E. (1995). Cellular and molecular biologic approaches for analyzing the in vivo development and maintenance of gut mucosal IgA responses. *Advances in Experimental Medicine & Biology* 371A: 429-34.
39. Craig, S. W., Cebra, J. J. (1971). Peyer's patches: an enriched source of precursors for IgA-producing immunocytes in the rabbit. *Journal of Experimental Medicine* 134: 188-200.

40. Cebra, J. J., Periwal, S. B., Lee, G., Lee, F., Shroff, K. E. (1998). Development and maintenance of the gut-associated lymphoid tissue (GALT): the roles of enteric bacteria and viruses. *Developmental Immunology* 6: 13-8.
41. Roux, M. E., McWilliams, M., Lamm, M. E., Phillips-Quagliata, J. M. (1979). Site maturation of IgA plasma cell precursors. *Fed. Proc.* 38: 1081.
42. Roux, M. E., McWilliams, M., Phillips-Quagliata, J. M., Lamm, M. E. (1981). Differentiation pathway of Peyer's patch precursors of IgA plasma cells in the secretory immune system. *Cellular Immunology* 61: 141-53.
43. Chen, H. T., Makover, S. D., Kabat, E. A. (1987). Immunochemical studies on monoclonal antibodies to stearyl-isomaltotetraose from C58/J and a C57BL/10 nude mouse. *Mol. Immunol.* 24: 333-338.
44. Wang, D., Stall, A. M., Kabat, E. A. 1995. *B Cell Responses to α(1->6)dextran in T Cell Deficient Mice.* Presented at the The 9th International Congress of Immunology, San Francisco, California, July 23-29, 1995.
45. Wang, D., Seidl, K. J., Liao, J., Gruezo, F., Kabat, E. A., Stall, A. 1995. Paradoxical IgA response to α(1->6)dextran in TCR β/δ KO mice: impaired serum IgA production and enhanced class switching of IgM to IgA in spleen. Presented at the 7th International Workshop of WACIID, Ikoino Mura, Yamagata, Japan., December 16-19 1995.
46. Wang, D., Erlanger, B. F. 1997. An alternative vaccination strategy against HIV: In vivo induction of antigen-specific IgA in the absence of T cell. (selected for a Travel Fund Award). Presented at the Symposium on HIV-1 Infection, Mucosal Immunity and Pathogenesis, The Natcher Conference Center, Bethesda, Maryland 1997.
47. Schuler, W., Schuler, A., Kolsch, E. (1984). Immune response against the T-independent antigen alpha (1--3) dextran. II. Occurrence of B gamma memory cells in the course of immunization with the native polysaccharide is T cell dependent. *European Journal of Immunology* 14: 578-85.
48. Tittle, T. V. (1989). Immunochemical analyses of C57BL/6J monoclonal anti-α(1->3) Dextran antibodies. *Mol. Immunol.* 26: 343-350.
49. Wang, D., Chen, H., Liao, J., Akolkar, P. N., Sikder, S. K., Gruezo, F., Kabat, E. A. (1990). Two families of monoclonal antibodies to α(1->6)dextran, V_H19.1.2 and V_H9.14.7, show distinct patterns of J_K and J_H minigene usage and amino acid substitutions in CDR3. *J. Immunol.* 145: 3002-3010.
50. Waldschmidt, T. J. (1996). B-cell subpopulations and microenvironments. In *B-cell subpopulations and microenvironments*, ed. C. M. Snapper, pp. 73-98. Chichester, New York, Brisbane,Toronto, Singapore: John Wiley & Sons Ltd.
51. Herzenberg, L. A., Kantor, A. B. (1993). B-cell lineages exist in the mouse. [Review]. *Immunology Today* 14: 79-83.
52. Kantor, A. B., Herzenberg, L. A. (1993). Origin of murine B cell lineages. *Ann. Rev. Immunol.* 11: 1-29.
53. Herzenberg, L. A., Stall, A. M., Lalor, P. A., Sidman, C., Moore, W. A., Parks, D. R., Herzenberg, L. A. (1986). The LY-1 B cell lineage. *Immunol. Rev.* 93: 81-102.
54. MacLennan, I. C. (1994). Germinal centers. *Annual Review of Immunology* 12: 117-39.
55. Hardy, R. R., Hayakawa, K., Haaijman, J., Herzenberg, L. A. (1982). B-cell subpopulations identifiable by two-color fluorescence analysis using a dual-laser FACS. *Annals of the New York Academy of Sciences* 399: 112-21.
56. Wells, S. M., Kantor, A. B., Stall, A. M. (1994). CD43 (S7) expression identifies peripheral B cell subsets. *J. Immunol.* 153: 5503-5515.
57. Hardy, R. R., Carmack, C. E., Li, Y. S., Hayakawa, K. (1994). Distinctive developmental origins and specificities of murine CD5+ B cells. *Immunological Reviews* 137: 91-118.
58. Hardy, R. R., Hayakawa, K. (1994). CD5 B cells, a fetal B cell lineage. *Advances in Immunology* 55: 297-339.

59. Herzenberg, L. A., Herzenberg, L. A. (1989). Toward a layered immune system. *Cell* 59: 953-954.
60. Timens, W., Boes, A., Poppema, S. (1989). Human marginal zone B cells are not an activated B cell subset: strong expression of CD21 as a putative mediator for rapid B cell activation. *European Journal of Immunology* 19: 2163-6.
61. Timens, W., Boes, A., Rozeboom-Uiterwijk, T., Poppema, S. (1989). Immaturity of the human splenic marginal zone in infancy. Possible contribution to the deficient infant immune response. *Journal of Immunology* 143: 3200-6.
62. Lane, P. J., Gray, D., Oldfield, S., MacLennan, I. C. (1986). Differences in the recruitment of virgin B cells into antibody responses to thymus-dependent and thymus-independent type-2 antigens. *European Journal of Immunology* 16: 1569-75.
63. Howard, J. (1987). T cell-independent responses to polysaccharides: their nature and delayed ontogeny. In *T cell-independent responses to polysaccharides: their nature and delayed ontogeny*, ed. R. Bell, G. Torrigiani, pp. 221-229. Chichester New York Brisbane Toronto Singapore: John Wiley & Sons.
64. Wells, S. M., Wang, D., Kabat, E. A., Stall, A. M. (2000). Involvement of B-1 and Conventional B Cell Populations in the T-independent Response to α(1->6)dextran. *J. Immunol.* (Submitted).
65. Fernandez, C., Moller, G. (1991). The influence of T cells on the immunoglobulin repertoire and the affinity maturation of the immune response agwinst dextran B512 in C57BL/6 mice. *Scand. J. Immunol.* 33: 307-317.
66. Förster, I., Rajewsky, K. (1987). Expansion and functional activity of Ly-1^+ B cells upon transfer of peritoneal cells into allotype-congenic, newborn mice. *Eur. J. Immunol.* 17: 521-528.
67. Kroese, F. G., Cebra, J. J., van der Cammen, J. F., Kantor, A. B., Bos, N. A. (1995). Contribution of B1 cells to intestinal IgA production in the mouse. *Methods* 8: 37-43.
68. Baker, P. J. (1992). T cell regulation of the antibody response to bacterial polysaccharide antigens: an examination of some general characteristics and their implications. *J. Infect. Dis.* 165(suppl 1): S44-48.
69. Baker, P. J., Stashak, P. W., Amsbaugh, D. F., Prescott, B. (1971). Characterization of the antibody response to type III *pneumococcal* polysaccharide at the cellular level. *Immunology* 20: 469-480.
70. Baker, P. J., Prescott, B. (1979). Regulation of the antibody response to pneumococcal polysaccharides by thymus-derived (T) cells: mode of action of suppressor and amplifier T cells. In *Regulation of the antibody response to pneumococcal polysaccharides by thymus-derived (T) cells: mode of action of suppressor and amplifier T cells*, ed. J. A. Rudbach, P. J. Baker, pp. 67-105. New York: Elsevier/North-Holland.
71. Baker, P. J. (1990). Regulation of magnitude of antibody response to bacterial polysaccharide antigens by thymus-derived lymphocytes. *Infect. Immun.* 58: 3465-3468.
72. Schrater, A. F., Goidl, E. A., Thorbecke, G. J., Siskind, G. W. (1979). Production of auto-anti-idiotypic antibody during the normal immune response to TNP-ficoll. III. Absence in nu/nu mice: evidence for T-cell dependence of the anti-idiotypic-antibody response. *J. Exp. Med.* 150: 808-817.
73. Hiernaux, J. R., Chiang, J., Baker, P. J., Delisi, C., Prescott, B. (1982). Lack of involvement of auto-anti-idiotypic antibody in the regulation of oscillations and tolerance in the antibody response to levan. *Cell. Immunol.* 67: 334-345.
74. Mosier, D. E., Feeney, A. J. (1984). Idiotype regulation in the antibody response to phosphocholine. Antigen selection of B-lymphocyte subsets with differential idiotype expression? In *Idiotype regulation in the antibody response to phosphocholine. Antigen selection of B-lymphocyte subsets with differential idiotype expression?*, ed. M. Green, A. Nisonoff, pp. 403-415. New York: Plenum Press.

75. Mongini, P. K. A., Stein, K. E., Paul, W. E. (1981). T-cell regulation of IgG subclass antibody production in response to T-independent antigens. *J. Exp. Med.* 153: 1-12.
76. Mongini, P. K., Paul, W. E. (1982). T cell regulation of the IgG2a response to TNP-Ficoll: evidence that allotype congenic mice contain both helper cells that preferentially enhance IgG2a synthesis and suppressor cells that specifically suppress IgG2 synthesis. *Journal of Immunology* 128: 2405-10.
77. Mongini, P. K., Paul, W. E., Metcalf, E. S. (1982). T cell regulation of immunoglobulin class expression in the antibody response to trinitrophenyl-ficoll. Evidence for T cell enhancement of the immunoglobulin class switch. *Journal of Experimental Medicine* 155: 884-902.
78. van den Eertwegh, A. J., Van Meurs, M., Foy, T. M., Noelle, R. J., Boersma, W. J., Claassen, E. (1994). In vivo gp39-CD40 interactions occur in the non-follicular compartments of the spleen and are essential for thymus dependent antibody responses and germinal center formation. *Advances in Experimental Medicine & Biology* 355: 75-80.
79. Kiyono, H., McGhee, J. R., Mosteller, L. M., Eldridge, J. H., Koopman, W. J., Kearney, J. F., Michalek, S. M. (1982). Murine Peyer's patch T cell clones. Characterization of antigen-specific helper T cells for immunoglobulin A responses. *Journal of Experimental Medicine* 156: 1115-30.
80. Mombaerts, P., Mizoguchi, E., Ljunggren, H. G., Iacomini, J., Ishikawa, H., Wang, L., Grusby, M. J., Glimcher, L. H., Winn, H. J., Bhan, A. K., Tonegawa, S. (1994). Peripheral lymphoid development and function in T cell receptor mutant mice. *Int. Immunol.* 6: 1061-1070.
81. Franco, M. A., Greenberg, H. B. (1997). Immunity to rotavirus in T cell deficient mice. *Virology* 238: 169-179.
82. Poussier, P., Edouard, P., Lee, C., Binnie, M., Julius, M. (1992). Thymus-independent development and negative selection of T cells expressing T cell receptor α/β in the intestinal epithelium: evidence for distinct circulation patterns of gut- and thymus-derived T lymphocytes. *J. Exp. Med.* 176: 187-199.
83. Saito, H., Kanamori, Y., Takemori, T., Nariuchi, H., Kubota, E., Takahashi-Iwanaga, H., Iwanaga, T., Ishikawa, H. (1998). Generation of intestinal T cells from progenitors residing in gut cryptopatches. *Science* 280: 275-278.
84. Mosmann, T. R., Sad, S. (1996). The expanding universe of T-cell subsets: Th1, Th2 and more [see comments]. *Immunology Today* 17: 138-46.
85. Romagnani, S. (1999). Th1/Th2 cells. *Inflammatory Bowel Diseases* 5: 285-94.
86. McGhee, J. R., Mestecky, J., Elson, C. O., Kiyono, H. (1989). Regulation of IgA synthesis and immune response by T cells and interleukins. [Review]. *Journal of Clinical Immunology* 9: 175-99.
87. Kagnoff, M. F., Murray, P. D. (1987). T dependent induction of an IgA and IgM anti-polysaccharide response. *Advances in Experimental Medicine & Biology* 216A: 155-67.
88. Ivars, F., Nyberg, G., Holmberg, D., Coutinho, A. (1983). Immune response to bacterial dextrans. II. T cell control of antibody isotypes. *Journal of Experimental Medicine* 158: 1498-510.
89. Xu-Amano, J., Aicher, W. K., Taguchi, T., Kiyono, H., McGhee, J. R. (1992). Selective induction of Th2 cells in murine Peyer's patches by oral immunization. *International Immunology* 4: 433-45.
90. Jackson, R. J., Fujihashi, K., Xu-Amano, J., Kiyono, H., Elson, C. O., McGhee, J. R. (1993). Optimizing oral vaccines: induction of systemic and mucosal B-cell and antibody responses to tetanus toxoid by use of cholera toxin as an adjuvant. *Infection & Immunity* 61: 4272-9.
91. Xu-Amano, J., Kiyono, H., Jackson, R. J., Staats, H. F., Fujihashi, K., Burrows, P. D., Elson, C. O., Pillai, S., McGhee, J. R. (1993). Helper T cell subsets for immunoglobulin A responses: oral immunization with tetanus toxoid and cholera toxin as adjuvant selectively induces Th2 cells in mucosa associated tissues. *Journal of Experimental Medicine* 178: 1309-20.

92. Vajdy, M., Kosco-Vilbois, M. H., Kopf, M., Kohler, G., Lycke, N. (1995). Impaired mucosal immune responses in interleukin 4-targeted mice. *Journal of Experimental Medicine* 181: 41-53.
93. Parish, C. R. (1971). Immune response to chemically modified flagellin. I. Induction of antibody tolerance to flagellin by acetoacetylated derivatives of the protein. *Journal of Experimental Medicine* 134: 1-20.
94. Parish, C. R. (1971). Immune response to chemically modified flagellin. II. Evidence for a fundamental relationship between humoral and cell-mediated immunity. *Journal of Experimental Medicine* 134: 21-47.
95. Parish, C. R. (1972). The relationship between humoral and cell-mediated immunity. *Transplantation Reviews* 13: 35-66.
96. Parish, C. R. (1996). Immune deviation: a historical perspective. *Immunology & Cell Biology* 74: 449-56.
97. Romagnani, S., Parronchi, P., D'Elios, M. M., Romagnani, P., Annunziato, F., Piccinni, M. P., Manetti, R., Sampognaro, S., Mavilia, C., De Carli, M., Maggi, E., Del Prete, G. F. (1997). An update on human Th1 and Th2 cells. *International Archives of Allergy & Immunology* 113: 153-6.
98. Mavalia, C., Scaletti, C., Romagnani, P., Carossino, A. M., Pignone, A., Emmi, L., Pupilli, C., Pizzolo, G., Maggi, E., Romagnani, S. (1997). Type 2 helper T-cell predominance and high CD30 expression in systemic sclerosis. *American Journal of Pathology* 151: 1751-8.
99. Parronchi, P., Romagnani, P., Annunziato, F., Sampognaro, S., Becchio, A., Giannarini, L., Maggi, E., Pupilli, C., Tonelli, F., Romagnani, S. (1997). Type 1 T-helper cell predominance and interleukin-12 expression in the gut of patients with Crohn's disease. *American Journal of Pathology* 150: 823-32.
100. Scott, H., Nilsen, E., Sollid, L. M., Lundin, K. E., Rugtveit, J., Molberg, O., Thorsby, E., Brandtzaeg, P. (1997). Immunopathology of gluten-sensitive enteropathy. *Springer Seminars in Immunopathology* 18: 535-53.
101. D'Elios, M. M., Manghetti, M., Almerigogna, F., Amedei, A., Costa, F., Burroni, D., Baldari, C. T., Romagnani, S., Telford, J. L., Del Prete, G. (1997). Different cytokine profile and antigen-specificity repertoire in Helicobacter pylori-specific T cell clones from the antrum of chronic gastritis patients with or without peptic ulcer. *European Journal of Immunology* 27: 1751-5.
102. D'Elios, M. M., Amedei, A., Manghetti, M., Costa, F., Baldari, C. T., Quazi, A. S., Telford, J. L., Romagnani, S., Del Prete, G. (1999). Impaired T-cell regulation of B-cell growth in Helicobacter pylori--related gastric low-grade MALT lymphoma. Gastroenterology 117: 1105-12.
103. Piccinni, M. P., Beloni, L., Livi, C., Maggi, E., Scarselli, G., Romagnani, S. (1998). Defective production of both leukemia inhibitory factor and type 2 T-helper cytokines by decidual T cells in unexplained recurrent abortions. *Nature Medicine* 4: 1020-4.
104. D'Elios, M. M., Romagnani, P., Scaletti, C., Annunziato, F., Manghetti, M., Mavilia, C., Parronchi, P., Pupilli, C., Pizzolo, G., Maggi, E., Del Prete, G. F., Romagnani, S. (1997). In vivo CD30 expression in human diseases with predominant activation of Th2-like T cells. *Journal of Leukocyte Biology* 61: 539-44.
105. Romagnani, S. (1997). The Th1/Th2 paradigm. *Immunology Today* 18: 263-6.
106. Galli, G., Annunziato, F., Mavilia, C., Romagnani, P., Cosmi, L., Manetti, R., Pupilli, C., Maggi, E., Romagnani, S. (1998). Enhanced HIV expression during Th2-oriented responses explained by the opposite regulatory effect of IL-4 and IFN-gamma of fusin/CXCR4. *European Journal of Immunology* 28: 3280-90.
107. Maggi, E., Manetti, R., Annunziato, F., Cosmi, L., Giudizi, M. G., Biagiotti, R., Galli, G., Zuccati, G., Romagnani, S. (1997). Functional characterization and modulation of cytokine production by CD8+ T cells from human immunodeficiency virus-infected individuals. *Blood* 89: 3672-81.

108. Stavnezer, J. (1996). Transforming growth factor beta. In *Transforming growth factor beta*, ed. C. M. Snapper, pp. 291-324. Chichester, New York, Brisbane,Toronto, Singapore: John Wiley & Sons Ltd.
109. Snapper, C. M. (1996). Interferon-Gamma. In *Interferon-Gamma*, ed. C. M. Snapper, pp. 325-346. Chichester, New York, Brisbane,Toronto, Singapore: John Wiley & Sons Ltd.
110. McIntyre, T. M., Kehry, M. R., Snapper, C. M. (1995). Novel in vitro model for high-rate IgA class switching. *Journal of Immunology* 154: 3156-61.
111. Snapper, C. M., Rosas, F., Moorman, M. A., Jin, L., Shanebeck, K., Klinman, D. M., Kehry, M. R., Mond, J. J., Maliszewski, C. R. (1996). IFN-gamma is a potent inducer of Ig secretion by sort-purified murine B cells activated through the mIg, but not the CD40, signaling pathway. *International Immunology* 8: 877-85.
112. Snapper, C. M., Mond, J. J. (1996). A model for induction of T cell-independent humoral immunity in response to polysaccharide antigens. *Journal of Immunology* 157: 2229-33.
113. Gruss, H. J., Brach, M. A., Drexler, H. G., Bross, K. J., Herrmann, F. (1992). Interleukin 9 is expressed by primary and cultured Hodgkin and Reed-Sternberg cells. *Cancer Research* 52: 1026-31.
114. Cebra, J. J., Gearhart, P. J., Halsey, J. F., Hurwitz, J. L., Shahin, R. D. (1980). Role of environmental antigens in the ontogeny of the secretory immune response. *Journal of the Reticuloendothelial Society* 28: 61s-71s.
115. Gett, A. V., Hodgkin, P. D. (1998). Cell division regulates the T cell cytokine repertoire, revealing a mechanism underlying immune class regulation. *Proceedings of the National Academy of Sciences of the United States of America* 95: 9488-93.
116. Hasbold, J., Lyons, A. B., Kehry, M. R., Hodgkin, P. D. (1998). Cell division number regulates IgG1 and IgE switching of B cells following stimulation by CD40 ligand and IL-4. *European Journal of Immunology* 28: 1040-51.
117. Deenick, E. K., Hasbold, J., Hodgkin, P. D. (1999). Switching to IgG3, IgG2b, and IgA is division linked and independent, revealing a stochastic framework for describing differentiation. *Journal of Immunology* 163: 4707-4714.
118. Honjo, T., Kataoka, T. (1978). Organization of immunoglobulin heavy chain genes and allelic deletion model. *Proceedings of the National Academy of Sciences of the United States of America* 75: 2140-4.
119. Humphrey, J. H., Grennan, D. (1981). Different macrophage populations distinguished by means of fluorescent polysaccharides. Recognition and properties of marginal-zone macrophages. *Eur. J. Immunol.* 11: 221-228.
120. Humphrey, J. H. (1981). Tolerogenic or immunogenic activity of hapten-conjugated polysaccharides correlated with cellular localization. *Eur. J. Immunol.* 11: 212-220.
121. Van den Eertwegh, A. J. M., Boersma, W. J. A., Claassen, E. (1992). Immunological functions and *in vivo* cell-cell interactions of T cells in the spleen. *Critic. Rev. Immunol.* 11: 337-380.
122. Jeanes, A. (1986). Immunochemical and related interactions with dextrans reviewed in terms of improved structural information. *Mol. Immunol.* 23: 999-1028.
123. Goldstein, I. J., Whelan, W. J. (1962). Structural studies of dextrans. Part I. A dextran containing α-1,3-glucosidic linkages. *J. Chem. Soc.* January: 170-175.
124. Tew, J. G., Burton, G. F., Kupp, L. I., Szakal, A. (1993). Follicular dendritic cells in germinal center reactions. *Advances in Experimental Medicine & Biology* 329: 461-5.
125. Tew, J. G., Burton, G. F., Helm, S., Wu, J., Qin, D., Hahn, E., Szakal, A. K. (1995). Murine follicular dendritic cells: accessory activities in vitro. *Current Topics in Microbiology & Immunology* 201: 93-104.

PROTECTIVE IMMUNITY AND GENE EXPRESSION RELATED TO PNEUMOCOCCAL GLYCOCONJUGATE

Chi-Jen Lee and Zhong-Ming Li

Center for Biologics Evaluation and Research, FDA, Bethesda, MD U.S.A.

Pneumococcal infection persists as a major cause of pneumonia, meningitis and otitis media. These diseases remain some of the most prevalent infections in the U.S. and other areas of the world, particularly in young children, the elderly, and immunodeficient patients. The World Health Organization estimates that there are one million deaths among children due to pneumococcal meningitis and sepsis each year world-wide, with 98% of these deaths in developing countries. The emergence of pneumococcal strains with antimicrobial resistance further emphasizes the importance of prevention of pneumococcal infection by immunization (1-4,18).

The current 23-valent pneumococcal polysaccharide (PS) vaccine induces a high antibody response in adults, but the immunogenicity of most PSs in the vaccine is low in children less than 2 years of age. One approach for enhancing the antibody response in children is development of PS-protein conjugate vaccines for selected types. Several such pneumococcal vaccines have been prepared and are currently in clinical trials. These conjugate vaccines have used tetanus toxoid or diphtheria toxoid as a carrier protein. Thus, these vaccines confer protection only to infection with the types included in the vaccine.

Several investigators have considered the possibility of using pneumococcal cell-surface proteins, such as pneumolysin, autolysin, pneumococcal surface protein A (PspA), or hemin-binding protein as carriers (19-22). Since these proteins are not used in childhood vaccination programs, the risk of inducing anti-protein antibodies causing suppression of subsequent antibody response to conjugate vaccines is low. Furthermore, the use of a cell-surface protein antigen as a carrier may provide a cross-protective antigen for prevention of pneumococcal infection.

Pneumococcal group 19 contains 4 types; 19F, 19A, 19B and 19C. Among these, types 19F and 19A account for almost all pneumococcal disease isolates from this group. They are more prevalent in young children. The structure of type 19F PS is shown in Figure 1 (5). The phosphodiester linkage is easily hydrolyzed by phosphodiesterase and chemical agents. Thus, it is less stable and less immunogenic than other PSs contained in the PS vaccine. Pneumolysin (*Ply*) is a potent pneumococcal virulence factor. It is a main protein toxin, causing cytotoxicity, hemolytic activity, inflammation and complement activation (6-8). Through its cytotoxicity to respiratory cells, pneumolysin disrupts pulmonary tissue barriers

The Molecular Immunology of Complex Carbohydrates–2.
Edited by Albert M. Wu. Kluwer Academic/Plenum Publishers, 2001.

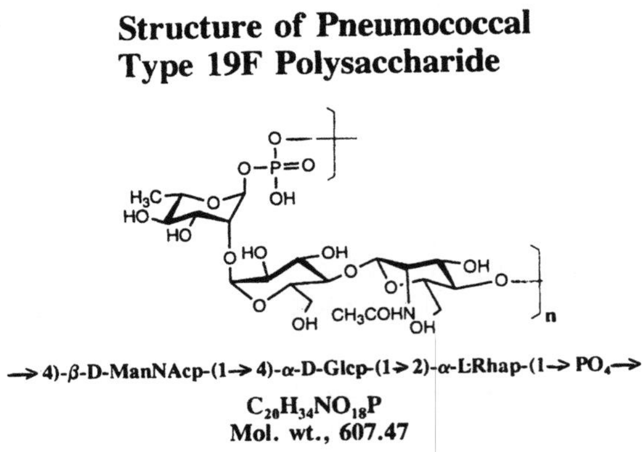

Fig. 1. Structure of pneumococcal type 19F polysaccharide.

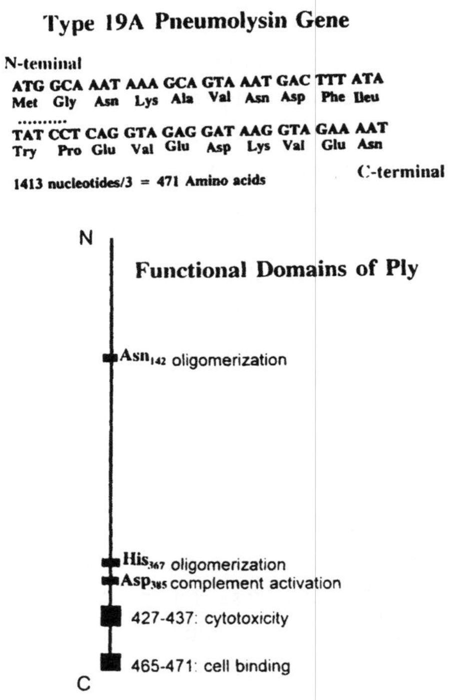

Fig. 2. Nucleotide and amino acid sequences of N-terminal, C-terminal ends of type 19A *ply* and functional domains of pneumolysin.

to facilitate pneumococcal growth and multiplication. Through direct inhibitory effects on immune system and complement response, it inhibits bacterial clearance from the blood and lung interstitial tissues. The critical domains involved in pneumolysin functions have been identified by site-directed mutagenesis (Figure 2). When pneumolysin-immunized mice were challenged with virulent pneumococci, they survived significantly longer than non-immunized control group (9).

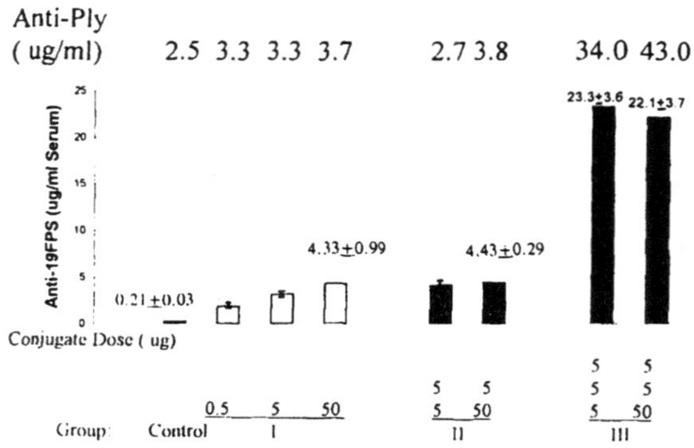

Groups of 6 Balb/c mice, 8 wks of age were injected IP at 10 days interval between injections.

Fig. 3. Antibody response of mice to type 19F PS-PdB conjugate.

I. Protective immunity of type 19F PS-protein conjugate

Studies were conducted to assess the protective capacity of a pneumococcal glycoconjugate. Conjugation of type 19F PS with mutant pneumolysin (pneumolysoid B, PdB, prepared from type 1 mutant *ply* gene) produced higher serum levels of specific antibodies in young mice and led to rapid bacterial clearance from the blood. Type 19F PS-PdB conjugate, 0.5 to 50 ug/dose mixed with aluminum hydroxide adjuvant (0.1 mg/dose), was injected intraperitoneally to first group of six Balb/c mice. The second group (5 and 50 ug/dose) received a booster injection of 5 ug conjugate, while the third group (5 and 50 ug/dose) received 2 booster injections at 10 day intervals between the injections. At 7 days after the final injection, the serum levels of anti-PS antibodies and anti-*Ply* titers were measured. For all groups receiving the conjugate, the anti-PS levels were significantly greater than the control ($P<0.001$, t-test). The highest anti-PS levels, 22-23 ug Ab/ml were observed in the group that received 3 doses. These anti-19F PS levels were 100-fold higher than those of the control group. The maximum anti-*Ply* response was also found in the mice that received 3 doses of the conjugate. These anti-*Ply* titers were 13.6-17.2 fold higher than the control group (Figure 3)(10). These results indicate that the effectiveness of pneumococcal vaccine can be improved by the conjugation of PS to a common cell-surface protein.

Furthermore, the antibody response, protective immunity, and bacterial clearance were examined in mice exposed to 19F PS-PdB conjugate during gestation and lactation. The experimental design for these studies is shown in Table 1. For group I (control group), female mice at 14 days of gestation were injected with saline containing adjuvant. For group II, pregnant mice at 14 days of gestation received 5 ug of conjugate. For group III, mothers received 5 ug of conjugate at one week.

Table I

Experimental design for challenge of young mice from immunized mothers with the glycoconjugate during gestation and lactation.

(Wks)	-2	-1	Birth	+1	+2	+3	+4
	(Gestation, G)			(Lactation, L)			
Treatment	**Mother**				**Young**		
							Challenge
19FPS-PdB	**5 ug**				**1 ug**		**T-19F 1.3 x 10^6 (cfu/0.1ml)**
Group:							
I, G-L-	-				-		+
II, G+L-	+				-		+
III, G-L+	-			+	+		+
IV, G+L+	+			+	+		+

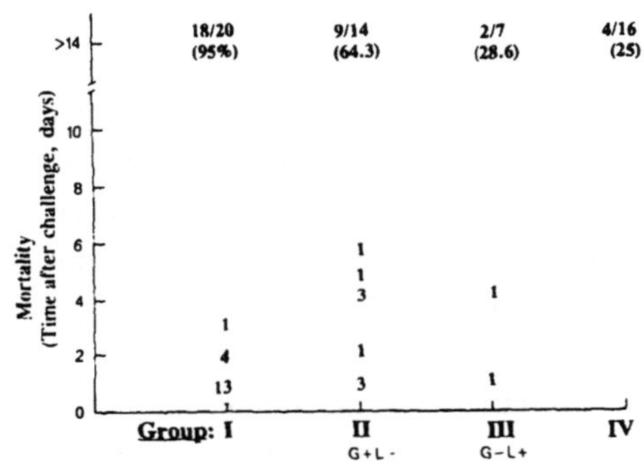

	I	II	III	IV
(19F Ab, µg/ml serum)	# 1.07 ± 0.07 (9)	**2.16 ± 0.22 (10)	*1.66 ± 0.14 (7)	**4.02 ± 0.62 (9)
Ply Ab (ug/ml)	**17.6±6**	****67±14**	****778±368**	****213±60**

 # : Mean ± Standard error.
*** : $P < 0.05$, ** : $P < 0.01$.**

Fig. 4. Protective immunity of young mice from immunized mothers against challenge with type 19F pneumococci. Mice in group I, II, III, and IV were immunized as described in the text and injected i.p. with 1.3 x 10 c.f.u. of type 19F pneumococci at 4 weeks of age. Mortality was observed for 14 days.

Pneumolysoid B (PdB) was prepared from expression of mutant type 1 pneumolysin gene in *E. coli*. Type 19F was conjugated with PdB by the activation of PS by cyanogen bromide and coupling reaction with the carrier protein by carbodiimide reagent. The ratio of PS to protein was 0.3 to 0.4. The serum levels of 19F PS antibody were measured by radioimmunoassay, whereas anti-Ply titers were measured by ELISA after delivery, and offspring received an additional dose of 1 ug conjugate at 2 weeks of age. For group IV, mothers received 5 ug of conjugate during gestation and lactation, and the offspring received an additional dose of conjugate at 2 weeks of age. At 4 weeks of age, young mice from each group were challenged intraperitoneally with 1.3 x 10^6 cfu of type 19F pneumococci. The mortality rate was observed for 14 days (Figure 4). Most young mice in the control group(95%)

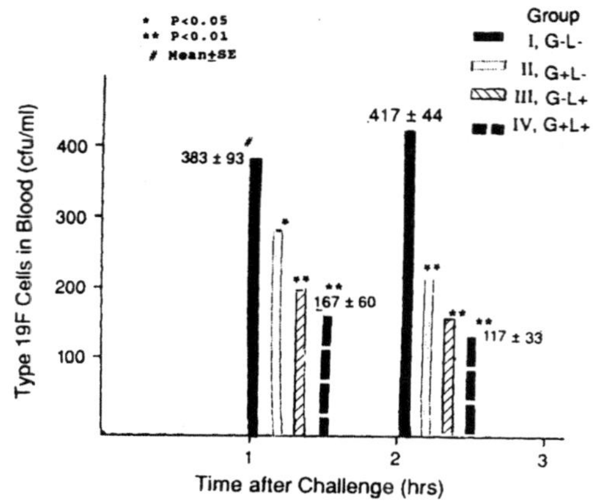

Fig. 5. Bacterial clearance in mice from immunized mothers during gestation and lactation. Mice at 8 weeks of age were injected i.p. with 5 x 10 c.f.u. of type 19F pneumococci. Blood samples were collected at various time after challenge, diluted in the culture medium and inoculated on blood agar plate for determination of bacterial colonies. Data shown are the average of triple samples $\pm$ standard error.

died within 3 days after challenge. The mortality rate was significantly lower in young mice from all immunized mothers (Figure 4). The mortality of group III which contained young mice from mothers immunized during gestation that had received an additional dose of conjugate after birth showed lower mortality than group II which included young mice from mothers immunized during gestation ($p<0.05$). Although the 19F PS Ab titer of group III was lower than that of group II, the anti-*Ply* titer of group III was more than 10 fold higher than that of group II. This result suggests that the antibodies induced by the pneumolysin component of the conjugate make an important contribution to the overall protective immunity.

In addition, the protective effect of immunization with conjugate was examined by testing the capability of mice to clear pneumococci from the blood (Figure 5). The mice which had not previously been challenged were injected IP at 8 weeks of age with 5 x 10^4 cfu/0.1 ml of type 19F pneumococci, and bacterial clearance from the blood at various times after challenge was determined. The clearance was greater at 1 and 2 hrs after the bacterial challenge in young mice from all immunized mothers than from the non-immunized controls (11). These results indicate that effective protective immunity against pneumococcal infection can be induced in young mice by maternal immunization with the PS-protein conjugate during gestation and at early infancy.

II. Expression of type 19A pneumolysin gene in mammalian cells

DNA vaccinations have been reported to induce specific immune response and protection in animal disease models (12-14). Furthermore, animal studies suggest that bacterial DNA can play a nonspecific role in host defense, both by stimulating B cells for antibody production as well as inducing cytokines (15, 16).

Pneumolysin, a protein toxin, is involved in the pathogenesis of pneumococcal infection and contributes to evasion of the host defense mechanisms. The type 19A pneumolysin gene (*ply*) consists of 1,413 nucleotides that encode 471 amino acids (17).

Type 19A *ply* gene was amplified by PCR using Vent Taq polymerase (New England BioLab). The PCR product was subcloned and transformed into competent *E. coli* cells using ZeroBlunt vector (Invitrogen Co.,San Diego, CA) (Figure 6).

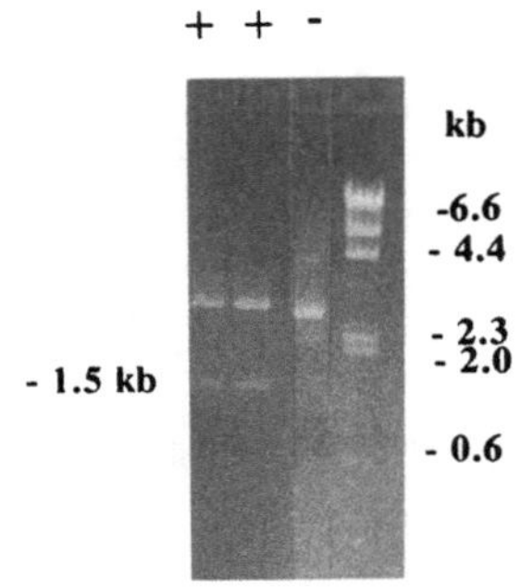

Fig. 6. Detection of inserted type 19A *ply* DNA after treatment with restriction enzymes. Type 19A *ply* DNA was amplified by PCR, inserted into competent *E. coli* cells. The inserted *ply* was checked by treatment with enzymes, NheI and XhoI, and analyzed by gel electrophoresis. The *ply* DNA was detected at 1.5 kb band.

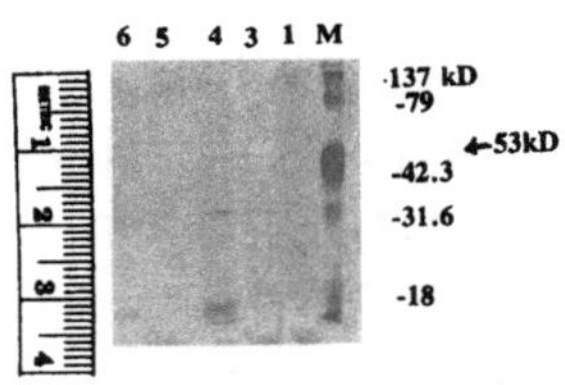

Sample	Transfection: Liposome	Ca++	Break cells: SDS	Detergent
1(Cont)	+		+	
rDNA				
3	+		+	
4	+			+
5		+	+	
6		+		+

Fig. 7. Transfection and expression of type 19A *ply* plasmid DNA in human rhabdomyosarcoma cells by immunoblot analysis. Plasmid *ply* DNA inserted into JW4303 vector was treated with cationic lipid (samples 3 and 4) or calcium phosphate (samples 5 and 6) reagent to form a DNA-reagent complex. The DNA complex was reacted with a layer of human rhabdomyosarcoma cells and incubated at 37° C for 72 hrs. The transfected cells were lysed by SDS or detergent and analyzed by immunoblot using rabbit pneumolysin antibody.

The 19A *ply* DNA fragment was cut with NheI and BamHI and inserted into the JW4303 vector containing a cytomegalovirus (CMV) promoter. The plasmid *ply* DNA was transformed into human rhabdomyosarcoma cells for gene expression in vitro. The *ply* plasmid DNA was first treated with cationic lipid or calcium phosphate reagent to form a DNA-reagent complex, and then incubated with a layer of human rhabdomyosarcoma cells at 37°C for 72 hrs. The transfected cells were extracted in 0.2M n-octylglucoside containing phenylmethyl sulfonylfluoride and 1 mM N-ethylmaleimide for 20 min at 25°C. Extracts were centrifuged at 12,000 x g for 10 min and the supernatants were examined for expression of the *ply* gene by immunoblot analysis. All samples of transfected cells demonstrated expression of pneumolysin. The control cells that received vector only did not show any pneumolysin expression (Figure 7).

III. Antibody response and bacterial clearance in mice immunized with *ply* plasmid DNA

Type 19A *ply* plasmid DNA was injected intramuscularly into mice at a dose of 50-100 ug DNA at a concentration of 1 ug/ul. A second group also received a booster injection of 50 ug DNA, while a third group received two booster injections at 6 week-intervals. At 7 days after the final injection, the serum levels of pneumolysin IgG and IgM antibodies were measured by ELISA method. The mice that received 2 and 3 injections produced 9.9 to 17.8 fold more anti-*Ply* IgG Ab than did the control group. The mice that received 3 doses of the 19A ply plasmid DNA produced 8.1 fold more Ply IgM Ab than the control group. These results indicate that the ply plasmid DNA can express pneumolysin antigen in vivo and stimulate the immune system to produce high levels of specific IgG and IgM antibodies in mice (Figure 8).

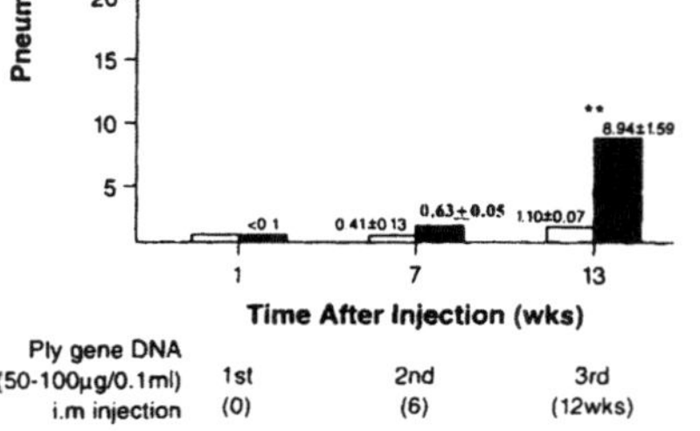

Fig. 8. Antibody response in mice injected with type 19A *ply* plasmid DNA. Experimental conditions are described in the text. Daa of antibody values shown are the average of 6 samples ± standard error.

In addition, bacterial clearance from the blood was examined in mice injected with *ply* plasmid DNA. Groups of 8 weeks old mice were injected with 3 doses of 50-100 ug *ply* plasmid DNA at 6 week intervals. At 7 days after the final injection, mice were challenged with 6.4 x 10^5 cfu of type 19F pneumococci. At various times after the challenge, 20 ul blood samples were collected, inoculated on blood agar plates and incubated at 37°C for 17 hrs. The mice immunized with *ply* plasmid DNA showed rapid bacterial clearance from the blood at 1, 2, 4 and 24 hrs after challenge with virulent type 19F pneumococci as compared to the non-immunized control group (Figure 9). These results indicate that the injection of mice with *ply* plasmid DNA conferred protective immunity against pneumococcal infection.

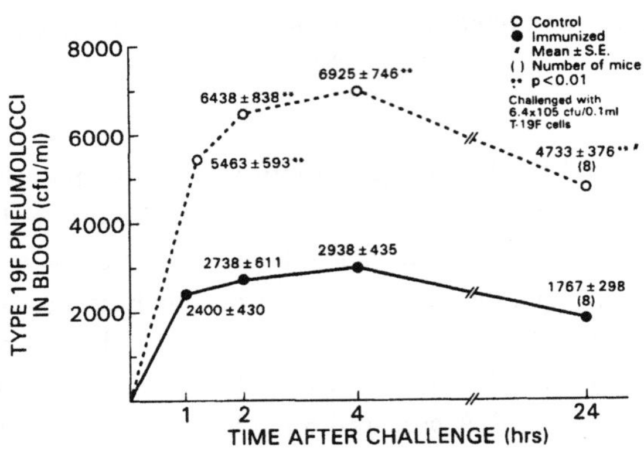

Fig. 9. Bacterial clearance of mice injected with type 19A *ply* plasmid DNA. Experimental conditions are described in the text. Data of c.f.u. values are the average of 8 samples ± standard error.

IV. Conclusions

The following conclusions can be reached from these studies:

(1). The development of new and broadly effective pneumococcal conjugate vaccines are an important approach for prevention of pneumococcal infection.

(2). The effectiveness of pneumococcal polysaccharide vaccines can be improved by supplementation with pneumococcal cell-surface proteins, such as pneumolysin and hemin binding protein used for the glycoconjugates.

(3). The mice that were immunized with *ply* plasmid DNA produced high serum levels of specific IgG and IgM Abs and showed rapid bacterial clearance from the blood. These results suggest that the efficacy of pneumococcal vaccine will also be improved by the immunization with *ply* plasmid DNA.

Several issues concerning the improvement of the effectiveness of pneumococcal vaccine and its immunization strategy need to be addressed: (a) New conjugate vaccines using cell-surface protein carriers such as inactivated pneumolysin and hemin binding protein may provide cross-reactive immunity for protection against pneumococcal infection; (b) the nucleotides that encode 5-10 amino acids from the C-terminal of *ply* gene can be deleted or cysteine 428 changed to other amino acids, e.g. glycine, serine or phenylalanine for the production of inactivated pneumolysin. The mutated pneumolysin can be used as a protein carrier for pneumococcal glycoconjugate; (c) maternal immunization during a critical period of perinatal development and vaccine delivery system through mucosal immunity will provide a more effective immunization approach.

SUMMARY

The immunogenicity of most polysaccharides (PSs) contained in the pneumococcal vaccine is low in children less than 2 years of age. Enhancement of immune response in early life can be induced by immunization of pneumococcal glycoconjugate as well as plasma DNA coding for cell-surface protein antigen.

Pneumococcal type 19F PS conjugated with inactivated pneumolysin (Ply) induced in mice remarkable antibody responses to both type 19F PS and the protein carrier. In addition, the conjugate was administered to pregnant mice during gestation and/or lactation, and to their offspring during early infancy. When the young mice were challenged with type 19F pneumococci, the bacteria were cleared more rapidly from the blood of immunized mice than from that of the control group. The mortality rate of young mice from immunized mothers was also significantly lower than the control group. These results indicate that the effective protective immunity against pneumococcal infection can be induced in young mice by the maternal immunization with the glycoconjugate during gestation and at early infancy.

Studies have been conducted to express type 19A pneumolysin gene (*ply*) in mammalian cells. *Ply* DNA was inserted into the cloning site of a vector containing CMV promoter. The recombinant plasmid DNA containing *ply* was transfected in human rhabdomyosarcoma cells and the gene expression was confirmed by immunoblot. Injection of mice three times 50-100 ug per dose *ply* DNA produced high serum levels of *Ply* IgG and IgM antibodies and showed rapid bacterial clearance from the blood.

REFERENCES

1. Lee, C.J., S.D. Bank and J.P. Li. Virulence, immunity and vaccine related to *Streptococcus pneumoniae*. Crit. Rev. Microbiol., 18(2):89-114, 1991.
2. Lee, C.J., M. Koizumi, K.T. Lin and K. Koizumi. Protective immunity to cell-surface antigens derived from *Streptococcus pneumoniae* and new conjugate vaccine development. Clin. Microbiol., 23(2):227-233, 1996.
3. Lee, C.J. and T.R. Wang. Pneumococcal infection and immunization in children. Crit. Rev. Microbiol., 20:1-12, 1994.
4. Butler, J.C., J. Hofmann, M.S. Cetron et al. The continued emergence of drug-resistant *Streptococcus pneumoniae* in the United States: an update from the Center for disease Control and Prevention's Pneumococcal sentinel Surveillance system. J. Infect. Dis., 174:986-993, 1996.
5. Lee, C.J. and B.A. Fraser. Structure of the cross-reactive type 19 (19F) and 57 (10A) pneumococcal capsular polysaccharides. J. Biol. Chem., 255:6847-6853, 1980.
6. Rubins, J.B. and E. N. Janoff. Pneumolysin: A multifunctional pneumococcal virulence factor. J. Lab. Clin. Med.,131:21-27, 1998.
7. Paton, J.C. The contribution of pneumolysin to the pathogenicity of *Streptococcus pneumoniae*. Trends Microbiol., 28:103-106, 1996.

8. Paton, J.C., P. W. Andrew, G.J. Boulnois and T.J. Mitchell. Molecular analysis of the pathogenicity of *Streptococcus pneumoniae*: the role of pneumococcal proteins. Ann. Rev. Microbiol., 47: 89-115, 1993.
9. Paton, J.C., R.A. Rock and D. Hansman. Effect of immunization with pneumolysin on survival time of mice challenged with *Streptococcus pneumoniae*. Infect. Immun., 40:548-552, 1983.
10. Paton, J.C., R.A. Lock, C.J. Lee, et. al. Purification and immunogenicity of genetically obtained pneumolysin toxoids and their conjugation to *Streptococcus pneumoniae* type 19F polysaccharide. Infect. Immun., 59:2297-2304, 1991.
11. Lee, C.J., R.A. Lock, P.W. Andrew, et al. Protection of infant mice from challenge with *Strep-tococcus pneumoniae* type 19F by immunization with a type 19F polysaccharide-pneumolysoid conjugate. Vaccine 12:875-878, 1994.
12. Liu, M. Overview of DNA vaccines, in Liu, M.A., Hilleman, M.R., Kurth, R. (eds): DNA Vaccines -- a new era in vaccinology. N.Y. Acad.Sci., New York City, pp 15-20, 1995.
13. Hassett, D.E. and J.L. Whitton. DNA immunization. Trends Microbiol., 4:307-312, 1966.
14. Fynan, E.F., R.G. Webster, D.H. Fuller et al. DNA vaccines: a novel approach to immunization. Int. J. Immunopharmacol., 17:79-83, 1995.
15. Pisetsky, D.S., C. Reich, S.D. Ctowley, and M.D. Halpern. Immunological properties of bacterial DNA, in Liu, M.A., Hilleman, M.R. Kurth, R. (eds): DNA vaccines - a new era in vaccinology. N.Y. Acad. Sci., New York City, pp 152-163, 1995.
16. Whitton, J.L., R. Fernando, J. Zhang et al. DNA immunization:mechanistic studies. Vaccine 17:1612-1619, 1999.
17. Lee, C.J., T.R. Wang and S.S. Tai. Immunologic epitope, gene, and immunity involved in pneumococcal glycoconjugate. Crit. Rev. Microbiol., 23(2):121-142, 1997.
18. Immunization Practice Advisory Committee: Prevention of pneumococcal disease: recommen-dation of the advisory committee on immunization practices (ACIP). Mort. Morb. Wk. Rep., 46 (RR-8):1-24, 1997.
19. Ogunniiyi, A.D., D.E. Briles and J.C. Paton. Development of a pneumococcal common protein vaccine based a combination of pneumolysin, PspA and PssA. Pneumococcal Vaccines for the World 1998 Conference, p. 29, 1998, Washington DC.
20. Lock, R.A., J.C. Paton and D. Hansman. Comparative efficacy of pneumococcal neuraminidase and pneumolysin as immunogens protective against *Streptococcus pneumoniae*. Microb. Pathog., 5:461-567, 1998.
21. Tart, R.C., L.S. McDaniel, B.A. Ralph et al. Truncated *Streptococcus pneumoniae* PspA molecules elicit cross-protective immunity against pneumococcal challenge in mice. J. Infect. Dis., 173:380-386, 1996.
22. Tai, S.S., T.R. Wang and C.J. Lee. Characterization of hemin binding activity of *Streptococcus pneumoniae*. Infect. Immun., 65:1083-1087, 1997.

STRUCTURE AND FUNCTIONAL GENOMICS OF LIPOPOLYSACCHARIDE EXPRESSION IN *HAEMOPHILUS INFLUENZAE*[1]

James C. Richards[a], Andrew D. Cox[a], Elke K.H. Schweda[b], Adele Martin[a], Derek W. Hood[c], and E. Richard Moxon[c]

[a]Institute for Biological Sciences, National Research Council, Ottawa, Ontario,da, K1A 0R6; [b]University College of South Stockholm, Clinical Research Centre, NOVUM, S-141 86, Huddinge, Sweden; and [c]Molecular Infectious Diseases Group, University of Oxford, Department of Paediatrics, John Radcliffe Hospital, Headington, Oxford, 0X3 3DU, United Kingdom

Diseases caused by *Haemophilus influenzae* remain a significant problem world-wide. Seven capsular serotypes and an indeterminant number of acapsular or non-typeable strains are known. Type b capsular strains are associated with invasive diseases, including meningitis and pneumonia, while non-typeable strains cause otitis media and respiratory tract infections. Lipopolysaccharide (LPS) is an essential and characteristic surface component of these pathogens and is implicated as a major virulence factor (1). *H. influenzae* express heterogeneous populations of short-chain LPS which exhibit extensive antigenic diversity among multiple oligosaccharide epitopes. This short-chain LPS is composed of a membrane-bound lipid A to which the oligosaccharide portion is attached by a single 3-deoxy-D-*manno*-octulosonic acid (KDO) residue. Molecular structural studies of *H. influenzae* type d (strain Rd), type b, and non-typeable LPS have resulted in a structural model consisting of a conserved tri-heptosyl inner-core moiety in which each of the heptose residues can provide a point for attachment by hexose containing oligosaccharide chains or non-carbohydrate substituents (2-13). Phosphate-containing substituents such as phosphoethanolamine (*P*Etn), pyrophosphoethanolamine (*PP*Etn) and phosphocholine (*P*Cho) have been observed (Table 1). Considerable intra-strain and strain-to-strain variation in LPS structure has been observed (6,8). Detailed structural studies of LPS from defined mutants of type b strain RM7004 (3,4,11), transformed variants of the type d

---

[1] Corresponding author: James C. Richards. Fax: (613) 941-1327. E-mail: jim.richards@nrc.ca

derived strain, RM118 (7,12,14) and transposon mutants of type b strain A2 (5) have provided further evidence for a common heptose containing trisaccharide inner-core moiety in *H. influenzae* LPS. Structural studies of LPS from an acapsular strain of *H. influenzae* Rd (RM118), derived from the strain on which the complete genome has been sequenced, has revealed three major populations of LPS glycoforms, all containing a *P*Cho→ 6)-β-D-Glc*p* residue substituted at *O*-4 of HepI, but differing in the length of the oligosaccharide chains attached to HepIII (8).

Table 1. Structural model of *H. influenzae* LPS showing major structural features of variable LPS oligosaccharides.

R_1→4)-β-D-Glc*p*-(1→4)-L-α-D-Hep*p*-(1→5)-α-KDO*p*-(2→
3 ↑ 1 (on Hep*p*); 4 ↑ P...*P*Etn (on KDO*p*)

R_2→4)-L-α-D-Hep*p*6←*P*etn
2 ↑ 1

R_3→2)-L-α-D-Hep*p*4←Y

Strain	R1	R2	R3	Y	Reference
Rd					
RM118[a]	*P*Cho[b]	H	β-D-Gal-(1→4)-β-D-Glc	H	8
	*P*Cho[b]	H	α-D-Gal-(1→4)-β-D-Gal-(1→4)-β-D-Glc	H	
	*P*Cho[b]	H	β-D-GalNAc-(1→3)-α-D-Gal-(1→4)-β-D- Gal-(1→4)-β-D-Glc		
Type b					
A2[c]	β-D-Gal	β-D-Gal-(1→4)-α-D-Glc	H	H	5
Eagan[d]	H	β-D-Glc-(1→4)-α-D-Glc	β-D-Gal	H	6
	H	β-D-Glc-(1→4)-α-D-Glc	β-D-Gal	H_2PO_3	
	H	β-D-Gal-(1→4)-β-D-Glc-(1→4)-α-D-Glc	β-D- Gal		
	H	α-D-Gal-(1→4)-β-D-Gal-(1→4)-β-D-Glc-(1→4)-α-D-Glc	β-D-Gal		
NTHi					
2019	β-D-Gal	H	H	PEtn[e]	2
375	H	H	β-D-Gal-(1→4)-β-D-Glc	H[f]	13
	H	H	α-D-Neu5Ac-(2→3)-β-D-Gal-(1→4)-β-D-Glc	H[f]	

[a] Capsule deficient *H. influenzae* type d strain derived from the strain used for genome sequencing (17)
[b] *P*Cho is substituted at *O*-6 position.
[c] This is the major LPS glycoform present in the parent strain. The structural details were determined on the Hex4 LPS glycoform in the mutant 281.25. The location of the *P*Etn was not determined, but is assumed to be located at *O*-6 of HepII.
[d] Glycoforms which contain an *O*-acetyl group on HepIII and *P*Cho on the β-D-Gal attached to Hep III have been identified.
[e] The linkage position of the *P*Etn substituent was not determined.
[f] A population of glycoforms contain *P*Etn at O-4 of Hep III.

In *H. influenzae* strain Rd, the inner-core HepIII residue was found to carry lactose (β-D-Gal*p*-(1→4)-β-D-Glc*p*), globoside (α-D-Gal*p*-(1→4)-β-D-Gal*p*-(1→4)-β-D-Glc*p*) or globotetraose (β-D-Gal*p*NAc-(1→3)-α-D-Gal*p*-(1→4)-β-D-Gal*p*-(1→4)-β-D-Glc*p*) oligosaccharide chains, mimics of the mammalian globo series of glycolipids (Table 1). Recently, we have shown that a lactose moiety attached to HepIII in a non-typeable strain NTHi 375 can carry a sialic acid residue (Neu5Ac) (13). Sialylation of *H. influenzae* LPS has been reported to be dependent upon an environmental source of sialic acid (15). We have found that strain RM118, when grown under appropriate conditions can express sialylated oligosaccharide epitopes (16).

Using high-field NMR techniques and sensitive MS-based methods we have determined that the globoside trisaccharide (also known as the pK blood group epitope) can be present on oligosaccharide chains that are attached to different heptose residues in different *H. influenzae* strains (6,8). Thus, in strain Eagan this LPS epitope is part of an oligosaccharide chain attached to HepII, while in strain Rd it is in a different molecular environment where it is attached to HepIII (Figure 1).

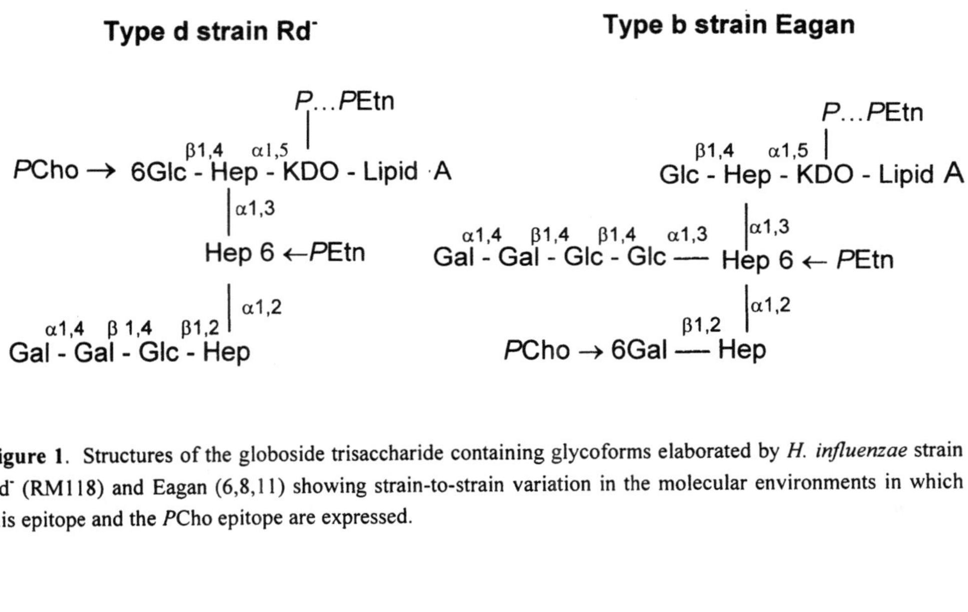

Figure 1. Structures of the globoside trisaccharide containing glycoforms elaborated by *H. influenzae* strain Rd⁻ (RM118) and Eagan (6,8,11) showing strain-to-strain variation in the molecular environments in which this epitope and the *P*Cho epitope are expressed.

Oligosaccharide expression of *H. influenzae* LPS is known to be subject to high-frequency phase variation (18). Phase variation among certain surface exposed LPS oligosaccharide epitopes is thought to provide an adaptive mechanism which is advantageous for the survival of this bacterium when it is confronted by the differing microenvironments and immune responses presented by the host during the pathogenesis of disease. The structural diversity of *H. influenzae* LPS arising from phase variation has complicated the study of the molecular features of these molecules and the understanding of the role they play in commensal and pathogenic interactions with the host. With the availability of the complete genome sequence of *H. influenzae* strain Rd (17), significant progress has been made in identifying the genes that are responsible for intra-strain and strain-to-strain variation of LPS oligosaccharide epitopes (19). A mechanism contributing to LPS phase variation has been established for a number of chromosomal loci, including *lic1*, *lic2* , *lic3* and *lgtC* (20). Each of these genetic loci contains multiple tandem tetranucleotide repeats in the coding region of the first open reading frame (orf). Loss or gain of one or more of these tetranucleotides through slipped-strand misspairing shifts potential initiation codons into or out of frame with the remainder of the orf, creating a translational switch. A schematic representation of how this occurs is illustrated in Figure 2 for expression of the phosphocholine epitope which is subject to phase variation mediated by genes in the *lic1* locus (21).

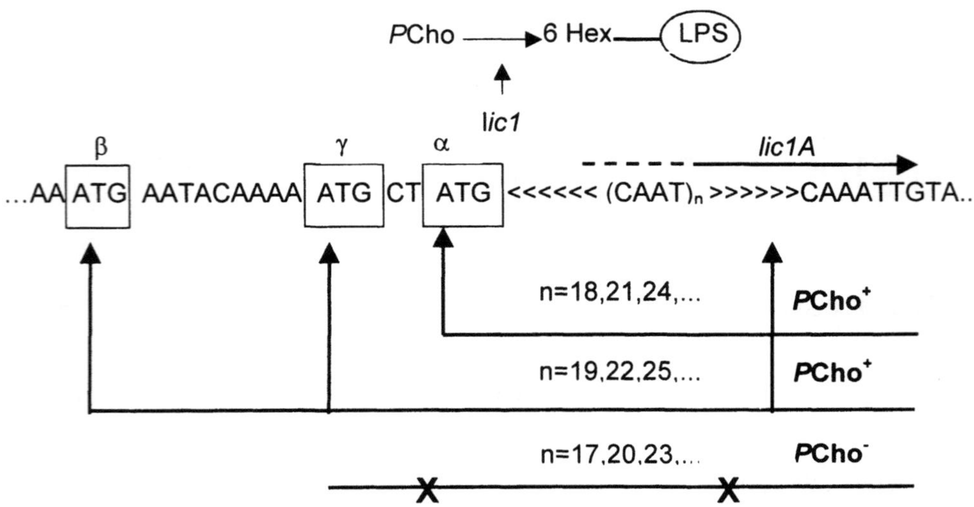

Figure 2. Correlation between expression of *P*Cho and the genotype of the *lic1* locus containing four genes, *lic1A-D*, which are required for the addition of *P*Cho to *H. influenzae* LPS (21). The molecular mechanism controlling phase variation is indicated in which, depending on the number of tandem repeats of 5'-(CAAT)-3', one of three initiation codons (α, β, or γ) is either positioned in or out of phase with the remainder of the *lic1A* open reading frame. In strains containing for example 18, 19 or 21 repeats an initiation codon is in phase with the remainder of the orf and *P*Cho is expressed, while in strains containing 17, 20 or 23 repeats no initiation codon is in phase with the remainder of the orf and *P*Cho is not expressed.

It is interesting to note that certain type b strains (Eagan and RM7004) generally exhibit low expression levels of the *P*Cho epitope (11). We have determined expression levels to be of the order 1-2% by ^{1}H NMR spectroscopy under a variety of growth conditions. By contrast, significant *P*Cho expression levels are observed in strain Rd and many of its derived mutants (ca 95%) (8). Genomic analysis shows that the *lic1* gene in these type b strains typically contains multiples of (CAAT) repeats which are incompatible with *P*Cho expression (Figure 2).

From our detailed structural analysis of LPS from strain Rd (8), oligosaccharide chain extension was found only to occur from HepI and HepIII. In type b strains RM7004 and Eagan, genes in the *lic2* locus are involved in chain extension from HepII. Thus, inactivation of the complete *lic2* locus by deletion-insertion mutation of these type b strains (11) leads to expression of LPS completely devoid of oligosaccharide chain extension from HepII. Analysis of LPS obtained from insertion mutants generated by inactivation of each of the four open reading frames of the *lic2* locus of strain Eagan demonstrated that the third (*orf3*) and perhaps the fourth (*lic2B*) gene in the locus play a role in determining chain extension from HepII, while the first gene (*lic2A*) is involved in the expression of the digalactoside epitope. Complete genome analysis indicated that the genes *lic2orf3* and *lic2B* are absent from the *lic2* locus in strain Rd which is consistent with a lack of chain extension from HepII in *H. influenzae* strain Rd LPS. The gene product, Lic2orf3 shows a high degree of homology to a glucosyltransferase in *Neisseria gonorrhoeae*. The latter gene encoded by *lgtG* is responsible for the tranfer of D-Glc*p* in an α1,3 linkage to HepII of the gonnococcal LPS (22). The structural implications of mutations in the *lic* loci of strain Eagan are summarized in Figure 3.

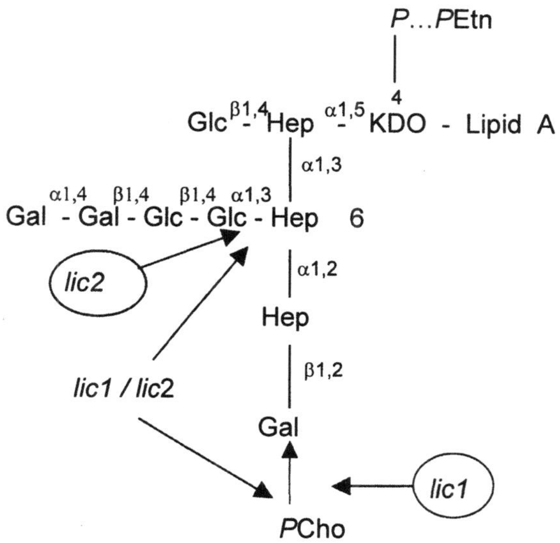

Figure 3. Structural model of the fully extended *P*Cho-containing LPS of *H. influenzae* strain Eagan (6,11). The sites at which structural alterations occur due to oligosaccharide truncations in type b mutant strains *lic1*, *lic2* and *lic1/lic2* are indicated.

In strain Rd, a terminal β-D-Glc*p* residue on HepI is substituted by a *P*Cho substituent at the *O*-6 position. It is noteworthy that *P*Cho occupies a different molecular environment in the LPS elaborated by the type b strains RM7004 and Eagan where it is attached at the *O*-6 position of a terminal β-D-Gal*p* residue on HepIII (Figure 1). In an isogenic mutant of strain Rd (RM118-28) which was transformed with a 0.45 kb restriction fragment from within the fourth open reading frame of *lic1* (*lic1D*) of strain RM7004, the *P*Cho substituent occupies the same molecular environment as the RM7004 LPS, except that it is attached to the *O*-6 position of a teminal β-D-Glc*p* residue on HepIII (7). In a complementary experiment, we have recently shown that LPS from strain Eagan transformed with *lic1D* from strain Rd expresses *P*Cho in the same molecular environment as strain Rd (14). *Lic1D* is thought to encode the *P*Cho transferase. A comparison of the gene products of *lic1D* of the type b strains, RM7004 and Eagan, and strain Rd shows differences in 29 of the 263 amino acid residues (14) which would account for the difference in their acceptor specificities (i.e., Hex-HepIII vs. Hex-HepI).

The sequence of a wide range of LPS biosynthetic genes from Gram-negative bacteria were used as probes to screen the *H. influenzae* strain Rd genome database for homologous genes by DNA or amino acid similarity. Mutations were constructed in putative *H. influenzae* LPS-associated genes which were identified from the strain Rd genome sequence and these were characterized for alterations in cell wall LPS (19,23). The proposed sites of action in LPS biosynthesis of genes involved in chain elongation from the tri-heptosyl inner-core element are shown in Figure 4.

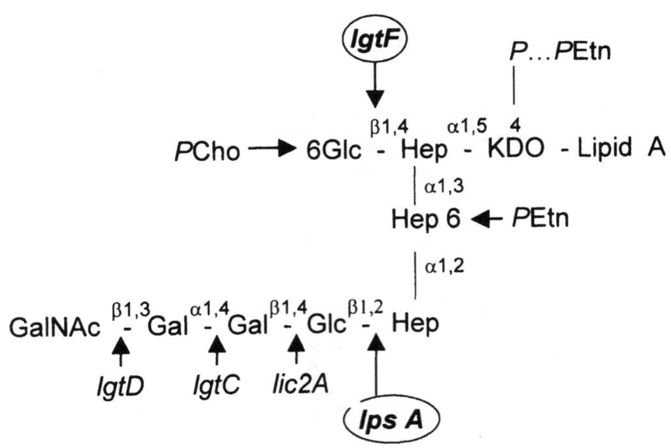

Figure 4. Structural model of the fully extended LPS of *H. influenzae* strain Rd showing the site of action of genes involved in controlling oligosaccharide extension from HepI and HepIII of the tri-heptosyl inner-core. Downstream genes in the *lic2* locus have been shown to initiate chain extension from HepII in type b strains (See Figure 3). These genes are absent from the strain Rd genome. The first orf in *lic2* (*lic2A*), *lgtC* and *lgtD* are involved in assembly of the globotetraose side chain (23).

A structural fingerprinting strategy involving electrospray mass spectrometry (ESI-MS), methylation analysis and 1D ^{1}H NMR analysis of *O*-deacylated LPS was employed to identify the effect of mutations generated in each of the genes suspected to be involved in LPS biosynthesis (23). In this way, the *lpsA* gene product was shown to be involved in controlling oligosaccharide chain extension from HepIII. A mutation in the *lpsA* gene afforded a truncated LPS in which HepIII is devoid of oligosaccharide chain extension. Thus, ESI-MS of the *H. influenzae* Rd *lpsA O*-deacylated LPS indicated a *P*Cho-containing Hex1 glycoform as the major LPS species (Figure 5). The Rd *lic2A, lgtC* and *lgtD* mutants which contain a functional *lpsA* gene are capable of adding a β-D-Glc*p* residue in a 1,2-linkage to initiate chain extension from HepIII of the respective LPS. These data are consistent with *lpsA* encoding the β-1,2-glucosyltransferase and this is supported by sequence homologies with other known bacterial transferases. It is noteworthy that the homologous gene from type b strains, Eagan and RM7004 mediates the addition of a β-D-Gal*p* residue instead of a β-D-Glc*p* residue to *O*-2 of HepIII (19). As in the case of *lic1D*, the observed difference in donor specificity may be due to polymorphisms between the respective *lpsA* genes.

Analysis of the Rd strain containing a mutation in *lgtF* was indicative of the role of this gene in LPS chain extension from HepI. The ESI-MS of *O*-deacylated LPS from this strain gave molecular ions corresponding to a mixture of glycoforms having chain extension from HepIII of a tri-heptosyl inner-core unit which lacked (*P*Cho→6)-β-D-Glc*p* at HepI. The gene product of *lgt*F shows sequence homology to an analogous β-D-glucosyltransferase identified in *Neisseria spp.*

The Rd strain containing a mutation in *lic2A* expresses LPS which contains a single β-D-Glc*p* residue attached in a 1,2-linkage of HepIII. The Rd *lgtC* mutant which contains a functional *lic2A* gene elaborates LPS containing a lactose side chain at HepIII. This is consistent with the involvement of the *lic2A* gene in addition of the β-D-Gal*p* unit in a 1,4-linkage to the terminal β-D-Glc*p* attached to HepIII.

Correspondingly, the *lgtD* mutant containing a functional *lgtC* gene is capable of adding an α-D-Gal*p* in 1,4-linkage to the terminal β-D-Gal*p* of the lactose epitope. This result is consistent with a function for *lgtC* in encoding the specific α-galactosyltransferase for the synthesis of the α-D-Glc*p*-(1→4)-β-D-Gal*p* unit of the globoside trisaccharide in the Rd Hex 4 LPS glycoform (Table 1). The parent strain (Rd) which contains a functional *lgtD* gene is capable of elaborating the complete globotetraose unit which is indicative of the role of *lgtD* in adding the terminal β-D-Gal*p*NAc. *LgtD* and its gene product show sequence homology to *N. gonorrhoeae lgtA* and *lgtD* genes which are known to have β-D-Glc*p*NAc and β-D-Gal*p*NA*c* transferase functions, respectively.

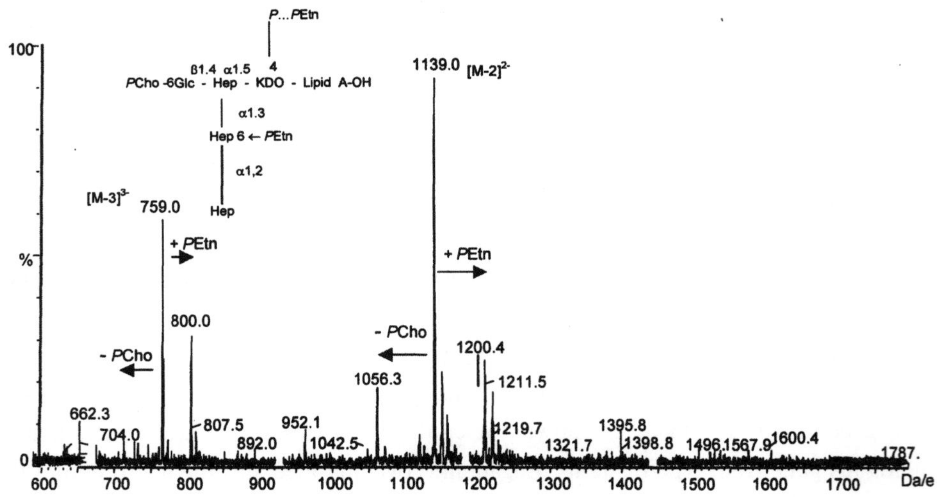

Figure 5. Negative ion electrospray ionization mass spectrum of *O*-deacylated LPS from *H. influenzae* strain Rd in which the *lpsA* gene was inactivated by a deletion-insertion mutation. The major LPS species (structure shown in inset) lacks chain extention from HepIII pointing to the importance of LpsA in initiating chain extension.

The *lic*2A homologues in type b strains, Eagan and RM7004, have been shown to be involved in expression of the globoside trisaccharide pK epitope. The role of *lic2A* in phase variable expression of this epitope has been previously demonstrated (24). The gene product from the *lgtC* homologue in type b strain Eagan has been shown to have α-galacosyltransferase activity (23). Moreover, mutations of this gene in Eagan and a non-typeable strain (NTHi 375) leads to expression of truncated LPS which lacks terminal α-D-Gal*p* units in the oligosaccharide chains attached to HepIII. *LgtC* contains a tetranucleotide repeat motif, GACA, at the extreme 5' end of the coding region. In a similar fashion to 5'-CAAT-3' repeats present in *lic2A* (24), multiple tandem repeats of the tetramer 5'-GACA-3' are also found in the homologous *lgtC* genes of type b strains, Eagan and RM7004, thus contributing to the mechanism for phase variable expression of the globoside trisaccharide epitope. The digalactoside unit, α-D-Gal*p*-(1→4)-β-D-Gal*p*, is found on many human cells, for example as a component of the globoseries of glycolipids, structures implicated as receptors for bacterial ligands including *E. coli* fimbriae and the toxin of *Shigella*. It is thought that this epitope is an important virulence determinant of *H. influenzae* and may be involved in molecular mimicry of epithelial cell structures. Although the molecular environment in which the digalactoside found on LPS of *H. influenzae* can differ, the biological implications of these structural differences are not known.

SUMMARY

The involvement of genes in the *lic* loci in *H. influenzae* LPS expression has been known for some time (20). However, it was not until recently that it was shown that the *lic1* locus contains genes required for phase variable expression of phosphocholine substituents (21), while genes in the *lic2* locus and *lgtC* are required for expression of the globoside trisaccharide, α-D-Gal*p*-(1→4)-β-D-Gal*p*-(1→4)-β-D-Glc*p* (i.e., the pK blood group epitope) (19,23). The availability of the complete sequence of the *H. influenzae* strain Rd genome has facilitated significant progress in understanding the role of these and other genes in the expression and biosynthesis of LPS. We have employed a comparative structural fingerprinting strategy to establish the structural relationships among LPS from *H. influenzae* mutant strains in which putative biosynthesis genes were inactivated. Using this functional genomics approach, we have gained considerable insight into the genetic basis for intra-strain and strain-to-strain variation in epitope expression.

REFERENCES

1. Preston, A., Mandrell, R.E., Gibson, B.W., & Apicella, M.A. (1996) The lipopolysaccharide of Gram negative bacteria. Crit. Rev. Microbiol. **22**: 139-1802.
2. Phillips, N.J., Apicella, M.A., Griffiss, J.M. & Gibson, B.W. (1992) Structural characterization of the cell surface lipooligosaccharides from a nontypable strain of *Haemophilus influenzae*. *Biochemistry* **31**: 4515-4526.
3. Schweda, E.K.H., Hegedus, O.E., Borelli, S., Lindberg, A.A., Weiser. J.N., Maskell, D.J. & Moxon, E.R. (1993) Structural studies of the cell envelope lipopolysacchaide from *Haemophilus influenzae* strain AH1-3 (lic3+). *Carbohydr. Res.* **246**: 319-330.
4. Schweda E.K.H., Jansson, P.-E., Moxon, E.R. & Lindberg, A.A. (1995) Structural studies of the saccharide part of the cell envelope lipopolysaccharide from *Haemophilus influenzae* strain galEgalK. *Carbohydr. Res.* **272**: 213-224.
5. Philips, N.J., McLaughlin, R., Miller, T.J., Apicella, M.A. & Gibson, B.W. (1996) Characterization of two transposon mutants from *Haemophilus influenzae* type b with altered lipooligosaccharide biosynthesis. *Biochemistry* **35**: 5937-5947.
6. Masoud, H., Moxon, E.R., Martin, A., Krajcarski, D. & Richards, J.C. (1997) Structure of the variable and conserved lipopolysaccharide oligosaccharide epitopes expressed by *Haemophilus influenzae* serotype b strain Eagan. *Biochemistry* **36**: 2091-2103.
7. Risberg, A., Schweda, E.K.H & Jansson, P.-E. (1997) Structural studies of the cell-envelope oligosaccharide from the lipopolysaccharide of *Haemophilus influenzae* strain RM.118-28. *Eur. J. Biochem.* **243**: 701-707.
8. Risberg, A., Masoud, H., Martin, A., Richards, J.C., Moxon, E.R. & Schweda E.K.H. (1999) Structural analysis of the lipopolysaccharide oligosaccharide epitopes expressed by a capsule-deficient strain of *Haemophilus influenzae* Rd. *Eur. J. Biochem.* **261**: 717-180.
9. Rahman, M.M., Gu, X.-X., Tsai, C.-M., Kolli, V.S.K & Carlson, R.W. (1999) The structural heterogeneity of lipooligosaccharide (LOS) expressed by pathogenic non-typeable *Haemophilus influenzae* strain NTHi 9274. *Glycobiology* **9**: 1371-1380.
10. Thibault, P. & Richards, J.C. (1999) Application of combined capillary electrophoresis electrospray mass spectrometry in the characterization of short-chain lipopolysaccharide: *Haemophilus influenzae*. In *Methods in Molecular Biology*, Vol. 145: Bacterial Toxins: Methods and Protocols (Holst, O., ed.), pp. 327-344. Humana Press Inc., Totowa, NJ, USA.
11. Schweda, E.K.H., Brisson, J.-R., Alvelius, G., Martin, A., Weisser, J.N., Hood, D.W., Moxon, E.R. & Richards, J.C. (2000) Characterization of the phosphocholine-substituted oligosaccharide in lipopolysaccharides of type b *Haemophilus influenzae*. *Eur. J. Biochem.* **267**: 3902-3913.

12. Risberg, A., Alvelius, G., and Schweda, E.K.H. (1999) Structural analysis of the lipopolysaccharide oligosaccharide epitopes expressed by *Haemophilus influenzae* strain RM.118-26. *Eur. J. Biochem.* **265**: 1067-1074.
13. Hood, D.W., Makepeace, K., Deadman, M.E., Rest, R.F., Thibault, P., Martin, A., Richards, J.C., and Moxon, E.R. (1999) Sialic acid in the lipopolysaccharide of *Haemophilus influenzae*: strain distribution, influence on serum resistance and structural characterisation. *Mol. Microbiol.* **33**: 679-692.
14. Lysenko, E., Richards, J.C., Cox, A.D., Stewart, S., Martin, A., Kapoor, M. & Weiser, J.N. (2000) The position of phosphocholine on the lipopolysaccharide of *Haemophilus influenzae* affects binding and sensitivity to C-reactive protein mediated killing. *Mol. Microbiol.* **35**: 234-245.
15. Vimr, E., Lichtensteiger, C., and Steenbergen, S. (2000) Sialic acid metabolism's dual function in *Haemophilus influenzae. Mol. Microbiol.* **36**: 1113-1123.
16. Hood, D.W., Cox, A.D., Gilbert, M., Makepeace, K., Walsh, S., Deadman, M.E., Cody, A., Martin, A., Månson, M., Schweda, E.K.H., Brisson, J.-R., Richards, J.C., Moxon, E.R. & Wakarchuk, W.W. (2000) Identification of lipopolysaccharide α-2,3-sialyltranferase from *Haemophilus influenzae. Mol Microbiol.* In Press.
17. Fleishmann, R.D., Adams, M.D., White, O., Clayton, R.A., Kirkness, E.F., Kerlavage, A.R., Butt, C.J., Tomb, F.-F., Doughtery, B.A, Merrick,, J.M., McKenney, K., Sutton, G., Fitzhugh, W., Fields, C., Gocaybe, J.D., Scott, J., Shirley, R., Liu, L.-I., Glodek, A., Kelly, J.M., Weildman, J.F., Phillips, C.A., Spiggs, T., Hedblom, E., Cotton, M.D., Utterbask, T.R., Hanna, M.C., Nguyen, D.T., Saudek, D.M., Brandon, R.C., Fine, L.D., Fritchmann, J.L., Georghagen, N.S.M., Gnehm, C.L., McDonald, L.A., Small, K.V., Fraser, C.M., Smith, H.O. & Venter, J.C. (1995) The genome of *Haemophilus influenzae* Rd. Science **269**: 496-512.
18. Kimura, A. & Hansen, E.J. (1986) Antigenic and phenotypic variations of *Haemophilus influenzae* type b lipopolysaccharide and their relationship to virulence. *Infect. Immun.* **51**: 69-79.
19. Hood, D.W., Deadman, M.E., Allen, T., Masoud, H., Martin, A., Brisson, J.-R., Fleischmann, R., Venter, J.C., Richards, J.C., and Moxon, E.R. (1996) Use of the complete genome sequence information of *Haemophilus influenzae* strain Rd to investigate lipopolysaccharide biosynthesis. *Mol Microbiol* **22**: 651-965.
20. Hood, D.W., and Moxon, E.R. (1999) Lipopolysaccharide phase variation in *Haemophilus* and *Neisseria.* In *Endotoxin in Health and Disease.* New York, USA: Marcel Dekker Inc,pp 39-54.
21. Weiser, J.N., Shchepetov, M. & Chong, S.T.H. (1997) Decoration of lipopolysaccharide with phosphocholine: a phase-variable characteristic of *Haemophilus influenzae. Infect. Immunity* **65**: 943-950.
22. Banerjee, A., Wang, R., Uljon, S.E., Rice, R.A., Gotschlich, E.C., Stein, D.C., (1998) Identification of the gene (*lgtG*) encoding the lipooligosaccharide β chain synthesizing glucosyltransferase from *Neisseria gonorrhoeae. Proc. Natl. Acad. Sci. USA 95*: 10872-10877.
23. Hood, D.W., Cox, A.D., Wakarchuk, W.W., Schur, M., Schweda, E.K.H., Walsh, S., Deadman, M.E., Martin, A., Moxon, E.R., Richards, J.C., (2000) Genetic basis for expression of the major globotetraose containing lipopolysaccharide in *Haemophilus influenzae* strain Rd (RM118). *Submitted.*
24. High, N.J., Deadman, M.E., and Moxon, E.R. (1993) The role of the repetitive DNA motif (5'-CAAT-3') in the variable expression of the *Haemophilus influenzae* lipopolysaccharide epitope αGal (1-4)βGal. *Mol. Microbiol.* **9**: 1275-1282.

MOLECULAR MIMICRY OF HOST STRUCTURES BY LIPOOLIGOSACCHARIDES OF *NEISSERIA MENINGITIDIS*: CHARACTERIZATION OF SIALYLATED AND NONSIALYLATED LACTO-N-NEOTETRAOSE (Galβ1-4GlcNAcβ1-3Galβ1-4Glc) STRUCTURES IN LIPOOLIGOSACCHARIDES USING MONOCLONAL ANTIBODIES AND SPECIFIC LECTINS

Chao-Ming Tsai
Division of Bacterial Products, Center for Biologics Evaluation and Research, FDA, Bethesda, MD 20892 USA

Neisseria meningitidis is a leading cause of bacterial meningitis in humans (Gotschlich, 1990). The mortality rate of meningococcal disease is about ten percent. Disease strains isolated from the blood and cerebrospinanl fluid of patients are capsulated. Besides the surface capsular polysaccharide, the other surface components on *N. meningitidis* are outer-membrane proteins, lipopolysaccharides (LPS), and pili. Both meningococcal capsular polysaccharides and outer-membrane proteins have been reviewed extensively (Jennings, 1983; Frasch, 1995; Zollinger, 1997). A review on meningococcal LPS has also been reported (Verheul *et al.*, 1993). In this article, I will focus only on the LPS of *N. meningitidis*. In particular, using monoclonal antibodies and specific lectins to characterize sialylated and nonsialylated lacto-N-neotetraose sequence, a precursor of human ABH blood group antigens, in the LPS. Molecular mimicry of host structures by the LPSs of *N. meningitidis* and some other Gram-negative pathogens will be presented.

Serological classification of *N. meningitidis*

N. meningitidis is serologically classified into serogroup, immunotype, and serotype based on the three surface antigens: capsular polysaccharide, lipopolysaccharide (LPS), and outer-membrane protein respectively. Serological classification based on its pilus has not been explored due to a very high degree of antigenic variability. For the capsular polysaccharide antigen, twelve serogroups are currently recognized: A, B, C, H, I, K, L, X, Y, Z, 29E, and

The Molecular Immunology of Complex Carbohydrates–2.
Edited by Albert M. Wu. Kluwer Academic/Plenum Publishers, 2001.

W-135 (Ashton *et al.,* 1983). Among the 12 serogroups, serogroup A was historically the main cause of epidemic meningococcal disease globally and still predominates in Africa and Asia (Peltola, 1983; Gotschlich, 1990; Tikhomirov *et al.*, 1997). In contrast, serogroups B and C have been the major cause of the disease in the U. S. and Europe since 1950; and incidences of serogroups Y and W-135 have increased in recent years (Jones, 1995; Zollinger, 1997). Based on the antigenicity of LPS, *N. meningitidis* can be divided into 12 immunotypes. Immunotypes L1 to L8 were found among serogroup B and C organisms (Mandrell and Zollinger, 1977; Zollinger and Mandrell, 1977) and L9 to L12 were in serogroup A organisms (Zollinger and Mandrell, 1980). For serotypes based on outer-membrane proteins, currently there are about 18 serotypes and 20 serosubtypes (Frasch *et al.*, 1985; Zollinger, 1997; Sacchi *et al.*, 1998).

Heterogeneity of *N. meningitidis* lipooligosaccharides (LOS)

In contrast to the LPS of enteric bacteria, *N. meningitidis* LPS consists of only a branched oligosaccharide (OS) up to about ten monosaccharide units linked to lipid A through 2-keto-3-deoxy-octonic acid (Kdo) but with no O-polysaccharide chain (Jennings *et al.*, 1980; Kogan *et al.*, 1997) as shown in Fig. 1. *N. meningitidis* LPS, therefore, is now called lipooligosaccharide (LOS) by most investigators (Schneider *et al.*, 1984; Tsai *et al.*, 1987). Most of *N. meningitidis* LOSs are heterogeneous in having both 4.0 and 3.6 kDa components and also other minor components as revealed by SDS-PAGE analysis in Fig. 2 (Tsai *et al.*, 1987; Tsai and Civin, 1991). The degree of heterogeneity depends on strains and growth conditions of the organism (Tsai *et al.*, 1983). The multiple components of an LOS seen on SDS-PAGE represent OSs of different sizes linked to lipid A (Tsai *et al.*, 1983; Schneider *et al.*, 1984). The shorter OSs are the biosynthetic precursors of the longest OS, which has a lacto-N-neotetraose structure at the nonreducing end, since these OSs carry the same structure at their Kdo reducing ends (Michon *et al.*, 1990). The structure of *N. meningitidis* lipid A has been elucidated and the lipid A is also heterogeneous (Kulshin *et al.*, 1992).

Presence of lacto-*N*-neotetraose sequence in *N. meningitidis* LOSs

The immunotype specificity of *N. meningitidis* LOS resides in the OS portion of the LOS (Jennings *et al.*, 1984). Structural studies on purified OSs from several LOS types have revealed common regions as well as variable regions, which presumably carry type-specific

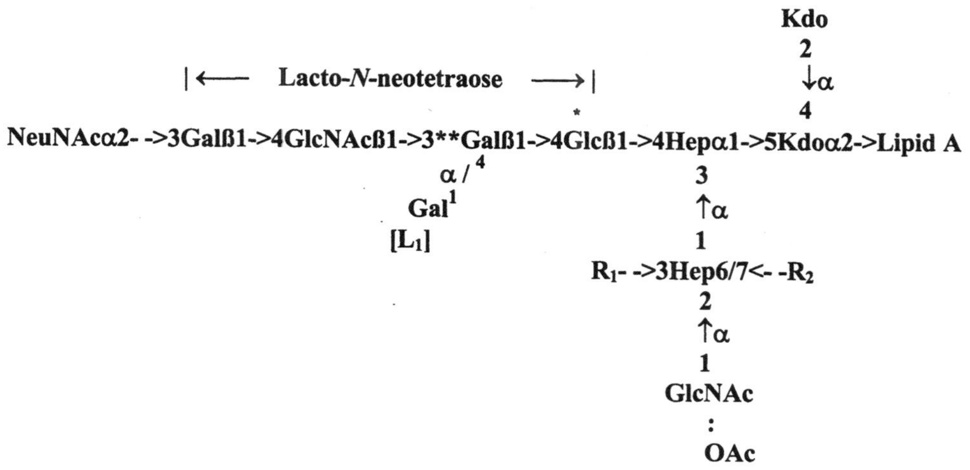

Fig.1. A generalized oligosaccharide structure of *Neisseria meningitidis* LOSs.
Double asterisks (**) indicate the penultimate Gal for L1 and L6 oligosaccharides with Galα1-4 for L1 and GlcNAcß1-3 for L6 at the nonreducing end. Dotted line (- -) indicates incomplete sialylation of LOS with NeuAc in L2, L3, L4, L5, L7, & L8. The single asterisk (*) indicates an additional internal ß1-4 linked Glc in the chain for L5. Hep is L-glycero-D-manno-heptose. R_1 is αGlc for L2 & L5, and phosphoethanolamine ($-P-EtNH_2$) for L1, L3, & L7; R_2, $-P-EtNH_2$ for L2 (6/7), & L6 (7); OAc, acetyl group in L2 (40%), L4 (50% at 3 position), & L5 (60%). Adapted from Tsai *et al.*, Glycobiology 8: 359-365, 1998, Fig. 1 and Kogan *et al.*, Carbohydrate Research 298: 191-199, 1997, Fig. 2.

epitopes (Jennings *et al.*, 1987). Lacto-N-neotetraose (LNnT, Galβ1-4GlcNAcβ1-3Galβ1-4Glc) sequence was found to be a common structure at the nonreducing end of branched OSs of L3, L5, and L9 immunotype LOSs (Jennings *et al.*, 1983; Dell *et al.*, 1990; Michon *et al.*, 1990). The presence of LNnT sequence in *N. meningitidis* LOSs is interesting because this structure is a precursor of human ABH blood group antigens (a type 2 chain) and exists in glycosphingolipids such as paragloboside (LNnT-ceramide) on human erythrocytes and granulocytes (Macher and Klock, 1980; Hakomori, 1981). Therefore, we (Tsai and Civin, 1991) used a monoclonal antibody, anti-My-28, to probe the LNnT sequence in all 12 immunotypes of *N. meningitidis* LOS. Anti-My-28 is a mouse monoclonal antibody to human granulocytes with specificity for LNnT as shown in Table 1 (Spitalnik *et al.*, 1990). The binding activities of *N. meningitidis* LOSs with the anti-My-28 monoclonal antibody were analyzed by immunoblot and ELISA.

Table 1

Specificity of anti-My-28 monoclonal antibody as measured by sugar inhibition of antibody binding to paragloboside containing granulocyte neutral glycolipids in radioimmunoassay

Sugar	Conc. mM 2	25	50% inhibition
Lacto-N-neotetraose (LNnT) Galß1-4GlcNAcß1-3Galß1-4Glc	60%	-	1.6 mM
Lacto-N-tetraose (LNT) Galß1-3GlcNAcß1-3Galß1-3Glc	0	0	-
N-acetyllactosamine Galß1-4GlcNAc	10%	90%	4.8 mM
Lactose Galß1-4Glc	2%	85%	5.1 mM
Galactose Gal	0	0	-

Note: No inhibition of the binding when the terminal Gal of LNnT, N-acetyllactosamine, or lactose was substituted with sialic acid. Adapted from Spitalnik *et al.*, (1985) Blood 66: 319-326, Fig. 5.

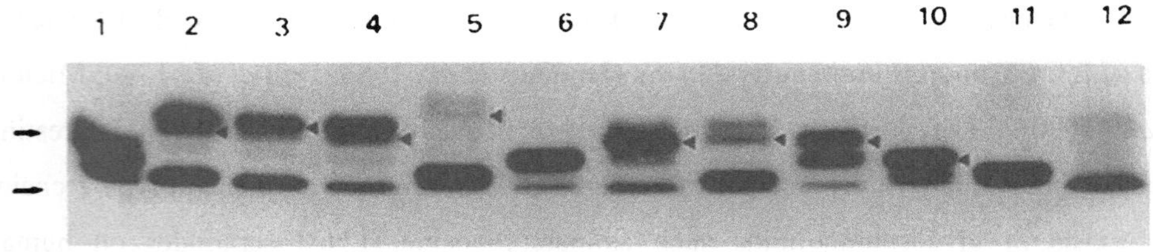

Fig. 2. Heterogeneity of *N. meningitidis* LOSs as revealed by SDS-PAGE and anti-My-28 antibody-binding components by immunoblot analysis. Lanes 1 through 12 designate the twelve immunotype LOSs from prototype strains grown in TSB as indicated in Table 2. The two arrows indicate the locations of 4.0 and 3.6 kDa components of LOSs. An arrowhead indicates the antibody-binding component in a multiple-component LOS on immunoblot. Sample loads: 0.2 μg of LOS for SDS-PAGE and silver stain and 0. 4 μg for immunoblot. Adapted from Tsai and Civin (1991), Infect. Immun. 59: 3604-3609, Fig. 2.

An initial immunodot analysis indicated that eight of 12 immunotype LOSs bound the antibody; they were L2, L3, L4, L5, L7, L8, L9, and L10. Since most *N. meningitidis* LOSs were heterogeneous and contained more than one component, immunoblot was used to analyze the antibody-binding component in a multiple-component LOS. Only one antibody-binding component was found in each of the reactive LOSs as indicated by an arrowhead in Fig. 2. The apparent molecular sizes of antibody-binding components on SDS-PAGE were the same, 4.0 kDa, except for the L5 and L10 LOSs. The L5 was slightly larger, 4.2 kDa, reflecting the additional Glc in the LNnT-containing OS chain as indicated by a star (*) in Fig. 1. The L10 had no 4.0 kDa component but its major 3.7 kDa component with a shorter OS had a weak binding with the antibody indicating a cross-reactivity of its terminal lactosyl sequence with the antibody (Kim *et al.,* 1994).

In an ELISA, only four LOSs had good binding with the antibody (Table 2). However, eight of 12 immunotype LOSs bound the antibody when the LOSs were treated with neuraminidase. The results indicate that the antibody binding with some LOSs was blocked if they were sialylated. The four non-binding LOSs even after neuraminidase treatment were L1, L6, L11 and L12. The binding of antibody to the LOSs was inhibited by N-acetyllactosamine. The binding of a representative LOS (L2) was best inhibited by LNnT followed by

Table 2

Effects of neuraminidase treatment on the binding of 12 LOSs of *N. meningitidis* with anti-My-28 antibody and inhibition of binding by 5 Mm N-acetyllactosamine as measured by ELISA at A_{405nm}

LOS	L1	L2	L3	L4	L5	L6	L7	L8	L9	L10	L11	L12
Strain	126E	35E	6275	89I	M981	M992	6155	M978	M120	7880	7889	7897
Serogroup	C	C	B	C	B	B	B	B	A	A	A	A
No enzyme	.01	.02	.02	.01	.05	0	.43	1.20	1.19	.49	.01	.01
Neuraminidase	.01	1.28	1.69	1.48	.35	0	.51	1.31	1.25	.57	.01	.01
% Inhibition 5 mM LacNAc	-	95	93	95	99	-	99	94	95	99	-	-

The LOSs were purified from the strains grown in TSB medium.
Modified from Tsai and Civin (1991), Infect. Immun. 59: 3604-3609, Tables 1 & 2 with permission.

N-acetyllactosamine and lactose but not by lacto-N-tetraose, Galβ1-3GlcNAcβ1-3Galβ1-4Glc (Tsai and Civin, 1991). The above results indicate the presence of the LNnT sequence in the L2, L4, L7, and L8 LOSs in addition to the L3, L5, and L9 LOSs which were previously shown to have the LNnT sequence in their LOSs by structure analysis. The presence of the LNnT sequence in the L2, L4, and L7 was later confirmed by structural analysis (Gamian *et al.*, 1992; Kogan *et al.*, 1997). Since the L10 LOS has a low activity in ELISA and its 3.7 kDa antibody-binding component has a shorter oligosaccharide than the 4.0 kDa component of other LOSs, the activity of L10 is probably due to a cross-reaction of its terminal lactosyl sequence with the antibody (Kim *et al.*, 1994). Thus, the possibility of cross-reactivity of an antibody with a similar structure should be assessed when using monoclonal antibodies to characterize the structure of an LOS. Based on the above results, we concluded that seven of the eight antibody-binding LOSs contained the LNnT sequence in their LOSs. This conclusion is consistent with the published structures of 6 immunotypes by Jennings and his colleagues in showing a common LNnT sequence in addition to their unique carbohydrate sequences (Kogan *et al.*, 1997). The structure of the L8 LOS (M978 L8 prototype strain) has yet to be determined. In a different study, other investigators have shown that certain monoclonal antibodies to Neisserial LOS bound to human erythrocytes and to the paragloboside (LNnT-ceramide) and other related higher glycoshingolipids purified from human erythrocytes (Mandrell *et al.*, 1988).

Sialylation of *N. meningitidis* LOSs

N. meningitidis LOS in Group B and C organisms, but not Group A, may be sialylated (Mandrell *et al.*, 1988; Tsai and Civin, 1991). This is due to the finding that the former two group organisms but not the later produce CMP-NeuNAc, which is the required substrate for the sialylation of the LPSs (Mandrell *et al.* 1993). The terminal Gal of LNnT in *N. meningitidis* LOSs could be either sialylated or non-sialylated depending on growth conditions (Mandrell *et al.*, 1988; Tsai and Civin, 1991; Mandrell *et al.*, 1993). The sialic acid in *N. meningitidis* LOSs has been shown to be N-acetylneuraminic acid (NeuNAc) (Mandrell *et al.*, 1991; Tsai and Civin, 1991). Structural analyses on the linkage between NeuNAc and Gal in *N. meningitidis* LOS have revealed a 2->3 linkage for the sialylated immunotype 3 and 4 LOSs as presented in figure 1 (Pavliak *et al.*, 1993; Yamasaki *et al.*, 1993; Kogan *et al.*, 1997). The sialic acid linkages in other immunotypes have been characterized only recently using specific lectins (Tsai *et al.*, 1995; Tsai *et al.*, 1998). This is because the OSs used in most previous structural studies were released from *N. meningitidis*

LOSs by acid hydrolysis (1% acetic acid at 100 °C for two hours), which also hydrolyzed sialic acid from the OSs when sialylated. Recently, the sialyltransferase gene for the biosynthesis of *N. meningitidis* LOS was cloned and the enzyme was characterized to be an α-2,3-sialyltransferase (Gilbert *et al.*, 1996). Elucidation of the biosynthesis genes of the sialyltransferase and three other glycosyltranferases for expression of the LNnT sequence has thus established the genetic basis for the biosynthesis of the terminal NeuNAcα2-3LNnT structure of LOS in *N. meningitidis* (Preston *et al.*, 1996; Wakarchuk *et al.*, 1996).

To examine the linkage of sialic acid in *N. meningitidis* LOS, we used the sialic acid-binding lectins, *Limax flavus* agglutinin, *Maackia amurensis* leukoagglutinin, and *Sambucus nigra* agglutinin to investigate the sialic acid linkage in *N. meningitidis* LOSs. The specificity of *Maackia amurensis* leukoagglutinin (MAL) is shown in Table 3 (Knibbs 1991). The MAL lectin is specific for NeuNAcα2-3Galβ1-4GlcNAc/Glc trisaccharide sequence. Neither NeuNAcα2-6Galβ1-4GlcNAc trisaccharide nor NeuNAcα2-3Gal disaccharide bind the lectin.

Table 3

Binding specificity of *Maackia amurensis* leukoagglutinin (MAL) as measured by precipitation reaction with BSA-glycoconjugates

Neu5Acα**2->3**Galß**1->4**GlcNAc	100%
Neu5Acα2->3Gal	≤ 5%
Galß1->4GlcNAc	≤10%
Neu5Acα2->3Galß**1->3**GlcNAc	≤ 5%
Neu5Acα**2->6**Galß1->4GlcNAc	≤ 5%

Adapted from Knibbs *et al.*, (1991) J. Biol. Chem. 266:83-88, Fig.s 1 and 2.

Using an ELISA, six of the above seven LNnT-containing LOSs (L2, L3, L4, L5, L7, L8 and L9) bound with MAL as shown in Table 4. The LNnT-containing, but MAL-non-binding, L9 LOS is from a serogroup A organism and it does not synthesize CMP-NeuAc required for the sialylation of the LOS. The LOS-lectin binding was abolished when the LOSs were treated with Newcastle disease viral neuraminidase which cleaves α2->3 linked sialic acid. Methylation analysis (York *et al.*, 1986) of a representative LOS (L2) confirmed that NeuNAc was α2-3 linked to the terminal Gal of LNnT in the LOS.

Table 4

Reactivities of 12 *N. meningitidis* LOSs with lectin MAL by ELISA at A_{405nm} and inhibition by 10 mM sialyllactose

LOS	L1	L2	L3	L4	L5	L6	L7	L8	L9	L10	L11	L12
Strain	126E	35E	6275	89I	M981	M992	6155	M978	M120	7880	7889	7897
Serogroup	C	C	B	C	B	B	B	B	A	A	A	A
No sugar	.03	2.07	2.03	1.59	.52	.05	.59	1.06	.12	.06	.05	.07
Sialyllactose	.03	.29	.24	.20	.11	.07	.09	.18	.10	.07	.06	.09

Modified from Tsai *et al.*, (1998), Glycobiology 8: 359-365, Table IV with permission

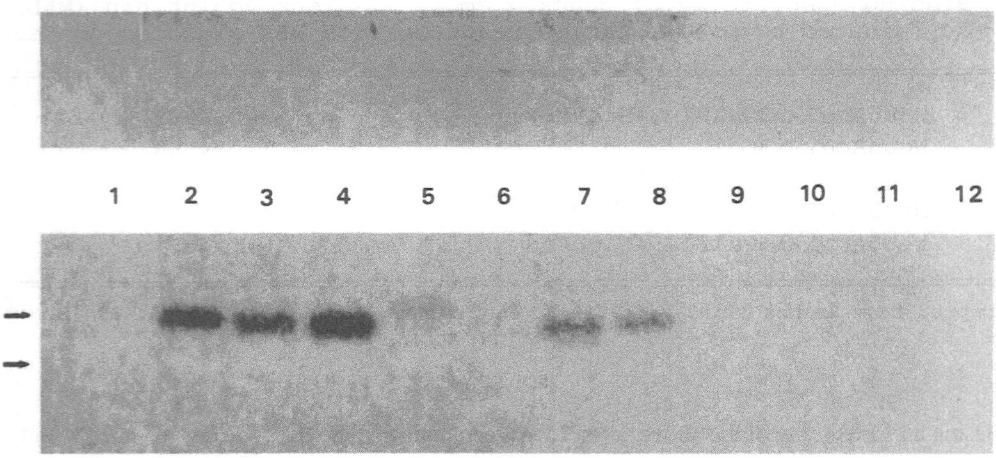

Fig. 3. MAL blot analysis of 12 *Neisseria meningitidis* immunotype LOSs.
The two arrows on the left of the panel indicate the locations of 4.1 and 3.6 kDa LOS components in an LOS. The numbers 1 through 12 designate immunotypes L1 through L12. The corresponding silver-stained pattern on SDS-PAGE was shown in Fig. 2. Sample loads: 0.2 μg of LOS for silver stain and 0.4 μg for lectin blot. Reproduced with permission from Tsai *et al.*, (1998), Glycobiology 8: 359-365, Fig. 3.

The combination of SDS-PAGE and MAL-blot analyses as shown in Figs. 2 and 3 revealed that the six MAL-binding LOSs contained an LOS component with an apparent MW of 4100. Neuraminidase treatment of *N. meningitidis* LOSs showed that the 4.1 kDa LOS component in an LOS was the sialylated form of the 4.0 kDa component which has a terminal LNnT structure as presented in the preceding section. The 4.0 and 4.1 kDa components appeared as a closely spaced doublet on SDS-PAGE (Tsai and Civin, 1991). For the L8 (M978) LOS seen on SDS-PAGE, it had a minor 4.0 kDa component and a major 3.6 kDa component which possesses the L8 epitope with a terminal lactose sequence (Schneider *et al.*, 1991; Chen *et al.*, 1995). However, the sialylation of the L8 LOS occurred only to the minor 4.0 kDa component but not to the major 3.6 kDa component as revealed by the lectin blot analysis (see Fig. 3). Thus, the terminal Gal at LNnT but not at lactose structure of the L8 LOS serves as the natural receptor for the sialyltransferase. These MAL-binding LOSs did not bind to *Sambucus nigra* agglutinin (SNA) which recognizes NeuNAcα2-6Gal sequence (Shibuya *et al.*, 1987) indicating no α2-6 linked NeuNAc in these LOSs. Thus, we concluded that NeuNAc is α2-3 linked to the terminal Gal of LNnT in *N. meningitidis* LOSs (Tsai *et al.*, 1998). This conclusion is in agreement with a previous structural analysis of the L3 LOS in that NeuNAc is α2-3 linked to the Gal of LNnT (Pavliak *et al.*, 1993; Yamasaki *et al.*, 1993).

N. meningitidis L1 and L6 LOSs did not bind the MAL lectin although they were isolated from Group B organisms that produce CMP-NeuNAc, the required substrate for the sialylation of LOS. It has been reported that the L1 LOS had a NeuNAc α2-6 linked to its Galα1-4Galβ1-4Glc terminal (Wakarchuk *et al.*, 1998). We did not observe any binding between the L1 LOS and the SNA lectin, which recognizes NeuNAcα2-6Gal sequence. However, our unpublished results showed that both the L1 (126E) LOS and the MAL-reactive LOSs bound *Limax flavus* agglutinin (LFA), a sialic acid-specific lectin (Miller *et al.*, 1982), and treatment of the L1 LOS with *V. cholera* neuraminidase eliminated the binding with concomitant release of NeuNAc from the LOS. Thus, the non-binding of 126E (L1) LOS with SNA may be due to steric hindrance by a neighboring group surrounding the NeuNAcα2-6Galα1-4Galβ1-4Glc moiety of the LOS. For the L6 LOS, the nonreducing terminal is GlcNAc, which is presumably not a receptor for the sialyltransferase, and no sialylation is expected. In fact, the L6 LOS did not bind LFA lectin and no sialic acid was released by neuraminidase treatment.

The binding of 12 *N. meningitidis* LOSs with the MAL lectin and the anti-My-28 antibody is summarized in Table 5.

Table 5

Reactivities and Binding Regions of 12 N. meningitidis LOSs from Cells Grown in TSB with Lectin MAL and Antibody My-28

LOS			Lectin MAL		Antibody My-28	
type	strain	Ref	Reactivity	Binding Region	Reactivity#	Binding Region
L1	126E(C)	1	-	none	-	none
L2	35E (C)	2	++	NeuNAcα2-3Galß1-4GlcNAc	++	LNnT
L3	6275(B)	3	++	"	++	"
L4	89I (C)	4	++	"	++	"
L5	M981(B)	5	+	"	+	"
L6	M992(B)	1	-	none	-	none
L7	6155(B)	4	+	NeuNAcα2-3Galß1-4GlcNAc	++	LNnT
L8	M978(B)		+	"	+	"
L9	M120(A)	6	-	none	++	"
L10	7880(A)	7	-	"	+	Lactose
L11	7889(A)		-	"	-	none
L12	7897(A)		-	"	-	"

* My-28, a mouse monoclonal antibody with specificity for LNnT
\# Neuraminidase treated LOSs
Ref: 1. Can. J. Chem. 68:1029 (1990)
2. J. Biol. Chem. 267:922 (1992)
3. J. Biol. Chem. 268:14146 (1993)
4. Carbohydr. Res. 298:191 (1997)
5. J. Biol. Chem. 265:7243 (1990)
6. Carbohydr. Res. 121: 233 (1983)
7. Infect. Immun. 64:1566 (1994)

The sialylation of LOS influences many biological effects. For *N. gonorrhoeae*, the sialylation of LOS blocked the bacteriocidal effects of serum antibodies to certain outer membrane proteins (Parsons *et al.*, 1989; Wetzler *et al.*, 1992, Rest and Mandrell, 1995). The sialylation of *N. meningitidis* LOS also conferred serum resistance by inhibiting both the classical and alternate complement pathways, but the effect is not as obvious as that in *N. gonorrhoeae* (Hammerschmidt *et al.*, 1994; Vogel *et al.*, 1997). Sialylation of LOS increased resistance of *N. gonorrhoeae* and some strains of *N. meningitidis* to opsonophagocytosis by human neutrophils (Estabrook *et al.*, 1992; Kim *et al.*, 1992). In addition, sialylated gonococci decreased both non-opsonic adherence and stimulation of oxidative burst in the neutrophils (Rest and Frangipane, 1992). Furthermore, the sialylated LNnT moiety in LOS may serve as a ligand for the binding of the LOS or whole organism to the lectin-like receptor on macrophages, i.e. sialoadhesin, which has a specificity for terminal NeuNAcα2-3Galβ1->3(4)GlcNAc or NeuNAcα2-3Galβ1->3GalNAc sequence (Crocker *et al.*, 1991; Kelm *et al.*, 1994). The LOS-sialoadhesin binding may subsequently facilitate transfer of the LOS to CD14 receptor and result in the production of proinflammatory cytokines by the macrophages. This may explain our recent observation that the sialylated meningococcal LOS induced 3-4 times more TNF-α and IL6 in mouse peritoneal macrophages than nonsialylated LOS (Quakyi *et al.*, 1997). Thus, the sialylation of LOS contributes many factors in the virulence of pathogenic Neisseria (Apicella and Mandrell, 1989; Smith *et al.*, 1992).

Molecular mimicry of host structures by LOS and LPS of some pathogenic bacteria

The LNnT and NeuNAcα2->3 LNnT structures found at the distal end of *N. meningitidis* LOSs are present in the glycosphingolipids, i.e., paragloboside and sialylparagloboside, of human erythrocytes, granulocytes, and lymphocytes (Macher *et al.*, 1980; Hakomori, 1981; Spitalnik *et al.*, 1985). The terminal N-acetyllactosamine sequence with or without 2->3 linked NeuNAc is also present in glycoproteins on human cells such as the predominant surface protein (Band 3) of erythrocytes (Fukuda *et al.*, 1984). *N. meningitidis* LOS therefore mimics certain glycosphingolipids and glycoproteins on human cells or tissues in having the same carbohydrate sequences. Consequently, molecular mimicry of *N. meningitidis* LOS may play a role in infection by assisting the pathogen to evade host recognition in mounting immune defense due to the presence of host-like structures in the LOS (Mandrell *et al.*, 1988; Tsai and Civin, 1991; Mandrell and Apicella, 1993; Preston *et al.*, 1996). In fact, epidemiological studies have shown that most of disease strains express L2, L3, L4, and L7 immunotypes among serogroups B and C organisms isolated in the United States and Europe (Zollinger and Mandrell, 1977; Jones *et al.*, 1992). All these immunotypes have the LNnT structure with or without sialylation in their LOSs. The LNnT sequence is also found in the LOSs of other pathogens such as *N. gonorrhoeae* and *Haemophilus ducreyi*, both sexually transmitted pathogens in humans (Mandrell *et al.*, 1988; John *et al.*, 1991; Yamasaki *et al.*, 1991; Melaugh *et al.*, 1992).

For *N. meningitidis* L1 LOS, a distal globotriaose (Galα1-4Galβ1-4Glc) instead of the LNnT structure is attached to the inner heptose-Kdo common core (Di Fabio *et al.*, 1990). This trisaccharide is presented in human blood group P^k antigen (globotriaosyl ceramide) and is also presented in a number of other human tissues including uroepithelial cells (Kasai *et al.*, 1985; Lund *et al.*, 1987). The globotriaosyl structure is also present in many strains of *H. influenzae*, *N. gonorrhoeae*, and *Moraxella (Branhamella) catarrhalis* (Virji *et al.*, 1990; Mandrell, 1992; Masoud *et al.*, 1994; Masoud *et al.*, 1997). More recently, molecular mimicry of host structures by the LPSs of *Campylobacter* and *Helicobacter* spp. has also been reported in that they have ganglioside-like structure and $Lewis^{x/y}$ blood group antigens respectively. It has been suggested that the molecular mimicry of host structures by the LOS or LPS may play a role in the pathogenesis of these pathogens in man (Moran *et al.*, 1996; Preston *et al.*, 1996). Table 6 summarizes the molecular mimicry of host structures by the LOS or LPS of the above Gram-negative pathogens.

Table 6

Molecular mimicry of host structures by bacterial LOS or LPS

Human pathogens	Glycolipids	Common structures in LOS and glycolipids
Neisseria meningitidis	Paragloboside (LNnT-ceramide) Sialylparagloboside P^k antigen (Globotriaosylceramide)	Galβ1-4GlcNAcβ1-3Galβ1-4Glc NeuNAcα2-3Galβ1-4GlcNAcβ1-3Galβ1-4Glc Galα1-4Galβ1-4Glc
Neisseria gonorrhoeae	Paragloboside P^k antigen	Galβ1-4GlcNAcβ1-3Galβ1-4Glc Galα1-4Galβ1-4Glc
Haemophilus ducreyi	Paragloboside	Galβ1-4GlcNAcβ1-3Galβ1-
Haemophilus influenaze	P^k antigen	Galα1-4Galβ1-4Glc
Moraxella catarrhalis	P^k antigen	Galα1-4Galβ1-4Glc
Campylobactor jejuni	Gangliosides (GM_1 and others)	Galβ1-3GalNAcβ1-4Galβ1- NeuNAcα2/3
Helicobacter pylori	Le^x and Le^y blood group antigens	Galβ1-4GlcNAcβ1-, Galβ1-4GlcNAcβ1- Fucα1/3 Fucα1/2 Fucα1/3

Ref: 1. Preston *et al*, Critical Rev. in Micro. **22:** 139-180 (1996)
2. Moran *et al*, J. Endotoxin Res. **3:** 521-531 (1996)

Summary

Neisseria meningitidis lipooligosaccharides (LOSs) are classified into 12 immunotypes. Most LOSs are heterogeneous in having a few components by SDS-PAGE analysis that differ antigenically and chemically. We have utilized a monoclonal antibody that recognizes lacto-*N*-neotetraose (LNnT) and the lectin, *Maackia amurensis* leukoagglutinin (MAL), which is specific for NeuNAcα2-3Galβ1-4GlcNAc trisacchride sequence to characterize the 12 *N. meningitidis* LOSs. Using the combination of ELISA, SDS-PAGE, Western blotting, and other chemical analyses, we have shown that the LNnT (Galβ1-4GlcNAcβ1-3Galβ1-4Glc) sequence was present in the 4.0-kDa LOS components of seven immunotype LOSs seen on SDS-PAGE. Six of the seven LNnT-containing LOSs also bound the MAL lectin indicating that N-acetylneuraminic acid (NeuNAc) was α2,3-linked to the LNnT sequence in the LOSs. Sialylation of the terminal Gal of LNnT-containing 4.0-kDa component caused only a slight increase in its apparent MW to 4100 on SDS-PAGE. The one LOS with the LNnT-containing component, but not MAL-binding, was from a Group A *N. meningitidis*, which does not synthesize CMP-NeuNAc, the substrate needed for LOS sialylation. Thus, it is concluded (1) a common LNnT sequence is present in seven immunotype LOSs in addition to their immunotype epitopes, and (2) NeuNAc is α2->3 linked to the terminal Gal of LNnT if a organism synthesizes CMP-NeuNAc such as Groups B and C organisms. The above conclusions are consistent with the published structures of *N. meningitidis* LOSs. The results also demonstrate that specific carbohydrate-binding lectins and monoclonal antibodies can be used as simple yet effective tools to characterize specific carbohydrate sequences in a bacterial LOS or LPS such as *N. meningitidis* LOS.

It is intriguing that *N. meningitidis* LOSs mimic certain glycosphingolipids, such as paragloboside (LNnT-ceramide) and sialylparagloboside, and some glycoproteins of the host in having LNnT and N-acetyllactosamine sequences respectively with or without α2->3 linked NeuNAc. Epidemiological studies of *N. meningitidis* suggest that the molecular mimicry of host structures by its LOS plays a role in the pathogenesis of *N. meningitidis* by helping the organism to evade host immune defenses in man. The molecular mimicry of host structures by LOS or LPS is also found in other human pathogens such as *N. gonorrhoeae*, *Haemophilus ducreyi*, *H. influenaze*, *Moraxella catarrhalis*, *Campylobacter jejuni*, and *Helicobacter pylori*.

References

Apicella,M.A. and Mandrell,R.E. (1989) Molecular mimicry as a factor in the pathogenesis of human neisserial infection: *in vitro* and *in vivo* modification of the lipooligosaccharide *Neisseria gonorrhoeae* by N-acetylneuraminic acid. *Pediatr. Infect. Dis. J.*, **8**, 901-902.

Ashton,F.E., Ryan,A., Diena,B., and Jennings,H.J. (1983) A new serogroup (L) of *Neisseria meningitidis*. *J. Clin. Microbiol.* **17**, 722-727.

Chen,W.H., Balakonis,P., and Tsai,C-M. (1995) Detection of lipopolysaccharides blotted on nylon membranes. *J. Endotoxin Res.*, **2**, 405-410.

Crocker, P.R., Kelm, S., Dubois, C., Martin, B., McWilliam, A.S., Shotton, D.M., Paulson, J.C., and Gordon, S. (1991) Purification and properties of sialoadhesin, a sialic acid-binding receptor of murine tissue macrophages. *EMBO J.* **10**, 1661-11669.

Dell,A., Azadi,P., Tiller,P., Thomas-Oates,J., Jennings,H.J., Beurret,M., and Michon,F. (1990) Analysis of oligosaccharide epitopes of meningococcal lipopolysaccharide by fast-atom-bombardment mass spectrometry. *Carbohydr. Res.*, **200**, 59-76.

Di Fabio,J.L., Michon,F., Brisson,J-R.,and Jennings,H.J. (1990) Structure of the L1 and L6 core oligosaccharide epitopes of *Neisseria meningitidis*. *Can. J. Chem.*, **68**, 1029-1034.

Estabrook,M.M., Christopher,N.C., Griffiss,J.M., Baker,C.J. and Mandrell,R.E. (1992) Sialylation and human neutrophil killing of group C *Neisseria meningitidis*. *J. Infect. Dis.*, **166**, 1079-1088.

Frasch,C.E., Zollinger,W.D., and Poolman,J.T. (1985) A proposed nomenclature for designation of serotypes within *Neisseria meningitidis*. *Rev. Infect. Dis.*, **7**, 504-510.

Frasch,C.E., (1995) Meningococcal vaccines: past, present and future, In: Cartwright,K., ed. Meningococcal disease, John Wiley & Sons, New York, pp. 245-284.

Fukuda,M., Dell,A. and Fukuda,M.N. (1984) Structure of fetal lactosaminoglycan. *J. Biol. Chem.*, **259**, 4782-4791.

Gamian,A., Beurret,M., Michon,F., Brisson,J-R. and Jennings,H.J. (1992) Structure of the L2 lipopolysaccharide core oligosaccharide of *Neisseria meningitidis*. *J. Biol.Chem.*, **267**, 922-925.

Gilbert, M., Watson, D.C., Cunningham, A-M., Jennings, M.P., Young, N.M., and Wakarchuk, W.W. (1996) Cloning of the lipooligosaccharide α-2,3-sialyltransferase from the bacterial pathogens *Neisseria meningitidis* and *Neisseria gonorrhoeae*. *J. Biol. Chem.*, **271**, 28271-28276.

Gill,M.J., McQuillen,D.P., Van Putten,J.P.M., Wetzler,L.M., Bramley,J., Crooke,H., Parsons,N.J., Cole,J.A. and Smith,H. (1996) Functional characterization of sialyltransferase-Deficient mutant of *Neisseria gonorrhoeae*. *Infect. and Immun.*, **64**, 3374-3378.

Gotschlich,E.C. (1990) *Neisseriae*. In Microbiology (Eds Davis,B.D., Dulbecco,R., Eisen,H.N., and Ginsberg,H.S.) 4th edn. J.B. Lippincott, Philadelphia, PA, pp. 551-560.

Gu,X.X., and Tsai,C.M. (1991) Purification of rough-type lipopolysaccharides of *Neisseria meningitidis* from cells and outer membrane vesicles in spent media. *Anal. Biochem.*, **196**, 311-318.

Hakomori,S. (1981) Blood groups ABH and Ii antigens of human erythrocytes. Chemistry, polymorphism, and their development change. *Semin. Hematol.*, **18**, 39-52.

Hammerschmidt,S., Birkholz,C., Zahringer,U., et al., (1994) Contribution of genes from the capsule gene complex (cps) to lipooligosaccharide biosynthesis and serum resistance in *Neisseria meningitidis*. *Mol. Microbiol.*, **11**, 885-896.

Jennings,H.J. (1983) Capsular polysaccharides as human vaccines. *Adv Carbohydr Chem Biocem*, **41**, 155-208.

Jennings,H.J., Beurret,M., Gamian,A. and Michon,F. (1987) Structure and immunochemistry of meningococcal lipopolysaccharides. *Antonie van Leeuwenhoek*, **53**, 519-522.

Jennings,H.J., Bhattacharjee,A.K., Kenne,L., Kenny,C.P. and Calver,G. (1980) The R-type lipopolysaccharides of *Neisseria meningitidis*. *Can. J. Biochem.*, **58**, 128-136.

Jennings,H.J., Johnson,K.G. and Kenne,L. (1983) The structure of an R-type oligosaccharide core obtained from some lipopolysaccharides of *Neisseria meningitidis*. *Carbohydr. Res.*, **121**, 233-241

Jennings,H.J., Lugowski,C. and Ashton,F.E. (1984) Conjugation of meningococcal lipopolysaccharide R-type oligosaccharides to tetanus toxoid as route to a potential vaccine against group B *Neisseria meningitidis*. *Infect. Immun.*, **43**, 407-412.

John,C.M., Griffiss,J.M., Apicella,M.A., Mandrell,R.E. and Gibson,B.W. (1991) The structural basis for pyocin-resistance in *Neisseria gonorrhoeae* lipooligosaccharides. *J. Biol. Chem.*, **266**,19303-19311.

Jones,D. (1995) Epidemiology of meningococcal disease in Europe and the USA in Meningococcal disease (ed Cartwright,K.) Johns Wiley & Sons, New York, NY, pp. 147-158

Jones,D M., Borrow,R., Fox,A.J., Gray,S., Cartwright,K.A. and Poolman,J.T. (1992) The lipooligosaccharide immunotype as a virulence determinant in *Neisseria meningitidis*. *Microb. Pathogen.*, **13**, 219-224.

Kasai, K., Galton, J., Terasaki, P.I., Wakisaka, A., Kawahara, M., Root, T., and Hakomori, S.I. (1985) Tissue distribution of the P^k antigen as determined by a monoclonal antibody. *J. Immunogenet*, **12**, 213-220.

Kelm, S., Schauer, R., Manuguerra, J.-C., Gross, H.-J., and Crocker, P.R. (1994) Modifications of cell surface sialic acids modulate cell adhesion mediated by sialoadhesin and CD22. *Glycoconj. J.* **11**, 576-585.

Kim,J.J., Zhou,D., Mandrell,R.E. and Griffiss,J.M. (1992) Effect of exogenous sialylation of the lipooligosaccharide of *Neisseria gonorrhoeae* on opsonophagocytosis. *Infect. Immun.*, **60**, 4439-4442.

Kim,J.J., Phillips,N.J., Gibson,B.W., Griffiss,J.M. and Yamasaki,R. (1994) Meningococcal group A lipooligosaccharides (LOS): preliminary structural studies and characterization of serotype-associated and conserved LOS epitopes. *Infect. Immun.*, **62**, 1566-1575.

Knibbs,R.N., Goldstein,I.J., Ratcliffe,R.M. and Shibuya,N. (1991) Characterization of the carbohydrate binding specificity of the leukoagglutinating lectin from *Maackia amurensis*. *J. Biol. Chem.*, **266**, 83-88.

Kogan,G., Uhrin,D., Brisson, J.-R., and Jennings, H. J. (1997) Structural basis of *Neisseria meningitidis* immunotypes including the L4 and L7 immunotypes. *Carbohydr. Res.*, **298**, 191-199.

Kulshin,V.A., Zahringer,U., Lindner,B., Frasch,C.E., Tsai,C.M., Dmitriev,B.A. and Rietschel,E.T. (1992) Structural characterization of the lipid A component of pathogenic *Neisseria meningitidis*. J. Bacteriol., **174**, 1793-1800.

Lund, B., Lindberg, F., Marklund, B.I., Normark, S. (1987) the PapG protein is the α-D-galactopyranosyl(1->4)β-D-galactopyranose-binding adhesin of uropathogenic *Escherchia coli*. *Proc. Nat'l. Acad. Sci. USA*, **84**, 5898-5902.

Macher,B.A. and Klock,J.C. (1980) Isolation and Chemical Characterization of Neutral Glycosphingolipids of Human Neutrophils. *J. Biol. Chem.*, **255**, 2092-2096.

Mandrell,R.E (1992) Further antigenic similarities of *Neisseria gonorrhoeae* lipooligosaccharides and human glycosphingolipids. *Infect. Immun.*, **60**, 3017-3020.

Mandrell,R.E., Griffiss,J.M. and Macher,B.A. (1988) Lipooligosaccharides (LOS) of *Neisseria gonorrhoeae* and *Neisseria meningitidis* have components that are

immunochemically similar to precursors of human blood group antigens. *J. Exp. Med.*, **168**, 107-126.
Mandrell,R.E. and Apicella,M.A. (1993) Lipo-oligosaccharides (LOS) of Mucosal Pathogens: Molecular Mimicry and Host-Modification of LOS. *Immunobiol.*, **187**, 382-402.
Mandrell,R.E., Griffiss,J.M., Smith,H. and Cole,J.A. (1993) Distribution of a lipooligosaccharide-specific sialyltransferase in pathogenic and non-pathogenic *Neisseria. Micro. Pathogen.*, **14**,315-327.
Mandrell,R.E., Kim,J.J., John,C.M., Gibon,B.W., Sugai,J.V., Apicella,M.A., Griffiss,J.M. and Yamasaki,R. (1991) Endogenous sialylation of the lipooligosaccharides of *Neisseria meningitidis*. *J. Bacteriol.*, **173**, 2823-2832.
Mandrell,R.E. and Zollinger,W.D. (1977) Lipopolysaccharide serotyping of *Neisseria meningitidis* by hemagglutination inhibition. *Infect. Immun.*, **16**, 471-475.
Masoud, H., Perry, M.B., Brisson, J-R., Uhrin, D., and Richards, J.C. (1994) Structural elucidation of the backbone oligosaccharide for the lipopolysaccharide of *Moraxella catarrhalis* serotype. *Can. J. Chem.*, **72**, 1466-1477.
Masoud, H., Moxon, E.R., Martin, A., Krajcarski, D.T., and Richards, J.C. (1997) Structural elucidation of the vaiable and conserved lipopolysaccharide oligosaccharide epitopes expressed by *Haemophilus influenzae* serotype b strain Eagan. *Biochemistry*, **36**, 2091-2103.
Melaugh,W., Phillips,N.J., Camagnari,A.A., Karalus,R. and Gibson,B.W. (1992) Partial characterization of major lipooligosaccharide from a strain of Haemophilus ducreyi, the causative agent of chancroid, a genital ulcer disease. *J. Biol. Chem.*, **267**, 13434-13439.
Michon,F., Beurret,M., Gamian,A., Brisson,J-R., and Jennings,J.H. (1990) Structure of the L5 lipopolysaccharide core oligosaccharides of *Neisseria meningitidis*. *J. Biol. Chem.*, **265**, 7243-7247.
Miller, R.L., Collawn, J.F., amd Fish W.W. (1982) Purification and macromolecular properties of a sialic acid-specific lectin from slug *Limax flavus*. *J. Biol. Chem.*, **257**, 7574-7580.
Moran,A.P., Appelmelk,B.J., and Aspinall,G.O. (1996) Molecular mimicry of host structures by lipopolysaccharides of *Campylobacter* and *Helicobacter* spp.: implications in pathogenesis. *J. Endotoxin Res.*, **3**, 521-531.
Moran,E.E., Brandt,B.L. and Zollinger,W.D. (1994) Expression of the L8 lipopolysaccharide determinant increases the sensitivity of *Neisseria meningitidis* to serum bactericidal activity. *Infect. Immun.*, **62**, 5290-5295.
Parsons,N.J., Andrade,J.R., Patel,P.V., Cole,J.A. and Smith,H. (1989) Sialylation of lipopolysaccharide and loss of absorption of bactericidal antibody during conversion of gonococci to serum resistance by cytidine 5'-monophospho-N-acetyl neuraminic acid. *Microb. Pathog.*, **7**, 63-72.
Pavliak,V., Brisson,J-R., Michon,F., Uhrin,D. and Jennings,H.J. (1993) Structure of the sialylated L3 lipopolysaccharide of *Neisseria meningitidis*. *J. Biol. Chem.*, **268**, 14146-14152.
Peltola,H. (1983) Meningococcal disease: still with us. *Rev. Infect. Dis.*, **5**, 71-91.
Preston,A., Mandrell,R.E., Gibson,B.W., and Apicella,M.A. (1996) The lipooligosaccharides of pathogenic gram-negative bacteria. *Critical Reviews in Micro.*, **22(3)**, 139-180
Quakyi,E., Hochstein, H.D., and Tsai,C.-M. (1997) Modulation of the biological activities of meningococcal endotoxins by association with outer membrane proteins is not inevitably linked to toxicity. *Infect. Immun.*, **65**, 1972-1979.
Rest,R.F. and Frangipane,J.V., (1992) Growth of *Neisseria gonorrhoeae* in CMP-N-acetylneuraminic acid inhibits nonopsonic (opacity-associated outer membrane protein-mediated) interactions with human neutrophils. *Infect. Immun.*, **60**, 989-997.

Rest,R.F., and Mandrell,R.E. (1995) *Neisseria* sialyltransferases and their role in pathogenesis. *Microb. Pathog.* **19**, 379-390.
Schneider,H., Hale,T.L., Zollinger,W.D., Seid,R.C., Hammack,C.A. and Griffiss,J.M. (1984) Heterogeneity of molecular size and antigenic expression within the lipooligosaccharides of individual strains of *Neisseria gonorrhoeae* and *Neisseria meningitidis*. *Infect. Immun.*, **45**, 544-549.
Schneider,H., Griffiss,J.M., Boslego,J.W., Hitchcock,P.J., Zahos,K.M., and Apicella,M.A. (1991) Expression of paragloboside-like lipooligosaccharides may be a necessary component of gonococcal pathogenesis in men. *J. Exp. Med.*, **174**, 1601-1605.
Shibuya,N., Goldstein,I.J., Broekaert,W.F., Nsimba-Lubaki,M., Peeters,B. and Peumans,W.J. (1987) The elderberry (*Sambucus nigra L.*) bark lectin recognizes the NeuNAc(α2-6)Gal/GalNAc sequence. *J. Biol. Chem.*, **262**, 1596-1601.
Smith,H., Parsons,N.J., and Cole,J.A., (1995) Sialylation of neisserial lipopolysaccharide: a major influence on pathogenicity. *Microb. Pathog.*, **19**, 365-377.
Spitalnik,S.L., Schwartz,J.F., Magnani,J.L., Roberts,D.D., Spitalnik,P.F., Civin,C.I. and Ginsberg,V. (1985) Anti-My-28, an antigranulocyte mouse monoclonal antibody, binds to a sugar sequence in lacto-N-neotetraose. *Blood.*, **66**, 319-326.
Tikhomirov, E., Santamaria, M., and Esteves, K. (1997) Meningococcal disease: public health burden and control. *World Health Stat. Quart.* **50**, 170-177.
Tsai,C.-M. and Frasch,C.E. (1982) A sensitive silver stain for detecting lipopolysaccharides in polyacrylamide gels. *Anal. Biochem.*, **119**, 115-119.
Tsai,C.-M., Boykins,R. and Frasch,C.E. (1983) Heterogeneity and variation among *Neisseria meningitidis* lipopolysaccharides. *J. Bacteriol.*, **155**, 498-504.
Tsai,C.-M., Mocca,L.F. and Frasch,C.E. (1987) Immunotype epitopes of *Neisseria meningitidis* lipooligosaccharide type 1 through 8. *Infect. Immun.*, **55**, 1652-1656.
Tsai,C.-M., and Civin,C.V. 1991. Eight lipooligosaccharides of *Neisseria meningitidis* react with monoclonal antibody which binds lacto-*N*-neotetraose (Galβ1-4GlcNAcβ1-3Galβ1-4Glc). *Infect. Immun.*, **59**, 3604-3609.
Tsai,C.M., Chen,W. and Balakonis,P.A. (1995) *Neisseria meningitidis* lipooligosaccharides mimic glycolipids and glycoproteins in having NeuNAcα2-3Galβ1-4GlcNAc sequence. *Glycoconjugate J.*, **12**, 562.
Tsai,C.M., Chen,W. and Balakonis,P.A. (1998) Characterization of terminal NeuNAcα2-3Galβ1-4GlcNAc sequence in lipooligosaccharides of *Neisseria meningitidis*. *Glycobiology*, **8**, 359-365.
Verheul,A.F.M., Snippe,H, and Poolman,J.T., (1993) Meningococcal lipopolysaccharides: virulence factor and potential vaccine component. *Microbiol Rev*, **57**, 34-49.
Virji, M., Weiner, J.N., Lindberg, A.A., and Moxon, E.R., (1990) Antigenic similarities in lipopolysaccharides of *Haemophilus* and *Neisseria* and expression of a digalactoside structure also present on human cells. *Microb. Pathog.*, **9**, 441-450.
Vogel, U., Weinberger, A., Frank, R., Muller, A., Kohl, J., Atkinson, J.P., and Frosch, M. (1997) Complement factor C3 deposition and serum resistance in isogenic capsule and lipooligosaccharide sialic acid mutants of serogroup B *Neisseria meningitidis*. *Infect. Immun.*, **65**, 4022-4029.
Wakarchuk, W.W., Gilbert, M., Martin, A., Wu,Y., Brisson, J-R., Thibault, P., and Richards, J.C. (1998) Structure of an α-2,6-sialylated lipooligosaccharide from *Neisseria meningitidis* immunotype L1. *Eur. J. Biochem.*, **254**, 626-633.
Wakarchuk, W.W., Martin, A., Jennings, M.P., Moxon, E.R., and Richards, J.C. (1996) Functional relationships of the genetic locus encoding the glycosyltransferase enzymes involved in expression of the lacto-N-neotetraose terminal lipopolysaccharide structure in *Neisseria meningitidis*. *J. Biol. Chem.*, **271**, 19166-19173.

Wang,W.-C., and Cummings,R.D., (1988) The immobilized leukoagglutinin from the seeds of *Maackia amurensis* binds with high affinity to complex-type Asn-linked oligosaccharides containing terminal sialic acid-linked α-2,3 to penultimate galactose residues. *J. Biol. Chem.*, **263**, 4576-4585.

Wetzler,L.M., Barry,K., Blake,M.S. and Gotschlich,E.C. (1992) Gonococcal lipooligosaccharide sialylation prevents complement-dependent killing by immune sera. *Infect. Immun.*, **60**, 39-43.

Yamasaki,R., Bacon,B.E., Nasholds,W., Schneider,H., and Griffiss,J.M. (1991) Structural determination of oligosaccharides derived from lipooligosaccharide of *Neisseria gonorrhoeae* F62 by chemical, enzymatic, and two-dimensional NMR methods. *Biochem.*, **30**, 10566-10575.

Yamasaki,R., Griffiss,J.M., Quinn,K.P. and Mandrell,R.E. (1993) Neuraminic acid is α2-3 linked in the lipooligosaccharide of *Neisseria meningitidis* serogroup B strain 6275. *J Bacteriol.*, **175**, 4565-4568.

York,W.S., Darvill,A.G., McNeil,M., Stevenson,T.T., and Albersheim,P. (1986) Isolation and characterization of plant cell walls and cell wall components. *Methods Enzymol.*, **118**, 3-40

Zollinger,W.D. and Mandrell,R.E. (1977) Outer membrane protein and lipopolysaccharide serotyping of *Neisseria meningitidis* by inhibition of a solid-phase radioimmunoassay. *Infect. Immun.*, **18**, 424-433.

Zollinger,W.D. and Mandrell,R.E. (1980) Type-specific antigens of group A *Neisseria meningitidis* : lipopolysaccharide and heat-modifiable outer membrane proteins. *Infect. Immun.*, **28**, 451-458.

Zollinger,W.D., (1997) New and improved vaccines against meningococcal disease. In: Levine, M.M., Woodrow, G.C., eds. New Generation Vaccines, 2nd edition. Marcel Dekker, Inc. New York, pp.469-488.

NMR AND MOLECULAR MODELING OF COMPLEX CARBOHYDRATES AND CARBOHYDRATE-PROTEIN INTERACTIONS. APPLICATIONS TO ANTI-BACTERIA VACCINES[1]

Jean-Robert Brisson and Harold J. Jennings

Institute for Biological Sciences, National Research Council of Canada, Ottawa, Ontario, Canada K1A 0R6

Group B *Neisseria meningitidis* and *Escherichia coli* K1 are major causes of meningitis. Both bacteria have identical capsular polysaccharide composed of α(2-8)-linked sialic acid residues. Although, vaccine development has been hampered by the poor immunogenicity of the group B meningococcal polysaccharide in humans, by *N*-propionylation of its sialic acid residues, antibodies are produced which are both bactericidal and protective. The α(2-8)-linked sialic acid homopolymer has many interesting immunological properties. The presence of a conformational epitope was hypothesized to explain the inability of oligosaccharides up to five residues to inhibit the binding of the group B polysaccharide to a homologous horse antibody[1]. Later, it was established that an unusually large decasaccharide was required to bind or inhibit group B polysaccharide specific antibodies. This is in contrast with a conventional sequential oligomer where only a maximum of 6 or 7 residues are necessary for binding [2,3].

For vaccine development the binding specificity of antibodies specific for the group B meningococcal polysaccharide and its *N*-propionylated derivatives must thus be understood in terms of their three-dimensional structure. The conformational properties of the α(2-8)NeuAc polymer, oligosaccharides, and *N*-acyl derivatives were evaluated by nuclear magnetic resonance (NMR) studies at 600 MHz, in conjunction with molecular modeling. The interpretation of nuclear Overhauser enhancements (NOE) which are highly dependent on conformation was based on a model which allows for the inherent flexibility of the molecule. The biological implications were rationalized in terms of a model where the random coil nature of a polysaccharide can be described by the presence of local helices.

[1] Corresponding author: Jean-Robert Brisson. Fax: (613) 952-9092. E-mail: jean-robert.brisson@nrc.ca

The Molecular Immunology of Complex Carbohydrates–2.
Edited by Albert M. Wu, Kluwer Academic/Plenum Publishers, 2001.

The capsule of the group B meningococcal polysaccharide (GBMP) is composed of a homopolymer of sialic acid, (→8NeuAcα2→). It order to determine the conformational properties of GBMP, the linkage conformation ϕ, ψ, ω_7, and ω_8, and the orientation of the pendant groups must be determined (Figure 1).

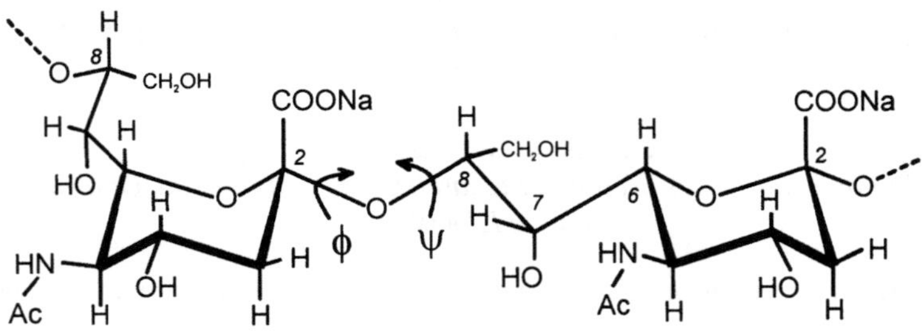

Figure 1. Structure of αNeuAc(2-8)αNeuAc (*b-a*) with ϕ = (*b*O6-*b*C2-*b*O2-*a*C8), ψ = (*b*C2-*b*O2-*a*C8-*a*C7), ω_8 = (O7-C7-C8-O8) and ω_7 = (O6-C6-C7-O7) and $\omega_7 = \omega_8 = 60°$. The orientation of the pendant groups for C1 and C9 is defined by ω_1 = (O6-C2-C1-O1) and ω_9 = (O8-C8-C9-O9).

Various NMR, circular dichroism, and molecular modeling studies all indicate that due to the high degree of flexibility about the glycosidic bonds and exocyclic bonds, GBMP exists as random coil in solution [4-7]. NOE measurements (Figure 2) which are dependent on interproton distances cannot be used to determine a unique conformation in solution due to high degree of flexibility of the molecule.

NOE simulations can only be done with motional averaging taking into account all possible conformations. Also, due to the high degree of flexibility of the molecule and systematic errors involved, interproton distances cannot be extracted from the NOE data of sugars, since they are meaningless[8,9]. For these reasons, NOE simulations were done taking into account the flexibility of the molecule using a potential suitable for carbohydrates (Figure 3), which adequately described the conformational properties of GBMP. Once the conformational properties of the polysaccharide in solution were described adequately, a unique conformer most suitable for binding to an antibody was selected using serological data in order to justify the choice of the epitope.

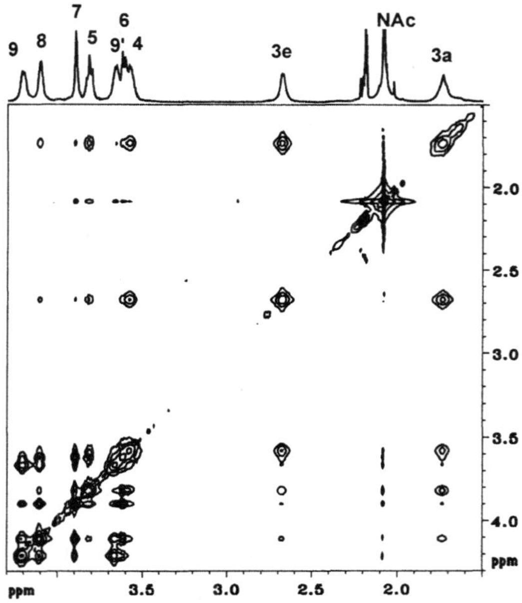

Figure 2. 2D NOE spectrum at 600 MHz of the →8NeuAcα2→ polysaccharide. The 1D spectrum is shown on top.

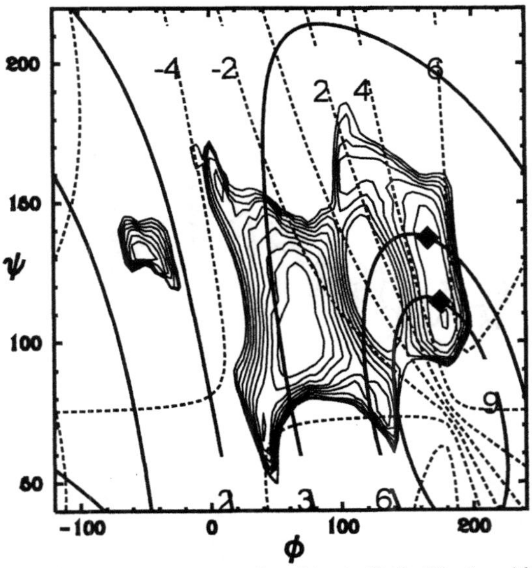

Figure 3. Potential energy contour map for αNeuAc(2-8)αNeuAc with (ω_7, ω_8) = (55°, 70°) and ω_1 and ω_9 adopting minimum energy conformers. With respect to the global minimum the 1 to 10 kcal/mol levels are drawn. The helical parameters, *n* (solid lines) and *h* (dotted line) are also displayed. The two helices with φ near 180°, located in the same energy well, that have (*n*,*h*) = (6, 5.3 Å) and (9, 5.8 Å) are marked. *n* is the number of residues per turn of the helix and *h* is the translation of the monosaccharide along the helix axis.

Upon binding, a well-defined conformation is selected. Binding can involve a short or an extended epitope. A short epitope (< 6 residues) is not affected by the length of the polysaccharide. An extended epitope (>9 residues) is dependent on the length of the PS and is due to extended conformational features present in the internal residues. For an extended epitope, the spatial organization of the polymer becomes important. For polysaccharide conformations, the spatial organization of the sugar chains can be defined by consecutive fragments which can have well defined helical parameters. For GBMP, a wide range of helical structures are probable (Figure 3). Serological data, X-ray crystallographic, thermodynamic and NMR evidence are consistent with the presence of an extended binding site of at least 9 residues in length for an antibody specific for (→8NeuAcα2→) [10]. Initial NMR studies revealed differences in conformation between the trisaccharide and polysaccharide as being the basis for the immunological properties of GBMP. Further NMR and molecular modeling studies performed for (→8NeuAcα2→) postulated that an extended helix with 9 turns per residue was the epitope (Figure 4). Xray crystallographic studies further defined this epitope has being an extended helix with similar conformational features but with 6 turns per residue (Figure 4). The postulated epitope for GBMP is thus made up of a high-order extended helix with a large pitch suitable for binding to a groove on the surface of the antibody. In order to elicit an antibacterial response to GBMP, its polysaccharide must also be *N*-propionylated. The conformational properties of these polysaccharides were not affected by replacing the NAc group with a NPr group, since chemical shifts and NOEs were similar in all related polysaccharides [11].

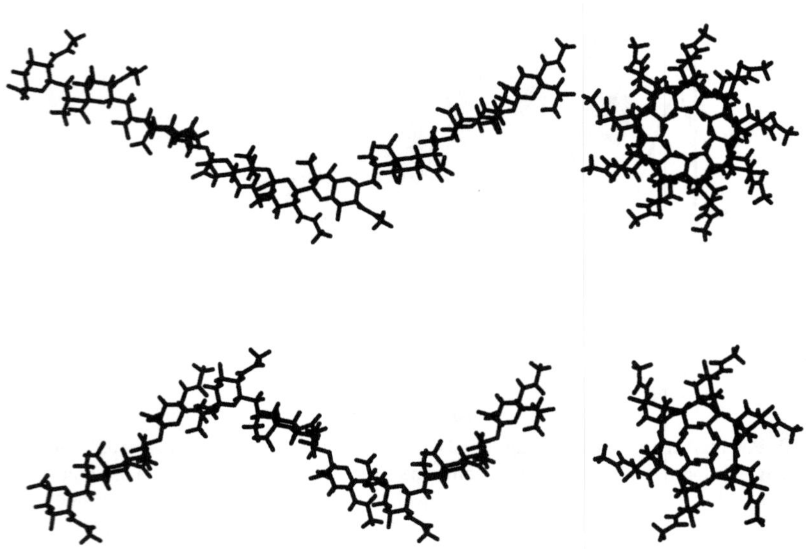

Figure 4. Side and top view for the $n = 6$ and $n = 9$ helices of (-8NeuAcα2-)$_9$ from Figure 3.

Group B *Streptococcus* is also a major cause of neonatal sepsis and meningitis. The conformational epitope of the type III group B *Streptococcus* capsular polysaccharide (GBSP III) exhibits unique properties which can be ascribed to the presence of sialic acid in its structure and the requirement for an extended binding site (decasaccharide) due to the length dependency of binding. The influence of sialic acid on the conformational properties of GBSP III was examined by performing conformational analysis on desialylated GBSP III which is identical to the polysaccharide of *Streptococcus pneumoniae* type 14, and also on oxidized and reduced GBSP III [12]. As compared to GBMP, the conformational features of GBSP III are much more complex (Figure 5). The conformational properties of GBSP III depend on the linkage conformation ϕ, ψ for the five linkages in the repeat unit, ω for the (1-6) linkage, and the orientation of the pendant groups.

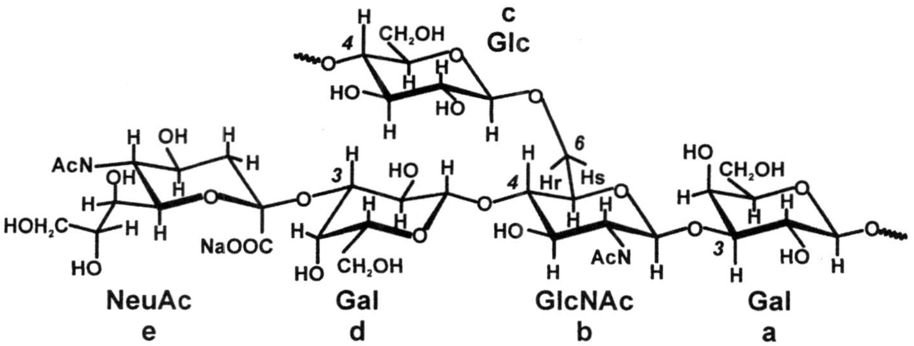

Figure 5. Structure for the repeat unit of GBSP III denoted as **[c(ed)ba]**$_n$ where n denotes the number of repeat units. The desialylated repeat units have the same structure as the *Streptococcus pneumoniae* type 14 polysaccharide.

Conformational changes were gauged by ^1H and ^{13}C chemical shift analysis, NOE, 1D selective TOCSY-NOESY experiments, J_{HH} and J_{CH} variations, and NOE of OH resonances. The NOE spectra were more complex due to the large number of resonances and consequent overlap of intra-residue and inter-residue NOEs of the anomeric resonances in the 3.5 to 4.0 ppm region (Figure 6). Also, the complexity of the NOE spectra will increase the already large error associated with NOE measurements [13]. Due to the high degree of flexibility, overlap of resonances, and inaccuracy of distances extrapolated from NOE data of sugars, interproton distances from NOE data cannot be used as constraints for molecular modeling of sugars.

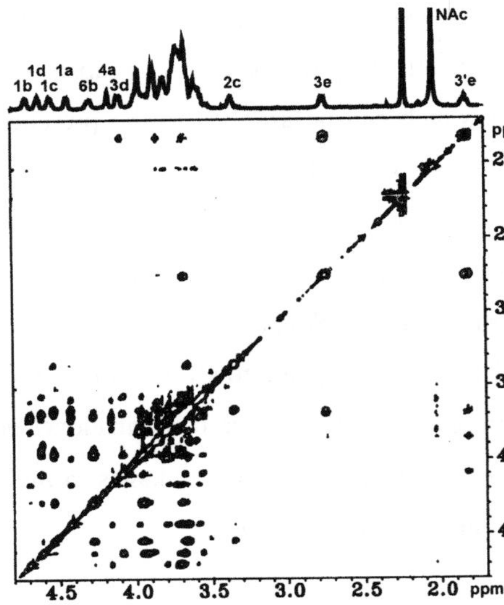

Figure 6. 2D NOE spectrum at 600 MHz of GBSP III with 14 repeat units **[c(ed)ba]**$_{14}$. The 1D spectrum is shown on top. The anomeric resonances are labelled 1a, 1b, 1c and 1d.

Also, due to the complexity of the repeating unit of GBSP III and the *S. pneumoniae* type 14 polysaccharide, unrestrained molecular dynamics with parameters suitable for carbohydrates were thus used to assess static and dynamic conformational models, simulate the observable NMR parameters, and calculate helical parameters (Figure 7). Changes in mobility and degree of flexibility were examined by ^{13}C T_1 and T_2 measurements. As expected, GBSP III and the *S. pneumoniae* type 14 polysaccharide are highly flexible random coils in solution and no unique conformation can adequately described their solution conformational properties.

From the calculation of helical parameters (Figure 7), GBSP III was found to be capable of forming extended helices with a large pitch suitable for binding to a groove type of surface on the antibody. Hence, as for GBMP, the length dependence of the conformational epitope for GBSP III could be explained by its location on extended helices within the random coil structure of the polysaccharide.

Interactions of sialic acid with the backbone of the PS were also found to be important in defining the conformational epitope of GBSP III. Removal of sialic acid affected slightly the conformation of the backbone, mainly at the branching point. Since the backbone for GBSP III and the *S. pneumoniae* type 14 polysaccharide is the same (Figure 5), the *S. pneumoniae* type 14 polysaccharide is also capable of forming extended helices with a large pitch suitable for binding to a groove type of surface of the antibody (Figure 7). Hence, as expected, the binding of antibodies to *S. pneumoniae* type 14 polysaccharide also depends on the length of the polysaccharide, consistent with the presence of a conformational epitope [14,15].

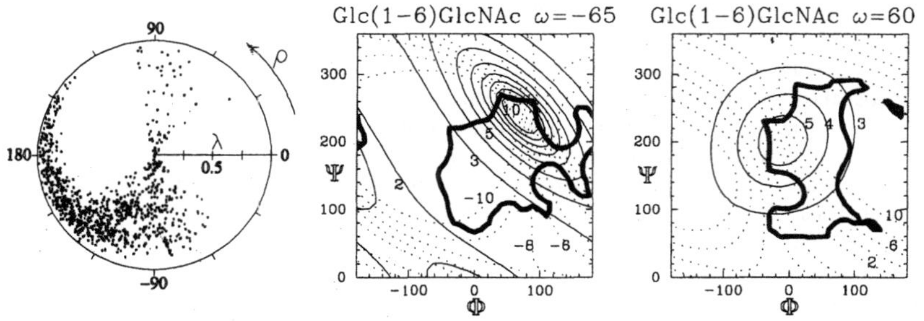

Figure 7. Polar (ρ,λ) representation of the distribution of (*n,h*) helical parameters calculated from the MD trajectory for **[ac(ed)ba]$_1$** with *n* being the number of residues per turn, ρ = 360°/*n*, and λ is the percentage of maximum elongation (0 to 1). The helical parameters *n* (solid lines) and *h* (dotted lines) (Å) are also plotted as a function (Φ,Ψ) for the (1-6) linkage in **[bac]$_2$** with ω, (O5, C5, C6, O6) = -65° and ω = 60°, along with the 6 kcal/mole contour level above the global energy minimum.

SUMMARY

In order to characterize the conformational epitope of the group B meningococcal polysaccharide and of the type III group B *Streptococcus* capsular polysaccharide NMR measurements were done on a wide variety of native and modified polysaccharides and oligosaccharides. Since these saccharides are highly mobile and exist as random coils in solution, the analysis of the NMR data and molecular modeling was done to take into account this inherent flexibility. The conformational model of extended high-order helices being selected upon binding to a protein, although still hypothetical at this stage, has proven useful in explaining the serology for the conformational epitopes for polysaccharides of group B *Neisseria meningitidis*, group B *Streptococcus* type III and *Streptococcus pneumoniae* type 14.

REFERENCES

1. H.J. Jennings, E. Katzenellenbogen, C. Lugowski, F. Michon, R. Roy, and D.L. Kasper, Structure, conformation, and immunology of sialic acid- containing polysaccharides of human pathogenic bacteria, *Pure Appl.Chem.* 56:893-905 (1984).
2. H.J. Jennings, J.R. Brisson, M. Kulakowska, and F. Michon, Polysialic acid vaccines against meningitis caused by *Neisseria meningitidis* and *Escherichia coli* K1, in: "Polysialic Acid", J. Roth et al., eds., Birkhauser Verlag, Basel, pp. 25-38 (1993).
3. R.A. Pon, L. Lussier, Q.L. Yang, and H.J. Jennings, *N*-propionylated group B meningococcal polysaccharide mimics a unique bactericidal capsular epitope in group B *Neisseria meningitidis*, *J.Exp.Med.* 185:1929-1938 (1997).
4. J.R. Brisson, H. Baumann, A. Imberty, S. Perez, and H.J. Jennings, Helical epitope of the group B meningococcal α(2-8)-linked sialic acid polysaccharide, *Biochem.* 31:4996-5004 (1992).
5. R.M. Venable, F. Bizik, T.J. Henderson, and W. Egan, Molecular dynamics simulations of an α-(2-8)-linked sialic acid tetramer in vacuum and solvent, *J.Mol.Struct.(Theochem)* 395:375-388 (1997).
6. S. Bystricky, V. Pavliak, and S.C. Szu, Characterization of colominic acid by circular dichroism and viscosity analysis, *Biophys.Chem.* 63:147-152 (1997).
7. M.K. Koenig and R.J. McLean, Influence of metal ions and temperature on the conformation of *Escherichia coli* K1 capsular polysaccharide, *BioMetals* 12:47-52 (1999).
8. D.A. Cumming and J.P. Carver, Virtual and solution conformations of oligosaccharides, *Biochem.* 26:6664-6676 (1987).
9. G. Widmalm, R.A. Byrd, and W. Egan, A conformational study of α-L-Rhap-(1-2)-α-L-Rhap-(1-OMe) by NMR nuclear Overhauser effect spectroscopy (NOESY) and molecular dynamics calculations, *Carbohydr.Res.* 229:195-211 (1992).
10. S.V. Evans, B.W. Sigurskjold, H.J. Jennings, J.R. Brisson, R. To, W.C. Tse, E. Altman, M. Frosch, C. Weisgerber, H.D. Kratzin, S. Klebert, M. Vaesen, D. Bitter-Suermann, D.R. Rose, N.M. Young, and D.R. Bundle, Evidence for the extended helical nature of polysaccharide epitopes. The 2.8 Å resolution structure and thermodynamics of ligand binding of an antigen binding fragment specific of α(2-8)- polysialic acid, *Biochem.* 34:6737-6744 (1995).
11. H. Baumann, J.R. Brisson, F. Michon, R. Pon, and H.J. Jennings, Comparison of the conformation of the epitope of α(2-8) polysialic acid with its reduced and *N*-acyl derivatives, *Biochem.* 32:4007-4013 (1993).
12. J.R. Brisson, S. Uhrinova, R.J. Woods, M. van der Zwan, H.C. Jarrell, L.C. Paoletti, D.L. Kasper, and H.J. Jennings, NMR and molecular dynamics studies of the type III group B *Streptococcus* capsular polysaccharide and derivatives, *Biochem.* 36:3278-3292 (1997).
13. G.H. Weiss, J.E. Kiefer, and J.A. Ferretti, Accuracy and precision in the estimation of internuclear distances for structure determination, *J.Magn.Reson.* 97:227-234 (1992).
14. C.A. Laferrière, R.K. Sood, J.-M. de Muys, F. Michon, and H.J. Jennings, *Streptococcus pneumoniae* type 14 polysaccharide-conjugate vaccines: Length stabilization of opsonophagocytic conformational polysaccharide epitopes., *Infect.Immun.* 66:2441-2446 (1998).
15. M.R. Wessels and D.L. Kasper, Antibody recognition of the type 14 pneumococcal capsule, *J.Exp.Med.* 169:2121-2131 (1989).

A GUIDE TO THE CARBOHYDRATE SPECIFICITIES OF APPLIED LECTINS-2[3-5]

(Updated in 2000)

Albert M. Wu[1], Shuh-chyung Song[2], Ming-sung Tsai and Anthony Herp

Glyco-Immunochemistry Research Laboratory, Chang-Gung Medical College, Kwei-San, Tao-Yuan 33332, Taiwan

The lectins, that can be used as tools to study glycobiological systems are defined as applied lectins (1-14). They are easily obtained , stable and have their own binding specificity extending beyond the monosaccharide(1-4,15-19). Their biochemical application has been reviewed extensively by Goldstein and Poretz (19), Sharon and Lis (8) and Goldstein *et al.* (11). Hundreds of lectins are used as applied lectins. Thus, organizing and grouping binding properties of these lectins should facilitate the selection of lectins as structural probes for studying glycans as well as the interpretation, distribution and properties of the carbohydrate chains on the cell surface. From the information provided by inhibition assays and binding properties, the carbohydrate specificities of applied lectins are classified into six groups according to their specificities to monosaccharides. The subgroups are based on lectin affinities to GalNAcα1→*O* to Ser(Thr) of the peptide chain, disaccharides, trisaccharides and, the number and the location of LFucα1→linked to oligosaccharides; all of these structures are found mainly in soluble glycoproteins and as cell surface glycoconjugates in mammals (1-4,18). Reviews concerning coding and classification of DGal, GalNAc and Galβ1→3/4GlcNAc specificities of applied lectins are given in Secion I-4 of this proceeding (18) together with references 1-4 . A scheme of the classification is shown as follows.

---

[1] Corresponding author (Fax: 886-3-328-6456; e-mail: amwu@mail.cgu.edu.tw); [2] Present address: Depatrment of Pathology and Lab. Medicine, University of Rochester Medical Center,601, Elmwood Avenue, Box 626 Rochester, New York, 14642, USA; [3] This work was supported by grants from the Chang-Gung Medical Research plan (CMRP No.676) Kwei-san ,Tao-Yuan, Taiwan, and the National Science Council (NSC 89-2320-B-182-003) Taipei, Taiwan; [4] The content of this article is an updated form of "A guide for carbohydrate specificities of lectins" published in "The molecular immunology of complex carbohydrates" 819-847, 1988(17). Adv. Exp. Med. & Biol. 228; [5] Abbreviations: Gal, D-galactopyanose; Glc, D-glucopyranose; Man, D-mannopyranose; LFuc or Fuc, L-fucopyranose; DFuc, D-fucopyranose; GalNAc, 2- acetamido-2-deoxy-D-galactopyranose; GlcNAc,2-acetamido-2-deoxy-D-glucopyranose; **C**, Chitin disaccharide, GlcNAcβ1→4GlcNAc, (GlcNAcβ1→4)n, repeat unit of GlcNAcβ1→4; NeuNAc or Neu5Ac, *N*-acetylneuraminic acid; NeuGly or NeuGc, *N*-glycolylneuraminic acid; R, carbohydrate residue; Melibiose, Galα1→6Glc; Raffinose, Galα1→6Glcβ1→2Fru$_f$; Stachyose, Galα1→6Galα1→6Glcβ1→2Fru$_f$; m, multivalent; c, cluster; **A$_h$**, GalNAcα1→2(LFuc α1→3)Gal; Lectin, agglutinin.

I. GalNAc-Specific agglutinins

1. **F/A**, GalNAcα1→3GalNAc (Forssman) and GalNAcα1→3Gal (Blood group A determinant disaccharide) specific agglutinins-*Dolichos biflorus* (DBA), and *Helix pomatia* (HPA).
2. **A**, GalNAcα1→3Gal (Blood Group A determinant disaccharide)- specific agglutinins- Soy bean (SBA), and Lima bean (LBA).
3. **Tn**, GalNAcα1→Ser/Thr specific agglutinins-*Vicia villosa* B_4 (VVA-B_4),*Salvia sclarea* (SSA).

II. Gal-specific agglutinins

1. **T**, Galβ1→3GalNAc (T_α, Galβ1→3GalNAcα1→ the mucin type sugar sequence on human erythrocyte membrane or T_β, Galβ1→3GalNAcβ1→ at the terminal nonreducing end of the gangliosides) specific agglutinins - Peanut (PNA),and *Bauhinia purpurea alba* (BPA).
2. **I/II**, Galβ1→3(4)GlcNAcβ1→ (Lacto-*N*-biose/*N*-acetyllactosamine) specific agglutinins (Human blood group type **I** (Galβ1→3GlcNAc) and type **II** (Galβ1→4GlcNAc) carbohydrate sequences (23, 25, 26), the disaccharide residues at the nonreducing end of the carbohydrate chains derived from either *N*-glycosidic or *O*-glycosidic linkages.) - *Ricinus communis* agglutinin (RCA_1), *Datura stramonium* (Thorn apple, TAA) and *Wheat germ* agglutinin (WGA).
3. **B**, Galα1→3Gal (Human blood group **B** disaccharide -*Griffonia (Bandeiraea) simplicifolia* B_4 (GSI-B_4).
4. **E**, Galα1→4Gal, Blood group P^k active disaccharide- abrin-a and Mistletoe lectin-I (ML-I).

III. Man and/or Glc-specific agglutinins

Man-linked oligosaccharide-specific agglutinins -*Concanavalin ensiformis* (Jack bean, ConA) and *Lens culinaris* (LCA).

IV. GlcNAc, and/or Galβ1→4GlcNAcβ1→linked specific agglutinins

Chitin oligosaccharide specific aggluinins - Wheat germ (WGA) agglutinin and *Griffonia (Bandeiraea) simplicifolia* II (GSA-II).

V. LFuc-specific agglutinins

1. Monofucosyl-specific agglutinins - *Ulex europaeus* I (UEA-I) and *Ulex europaeus* II (UEA-II).
2. Difucosyl-specific agglutinins - *Griffonia (Bandeiraea) simplicifolia* IV (GSA-IV).

VI. Sialic acid specific agglutinins

Limulus polyphemus agglutinin (LPA).

The source and the structural relationship of the proposed glycoconjugate structural units (lectin determinants) are shown in Fig.1 and Table 1 of Section I-4 (18). Most of the lectin $\mathbf{F}_\beta$ and $\mathbf{T}_\beta$ determinants are found in glycosphingolipids (20-22), and the other determinants are present in mammalian glycoproteins (23-27), especially in the human blood group A,B,H,Le^a,Le^b, and Ii active glycoproteins prepared from human ovarian cyst fluid (23,24) and salivary glycoproteins (28). The reactivities of lectin determinants represent a combination result of the binding of individual sugars. The contribution of each sugar to the binding is not necessarily equal, and is different among lectins such as both *Maclura pomifera*(MPL) and peanut (PNA) are Galβ1→3GalNAc specific (2, 18). However, the inhibitory profile of the monosaccharides with these two lectins is quite different; GalNAc>Gal in MPL and Gal>>GalNAc (inactive) in PNA. Both *Ricinus communis* agglutinin (RCA_{120}) and *Wheat germ* agglutinins (WGA) are Galβ1→4GlcNAcβ1→6 specific, and the inhibitory profile of the monosaccharide residues with these two lectins is also very different: Gal>>>GlcNAc in RCA_{120} and GlcNAcβ1→ >> Gal in *Wheat germ* (1,2,18). From the data available, the current concepts of the carbohydrate specificities of applied lectins are expressed as follows.

I. GalNAc-specific lectins

Three lectin determinants, **F, A and Tn**, are proposed. Subgrouping can be classified according to individual or combination of several cross-reactive determinants (1-3, 18).

I-1. F/A, GalNAcα1→3GalNAc(Forssman)/GalNAcα1→3Gal (Human blood Group A) specific agglutinins

ABrA, *Amphicarpaea bracteata* agglutinin (Hog-peanut) (**F > A > Tn**, 29)

GalNAcα1→3GalNAc > GalNAcα1→3Gal > GalNAc.

CFT, *Codium fragile* subspecies *tomentosoides* ($\mathbf{F_p}$, **A**, $\mathbf{T_\alpha}$ and **Tn**, 30, 31).

GalNAcα1→3GalNAc, 3Galα1→4Galβ1→4Glc (F_p,Forssman specific pentasaccharide), GalNAcα1→3Gal, Galβ1→3GalNAcα1→Ser/Thr(T_α) and GalNAcα1→Ser/Thr(**Tn**) >>> Galβ1→3/4GlcNAc.(I/II)

DBA, *Dolichos biflorus* agglutinin (Horse Gram) ($\mathbf{F_p}$ > $\mathbf{A_h}$ > **A** > **Tn**, 32-34)

GalNAcα1→3GalNAcβ1→3Galα1→4Galβ1→4Glc (F_p) >> GalNAcα1→3GalNAc (F)>> GalNAcα1→3[LFucα1→2]Galβ1→4GlcNAcβ1→6R(Blood group A active pentasaccharide) > GalNAcα1→3Galβ1→3GlcNAc and GalNAcα1→3Gal > GalNAc.

GCL, *Geodia cydonium* lectin I (sponge) (Tri-**II**, **F**, Fp, $\mathbf{A_h}$ >> **L** > **II** ≥ **T** > **I**, 35-37)

GalNAcα1→3GalNAcβ1→3Galα1→4Galβ1→4Glc,($\mathbf{F_p}$),
Tri-antennary Galβ1→4GlcNAc, GalNAcα1→3GalNAc,
GalNAcα1→3[Fucα1→2]Gal($\mathbf{A_h}$) >> Lac > Galβ1→4GlcNAc ≥
Galβ1→3GalNAc > Galβ1→3GlcNAc > Melibiose > Gal, GalNAc.(35, 36)

Galβ1→4GlcNAc, Galβ1→3GlcNAc and Lactose > GalNAc (37).

GSI-A_4, *Griffonia (Bandeiraea) simplicifolia* A_4 (**F** >**A**, and **Tn** cluster >**E** >**B** >**I** >**T** >>**L**, 38-40a)

GalNAcα1→3Galβ1→3GlcNAc, GalNAcα1→3Gal,
GalNAcα1→6Gal and GalNAcα1→R >
Galα1→3Gal, Galα1→6Glc and Raffinose > Gal. (38, 39)

p-NO_2-phenyl αGalNAc > GalNAcα1→3GalNAcβ1→3Galα1→4
Galβ1→4Glc (F_p)> p-NO_2-phenyl αGal > GalNAcα1→3[Fucα1→2]
Gal ($\mathbf{A_h}$)> GalNAc, p-NO_2-phenyl βGalNAc > Melibiose > Me αGal,
Galα1→4Gal > Galα1→3Gal, Raffinose, Stachyose > *p*-NO_2-phenyl βGal > Gal >
Galβ1→3GlcNAc > Me βGal > Galβ1→3GalNAc >> Galβ1→4Glc. (32, 40, 4a)

HPA, *Helix pomatia* agglutinin (Edible snail) ($\mathbf{F_p}$ > **A** > $\mathbf{A_h}$ ≥ **Tn**, **T**, 32, 34, 41)

GalNAcα1→3GalNAcβ1→3Galα1→4Galβ1→4Glc(F_p) >
GalNAcα1→3GalNAc > GalNAcα1→3Galβ1→3GlcNAc > Galβ1→3GalNAc (32) and GalNAc.

MDL, *Mucuna derringiana* lectin (velvet bean seed) (**F** > **A** >>> **L**, 42).

Phenyl αGalNAc, *p*-nitrophenyl αGalNAc ≥
GalNAcα1→3GalNAcβ-*O*-$(CH_2)_8COOCH_3$ > Me αGalNAc, o-nitrophenyl αGalNAc,
GalNAcα1→3Gal > GalNAc, GalNAcα1→6Gal > Melibiose > Me αGal, Me βGalNAc >
Gal > Me βGal > DFuc > Galβ1→4Glc, LAra

PTA, *Psophocarpus tetragonolobus* agglutinin (**A/A$_h$, F, Tn > B**, 43)

GalNAcα1→3[LFucα1→2]Galβ1→4[LFucα1→3]Glc (A-penta)
GalNAcα1→3[LFucα1→2]Galβ1→4Glc (A-tetra), and
GalNAcα1→3Galβ1→3GlcNAc >
GalNAcα1→3[LFucα1→2]Gal (A-tri), GalNAcα1→3Gal (A-di),
Galα1→3[LFucα1→2]Galβ1→4[LFucα1→2]Glc (B-penta)
GalNAcα1→3GalNAcβ1→3GalαO (CH_2)$_8$COOMe (Forssman trisacchride), and GalNAc >
Galα1→3Gal(B-Di) > Gal.

WFA, *Wistaria floribunda* agglutinin (**A** (> **A$_h$**), **F** > **Tn**, **I(II)**, 32, 34, 44)

GalNAcα1→6Gal > GalNAcα1→3Gal,GalNAcα1→3GalNAcβ1→
3Galα1→4Galβ1→4Glc (**F$_p$**) and GalNAcα1→3Galβ1→3GlcNAc >
GalNAcβ1→3Galα1→4Galβ1→4Glc(Globoside) >
GalNAcα1→3GalNAc > GalNAc and
GalNAcα1→3[LFucα1→2]Galβ1→4GlcNAcβ1→6R >
Galβ1→4GlcNAc(*N*-acetyllactosamine) > Lactose.

WSA, *Wistaria sinensis* lectin (**F** ≥ **Tri-II** > **II** ≥ **Tn**, **I** and **A$_h$** ≥ **E** and **L** > **T** > Gal, 45, 46)

p-NO_2-phenyl βGalNAc > *p*-NO_2-phenyl αGalNAc > GalNAc,
GalNAcα1→3GalNAcβ1→3Galα1→4Galβ1→4Glc (Fp) > Triantennary
Galβ1→4GlcNAc > GalNAcβ1→3Gal-O-Me > Galβ1→4GlcNAc > Tn mixture,
Galβ1→3GlcNAc, GalNAcα1→3[Fucα1→2]Gal, *p*-NO_2-phenyl βGal > Galα1→4Gal >
Galβ1→4Glc, Galβ1→4Man, Galα1→3Galα1-Me, Galβ1→3GalNAcα1-benzyl >
Galβ1→3GalNAc > Me αGal > *p*-NO_2-phenyl αGal > Galβ1→3Arb, Galβ1→6GlcNAc >
Melibiose > Stachyose > DFuc > Me βGal > Gal. (45)

Galβ1→4GlcNAc > GalNAc > Galβ1→4Glc. (46)

I-2. GalNAcα1→3Gal (Human blood group A) specific agglutinins

DIA, Dutch Iris (*Iris* x *hollandica*) lectin (**A** > **L**, 47)

GalNAcα1→3Gal > GalNAcα1→6Gal > Me α,βGalNAc >
GalNAc, *p*-NO_2-phenyl βGal > *p*-NO_2-phenyl αGal > Me αGal, Me βGal,
Galα1→6Gal, Galβ1→4Glc, Gal >> Rha.

LBA, *Phaseolus lunatus agglutinin*(Lima bean) (Hexa-**A$_h$** > **A$_h$** >> **B**, 48-50)

GalNAcα1→3[LFucα1→2]Galβ1→3GlcNAcβ1→3Galβ1→4Glc >
GalNAcα1→3[LFucα1→2]Galβ1→4Glc >
GalNAcα1→3[LFucα1→2]Galβ1→4GlcNAcβ1→6R >
GalNAcα1→3[LFucα1→2]Gal >> GalNAc >> Gal
LFucα1→2 to subterminal Gal is an important factor for binding.

SBA, *Glycine max* agglutinin (Soy bean) (**F**, **A** > **A$_h$**, **Tn** and **I(II)**, 51, 52)

GalNAcα or β1→linked, GalNAcα1→3Galβ1→3GlcNAc,
GalNAcα1→3GalNAcβ1→R, **F** (28), and GalNAcα1→3Gal >>
GalNAcα1→3[LFucα1→2]Galβ1→4GlcNAcβ1→6R,
Galα1→6Glc (melibiose), Galβ1→6Glc, Raffinose,
Stachyose, Galβ1→4Glc (Lactose) and LFucα1→2Galβ1→4GlcNAcβ1→6R
(Human blood H active oligosaccharide)> Gal.
LFucα1→2 to subterminal Gal blocks A determinant for binding.

SNA-II, *Sambucus nigra* agglutinin (**A** > **T** >> **II**, 53)

GalNAcα1→3Gal and GalNAcα1→6Gal > GalNAcβ1→6Gal >
GalNAc > di-(Penta 2,6[5a], Hex 3,6[5b]), tri-(Nona I[5c]), tetra-(Undeca[5d]),
antennary oligosaccharides containing Galβ1→4GlcNAc at nonreducing end >
Galβ1→3GalNAc (**T**) >> Galβ1→4GlcNAc > Gal.

VVA, *Vicia villosa* agglutinin (Hairy Vetch, it is assumed that the agglutinin used to study the following information is a mixture of three isolectins; A_4-minor,B_4-major, and A_2B_2-minor) (**A** (> $\mathbf{A_h}$) and **Tn** mainly, 54, 55)

> GalNAcα1→3Gal > GalNAcα1→6Gal >
> GalNAcα1→3Galβ1→3GlcNAc and GalNAcα1→3GalNAc >>
> GalNAcα1→3[LFucα1→2]Galβ1→4GlcNAcβ1→6R > Galα1→3Gal.

CFT, *Codium fragile* subspecies *tomentosoides* (see **F/A** specific lectins, Section **I-1**).

DBA, *Dolichos biflorus* agglutinin (see **F/A** specific agglutinins, Section **I-1**).

HPA, *Helix pomatia* agglutinin (see **F/A** specific agglutinins, Section **I-1**).

PTA, *Psophocarpus tetragonolobus* agglutinin (see **F/A** specific agglutinins, Section **I-1**).

WFA, *Wistaria floribunda* agglutinin (see **F/A** specific agglutinins, Section **I-1**).

I-3. GalNAcα1-Ser or Thr (Tn) specific agglutinins

MLL, *Moluccella laevis* lectin (**Tn** >> $\mathbf{A_h}$, **F** , 56, 57)

> GalNAcα-OMe > *p*-NO_2-Phenyl αGalNAc, *p*-NO_2-Phenyl βGalNAc > GalNAc >> Gal–terminated compounds.
>
> the reaction with **T** and other Gal-terminated structures is negligible or not detectable in glycoproteins
>
> On thin-layer chromatograms or adsorbed micotiter wells, MLL showed the highest binding activity for glycosphingolipids with terminal α-linked GalNAc, both in linear structures, as Forssman glycosphingolipids, and in branched structures, as glycosphingolipids with the blood group A determinant. It bound weakly to linear glycosphingolipids with terminal α–linked Gal. Glycosphingolipids with the Gal in the α4-linkage had a higher binding activity than those with the sugar in an α3-linkage, and the presence of an α-linked fucose at the C2 position of the pentultimate galactose abolished the binding (57).

SSA, *Salvia sclarea* agglutinin (Two **Tn** > single or three sequential **Tn** structure, 32, 58)

> Glycopeptides containing two Tn structures > Glycopeptides containing one Tn structures or Glycopeptides containing three sequential Tn structures. >>
> Galβ1→3GalNAcα1→R (inactive) (58).
> SSA is also specific for **F** determinant (32).

5a. Galβ1→4GlcNAcβ1→2Man
Galβ1→4GlcNAcβ1, 6↑

Penta 2,6

5b. Galβ1→4GlcNAcβ1→3Manα1→2Man
Galβ1→4GlcNAcβ1, 6↑

Hexa 3,6

5c. Galβ1→4GlcNAcβ1, 4↓
Galβ1→4GlcNAcβ1→2Manα1, 3↓
Man
Manα1, 6↑
Galβ1→4GlcNAcβ1, 2↑

Nona I

5d. Galβ1→4GlcNAcβ1, 4↓
Galβ1→4GlcNAcβ1→2Manα1, 3↓
Man
Galβ1→4GlcNAcβ1→6Manα1, 6↑
Galβ1→4GlcNAcβ1, 2↑

Undeca

VVA-B_4 ·, *Vicia villosa* B_4 agglutinin (Hairy Vetch) (Two **Tn** > > one **Tn** >> one or two **T**, 32, 55, 59)

A fetuin glycopeptide containing two GalNAcα1→*O* to Ser(Thr) of the peptide chains (two Tn structures) >> single Tn structure >> a fetuin glycopeptide containing one or two Galβ1→3GalNAcα1→*O* to Ser (Thr) of the peptide chain(s) (T structure) > NeuNAcα2→3 Galβ1→3GalNAcα1→Ser(Thr) of the peptide chain >> Galβ1→3GalNAcα1→R (inactive).

I-4. Other GalNAc specific lectins required for further characterization

ABl A, *Agaricus blazei* agglutinin (basidiomycetous fungi) (60)

Methyl αGalNAc > GalNAc

ALeuL, *Amaranthus leucocarpus* lectin (60a); GalNAc, Galβ1→4GlcNAc

BPA, *Bauhinia purpurea alba* agglutinins; (see **T** specific agglutinins,Section **II-1**).

CGL, *Crenomytilus grayanus* lectin (sea mussel) (61)

p-NO_2-phenyl βGal > GalNAc > Raffinose ≥ Melibiose > *p*-NO_2-phenyl βGalNAc > Me αGal > Gal, D-Talose > Me βGal

CStrL, *Crotalaria striata* lectin (62); GalNAc > Gal >GalN

FJL, *Falcata japonica* (62a, 62b)

Phenyl βGalNAc, phenyl αGalNAc > GalNAc > Gal > GlcNAc, GalN > GlcN, Man , Glc, LFuc.

GFL, *Grifola frondosa* (62c)

Methyl βGalNAc > Methyl αGalNAc > GalNAc, $(GalNAc\alpha1\rightarrow4GalNAc)_{n=1-6}$ > Galβ1→4Fru > Methyl βGal > Galβ1→4GlcNAc > Galβ1→4Glc > Galα1→6Glc, Gal and Fuc > Methyl αGal > Galα1→6Glcβ1→2Fru > GalN > L-Ara (L-Arabinose) > Methyl αGlcNAc > GlcNAc > GlcN > Methyl βGlc > L-Rha (L-Rhamnose) > Xylose

GTL, *Galactia tashiroi* lectin (63)

GalNAc >> Gal, Me αGal, Me βGal, *p*-nitrophenyl αGal, *p*-nitrophenyl βGal, Galβ1→4Glc, GlcNAc > Glc, Maltose, Melibiose, Raffinose

ML-II, Mistletoe toxic lectin-II (63a)

GalNAc and Galβ1→

ML-III, Mistletoe toxic lectin-III (63a)

GalNAc > Galβ1→

MPA, *Maclura pomifera* agglutinin; (see **T** specific agglutinins,Section **II-1**)

PCA-I, *Phaseolus coccineus* L. phytohemagglutinin-I (scarlet runner seeds) (64); GalNAc

Tachylectin-3, *Tachypleus tridentatus* lectin-3 (Japanese horseshoe crab) (65); A-pentasaccharide

TCA-I, *Tetracarpidium conophorum* agglutinin-I (Nigerian walnut) (66)

GalNAc > Me βGal > Galβ1→4Glc > Gal, Me αGal

II. Gal-specific lectins

II-1. Galβ1→3GalNAc(T) specific agglutinins

ABiA, *Agaricus bisporus* agglutinin (Mushroom) (Tα > T > L, 67-69)

Galβ1→3GalNAcα→*N*-tosyl-L-serine.
Galβ1→3GalNAcα*P*–NO_2-phenyl, Galβ1→3GalNAc >>> Galβ1→6GlcNAc >
Galβ1→4Glc > *P*–NO_2-phenyl αGalNAc, *P*–NO_2-phenyl βGalNAc, GalNAc >
Gal, Me αGal, Me βGal.

ACA, *Amaranthus caudatus* agglutinin (Tα > T >> $T_β$, 70)

Galβ1→3GalNAcα1→Ser/Thr >> GalNAc >>> Gal,Galβ1→3(4)GlcNAc (inactive).
Galβ1→3GalNAcα-*O*-benzyl, Galβ1→3[6-F]GalNAcα-*O*-benzyl,
Galβ1→3GalNAcα-O(CH_2)$_8$-CO_2CH_3, Neuα2→3Galβ1→3GalNAcα-O(CH_2)$_8$-CO_2CH_3,
DFucβ1→3GalNAcα-*O*-benzyl, 3-*O*-MeGalβ1→3GalNAcα-*O*-benzyl,
Galβ1→3GalNAcα-*O*-*o*-nitrophenyl, Galβ1→3GalNAcα-*O*-$CH_2CH{=}CH_2$>
Galβ1→3GalNAcα-*O*-CH_3 > Galβ1→3GalNAc > Galβ1→3 (2-acetamido-2-deoxy-1,5-anhydrogalactitol) > GlcNAcβ1→3Galβ1→3GalNAcα-*O*-benzyl >
Galβ1→3GalNAcβ-*O*-$CH_2CH{=}CH_2$, Glcβ1→3GalNAcα-*O*-CH_3 >
3-*O*-(2-*N*-acetyl-α-D-galactopyranosyl)-*N*-(benzyloxy-carbonyl)-L-Serine methyl ester >
Methyl (2-*N*-nitro-benzamido)α-galactosaminide > GalNAcα1→6Gal > Me
αgalactosaminide > GalNAcα1→3Gal > *P*-NO_2-phenyl αgalacto-saminide > GalNAc >
Methyl (2-*N*-benzamido)α-galactosaminide > Methyl (2-*O*-acetyl)α-Gal >
Methyl *N*-acetyl-β-galactosaminide.

AIA, *Artocarpus integrifolia* agglutinin, Jacalin (**T** > **Tn** >> **I(II)**, 71-73)

Galβ1→3GalNAc > GalNAcα1→Ser/Thr >> Galβ1→3GlcNAc;
Galβ1→4GlcNAc and Galβ1→4Glc.

ALA, *Artocarpus lakoocha* agglutinin (**T** > **Tn**, 74, 75)

Galβ1→3GalNAc > GalNAcα1→*O* Ser/Thr > Melibiose > Raffinose > Gal.

APA, *Abrus precatorius* agglutinin (**T** > **I/II** > **E** > **B** > **Tn**, 76).

Galβ1→3GalNAc > Galβ1→3/4GlcNAc > Galα1→4Gal > Galα1→3Gal >>
GalNAcα1→Ser(Thr).

ARA, *Agropyrum repens* agglutinin (Couch grass) (**T**, 77)
Galβ1→3GalNAc > GalNAc and chitin oligosaccharide (with type A cells).

BPA, *Bauhinia purpurea alba* agglutinin (**T** > **I(II)** and **Tn**, 78, 79)

Galβ1→3GalNAc > GalNacα1→6Gal > GalNAc and
Galβ1→3GlcNAcβ1→3Galβ1→4Glc(Lacto-*N*-tetraose) >
Galβ1→3GlcNAc, Galβ1→6GlcNAc and
GalNAcα1→3Galβ1→3GlcNAc > GalNAcα1→3Gal >
Galβ1→4GlcNAc (*N*-acetyllactosamine) and
Galβ1→3[LFucα1→4]GlcNAcβ1→3Galβ1→4Glc
(Lacto-*N*-fucopentaose II) > Galβ1→4GlcNAcβ1→6Gal and Lactose >Gal.

LDelL, *Lactarius deliciosus* lectin (edible mushroom) (80)

Galβ1→3GalNAc > GalNAc 1P > Galβ1→3GlcNAc

LDetL, *Lactarius deterrimus* lectin (mushroom) (81)

Galβ1→3GalNAc

MPA, *Maclura pomifera* agglutinin (Osage orange tree, Hedge apple tree) (**T** > **Tn**, 82, 83)

Galβ1→3GalNAc >> GalNAcα1→6Gal > Galα1→6Glc

(Melibiose), Stachyose > GalNAc > Gal.

PNA, *Arachis hypogaea* agglutinin (Peanut) (**T** >> **I(II)**, 84, 85)

Galβ1→3GalNAc >> Galβ1→4GlcNAc(*N*-acetyllactosamine)>
Gal and Galβ1→3GlcNAc >> GalNAc (inactive).

RCA2, Ricin (**T** > **I(II)** > **E** > **B**> **Tn**, 52, 86, 87)

[Galβ1→3GalNAcα1→Ser or Thr]4 glycopeptide >> Triantennary oligosaccharides containing Galβ1→4GlcNAc at nonreducing end > [Galβ1→3GalNAcα1→Ser or Thr]2 glycopeptide with space between carbohydrate side chains, Galβ1→3GalNAcα1→Ser or Thr >> GalNAcα1→Ser or Thr.

The ability of RCA2 to bind to Galβ1→3GalNAc disaccharides appears to be markedly influenced by the relative locations of the disaccharides along the peptide backbone.

From Debray *et al.* (52) and Baenziger *et al.* (86)
Biantennary oligosaccharides containing Galβ1→4GlcNAc at nonreducing end > Lac > GalNAc > Melibiose and Gal (≥ GalNAc, see ref. 56).

SJA, *Sophora japonica* agglutinin (Japanese pagoda tree) (**T** and **I(II)**, 88, 89)

Galβ1→3GalNAcβ1→*N*-tosyl-L-Serine > Galβ1→3GlcNAc >
Galβ1→3GlcNAcβ1→3Galβ1→4Glc(Lacto-*N*-tetraose) >
Galβ1→3GlcNAcβ1→6Gal > Galβ1→4GlcNAcβ1→6Gal
(Blood group I Ma-active trisaccharide) and Galβ1→4GlcNAc > GalNAc >>
Lactose > Gal and Melibiose.

Some unknown, mild acid sensitive compound near the terminal group at the nonreducing end plays an important role in binding or precipitating this agglutinin.

VGA, *Vicia graminea* agglutinin(Blood group N specific) (**T**, 90-92)

clusters of Galβ1→3GalNAcα1→linked >> cluster of sialylated Galβ1→3GalNAcα1→linked; reaction enhanced by adjacent hydrophobic group(s), and for this reason stronger with N (containing Leu1), than with M (containing Ser1).
Glycopeptides: asialoN > N > asialoM > M.

II-2. Galβ1→3(4)GlcNAcβ1→linked (Human blood group type I and Type II carbohydrate sequences, I & II) specific agglutinins (25, 26)

ADA-II, Allo A-II, *Allomyrina dichotoma* agglutinin II (hemolymph of a beetle) (93, 94)

Galβ1→4GlcNAc, substitution of the terminal Gal residues by Neu5Acα1→6 will enhance their affinity for this lectin.

BFL, *Butea frondosa* lectin (**L** > **I/II** > **E** > **T**, 95, 96)

Lactose > DFuc > Gal > GalNAc > Arabinose

GalNAc, Fucα1→2Galβ1→4Glc(2'-fucosyllactose) > *p*-NO_2-phenyl αGalNAc, *p*-NO_2-phenyl βGalNAc > Lac, Lactulose > Me αGal, Me βGal, *p*-NO_2-phenyl αGal, *p*-NO_2-phenyl βGal, MurNAc, Galβ1→3GlcNAc, Galβ1→4GlcNAc, Galβ1→6Galβ(1→*O*)-*p*-$NO_2C_6H_4$ > Galβ1→6GlcNAc, GlcNAcβ1→6Gal, GlcNAcβ1→6Galβ1→4Glc, Raffinose > Gal, GalN,*p*-NO_2-phenyl β-thiogalactoside, NeuAc, Muramic acid, Galβ1→6Gal, Galβ1→4Man, Galβ1→4Gal, Galβ1→3GlcNAc, Galβ1→3GlcNAcβ(1→*O*)-*p*-$NO_2C_6H_4$,Galβ1→4Galβ1→4GlcNAc, NeuAcα2→3Galβ1→4Glc, Melibiose > Galβ1→3GalNAc > GalU > GlcU.

BPA, *Bauhinia purpurea alba* agglutinin(see **T** specific agglutinins, Section **II-1**).

CAu, *Colchicum autumnale* agglutinin (meadow saffron tuber) (97)
Galβ1→4Glc > GalN > GalNAc > Gal > Melibiose, Raffinose, DFuc

ECA, *Erythrina cristagalli* agglutinin (Coral tree) (Multiple antennary **I(II)**, **A** and **mII** clusters, **F** > **A** > **mII** >> **II**, 98)

Tetra- and tri-antennary oligosaccharides containing
Galβ1→4GlcNAc linked units at nonreducing end >
Galβ1→4GlcNAcβ1→6Gal(Blood group I Ma specific) >
Galβ1→4GlcNAc > Galβ1→3GlcNAc >> Lactose >>>
Galβ1→6Gal and,Galα1→3Gal > Raffinose.

ECosL, *Erythrina costaricensis* lectin (99)

Nitrophenyl βGal > GalNAc, Galβ1→4Glc ≥ Me αGal > Gal > Me βGal, Raffinose > GalN.

EHA, *Eranthis hyemalis* agglutinin (from Winter-aconite root tubers) (100)

Galβ1→4GlcNAc > Galβ1→4Glc > Gal, Melibiose > Raffinose.

EVA, *Erythrina variegata* agglutinin (101)

Galβ1→4GlcNAcβ1→3[Galβ1→4GlcNAcβ1→6]Gal sugar sequence is the active one.

Other *Erythrina* agglutinins (102)

E. caffra^{+} (ECaf,L),*E. corallodendron* (ECor,A), *E. flabelliformis* (EFA), *E.humeana* (EHA), *E.latissima* (ELat,A), *E. lysistemon* (ELys,A), *E. perrieri*$^{+}$ (EPA), *E.stricta* (ESA), *E. zeyheri* (EZA)

Triantennary or biantennary oligosaccharides containing Galβ1→4GalNAc at nonreducing end > *N*-acetyllactosamine >Lactose ≥ GalNAc > Gal(GalNAc ≥ Lactose ≥ LGal)$^{+}$.

LEA, *Lycopersicon esculentum* agglutinin (Tomato) (95, 103, 104)

In addition to (GlcNAcβ1→4)$_{2\text{-}4}$, Galβ1→4GlcNAcβ1→3-
Galβ1→4GlcNAcβ1→6Gal-ol sequence is also one of active oligosaccharides.

PAA, *Phytolacca americana* agglutinin (pokeweed mitogen) (95, 105-107)

Four kinds of carbohydrate specificties have been reported. Their relationship has to be defined.
Galβ1→4GlcNAcβ1→6Gal-ol and
Galβ1→4GlcNAcβ1→6[Galβ1→4GlcNAcβ1→3]Gal-ol [78);
Branched poly (Galβ1→4GlcNAc) complex (86);
(GlcNAcβ1→4)$_{n>5}$-units(87) or Manα1→2Manα1→2-
Manα1→3Manβ1→4GlcNAcβ1→4GlcNAcβ-AsN(88).

PHA-E, *Phaseolus vulgaris* (isolectin E$_4$) erythroagglutinin (Red kiney bean) (108-110)

Essential structural unit:

```
Galβ1→4GlcNAcβ1→2Manα1,6↓
                          GlcNAcβ1→4Manβ1→4GlcNAcβ1→4R3
GlcNAcβ1→2Manα1,3↑
     ↑          ↑
     R1         R2
```

in which **R**$_1$ and **R**$_2$ represent either hydrogen atoms or sugars and **R**$_3$ represents either GlcNAc or (Fucα1→6) GlcNAc.

PHA-L, *Phaseolus vulgaris* (isolectin L_4) leukoagglutinin (108, 111)

Tri- and tetra-antennary oligosaccharides containing
Galβ1→4GlcNAc→ linked units at nonreducing end >>
Penta-2,6[5e] > Tri-2,6[5f] >

```
Galβ1→4GlcNAcβ1→2Manα1,3↓
                                 Man (hepta-) >
Galβ1→4GlcNAcβ1→2Manα1,6↑
Galβ1→4GlcNAcβ1→2Man > GlcNAcβ1→2Man.
```

PNA, *Arachis hypogaea* agglutinin (see **T** specific agglutinins, Section **II-1**).

RCA1, *Ricinus communis* agglutinin (Castor bean) (**II** > **I** > **B** > **T**, 52, 112, 113)

Biantennary oligosaccharides containing
Galβ1→4GlcNAc-linked units at nonreducing end >
Galβ1→4GlcNAcβ1→6Gal(Blood group I Ma specific) >
Galβ1→4GlcNAc > Galβ1→3GlcNAc > Lactose > Galα1→3Gal > Raffinose.
RCA1 is also specific for Galβ1→3GalNAcα1→Ser (Thr). For further information, see reference 86.

Ricin, (see **T** specific agglutinins, Section **II-1**).

SNA-I, *Sambucus nigra* agglutinin (114, 115) (see sialic acid specific agglutinins Section **VI**).

TAA, *Datura stramonium* agglutinin (DSA,Thorn apple) (Biantennary **I**(**II**, penta-2,6[5e]) >> C , 52, 116-119)

Penta-2,6[5e] >> Penta-2,4[5f] > $[\mathrm{GlcNAc}\beta1\rightarrow4]_4$ >
$[\mathrm{GlcNAc}\beta1\rightarrow4]_3$ > Galβ1→4GlcNAc(*N*-acetyllactosamine),
GlcNAcβ1→4GlcNAc, and Galβ1→4GlcNAcβ1→6Man >
Manα1→4Manβ1→2GlcNAc > GlcNAcβ1→4Glc. Information
about Galβ1→4GlcNAcβ1→3Galβ1→4GlcNAcβ1→6Gal-ol and
Galβ1→4GlcNAcβ1→3Galβ1→4GlcNAcβ1→3Galβ1→4GlcNAc-ol
as active oligosaccharides is given in reference 95 and additional information is also available in reference 117.

TCA-II, *Tetracarpidium conophorum* agglutinin (**II** > **I**, 120, 121).

Galβ1→4GlcNAcβ1→3Galβ1→4Glc, Galβ1→4GlcNAc > Galβ1→6Gal > Galβ1→4Glc >
Galβ1→3GlcNAcβ1→3Galβ1→4Glc > Galβ1→6GlcNAc, Gal.

UDA, *Urtica dioca* agglutinin (Stinging nettle)(see GlcNAcβ1→4GlcNAc specific agglutinin,Section **IV**).

WGA, *Triticum vulgaris* agglutinin (Wheat germ) (52, 95, 122-127)

Galβ1→4GlcNAcβ1→6Gal, (Galβ1→4)GlcNAcβ1→6GalNAc and
$(\mathrm{GlcNAc}\beta1\rightarrow4)_{2\ \mathrm{or}\ 3}$ > Galβ1→4GlcNAcβ1→3Gal >>>
Galβ1→4GlcNAc (*N*-acetyllactosamine) > Galβ1→3GalNAc (inactive).

```
5e. Galβ1→4GlcNAcβ1→2Man          5f. ···GlcNAcβ1→2Man
       Galβ1→4GlcNAcβ1,4↑                ···GlcNAcβ1,6↑
```

Penta-2,4 **Tri-2,6**

In addition to GlcNAcβ1→4GlcNAc, the GlcNAcβ1→6 sequence is the most important structure for binding. GlcNAcβ1→4GlcNAc sequence, the lectin C determinant, can be found in mammalian *N*-linked carbohydrate chains as part of a core trisaccharide (27) and Galβ1→4GlcNAcβ1→6Gal frequently found in mammalian *O*- and *N*-linked glycoproteins (23).

Recently report indicated that cluster (multivalent) effects through the high density of weak inhibitory determinants on glycans, such as GalNAcα1→Ser/Thr (**Tn**), GalNAc at nonreducing terminal, GlcNAcβ1→ at nonreducing end and/or as internal residue, play important roles in precipitation, while the GlcNAcβ1→4GlcNAc disaccharide may play a minor role in the precipitation of mammalian glycan-WGA complexes. (127)

II-3. Galα1→3Gal(B) specific agglutinin

CJA-B, *Charybdis japonica* agglutinin (128)

I. Galα1→3Galβ-*O*-R > Galα1→3[LFucα1→2]Galβ-O-R >> Galα1→4Galβ1→4GlcNAcβ-O-R

II. Glcα1→6Fru (Palatinose), Stachyose >> Me αGal > Raffinose, Me αGlc, Sucrose > Glc, Glcβ1→6Glc, Ara

GSI-B4, *Griffonia (Bandeiraea) simplicifolia* I-B4 agglutinin (38, 39, 129)

Galα1→3Gal > Melibiose > Raffinose > Gal > GalNAc (inactive).

PTL, *Psophocarpus tetragonolobus* (See **F/A** specific agglutinins Section **I-1**).

II-4 Galα1→4Gal (E) specific lectins

Abrin-a (130, 131).

Galα1→4Gal (**E**) > Galβ1→3(4)GlcNAc (**I/II**) and Galβ1→4Glc (**L**).

ML-I, Mistletoe toxic lectin-I (130-137).

Galα1→4Gal > Galβ1→3(4)GlcNAc and Galβ1→4Glc.

ML-I is specific for sialic acid as human α_1 acid gp interacts strongly with ML-I. Sialic acid of some *O*-glycans also acts as masking molecule. (134)

PA-IL, *Pseudomonas aeruginosa*-I lectin (E > B > L, 138)

Phenyl βGal > *p*-Nitrophenyl βGal > Galα1→6Glc (Melibiose)> Stachyose > Galα1→3Galα→1Me > Raffinose > *p*-Nitrophenyl αGal > Me αGal > Me βGal > Galα1→4Gal (**E**)> Gal > Galα1→3Gal (**B**)> GalNAc, Galβ1→4Glc (**L**) > Galβ1→3Ara >> DFuc

II-5 Other Gal-specific lectins required for further characterization

AAltL, *Artocarpus altilis* lectin (139); Galβ1→3GalNAc

EML, *Echinocytis lobata* lectin (wild cucumber) (140)

Gal, GalNAc

EMaL, *Euphobia marginata* latex lectin (141)

Glcβ1→6Glc, Galβ1→4Glc, Talose, Me βGal > Me αGal, Gal > GalN.

ENLA, *Euphorbia neriifolia* latex agglutinin (142); Lac > Gal ≥ Melibiose ≥ Raffinose

MeA, *Machaerocereus eurca* agglutinin (143)

MeA_1, *Machaerocereus eurca* agglutinin I

Nitrophenyl βGal ≥ Nitrophenyl αGal >> Me αGal, Me βGal > Gal, GalNAc, Galβ1→4Glc, Galβ1→4GlcNAc, L-Fuc.

MeA_2, *Machaerocereus eurca* agglutinin II

Nitrophenyl βGal > Nitrophenyl αGal, Me αGal, Me βGal > Gal, GalNAc, GalN, Galβ1→4Glc, Galβ1→4GlcNAc, L-Fuc.

PFML, *Pinctada fucata martensii* lectin (pearl oyster) (144)

Gal > GalNAc > NeuAc > ManNAc > D-Fuc > GalN > GlcNAc

PML, *Phaseolus mungo* lectin (145)

Me α Gal > *p*-Nitrophenyl αGal > Melibiose > Gal > GalNAc

PSL, *Pellina semitubulosa* lectin (sponge) (146)

Gal, Ara, Lac > D-Fuc, Melibiose > Glc > Cellobiose

TAngL, *Trichosanthes anguina* lectin (snake gourd) (147)

Lac, Me βGal > Gal GalN, GalNAc, Me αGal > Fuc, Melibiose > Ara

TCA-II, *Tetracarpidium conophorum* agglutinin-II (Nigerian walnut) (148)

Galβ1→4Glc > Me βGal > Me αGal, Gal, GalNAc

III. Man and/or Glc specific lectins

III-1. Manα1→linked oligosaccharide-specific agglutinins

AAscL, *Allium ascalonicum* lectin (shallot) (149)

Manα1→6[Manα1→3]Manα-*O*-Me > Manα1→3Manα-*O*-Me > Manα1→6Manα1→6Man > Manα1→6Manα-*O*-Me > Manα1→2Manα-*O*-Me > Manα-*O*-Me > Manβ-*O*-Me > Man

ASA, *Allium sativum* lectin (garlic) (150)

Manα1→3Manα-*O*-Me ≥ Manα1→3Man, Manα1→3Manα-*O*-allyl > Manα-*O*-Me > Man

ASA-I (25kDa) (151)

Man_9Gn_2Asn > Man_8Gn_2Asn > Man_7Gn_2Asn > Manα1→6[Manα1→3]-Manβ1→4[Fucα1→6]GlcNAcβ1→4GlcNAc > Man_3Gn_2 > Manα1→6-[Manα1→3]Manβ1→4[Fucα1→3]GlcNAcβ1→4GlcNAc > Man_5Gn_2 > Xylβ1→2[Manα1→6(Manα1→3)Man]Manβ1→4GlcNAcβ1→4GlcNAc > Gn_2Man_3 > Manα1→3[Manα1→6]Man ≥ Mannopentose > Xylβ1→2 [Manα1→6(Manα1→3)Man]Manβ1→4[Fucα1→3]GlcNAcβ1→4GlcNAc > Manα1→3Manβ1→4GlcNAc > Manα1→3Man > 4-Methyl-umbelliferyl α-D-Man ≥ Manα1→2Man > Manα1→6Man > Me αMan > Man > *p*-NO_2-Phenyl αMan

ASA-III (48kDa) (151)

Man_9Gn_2Asn > Man_8Gn_2Asn > Man_7Gn_2Asn > Manα1→6[Manα1→3]Manβ1→4[Fucα1→6]GlcNAcβ1→4GlcNAc > Man_3Gn_2 > Manα1→6[Manα1→3]Manβ1→4[Fucα1→3]GlcNAcβ1→4GlcNAc > Man_5Gn_2 > Gn_2Man_3 > Xylβ1→2[Manα1→6(Manα1→3) Man]Manβ1→GlcNAcβ1→4GlcNAc > Manα1→3[Manα1→6]Man > Mannopentose ≥ Xylβ1→2[Manα1→6(Manα1→3)Man]Manβ1→4 [Fucα1→3]GlcNAcβ1→4GlcNAc > Manα1→3Manβ1→4GlcNAc > Manα1→3Man > Manα1→2Man > Manα1→6Man > 4-Methyl-umbelliferyl α-DMan > Me αMan > Man > *p*-NO_2Ph αMan

AUA, *Allium ursinum* lectin (ramsons) (150)

Manα1→6[Manα1→3]Manα-*O*-Me > Manα1→3Manα-*O*-Me > Manα1→6Manα1→6Man > Manα1→6Manα-*O*-Me > Manα-*O*-Me > Man

BMA, *Bowringia milbraedii* agglutinin (West African legume) (152)

Manα1→2Manα1→2Manα1→3Manβ1→4GlcNAc-ol, Manα1→2Manα1→3Manβ1→4Glc-ol > Manα1→2Manα1→6Manα1→6Man,and Manα1→2Man > Manα1→3Man > Man >> GlcNAc and Glc.

B-SJA-II, *Sophora japonica* B-II lectin (Japanese pagoda) (153)

Me αMan ≥ Glcα1→4Glcα1→4Glc > Man, *p*-NO_2-Phenyl αGlc > GlcNAc, Ph αGlc > Glcα1→4Glc ≥ Me αGlc > GlcN > Glc, Man-6-PO_4 > Glcα1→6Glc > 3-*O*-Me βGlc > Ph βGlc > Glcα1→2Fru > Me βGlc > Me αGlcNAc

Calsepa, *Calystegia sepium* agglutinin (154)

Me αMan > Palatinose, Panose ≥ Me αGlc, Me α-deoxy-Glc, Me αGlcNAc, Maltose, Isomaltose > Man > Fru ≥ Melezitose ≥ GlcNAc > Glc

CCL, *Cajanus cajan* lectin (155)

Me αMan > Maltose > Me αGlc > Man > Fru ≥ Glc, Sucrose

CGA, *Canavalia gladiata* agglutinin (Japanese jack bean) (156)

Me αMan > Maltose > Trehalose > Man >> Glc

Champedak lectin-M, *Artocarpus integer* lectin (157)

Manα1→3Man > Me αMan > Manα1→3Manα-*O*-Me ≥
Manα1→6Manα-*O*-Me > Manα1→6Man > Manα1→4Manα-*O*-Me ≥
Man > Manα1→2Manα-*O*-Me > NeuAc > Glc ≥ GlcNAc

ConA, *Concanavalia ensiformis* agglutinin(Jack Bean) (52,158-161)

Manα1→linked > Glcα1→linked > GlcNAcα1→linked.

The binding sugar requires the C-3, C-4 and C-5 hydroxyl groups of the Man or Glc ring (158). One of the essential core structures in glycopeptides reactive with this agglutinin is shown below (52,159).

```
R1-Manα1,3↓
          R2→Manβ1→4GlcNAcβ1→4GlcNAcβ1-Asn
               ↑
R3-Manα1,6
```

ConA has a great affinity for the trimannosidic core structure (Lectin **M** determinant Manα1,3[Manα1,6]→Manβ→) substituted by two GlcNAc residues (**R**1) at the nonreducing end; the affinity is not affected when this structure is modified by addition of a β1→4 linked GlcNAc on the β→linked Man (**R**2) (52),but its affinity is reduced when these GlcNAc residues are substituted by Gal residues. **LFuc** is not as important for the binding of ConA as for LCA, PSA, and VFA (52).

CVA, *Crocus vernus* lectin (crocus) (162)

Manα1→6[Manα1→3]Manα-*O*-Me > Manα1→3Manα-*O*-Me, >
Manα1→6[Manα1→3]Manα1→6[Manα1→3]Manα-*O*-Me >
Manα1→3[Glcα1→6]Manα-*O*-Me > Manα1→3[Galα1→6]Manα-*O*- Me

DLL, *Dolichos lablab* lectin (163)

Trehalosamine > Manα1→3[Manα1→6]ManαMe > α, αTrehalose, Me αMan > *p*-NO_2-phenyl αMan >Manα1→6ManαMe, Me αGlcNAc ≥ GlcNAcα1→6Gal, Me αGlc, Manα1→3ManαMe ≥ GlcNAc, *p*-NO_2-phenyl βMan ≥ Kojibiose ≥ Man > Isomaltose > 1,5-anhydro-D-Mannitol ≥ Glc, Me βGlcNAc, Maltose, α, β-Trehalose ≥ 2-*O*-Me-Glc ≥ Me βMan, 2-Deoxy-D-Glc ≥ Sucrose, 1,5-anhydro-D-Glucitol, Me βGlc ≥ Gentiobiose ≥ β, βTrehalose

GNA, *Galanthus nivalis* agglutinins (Snowdrop blue) (164)

```
  Manα1,3↓
Manα1→6Manα1→6Manβ1→4GlcNAcβ1→4GlcNAc-AsN,
(Manα1→2)Manα1,3↑

  Manα1,3↓                        Manα1,3↓
Manα1→6Manα... ,and           Manβ1→4GlcNAc >

Manα1→3Manα... ,and Manα1→3Manα1→2Manα1→2Man >

Manα1→6Man,  ...Manα1→6Manα1→6Man, and

Manα1→2Manα1→2(3 or 6)Man > Manα1→2Man, and

Manα1→6Glc > Manβ1→2Man, and Manβ1→2Manβ1→2Man >>>

 Manα1→2Manα 1,3↓                Manα1,4↓
Manα1→2Manα1→6Man...   ,and  Manα1→2Man   (poor).
```

A nonreducing terminal residue is necessary for the interaction of oligosaccharides, and oligosaccharides with terminal Manα1→3Man units showed the highest inhibitory potency (10-30 times greater than Man) among oligosaccharides tested.

Heltuba, *Helianthus tuberosus* tuber agglutinin (165)
Me αMan > Man, Maltose > Me αGlc > Glc

HHA, *Hippeastrum hybr.* lectin (amaryllis) (166)

Manα1→6Manα1→6Man > Manα1→6[Manα1→3]Manα-*O*-Me > Manα1→3Manα-*O*-Me > Manα1→6Manα-*O*-Me > Manα1→3Man > Manα1→2Manα1→2Man > Manα1→2Man > *p*-NO_2-Phenyl αMan > Me αMan > Man > Me βMan

LCA, *Lens culinaris* agglutinin(Lentil) (52,167-169)

Manα1→linked > Glcα1→linked >> GlcNAc.

For comparison, see PSA, and VFA. LFucα1→6 linked to GlcNAc at the **reducing end** in the following structure is an important binding determinant (as shown in glycopeptide GP-hLTF-1 in Fig.2, reference 52),but not LFucα1→3 linked to GlcNAc near the nonreducing end (Glycopeptide,GP-hLTF-2 in Fig.2, reference 52).

```
NeuAcα2→6Galβ1→4GlcNAcβ1→2Manα1,3↓
                                   Manβ1→4GlcNAcβ1→4GlcNAcβ1-Asn
NeuAcα2→6Galβ1→4GlcNAcβ1→2Manα1,6↑          LFucα1,6↑
           LFucα1,3↑                (Important factor for binding)
```

Removal of LFucα1→6 linked from the above glycopeptide with an αLfucosidase yields a structure that is about 10-fold less active as compared to the untreated one.

LOA, *Listera ovata* agglutinin (orchid twayblade) (170)

Manα1→3Manα1→3Manα1→3Manα1→3Man > Manα1→3Manα1→3 Manα1→3Man > Manα1→3Manα-*O*-Me > Manα1→3Manα1→3Man > Manα1→6[Manα1→3]Manα-*O*-Me > Manα1→3Man > Manα1→6Man > Manα1→2Man > Me αMan > Man > Me βMan

NPA, *Narcissus pseudonarcissus* lectin (daffodil) (171)

Manα1→6Manα1→6Man > Manα1→6Manα-*O*-Me > Manα1→6[Manα1→3]Manα-*O*-Me, Manα1→2Man ≥ Manα1→3Manα-*O*-Me > Manα1→3Man > *p*-NO_2-Phenyl αMan > Manα1→2Manα1→2Man > Me αMan > Man > Manα1→6Glc > Man1-Pi > Me βMan > *p*-NO_2-Phenyl βMan

NFL, *Neoregelia flandria* lectin (172)

Manα1→3Man > Manα1→6Man, Sucrose, α, α-Trehalose > Manα1→6[Manα1→3]Man, Man > GlcNAc, Me αMan, 3-*O*-Me-Glc, Nigerose, Laminaribiose, Isomaltose > Fru, Glc, Me αGlc, Gentiobiose > GlcN, Me βMan, 2-Deoxy-Glc, Sophorose > Me βGlc, Maltose, *N*-Acetylchitobiose > Cellobiose > Lyxose

PAA, *Phytolacca americana* Pa-2 agglutinin (pokeweed) (105-107)

See Galβ1→3 or 4GlcNAc specific agglutinin (II-2) and chitin oligosaccharide specific lectin, Section **IV**.

PSA, *Pisum sativum* agglutinin (Pea) (52, 167-169)

Manα1→linked > Glcα1→linked >> GlcNAc.

For comparison, see LCA and VFA. LFucα1→6 linked to GlcNAc at the **reducing** end is also an important binding determinant as described for Lentil in this Section. The removal of the α1→6 linked fucose residue produces an αLFuc glycopetide 17-fold less inhibitory. The biantennary structure is also an essential feature (173).

SLA, *Sternbergia lutea* agglutinin (174)

Manα1→6[Manα1→3]Manα1→6[Manα1→3]Man >> Manα1→3 Manα-*O*-Me, Manα1→6[Manα1→3]Man > Manα1→6Manα-*O*-Me > Manα1→3Man > Manα1→2Manα-*O*-Me > Manα1→6Man ≥ Manα1→2Man > Bn αMan ≥ Me αMan > Man > Me βMan

TGA, *Tulipa gesneriana* (Tulip bulbs) agglutinin (175)

Manα1→6Manα1→6Manα1→6Man > Manα1→6Manα1→6Man, Manα1→3Manα1→2Manα1→2Man > Manα1→2Manα1→2Man, Manα1→2 or 6Man > Man, LFuc.

VFA, *Vicia faba* agglutinin (Fava bean) (52, 167, 169, 173)

Manα1→linked > Glcα1→linked >> GlcNAc.

For comparison see LCA and PSA. It is the same as Lentil and PSA, i.e. the binding properties of Lentil, *Vicia* and *Pisum* are similar in many aspects, but quite different fromConA. Specifically, LFucα1→6 linked to GlcNAc at the reducing end is not required for ConA binding (52). The fine affinity of these three lectins can be differentiated by some of the mannose containing oligosaccharides such as sialoside 9 and asialooligsaccharides derived from sialoside 9, as shown in Table 2, reference 52.

Diocleinae lectins (176)

Cratylia brasiliensis (Cbr, L), *C. bonariensis* (Cbon, L), *C. floribunda* (Cflo, L) *Dioclea rostrata* (Dros, L), *D.guianensis* (Dgui, L), *D. violacea* (Dviol, L), *D. virgata* (Dvir, L), *D. grandiflora* (Dgr, L)

Manα1→ linked > Me αMan > Glcα1→ linked > Glc

These lectins from the Diocleinae subtribe are Man/Glc-binding proteins with conserved binding specificities for the core trimannoside of *N*-linked carbohydrates. Using deoxy analogs of the trimannoside, all lectins were shown to possess conserved binding sites that recognize the 3-, 4- and 6-hydroxyl groups on the α1→6Man, the 3- and 4-hydroxyl group on the α1→3Man, and the 2- and 4-hydroxyl groups of the control Man of the trimannoside. The fine affinity of these lectins and the comparison with ConA were shown in references Gupta *et al.* (177) and Dam *et al.*. (176)

III-2. Other Man and/or Glc-specific lectins required for further characterization

AloeL, Kidachi aloe (*Aloe arborescens Miller var. natalensis Berger*) (178a), Man

Boonins, A, B, C and D, *Boodlea coacta* iso agglutinins (Marin green alga) (178a), Man

CMinL, *Clivia miniata* lectin (179), Man

Cymbidium hybrid lectin (180), Man

DLehL, *Dioclea lehmanni* lectin (181)

p-NO_2-Phenyl αGlc >>>> Me αMan > Me αGlc, Trehalose > Palatinose, Melezitose ≥ Sorbose ≥ Glc, Man, Fru > 2-Deoxy-Glc, GlcNAc, Sucrose > Me βGlc

EHA, *Epipactis helleborine* agglutinin (broad-leaved helleborine) (180), Man

LOL, Lathyrus ochrus, LaOL-1, -2 (180a, 180b)

Lathyrus ochrus lectin is a well-defined saccharide sequence on a bi-antennary *N*-acetyllactosamine-type glycan. The best inhibitor is a glycopeptide from human lactotransferrin with an a-L-fucose residue at the C-6 position on the *N*-acetylglucosamine residue involved in the *N*-glycosylamine bond.

Human lactotransferrin-1

```
NeuAcα2→6Galβ1→4GlcNAcβ1→2Man↓α1,3
                                  Manβ1→4GlcNAcβ1→4GlcNAcβ1-Asn  >
                                   ↑α1, 6              ↑α1,6
  NeuAcα2→6Galβ1→4GlcNAcβ1→2Man                        Fuc
```

LaOL, *Lathyrus odoratus* lectin (Sweet pea) (19), αMan > αGlc > GlcNAc

OML, *Oncorhynchus mykiss* lectin (rainbow trout) (182)

Glc, GlcNAc, Man, ManNAc, Fuc, Maltose, Me αMan

OEMLs, *Osmerus eperlanus mordax* lectins, OEML-1, -2, -3 (Olive rainbow smelt) (183)

OEML-1

Rha > Neohesperidose > Man > LLyxose, Melibiose > Me αGal, Raffinose > Me βGal > Gal, DFuc >> Lac > Lactulose > LFuc, DLyxose > DXylose > LXylose

OEML-2

Neohesperidose > Rha > Man, Melibiose > LLyxose ≥ Me αGal > Me βGal, Raffinose > Gal, DFuc > Lac > LFuc > DLyxose > Lactulose

OEML-3

Rha > Neohesperidose > Man > Raffinose > LLyxose > Melibiose ≥ Me αGal, Me βGal > Gal > DFuc >> Lac > DLyxose > Lactulose > LFuc > DLyxose

OVL, *Onobrychis viciifolia* lectin (Sainfoin) (19), αMan > αGlc

SAL, *Silurus asotus* lectin (catfish) (184)

Me αRha > Rha > Man > Thiodigalactoside > Stachyose > Phenyl αGal, Neohesperidose, Melibiose, Raffinose > LLyxose, Me αGal, Me βGal > Gal, DFuc > Ara > Lac > Lactulose > Phenyl βGal > Fru > Glc > Sorbose, Xylose

VCL, Vicia cracca lectin (Man) (19), αMan > αGlc

VEL, *Vicia ervilia* lectin (19), Man, Glc, Me αMan, maltose, α, α′ trehalose

VSL, *Vicia sativa* lectin (19), Melezitose > Me αMan > Man> Glc

VVL_M, *Vicia villosa* mannose-binding lectin (185) Me αMan > Glc, GlcNAc

IV. GlcNAcβ1→4 specific lectins (chitin oligosaccharide-specific agglutinins)

GlcNAcβ1→4GlcNAc sequence, the lectin C determinant, can be found in mammalian *N*-linked carbohydrate chains as part of core trisaccharide at the reducing end (27).

APA-I, *Aaptos papillata* I agglutinin (186)

(GlcNAcβ1→4)$_4$ > (GlcNAcβ1→4)$_3$ > GlcNAcβ1→4GlcNAc.

APA-II and III, *Aaptos papillata* II and III agglutinins (186)

(GlcNAcβ1→4)$_3$ > GlcNAcβ1→4GlcNAc.

CIA, *Coccinia indica* agglutinin (187)

Umb (GlcNAc)$_4$, Umb (GlcNAc)$_3$, Umb: 4-methylumbelliferyl

CMA, *Chelidonium majus* agglutinin (greater celandine seed) (188)

(GlcNAc)$_4$ > (GlcNAc)$_3$ > (GlcNAc)$_2$ > GlcNAc

CMA, *Cucurbita maxima* agglutinin (Pumpkin) (189)

(GlcNAcβ1→4)$_4$ > (GlcNAcβ1→4)$_3$ > GlcNAcβ1→4GlcNAc.

CMA-II, *Cytisus multiflorus-II* agglutinin (190)

(GlcNAcβ1→4)$_3$ and GlcNAcβ1→4GlcNAc > Cellobiose (Glcβ1→4Glc) and Salicin(*O*-hydroxymethylphenyl-β-D-Glucopyranoside) > Lactose.

CPA, *Cucurbita pepo* agglutinin (191)

(GlcNAcβ1→4)$_4$, (GlcNAcβ1→4)$_3$ and GlcNAcβ1→4GlcNAc > GlcNAc→ NeuNAc > GlcNAc.

DSA, *Datura stramonium* agglutinin (Thorn-apple)

see DGalβ1→3(4)GlcNAc specific agglutinin, section **II-2**.

FCL, *Ficus cunia* lectin (192)

N,N'N"-tetraacetylchitotetraose > GlcNAcβ1→6GlcNAc >Galβ1→4GlcNAc, N,N',N"-triacetylchitotriose > Me βGlcNAc > Me αGlcNAc > Galβ1→4Glc

GSA-II, *Griffonia (Bandeiraea) simplicifolia II* agglutinin (193-195)

(GlcNAcβ1→4)$_4$, (GlcNAcβ1→4)$_3$, and GlcNAcβ1→4Glc > GlcNAc.

LAA, *Luffa acutangula* agglutinin (Ridge gourd) (196)

(GlcNAcβ1→4)$_5$ > (GlcNAcβ1→4)$_3$ > GlcNAcβ1→4GlcNAc > GlcNAc.

LEA, *Lycopersicon esculentum* agglutinin (Tomato)(104, 197)

(GlcNAcβ1→4)$_4$ > (GlcNAcβ1→4)$_3$ > GlcNAcβ1→4GlcNAc. *N*-acetyllactosamine as active oligosaccharide, see Section **II-2**.(95, 103)

OSA, *Oryza sativa* agglutinin, *Rice* agglutinin.(198)

Fetuin >> Desialized fetuin >> (GlcNAcβ1→4)$_3$ > GlcNAcβ1→4GlcNAc >> GlcNAc > GalNAc.

PAA, *Phytolacca americana Pa-2* agglutinin (Pokeweed) (105-107)

See Galβ1→4(3)GlcNAc specific agglutinin.(**II-2**).

PVA, *Psathyrella velutina* agglutinin (Mushroom) (199)

GlcNAcβ1→3Galβ1→4GlcNAc > GlcNAcβ1→6Galβ-*p*-nitro-phenyl >GlcNAc
and GlcNAcβ1→3Galβ1→3GalNAcα-benzyl > GlcNAcβ1→6[Galβ1→3]GalNAcα-benzyl
and GlcNAcβ1→6-GalNAcα-benzyl > (GlcNAcβ1→4)$_{2 \text{ or } 3}$.

STA, *Solanum tubersum* agglutinin (Potato) (52, 200)

(GlcNAcβ1→4)$_4$ > (GlcNAcβ1→4)$_3$ > GlcNAcβ1→4GlcNAc and
(GlcNAc→MurNAc)$_2$ >>> GlcNAc.

Tomentine, *Codium tomentosum* (Huds.) Stackh. lectin (green alga) (201)

GlcNAc > Melibiose, Raffinose > ManNAc

UDA, *Urtica dioica* agglutinin (Stinging nettle) (202)

(GlcNAcβ1→4)$_4$ > (GlcNAcβ1→4)$_3$ >> GlcNAcβ1→4GlcNAc and
Galβ1→4GlcNAcβ1→6Man ≥ Manα1→3Manβ1→4GlcNAc >
Galβ1→4GlcNAc(*N*-acetyllactosamine) >> GlcNAc >> GalNAc.

UEA-II, *Ulex europaeus* II agglutinin (see LFuc specific agglutinins, **V-I**).

VisalbcBA, *Viscum album* chitin binding agglutinin (European mistletoe) (203)

GlcNAcβ1→4GlcNAcβ1→4GlcNAcβ1→4GlcNAc >>
GlcNAcβ1→4GlcNAcβ1→4GlcNAc >> GlcNAcβ1→4GlcNAc
> GlcNAc

VPL, Chitovibrin: *Vibrio parahemolyticus* (203a)

chitin, chitin-ologomers > dp9 (dp, degree of polymer)

WGA, *Triticum vulgaris* agglutinin (Wheat germ)(52, 122-127)

(GlcNAcβ1→4)$_{3\text{-}5}$ > GlcNAcβ1→4GlcNAc and sialic acid.
As *N*-acetyllactosamineβ1→linked specific lectin,
see Section **II-2**.

HBL, Hevein, *Hevea brasiliensis* lectin (rubber tree) GlcNAc (204)

V. LFuc specific lectins

V-1. Monofucosyl specific agglutinins

AAnA, *Anguilla anguilla* agglutinin (Eel serum) (19, 205-207)

LFucα1→3Gal > LFuc > LFucα1→3Galβ1→4GlcNAc.

AAurA, *Aleuria aurantia* agglutinin (Orange peel fungus) (208, 209)

LFuc > LFucα1→2Galβ1→4Glc(2'-Fucosyllactose) >
Galβ1→3[LFucα1→2]GlcNAcβ1→3Galβ1→4Glc
(Lacto-*N*-fuçopentaose II, LNFP-II)
Galβ1→4[LFucα1→3]Glc(3'-fucosyllactose) >
LFucα1→2Galβ1→3GlcNAcβ1→3Galβ1→4Glc
(Lacto-*N*-fucopentaose I, LNFP-I).

```
It binds to   GlcNAcβ1→4GlcNAc-Asn-peptide/protein and/or
                        LFucα1,6↑
LFucα1→6GlcNAc Asn-peptide/protein.
```

Debray and Montreuil (1989) (210)

```
┌ Galβ1→4GlcNAcβ1→  ┐2Manα1→3Manβ1→4GlcNAcβ1→4GlcNAcβ1→4Asn >
│        ↑ α1,3     │           ↑ α1,6              ↑ α1,6
└       LFuc        ┘4         LFuc                LFuc

NeuAcα1→6Galβ1→4GlcNAcβ1→2Manα1→3Manβ1→4GlcNAcβ1→4GlcNAcβ1→4Asn >
                                   ↑ α1,6             ↑ α1,6
                   Galβ1→4GlcNAcβ1→2Man               LFuc
                           ↑ α1,6
                          LFuc

     LFucα1→6GlcNAcβ1→4Asn, LFucα1→2Galβ1→4Glc          >
                                       ↑ α1,3
                                      LFuc

NeuAcα1→6Galβ1→4GlcNAcβ1→2Manα1→3Manβ1→4GlcNAcβ1→4GlcNAcβ1→4Asn,
                                   ↑ α1,6               ↑ α1,6
          NeuAcα1→6Galβ1→4GlcNAcβ1→2Man                LFuc

GalNAcα1→3[LFucα1→2]Gal > Manα1→3Manβ1→4GlcNAcβ1→4GlcNAcβ1→4Asn,
                                                         ↑ α1,6
                                                        LFuc

     GalNAcα1→3Galβ1→4Glc    > LFuc > Galα1→3Galβ1→4Glc      >
                 ↑ α1,2 ↑ α1,3                  ↑ α1,2 ↑ α1,3
                LFuc    LFuc                   LFuc    LFuc

NeuAcα1→6Galβ1→4GlcNAcβ1→2Manα1→3Manβ1→4GlcNAcβ1→4GlcNAcβ1→4Asn >>>
                                   ↑ α1,6
        NeuAcα1→6Galβ1→4GlcNAcβ1→2Man

     Me αMan
```

BFL, *Butea frondosa* lectin, (see Gal/GalNAc specific lectin)

CSA, *Cytisus sessilifolius* agglutinin (211, 206)

LFucα1→2Galβ1→4Glc (2'-fucosyllactose) >
LFucα1→2Galβ1→3GlcNAcβ1→3Galβ1→4Glc (LNFP-I) >
Galβ1→3GlcNAcβ1→3Galβ1→4Glc (Lacto-*N*-tetraose) and
Galβ1→4GlcNAcβ1→3Galβ1→4Glc(Lacto-*N*-neotetraose).

EEA, *Euonymus europaeus* agglutinin (Spindle tree) (212, 213)

Galα1→3[LFucα1→2]Galβ1→4GlcNAcβ1→6[Galβ1→3]GalNAc-ol,
and Galα1→3[LFucα1→2]Galβ1→4GlcNAcβ1→6R >
Galα1→3[LFucα1→2]Galβ1→3GlcNAcβ1→3(4)Gal-ol >
LFucα1→2Galβ1→3GlcNAcβ1→3Galβ1→4Glc(LNFP-I) >
Galα1→3[LFucα1→2]Galβ1→4[LFucα1→3]Glc (Urine B
pentasaccharide) > Galα1→3Gal > Galβ1→4GlcNAc.

The extent of specificity can be summarized in the following order: (213)

1. best, tetrasaccharides with blood group B activity.Galα1→3[LFucα1→2]Galβ1→4(3)GlcNAc.
2. good, trisaccharides derived from blood group B tetrasaccharides.
   LFucα1→2Galβ1→4(3)GlcNAc, Galα1→3Galβ1→4GlcNAc or Galα1→3[LFucα1→2]Gal.
3. poor, disaccharides derived from blood group B tetrasaccharides.
   Fucα1→2Gal, Galβ1→4GlcNAc, or Galα1→3Gal.

LAA, *Laburnum alpinum* agglutinin (206)

LFucα1→2Galβ1→4Glc (2'-fucosyllactose) >
LFucα1→2Galβ1→3GlcNAcβ1→3Galβ1→4Glu (LNFP-I)
Galβ1→3GlcNAcβ1→3Galβ1→4Glc(Lacto-*N*-tetraose) and
Galβ1→4GlcNAcβ1→3Galβ1→4Glc(Lacto-*N*-neotetraose).

LTA, *Lotus tetragonolobus* agglutinin[6] (Asparagus pea) (52, 214)

LFucα1→2Galβ1→4[LFucα1→3]GlcNAcβ1→6R >
LFucα1→2Galβ1→4[LFucα1→3]GlcNAc > LFucα1→2Galβ1→4GlcNAc.

TJA-II, *Trichosanthes japonica* root tuber agglutinin-II (215)

```
LFucα1→2Galβ1→4Glc (2'-fucosyllactose) >>

LFucα1→2Galβ1→3GlcNAcβ1→3Galβ1→4Glc (Lacto-N-fucopentaose I)

  >> Galβ1→4GlcNAcβ1→6[Galβ1→4GlcNAcβ1→2]Man
                                          ↓ α1,6
                                         Manβ1→4GlcNAc >
                                          ↑ α1,3
      Galβ1→4GlcNAcβ1→2[Galβ1→4GlcNAcβ1→4]Man

                              Galβ1→4GlcNAcβ1→2Man
                                          ↓ α1,6
                                         Manβ1→4GlcNAc >
                                          ↑ α1,3
      Galβ1→4GlcNAcβ1→2[Galβ1→4GlcNAcβ1→4]Man

                              Galβ1→4GlcNAcβ1→2Man
                                          ↓ α1,6
                                         Manβ1→4GlcNAc,
                                          ↑ α1,3
                              Galβ1→4GlcNAcβ1→2Man
```

GalNAc, Me αGalNAc, Me βGalNAc >
Galβ1→4GlcNAcβ1→2Manα1→3Manβ1→4GlcNA,
Galβ1→3GlcNAcβ1→3Galβ1→4Glc (Lacto-*N*-tetraose), Galβ1→4GlcNAc,
Galβ1→4Glc > Gal, Me αGal, Me βGal, GalNAc-HCl DFuc > DTalose >
2-Deoxy-D-Gal

UEA-I, *Ulex europaeus*-I agglutinin[6] (Gorse or Furz seeds) (52, 216, 217)

LFucα1→2Galβ1→4GlcNAcβ1→6R >
LFucα1→2Galβ1→4[LFucα1→3]GlcNAcβ1→6R >
LFucα1→2Galβ1→4GlcNAc (2'-fucosyllactosamine) >
LFucα1→2Galβ1→4[LFucα1→3]Glc (Lacto-difucotetraose).

---

6. For differential affinities among these agglutinins, see reference 213.

UEA-II, *Ulex europaeus*-II agglutinin (52, 218, 219)

LFucα1→2Galβ1→4GlcNAcβ1→6R >
LFucα1→2Galβ1→4Glc (2'-fucosyllactose) >
(GlcNAcβ1→4)4 > (GlcNAcβ1→4)3 > GlcNAcβ1→4GlcNAc >
GalNAcα1→3Galβ1→3GlcNAc > GlcNAcβ1→6Gal
Galβ1→3GlcNAcβ1→3Galβ1→4Glc (Lacto-*N*-tetraose) >
LFucα1→2Galβ1→3GlcNAcβ1→3Galβ1→4Glc (L*N*FP-I) >
GalNAcα1→3Gal.

V-2. Difucosyl specific agglutinins

GSA-IV, *Griffonia (Bandeiraea) simplicifolia* IV agglutinin (220)

LFucα1→2Galβ1→3[LFucα1→4]GlcNAcβ1→3Galβ1→4Glc
(Lacto-*N*-difucohexaose I) >
Galβ1→3[LFucα1→4]GlcNAcβ1→3Galβ1→4[LFucα1→3]Glc
(Lacto-*N*-difucohexaose II) >
LFucα1→2Galβ1→4[LFucα1→3]Glc (Lacto-difucotetraose)
and LFucα1→2Galβ1→4[LFucα1→3]GlcNAc >>
(Lacto-*N*-difucotetraose)
Galβ1→3[LFucα1→4]GlcNAcβ1→3Galβ1→4Glc (L*N*FP-II).

V-3. Other LFuc-specific lectins required for further characterization

STA, *Salmonella typhimurium* agglutinin (221)

LFuc > DGal > DGlu > DMan

Ulva-I, *Ulva lactuca* lectin (green marine alga) (222)

LFuc and LGal
The Ulva-I lectin reacted not only with soluble H substance, but also reacted with soluble Lewis substances present in the saliva of ABH 'secretors' Le(a-b+) and 'non-secretors' Le(a+b-).

VI. Sialic acid specific lectins (223-225)

AAusA, *Androctonus australis* agglutinin (Saharan scorpion) (226)
NeuAc.

ADA-II, Allomyrina dichtroma agglutinin II (Hemolymph of a beetle) (See **I/II** specific lectin in Section **II-2**)

AFA, *Achatina fulica* agglutinin, Achatinin$_H$, the hemolymph of a land snail. (227)

NeuNAcα2→3Galβ1→4Glc > NeuNAcα2→6Gal-OH > 9-*O*-AcNeuAc > NeuNAc > NeuNGc.

Agrocybe lectin, *Agrocybe cylindracea* lectin (edible fungus) (228)

NeuAcα2→3Galβ1→4GlcNAc > NeuAcα2→3Galβ1→4Glc > NeuAc, NeuGc > Galβ1→4GlcNAc, Galβ1→4Glc, Galβ1→3GlcNAc

AGL, Anadara granosa (228a)

NeuGcα2→6GalNAc-ol > NeuGc > > NeuAc

ATN$_H$, Achatinin$_H$, *Achatina fulica* lectin (snail) (229)

GlcNAcβ1→3(Neu5Gcα2→6)GalNAc-ol > Galβ1→3-(Neu5Gcα2→6)GalNAc-ol > Neu5,9Ac$_2$ > colominic acid >Neu5Gcα2→6GalNAc-ol, Neu5Acα2→3Galβ1→3(Neu5Acα2→6)-GalNAc-ol > Neu5Acα2→6GalNAc-ol > GlcNAcβ1→3(Neu5Acα2→6)GalNAc-ol > Ben αNeu5Ac9Pr > Neu5Gc > Neu5Acα2→3Galβ1→3GalNAc-ol > Me 9-acetamido-α–Neu5Ac > Neu5Acα2→3Galβ1→4Glc > Neu5Ac > Me 9-thioacetamido αNeu5Ac

Avocado agglutinin (230)

NeuAc > Melibiose > Stachyose.

BGA, *Biomphalaria glabrata* agglutinin (231)

NeuAc > Fru > Gal > Man > LFuc > Glc.

BLA, *Birgus latro* agglutinin (Coconut crab) (232)

NeuAc and NeuGly (NeuGlc).

CAL, *Cancer antennarius* lectin (California crab) (233)

9-*O*-Ac-NeuAc > 4-*O*-Ac-NeuAc > NeuAc

CHA-I, *Cepaea hortensis* agglutinin (snail) (234)

2-α-Me-9-*O*-Ac-NeuAc > 2-α-Me-NeuAc, 2-α-Bzl-9-*O*-prop-NeuAc > 2-α-Me-Bzl-NeuAc > 2-α-Me-9-*O*-tos-NeuAc, NeuAcα2→3Galβ1→4Glc, NeuAcα2→6Galβ1→4Glc, NeuAcα2→8NeuAcα2→3Galβ1→4Glc > 2-α-Bzl-8, 9-*O*-isoprop-NeuAc > NeuAc, 2,3-Dehydro-2-deoxy-NeuAc, 4-amino- NeuAc > 2-α-Me-4-deoxy-NeuAc, NeuAcα2→8$_{n=10-12}$ (colominic acid) > 2-β-Me-NeuAc, NeuAc-Me-ester, 2-α-N$_3$-NeuAc > 2-α-Me-NeuAc-Me-ester, 2-α-Me-NeuAc-amide > 2-α-N$_3$-6-thio-NeuAc > 2-α-Bzl-*N*-Ac-heptulosaminic acid > NeuGc

CRCA, *Carcinoscorpius rotunda cauda* agglutinin (Indian horseshoe crab) (235)

NeuAcα2→6Gal > NeuAcα2→3 or α2→6Galβ1→4Gal >Glucuronic acid > 3-deoxy-2-oxo-octonate (KDO) > GalNAc and GlcNAc.

GTA, *Gracilaria tikvahiae* agglutinin (red alga) (236)

NeuAcα2→3Galβ1→4Glc > NeuAc

HAA, *Homarus americanas* agglutinin (Lobster) (237)

NeuNAc, NeuGly (NeuGc) and ManNAc.

HEL, Hericium erinaceum (237a)

NeuGc > NeuAc > 3'-NeuAc-lactose > galacturonic acid

LFA, *Limax flavus* agglutinin (Slug) (238, 239)

NeuNAc and NeuGly (NeuGc) (238)

CMP-Neu5Ac > Me α-4-*O*-Ac-Neu5Ac > Neu5Acα2→8Neu5Ac > Me α-Neu5Ac > Bnz α-Neu5Ac ≥ Me α-4-*O*-Bnz-Neu5Ac > Me α-4-*O*-deoxy-Neu5Ac > Divalent sialoside G(4,4) > 2,3-Dehydro-2-deoxy-Neu5Ac > Neu5Ac ≥ Me α-Neu5Ac-CONHOH ≥ Me α-Neu5AcCOOMe > Me α-Neu5AcCONH$_2$ > Me α-4-epi-Neu5Ac > Me α-9-*O*-Ac-Neu5Ac > Me β-Neu5AcCOOMe > Me β-Neu5Ac ≥ Me α-4,9-di-*O*-Ac-Neu5Ac > Bnz α-5-formyl-NeuN ≥ Me α-C-7-Neu5 > Neu5Gly (NeuGlc) > GlcU > GlcNAc (239)

LPA, *Limulus polyphemus* agglutinin (Horseshoe crab) (240, 241)

Sialic acid, GlcNAc, Glucuronic acid, and Teichoic acid.

MAA, *Maackia amurensis* agglutinin (242, 243)

2,3-sialyllactose (NeuNAcα2→3Galβ1→4Glc) >>2,6 sialyllactose, lactose, and free NeuAc (inactive) (148)

Neu5Acα2→3Galβ1→4Glc > Galβ1→4GlcNAc > Galβ1→4Glc (243)

ML-I, Mistletoe toxic lectins (130-137)

ML-I is specific for sialic acid as human α_1 acid gp interacts strongly with ML-I. Sialic acid of some *O*-glycans also acts as masking molecule. (134), see II-4

PAL, *Pila globosa* albumin gland lectin (albumin gland of the apple snail) (244)

NeuGcα2→6GalNAc-ol > NeuGc > Neuraminic acid > 2-*O*-Me-β-glycoside of NeuGc > > NeuAc ≥ Methylester-2-*O*-Me-β-glycoside of NeuGc > 9-OAc- NeuGc > pyruvic acid > glycolic acid

SCA, *Sambucus canadensis* agglutinin. (North America) (245)

See SSA.

SNA-I, *Sambucus nigra* agglutinin (114, 115)

Bovine fetuin triantennary carbohydrate chains[5g] or porcine thyroglobulin biantennary carbonhydrate chains[5h] > Neu5Acα2→6Galβ1→4GlcNAcβ1→3Galβ1→4Glc > Neu5Acα2→6Galβ1→4Glc > Neu5Acα2→6Galβ1→4Glc-ol >> Neu5Acα2→3Galβ1→4Glc > Neu5Acα2→3Galβ1→3GlcNAcβ1→3Galβ1→4Glc > Galβ1→4GlcNAc and GalNAcβ1→Gal > Galβ1→4Glc and GalNAcα1→6Gal > GalNAc and GalNAcα1→3-Gal ≥ Gal > Melibiose, Raffinose and Stachyose.

```
5g. Neu5Acα2→3 (60%)
      or                  }Galβ1→4GlcNAcβ1→2Manα1,6↓
    Neu5Acα2→6 (40%)                             Manβ1→4GlcNAcβ1→4GlcNAc
          Neu5Acα2→6Galβ1→4GlcNAcβ1→2Manα1,3↑
            Neu5Acα2→3Galβ1→4GlcNAcβ1,4↑

5h.   [Galα1→3] ± Galβ1→4GlcNAcβ1→2Man1,6↓
                                             Manβ1→4GlcNAcβ1→4GlcNAc
        Neu5Acα2→6Galβ1→4GlcNAcβ1→2Man1,3↑
```

SSL, *Scylla serrata* lectin (marine crab) (246)

NeuGc, NeuGcα2→6Gal, NeuGcα2→6GalNAc, NeuGcα2→3Gal/GalNAc, NeuGcα2→3Gal

SSA, *Sambusus sieboldiana* agglutinin (Japan) (245)

SNA-I (Europe), SCA (North America) and SSA (Japan) showed a very high specificity for the Neu5Acα2→6Gal/GalNAc sequence. Difference in the detailed structure of the carbohydrate binding sites of three lectins is described in ref. 245.

TGL, *Tachypleus gigas* lectin (Southeast Asian horseshoe crab) (247)

NeuAc, NeuGc >> NeuAcα2→6Galβ1→4Glc > NeuAcα2→3Galβ1→4Glc

TJA-I, *Trichosanthes japonica* agglutinin I (248)

Neu5Acα2→6Galβ1→4GlcNAcβ1→3Galβ1→4Glc (6'-sialyllacto-*N*-neotetraose, LST-c) >>> Neu5Acα2→6Galβ1→4Glc (6'-sialyllactose) >> Galβ1→4GlcNAc > Galβ1→4Glc > Me αGal, Me βGal > Gal, D-Fuc, Me βGalNAc > D-Talose, GalNAc, Me αGalNAc > GalN-HCl

TML, *Tritrichomonas mobilensis* lectin (protozoan) (249)

NeuAcα2→6Galβ1→4Glc (6'-sialyllactose) > NeuAcα2→3Galβ1→4Glc > NeuAc > NeuAcα2→8NeuAc >> NeuGc, NeuAc-1Me (NeuAc methyl ester)

TTA, *Tachypleus tridentatus* agglutinin (Asian horseshoe crab) (250)

NeuAcα(2-8)$_{n=10\text{-}12}$ (colominic acid) > NeuAcα2→3Galβ1→4Glc, NeuAcα2→6Galβ1→4Glc > 2-α-Me-9-*O*-Ac-NeuAc, 2-α-Bzl-NeuAc, 2-β-Me-NeuAc, NeuAc-Me-ester, 2-α-Me-NeuAc-amide > 2-α-Bzl-8,9-*O*-isoprop-NeuAc ≥ 2-α-Me-NeuAc > 2-α-Me-NeuAc-Me-ester, 2-α-Me-9-*O*-tos-NeuAc > 2,3-Deoxy-NeuAc, 2,4,7,8,9-penta-*O*-Ac-NeuAc > NeuAc > 2-α-Bzl-5-*N*-fluoroacetyl-Neu, GlcNAc ≥ AllNAc > ManNAc > GalNAc, MurAc

WGA, *Triticum vulgaris* agglutinin, (see GlcNAcβ1→4 specific agglutinins Section **IV**).

REFERENCES

1. Wu, A.M., and Sugii, S. (1991) Coding and classification of D-galactose, *N*-acetyl-Dgalactosamine, and β-D-Galp-[1→3(4)-β-D-GlcpNAc, specificities of applied lectins. *Carbohydr. Res.*, 213: 127-143.
2. Wu, A.M. and Sugii, S. (1988) Differential binding properties of GalNAc and/or Gal specific lectins. *Adv. Exp. Med. Biol.*, 228: 205-263.
3. Wu, A.M., Song, S.C., Sugii, S. and Herp, A. (1997) Differential binding properties of Gal/GalNAc specific lectins available for characterization of glycoreceptors. *Ind. J. Biochem. Biophys.*, 34: 61-71.
4. Wu, A.M. (1984) Differential binding characteristics and application of DGalβ1→3DGalNAc specific lectins. *Mol. Cell Biochem.*, 61: 131-141.
5. Lis, H. and Sharon, N. (1986) Lectins as molecules and as tools. Ann. Rev. Biochem., 55:35-67.
6. Lis, H. and Sharon, N. (1986) Applications of lectin. In "The Lectins, Properties, Functions and Applications in biology and Medicine." (Liener, I.E. Sharon, N. and Goldstein, I.J. eds.) pp.33-247 Academic Press, New York.
7. Lis, H. and Sharon, N. (1974) Lectins: their chemistry and application to immunology. in "The Antigens", (Sela, M. ed) Vol 4., pp. 429-529, Academic Press, New York.
8. Lis, H. and Sharon, N. (1989) Lectin., Chapman and Hall, New York and London.
9. Gabius, H.-J., and Gabius, S., (Eds) (1993) Lectins and Glycobiology ., Springer-Verlag, Berlin, New York, London and Paris.
10. Nicolson, G.L. (1974) The interaction of lectins with animal cell surfaces. Int. Rev. Cytol., 39:89-190.
11. Goldstein, I.J., Winter, H.C. and Poretz, R. (1997) Plant lectins: tools for of complex carbohydrates. In "Glycoproteins II". (Montreuil, J., Vliegenthart, J.F.G. and Schachter, H. eds.) Elsevier, Amsterdam, New York, Oxford, Tokyo.

12. Stanley, W.S., Peters, B.P., Blake, D.A., Yep, D., Chu, E.H.Y. and Goldstein, I.J., (1979) Interaction of wild-type and variant mouse 3T3 cells with lectins from *Bandeiraea simplicifolia* seeds. Proc. Natl. Acad. Sci. U.S.A., 76: 303-307.
13. Hakomori, S. and Knnagi, (1986) in Handbook of Experimental Immunology (Weir, D.M.ed.) Vol.1, Immunochemistry, pp. 9.1-9.39 Blackwell Scientific, Oxford, London and Boston.
14. Merkle, R.K. and Cumming, R.D., (1987) Lectin affinity chromatography of glycopeptides. Meth. Enzymol., 138:232-259.
15. Gallagher, J.T., (1984) Carbohydrate-binding properties of lectins: a possible approach to lectin nomeclature and classification. Biosci. Rep., 4: 621-632.
16. Mäkelä, O., (1957) Studies in hemagglutinins of leguminosae seeds. Ann. Med. Exp. Biol. Fenn. Suppl., 11 35:1-156.
17. Wu, A.M. and Sugii, S., and Herp, A. (1988) A guide for carbohydrate specificities of lectins. In "The Molecular Immunology of Complex Carbohydrates" (Wu, A.M. ed.), Plenum Press, New York and London. Adv. Exp. Med. Biol. 228:819-847.
18. Wu, A.M. (2000) Expression of binding properties of Gal/GalNAc reactive lectins by mammalian glycotopes (I-4). In "The Molecular Immunology of Complex Carbohydrates-2" (Wu, A.M. ed) Adv. Exp. Med. & Biol. in press.
19. Goldstein, I.J. and Poretz, R.D.(1986) Isolation physicochemical characterization and carbohydrate-binding specificity of lectins. In "The Lectins, Properties, Functions and Applications in biology and Medicine." (Liener, I.E. Sharon, N. and Goldstein, I.J. eds.) pp.33-247 Academic Press, New York.
20. Springer, G.F., (1971) Blood-group and Forssman antigenic determinants shared between microbes and mammalian cells. Prog. Allergy 15:9-77.
21. Makita, A., Suzuki, C., Yoshizawa, Z., and Konno, T. (1966) Glycolipids isolated from the spleen of Gaucher's disease. Tohoku J. Exp. Med., 88:277-288.
22. Hakomori, S-I. and Kannagi, R., (1983) Glycosphingolipids as tumor-associated and differentiation markers. J. Nat. Cancer Inst., 71:231-351.
23. Wu, A.M., (1988) in Wu, A.M. (ed.) Structure concepts of the human blood group A, Le^a, Le^b, I, and i active glycoproteins purified from human ovarian cyst fluid. The Molecular Immunology of Complex Carbohydrates. Adv. Exp. Med. Biol. 228:351-394, Plenum Press. New York and London.
24. Maisonrouge-McAuliffe, F. and Kabat, E.A., (1976) Immunochemical studies on blood groups. Heterogeneity of oligosaccharides liberated by degradation with alkaline borohydride of two human ovarian cyst fractions differing in B, I, and i active and in reactivity toward concanavalin A. Arch. Biochem. Biophys., 175:90-113.
25. Watkins, W.M. (1972) Blood-group specific substances. in Glycoproteins (Gottschalk, A.ed.) 2nd Ed. pp.830-890. Elsevier, Amsterdam.
26. Watkins, W.M., (1980) Biochemistry and genetics of the ABO, lewis, and P blood group system. advan. Hum. Genet. 10:1-136.
27. Fukuda, M. and Fukuda, M.N., (1984) in (Ivatt, R.J. ed)the Biology of Glycoproteins pp.183-234. Plenum Press, New York and London.
28. Wieruszeski, J.-M., Michalski, J.-C., Montreuil, J., Strecker, G., Peter-Katalinic J., Egge, H., van Halbeek, H., Mutsaers, J.H.G.M. and Vliegenthart, J.F.G. (1987) Structure of the monosialyl oligosaccharides derived from salivary gland mucin glycoproteins of the Chinese swiftlet (Genus Collocalia). J. Biol. Chem., 262: 6650-6657.
29. Maliarik, M.J., Roberts, D.D. and Goldstein, I.J. (1987) Properties of the lectin from the Hog peánut (*Amphicarpaea bracteata*). Arch. Biochem. Biophy., 255:194-200.
30. Wu, A.M., Song, S.-C., Wu, J.H., Chang, K.S.S. and Kabat, E.A. (1997) Further characterization of the binding properties of a GalNAc specific lectin from Codium Fragile subspecies tomentosoides. Glycobiology, 7, 1061-1066.
31. Wu, A.M., Song, S.-C., Hwang, P.-Y., Wu, J.H. and Chang, K.S.S. (1995) Binding studies on the combining sites of a GalNAcα1→ specific lectin with Thomsen-Friedenreich activity prepared from Green marine alge *Codium fragile* subspecies *tomentosoides*. Eur. J. Biochem., 233, 145-151.
32. Piller, V., Piller, F. and Cartron. J.-P.,(1990) Comparison of the carbohydrate-binding specificities of seven *N*-acetyl-D-galactosamine-recognizing lectins. Eur.J. Biochem. 191:461-466.
33. Etzler, M.E. and Kabat, E.A. (1970) Purification and characterization of a lectin (plant hemaggutinin) with blood group A specificity from *Dolichos biflorus*. Biochemistry, 9:869-877.
34. Baker, D.A., Sugii, S., Kabat, E.A., Ratcliffe, R. M., Hermentin, P. and Lemieux, R.U. (1983) Immunochemical studies on the combining sites of Forssman hapten reactive hemagglutinins from *Dolichos biflorus*, *Helix pomatia* and *Wistaria floribunda*. Biochemistry, 22: 2741-2750.
35. Wu, J.H., Song, S.C., Chen, Y.Y., Tsai, M.C., Kabat, E.A. and Wu, A.M. (1998) Muti-antennary Galβ1→4GlcNAc and Galβ1→3GalNAc clusters as important ligands for a lectin isolated from the sponge *Geodia cydonium*. FEBS Lett. 427: 134-138.
36. Hanishch, F.-G., Baldus, S.-E., and Kummel, A.T. (1996) Forssman disaccharide is the specific ligand of a galectin from the sponge *Geodia cydonium* but does not mediate its binding to nuclear protein np56. Glycobiology, 6, 321-336.

37. Bretting, H., Phillips, S.G., Klumpart, H.J. and Kabat, E.A. (1981) A mitogenic lactose-binding lectin from the sponge *Geodia cydonium*. J. Immunol., 127:1652-1658.
38. Murphy, L.A. and Goldstein, I.J. (1977) Five α-D-galactopyranosyl-binding isolectins from *Bandeiraea simplicifolia* seeds. J. Biol. Chem., 252: 4739-4742.
39. Wood, C., Kabat, E.A., Murphy, L.A. and Goldstein, I.J. (1979) Immunochemical studies on the combining sites of the two isolectins, A_4 and B_4, isolated from *Bandeiraea simplicifolia*. Arch, Biochem. Biophys., 198: 1-11.
40. Wu, A.M., Wu, J.H., Chen, Y.Y., Song, S.C. and Kabat, E.A. (1999) Further characterization of the combining sites of *Bandeiraea* (*Griffonia*) *simplicifolia* lectin-I, isolectin A_4. Glycobiology 9: 1161-1170.
40a. Wu, A.M., Wu, J.H., Song, S.-C. and Herp, A. (1996) *Bandeiraea* (*Griffonia*) *simplicifolia* lectin-I, isolectin A_4, reacting with Tn (GalNAcα1→Ser/Thr) or galabiose (Galα1→4Gal) containing ligands. FEBS lletters, 398: 183-186.
41. Hammarström, S. and Kabat, E.A.(1969) Purification and characterization of a blood-group A reactive hemagglutinin from the snail *Helix pomatia* and a study of its combining site. Biochemistry, 8:2696-2705.
42. Mo, H. and Goldstein, I.J. (1994) Isolation and characterization of a Forssman antigen-binding lectin from velvet bean (*Mucuna derringiana*) seeds. Glycoconjugate J. 11:424-431.
43. Matsuda, T. Kabat, E.A. and Surolia, A (1989) Carbohydrate binding specificity of the basic lectin from winged bean (*Psophocarpus tetragonolobus*). Mol. Immunol. 26:189-195.
44. Sugii, S. and Kabat,E.A. (1980) Immunochemical specificity of the combining site of *Wistaria floribunda* hemagglutinin. Biochemistry, 19:1192-1199.
45. Song, S.C., Chen, Y.Y., Herp, A., Tsai, M.S. and Wu, A.M. (1999) Lectinochemical characterization of a GalNAc and muti-Galβ1→4GlcNAc reactive lectin from *Wistaria sinensis* seeds. Eur. J. Biochem. 266: 778-788.
46. Ahmed, H. and Chatterjee, B.P. (1988)*Wistaria sinensis* agglutinin: purification, carbohydrate specificity, and characterization of the combining site. Carbohydr. Res., 177: 173-183.
47. Mo, H., Van Damme, E.J.M., Peumans, W.J. and Goldstein, I.J. (1994) Isolation and characterization of an N-acetyl-D-galactosamine-binding lectin from Dutch Iris bulbs which recognizes the blood group A dissacharide (GalNAcα1→3Gal). J. Biol. Chem. 269: 7666-7673.
48. Hammarström, S., Murpfy, L.A., Goldstein, I.J. and Etzler, M.E. (1977) Carbohydrate binding specificity of *N*-acetyl-D-galactosamine- "specific" lectins: *Helix pomatia* A hemagglutinin, soy bean agglutinin, lima bean lectin and *Dolichos biflorus* lectin. Biochemistry, 16: 2750-2755.
49. Roberts, D.D. and Goldstein, I.J. (1984) Effect of carbohydrate and metal ion binding on the reactivity of the essential thiol groups of lima bean lectin. J. Biol. Chem., 259: 903-908.
50. Sikder, S.K., Kabat, E.A., Roberts, D.D. and Goldstein, I.J. (1986) Immunochemical studies on the combining site of the blood group A-specific lima bean lectin. Carbohydr. Res., 151:247-260.
51. Pereira, M.E.A., Kabat, E.A. and Sharon, N. (1974) Immunochemical studies on the specificty of soybean agglutinin.Carbohydr. Res., 37:89-102.
52. Debray, H., Decout, D., Strecker, G., Spik, G. and Montreuil, J. (1981) Specificty of twelve lectins towards oligosaccharides and glycopeptides related to *N*-glycosylproteins. Eur. J. Biochem., 117:41-55.
53. Kaku, H., Peumans, W.J., and Goldstein, I.J. (1990) Isolation and characterization of a second lectin SNA-II) present in elderberry (*Sambucus nigra L.*) bark. Arch. Biochem. Biophys. 277:255-262.
54. Kaladas, P.M., Kabat, E.A., Kimura, A. and Ersson, B. (1981) The specificity of the combining site of the lectin from *Vicia villosa* seeds which reacts with cytotoxic T-lymphocytes. Mol. Immunol., 18:969-977.
55. Tollefsen, S.E. and Kornfeld, R. (1983) Isolation and characterization of lectins from *Vicia villosa*. Two distinct carbohydrate binding activities are present in seed extracts. J. Biol. Chem., 258:5166-5171.
56. Duk,M., Mitra, D., Lisowska, E., Kabat, E.A., Sharon, N. and Lis, H. (1992) Immunochemical studies on the combining site of the A+N blood type specific *Moluccella laevis* lectin. Carbohydr. Res. 236: 245-258.
57. Teneberg, S., Leonardsson, I., Angstrom, J., Ehrlich-Rogozinski, S. and Sharon, N. (1994) Characterization of the specificity of binding of *Moluccella laevis* lectin to glycosphingolipids. Glycoconjugate J. 11: 418-423.
58. Piller, V., Piller, F. and Cartron, J-P. (1986) Isolation and characterization of an *N*-acetylgalactosamine specific lectin from *Salvia sclarea* seeds. J. Biol. Chem., 261: 14069-14075.
59. Tollefsen, S.E. and Kornfeld, R. (1983) The B4 lectin from Vicia villosa seeds interact with *N*-acetyl-galactosamine residues α-linked to serine or threonine residues in cell surface glycoproteins. J. Biol. Chem., 258: 5172-5176.
60. Kawagishi, H., Nomura, A., Yumen, T. and Mizuno, T. (1988) Isolation and properties of a lectin from the fruiting bodies of *Agaricus blazei.*, Carbohydr. Res., 183: 150-154.
60a. Zenteno, E. and Ochoa, J.L. (1988) Purification of a lectin from *Amaranthus leucocarpus* by affinity chromatography. Phytochemistry 27: 313-317.

61. Belogortseva, N.I., Molchanova, V.I., Kurika, A.V., Skobun, A.S. and Glazkova, V.E. (1998) Isolation and characteriz-ation of new GalNAc/Gal-specific lectin from the sea mussel *Crenomytilus grayanus*. Comp. Biochem. Physiol. 119C: 45-50.
62. Khang, N.Q., Jean-Luc, G. and Johan, H. (1990) A blood group A specific lectin from the seeds of *Crotalaria striata*. Biochim. Biophys. Acta 1033: 210-213.
62a. Nakajima, T., Yazawa, S., Kogure, T. and Furukawa, K. (1988) Purification and characterization of *N*-acetyl-D-galactosamine-binding lectin from *Falcata japonica*., Biochim. Biophys. Acta., 964: 207-212.
62b. Nakajima, T., Kogure, T. and Furukawa, K. (1986) Specificity of hemagglutinin of *Falcata japonica* which reacts with blood group active *N*-acetyl-D-Galactosamine residues., Expl. Clin. Immunogenet., 3: 187-194.
62c. Kawagishi, H., Nomura, A., Mizuno, T., Kimura, A. and Chiba S. (1990) Isolation and characterization of a lectin from *Grifola frondosa* fruiting bodies., Biochim. Biophys. Acta., 1034, 247-252.
63. Fukuda, N., Yoshimaru, A., Hidaka, T., Ohta, H., Yamamoto, K. and Yomo, H. (1994) Isolation a nd characterization of *N*-acetylgalactosamine-specific lectin from *Galactia tashiroi* seeds. Biosci. Biotech. Biochem. 58: 423-424.
63a. Franz, H. (1990) Mistletose lectins (2). In "Advances in Lectin Research" (Franz H. ed.) pp. 33-50 Springer-Verlag, Berlin, Heidelberg, New York, London and Paris.
64. Nowaková, N. and Kocourek, J. (1974) Studies on phytohemagglutinins. XX. Isolation and characterization of hemagglutinins from scarlet runner seeds (*Phaseolus coccineus* L.). Biochim. Biophys. Acta 359: 320-333.
65. Inamori, K., Saito, T., Iwaki, D., Nagira, T., Iwanaga, S., Arisaka, F. and Kawabata, S. (1999) A newly identified horsshoe crab lectin with specificity for blood group A antigen recognizes specific O-antigens of bacterial lipopolysaccharides. J. Biol. Chem. 274: 3272-3278.
66. Animashaun, T., Togun, R.A. and Hughes, R.C. (1994) Characterization of isolectins in *Tetracarpidium conophorum* seeds (Nigerian walnut). Glycoconjugate J. 11: 299-303.
67. Presant, C.A. and Kornfeld, S. (1972) Characterization of the cell surface receptor for the *Agaricus bisporus* hemagglutinin. J. Biol. Chem., 247:6937-6945.
68. Sueyoshi, S., Tsuji, T. and Osawa, T. (1985) Purification and characterization of four isolectins of mushroom (*Agaricus bisporus*). Biol. Chem., Hoppe-Seyler, 366: 213-221.
68. Sueyoshi, S., Tsuji, T. and Osawa, T. (1985) Purification and characterization of four isolectins of mushroom (*Agaricus bisporus*). Biol. Chem., Hoppe-Seyler, 366: 213-221.
69. Irazoqui, F., Vides, M.A. and Nores, G.A. (1999) Structural requirements of carbohydrates to bind *Agaricus bisporus* lectin. Glycobiology 9: 59-64.
70. Rinderle, S.J., Goldstein, I.J., Matta, K.L. and Ratcliffe, R.M. (1989) Isolation and characterization of amaranthin, a lectin present in the seeds of *Amaranthus caudatus*, that recognizes the T- (or cryto T)-antigen. J. Biol. Chem. 264: 16123-16131.
71. Sastry, M.V.K., Banarjee, P. Patanjali, S.R., Swamy, M.J., Swarnalatha, G.V. and Surolia, A. (1986) Analysis of saccharide binding to *Artocarpus integrifolia* lectin reveals specific recognition of T-antigen (β-D-Gal(1→3)-D-GalNAc). J. Biol. Chem., 261: 11726-11733.
72. Mahanta,S.K., Sastry, M.V.K., and Surolia, A. (1990) Topography of the combining region of a Thomsen-Friedenreich-antigen-specific lectin jacalin (*Artocarpus integrifolia* agglutinin). A thermodynamic and circular-dichroism spectroscopic study. Biochem. J. 265: 831-840.
73. Ahmed, H., and Chatterjee B.P., (1989) Further characterization and immunochemical studies on the carbohydrate specificity of jackfruit (*Artocarpus integrifolia*) lectin. J. Biol. Chem. 264:9365-9372.
74. Chowdhury, S., Ahmed, H. and Chatterjee, B.P.,(1991) Chemical modification studies of *Artocarpus lakoocha* lectinartocarpin. Biochimie 73: 563-571.
75. Chatterjee, B.P., Chowdhury, S., and Ahmed, H. Further characterization of *Artocarpus lakoocha* lectin purified by rivanol. In: Abstract (B-8) 4th European Carbohydrate Symposium July 12-17,1987, Darmstadt. FRG.
76. Wu, A.M., Lin, S.R., Chin, L.K., Chow, L.P. & Lin, J. Y. (1992) Defining the carbohydrate specificities of Abrus precatorius agglutinin as T (Galβ1→3GalNAc) > **I/II** (Galβ1→3/4GlcNAc). *J.* Biol. Chem. 267, 19130-19139.
77. Cammue, B., Stinissen, H.M. and Peumans, W.J. (1985) A new type of cereal lectin from leaves of cough grass (*Agropyrum repens*). Eur. J. Biochem., 148: 315-322.
78. Osawa, T., Irimura, T. and Kawaguchi, T.(1978) *Bauhinia purpurea* agglutinin. Method.Enzymol., 50:367-372.
79. Wu, A.M., Kabat, E.A., Gruezo, F.G. and Allen, H.J. (1980) Immunochemical studies on the combining site of the D-galactopyranose and 2-acetamido-2-deoxy-D-galactopyranose specific lectin isolated from *Bauhinia purpurea alba* seeds. Arch. Biochem. Biophys., 204:622-639.
80. Guillot, J., Giollant, M., Damez, M. and Dusser, M. (1991) Isolation and characterization of a lectin from the mushroom, *Lactarius deliciosus*. J. Biochem. 109: 840-845.

81. Giollant, M., Guillot, J., Damez, M., Dusser, M., Didier, P. and Didier, E. (1993) Characterization of a lectin from *Lactarius deterrimus*. Research on the possible involvement of the fungal lectin in recognition between mushroom and spruce during the early stages of mycorrhizae formation. Plant Physiol. 101: 513-522.
82. Bausch, J.N. and Poretz, R.D. (1977) Purification and properties of the hemagglutinin from *Maclura pomifera* seeds. Biochemistry, 16:5790-5794.
83. Sarkar, M., Wu, A.M. and Kabat, E.A. (1981) Immunochemical studies on the carbohydrate specificity of *Maclura pomifera* lectin. Arch. Biochem. Biophys., 209:204-218.
84. Pereira, M.E.A., Kabat, E.A., Lotan, R. and Sharon, N.(1976) Immunochemical studies on the specificity of the peanut (*Arachis hypogaea*) agglutinin. Carbohydr. Res., 51:107-118.
85. Lotan, R. and Sharon, N. (1978) Peanut (*Arachis hypogaea*) agglutinin. Method. Emzymol., 50:361-367.
86. Baenziger, J.U. and Fiete, D. (1979) Structural determinants of *Ricinus communis* agglutinin and toxin specificity for oligosaccharides. J. Biol. Chem., 254:9795-9799.
87. Nicolson, G.L., Blaustein, J. and Etzler, M.E. (1979) Characterization of two plant lectins from *Ricinus communis* and their quantitative interaction with a murine lymphoma. Biochemistry, 13:196-204.
88. Poretz, R.D., Riss, H., Timberlake, J.W. amd Chien, S-M. (1974) Purification and properties of the hemagglutinin from *Sophora japonica* seeds. Biochemistry, 13:250-256.
89. Wu, A.M., Kabat, E.A., Gruezo, F.G. and Poretz, R.D. (1981) Immunochemical studies on the reactivities and combining sites of the d-galactopyranose- and 2-acetamido-2-deoxy-D-galactopyranose specific lectin purified from *Sophora japonica* seeds. Arch. Biochem. Biophys., 209:191-203.
90. Uhlenbruck, G. and Dahr, W. (1971) Studies on lectins with a broad agglutination spectrum. XII. *N*-acetyl-D-galactosamine specific lectins from the seeds of *Soja hispida*, *Bauhinia purpurea*, Iberis amara, *Moluccella laevis* and *Vicia graminea*. Vox Sang, 21:338-351.
91. Duk, M. and Lisowska, E. (1981) *Vicia graminea* anti-N lectin: partial characterization of the purified and its binding to erythrocytes. Eur. J. Biochem., 118: 131-136.
92. Duk, M., Lisowska, E., Kordowicz, M. and Warniowska, K. (1982) Studies on the specificity of the binding site of *Vicia graminea* anti-N lectin. Eur. J. Biochem., 123:105-112.
93. Yamashita, K., Umetsu, K., Suzuki, T., Iwaki, Y., Endo, T. and Kobata, A. (1988) Carbohydrate binding specificity of immobilized *Allomyrina dichotoma* lectin II. J. Biol. Chem. 263:17482-17489.
94. Sueyoshi, S., Yamamoto, K., and Osawa, T., (1988) Carbohydrate binding specificity of a beetle (*Allomyrina dichotoma*) lectin. J. Biochem. 103:894-899.
95. Kawashima, H. Sueyoshi, S. Li, H. Yamamoto, K., and Osawa T., (1990) Carbohydrate binding specificities of several poly-*N*-acetyllactosamine-binding lectins. Glycoconjugate J. 7:323-334.
96. Ayouba, A., Debray, H. and Rouge, P. (1992) Fine sugar specificity of the *Butea frondosa* seed lectin. Glycoconjugate J. 9: 141-147.
97. Peumans, W.J., Allen, A.K. and Cammue, B.P.A. (1986) A new lectin from meadow saffron (*Colchicum autumnale*). Plant Physiol. 82: 1036-1039.
98. Kaladas, P.M., Kabat, E.A., Iglesias, J.L., Lis, H. and Sharon, N. (1982) Immunochemical studies on the combining site of the D-galactose/*N*-acetyl-D-galacto-samine specific lectin from *Erythrina cristagalli* seeds. Arch. Biochem. Biophys., 217:624-637.
99. Perez, G. (1995) Purification and characterization of a lectin from the seeds of *Erythrina costaricensis*. Int. J. Biochem. Cell. Biol. 27: 857-863.
100. Cammue, B.P., Peeters, B. and Peumans, W.J. (1985) Isolation and partial characterization of an *N*-acetylgalactosamine-specific lectin from winter-aconite (Eranthis hyemalis) root tubers. Biochem. J., 227:949-955.
101. Li, H. Yamamoto, K., Kawashima, H. and Osawa, T. (1990) Structural requirement for the binding of oligosaccharides to immobilized lectin of *Erythrina variegata* (Linn) var. orientalis. Glycoconjugate J., 7:311-322.
102. Lis, H., Joubert, F.J. and Sharon, N. (1985) Isolation and properties of *N*-acetyllactosamine-specific lectins from nine *Erythrina* species. Phytochemistry. 24:2803-2809.
103. Zhu, B.C.R., and Laine, R.A. (1990) Novel hyperglycosylated weak gelatin-binding fibronectin from human fetal placenta. Fractionation of a high poly (*N*-acetyllactosamine) fragment by tomato lectin affinity chromatography. Eur. J. Biochem. 188:67-71
104. Kilpatrick, D.C. (1980) Purification and some properties of a lectin from the fruit juice of the tomato (*Lycopersicon esculentum*). Biochem. J., 185:262-272.
105. Irimura, T. and Nicolson, G.L. (1983) Interaction of Pokeweed mitogen with poly(*N*-acetyllactosamine)-type carbohydrate chains. Carbohydr. Res., 120:187-195.
106. Yokoyama, K., Terao, T. and Osawa, T. (1978) Carbohydrate-binding specificity of pokeweed mitogens. Biochim. Biophys. Acta, 538:384-396.
107. Katagiri, Y., Yamamoto, K., Tsuji, T. and Osawa, T. (1983) Structural requirements for the binding of high-mannose-type glycopeptide to immobilized pokeweed Pa-2 lectin. Carbohydr. Res., 120:283-292.

108. Cummings, R.D. and Kornfeld, S. (1982) Characterization of the structural determinants required for the high affinity interaction of asparagine-linked oligosaccharides with immobilized *phaseolus vulgaris* leukoagglutinin and erythroagglutinin lectins. J. Biol. Chem., 257: 11230-11234.
109. Irimura, T., Tsuji, T., Tagami, S., Yamamoto, K. and Osawa, T. (1981) Structural of a complex-type surgar chain of human glycophorin A. Biochemistry, 20:560-566.
110. Yamashita, K., Hitoi, A. and Kobata, A. (1983) Structural determinants of *phaseolus vulgaris* erythroagglutinin lectin for oligosaccharides. J. Biol. Chem., 258:14753-14755.
111. Hammarström, S., Hammarström, M.L., Sundblad, G., Arnarp, J. and Lönngren, J. (1982) Mitogenic leukoagglutinin from *Phaseolus vulgaris* binds to a pentasaccharide unit in N-acetyllactosamine-type glycoprotein glycans. Proc. Natl. Acad. Sci., 79:1611-1615.
112. Wu, A.M., Sugii, S., Gruezo, F.G. and Kabat, E.A. (1988) Immunochemical studies on the Galβ1→4 or 3GlcNAc specificity of *Ricinus communis* agglutinin. Carbohydr. Res., 178:243-257.
113. Wu, A.M., Herp, A., and Wu, J.H. (1993) Defining carbohydrate specificity of *Ricinus communis* agglutinin as Galβ1→4GlcNAc (II) > Galβ1→3GlcNAc (I) > Galα1→3GalNAc (T). Mol. Immunol., 30: 333-339.
114. Broekaert, W.F., Nsimba-Lubaki, M., Peeters, B.and Peumans,W.J. (1984) A lectin from elder (*Sambucus nigra* L.) bark. Biochem. J., 221:163-169.
115. Shibuya, N. Goldstein, I.J., Broekaert, W.F., Nsimba-Lubaki, M., Peeters, B. and Peumans, W.J. (1987) The elderberry (*Sambucus nigra* L.) bark lectin recogizes the Neu5Ac (α2→6)Gal/GalNAc sequence. J. Biol. Chem., 262: 1596-1601.
116. Desai, N.N., Allen, A.K. and Neuberger, A. (1981) Some properties of the lectin from *Datura stramonium* (thornapple) and the nature of its glycoprorein linkage. Biochem. J., 197:345-353.
117. Crowley, J.F., Goldstein, I.J., Arnarp, J. and Lonngren, J. (1984) Carbohydrate binding studies on the lectin from *Datura stramonium* seeds. Arch. Biochem. Biophys., 231:524-533.
118. Yamashita, K., Totani, K., Ohkura, T., Takasaki, S., Goldstein, I.J. and Kobata, A. (1987) Carbohydrate binding properties of complex-type oligosaccharides on immobilized *Datura stramonium* lectin. J. Biol. Chem., 262:1602-1607.
119. Cummings, R.D. and Kornfeld, S. (1984) The distribution of repeating [Galβ1,4GlcNAcβ1,3] sequences in asparagine-linked oligosaccharides of the mouse lymphoma cell lines BW5147 and PHAR2.1. Finding of oligosaccharides containing these sequences to immobilized *Datura stramonium* agglutinin. J. Biol. Chem., 259:6253-6260.
120. Sato, S., Aninashaum, S. and Hughes, R.C. (1991) Carbohydrate-binding specificity of *Tetracarpidium conophrum* lectin. J. Biol. CHem., 266: 11485-11494.
121. Animashaum, S. Togun, R.A. and Hughes, R.C. (1994) Characterization of isolectins in *Tetracarpidium conophrum seeds.* (Nigerian walnut). Glycoconj. J., 11: 299-303.
122. Goldstein, I.J., Hammarström, S., and Sundblad, G. (1975) Precipitation and carbohydrate binding specificity studies on wheat germ agglutinin. Biochim. Biophys. Acta, 405: 53-61.
123. Allen, A.K., Neuberger, A. and Sharon, N. (1973) The purification, composition and specificity of wheat-germ agglutinin. Biochem.J., 131: 155-162.
124. Gallagher, J.T., Morris, A. and Dexter, T.M. (1985) Identification of two binding sites for wheat-germ agglutinin on polylactosamine-type oligosaccharides. Biochem. J., 231:115-122.
125. Renkonen, O., Makinen, P., Hard, K., Helin, J., Penttila, L. (1988) Immobilized wheat germ agglutinin separates small oligosaccharides derived from poly-*N*-acetyllactosamino-glycans of embryonal carcinoma cells. Biochem. & Cell Biol., 66: 449-453.
126. Renkonen, O. Pettila, L. Niemela, R., Vainio. A., Leppänen, A., Helin, J., Seppo. A., Makkonen, A., and Maaheimo, H. (1991) *N*-acetyllactosaminooligosaccharides that contain the β-D-GlcpNAc-(1→6)-D-Gal or β-D-GlcpNAc-(1→6)-D-GalNAc sequences reveal reduction-sensitive affinities for wheat germ agglutinin. Carbohydr. Res., 213:169-183.
127. Wu, A.M., Wu, J.H., Song, S.-C., Tsai, M.-S., and Herp, A. (1998) Studies on the binding of wheat germ agglutinin (*Triticum vulgaris*) to *O*-glycans. FEBS Letters, 440: 315-319.
128. Umetsu, K., Yamashita, K. and Suzuki, T. (1991) Purification and carbohydrate-binding specificities of a blood type B binding lectin from hemolymph of a crab (*Charybdis japonica*). J. Biochem. 109: 718-721.
129. Wu, A.M., Song, S.-C., Wu, J.H. and Kabat, E.A. (1995) Affinity of *Bandeiraea* (*Griffonia*) *simplicifolia* lectin-I, Isolectin B4 (BSI-B4) for Gala1Æ4Gal ligand. Biochem, Biophys. Res. Commun., 216: 814-820.
130. Wu, A.M., Song, S.-C., Wu, J.H., Pfüeller, U., Chow, L.P. and Lin, J.Y. (1995) A sheep hydatid cyst glycoprotein as receptor for three toxic lectins, as well as *Abrus Precatorius* and *Ricinus communis* agglutinins. Biochim. Biophys. Acta., 1243: 124-128.
131. Wu, A.M., Watkins, W.M., Chen, C.-P., Song, S.-C., Chow, L.-P. and Lin, J.-Y. (1995) Native and/or asialo-Tamm-Horsfall glycoproteins Sd(a+) are important receptors for *Triticum vulgaris* (Whesat germ) agglutinin and for three toxic lectins (abrin-a and mistletoe) toxic lectin-I). FEBS letters, 371: 32-34.
132. Wu, A.M., Chin, L.K., Franz, H., Pfüeller, U. and Herp, A. (1992) Carbohydrate specificity of the receptor sites of Mistletoe toxic lectin-I. Biochim. Biophys. Acta., 1117: 232-234.

133. Pfüeller, U., Weyhenmeyer, R., Witthohn, K. and Pfüeller, K. (1999) Sugar recognition by mistletoe lectins: comparison of native glycoforms and the recombinant lectin. In abstract 122, INTERLEC 18, University of Portsmouth, Portsmouth, UK, 27-31 July 1999.
134. Wu, A.M., Song, S.-C., Hwang, P.-Y., Wu, J.H. and Pfüller, U. (1995) Interaction of Mistletoe toxic lectin-I with sialoglycoproteins. Biochem. Biophys. Res. Commun., 214: 396-402.
135. Debray, H., Montreuil, J. and Franz H. (1994) Fine sugar specificity of the mistletoe (*Viscum album*) lectin I. Glycoconjugate J., 11: 550-557.
136. Lee, R.T., Gabius, H.-J. and Lee, Y.C. (1994) The sugar-combining area of the galactose-specific toxic lectin of mistletoe extends beyond the terminal sugar residue: comparison with a homologous toxic lectin, ricin. Carbohydr. Res., 254: 269-276.
137. Lee, R.T., Gabius, H.-J. and Lee, Y.C. (1992) Ligand binding characteristics of the major mistletoe lectin. J. Biol. Chem., 267: 23722-23727.
138. Chen, C.P., Song, S.C., Gilboa-Garber, N., Chang, K.S.S. and Wu, A.M. (1998) Studies on the binding site of the galactose-specific agglutinin PA-IL from *Pseudomonas aeruginosa*. Glycobiology 8: 7-16.
139. Pekelharing, J.M. and Animashaun, T. (1989) Binding of *Artocarpus* lectins to plasma glycoproteins. Biochem. Soc. Trans. 17: 131-132.
140. Lönngren, J., Goldstein, I.J. and Bywater, R. (1976) Cross-linked guaran: a versatile immunosorbent for D-galactopyranosyl binding lectins. FEBS Lett. 68: 31-34.
141. Stripe, F., Licastro, F., Morini, M.C., Parente, A., Savino, G., Abbondanza, A., Bolognesi, A., Falasca, A.I. and Rossi, C.A. (1993) Purification and partial characterization of a mitogenic lectin from the latex of *Euphobia marginata*. Biochim. Biophs. Acta 1158: 33-39.
142. Seshagirirao, K. and Prasad, M.N. (1995) Purification and partial characterization of a lectin from *Euphorbia neriifolia* latex. Biochem. Mol. Biol. Int. 35: 1199-1204.
143. Zenteno, E., Vazquez, L., Chavez, R., Cordoba, F., Wieruszeski, J.M., Montreuil, J. and Debray, H. (1995) Specificity of the isolectins from the plant cactus *Machaerocereus eurca* for oligosaccharides from porcine stomach mucin. Glycoconjugate J. 12: 699-706.
144. Suzuki, T. and Mori, K. (1989) A galactose-specific lectin from the hemolymph of the pearl oyster *Pinctada fucata martensii*. Comp. Bioch. Physiol. 92B: 455-462.
145. Sharma, S. and Salahuddin, A. (1993) Purification and some properties of *Phaseolus mungo* lectin. J. Agri. Food Chem. 41: 700-703.
146. Engel, M., Bachmann, M., Schröder, H.C., Rinkevich, B., Klyajic, Z., Uhlenbruck, G. and Müller, W.E.G. (1992) A new galactose- and arabinose-specific lectin from the sponge *Pellina semitubulosa*: isolation, characterization and immunobiological properties. Biochimie 74: 527-537.
147. Komath, S.S., Nadimpalli, S.K. and Swamy, M.J. (1996) Purification in high yield and characterization of the galactose-specific lectin from the seeds of snake gourd (*Trichosanthes anguina*). Biochem. Mol. Biol. Int. 39: 243-252.
148. Animashaun, T., Togun, R.A. and Hughes, R.C. (1994) Characterization of isolectins in *Tetracarpidium conophorum* seeds (Nigerian walnut). Glycoconjugate J. 11: 299-303.
149. Mo, H., Van Damme, E.J.M., Peumans, W.J. and Goldstein, I.J. (1993) Purification and characterization of a mannose-specific lectin from shallot (*Allium ascalonicum*) bulbs. Arch. Biochem. Biophys. 306: 431-438.
150. KaKu, H., Goldstein, I.J. Van Damme, E.J.M. and Peumans, W.J. (1992) New mannose-specific lectin from garlic (*Allium sativum*) and ramsons (*Allium ursinum*) bulbs. Carbohydr. Res. 229: 347-353.
151. Dam, T.K., Bachhawat, K., Rani, P.G. and Surolia, A. (1998) Garlic (*Allium ursinum*) lectins bind to high mannose oligosaccharide chains. J. Biol. Chem. 273: 5528-5535.
152. Animashaun, T. and Hughes, R.C. (1989) *Bowringia mibraedii* agglutinin. Specificity of binding to early processing intermediates of asparagine-linked oligosaccharide and use as a marker of endoplasmic reticulum glycoproteins. J. Biol. Chem., 264:4657-4663.
153. Ueno, M., Ogawa, H., Matsumoto, I. and Seno, N. (1991) A novel mannose- specific and sugar specifically aggregatable lectin from the bark of the Japanese pagoda tree (*Sophora japonica*). J. Biol. Chem. 266: 3146-3153.
154. Peumans, W.J., Winter, H.C., Bemer, V., Van Leuven, F., Goldstein, I.J., Truffa-Bachi, P. and Van Damme, E.J.M. (1997) Isolation of a novel plant lectin with an unusual specificity from *Calystegia sepium*. Glycoconjugate J. 14: 259-265.
155. Siddiqui, S., Hasan, S. and Salahuddin, A. (1995) Isolation and characterization of *Cajanus cajan* lectin. Arch. Biochem. Biophys. 319: 426-431.
156. Kojima, K., Ogawa, H., Seno, N. and Matsumoto, I. (1991) Purification and characterization of *Canavalia gladiata* agglutinin. Carbohydr. Res. 213: 275-282.
157. Lim, S.B., Chua, C.T. and Hashim, O.H. (1997) Isolation of a mannose-binding and IgE- and IgM-reactive lectin from the seeds of *Artocarpus integer*. J. Immunol. Methods 209: 177-186.
158. Goldstein, I.J., Hollerman, C.E. and Smith, E.K.(1965) Protein-carbohydrate interaction of concanavalin A with polysaccharides. Biochemistry, 4:876-883.

159. Baenziger, J.U. and Fiete, D. (1979) Structural determinants of concanavalin A specificity for oligosaccharides. J. Biol. Chem., 254:2400-2407.
160. Allen, A.K., Desai, N.N. and Neuberger, A. (1976) The purification of the glycoprotein lectin from the broad bean (*Vicia faba*) and a comparison of its properties with lectins of similar specificity. Biochem. J., 155:127-135.
161. Hardman, K.D., and Goldstein, in I.J. (1977) The Structure and Activity of Concanavalin A. In Immunochemistry of proteins. (Atassi, M.Z. ed.) 2:373-416. Plenum Press, New York.
162. Misaki, A., Kakuta, M., Meah, Y. and Goldstein, I.J. (1997) Purification and characterization of the -1,3-mannosyl-mannose-recognizing lectin of *Crocus vernus* bulbs. J. Biol. Chem. 272: 25455-25461.
163. Mo, H., Meah, Y., Moore, J.G. and Goldstein, I.J. (1999) Purification and characterization of *Dolichos lablab* lectin. Glycobiology 9: 173-179.
164. Shibuya, N., Goldstein, I.J., Van Damme, E.J.M., and Peumans, W.J. (1988) Binding properties of a mannose-specific lectin from the snowdrop (*Galanthus nivalis*) bulb. J. Biol. Chem. 263:728-734.
165. Van Damme, E.J.M., Barre, A., Mazard, A.M., Verhaert, P., Horman, A., Debray, H., Rouge, P. and Peumans, W.J. (1999) Characterization and molecullar cloning of the lectin from *Helianthus tuberosus*. Eur. J. Biochem. 259: 135-142.
166. Kaku, H., Van Damme, E.J.M., Peumans, W.J. and Goldstein, I.J. (1990) Carbohydrate-binding specificity of the daffodil (*Narcissus pseudonarcissus*) and amaryllis (*Hippeastrum hybr.*) bulb lectins. Arch. Biochem. Biophys. 279: 298-304.
167. Kornfeld, K., Reitman, M.L. and Kornfeld, R. (1981) The carbohydrate-binding specificty of pea and lentil lectins. Fucose is an important determinant. J. Biol. Chem., 256: 6633-6640.
168. Yamamoto, K., Tsuji, T. and Osawa, T. (1982) Requirement of the core structure of a complex-type glycopeptide for the binding to immobilized lentil- and pea lectins. Carbohydr. Res., 110:283-289.
169. Matsumoto, I., Uehara, Y.,Jimbo, A. and Seno, N. (1983) Immunochemical and spectral studies on *Vicia faba* agglutinin. J. Biochem., 93:763-769.
170. Saito, K., Komae, A., Kakuta, M., Van Damme, E.J.M., Peumans, W.J., Goldstein, I.J. and Misaki, A. (1993) The α-mannosyl-binding lectin from leaves of the orchid twayblade (*Listera ovata*) application to separation of α-D-mannans from α-D-glucans. Eur. J. Biochem. 217: 677-681.
171. Kaku, H., Van Damme, E.J.M., Peumans, W.J. and Goldstein, I.J. (1990) Carbohydrate-binding specificity of the daffodil (*Narcissus pseudonarcissus*) and amaryllis (*Hippeastrum hybr.*) bulb lectin. Arch. Biochem. Biophys., 279: 298-304.
172. Yagi, F., Hidaka, M., Minami, Y., ans Tadera, K. (1996) Alectin from leaves of *Neoregelia flandria* recognizes D-glucose, D-mannose and N-acetyl D-glucosamine, differing from the mannose-specific lectins of other monocotyledonous plants. Plant Cell Physiol. 37: 1007-1012.
173. Katagiri, Y., Yamamoto, K., Tsuji, T. and Osawa, T. (1984) Structural requirements for the binding of glycopeptides to immobilized *Vicia faba* (fava) lectin. Carbohydr. Res., 129:257-265.
174. Saito, K., Misaki, A. and Goldstein, I.J. (1997) Purification and characterization of a new mannose-specific lectin from *Sternbergia lutea* bulbs. Glycoconjugate J. 14: 889-896.
175. Oda, Y., and Minami, K., (1986) Isolation and characterization of a lectin from tulip bulbs, *Tulipa gesneriana*. Eur. J. Biochem. 159:239-245.
176. Dam, T.K., Cavada, B.S., Grangeiro, T.B., Santos, C.F., de Sousa, F.A.M., Oscarson, S. and Brewer, C.F. (1998) Diocleinae lectins are a group of proteins with conserved binding sites for the core trimannoside of asparagine-linked oligosaccharides and differential specificities for complex carbohydrates. J. Biol. Chem. 273: 12082-12088.
177. Gupta, D., Oscarson, S., Raju, T.S., Stanley, P., Toone, E.J. and Brewer, C.F. (1996) A comparison of the fine saccharide-binding specificity of *D. grandiflora* lectin and concanavalin A. Eur. J. Biochem. 242: 320-326.
178. Koike, T, Titani, K., Suzuki, M., Beppu, H., Kuzuya, H., Maruta, K., Shimpo, K. and Fujita, K. (1995) The complete amino acid sequence of a mannose-binding lectin from "kidachi aloe" (Aloe arborescens Miller var. natalensis Berger). Biochem. Biophys. Res. Commun., 214: 163-170.
178a. Hori, K., Miyazawa, K. and Ito, K. (1986) Isolation and characterization of glycoconjugate-specific isoagglutinins from a marine green alga *Boodlea coacta* (Dickie) Murray et De Toni. Bot. Mar. 29: 323-328.
179. Van Damme, E.J.M., Goldstein, I.J. and Peumans, W.J. (1991) A comparative study of mannose-binding lectins from the Amaryllidaceae and Alliaceae. Phytochemistry 30: 509-514.
180. Van Damme, E.J.M., Smeets, K., Torrekens, S., Van Leuven, F. and Peumans, W.J. (1994) Characterization and molecular cloning of mannose-binding lectins from the Orchidaceae species *Listera ovata*, *Epiactis helleborine* and *Cymbidium* hybrid. Eur. J. Biochem. 221: 769-777.
180a. Debray, H. and Rouge, P. (1984) The fine sugar specificity of the Lathyrus ochrus seed lectin and isolations., FEBS letters, 176: 120-124.
180b. Imberty, A. and Pérez (1994) Molecular modelling of protein-carbohydrate interaction. Understanding the specificities of two legume lectins towards oligosaccharides., Glycobiology, 7: 351-366.

181. Perez, G., Hernandez, M. and Mora, E. (1990) Isolation and characterization of a lectin from the seeds of *Dioclea lehmanni*. Phytochemistry 29: 1745-1749.

182. Jensen, L.E., Thiel, S., Petersen, T.E. and Jensenius, J.C. (1997) A rainbow trout lectin with multimeric structure. Comp. Biochem. Physiol. 16B: 385-390.

183. Hosono, M., Kawauchi, H., Nitta, K., Takayanagi, Y., Shiokawa, H., Mineki, R. and Murayama, K. (1993) Three rhamnose-binding lectins from *Osmerus eperlanus mordax* (Olive rainbow smelt) roe. Biol. Pharm. Bull. 16: 239-243.

184. Hosono, M., Kawauchi, H., Nitta, K., Takayanagi, Y., Shiokawa, H., Mineki, R. and Murayama, K. (1993) Purification and characterization of *Silurus asotus* (catfish) roe lectin. Biol. Pharm. Bull. 16: 1-5.

185. Qian, R., Shi, W.X., Shen, Z.M. and Sun, C. (1994) A mannose-specific lectin from *Vicia villosa* seeds. Biochim. Biophys. Acta 1201: 61-68.

186. Bretting, H., Kabat, E.A., Liao, J. and Pereira, M.E.A. (1976) Purification and characterization of the agglutinins from the sponge *Aaptos papillata* and a study of their combining sites. Biochemistry, 15:5029-5038.

187. Sanadi, A.R. and Surolia, A. (1994) Studies on a chitooligosaccharide-specific lectin from *Coccinia indica*. Thermodynamic and kinetics of umbelliferyl glycoside binding. *J. Biol. Chem.*, 269: 5072-5077.

188. Peumans, W.J., De Ley, M., Stinissen, H.M. and Broekaert, W.F. (1985) Isolation and partial characterization of a new lectin from seeds of the greater celandine (*Chelidonium majus*). Plant Phyiol., 78: 379-383.

189. Read, S.M. and Northcote, D.H. (1983) Subunit structure and interactions of the phloem proteins of *Cucurbita maxima* (pumpkin). Eur. J. Biochem.,134:561-569.

190 Konami, Y., Yamamoto, K., Tsuji, T., Matsumoto, I. and Osawa,T. (1983) Purification and characterization of two types of *Cytisus multiflorus* hemagglutinin by affinity chromatography. J. Pharmacobiodyn., 6:737-747.

191. Allen, A.K. (1979) A lectin from the exudate of the fruit of the vegetable marrow (*Cucurbita pepo*) that has a specificity for β-1,4-linked *N*-acetylglucosamine oligosaccharides. Biochem. J., 183:133-137.

192. Ray, S., Ahmed, H., Basu, S. and Chatterjee, B.P. (1992) Purification, characterization, and carbohydrate specificity of the lectin of *Ficus cunia*. Carbohydr. Res., 242: 247-263.

193. Iyer, P.N.S., Wilkinson, K.D. and Goldstein, I.J. (1976) An *N*-acetyl-D-glucosamine binding lectin from *Bandeiraea simplicifolia* seeds. Arch. Biochem. Biophys., 177:330-333.

194. Ebisu, S., Iyer, P.N.S. and Goldstein, I.J. (1978) Equilibrium dialysis and carbohydrate-binding studies on the 2-acetamido-2-deoxy-D-glucopyranosyl-binding lectin from *Bandeiraea simplicifolia* seeds. Carbohydr. Res., 61:129-138.

195. Wood, C., Kabat, E.A., Ebisu, S. and Goldstein, I.J. (1978) An immunochemical study on the combining sites of the second lectin isolated from *Bandeiraea simplicifolia* (BS-II). Ann. Immunol., 129C:143-158.

196. Anantharam, V., Patanjali, S.R., Swamy, M.J., Sanadi, A.R., Goldstein, I.J. and Surolia, A. (1986) Isolation, macro-molecular properties and combining site of a chitooligosaccharide-specific lectin from the exudate of ridge gourd (*Luffa acutangula*). J. Biol. Chem., 261: 14621-14627.

197. Nachbar, M.S., Oppenheim, J.D. and Thomas, J.O. (1980) Lectins in the U.S. diet. Isolation and characterization of a lectin from the tomato (*Lycopersicon esculentum*). J. Biol. Chem., 255:2056-2061.

198. Tabary, F., Font, J. and Bourrillon, R. (1987) Isolation, molecular and biological properties of a lectin from rice embryo: relationship with wheat germ agglutinin properties. Arch. Biochem. Biophys. 259:79-88.

199. Kochibe, N. and Matta, K.L. (1988) Purification and properties of an *N*-acetylglucosamine-specific lectin from *Psathyrella velutina* mushroom. J. Biol. Chem., 264:173-177.

200. Allen, A.K. and Neuberger, A. (1973) The purification and properties of the lectin from potato tubers, a hydroxy-proline-containing glycoprotein. Biochem. J., 135:307-314.

201. Fabregas, J., Muñoz, A., Llovo, J. and Carracedo, A. (1988) Purification and partial characterization of tomentine. An N-acetylglucosamine-specific lectin from the green alga *Codium tomentosum* (Huds.) Stackh. J. Exp. Mar. Biol. Ecol., 124: 21-30.

202. Shibuya, N., Godstein, I.J., Shafer, J.A., Peumans, W.J. and Broekaert, W.F. (1986) Carbohydrate binding properties of the stinging nettle (*Urtica dioica*) rhizsome lectin. Arch. Biochem. Biophys., 249:215-224.

203. Peumans, W.J., Verhaert, P., Pfüller, U. and Van Damme, E.J.M. (1996) Isolation and partial characterization of a small chitin-binding lectin from mistletoe (*Viscum album*). FEBS Lett. 396: 261-265.

203a. Gildemeister, O., Zhu, B.C. and Laine, R.A. (1994) Chitovibrin: a chitin-binding lectin from *Vibrio parahemolyticus*., Glycoconjugate J., 11: 518-526.

204. Gidrol, X., Chrestin, H., Tan, H.L. and Kush, A. (1994) Hevein, a lectin-like protein from *Hevea brasiliensis* (rubber tree) is involved in the coagulation of latex. J. Biol. Chem. 269: 9278-9283.

205. Rege, Y.P., Painter, T.J., Watkins, W.M. amd Morgan, W.T.J. (1964) Isolation of serologically active fucose-containing oligosaccharides from human blood-group H substance. Nature, 203:360-363.

206. Matsumoto, I. and Osawa, T. (1971) On the specificity of various sugars heterologus anti-H hemagglutinin. Vox Sang., 21:548-557.

207. Baldus S.E., Thiele J., Park, Y.-O., Hanisch, F.-G., Bara, J. and Fischer, R. (1996) Characterization of the binding specificity of *Anguilla anguilla* agglutinin (AAA) in comparison to *Ulex europaeus* agglutunun I (UEA-I). Glycoconjugate J., 13: 585-590.

208. Kochibe, N. and Furukawa, K. (1980) Purification and properties of a novel fucose-specific hemagglutinin of *Aleuria aurantia*. Biochemistry, 19:2841-2846.

209. Yamashita, K., Kochibe, N., Ohkura, T., Ueda, I. and Kobata, A. (1985) Fractionation of L-fucose-containing oligosaccharides on immobilized *Aleuria aurantia* lectin. J. Biol. Chem., 260:4688-4693.

210. Debray, H. and Montreuil, J. (1989) *Aleuria aurantia* agglutinin. A new isolation procedure and further study of its specificity towards various glycopeptides and oligosaccharides. Carbohydr. Res. 185: 15-26.

211. Osawa, T. (1966) Inhibition tests of plant agglutinin from *Laburnum alpinum* and *Cytisus sessifolius* with various sugars and sugar derivatives. Indication of the type of linkage in certain disaccharides. Biochim. Biophys. Acta, 115:507-510.

212. Petryniak, J., Pereira, M.E.A. and Kabat, E.A. (1977) The lectin of Euonymus europeus: purification, characterization and an immunochemical study of its combining site. Arch. Biochem. Biophys., 178:118-134.

213. Petryniak, J. and Goldstein, I.J.(1986) Immunochemical studies on the interaction between synthetic glycoconjugates and α-L-fucosyl binding lectins. Biochemistry, 25:2829-2838.

214. Pereira, M.E.A. and Kabat, E.A. (1974) Specificity of purified hemagglutinin (lectin) from Lotus tetragonolobus. Biochemistry, 13:3184-3192.

215. Yamashita, K., Ohkura, T., Umetsu, K. and Suzuki, T. (1992) Purification and characterization of a Fucα1→2Galβ1→ and GalNAcβ1→ -specific lectin in root tubers of *Trichosanthes japonica*. J. Biol. Chem. 267:25414-25422.

216. Matsumoto, I. and Osawa, T. (1969) Purification and characterization of an anti-H(O) phytohemagglutinin of *Ulex europeus*. Biochim. Biophys. Acta., 194:180-189.

217. Pereira, M.E.A., Kisailus, E.C., Gruezo, F.G. and Kabat, E.A. (1978) Immunochemical studies on the combining site of the blood group H-specific lectin I from *Ulex europeus* seeds. Arch. Biochem. Biophys., 185:108-115.

218. Matsumoto, I. and Osawa, T. (1970) Purification and characterization of a cytisus-type anti-H(O) phytohemagglutinin from *Ulex europeus* seeds. Arch. Biochem. Biophys., 140:484-491.

219. Pereira, M.E.A., Gruzo, F.G. and Kabat, E.A. (1979) Purification and characterization of lectin II from *Ulex europeus* seeds and an immunochemical study of its combining site. Arch. Biochem. Biophys., 194:511-525.

220. Kaladas, P.M., Kabat, E.A., Shibata, S. amd Goldstein, I.J. (1983) Immunochemical studies on the binding specificity of the blood group Leb specific lectin *Griffonia simplicifolia* IV. Arch. Biochem. Biophys., 223:309-318.

221. Ensgraber, M., Genitsariotis, R., Störkel, S. and Loos, M. (1992) Purification and characterization of a *Salmonella typhimurium* agglutinin from gut mucus secretions. Microb. Pathogen. 12: 255-266.

222. Gilboa-Garber, N., Citronbaum, R., Levene, C. and Sela, R. (1988) H blood group detection by the L-fucose binding lectin of the green marine alga *Ulva lactuca*. Development. Comparat. Immunol. 12: 695-705.

223. Mandal, C., and Mandal, C. (1990) Sialic acid binding lectins. Experientia 46:433-441.

224. Goldstein, I.J. (1999) Sialic acid-binding plant lectins. In "Sialobiology and Other Novel Forms of Glycosyation." (Inoue, Y., Lee, Y.C. and Troy II, F.A. eds.) pp. 95-98 Gakushin Publishing Company, Osaka.

225. Schauer, R., Kelm, S., Reuter, G., Roggentin, P. and Shaw, L. (1995) Biochemistry and role of sialic acids. In "Biology of the Sialic Acids" (Rosenberg, A. ed.) pp. 7-37 Plenum Press, New York.

226. Vasta, G.R., Ilodi, G.H.U., Cohen, E. and Brahmi, Z. (1982) A comparative study on the specificity of *Androctonus australis* (Saharan scorpion) and *Limulus polyphemus* (horseshoe crab) agglutinins. Dev. Comp. Immunol., 6:625-634.

227. Mandal, C., and Bas, S., and Mandal, C. (1989) Physiochemical studies on Achatinin$_H$, a novel sialic acid-binding lectin. Biochem. J., 257:65-71.

228. Yagi, F., Miyamoto, M., Abe, T., Minami, Y., Tadera, K. and Goldstein, I.J. (1997) Purification and carbohydrate binding specificity of *Agrocybe cylindracea* lectin. Glycoconjugate J. 14: 281-288.

228a. Dam, T.K., Sarkar, M., Ghosal, J. and Choudhury, A. (1993) Purification and partial characterization of a *N*-glycolylneuraminic acid specific lectin from the clam Anadara granosa (L)., Biochem. Biophys. Res. Commun., 196: 422-429.

229. Sen, G. and Mandal, C. (1995) The specificity of the binding site of Achatinin$_H$, a sialic acid-binding lectin from *Achatina fulica*. Carbohydr. Res. 268: 115-125.

230. Yaakobovich, Y. and Neeman, I. (1983) Partial isolation and characterization of a hemagglutinating factor from avocado seed. Arch. Toxicol. Suppl., 6:52-57.

231. Bretting, H., Stanislawski, E., Jacob, G. and Becker, W.(1983) Isolation and characterization of a lectin from the snail. *Biomphalaria glabrata* and a study of its combining site. Biochim. Biophys. Acta, 749:143-152.

232. Vasta, G.R. and Cohen, E. (1984) Carbohydrate specificities of *Birgus latro* (coconut crab) serum lectins. Dev. Comp. Immunol., 8:197-202.

233. Ravindranath, M.H., Higa, H.H., Cooper, E.L. and Paulson, J.C. (1985) Purification and characterization of an O-acetylsialic acid-specific lectin from a marine crab *Cancer antennarius*. J. Biol. Chem. 260: 8850-8856.

234. Brossmer, R., Wagner, M. and Fischer, E. (1992) Specificity of the sialic acid- binding lectin from the snail *Cepaea hortensis*.. J. Biol. Chem. 267: 8752-8756.

235. Mohan, S., Dorai, D.T., Srimal, S. and Buchhawat, B.K. (1982) Binding studies of a sialic acid-specific lectin from horseshoe crab *Carcinoscorpius rotunda cauda* with various sialoglycoproteins. Biochim. J., 203:253-261.

236. Chiles, T.C. and Bird, K.T. (1990) *Gracilaria tikvahiae* agglutinin. Partial purification and preliminary characterization of its carbohydrate specificity. Carbohydr. Res. 207: 319-326.

237. Hall, J.L. and Rowlands, D.T. Jr. (1974) Heterogeneity of lobster agglutinins. II. Specificity of agglutinin-erythrocyte binding. Biochemistry, 13:828-832.

237a. Kawagishi, H., Mori, H., Uno, A., Kimura, A. and Chiba, S. (1994) A sialic acid-binding lectin from the mushroom *Hericium erinaceum*., FEBS Letters, 340: 56-58.

238. Miller, R.L., Collawn, J.F. Jr. and Fish, W.W. (1982) Purification and macromolecular properties of a sialic acid-specific lectin from the slug *Limax flavus*. J. Biol. Chem., 257:7574-7580.

239. Knibbs, R.N., Osborne, S.E., Glick, G.D. and Goldstein, I.J. (1993) Binding determinants of the sialic acid-specific lectin from the slug *Limax flavus*. J. Biol. Chem. 268: 18524-18531.

240. Nowak, T.P. and Barondes, S.H. (1975) Agglutinin from *imulus polyphemus*. Purification with formalinized horse erythrocytes as the affinity adsorbent. Biochim. Biophys. Acta, 393:115-123.

241. Brandin, E.R. and Pistole, T.G. (1983) Polyphemin: a teichoic acid-binding lectin from the horseshoe crab, *Limulus polyphemus*. Biochem. Biophys. Res. Comm., 113: 611-617.

242. Wang, W-C. and Cummings, R.D. (1988) The immobilized leukoagglutinin from the seeds of *Maackia amurensis* binds with high affinity to complex-type Asn-linked oligo-saccharides containing terminal sialic acid-linked α-2,3 to penultimate galactose residues. J. Biol. Chem., 263: 4576-4585.

243. Knibbs, R.N., Goldstein, I.J., Ratcliffe, R.M. and Shibuya, N. (1991) Characterization of the carbohydrate binding specificity of the leukoagglutinating lectin from *Maackia amurensis*. Comparison with other sialic acid-specific lectins. J. Biol. Chem. 266: 83-88.

244. Swarnakar, S., Chowdhury, P.S. and Sarkar, M. (1991) N-glycolylneuraminic acid specific lectin from *Pila globosa* snail. Biochem. Biophys. Res. Commun. 178: 85-94.

245. Shibuya, N., Tazak, K., Song, Z., Tarr, G.E., Goldstein, I.J. and Peumans, W.J.(1989) A comparative study of bark lectins from three elderberry (*Sambucus*) species. J. Biochem., 106:1098-1103.

246. Mercy, P.D. and Ravindranath, M.H. (1993) Purification and characterization of *N*-glycolylneuraminic-acid-specific lectin from *Scylla serrata*. Eur. J. Biochem. 215: 697-704.

247. Tsuboi, I., Matsukawa, M. and Sato, N. (1993) Isolation and characterization of a sialic acid-specific lectin from hemolymph of the southeast Asian horseshoe crab *Tachypleus gigas*. Biosci. Biotech. Biochem. 57: 1237-1242.

248. Yamashita, K., Umetsu, K., Suzuki, T. and Ohkura, T. (1992) Purification and characterization of a Neu5Acα2→6Galβ1→4GlcNAc and HSO_3^-→6Galβ1→4GlcNAc specific lectin in tuberous roots of *Trichosanthes japonica*. Biochemistry 31: 11647-11650.

249. Babál, P., Pindak, F.F., Wells, D.J. and Gardner, W.A.Jr. (1994) Purification and characterization of a sialic acid-specific lectin from *Tritrichomonas mobilensis*. *Biochem. J.* 299: 341-346.

250. Fischer, E., Khang, N.Q., Letendre, G. and Brossmer, R. (1994) A lectin from the Asian horseshoe crab *Tachypleus tridentatus*: purification, specificity and interaction with tumor cells. Glycoconjugate J. 11:51-58.

A GUIDE TO MONOCLONAL ANTIBODIES DIRECTED TO GLYCOTOPES[a,b]

Reiji Kannagi and Sen-itiroh Hakomori

Program of Molecular Pathology, Aichi Cancer Center, Nagoya 464-8681, Japan, and Division of Biomembrane Research, Pacific Northwestern Research Foundation, Seattle, WA 98122 USA

Monoclonal antibodies specific to glycotopes, the antigenic epitopes carried by glycoconjugates, have been widely utilized for study of the expression, distribution and function of carbohydrate determinants on a variety of cells and tissues (1,2). It is increasingly clear that cell surface glycoconjugates play important roles in cell-to-cell recognition and interaction (3,4). Functional analysis of glycotopes in such interaction has been greatly facilitated by the application of specific monoclonal antibodies. This chapter provides an introduction to the application of anti-glycotope antibodies in such analyses. For information regarding how to generate anti-carbohydrate antibodies, see refs. 1, 2 or 5.

New fields for application of anti-carbohydrate antibodies

A recently developed field for the application of anti-carbohydrate monoclonal antibodies is the study of cell adhesion. Monoclonal antibodies directed to these glycotopes serve as powerful tools in the identification of specific carbohydrate ligand, dissection and purification of ligand molecules, or experimental anti-adhesion therapy.

Cell adhesion and recognition mediated by selectins. Selectins, the family of cell adhesion molecules with a C-type lectin domain at the outer terminus of each molecule, have been shown to recognize specific carbohydrate ligands such as sialyl Lewis X (6-8) and sialyl Lewis A (Fig. 1) (9-11). The cell adhesion mediated by selectin and their

Correspondence: Reiji Kannagi, M.D., Ph.D. Program of Experimental Pathology, Research Institute, Aichi Cancer Center, 1-1 Kanokoden, Chikusaku, Nagoya 464-8681, Japan. E-mail <rkannagi@aichi-cc.pref.aichi.jp>.

[a] This work was supported in part by grants-in-aid from the Ministry of Education, Science, Sports and Culture, Japan (11680648 and on priority areas 10178104), grants-in-aid for Cancer Research (10-27) and the Second Term Comprehensive Ten-year Strategy for Cancer Control from the Ministry of Health and Welfare, Japan, and a grant from the Princess Takamatsu Foundation for the Promotion of Cancer Research.

[b] Abbreviations: DSGG, disialosyl galactosylgloboside; KDN, 3-deoxy-D-glycero-D-galacto-nonulosonic acid; MAG, myelin-associated glycoprotein; MSGG, monosialosyl galactosylgloboside; OB-BP, leptin-binding protein; PSGL-1, P-selectin glycoprotein ligand-1; Siglec, sialic acid-binding lectin of the immunoglobulin superfamily; SSEA-3, stage-specific embryonic antigen-3; SSEA-4, stage-specific embryonic antigen-4; SMP, Schwann cell myelin protein.

The Molecular Immunology of Complex Carbohydrates–2.
Edited by Albert M. Wu, Kluwer Academic/Plenum Publishers, 2001.

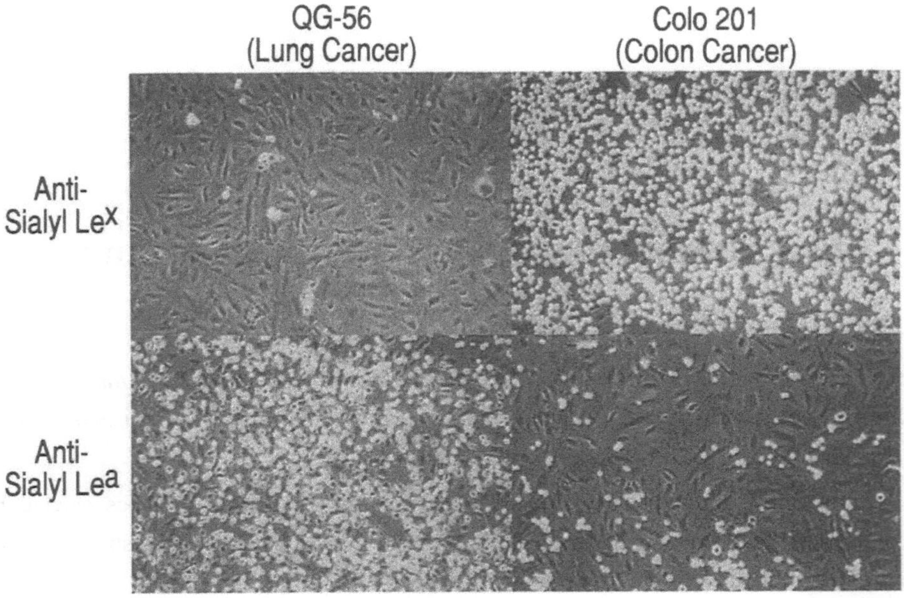

Fig. 1. Typical inhibition patterns in the adhesion of human cancer cells to rIL-1β-activated HUVECs by treatment with monoclonal anti-sialyl Lewis X and anti-sialyl Lewis A antibodies. Cancer cells were treated for 30 min with the respective antibodies (20 μg/ml) prior to the adhesion experiment. Results for QG-56 (lung cancer) and Colo201 (colon cancer) cells are shown. Note that adhesion of QG-56 is inhibited by anti-sialyl Lewis X antibody (SNH-3), while that of Colo201 cells is inhibited by anti-sialyl Lewis A antibody (2D3). Adopted from ref. 11.

carbohydrate ligands has been proposed to be implicated in the recruitment of leukocytes in inflammation (12,13), hematogenous metastasis of cancer cells (9,11,14) and tissue infiltration of leukemic cells (15,16). Anti-carbohydrate antibodies have been proved to be very useful in the study of cell-to-cell interaction involving cell surface glycoconjugates, a typical example of which is selectin-mediated cell adhesion.

There is a remarkable heterogeneity of carbohydrate determinant involved in cell adhesion. Table 1 indicates a list of putative ligands for selectins having sialyl Lewis X-related structures. Each determinant is synthesized through a distinct set of glycosyltransferases, the combination of which determines the molecular species of selectin ligand to be expressed on a given lineage of cells, and even a subtle difference in the carbohydrate structures may have biological significance in cell-to-cell recognition. It should be noted that there is considerable heterogeneity also in the carbohydrate specificity of anti-sialyl Lewis X antibodies which are used in the study of selectin-mediated cell adhesion, including cell adhesion inhibition experiments, and this adds more complexity to the analysis of cell adhesion mediated by selectins. The specificities of these antibodies are shown in Table 2.

Detection of carbohydrate ligands for P-Selectin. Recently, it has become increasingly clear that each member of the selectin family cell adhesion molecules has distinct ligand

Table 1. Heterogeneity of sialyl Lewis X-related glycotopes.

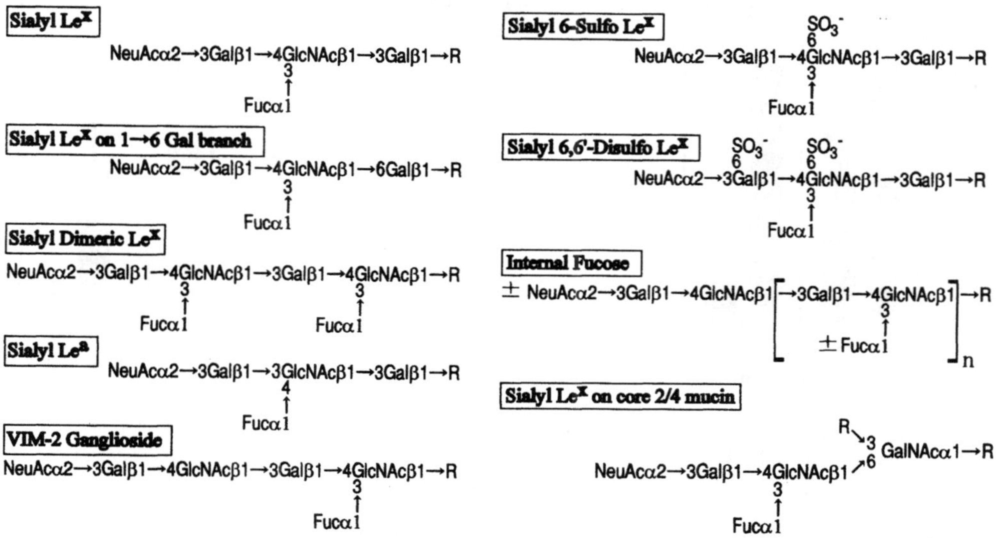

specificities. Since sialyl Lewis X and sialyl Lewis A were found to serve as a ligand for E-selectin in the early 1990's, it has been proposed that they would also serve as ligands for P- and L-selectins, because of the considerable homology in their C-type lectin domains with that of E-selectin. Subsequent studies, however, disclosed that P-selectin specifically recognizes sialyl Lewis X-like carbohydrate determinants carried by the *O*-glycans on the particular glycoprotein, PSGL-1 (P-selectin glycoprotein ligand-1). On the other hand, E-selectin recognizes sialyl Lewis X, irrespective of the core structures; *i.e.*, it binds sialyl Lewis X carried by *N*-glycans and/or glycolipids as well as *O*-glycans.

For example, a cultured human lymphoid cell line expressing sialyl Lewis X-like determinant significantly binds to both P- and E-selectin, but a treatment with *O*-

Table 2. Summary of reactivity of anti-sialyl Lewis X antibodies against pure carbohydrate determinants as ascertained by solid-phase enzymoimmunoassay.

Name of MoAb	Sialyl Le^{X}	Sialyl Le^{X} on 1→6 branch[1]	Sialyl Dimeric Le^{X}	Sialyl Le^{a}	VIM-2 Ganglio-side	Sialyl 6-sulfo Le^{X}	*N*-glycolyl sialyl Le^{X}	KDN-Le^{X}	8-*epi*-sialyl Le^{X}
Classical anti-sialyl Le^{X}									
CSLEX-1	++	+	+	−	−	−	+	*n.t.*	*n.t.*
SNH-3	++	++	++	−	++	−	+	*n.t.*	*n.t.*
FH-6	++	+	++	−	++	−	−	*n.t.*	*n.t.*
Variant type anti-sialyl Le^{X}									
2F3	++	+	+	+	−	++	−	−	+
HECA-452	++	++	+	++	−	++	−	−	+
2H5	++	++	+	+	±	++	++	++	−

[1]Sialyl Lewis X having the NeuAcα2→3Galβ1→4(Fucα1→3)GlaNAcβ1→6Galβ1→R sequence (see Table 1).

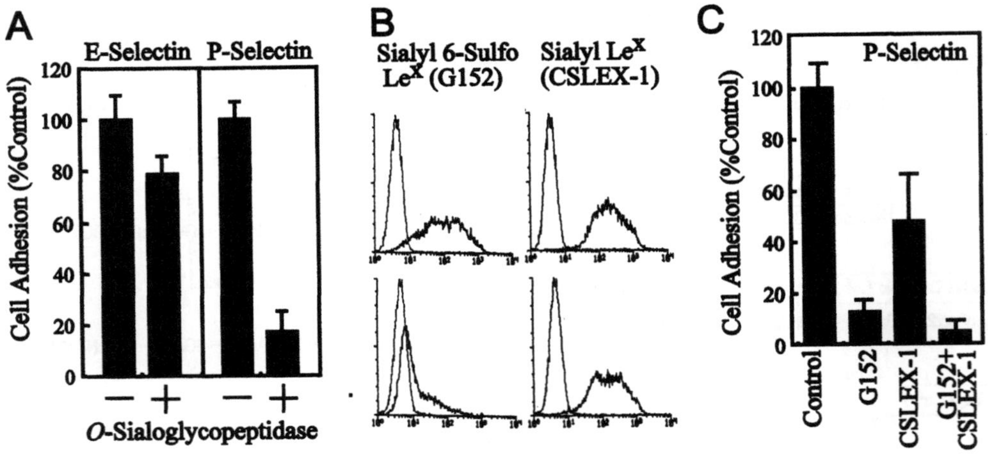

Fig. 2. Heterogeneity of sialyl Lewis X-related determinants involved in adhesion of a human lymphoid cell line to P-selectin. Panel A, effect of *O*-sialoglycoproteinase treatment of a human lymphoid cell line, Namalwa cells transfected with fucosyltransferase VII cDNA on the adhesion to E- and P-selectin/CHO cells. Note that adhesion of cells to P-selectin was strongly inhibited by the enzyme treatment, whereas adhesion to E-selectin was not seriously impaired, corroborating the significant difference in the ligand specificity of E- and P-selectins. Panel B, Expression of sialyl Lewis X-related glycotopes on cells treated with the enzyme. Expression of sialyl 6-sulfo Lewis X defined by G152 antibody was markedly reduced, while that of convemtional sialyl Lewis X defined by CSLEX-1 antibody was only minimally affected, indicating sialyl 6-sulfo Lewis X is preferentially carried by *O*-glycans. Panel C, inhibition of adhesion of cells to P-selectin/CHO by anti-glycotope antibodies. The adhesion is strongly inhibited by G152, but only moderately by CSLEX-1, showing that sialyl 6-sulfo Lewis X serves as a major ligand for P-selectin on these cells.

sialoglycoproteinase which specifically cleaves sialylated *O*-glycans, totally abrogates P-selectin binding while affecting binding of E-selectin only minimally (Fig. 2A). This indicates a significant difference between the carbohydrate ligands recognized by P- and E-selectins. This would suggest that P-selectin mediated cell adhesion will generally be effectively inhibited by the antibodies directed to the sialyl Lewis X-like determinant carried by *O*-glycans. Use of antibodies with a wider specificity would be required for the effective inhibition of E-selectin-mediated cell adhesion, when a non-negligible amount of sialyl Lewis X-like determinants carried by core molecules other than *O*-glycans serves as ligands for E-selectin on the cells under study.

Interestingly, there was a peculiar deviation in the distribution of molecular species of sialyl Lewis X-like determinants in the cultured lymphoid cells shown in Fig. 2B. In these cells, sialyl 6-sulfo Lewis X defined by the G152 antibody was almost exclusively carried by *O*-glycans, while a majority of conventional sialyl Lewis X determinants defined by the CSLEX-1 antibody was carried by core molecules other than *O*-glycans. Sialyl 6-sulfo Lewis X as well as conventional Lewis X serves as a ligand for P-selectin in these cells, and antibodies to be used in adhesion inhibition experiments should be properly selected with due care. Use of conventional anti-sialyl Lewis X antibody is insufficient for maximal inhibition of P-selectin-mediated adhesion of these cells (Fig. 2C). This is most probably because the 6-sulfotransferases required for the synthesis of sialyl 6-sulfo Lewis X prefer *O*-glycans as substrates. Both of two known 6-sulfotransferase isoenzymes (17,18) are acknowledged to favor *O*-glycans much more than *N*-glycans as substrates. This does not necessarily mean, however, that P-selectin prefers sialyl 6-sulfo Lewis X to conventional sialyl Lewis X as its ligand. Conventional sialyl Lewis X may well serve as a major ligand for P-selectin, when the cells have a reduced or no 6-sulfotransferase activity. There are lines of evidence clearly indicating that conventional sialyl Lewis X has significant binding activity to P-selectin, provided it is presented on *O*-glycans carried by PSGL-1. Use of conventional anti-sialyl Lewis X antibody would be quite inhibitory in such cases.

Detection of carbohydrate ligands for E-Selectin. Since E-selectin requires only sialyl Lewis X-like determinants on the ligand molecules, it binds to a wider range of cell lineages including epithelial cancer cells as well as leukocytes, and is involved not only in the process of inflammatory response but also in the extravasation and hematogenous metastasis of cancer cells. We have learned from experience that there is a considerable difference in the inhibitory activity of anti-sialyl Lewis X antibodies to E-selectin mediated cell adhesion. Adhesion of epithelial cancer cells such as lung cancer cells to E-selectin tends to be most effectively inhibited by the anti-sialyl Lewis X antibody, for example, SNH-3 (9,11), while that of lymphoid cells is usually inhibited by the anti-sialyl Lewis X antibody, exemplified by 2F3 (15). SNH-3 was much less effective in the inhibition of lymphoid cell adhesion,

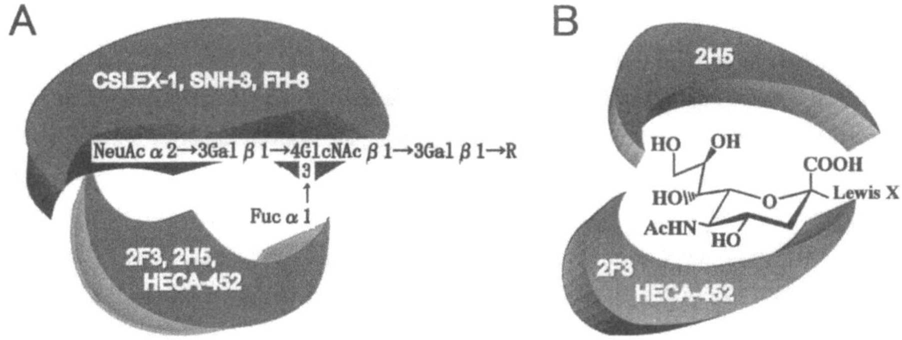

Fig. 3. Schematic illustration of reactivity of anti-sialyl Lewis X antibodies against sialyl Lewis X determinant. Left panel, reactivity of six different anti-sialyl Lewis X antibodies against sialyl Lewis X determinant. The antibodies CSLEX-1, SNH-3 and FH-6 rigorously require NeuAcα and the Galβ1→4 GlcNAcβ core for binding, and are therefore sensitive to the modifications at the GlcNAcβ moiety. The SNH-3 and FH-6 antibodies are cross-reactive to the structures devoid of Fucα. Antibodies 2F3, 2H5, and HECA-452 recognize both NeuAcα and Fucα residues, but their specificity to the Galβ1→ 4 GlcNAcβ core is not strong. The antibodies are sensitive to the deprivation of the Fucα moiety, but cross-reactive to structures having the Galβ1→3GlcNAcβ core (i.e., sialyl Le^{a}). Right panel, recognition of sialic acid moiety by three representative anti-sialyl Lewis X antibodies. The 2H5 antibody recognizes the tail portion of sialic acid, and is therefore sensitive to the *O*-acetyl modifications at this portion, while being cross-reactive to the determinants modified at *N*-aetyl residue, such as NeuGc or even KDN. On the other hand, 2F3 and HECA-452 antibodies require *N*-acetyl portion of sialic acid be intact for recognition, and are sensitive to the modification at this portion, while being cross-reactive to the structures modified at the tail portion of the sialic acid moiety, such as 9-*O*-acetyl sialic acid. See also Table 2.

while 2F3 frequently fails to inhibit cancer cell adhesion, which is clearly inhibited by the SNH-3 antibody.

The exact reason for this discrepancy remains unclear. These antibodies react equally to pure sialyl Lewis X determinant, but the ranges of their cross-reaction to sialyl Lewis X-related determinants are quite different (Table 2 and Fig. 3, left panel). SNH-3 is cross-reactive to the sialyl Lewis X-like determinants having fucosylation at internal GlcNAc, even when the penultimate GlcNAc was not fucosylated, as exemplified by the so-called VIM-2 determinant (19). Therefore, the carbohydrate determinants defined by the SNH-3 antibody form a set of such heterogenous sialyl Lewis X-like determinants. Sialyl Lewis X-like determinants in epithelial cancer cells are synthesized by the action of fucosyltransferase III and/or VI, which is capable of adding fucose either to penultimate or internal GlcNAc residues (6,20). Fucosyltransferase IV activity is known to be considerably increased in cancer cells (21,22), and this enzyme is also capable of transferring fucose to internal GlcNAc residues (23). Some internally fucosylated structures are reportedly capable of binding E-selectin (24-26), and the E-selectin ligands carried by these cells are estimated to be a set of such heterogenous sialyl Lewis X-like determinants, which would heavily overlap with the SNH-3 defined set of determinants.

On the other hand, sialyl Lewis X-like determinants in lymphoid cells are synthesized through the action of fucosyltransferase VII, which can transfer fucose to only penultimate GlcNAc, but not to internal GlcNAc residues (27,28). Fucosyltransferase IV is abundantly present in granulocytes and monocytes, but much less in lymphoid cells, further limiting the availability of internal fucosylation in these cells. Instead, the sialyl Lewis X determinant in lymphoid cells is frequently modified by *O*-acetylation in the sialic acid residue and 6-sulfation at the GlcNAc residues, which gives rise to a set of sialyl Lewis X-like determinants having considerable heterogeneity. The carbohydrate determinants defined by the 2F3 antibody, which invariably requires the fucosylation at penultimate GlcNAc but allows *O*-acetylation and/or 6-sulfation in sialic acid or GlcNAc (Table 2 and Fig. 3, right panel) (19,29,30), may well fit the range of determinants expressed on lymphoid cells.

This kind of ambiguity seems almost inherent in every carbohydrate-mediated cell recognition system, and this ambiguity *per se* may have some physiological significance. Cell-to-cell recognition mediated by glycotopes is thus sometimes far from key-and-key hole type interaction. Despite the ambiguity of the system, this difference results in surprisingly clear discrimination in the activity to inhibit cell adhesion, eventually in an almost all-or-none fashion. In any case, the key is the careful choice of an antibody effective for the cell adhesion inhibition experiments.

Detection of carbohydrate ligands for L-Selectin. L-selectin was first described as a lymph node receptor involved in the homing of lymphocytes to peripheral lymph nodes (31,32). High endothelial venules are known to serve as an entrance for the lymphocytes which home into peripheral lymph nodes. Since L-selectin is expressed on leukocytes, the

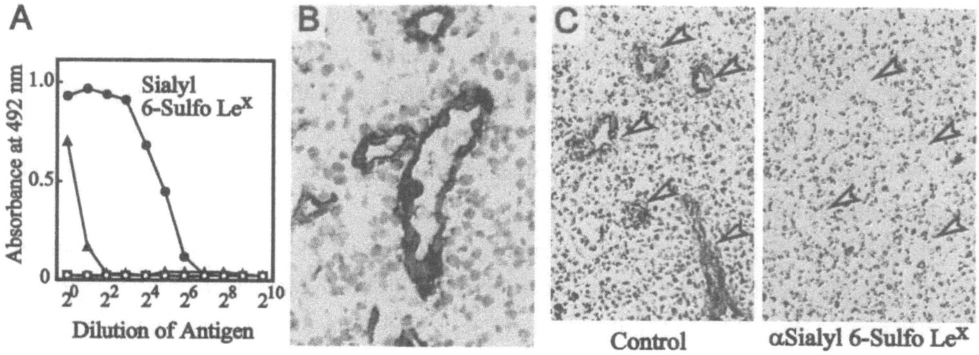

Fig. 4. Reactivity of anti-sialyl 6-sulfo Lewis X antibody G152 against sulfated and non-sulfated sialyl Lewis X determinants as ascertained by ELISA (Panel A), against human high endothelial cells in immunohistochemical staining (Panel B), and inhibition of binding of human recombinant L-selectin to the high endothelial cells. Panel A shows reactivity of the G152 antibody against sialyl 6-sulfo Lewis X (●), sialyl 6,6'-bis-sulfo Lewis X (▲), sialyl 6'-sulfo Lewis X (○), or sialyl Lewis X (■). Freshly obtained frozen sections of human lymph node were used in panels B and C. Adopted from ref. 30.

endothelial cells of high endothelial venules should express corresponding carbohydrate ligands for L-selectin, such as sialyl Lewis X. However, attempts to identify the L-selectin ligand on high endothelial cells using monoclonal anti-sialyl Lewis X antibodies led to complicated results, which were difficult to interpret (33). The high endothelial cells in human lymph nodes were stained by only a limited group of anti-sialyl Lewis X antibodies, but not by another group of anti-sialyl Lewis X antibodies (29,33). The group of antibodies that stained high endothelial cells included 2H5, 2F3 and HECA-452. Some of these antibodies inhibited the L-selectin-dependent cell adhesion to high endothelial cells (33). The group of antibodies which did not stain high endothelial cells included CSLEX-1, SNH-3 and FH-6, the three well-known classical anti-sialyl Lewis X antibodies. These findings indicated that a unique sialyl Lewis X-like determinant, but not the authentic sialyl Lewis X, was present on high endothelial cells, and served as a ligand for L-selectin. This was further supported by the finding that most of the anti-Lewis X antibodies failed to stain the sialidase-treated high endothelial cells. The genuine sialyl Lewis X, after sialidase treatment, should have yielded conventional Lewis X, which would have been readily detected by the usual anti-Lewis X antibodies.

Subsequently, the L-selectin ligand detected on human high endothelial venules by 2H5, 2F3 and HECA-452 turned out to be mainly sialyl 6-sulfo Lewis X (29,30,34). The reason for the lack of reactivity of the other three antibodies was that the addition of 6-sulfate residue to the determinant blocked the binding of the antibodies. The G152 antibody specific to sialyl 6-sulfo Lewis X strongly reacted with human high endothelial cells, and significantly inhibited the binding of L-selectin to high endothelial cells (Fig. 4) (30). Moreover, transfection of cells with fucosyltransferase VII and 6-sulfotransferase cDNAs conferred expression of sialyl 6-sulfo Lewis X and binding activity to L-selectin (Fig. 5) (34).

The distinct ligand specificity of three members of the selectin family was disclosed from the results of the transfection experiments. When the cells were transfected only with fucosyltransferase VII, conventional sialyl Lewis X appeared at the cell surface (Fig. 5), but the transfectant cells did not adhere to the cells expressing L-selectin. This indicated that conventional sialyl Lewis X was not sufficient in conferring L-selectin mediated cell adhesion, while it was sufficient for E-selectin (34).

The sialyl 6-sulfo Lewis X determinant detected by the G152 antibody appeared only when the cells were doubly transfected with fucosyltransferase VII and 6-aulfotransferase (Fig. 5), and only these double transfectant cells showed significant adhesion of cells expressing L-selectin (34). This transfectant also had an E-selectin binding activity. Interestingly, none of these transfectants had any adhesive activity to P-selectin, indicating that the requirements for P-selectin ligand are quite different from those for E- and L-selectin. Further transfection of PSGL-1 cDNA to the cells was necessary to confer significant binding to P-selectin, indicating that PSGL-1 is essential for P-selectin-mediated cell adhesion (Kanamori, A., et al., manuscript in preparation).

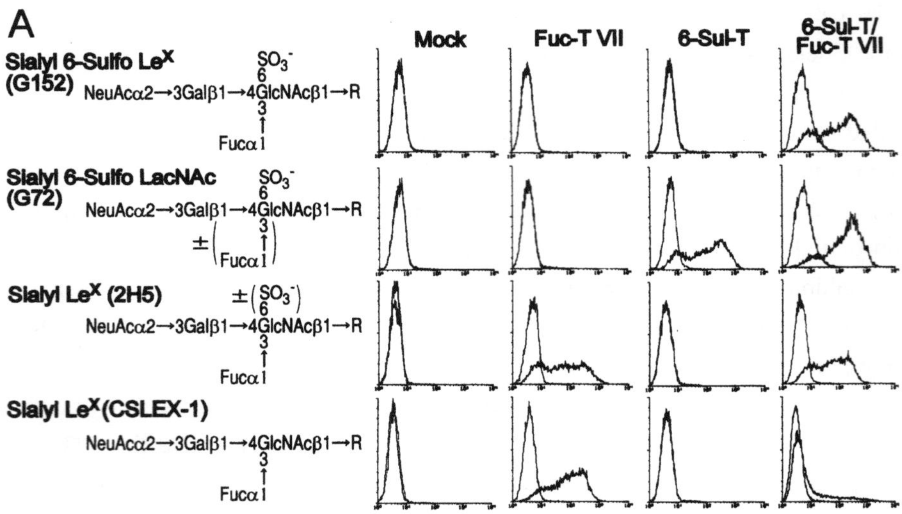

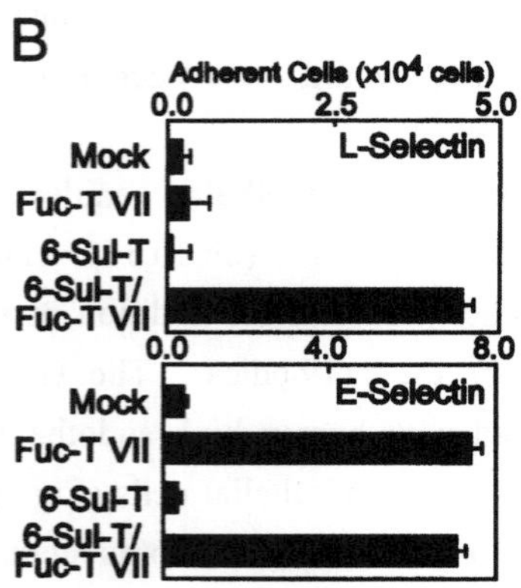

Fig. 5. Expression of glycotopes (panel A) and binding of L- and E-selectin-expressing cells to ECV304 cells transfected with fucosyltransferase VII (Fuc-T VII) and/or GlcNAcβ: 6-sulfotransferase (6-Sul-T) cDNAs. Panel A, only the cells co-transfected with both Fuc-T VII and 6-Sul-T express sialyl 6-sulfo Lewis X as defined by the G152 antibody, while cells transfected with 6-Sul-T express sialyl 6-sulfo *N*-acetyl lactosamine (LacNAc) as defined by the G72 antibody. Sialyl Lewis X as defined by 2H5 or CSLEX-1 antibody is expressed on the cells transfected with Fuc-T VII. Expression of the glycotope defined by the CSLEX-1 antibody is suppressed in the doubly transfected cells, since the 6-sulfation at GlcNAc blocks the binding of the CSLEX-1 antibody, but not the binding of 2H5. Adopted from ref. 34.

Many problems remain regarding the carbohydrate ligand for L-selectin, one of which would be the ligand in high endothelial cells of murine and rat lymph nodes. The anti-sialyl 6-sulfo Lewis X antibody G152 is not reactive to high endothelial cells of rodents. Instead, MECA-79 antibody, originally raised against murine lymph node stroma, detects L-selectin ligand on murine high endothelial cells, since the antibody strongly stains the cells and inhibits L-selectin-mediated adhesion on high endothelial cells. The antigen recognized by this antibody has been suggested to be sulfation-dependent, but not dependent on sialic acid or fucose. A significant reactivity to the MECA-79 antibody appeared when cells were transfected with 6-sulfotransferase, indicating that this antibody recognizes products of 6-sulfotransferase (34). MECA-79 reacts also to high endothelial cells of humans, suggesting a certain heterogeneity in the carbohydrate ligands for L-selectin.

Cell adhesion mediated by disialosyl galactosylgloboside (DSGG), $G_{D1\alpha}$ and related O-glycan structures. A sialylated glycolipid called disialosyl galactosylgloboside (Table 3) expressed on renal carcinoma cells was found to serve as a specific ligand for yet-uncharacterized receptor on endothelial cells of microvasculature in the lung, and to confer significant cell adhesion (35). The adhesion was specifically inhibited by the antibody directed to disialosyl galactosylgloboside. The receptor was not selectin, and was not

Table 3. Disialosyl galactosylgloboside (DSGG), $G_{D1\alpha}$ and related *O*-glycan structures.

Disialosyl galactosylgloboside (DSGG)

NeuAcα2→3Galβ1→3GalNAcβ1→3 Galα1→4Galβ1→4Glc1→Cer
(NeuAcα2→6 on GalNAc)

Monosialosyl galactosylgloboside (MSGG, SSEA-4)

NeuAcα2→3Galβ1→3GalNAcβ1→3 Galα1→4Galβ1→4Glc1→Cer

$G_{D1\alpha}$

NeuAcα2→3Galβ1→3GalNAcβ1→4Galβ1→4Glc1→Cer
(NeuAcα2→6 on GalNAc)

***O*-glycan (glycophorin)**

NeuAcα2→3Galβ1→3GalNAcα1→Ser/Thr
(NeuAcα2→6 on GalNAc)

expressed on TNFα-stimulated umbilical vein endothelial cells. Nor was the adhesion dependent on calcium ion, an absolute requirement for selectin-mediated cell adhesion. The degree of expression of sialylated galactosylgloboside proved to be correlated with the potential of renal cancer cells to metastasize to the lung and lymph nodes, and inversely correlated with patient survival. Later, the ganglioside $G_{D1\alpha}$, carrying the same terminal structure as disialosyl galactosyl-globoside (Table 3), was shown to mediate adhesion of murine lymphosarcoma cells to hepatic sinusoidal endothelial cells (36). The adhesion was specifically inhibit-ed by a anti-$G_{D1\alpha}$ antibody, $G_{D1\alpha}$-liposomes, and mimicking peptides of $G_{D1\alpha}$ (37). It is noteworthy that a similar structure is also carried by some *O*-glycans (Table 3), the cell adhesive activity of which awaits further investigation.

Sialoadhesins. Sialic acid binding Ig-like lectins (siglecs) are type 1 membrane proteins that constitute a distinct subset of the Ig super-family and are characterized by their sequence similarities and abilities to bind sialic acids in glycotopes (I-type lectin) (38). The group currently comprises 8 members including sialoadhesin/siglec-1, CD22/siglec-2, CD33/siglec-3, myelin-associated glycoprotein (MAG)/siglec-4A and its putative avian homologue SMP/siglec-4B, OB-BP1/siglec-5, OB-BP1/siglec-6 (CD33L1), and siglec-7, which latter is also characterized independently as the inhibitory NK cell receptor, p75/AIRM1. Each of these proteins has an extracellular region made up of a membrane distal V-set domain to which the sialic acid binding site has been mapped, followed by varying numbers of C2-set domains, which range from 1 in CD33 to 16 in sialoadhesin. Apart from MAG/Siglec-4, which is expressed exclusively on oligodendrocytes in the central nervous system and on Schwann cells in the peripheral nervous system, all other siglecs are expressed on discrete subpopulations of leukocytes. CD22 is present only on mature B cells, sialoadhesin is found on limited macrophage subsets, CD33 is a marker of early committed myeloid progenitor cells, siglec-5 is expressed by monocytes and mature neutrophils, siglec-6 is on B cells, siglec-7 is expressed by NK cells, a part of CD8+ T cells and monocytes, and siglec-8 is reportedly specific to eosinophils.

These siglecs are known to display significant differences in sialic acid linkage specificity. CD22 recognizes solely α2,6-linked sialic acids, MAG/SMP and CD33 require α2,3-linked sialic acids, and sialoadhesin binds terminal α2,3- or α2,8-linked sialic acids. Siglec-5 and siglec-7 are reported to be capable of binding glycoconjugates carrying sialic acid in either α2,3 or α2,6 linkages (39), and the most recently described siglec-8 recognizes α2,3-linked sialic acids (40). At this moment, anti-carbohydrate antibodies have not been actively applied in the elucidation of the ligands for siglecs, since the structure of the proposed ligands is so simple that does not await dissection using specific antibodies. All siglecs, however, were demonstrated to have considerable sialic acid substructural specificity, with differing requirements for the sialic acid *N*-acyl moiety as well as *O*-acetylation. In addition, G_{D1a}-like disialosyl structures or their sulfated analogs, resembling those described in the previous sections, were recently suggested to be more active ligands for MAG, SMP and sialoadhesin (41). OB-BP1/siglec-6 is known to selectively bind Neu5Acα2–6GalNAcα (sialyl-Tn). If these observations are physiologically significant, anti-carbohydrate antibodies discriminating these determinants would serve as useful tools for elucidation of endogenous ligands in cell-to-cell interactions.

Carbohydrate-carbohydrate interactions. Cell-to-cell recognition and adhesion are also mediated by carbohydrate-carbohydrate interactions. Since the Lewis X-Lewis X

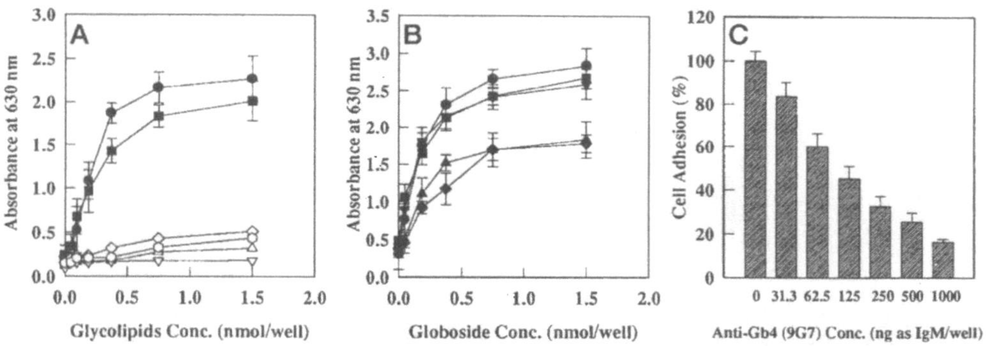

Fig. 6. Binding of undifferentiated human embryonal carcinoma 2102Ep cells to plates coated with various glycolipids, and the effect of anti-glycolipid antibodies on Gb4-dependent 2102 cell adhesion (45). Panel A, 2102 cell binding to glycolipid-coated plates. 2102 cells (1×10^5 /well) were incubated in wells coated with Gg_3 (●), Gb_4 (■), G_{M1} (△), G_{M3} (▽), Le^a (□), and nLc_3 (◇) (0–1.5 nmol of glycolipid/well). Adherent cells were stained with 0.1% Toluidine Blue O and measured at 630 nm. Panel B, effect of anti-glycolipid antibodies on Gb_4-dependent adhesion. 2102 cells (1×10^5) were incubated in control medium or in medium containing mAbs 2D4 (anti-Gg_3) (5 μg/ml), 9G7 (anti-Gb_4) (20 μg/ml), MC631 (anti-SSEA-3) (20 μg/ml), or 1B2 (anti-nLc_4)(2 μg/ml) for 1 h at 37°C. Cells were washed and plated in wells coated with various quantities of Gb_4 (abscissa). l, control (no treatment of cells); ●, 2D4; ▼, 9G7; ◆, MC631; ▲, 1B2. Adherent cells were stained and measured as in panel A. Each point represents the mean of triplicate experiments. Panel C, effect of anti-Gb_4 mAb 9G7 on 2102 cell adhesion to Gb_4-coated plates preincubated with various quantities of 9G7. Standard variation is indicated by bar.

interaction (42) was first reported, significant interaction has been observed between Gg_3Cer and G_{M3} ganglioside (43,44), between Gb_4Cer and nLc_4Cer or GalGb_4Cer (SSEA-3) (45), and between sulfatide and GalCer (46,47). The interaction between KDN G_{M3} and Gg_3Cer, or between Gb_4Cer and nLc_4Cer or GalGb_4Cer did not require a bivalent cation, while that between sulfatide and GalCer was dependent on calcium ion. On the other hand, some combination of glycotopes are anti-adhesive; for instance, G_{M3}-G_{M3} and Lewis Y-Lewis Y are highly repellent (48). Specific anti-carbohydrate antibodies such as anti-SSEA-3, anti-nLc4 or anti-Lewis Y, have been frequently utilized to elucidate the contribution of carbohydrate-carbohydrate interaction in cell-to-cell interaction of teratocarcinoma cells (Fig. 6) (45,49). Some of these interactions are known to initiate signal transduction in cells by activating focal adhesion kinase and/or transcription factors such as AP1 and CREB (43,45). The implications of the interaction among sulfated carbohydrates in the cell aggregation of the marine sponge or sea urchin embryos have been closely investigated (50,51). Cell aggregation blocking monoclonal antibodies directed to such acidic glycans isolated from marine sponges or sea urchin embryos was shown to detect a similar glycotopes present in human colon cancer cells (47).

Expression cloning of glycosyltransferases and related genes. Another recently developing area is application of specific monoclonal antibodies for the cloning of genes encoding glycosyltransferases and/or related genes. An anti-G_{D2} monoclonal antibody has been successfully utilized for the expression cloning of β-1,4 *N*-acetylgalactosaminyl-

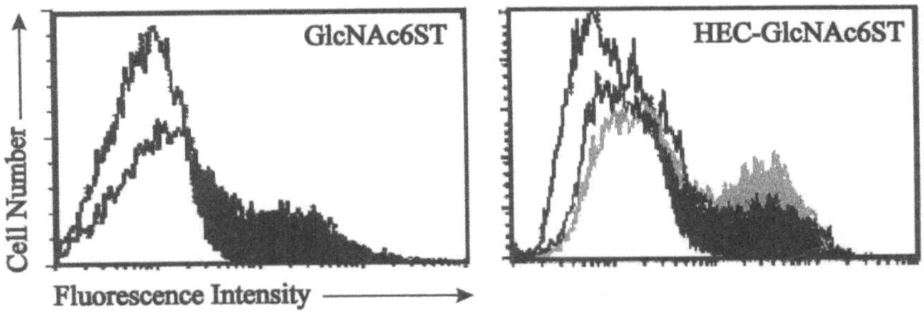

Fig. 7. Detection of synthetic products on the COS7 or CHO cells transfected with 6-sulfotransferase cDNA candidates obtained by est cloning. Left panel, induction of sialyl 6-sulfo LacNAc as defined by G72 antibody on the COS7 cells transfected with cDNA encoding putative first isoenzyme of GlcNAcβ: 6-*O*-sulfotransferase (GlcNAc6ST). Right panel, induction of the same glycotope on the CHO cells transfected with cDNA encoding putative second isoenzyme of GlcNAcβ: 6-*O*-sulfotransferase (HEC-GlcNAc6ST), with (gray) or without (black) CD34 cDNA. Adopted from refs. 17 and 18.

transferase cDNA involved in the synthesis of G_{M2} and G_{D2} gangliosides (52). More recently, an antibody specific to the type III mucin of gastric gland, HIK1083, enabled expression cloning of a human α1,4-*N*-acetylglucosaminyltransferase that forms GlcNAcα1→4Galβ1→R, a unique terminal structure on *O*-glycan specifically expressed in the gastric gland mucous cell-type mucin (type III mucin) (53). Expression cloning of the genes related to *O*-acetylated sialic acid using a specific antibody resulted in the cDNA cloning not of acetyltransferases, but of an acetyl-CoA transporter-like molecule (54). Anti-carbohydrate antibodies that were beneficial in innovative study of carbohydrate-mediated cell-cell recognition and interaction were also quite useful in the cloning of genes related to expression of the carbohydrate determinants involved in such interaction. Specific antibodies were successfully applied in the cloning of 6-*O*-sulfotransferases in the screening of cDNA clones obtained by *BLAST* search of expressed sequence tag (est) databases (Fig. 7) (17,18), which was involved in the synthesis of sialyl 6-sulfo Lewis X, the putative L-selectin ligand expressed on high endothelial cells in lymph nodes. The antibody used in the dissection of cell adhesive activity of G_{D1a} ganglioside also enabled est-based cloning of a sialyltransferase involved in its synthesis (55).

Diverse monoclonal antibodies directed to glycotopes

The properties of monoclonal antibodies directed to each species of glycoconjutgates are described in this section. A list of hitherto-established monoclonal antibodies directed to glycotopes and their properties are listed at the end of this chapter in Table 4.

Monoclonal antibodies directed to glycolipids. As shown in Table 4, glycolipids have been the most well-studied glycoconjugates utilizing monoclonal antibodies. Antibodies directed to almost every basic glycolipid structures were already established in the 1980's. This was because methods for purification of each molecular species, for generation of specific antibody and for detection of antigen-antibody reaction for glycolipids, were well-developed much earlier than those for other glycoconjugates. Still the list of anti-glycolipid antibodies is increasing, and antibodies which discriminate subtle differences in the intramolecular structure, such as *N*-acetyl, de-*N*-acetyl or *N*-glycolyl sialic acid, and its *O*-acetylated derivatives, have recently been generated. The list also contains antibodies which discriminate lactone or lactam ring formation in sialic acid, and even a specific antibody to KDN, the deaminated sialic acids.

Monoclonal antibodies directed to proteoglycans. Antibodies directed to glycosaminoglycans have long been frequently used as well. By now, antibodies directed to almost all members of glycosaminoglycans have been obtained, and some of them are commercially available. Although these antibodies have been used for localization and rough estimation of the amount of glycosaminoglycans in cells and tissues, they have not been frequently applied for functional studies. Most of the antigenic glycotopes of the glycosaminoglycan chain are located on relatively longer carbohydrate chains than in glycolipids, which are multiply modified by various types of sulfation. This makes it difficult to identify the exact structures of antigenic epitopes defined by the antibodies.

Monoclonal antibodies directed to N-glycans. It is not easy to obtain anti-carbohydrate antibody using glycoproteins as immunization antigen. Especially, *N*-linked glycans are generally poor immunogens, and there are only a few antibodies available for the carbohydrate determinants proper to *N*-glycans. Among the carbohydrate determinants proper to *N*-glycans, so-called bisecting GlcNAc, $\alpha 1 \rightarrow 6$ fucosylation at the chitobiose portion, and a high mannose structure, have caught the eye of many investigators, since these determinants are reportedly associated with malignant transformation of cells. Among these structures, only the very recently obtained antibody directed to GlcNAc$\beta 1 \rightarrow 4$[Fuc $\alpha 1 \rightarrow 6$GlcNAcβ] sequence is available. This antibody was originally raised against *N*-glycans carried by the glycoproteins of *Dictyostelium discoideum*, but reacts well with the same carbohydrate determinant present in human cancers.

Apart from the core portion of *N*-glycans, the more peripheral portion of *N*-glycan carries carbohydrate determinants which are common to glycolipids and/or *O*-glycans. Antibodies raised against such a determinant carried by glycolipids, for example, usually detect the same terminal determinant carried by *N*-glycans. These include various important cancer-associated determinants, such as sialyl Lewis X, which are frequently carried by a GlcNAc$\beta 1 \rightarrow 6$ branch in *N*-glycans.

Monoclonal antibodies directed to O-glycans. These groups classically contained antibodies directed to important carcinoma-associated antigens such as T, Tn and sialyl Tn determinants. Recently, antibodies directed to more complex mucin-type determinants were described, making this the most rapidly-growing area in terms of antibodies directed to glycotopes. The determinants defined by these new antibodies include disialosyl T antigen, $\alpha 1 \rightarrow 4$-linked GlcNAc, and sialyl Le^{X} linked to mucin core 2 or core 4 structure. The list of antibodies directed to the epitopes carried by *O*-glycans continues to grow, together with the importance of the *O*-glycan epitopes in cell-to-cell interaction.

REFERENCES

1. Kannagi,R. and Hakomori,S. (1986) Monoclonal antibodies directed to carbohydrate antigens, in *Handbook of experimental immunology, Vol.4, Applications of immunological methods in biomedical sciences* (Weir,D.M., Herzenberg,L., Blackwell,C. and Herzenberg,L.A. eds.), pp. 117.1-117.20, Blackwell Scientific Pub. Inc., Boston.
2. Kannagi,R. (1999) Monoclonal anti-glycosphingolipid antibodies, in *Methods in Enzymology, Sphingolipid Metabolism and Cell Signaling, Part B* (Hannun,Y.A. and Merrill,A.H.,Jr. eds.), in press, Academic Press, New York.
3. Hakomori, S. (1996) Tumor malignancy defined by aberrant glycosylation and sphingo-(glyco)lipid metabolism. *Cancer Res.* **56**: 5309-5318.
4. Hakomori, S. (1993) Structure and function of sphingoglycolipids in transmembrane signalling and cell-cell interactions. *Biochem. Soc. Trans.* **21**: 583-595.
5. Kannagi,R. (1999) Production of anti-carbohydrate monoclonal antibodies, in *Methods in Molecular Biology: Liposomes, Methods and Protocols* (Basu,S. and Basu,M. eds.), in press, Humana Press, Totowa.
6. Lowe, J. B., Stoolman, L. M., Nair, R. P., Larsen, R. D., Berhend, T. L. and Marks, R. M. (1990) ELAM-1-dependent cell adhesion to vascular endothelium determined by a transfected human fucosyltransferase cDNA. *Cell* **63**: 475-484.
7. Phillips, M. L., Nudelman, E., Gaeta, F. C. A., Perez, M., Singhal, A. K., Hakomori, S. and Paulson, J. C. (1990) ELAM-1 mediates cell adhesion by recognition of a carbohydrate ligand, sialyl-Le^x. *Science* **250**: 1130-1132.
8. Walz, G., Aruffo, A., Kolanus, W., Bevilacqua, M. and Seed, B. (1990) Recognition by ELAM-1 of the sialyl-Le^x determinant on myeloid and tumor cells. *Science* **250**: 1132-1135.
9. Takada, A., Ohmori, K., Takahashi, N., Tsuyuoka, K., Yago, K., Zenita, K., Hasegawa, A. and Kannagi, R. (1991) Adhesion of human cancer cells to vascular endothelium mediated by a carbohydrate antigen, sialyl Lewis A. *Biochem. Biophys. Res. Commun.* **179**: 713-719.
10. Berg, E. L., Robinson, M. K., Mansson, O., Butcher, E. C. and Magnani, J. L. (1991) A carbohydrate domain common to both sialyl Le^a and sialyl Le^x is recognized by the endothelial cell leukocyte adhesion molecule ELAM-1. *J. Biol. Chem.* **266**: 14869-14872.
11. Takada, A., Ohmori, K., Yoneda, T., Tsuyuoka, K., Hasegawa, A., Kiso, M. and Kannagi, R. (1993) Contribution of carbohydrate antigens sialyl Lewis A and sialyl Lewis X to adhesion of human cancer cells to vascular endothelium. *Cancer Res.* **53**: 354-361.
12. Springer, T. A. (1995) Traffic signals on endothelium for lymphocyte recirculation and leukocyte emigration. *Annu. Rev. Physiol.* **57**: 827-872.
13. Rosen, S. D. and Bertozzi, C. R. (1994) The selectins and their ligands. *Curr. Opin. Cell Biol.* **6**: 663-673.

14. Kannagi, R. (1997) Carbohydrate-mediated cell adhesion involved in hematogenous metastasis of cancer. *Glycoconjugate J.* **14**: 577-584.
15. Ohmori, K., Takada, A., Ohwaki, I., Takahashi, N., Furukawa, Y., Maeda, M., Kiso, M., Hasegawa, A., Kannagi, M. and Kannagi, R. (1993) A distinct type of sialyl Lewis X antigen defined by a novel monoclonal antibody is selectively expressed on helper memory T cells. *Blood* **82**: 2797-2805.
16. Furukawa, Y., Tara, M., Ohmori, K. and Kannagi, R. (1994) Variant type of sialyl Lewis X antigen expressed on adult T cell leukemia cells is associated with skin involvement. *Cancer Res.* **54**: 6533-6538.
17. Uchimura, K., Muramatsu, H., Kadomatsu, K., Fan, Q. W., Kurosawa, N., Mitsuoka, C., Kannagi, R., Habuchi, O. and Muramatsu, T. (1998) Molecular cloning and characterization of an *N*-acetylglucosamine-6-*O*-sulfotransferase. *J. Biol. Chem.* **273**: 22577-22583.
18. Bistrup, A., Bhakta, S., Lee, J. K., Belov, Y. Y., Gunn, M. D., Zuo, F. -R., Huang, C. -C., Kannagi, R., Rosen, S. D. and Hemmerich, S. (1999) Sulfotransferases of two specificities function in the reconstitution of high-endothelial-cell ligands for L-selectin. *J. Cell Biol.* **145**: 899-910.
19. Kannagi,R. (1998) CD15s cluster report, in *Leukocyte Typing VI* (Kishimoto,T., Kikutani,H., von dem Borne,E.G.Kr., Goyert,S.M., Mason,D.Y., Miyasaka,M., Moretta,L., Okumura,K., Shaw,S., Springer,T.A., Sugamura,K. and Zola,H. eds.), pp. 352-355, Garland Publishing Inc., New York.
20. Weston, B. W., Smith, P. L., Kelly, R. J. and Lowe, J. B. (1992) Molecular cloning of a fourth member of a human α(1, 3)fucosyltransferase gene family. Multiple homologous sequences that determine expression of the Lewis x, sialyl Lewis x, and difucosyl sialyl Lewis x epitopes. *J. Biol. Chem.* **267**: 24575-24584.
21. Yago, K., Zenita, K., Ginya, H., Sawada, M., Ohmori, K., Okuma, M., Kannagi, R. and Lowe, J. B. (1993) Expression of α(1,3)fucosyltransferases which synthesize sialyl Le^x and sialyl Le^a, the carbohydrate ligands for E- and P-selectins, in human malignant cell lines. *Cancer Res.* **53**: 5559-5565.
22. Ito, H., Hiraiwa, N., Sawada-Kasugai, M., Akamatsu, S., Tachikawa, T., Kasai, Y., Akiyama, S., Ito, K., Takagi, H. and Kannagi, R. (1997) Altered mRNA expression of specific molecular species of fucosyl- and sialyltransferases in human colorectal cancer tissues. *Int. J. Cancer* **71**: 556-564.
23. Lowe, J. B., Kukowska-Latallo, J. F., Nair, R. P., Larsen, R. D., Marks, R. M., Macher, B. A., Kelly, R. J. and Ernst, L. K. (1991) Molecular cloning of a human fucosyltransferase gene that determines expression of the Lewis x and VIM-2 epitopes but not ELAM-1-dependent cell adhesion. *J. Biol. Chem.* **266**: 17467-17477.
24. Stroud, M. R., Handa, K., Salyan, M. E., Ito, K., Levery, S. B., Hakomori, S., Reinhold, B. B. and Reinhold, V. N. (1996) Monosialogangliosides of human myelogenous

leukemia HL60 cells and normal human leukocytes. 2. Characterization of E-selectin binding fractions, and structural requirements for physiological binding to E-selectin. *Biochemistry* **35**: 770-778.

25. Stroud, M. R., Handa, K., Salyan, M. E., Ito, K., Levery, S. B., Hakomori, S., Reinhold, B. B. and Reinhold, W. N. (1996) Monosialogangliosides of human myelogenous leukemia HL60 cells and normal human leukocytes. 1. Separation of E-selectin binding from nonbinding gangliosides, and absence of sialosyl-Le^x having tetraosyl to octaosyl core. *Biochemistry* **35**: 758-769.
26. Handa, K., Stroud, M. R. and Hakomori, S. (1997) Sialosyl-fucosyl poly-LacNAc without the sialosyl-Le^x epitope as the physiological myeloid cell ligand in E-selectin-dependent adhesion: Studies under static and dynamic flow conditions. *Biochemistry* **36**: 12412-12420.
27. Natsuka, S., Gersten, K. M., Zenita, K., Kannagi, R. and Lowe, J. B. (1994) Molecular cloning of a cDNA encoding a novel human leukocyte α-1, 3-fucosyltransferase capable of synthesizing the sialyl Lewis x determinant. *J. Biol. Chem.* **269**: 16789-16794.
28. Sasaki, K., Kurata, K., Funayama, K., Nagata, M., Watanabe, E., Ohta, S., Hanai, N. and Nishi, T. (1994) Expression cloning of a novel α1,3-fucosyltransferase that is involved in biosynthesis of the sialyl Lewis x carbohydrate determinants in leukocytes. *J. Biol. Chem.* **269**: 14730-14737.
29. Mitsuoka, C., Kawakami-Kimura, N., Kasugai-Sawada, M., Hiraiwa, N., Toda, K., Ishida, H., Kiso, M., Hasegawa, A. and Kannagi, R. (1997) Sulfated sialyl Lewis X, the putative L-selectin ligand, detected on endothelial cells of high endothelial venules by a distinct set of anti-sialyl Lewis X antibodies. *Biochem. Biophys. Res. Commun.* **230**: 546-551.
30. Mitsuoka, C., Sawada-Kasugai, M., Ando-Furui, K., Izawa, M., Nakanishi, H., Nakamura, S., Ishida, H., Kiso, M. and Kannagi, R. (1998) Identification of a major carbohydrate capping group of the L-selectin ligand on high endothelial venules in human lymph nodes as 6-sulfo sialyl Lewis x. *J. Biol. Chem.* **273**: 11225-11233.
31. Gallatin, W. M., Weissman, I. L. and Butcher, E. C. (1983) A cell-surface molecule involved in organ-specific homing of lymphocytes. *Nature* **304**: 30-34.
32. Geoffroy, J. S. and Rosen, S. D. (1989) Demonstration that a lectin-like receptor ($gp90^{MEL}$) directly mediates adhesion of lymphocytes to high endothelial venules of lymph nodes. *J. Cell Biol.* **109**: 2463-2469.
33. Sawada, M., Takada, A., Ohwaki, I., Takahashi, N., Tateno, H., Sakamoto, J. and Kannagi, R. (1993) Specific expression of a complex sialyl Lewis X antigen on high endothelial venules of human lymph nodes: Possible candidate for L-selectin ligand. *Biochem. Biophys. Res. Commun.* **193**: 337-347.
34. Kimura, N., Mitsuoka, C., Kanamori, A., Hiraiwa, N., Uchimura, K., Muramatsu, T., Tamatani, T., Kansas, G. S. and Kannagi, R. (1999) Reconstitution of functional L-selectin ligands on a cultured human endothelial cell line by co-transfection of α1→3

fucosyltransferase VII and newly cloned GlcNAcβ: 6-sulfotransferase cDNA. *Proc. Natl. Acad. Sci. USA* **96**: 4530-4535.

35. Satoh, M., Handa, K., Saito, S., Tokuyama, S., Ito, A., Miyao, N., Orikasa, S. and Hakomori, S. (1996) Disialosyl galactosylgloboside as an adhesion molecule expressed on renal cell carcinoma and its relationship to metastatic potential. *Cancer Res.* **56**: 1932-1938.
36. Taki, T., Ishikawa, D., Ogura, M., Nakajima, M. and Handa, S. (1997) Ganglioside $G_{D1\alpha}$functions in the adhesion of metastatic tumor cells to endothelial cells of the target tissue. *Cancer Res.* **57**: 1882-1888.
37. Ishikawa, D., Kikkawa, H., Ogino, K., Hirabayashi, Y., Oku, N. and Taki, T. (1998) $G_{D1\alpha}$-replica peptides functionally mimic $G_{D1\alpha}$, an adhesion molecule of metastatic tumor cells, and suppress the tumor metastasis. *FEBS Lett.* **441**: 20-24.
38. Crocker, P. R., Clark, E. A., Filbin, M., Gordon, S., Jones, Y., Kehrl, J. H., Kelm, S., Le Douarin, N., Powell, L., Roder, J., Schnaar, R. L., Sgroi, D. C., Stamenkovic, K., Schauer, R., Schachner, M., Van den Berg, T. K., Van der Merwe, P. A., Watt, S. M. and Varki, A. (1998) Siglecs: a family of sialic-acid binding lectins. *Glycobiology* **8**: v.
39. Nicoll, G., Ni, J., Liu, D., Klenerman, P., Munday, J., Dubock, S., Mattei, M. G. and Crocker, P. R. (1999) Identification and characterization of a novel siglec, siglec-7, expressed by human natural killer cells and monocytes. *J. Biol. Chem.* **274**: 34089-34095.
40. Floyd, H., Ni, J., Cornish, A. L., Zeng, Z., Liu, D., Carter, K. C., Steel, J. and Crocker, P. R. (2000) Siglec-8. A novel eosinophil-specific member of the immunoglobulin superfamily. *J. Biol. Chem.* **275**: 861-866.
41. Collins, B. E., Ito, H., Sawada, N., Ishida, H., Kiso, M. and Schnaar, R. L. (1999) Enhanced binding of the neural siglecs, myelin-associated glycoprotein and schwann cell myelin protein, to Chol-1 (α-series) gangliosides and novel sulfated Chol-1 analogs. *J. Biol. Chem.* **274**: 37637-37643.
42. Eggens, I., Fenderson, B., Toyokuni, T., Dean, B., Stroud, M. and Hakomori, S. (1989) Specific interaction between Le^x and Le^x determinants. A possible basis for cell recognition in preimplantation embryos and in embryonal carcinoma cells. *J. Biol. Chem.* **264**: 9476-9484.
43. Iwabuchi, K., Yamamura, S., Prinetti, A., Handa, K. and Hakomori, S. (1998) GM3-enriched microdomain involved in cell adhesion and signal transduction through carbohydrate-carbohydrate interaction in mouse melanoma B16 cells. *J. Biol. Chem.* **273**: 9130-9138.
44. Kojima, N. and Hakomori, S. (1989) Specific interaction between gangliotriaosylceramide (Gg_3) and sialosyllactosylceramide (G_{M3}) as a basis for specific cellular recognition between lymphoma and melanoma cells. *J. Biol. Chem.* **264**: 20159-20162.
45. Song, Y., Withers, D. A. and Hakomori, S. (1998) Globoside-dependent adhesion of

human embryonal carcinoma cells, based on carbohydrate-carbohydrate interaction, initiates signal transduction and induces enhanced activity of transcription factors AP1 and CREB. *J. Biol. Chem.* **273**: 2517-2525.

46. Stewart, R. J. and Boggs, J. M. (1993) A carbohydrate-carbohydrate interaction between galactosylceramide-containing liposomes and cerebroside sulfate-containing liposomes: dependence on the glycolipid ceramide composition. *Biochemistry* **32**: 10666-10674.
47. Koshy, K. M. and Boggs, J. M. (1996) Investigation of the calcium-mediated association between the carbohydrate head groups of galactosylceramide and galactosylceramide I^3 sulfate by electrospray ionization mass spectrometry. *J. Biol. Chem.* **271**: 3496-3499.
48. Kojima, N. and Hakomori, S. (1991) Cell adhesion, spreading, and motility of G_{M3}-expressing cells based on glycolipid-glycolipid interaction. *J. Biol. Chem.* **266**: 17552-17558.
49. Zhu, Z. M., Kojima, N., Stroud, M. R., Hakomori, S. and Fenderson, B. A. (1995) Monoclonal antibody directed to Le^y oligosaccharide inhibits implantation in the mouse. *Biol. Reprod.* **52**: 903-912.
50. Misevic, G. N. and Burger, M. M. (1993) Carbohydrate-carbohydrate interactions of a novel acidic glycan can mediate sponge cell adhesion. *J. Biol. Chem.* **268**: 4922-4929.
51. Spillmann, D., Thomas-Oates, J. E., van Kuik, J. A., Vliegenthart, J. F., Misevic, G., Burger, M. M. and Finne, J. (1995) Characterization of a novel sulfated carbohydrate unit implicated in the carbohydrate-carbohydrate-mediated cell aggregation of the marine sponge, *Microciona prolifera*. *J. Biol. Chem.* **270**: 5089-5097.
52. Nagata, Y., Yamashiro, S., Yodoi, J., Lloyd, K. O., Shiku, H. and Furukawa, K. (1992) Expression cloning of β-1,4 *N*-acetylgalactosaminyltransferase cDNAs that determine the expression of G_{M2} and G_{D2} gangliosides. *J. Biol. Chem.* **267**: 12082-12089.
53. Nakayama, J., Yeh, J. C., Misra, A. K., Ito, S., Katsuyama, T. and Fukuda, M. (1999) Expression cloning of a human α1,4-*N*-acetylglucosaminyltransferase that forms GlcNAc α1→4Galβ1→R, a glycan specifically expressed in the gastric gland mucous cell-type mucin. *Proc. Natl. Acad. Sci. U. S. A.* **96**: 8991-8996.
54. Kanamori, A., Nakayama, J., Fukuda, M. N., Stallcup, W. B., Sasaki, K., Fukuda, M. and Hirabayashi, Y. (1997) Expression cloning and characterization of a cDNA encoding a novel membrane protein required for the formation of *O*-acetylated ganglioside: A putative acetyl-CoA transporter. *Proc. Natl. Acad. Sci. USA* **94**: 2897-2902.
55. Okajima, T., Fukumoto, S., Ito, H., Kiso, M., Hirabayashi, Y., Urano, T. and Furukawa, K. (1999) Molecular cloning of brain-specific $G_{D1\alpha}$ synthase (ST6GalNAc V) containing CAG/Glutamine repeats. *J. Biol. Chem.* **274**: 30557-30562.

Table 4. Monoclonal Antibodies Directed to Glycotopes.

Recognized glycotopes		Immunogens	Clone Name (Isotype)[a]	References [b]
Ganglio-series glycolipids				
asialo G_{M2}	GalNAcβ1→4Galβ1→4Glcβ1→Cer	pure lipid	2D4 (IgM)*	(1)
asialo G_{M1}		pure lipid	SH-34 (IgM)	(2)
		pure lipid	103HT30 (IgM)	(3)
	Galβ1→3GalNAcβ1→4Galβ1→4Glcβ1→Cer	pure lipid	MW-1, -2, -3, -4, -5 (IgM)	(4)
		pure lipid	H2G10 (IgM)	(5)
		pure lipid	AG-1 (IgM)§	(6)
G_{M4}	NeuAcα2→3Galβ1→Cer	pure lipid	AMR10 (IgM)	(7)
G_{M3}		mouse melanoma	M2590 (IgM)	(8)
	NeuAcα2→3Galβ1→4Glcβ1→Cer	melanoma patient	FCM1 (HIgM)	(9,10)
		G_{M3} lactone	DH2 (IgG_3)	(11)
G_{M2}		melanoma	5-3, 10-11 (IgM)	(12)
	NeuAcα2↘3	pure lipid	MK1-16§,1-17, 1-8 (IgM)	(13)
	GalNAcβ1→4Galβ1→4Glcβ1→Cer	melanoma patient	OFAI-1 (L55, HIgM)	(14)
		pure lipid	GMB28 (IgM)	(15)
G_{M2} (NeuGc)		melanoma patient	2-39M (HIgM)	(16)
	NeuGcα2↘3	rabbit thymocyte	YHD-07 (IgM)	(17)
	GalNAcβ1→4Galβ1→4Glcβ1→Cer	pure lipid	MK2-34 (IgM) §	(13)
G_{M2} and G_{D2}		melanoma patient	3-207 (HIgM)	(18)
G_{M1}	NeuAcα2↘3	pure lipid	GMB16 (IgM)§	(15)
	Galβ1→3GalNAcβ1→4Galβ1→4Glcβ1→Cer	pure lipid	AGM-1, -2, -3 (IgM)	(19)
G_{M1b}		pure lipid	NA-6 (IgM)	(20)
	NeuAcα2→3Galβ1→3GalNAcβ1→4Galβ1→4Glcβ1→Cer			
GalNAcG_{D1a}		liver cancer	2A3D2, 2D11E2 (IgM) §	(21)
	NeuAcα2↘3 NeuAcα2↘3			
	GalNAcβ1→4Galβ1→3GalNAcβ1→4Galβ1→4Glcβ1→Cer			

G_{D3}	human melanoma	R24 (IgG_3)*	(22)
NeuAcα2→8NeuAcα2→3Galβ1→4Glcβ1→Cer	human melanoma	4.2. (IgM)	(23)
	human melanoma	LeoMel3 (IgM κ)	(24)
	melanoma patient	HJM1 (HIgM)	(9,10)
	pure lipid	GMR19 (IgM)§	(25)
	pure lipid and SK-MEL-28	KM641 (IgG_3)	(26)
G_{D3} (NeuGc)$_2$	melanoma patient	32-27M (HIgM)	(16)
G_{D2}	neuroblastoma	14.18 (IgG_3), 126 (IgM)*	(27)
	neuroblastoma	2F7,3A7,3G6 (IgM),	(28)
NeuAcα2→8NeuGcα2↘3		3F8 (IgG_3)	
GalNAcβ1→4Galβ1→4Glcβ1→Cer	human glioma	BW625 (IgG_3),	(29)
		BW704 (IgG_3)	
	melanoma patient	OFA-I-2 (L72, HIgM)	(30)
	pure lipid	GMR7 (IgM)§	(25)
G_{D3} and G_{D2}	human melanoma	ME361 (IgG_3)*	(31)
9-*O*-acetyl G_{D3}	neuroepithelial tumor (B49)	antiD1 · 1 (IgM)	(32,33)
	retinal epithelia	JONES (IgM)	(34)
	rat embryonic brain cells	RB13-2 (IgM)	(35)
	pure lipid	GMR2 (IgM)§	(25)
	human melanoma	ME311 (IgG_3)	(36)
O -acetyl disialoganglioside	fetal rat brain	3G5 (IgM)*	(37,38)
G_{D1a}	pure lipid	anti-G_{D1a} (IgM)	(39)
NeuAcα2↘3			
NeuAcα2→3Galβ1→3GalNAcβ1→4Galβ1→4Glcβ1→Cer			
G_{D1b}	pure lipid	anti-G_{D1b} (IgM)	(40)
NeuAcα2→8NeuGcα2↘3	pure lipid	GGR12 (IgG_{2b})§	(25)
Galβ1→3GalNAcβ1→4Galβ1→4Glcβ1→Cer	pure lipid	MAb-5G6 (IgM)	(41)
G_{D1c}	pure lipid	YK-3 (IgM)	(42)
NeuAcα2→8NeuAcα2→3Galβ1→3GalNAcβ1→4Galβ1→4Glcβ1→Cer			
$G_{D1\alpha}$ NeuAcα2→3Galβ1→3GalNAcβ1→4Galβ1→4Glcβ1→Cer NeuAcα2↗6	pure lipid	KA-17 (IgM)	(43)

G_{T3} NeuAcα2→8NeuAcα2→8NeuAcα2→3Galβ1→4Glcβ1→Cer	chick embryo retinal cells neural tissue of 3-day chick embryos	18B8 (IgM), A_2B_5 (IgM) M6703 (IgG_3)	(44,45) (46)
9- *O* -acetyl G_{T3}	optic nerve of 10-day chick embryos	493D4 (IgM)	(47)
IV^2FucG_{M1} (fucosyl G_{M1}) NeuAcα2↘3 Fucα1→2Galβ1→3GalNAcβ1→4Galβ1→4Glcβ1→Cer	pure lipid	F3,F12,F14 (IgG) F-1,F-2,F4 (IgM)	(48)
G_{T1a} NeuAcα2↘3 NeuAcα2→8NeuAcα2→3Galβ1→3GalNAcβ1→4Galβ1→4Glcβ1→Cer	pure lipid	GMR11 (IgM)	(15)
G_{T1b} NeuAcα2→8NeuAcα2↘3 NeuAcα2→3Galβ1→3GalNAcβ1→4Galβ1→4Glcβ1→Cer	pure lipid	GMR5 (IgM)	(25)
G_{Q1b} NeuAcα2→8NeuAcα2↘3 NeuAcα2→8NeuAcα2→3Galβ1→3GalNAcβ1→4Galβ1→4Glcβ1→Cer	pure lipid pure lipid	4F10 (IgG_3), 7F5 (IgG_{2a}), 4E7 (IgM) GMR13 (IgM)§	(49) (25)
$G_{T1a\alpha}$(Chol-1_α-a), $G_{Q1b\alpha}$ (Chol-1_α-b) NeuAcα2↘3 NeuAcα2→3Galβ1→3GalNAcβ1→4Galβ1→4Glcβ1→Cer NeuAcα2↗6 NeuAcα2→8NeuAcα2↘3 NeuAcα2→3Galβ1→3GalNAcβ1→4Galβ1→4Glcβ1→Cer NeuAcα2↗6	pure lipid	GGR41 (IgG_{2a})	(50)
G_{T1c}, G_{Q1c}, G_{P1c}	chick brain	Q211 (IgM)	(51,52)
Lacto- and neolacto-series glycolipids			
LacCer Galβ1→4Glcβ1→Cer	ANLL	T5A7 (IgM)	(53)
Lc_3 GlcNAcβ1→3Galβ1→4Glcβ1→Cer	syngenic testis pure glycolipid	J-1 (IgM) JF12 (IgM)	(54,55) (56)

Lc_4 (Le^c, type 1 chain paragloboside)	teratocarcinoma	Fc-10.2 (IgM)	(57)
Galβ1→3GlcNAcβ1→3Galβ1→4Glcβ1→Cer	embryonal carcinoma cells (PA-1)	K21 (IgM)	(58,59)
	uterine cancer	HMST-1 (HIgM)	(60)
	A375 cells	11-50 (HIgM)	(61)
nLc_4 (paragloboside, p antigen)	pure lipid	1B2 (IgM)	(62)
Galβ1→4GlcNAcβ1→3Galβ1→4Glcβ1→Cer	granulocytes	My28 (IgM)	(63)
	germ cell tumor	2G10 (IgM)	(64)
	lipid mixture	α PGF1H11(IgM)	(65)
nLc_5	pure lipid	TE-4,TE-5,TE-7 (IgM), TE-6 (IgG_3)	(66)
GlcNAcβ1→3Galβ1→4GlcNAcβ1→3Galβ1→4Glcβ1→Cer			
nLc_6 (i-antigen)	lung cancer	MH21-134 ($HIgG_3$)	(67)
Galβ1→4GlcNAcβ1→3Galβ1→4GlcNAcβ1→3Galβ1→4Glcβ1→Cer	lung cancer	NCC-1004 (HIgM)	(68)
	EBV	GL-1, GL-2 (HIgM)	(69)
isonLc_8 (I-antigen)	mouse spermatogenic cells	C6 (IgM)	(70,71)
Galβ1→4GlcNAcβ1↘6 Galβ1→4GlcNAcβ1→3Galβ1→4GlcNAcβ1→3Galβ1→4Glcβ1→Cer			
IV^2FucLc_4 (Le^d, type 1 chain H)	epidermoid cancer	101 (IgM)	(72)
Fucα1→2Galβ1→3GlcNAcβ1→3Galβ1→4Glcβ1→Cer			
$IV^2FucnLc_4$ (type 2 chain H)	pure lipid	BE2 (IgM)	(62)
Fucα1→2Galβ1→4GlcNAcβ1→3Galβ1→4Glcβ1→Cer	hepatocellular carcinoma cells	S1 (IgG_{2a}), S3 (IgG_3)	(73)
	endometrial carcinoma cells	C12 (IgM)	(74)
	MCF-7 cells	A46-B/B10 (IgM)	(75)
III^4FucLc_4 (Le^a)	lipid mixture	CA37-4 (IgG_1),CF4-C4 (IgG_{2a})* BC9-ES (IgG_3),DG4-1 (IgG_1)	(76)
Fucα1↘4 Galβ1→3GlcNAcβ1→3Galβ1→4Glcβ1→Cer	pure lipid	3E8 (IgM)	(77)
	ovarial cyst mucin	7LE (IgG_1)§	(78)

III⁴V⁴Fuc₂Lc₆ (dimeric Le^a, Le^a/Le^a) Fucα1↘4 Fucα1↘4 Galβ1→3GlcNAcβ1→3Galβ1→3GlcNAcβ1→3Galβ1→4Glcβ1→Cer	gastric cancer	NCC-ST-421 (IgG_3)	(79)
III³FucnLc₄ (Le^x) Fucα1↘3 Galβ1→4GlcNAcβ1→3Galβ1→4Glcβ1→Cer	gastric cancer	WGHS29 (IgM)	(80)
	colon cancer	ZWG13,14,III (IgM)	(81)
	lung cancer	J525,34,38 (IgM)	(82)
	leukemia cell	My-1 (IgM)	(83)
	leukemia cell	VEP8,9 (IgM)	(84)
	leukemia cell	1G10 (IgM)	(85)
	pure lipid	FH-2 (IgM)	(86)
III³FucnLc₆ (Le^x-i) Fucα1↘3 Galβ1→4GlcNAcβ1→3Galβ1→4GlcNAcβ1→3Galβ1→4Glcβ1→Cer	teratocarcinoma	α SSEA-1 (IgM)	(87-89)
III³V³Fuc₂nLc₆ (dimeric Le^x, Le^x/Le^x) Fucα1↘3 Fucα1↘3 Galβ1→4GlcNAcβ1→3Galβ1→4GlcNAcβ1→3Galβ1→4Glcβ1→Cer	pure lipid	SH-2 (IgG)	(90)
III³V³VII³Fuc₃nLc₁₀ (internal Le^x)	gastric cancer	ACFH-18 (IgM)	(91)
III³ IV²Fuc₂Lc₄ (Le^b) Fucα1↘4 Fucα1→2Galβ1→3GlcNAcβ1→3Galβ1→4Glcβ1→Cer	colon cancer	NS-10 (IgM)	(92)
	ovarial cyst mucin	2-25LE (IgG_1)§	(78)
III³IV²Fuc₂nLc₄ (Le^y) Fucα1↘3 Fucα1→2Galβ1→4GlcNAcβ1→3Galβ1→4Glcβ1→Cer	colon adenoma	C14/1/16/10 (IgM)	(93)
	lung cancer	F_3 (IgM)	(94)
	teratocarcinoma	75.12 (IgM)	(95)
	gastric cancer	AH-6 (IgM)	(96)
	lung cancer	NCC-ST-433 (IgM)	(97)
	lung cancer	Luca 6 (IgM)	(97)
III³V³VI²Fuc₃nLc₆ (extended Le^y) Fucα1↘3 Fucα1↘3 Fucα1→2Galβ1→4GlcNAcβ1→3Galβ1→4GlcNAcβ1→3Galβ1→4Glcβ1→Cer	pure lipid	KH1 (IgM)	(98)
	colon cancer	CC-1 (IgM), CC-2 (IgM)	(99)

IV^{3}NeuAcIII^{4}FucLc$_4$ (sialyl Le^a)	colon cancer	N-19-9 (IgG)	(100,101)
Fucα1↘4	colon cancer	CC3C195 (IgM)	(102)
NeuAcα2→3Galβ1→3GlcNAcβ1→3Galβ1→4Glcβ1→Cer	pancreas cancer	SPan1 (IgM, κ)	(103)
	colon cancer	KM01 (IgG$_1$)	(104)
	colon cancer	MSW113 (IgG$_3$)	(105)
	pure lipid	2D3§, 4E2 (IgM), 1H4§ (IgG$_3$)	(106)
	colon cancer	C-50 (IgM)	(107)
IV^{3}NeuAcLc$_4$ (sialyl Le^c)	pancreas cancer	DU-PAN-2 (IgM)	(108,109)
NeuAcα2→3Galβ1→3GlcNAcβ1→3Galβ1→4Glcβ1→Cer	embryonal carcinoma cells (PA-1)	K4 (IgM)	(58,59)
	unknown	SL-50 (IgM)	(110)
IV^{3}NeuAcnLc$_4$ (sialyl paragloboside)	pure lipid	NS24 (IgM)	(111)
NeuAcα2→3Galβ1→4GlcNAcβ1→3Galβ1→4Glcβ1→Cer			
IV^{6}NeuAcnLc$_4$	pure lipid	1B9 (IgG$_3$)	(112)
NeuAcα2→6Galβ1→4GlcNAcβ1→3Galβ1→4Glcβ1→Cer	pure lipid	LM-4 (IgG$_{2a}$)	(113)
	meconium glycolipid mixture	MSG15 (IgM)	(114)
	lipid mixture	1-15B (IgG$_3$)	(77)
IV^{3}NeuAcIII^{3}FucnLc$_4$ (sialyl Le^X)	colon cancer	CSLEX-1 (IgM)*	(115)
Fucα1↘3	pure lipid	SNH-3 (1-12F,IgM)	(116)
NeuAcα2→3Galβ1→4GlcNAcβ1→3Galβ1→4Glcβ1→Cer	pure lipid	SNH-4 (IgG$_3$)	(117)
	lung cancer	KM-93 (IgM) §	(118)
	lipid mixture	2H5 (IgM) †	(119)
	pure lipid	2F3 (IgM) †	(120)
IV^{3}NeuAc III$^{3/4}$Fuc(n)Lc$_4$	tonsil stroma	HECA-452 (RIgM)*	(121)
VI^{3}NeuAcV^3 III^{3}Fuc$_2$nLc$_6$ (sialyl dimeric Le^X), VI^{3}NeuAcV^{3}FucnLc$_6$ (sialyl Le^X-i)	pure lipid	FH-6 (IgM)*	(122)
Fucα1↘3 ±Fucα1↘3		3-5E (IgG$_3$)	(116)
NeuAcα2→3Galβ1→4GlcNAcβ1→3Galβ1→4GlcNAcβ1→3Galβ1→4Glcβ1→Cer			
VI^{3}NeuAcIII^{3}FucnLc$_6$ (VIM-2)	leukocytes	VIM-2 (IgM)	(123)
±Fucα1↘3			
NeuAcα2→3Galβ1→4GlcNAcβ1→3Galβ1→4GlcNAcβ1→3Galβ1→4Glcβ1→Cer			

IV^3 III^{6}NeuAc$_2$ III^{4}FucLc$_4$ Fucα1↘4 NeuAcα2→3Galβ1→3GlcNAcβ1→3Galβ1→4Glcβ1→Cer NeuAcα2↗6	lipid mixture	FH-7 (IgG$_3$)	(124)
IV^3 III^{6}NeuAc$_2$Lc$_4$ NeuAcα2→3Galβ1→3GlcNAcβ1→3Galβ1→4Glcβ1→Cer NeuAcα2↗6	lipid mixture	FH-9 (IgG$_{2a}$)	(125)
IV^{3}Gal α nLc$_4$ Galα1→3Galβ1→4GlcNAcβ1→3Galβ1→4Glcβ1→Cer	rabbit erythrocyte	Gal-13 (IgG$_1$)	(126)
III^{3}V^{4}Fuc$_2$LcnLc$_4$(Le^{a}Le^x) Fucα1↘4 Fucα1↘3 Galβ1→3GlcNAcβ1→3Galβ1→4GlcNAcβ1→3Galβ1→4Glcβ1→Cer	milk-fat membrane lung cancer cells	115C2, 115G3 (IgG$_1$) 43-9F (IgM)	(127,128) (129,130)
IV^{3}VIII^{3}NeuAc$_2$*iso*nLc$_8$ NeuAcα2→3Galβ1→4GlcNAcβ1↘6 NeuAcα2→3Galβ1→4GlcNAcβ1→3Galβ1→4GlcNAcβ1→3Galβ1→4Glcβ1→Cer	human placenta	NUH-2 (IgM)*	(131)
VI^{3}NeuGcnLc$_6$ NeuGcα2→3Galβ1→4GlcNAcβ1→3Galβ1→4Glcβ1→Cer	pure lipid	SHS-1 (IgM)	(132)
Type 1 chain A Fucα1↘2 GalNAcα1→3Galβ1→3GlcNAcβ1→3Galβ1→4Glcβ1→Cer	pure lipid colon cancer cells HT-29	AH21(IgM) CLH6(IgM)	(133) (134)
Type 2 chain A Fucα1↘2 GalNAcα1→3Galβ1→4GlcNAcβ1→3Galβ1→4Glcβ1→Cer	pure lipid	HH4(IgG$_3$)	(135)
Type 1 and type 2 chain A	MKN-45 colon cancer cells HT-29 pure lipid	AH16(IgG$_3$) MH2/6D4 (IgM) HT29-36(IgG$_3$) HH1(IgG$_{2a}$)	(133) (136) (134) (137)

Type 3 chain A (A_1 substance)		TH1(IgG_{2a})	(138)
Fucα1↘2 Fucα1↘2 GalNAcα1→3Galβ1→3GalNAcα1→3Galβ1→4GlcNAcβ1→3Galβ1→4Glcβ1→Cer			
Type 3 and type 4 chain A	Renal cancer cells SK-RC-28	M2 (IgM)	(134)
Galactosyl-A (A_2 substance)	Galactosyl-A glycolipid	HH9 (IgG_1)	(139)
Fucα1↘2 Galβ1→3GalNAcα1→3Galβ1→4GlcNAcβ1→3Galβ1→4Glcβ1→Cer			
ALe[b]	pure lipid	HH3(IgG_{2a})	(137)
Fucα1↘2 Fucα1↘4 GalNAcα1→3Galβ1→3GlcNAcβ1→3Galβ1→4Glcβ1→Cer			
ALe[y]	pure lipid	HH2(IgG_3)	(137)
Fucα1↘2 Fucα1↘3 GalNAcα1→3Galβ1→4GlcNAcβ1→3Galβ1→4Glcβ1→Cer			
Type 1 and type 2 chain B	synthetic oligosaccharide-BSA	1-3E-4, 13-E-7, 1-1A-12 (IgM)	(140)
	Capan-2 cells	$E_2$83-52 (IgM)	(141)
Type 2 chain B	Capan-2 cells	$E_1$15-2 (IgM)	(141)
	dorsal root ganglion neurons	KH-10, LD2, TC6, TD10, LA4 (IgM)	(142)
Fucα1↘2 Galα1→3Galβ1→4GlcNAcβ1→3Galβ1→4Glcβ1→Cer			
BLe[y]	KATO-III cells	GOM-2(IgG_3)	(143)
Fucα1↘2 Fucα1↘3 Galα1→3Galβ1→4GlcNAcβ1→3Galβ1→4Glcβ1→Cer			
Globo- and extended globo-series glycolipids			
Gb_3 (CTH, P^k antigen)	Burkitt lymphoma	38-13 (RIgM)	(144)
Galα1→4Galβ1→4Glcβ1→Cer	teratocarcinoma	1A4 (IgM)	(145)
	lipid mixture	H-8 (IgM)	(146)
	pure lipid	BGR23 (IgM)§	(147)

III^{3}GalαGb$_3$ Galα1→3Galα1→4Galβ1→4Glcβ1→Cer	pure lipid	BGR47 (IgG$_3$)	(147)
Gb$_4$ (globoside, P antigen)	PC12 cells	CC1 (IgM)	(148)
GalNAcβ1→3Galα1→4Galβ1→4Glcβ1→Cer	tonsillar lymphocytes	HJ6 (IgM)	(149)
	pure lipid (*N*-glycolyl G$_{M1}$)	12D4 (IgM)	(149)
	pure lipid	BMR26 (IgM)	(147)
*iso*Gb$_4$ and an isoglobo-lacto series glycolipid	pure lipid	14.2, 14.10, 14.3 (IgM)	(150)
Gb$_5$ (SSEA-3, galactosylgloboside)	early mouse embryo	α SSEA-3 (MC631,RIgM)	(151)
Galβ1→3GalNAcβ1→3Galα1→4Galβ1→4Glcβ1→Cer	pure glycolipid (GLY)	5A3 (IgM)	(152)
	lung cancer patient	J309, D579 (HIgM)	(153)
V^{3}NeuAcGb$_5$ (SSEA-4, monosialosyl galactosylgloboside, MSGG)	teratocarcinoma	α SSEA-4 (IgM)	(154)
NeuAcα2→3Galβ1→3GalNAcβ1→3Galα1→4Galβ1→4Glcβ1→Cer	renal cancer cells (TOS1)	RM1 (IgM)	(154)
V^{3}IV^{6}NeuAc$_2$Gb$_5$ (disialosyl galactosylgloboside, DSGG) NeuAcα2→3Galβ1→3GalNAcβ1→3Galα1→4Galβ1→4Glcβ1→Cer NeuAcα2↗6	renal cancer cells (TOS1)	RM2 (IgM)	(155,156)
IV^{3}GalNAcGb$_4$ (Forssman)	mouse spleen cells	M1/22.25 (RIgM)	(157,158)
GalNAcα1→3GalNAcβ1→3Galα1→4Galβ1→4Glcβ1→Cer	influenza patient	H1-C4 (HIgM)	(159)
	mouse mammary tumor	33B12 (RIgG$_1$ or $_2$c), 117C9 (RIgM)	(160)
	pure lipid (*in vitro* immunization)	4C3 (RIgM)	(161)
V^{2}FucGb$_5$ (Globo-H) Fucα1→2Galβ1→3GalNAcβ1→3Galα1→4Galβ1→4Glcβ1→Cer	breast cancer cell	MBr-1 (IgM)	(162,163)
Lacto-ganglioseries glycolipids			
II^{3}GlcNAcGg$_3$ (II^{4}GalNAcLc$_3$) GalNAcβ1↘4 GlcNAcβ1→3Galβ1→4Glcβ1→Cer	synthetic lipid	YI328-18 (IgG$_3$), YI328-51 (IgG$_3$)	(164)

Le[x]-GM₁ Fucα1↘3 NeuAcucα2↘3 Galβ1→4GlcNAcβ1→3Galβ1→3GalNAcβ1→4Galβ1→4Glcβ1→Cer	frog skin extract	188C1 (IgM)	(165)
Sulfated glycotopes			
SM_4g (seminolipid) SO_4-3Galβ1→AAG,DAG[c]	pure lipid (*in vitro* immunization)	CA10 (IgM)	(166)
SM_4s (sulfatide) SO_4-3Galβ1→Cer	pure lipid	AIC3IA2 (IgG_3)	(167)
	PWM	M14-376 (HIgM)	(168)
GlcUA-PG-sulfate (HNK-1 epitope)	human lymphocytes	HNK-1 (IgM)	(169-171)
SO_4-3GlcUAβ1→3Galβ1→4GlcNAcβ1→3Galβ1→4Glcβ1→Cer	glycolipid mixture	L9 (IgM)	(172)
SB_1a	lipid mixture	2H11E5 (IgM) §	(173)
SO_4-3Galβ1→3GalNAcβ1→4(SO_4-3)Galβ1→4Glcβ1→Cer	liver cancer cells	4A9E10 (IgG_3) §	(173)
SM_3 SO_4-3Galβ1→4Glcβ1→Cer	liver cancer cells	49-D6 (IgM) §, 7-E10 (IgM) §	(174)
$IV^3SO_4III^4FucLc_4$ (3'-sulfo Le[a])	human salivary mucin	F2 (IgM)	(175)
Fucα1↘4	synthetic glycolipid	SU-59 (IgM)	(176)
SO_4-3Galβ1→3GlcNAcβ1→3Galβ1→4Glcβ1→Cer	human colonic mucin	91.9H (IgG_1)	(177)
$III^6SO_4III^3FucnLc_4$ (6-sulfo Le[x]) Fucα1↘3 Galβ1→4GlcNAcβ1→3Galβ1→4Glcβ1→Cer SO_4↗6	synthetic glycolipid	AG223, AG107 (IgM)	(178)
$IV^3NeuAcIII^6SO_4nLc_4$ (Sialyl 6-sulfo paragloboside) NeuAcα2→3Galβ1→4GlcNAcβ1→3Galβ1→4Glcβ1→Cer SO_4↗6	synthetic glycolipid	G72 (IgM)	(179)
$IV^3NeuAcIII^3Fuc\ III^6SO_4nLc_4$ (Sialyl 6-sulfo Le[x]) Fucα1↘3 NeuAcα2→3Galβ1→4GlcNAcβ1→3Galβ1→4Glcβ1→Cer SO_4↗6	synthetic glycolipid	G152 (IgM)	(179)

$IV^3NeuAcIII^3Fuc\ IV^6III^6(SO_4)_2nLc_4$ (Sialyl 6,6'-disulfo Le^x) Fucα1↘3 NeuAcα2→3Galβ1→4GlcNAcβ1→3Galβ1→4Glcβ1→Cer SO_4↗6 SO_4↗6	synthetic glycolipid	G2706, G27011, G27037, G27039 (IgM)	(179)
***O*-linked glycans**			
T antigen	desialylated erythrocytes	49H-24 (IgM)	(180)
Galβ1→3GalNAcα1→Ser/Thr		49H-8 (IgM)	(181)
	asialoglycophorin	A78-G/A7 (IgM)	(182)
	breast cancer cells	BW835 (IgG_1)	(183)
	galactosyl-A glycolipid	HH8 (IgM)	(139)
Tn antigen	synthetic antigen	164H.1	(184)
GalNAcα1→Ser/Thr	lung cancer	NCC-LU-35, -81 (IgG)	(185)
	Tn erythrocytes	FBT3 (IgM)	(186)
	Tn erythrocytes	BRIC111 (IgG_1)	(187)
	deglycosylated bovine submaxillary mucin	CU-1 (IgG_3)	(188)
	deglycosylated bovine submaxillary mucin	5F4 (IgM)	(189)
	colon cancer (LS180 cells)	MLS128 (IgG_3)	(190)
	deglycosylated colon cancer cells (LS174T cells)	91S8 (IgM), 10F4 (IgG_3), 15D3a (IgM)	(191)
Sialyl Tn antigen NeuAcα2↘6	colon cancer	MLS102 (IgM)	(192)
GalNAcα1→Ser/Thr	mucin	TKH-1 (IgM), TKH-2 (IgG)	(193)
	breast cancer	B72.3 (IgG_1)	(194,195)
T-Sialyl Tn NeuAcα2↘6	glycoprotein	CC49(IgM)	(196)
Galβ1→3GalNAcα1→Ser/Thr	glycophorin A	QSH1(IgG_3)	(155,197)

Disialyl T	NeuAcα2↘6 NeuAc α 2→3Galβ1→3GalNAcα 1→Ser/Thr	glycophorin A	QSH2(IgG$_{2a}$)	(155,197)
Truncated *O*-glycan	GlcNAc β1→3Galβ1→4GlcNAc β1→6GalNAcα1→Ser/Thr	gastric mucosa	2B5	(198)
α1→4-linked GlcNAc	(GlcNAcα1→4Galβ1→3GalNAcα)	rat gastric mucin	HIK1083 (IgM)	(199)
Core 2/4 structure	R↘3 Galβ1→4GlcNAc β1→6GalNAcα1→Ser/Thr	synthetic glycolipid	F1α79 (IgM)	(200)
Sialyl Le^x on core 2/4	Fucα1↘3 R↘3 NeuAc α 2→3Galβ1→4GlcNAc β1→6GalNAcα1→Ser/Thr	gastric cancer	NCC-ST-439 (IgM)	(201)
O-GlcNAc		streptococcal groupA carbohydrate	HGAC86, HGAC39 (IgG$_3$)	(202)
***N*-linked glycans**				
Fucα 1→6GlcNAcβ core		*Dictyostelium discoideum* glycoprotein	CAB2, CAB4 (IgG$_3$)	(203)
Glycosaminoglycans				
Keratan sulfate		human articular cartilage proteoglycan	1/20/5-D-4(IgG$_1$)§	(204)
Chondroitin sulfate and dermatan sulfate		Human dermal fibroblast	PG-4 (IgM)	(205)
Heparan sulfate		Meth-A cells	HepSS-1(IgM)§	(206)
Chondroitin Sulfate D		chick embryo limb bud proteoglycan (PG-M)	MO-225 (IgM)§	(207)
Heparan sulfate (*N*-sulfo GlcNAc)		human heparan sulfate proteoglycan (F58)	10E4 (IgM)§	(208)

Δ–Heparan sulfate	heparitinase-treated human heparan sulfate proteoglycan (F69)	3G10 (IgG_{2b}) §	(208)
Proteoglycan Δ–Di0S, Δ–Di4S, Δ–Di6S	chondroitinase-treated proteoglycans	1-B-5 (IgG_1), 2-B-6 (IgG_1), 3-B-3 (IgM) §	(209)
Lactones, Lactams and other modified forms of sialic acid			
Disialoganglioside lactone (G_{D1b})	lactonized ganglioside mixture	BBH5 (IgM)	(210)
G_{M3} lactam	synthetic glycolipid-BSA	P3, P5-1, P5-3 (IgG)	(211)
De-*N*-acetyl G_{M3}/G_{D3}	de-*N*-acetyl G_{M3}/G_{D3}	SMR36 (IgM), SGR37(IgG_3)	(212)
$IV^3NeuNIII^3Fuc\ III^6SO_4nLc_4$ -lactam (cyclic sialyl 6-sulfo Le^x)	synthetic glycolipid	G159 (IgG_1)	(176)
$IV^3NeuNIII^3Fuc\ III^6SO_4nLc_4$ (de-*N*-acetyl sialyl 6-sulfo Le^x)	synthetic glycolipid	NDA7 (IgG_3), NDA47 (IgG_1)	(176)
G_{M3} and G_{M3} lactone	G_{M3} lactone	DH2 (IgG_3)	(11)
G_{D2} and G_{D2} lactone	EL4 lymphoma cells	A1-201(IgM), A1-410 (IgG_3)	(213)
PolyKDN (→8KDNα2→)n	KDN-rich glycoprotein	kdn8kdn (IgM)	(214)

* Listed in ATCC catalogue. † Listed in Pharmingen catalogue. ‡ Listed in Becton-Dickinson catalogue. § Listed in Seikagaku Kogyo catalogue, available through Calbiochem outside Japan.

[a] Isotype refers to that of mice unless otherwise indicated. R, rat; H, human. [b] Selected references describing specificity of antibodies. [c] AAG, alkylacylglycerol; DAG, diacylglycerol

References for Table 4

1. Young, W. W.,Jr., MacDonald, E. M., Nowinski, R. C. and Hakomori, S. I. (1979) Production of monoclonal antibodies specific for two distinct steric portions of the glycolipid ganglio-*N*-triosylceramide (asialo G_{M2}). *J. Exp. Med.* **150**: 1008-1019.
2. Solomon, F. R. and Higgins, T. J. (1987) A monoclonal antibody with reactivity to asialo GM1 and murine natural killer cells. *Mol. Immunol.* **24**: 57-65.
3. Jacquemart, F., Millot, G., Goujet-Zalc, C., Mahouy, G. and Zalc, B. (1988) Production and characterization of a mouse monoclonal antibody to the glycolipid asialo-GM1. *Hybridoma* **7**: 323-331.
4. Shimada, S. and Iwata, D. (1987) Preparation of monoclonal antibodies.against a glycolipid asialo GM1. *Microbiol. Immunol.* **31**: 923-933.
5. Harrison, B. A., MacKenzie, R., Hirama, T., Lee, K. K. and Altman, E. (1998) A kinetics approach to the characterization of an IgM specific for the glycolipid asialo-GM1. *J. Immunol. Methods* **212**: 29-39.
6. Watarai, S., Handa, S., Tadakuma, T. and Yasuda, T. (1987) Application of liposomes to generation of monoclonal antibody to glycosphingolipid: production of monoclonal antibody to $GgOse_4Cer$. *J. Biochem. (Tokyo)* **102**: 59-67.
7. Ozawa, H., Kotani, M., Kawashima, I., Numata, M., Ogawa, T., Terashima, T. and Tai, T. (1993) Generation of a monoclonal antibody specific for ganglioside GM4: Evidence for GM4 expression on astrocytes in chicken cerebellum. *J. Biochem. (Tokyo)* **114**: 5-8.
8. Hirabayashi, Y., Hamaoka, A., Matsumoto, M., Matsubara, T., Tagawa, M., Wakabayashi, S. and Taniguchi, M. (1985) Syngeneic monoclonal antibody against melanoma antigen with interspecies cross-reactivity recognizes G_{M3}, a prominent ganglioside of B16 melanoma. *J. Biol. Chem.* **260**: 13328-13333.
9. Yamaguchi, H., Furukawa, K., Fortunato, S. R., Livingston, P. O., Lloyd, K. O., Oettgen, H. F. and Old, L. J. (1987) Cell-surface antigens of melanoma recognized by human monoclonal antibodies. *Proc. Natl. Acad. Sci. U. S. A.* **84**: 2416-2420.
10. Furukawa, K., Yamaguchi, H., Oettgen, H. F., Old, L. J. and Lloyd, K. O. (1989) Two human monoclonal antibodies reacting with the major gangliosides of human melanomas and comparison with corresponding mouse monoclonal antibodies. *Cancer Res.* **49**: 191-196.
11. Dohi, T., Nores, G. and Hakomori, S. (1988) An IgG3 monoclonal antibody established after immunization with GM3 lactone: immunochemical specificity and inhibition of melanoma cell growth in vitro and in vivo. *Cancer Res.* **48**: 5680-5685.
12. Natoli, E. J.,Jr., Livingston, P. O., Pukel, C. S., Lloyd, K. O., Wiegandt, H., Szalay, J., Oettgen, H. F. and Old, L. J. (1986) A murine monoclonal antibody detecting *N*-acetyl- and *N*-glycolyl-G_{M2}: characterization of cell surface reactivity. *Cancer Res.* **46**: 4116-4120.
13. Miyake, M., Ito, M., Hitomi, S., Ikeda, S., Taki, T., Kurata, M., Hino, A., Miyake, N. and Kannagi, R. (1988) Generation of two monoclonal antibodies that can discriminate *N*-glycolyl and *N*- acetyl neuraminic acid residues of G_{M2} gangliosides and distribution of the antigens in human lung cancer. *Cancer Res.* **48**: 6154-6160.
14. Cahan, L. D., Irie, R. F., Singh, R., Cassidenti, A. and Paulson, J. C. (1982) Identification of a human neuroectodermal tumor antigen (OFA-I-2) as ganglioside G_{D2}. *Proc. Natl. Acad. Sci. U. S. A.* **79**: 7629-7633.
15. Kotani, M., Ozawa, H., Kawashima, I., Ando, S. and Tai, T. (1992) Generation of one set of monoclonal antibodies specific for a-pathway ganglio-series gangliosides. *Biochim. Biophys. Acta Gen. Subj.* **1117**: 97-103.
16. Furukawa, K., Yamaguchi, H., Oettgen, H. F., Old, L. J. and Lloyd, K. O. (1988) Analysis of the expression of *N*-glycolylneuraminic acid-containing gangliosides in cells and tissues using two human monoclonal antibodies. *J. Biol. Chem.* **263**: 18507-18512.

17. Sanai, Y., Yamasaki, M. and Nagai, Y. (1988) Monoclonal antibody directed to a Hanganutziu-Deicher active ganglioside, GM2 (NeuGc). *Biochim. Biophys. Acta* **958**: 368-374.

18. Yamaguchi, H., Furukawa, K., Fortunato, S. R., Livingston, P. O., Lloyd, K. O., Oettgen, H. F. and Old, L. J. (1990) Human monoclonal antibody with dual GM2/GD2 specificity derived from an immunized melanoma patient. *Proc. Natl. Acad. Sci. U. S. A.* **87**: 3333-3337.

19. Watarai, S., Kiura, K., Shigeto, R., Shibayama, T., Kimura, I. and Yasuda, T. (1994) Establishment of monoclonal antibodies specific for ganglioside GM1: Detection of ganglioside GM1 in small cell lung carcinoma cell lines and tissues. *J. Biochem. (Tokyo)* **116**: 948-954.

20. Furuya, S., Hashikawa, T., Irie, F., Hasegawa, A., Nakao, T. and Hirabayashi, Y. (1995) Neuronal expression of a minor monosialosyl ganglioside GM1b in rat brain: Immunochemical characterization using a specific monoclonal antibody. *Neurosci. Res.* **22**: 411-421.

21. Hiraiwa, N., Tsuyuoka, K., Li, Y. T., Tanaka, M., Seno, T., Okubo, Y., Fukuda, Y., Imura, H. and Kannagi, R. (1990) Gangliosides and sialoglycoproteins carrying a rare blood group antigen determinant, Cad, associated with human cancers as detected by specific monoclonal antibodies. *Cancer Res.* **50**: 5497-5503.

22. Pukel, C. S., Lloyd, K. O., Travassos, L. R., Dippold, W. G., Oettgen, H. F. and Old, L. J. (1982) G_{D3}, a prominent ganglioside of human melanoma. Detection and characterisation by mouse monoclonal antibody. *J. Exp. Med.* **155**: 1133-1147.

23. Nudelman, E., Hakomori, S., Kannagi, R., Levery, S. B., Yeh, M. Y., Hellstr\m, K. E. and Hellstr\m, I. (1982) Characterization of a human melanoma-associated ganglioside antigen defined by a monoclonal antibody, 4.2. *J. Biol. Chem.* **257**: 12752-12756.

24. Werkmeister, J. A., Triglia, T., Mackay, I. R., Dowling, J. P., Varigos, G. A., Morstyn, G. and Burns, G. F. (1987) Fluctuations in the expression of a glycolipid antigen associated with differentiation of melanoma cells monitored by a monoclonal antibody, Leo Mel 3. *Cancer Res.* **47**: 225-230.

25. Ozawa, H., Kotani, M., Kawashima, I. and Tai, T. (1992) Generation of one set of monoclonal antibodies specific for b-pathway ganglio-series gangliosides. *Biochim. Biophys. Acta Lipids Lipid Metab.* **1123**: 184-190.

26. Ohta, S., Honda, A., Tokutake, Y., Yoshida, H. and Hanai, N. (1993) Antitumor effects of a novel monoclonal antibody with high binding affinity to ganglioside GD3. *Cancer Immunol. Immunother.* **36**: 260-266.

27. Mujoo, K., Cheresh, D. A., Yang, H. M. and Reisfeld, R. A. (1987) Disialoganglioside G_{D2} on human neuroblastoma cells: target antigen for monoclonal antibody-mediated cytolysis and suppression of tumor growth. *Cancer Res.* **47**: 1098-1104.

28. Cheung, N. K., Saarinen, U. M., Neely, J. E., Landmeier, B., Donovan, D. and Coccia, P. F. (1985) Monoclonal antibodies to a glycolipid antigen on human neuroblastoma cells. *Cancer Res.* **45**: 2642-2649.

29. Bosslet, K., Mennel, H. D., Rodden, F., Bauer, B. L., Wagner, F., Altmannsberger, A., Sedlacek, H. H. and Wiegandt, H. (1989) Monoclonal antibodies against epitopes on ganglioside G_{D2} and its lactones. Markers for gliomas and neuroblastomas. *Cancer Immunol. Immunother.* **29**: 171-178.

30. Tai, T., Paulson, J. C., Cahan, L. D. and Irie, R. F. (1983) Ganglioside G_{M2} as a human tumor antigen (OFA-I-1). *Proc. Natl. Acad. Sci. U. S. A.* **80**: 5392-5396.

31. Thurin, J., Thurin, M., Herlyn, M., Elder, D. E., Steplewski, Z., Clark, W. H.,Jr. and Koprowski, H. (1986) GD2 ganglioside biosynthesis is a distinct biochemical event in human melanoma tumor progression. *FEBS Lett.* **208**: 17-22.

32. Levine, J. M., Beasley, L. and Stallcup, W. B. (1984) The D1.1 antigen: a cell surface marker for germinal cells of the central nervous system. *J. Neurosci.* **4**: 820-831.

33. Cheresh, D. A., Reisfeld, R. A. and Varki, A. P. (1984) *O*-acetylation of disialoganglioside G_{D3} by human melanoma cells creates a unique antigenic determinant. *Science* **225**: 844-846.

34. Bonafede, D. M., Macala, L. J., Constantine-Paton, M. and Yu, R. K. (1989) Isolation and characterization of ganglioside 9-*O*-acetyl-G_{D3} from bovine buttermilk. *Lipids* **24**:

680-684.

35. Reinhardt-Maelicke, S., Cleeves, V., Kindler-Rohrborn, A. and Rajewsky, M. F. (1990) Differential recognition of a set of *O*-acetylated gangliosides by monoclonal antibodies RB13-2, D1.1, and JONES during rat brain development. *Dev. Brain Res.* **51**: 279-282.
36. Thurin, J., Herlyn, M., Hindsgaul, O., Stromberg, N., Karlsson, K. A., Elder, D., Steplewski, Z. and Koprowski, H. (1985) Proton NMR and fast-atom bombardment mass spectrometry analysis of the melanoma-associated ganglioside 9-*O*-acetyl-G_{D3}. *J. Biol. Chem.* **260**: 14556-14563.
37. Nayak, R. C., Berman, A. B., George, K. L., Eisenbarth, G. S. and King, G. L. (1988) A monoclonal antibody (3G5)-defined ganglioside antigen is expressed on the cell surface of microvascular pericytes. *J. Exp. Med.* **167**: 1003-1015.
38. Chou, D. K., Flores, S. and Jungalwala, F. B. (1990) Identification of disialosyl paragloboside and *O*-acetyldisialosyl paragloboside in cerebellum and embryonic cerebrum. *J. Neurochem.* **54**: 1598-1607.
39. Nishinaka, T., Iwata, D., Shimada, S., Kosaka, K. and Suzuki, Y. (1993) Anti-ganglioside GD1a monoclonal antibody recognizes senile plaques in the brains of patients with Alzheimer-type dementia. *Neurosci. Res.* **17**: 171-176.
40. Fredman, P., Jeansson, S., Lycke, E. and Svennerholm, L. (1985) A monoclonal antibody reacting specifically with ganglioside G_{D1b} in human brain. *FEBS Lett.* **189**: 23-26.
41. Maehara, T., Ono, K., Tsutsui, K., Watarai, S., Yasuda, T., Inoue, H. and Tokunaga, A. (1997) A monoclonal antibody that recognizes ganglioside GD_{1b} in the rat central nervous system. *Neurosci. Res.* **29**: 9-16.
42. Nakamura, K., Suzuki, H., Hirabayashi, Y. and Suzuki, A. (1995) $IV^3\alpha$(NeuGcα2-8NeuGc)-Gg_4Cer is restricted to $CD4^+$ T cells producing interleukin-2 and a small population of mature thymocytes in mice. *J. Biol. Chem.* **270**: 3876-3881.
43. Furuya, S., Irie, F., Hashikawa, T., Nakazawa, K., Kozakai, A., Hasegawa, A., Sudo, K. and Hirabayashi, Y. (1994) Ganglioside $G_{D1\alpha}$ in cerebellar Purkinje cells. Its specific absence in mouse mutants with Purkinje cell abnormality and altered immunoreactivity in response to conjunctive stimuli causing long-term desensitization. *J. Biol. Chem.* **269**: 32418-32425.
44. Dubois, C., Manuguerra, J. C., Hauttecoeur, B. and Maze, J. (1990) Monoclonal antibody A2B5, which detects cell surface antigens, binds to ganglioside G_{T3} (II^3NeuAc_3LacCer) and to its 9-*O*-acetylated derivative. *J. Biol. Chem.* **265**: 2797-2803.
45. Dubois, C., Magnani, J. L., Grunwald, G. B., Spitalnik, S. L., Trisler, G. D., Nirenberg, M. and Ginsburg, V. (1986) Monoclonal antibody 18B8, which detects synapse-associated antigens, binds to ganglioside G_{T3} (II^3NeuAc_3LacCer). *J. Biol. Chem.* **261**: 3826-3830.
46. Hirabayashi, Y., Hirota, M., Matsumoto, M., Tanaka, H., Obata, K. and Ando, S. (1988) Developmental changes of C-series polysialogangliosides in chick brains revealed by mouse monoclonal antibodies M6704 and M7103 with different epitope specificities. *J. Biochem. (Tokyo)* **104**: 973-979.
47. Zhang, G., Ji, L., Kurono, S., Fujita, S. C., Furuya, S. and Hirabayashi, Y. (1997) Developmentally regulated *O*-acetylated sialoglycans in the central nervous system revealed by a new monoclonal antibody 493D4 recognizing a wide range of *O*-acetylated glycoconjugates. *Glycoconjugate J.* **14**: 847-857.
48. Nilsson, O., Brezicka, F. T., Holmgren, J., Sorenson, S., Svennerholm, L., Yngvason, F. and Lindholm, L. (1986) Detection of a ganglioside antigen associated with small cell lung carcinomas using monoclonal antibodies directed against fucosyl-GM1. *Cancer Res.* **46**: 1403-1407.
49. Yamamoto, H., Tsuji, S. and Nagai, Y. (1990) Tetrasialoganglioside GQ1b reactive monoclonal antibodies: Their characterization and application for quantification of GQ1b in some cell lines of neuronal and adrenal origin(s). *J. Neurochem.* **54**: 513-517.
50. Kusunoki, S., Chiba, A., Hirabayashi, Y., Irie, F., Kotani, M., Kawashima, I., Tai, T. and Nagai, Y. (1993) Generation of a monoclonal antibody specific for a new class of minor ganglioside antigens, GQ1b-α and GT1a-α: Its binding to dorsal and lateral horn of human thoracic cord. *Brain Res.* **623**: 83-88.

51. Greis, C. and Rösner, H. (1990) C-pathway polysialogangliosides in the nervous tissue of vertebrates, reacting with the monoclonal antibody Q211. *Brain Res.* **517**: 105-110.

52. Rösner, H., Greis, C. and Henke-Fahle, S. (1988) Developmental expression in embryonic rat and chicken brain of a polysialoganglioside-antigen reacting with the monoclonal antibody Q 211. *Brain Res.* **470**: 161-171.

53. Symington, F. W., Bernstein, I. D. and Hakomori, S. (1984) Monoclonal antibody specific for lactosylceramide (LacCer). *J. Biol. Chem.* **259**: 6008-6012.

54. Fenderson, B. A., O'Brien, D. A., Millette, C. F. and Eddy, E. M. (1984) Stage-specific expression of three cell surface carbohydrate antigens during murine spermatogenesis detected with monoclonal antibodies. *Dev. Biol.* **103**: 117-128.

55. Symington, F. W., Fenderson, B. A. and Hakomori, S. (1984) Fine specificity of a monoclonal anti-testicular cell antibody for glycolipids with terminal *N*-acetyl-D-glucosamine structure. *Mol. Immunol.* **21**: 877-882.

56. Nakamura, M., Tsunoda, A., Sakoe, K. and Saito, M. (1993) A monoclonal antibody reactive with terminal lactotriaosyl residue- containing oligosaccharides and its application to characterizing cell surface expression of the glyco-epitopes in COS-1 cells. *Biochem. Biophys. Res. Commun.* **197**: 1025-1033.

57. Gooi, H. C., Williams, L. K., Uemura, K., Hounsell, E. F., McIlhinney, R. A. and Feizi, T. (1983) A marker of human foetal endoderm defined by a monoclonal antibody involves Type 1 blood group chains. *Mol. Immunol.* **20**: 607-613.

58. Wilkins, P. P., Moore, K. L., McEver, R. P. and Cummings, R. D. (1995) Tyrosine sulfation of P-selectin glycoprotein ligand-1 is required for high affinity binding to P-selectin. *J. Biol. Chem.* **270**: 22677-22680.

59. Fukuda, M. N., Bothner, B., Lloyd, K. O., Rettig, W. J., Tiller, P. R. and Dell, A. (1986) Structures of glycosphingolipids isolated from human embryonal carcinoma cells. The presence of mono- and disialosyl glycolipids with blood group type 1 sequence. *J. Biol. Chem.* **261**: 5145-5153.

60. Nozawa, S., Narisawa, S., Kojima, K., Sakayori, M., Iizuka, R., Mochizuki, H., Yamauchi, T., Iwamori, M. and Nagai, Y. (1989) Human monoclonal antibody (HMST-1) against lacto-series type 1 chain and expression of the chain in uterine endometrial cancers. *Cancer Res.* **49**: 6401-6406.

61. Yago, K., Zenita, K., Ohwaki, I., Harada, R., Nozawa, S., Tsukazaki, K., Iwamori, M., Endo, N., Yasuda, N., Okuma, M. and Kannagi, R. (1993) Immunoglobulin variable region sequences of two human monoclonal antibodies directed to an onco-developmental carbohydrate antigen, lactotetraosylceramide ($LcOse_4Cer$). *Mol. Immunol.* **30**: 1481-1489.

62. Young, W. W.,Jr., Portoukalian, J. and Hakomori, S. (1981) Two monoclonal anticarbohydrate antibodies directed to glycosphingolipids with a lacto-N-glycosyl type II chain. *J. Biol. Chem.* **256**: 10967-10972.

63. Spitalnik, S. L., Schwartz, J. F., Magnani, J. L., Roberts, D. D., Spitalnik, P. F., Civin, C. I. and Ginsburg, V. (1985) Anti-My-28, an antigranulocyte mouse monoclonal antibody, binds to a sugar sequence in lacto-*N*-neotetraose. *Blood* **66**: 319-326.

64. Fujimoto, J., Hata, J., Ishii, E., Tanaka, S., Kannagi, R., Ueyama, Y. and Tamaoki, N. (1987) Differentation antigens defined by mouse monoclonal antibodies against human germ cell tumors. *Lab. Invest.* **57**: 350-358.

65. Myoga, A., Taki, T., Arai, K., Sekiguchi, K., Ikeda, I., Kurata, K. and Matsumoto, M. (1988) Detection of patients with cancer by monoclonal antibody directed to lactoneotetraosylceramide (paragloboside). *Cancer Res.* **48**: 1512-1516.

66. Holmes, E. H. and Greene, T. G. (1991) Isolation and fine-structure characterization of four monoclonal antibodies reactive with glycoconjugates containing terminal GlcNAc residues: application to aspects of lacto-series tumor antigen biosynthesis. *Arch. Biochem. Biophys.* **288**: 87-96.

67. Miyake, M., Kohno, N., Nudelman, E. D. and Hakomori, S. (1989) Human IgG_3 monoclonal antibody directed to an unbranched repeating type 2 chain (Galβ1→4GlcNAcβ1→3Galβ1→4GlcNAcβ1→3Galβ1→R) which is highly expressed in colonic and hepatocellular carcinoma. *Cancer Res.* **49**: 5689-5695.

68. Hirohashi, S., Clausen, H., Nudelman, E., Inoue, H., Shimosato, Y. and Hakomori, S. (1986) A human monoclonal antibody directed to blood group i antigen: heterohybridoma between human lymphocytes from regional lymph nodes of a lung cancer patient and mouse myeloma. *J. Immunol.* **136**: 4163-4168.

69. Nagatsuka, Y., Watarai, S., Yasuda, T., Higashi, H., Yamagata, T. and Ono, Y. (1995) Production of human monoclonal antibodies to i blood group by EBV-induced transformation: Possible presence of a new glycolipid in cord red cell membranes and human hematopoietic cell lines. *Immunol. Lett.* **46**: 93-100.

70. Fenderson, B. A., Nichols, E. J., Clausen, H. and Hakomori, S. I. (1986) A monoclonal antibody defining a binary *N*-acetyllactosaminyl structure in lactoisooctaosylceramide ($IV^6Gal\beta 1 \rightarrow 4GlcNAcnLc_6$): a useful probe for determining differential glycosylation patterns between normal and transformed human fibroblasts. *Mol. Immunol.* **23**: 747-754.

71. Nichols, E. J., Fenderson, B. A., Carter, W. G. and Hakomori, S. (1986) Domain-specific distribution of carbohydrates in human fibronectins and the transformation-dependent translocation of branched type 2 chain defined by monoclonal antibody C6. *J. Biol. Chem.* **261**: 11295-11301.

72. Fredman, P., Richert, N. D., Magnani, J. L., Willingham, M. C., Pastan, I. and Ginsburg, V. (1983) A monoclonal antibody that precipitates the glycoprotein receptor for epidermal growth factor is directed against the human blood group H type 1 antigen. *J. Biol. Chem.* **258**: 11206-11210.

73. Imai, K., Sasanami, T., Nakanishi, T., Noguchi, T. and Yachi, A. (1985) Circulating blood group-related antigen(s) in cancer patients detected by the monoclonal antibodies produced against hepatocellular carcinoma cell line. *Tumour. Biol.* **6**: 257-272.

74. Tsuji, Y., Yoshioka, M., Ogasawara, T., Takemura, T. and Isojima, S. (1987) Identification of an H antigen-like blood group antigen in sera of cancer patients using a novel monoclonal antibody raised against endometrial carcinoma. *Cancer Res.* **47**: 3543-3550.

75. Karsten, U., Pilgrim, G., Hanisch, F. G., Uhlenbruck, G., Kasper, M., Stosiek, P., Papsdorf, G. and Pasternak, G. (1988) A new monoclonal antibody (A46-B/B10) highly specific for the blood group H type 2 epitope: generation, epitope analysis, serological and histological evaluation. *Br. J. Cancer* **58**: 176-181.

76. Young, W. W.,Jr., Johnson, H. S., Tamura, Y., Karlsson, K. A., Larson, G., Parker, J. M., Khare, D. P., Spohr, U., Baker, D. A., Hindsgaul, O. and Lemieux, R. U. (1983) Characterization of monoclonal antibodies specific for the Lewis a human blood group determinant. *J. Biol. Chem.* **258**: 4890-4894.

77. Zenita, K., Hirashima, K., Shigeta, K., Hiraiwa, N., Takada, A., Fujimoto, E., Hashimoto, K., Hakomori, S. and Kannagi, R. (1990) Northern hybridization analysis of V_H gene expression in murine monoclonal antibodies directed to cancer-associated ganglioside antigens having various sialic acid linkages. *J. Immunol.* **144**: 4442-4451.

78. Rouger, P., Tsikas, G., Gane, P., Oriol, R. and Salmon, C. (1987) Immunological approach of anti-H (9), anti-Lewis (6), anti-P (3) and anti-Pr (1) monoclonal antibodies. *Rev. Fr. Transfus. Immunohematol.* **30**: 663-669.

79. Stroud, M. R., Levery, S. B., Nudelman, E. D., Salyan, M. E., Towell, J. A., Roberts, C. E., Watanabe, M. and Hakomori, S. (1991) Extended type 1 chain glycosphingolipids: dimeric Le^a ($III^4V^4Fuc_2Lc_6$) as human tumor-associated antigen. *J. Biol. Chem.* **266**: 8439-8446.

80. Brockhaus, M., Magnani, J. L., Herlyn, M., Blaszczyk, M., Steplewski, Z., Koprowski, H. and Ginsburg, V. (1982) Monoclonal antibodies directed against the sugar sequence of lacto-*N*-fucopentaose III are obtained from mice immunized with human tumors. *Arch. Biochem. Biophys.* **217**: 647-651.

81. Huang, L. C., Brockhaus, M., Magnani, J. L., Cuttitta, F., Rosen, S., Minna, J. D. and Ginsburg, V. (1983) Many monoclonal antibodies with an apparent specificity for certain lung cancers are directed against a sugar sequence found in lacto-*N*-fucopentaose III. *Arch. Biochem. Biophys.* **220**: 318-320.

82. Cuttitta, F., Rosen, S., Gazdar, A. F. and Minna, J. D. (1981) Monoclonal antibodies that demonstrate specificity for several types of human lung cancer. *Proc. Natl. Acad. Sci. U. S. A.* **78**: 4591-4595.

83. Huang, L. C., Civin, C. I., Magnani, J. L., Shaper, J. H. and Ginsburg, V. (1983) My-1, the human myeloid-specific antigen detected by mouse monoclonal antibodies, is a sugar sequence found in lacto-*N*-fucopentaose III. *Blood* **61**: 1020-1023.

84. Gooi, H. C., Thorpe, S. J., Hounsell, E. F., Rumpold, H., Kraft, D., Forster, O. and Feizi, T. (1983) Marker of peripheral blood granulocytes and monocytes of man recognized by two monoclonal antibodies VEP8 and VEP9 involves the trisaccharide 3-fucosyl-N-acetyllactosamine. *Eur. J. Immunol.* **13**: 306-312.

85. Urdal, D. L., Brentnall, T. A., Bernstein, I. D. and Hakomori, S. I. (1983) A granulocyte reactive monoclonal antibody, 1G10, identifies the Galβ1-4(Fucα1-3)GlcNAc (X determinant) expressed in HL-60 cells on both glycolipid and glycoprotein molecules. *Blood* **62**: 1022-1026.

86. Fukushi, Y., Hakomori, S., Nudelman, E. and Cochran, N. (1984) Novel fucolipids accumulating in human adenocarcinoma. II. Selective isolation of hybridoma antibodies that differentially recognize mono-, di-, and trifucosylated type 2 chain. *J. Biol. Chem.* **259**: 4681-4685.

87. Gooi, H. C., Feizi, T., Kapadia, A., Knowles, B. B., Solter, D. and Evans, M. J. (1981) Stage-specific embryonic antigen involves α1→3 fucosylated type 2 blood group chains. *Nature* **292**: 156-158.

88. Kannagi, R., Nudelman, E., Levery, S. B. and Hakomori, S. (1982) A series of human erythrocyte glycosphingolipids reacting to the monoclonal antibody directed to developmentally-regulated antigen, SSEA-1. *J. Biol. Chem.* **257**: 14865-14874.

89. Kannagi, R., Nudelman, E. and Hakomori, S. (1982) Possible role of ceramide in defining the structure and function of membrane glycolipids. *Proc. Natl. Acad. Sci. USA* **79**: 3470-3474.

90. Singhal, A. K., Orntoft, T. F., Nudelman, E., Nance, S., Schibig, L., Stroud, M. R., Clausen, H. and Hakomori, S. (1990) Profiles of Lewis^x-containing glycoproteins and glycolipids in sera of patients with adenocarcinoma. *Cancer Res.* **50**: 1375-1380.

91. Nudelman, E., Levery, S. B., Stroud, M. R., Salyan, M. E., Abe, K. and Hakomori, S. (1988) A novel tumor-associated, developmentally regulated glycolipid antigen defined by monoclonal antibody ACFH-18. *J. Biol. Chem.* **263**: 13942-13951.

92. Brockhaus, M., Magnani, J. L., Blaszczyk, M., Steplewski, Z., Koprowski, H., Karlsson, K. A., Larson, G. and Ginsburg, V. (1981) Monoclonal antibodies directed against the human Le^b blood group antigen. *J. Biol. Chem.* **256**: 13223-13225.

93. Brown, A., Feizi, T., Gooi, H. C., Embleton, M. J., Picard, J. K. and Baldwin, R. W. (1983) A monoclonal antibody against human colonic adenoma recognizes difucosylated Type-2-blood-group chains. *Biosci. Rep.* **3**: 163-170.

94. Lloyd, K. O., Larson, G., Stromberg, N., Thurin, J. and Karlsson, K. A. (1983) Mouse monoclonal antibody F-3 recognizes the difucosyl type-2 blood group structure. *Immunogenetics* **17**: 537-541.

95. Blaineau, C., Le Pendu, J., Arnaud, D., Connan, F. and Avner, P. (1983) The glycosidic antigen recognised by a novel monoclonal antibody, 75.12, is developmentally regulated on mouse embryonal carcinoma cells. *EMBO J.* **2**: 2217-2222.

96. Abe, K., McKibbin, J. M. and Hakomori, S. (1983) The monoclonal antibody directed to difucosylated type 2 chian (Fucα1→2Galβ1→4[Fucα1→3]GlcNAc; Y determinant). *J. Biol. Chem.* **258**: 11793-11797.

97. Hirashima, K., Zenita, K., Kitahara, A., Ishihara, G., Harada, R., Ohmori, K., Hirohashi, S., Akiyama, S. and Kannagi, R. (1990) High idiotypic connectivity of the V_H7183-encoded antibodies directed to a murine embryonic carbohydrate antigen, Lewis Y, as ascertained by syngenic anti-idiotype monoclonal antibodies. *J. Immunol.* **145**: 224-232.

98. Kaizu, T., Levery, S. B., Nudelman, E., Stenkamp, R. E. and Hakomori, S. (1986) Novel fucolipids of human adenocarcinoma: monoclonal antibody specific for trifucosyl Le^y (III^{3}FucV^{3}FucVI^{2}FucnLc$_6$) and a possible three-dimensional epitope structure. *J. Biol. Chem.* **261**: 11254-11258.

99. Sun, Q., Siddiqui, B., Nudelman, E., Hakomori, S., Ho, J. J. L. and Kim, Y. S. (1987) New murine monoclonal antibodies to a human colonic cancer-associated glycolipid, extended difucosylated Le^y glycolipid. *Cancer J.* **1**: 213.

100. Koprowski, H., Herlyn, M., Steplewski, Z. and Sears, H. F. (1981) Specific antigen in serum of patients with colon carcinoma. *Science* **212**: 53-55.

101. Magnani, J. L., Nilsson, B., Brockhaus, M., Zopf, D., Steplewski, Z., Koprowski, H. and Ginsburg, V. (1982) A monoclonal antibody-defined antigen associated with gastrointestinal cancer is a ganglioside containing sialylated lacto-*N*-fucopentaose II. *J. Biol. Chem.* **257**: 14365-14369.

102. Fukuta, S., Magnani, J. L., Gaur, P. K. and Ginsburg, V. (1987) Monoclonal antibody CC3C195, which detects cancer-associated antigens in serum, binds to the human Le^a blood group antigen and to its sialylated derivative. *Arch. Biochem. Biophys.* **255**: 214-216.

103. Chung, Y. S., Ho, J. J., Kim, Y. S., Tanaka, H., Nakata, B., Hiura, A., Motoyoshi, H., Satake, K. and Umeyama, K. (1987) The detection of human pancreatic cancer-associated antigen in the serum of cancer patients. *Cancer* **60**: 1636-1643.

104. Kano, Y., Taniguchi, T., Uemura, Y., Yokoyama, K., Uesaka, K., Yamamoto, M., Ohyanagi, H. and Saitoh, Y. (1990) Characterization of an antigen defined by monoclonal antibody KMO1. *Hybridoma* **9**: 363-375.

105. Kitagawa, H., Nakada, H., Numata, Y., Kurosaka, A., Fukui, S., Funakoshi, I., Kawasaki, T. and Yamashina, I. (1988) A monoclonal antibody that recognizes sialyl-Le^a oligosaccharide, but is distinct from NS 19-9 as to epitope recognition. *J. Biochem. (Tokyo)* **104**: 817-821.

106. Takada, A., Ohmori, K., Takahashi, N., Tsuyuoka, K., Yago, K., Zenita, K., Hasegawa, A. and Kannagi, R. (1991) Adhesion of human cancer cells to vascular endothelium mediated by a carbohydrate antigen, sialyl Lewis A. *Biochem. Biophys. Res. Commun.* **179**: 713-719.

107. Nilsson, O., Månsson, J-E., Lindholm, L., Holmgren, J. and Svennerholm, L. (1985) Sialosyllactotetraosylceramide, a novel ganglioside antigen detected in human carcinomas by a monoclonal antibody. *FEBS Lett.* **182**: 398-402.

108. Hamanaka, Y., Hamanaka, S., Shinagawa, Y., Suzuki, T., Inagaki, F., Suzuki, M. and Suzuki, A. (1994) Ganglioside antigen of DU-PAN-2 in a human pancreatic cancer. *FEBS Lett.* **353**: 48-52.

109. Kawa, S., Tokoo, M., Oguchi, H., Furuta, S., Homma, T., Hasegawa, Y., Ogata, H. and Sakata, K. (1994) Epitope analysis of SPan-1 and DUPAN-2 using synthesized glycoconjugates sialyllacto-*N*-fucopentaose II and sialyllacto-*N*-tetraose. *Pancreas* **9**: 692-697.

110. Svennerholm, L., Bostrom, K., Fredman, P., Mansson, J. E., Rosengren, B. and Rynmark, B. M. (1989) Human brain gangliosides: developmental changes from early fetal stage to advanced age. *Biochim. Biophys. Acta* **1005**: 109-117.

111. Suzuki, Y., Nishi, H., Hidari, K., Hirabayashi, Y., Matsumoto, M., Kobayashi, T., Watarai, S., Yasuda, T., Nakayama, J. and Maeda, H. (1991) A new monoclonal antibody directed to sialyl α2-3lacto*neo*tetraosylceramide and its application for detection of human gastrointestinal neoplasms. *J. Biochem. (Tokyo)* **109**: 354-360.

112. Hakomori, S., Patterson, C. M., Nudelman, E. and Sekiguchi, K. (1983) A monoclonal antibody directed to *N*-acetylneuraminosyl α2→6 galactosyl residue in gangliosides and glycoproteins. *J. Biol. Chem.* **258**: 11819-11822.

113. Nilsson, O., Lindholm, L., Holmgren, J. and Svennerholm, L. (1985) Monoclonal antibodies raised against NeuAcα2-6 neolactotetraosylceramide detect carcinoma-associated gangliosides. *Biochim. Biophys. Acta* **835**: 577-583.

114. Taki, T., Yamamoto, K., Takamatsu, M., Ishii, K., Myoga, A., Sekiguchi, K., Ikeda, I., Kurata, K., Nakayama, J. and Handa, S. (1990) Accumulation of gangliosides with *N*-acetylneuraminosyl (α2-6) lactosamine structure in primary human hepatoma. *Cancer Res.* **50**: 1284-1290.

115. Fukushima, K., Hirota, M., Terasaki, P. I., Wakisaka, A., Togashi, H., Chia, D., Suyama, N., Fukushi, Y., Nudelman, E. and Hakomori, S. (1984) Characterization of sialosylated $Lewis^x$ as a new tumor-associated antigen. *Cancer Res.* **44**: 5279-5285.

116. Phillips, M. L., Nudelman, E., Gaeta, F. C. A., Perez, M., Singhal, A. K., Hakomori, S. and Paulson, J. C. (1990) ELAM-1 mediates cell adhesion by recognition of a carbohydrate ligand, sialyl-Le^x. *Science* **250**: 1130-1132.

117. Polley, M. J., Phillips, M. L., Wayner, E., Nudelman, E., Singhal, A. K., Hakomori, S. and Paulson, J. C. (1991) CD62 and endothelial cell-leukocyte adhesion molecule 1

(ELAM-1) recognize the same carbohydrate ligand, sialyl-Lewis x. *Proc. Natl. Acad. Sci. USA* **88**: 6224-6228.

118. Shitara, K., Hanai, N. and Yoshida, H. (1987) Distribution of lung adenocarcinoma-associated antigens in human tissues and sera defined by monoclonal antibodies KM-52 and KM-93. *Cancer Res.* **47**: 1267-1272.
119. Sawada, M., Takada, A., Ohwaki, I., Takahashi, N., Tateno, H., Sakamoto, J. and Kannagi, R. (1993) Specific expression of a complex sialyl Lewis X antigen on high endothelial venules of human lymph nodes: Possible candidate for L-selectin ligand. *Biochem. Biophys. Res. Commun.* **193**: 337-347.
120. Ohmori, K., Takada, A., Ohwaki, I., Takahashi, N., Furukawa, Y., Maeda, M., Kiso, M., Hasegawa, A., Kannagi, M. and Kannagi, R. (1993) A distinct type of sialyl Lewis X antigen defined by a novel monoclonal antibody is selectively expressed on helper memory T cells. *Blood* **82**: 2797-2805.
121. Berg, E. L., Robinson, M. K., Mansson, O., Butcher, E. C. and Magnani, J. L. (1991) A carbohydrate domain common to both sialyl Le^a and sialyl Le^x is recognized by the endothelial cell leukocyte adhesion molecule ELAM-1. *J. Biol. Chem.* **266**: 14869-14872.
122. Fukushi, Y., Kannagi, R., Hakomori, S., Shepard, T., Kulander, B. G. and Singer, J. W. (1985) Localization and distribution of difucoganglioside ($VI^3NeuAcIII^3V^3Fuc_2nLc_6$) in normal and tumor tissues defined by its monoclonal antibody FH6. *Cancer Res.* **45**: 3711-3717.
123. Macher, B. A., Buehler, J., Scudder, P., Knapp, W. and Feizi, T. (1988) A novel carbohydrate, differentiation antigen on fucogangliosides of human myeloid cells recognized by monoclonal antibody VIM-2. *J. Biol. Chem.* **263**: 10186-10191.
124. Nudelman, E., Fukushi, Y., Levery, S. B., Higuchi, T. and Hakomori, S. (1986) Novel fucolipids of human adenocarcinoma: IV. Disialosyl Le^a antigen ($III^4FucIII^6NeuAcIV^3NeuAcLc_6$) of human colonic adenocarcinoma and the monoclonal antibody (FH7) defining this structure. *J. Biol. Chem.* **261**: 5487-5495.
125. Fukushi, Y., Nudelman, E., Levery, S. B., Higuchi, T. and Hakomori, S. (1986) A novel disialoganglioside ($IV^3NeuAcIII^6NeuAcLc_4$) of human adenocarcinoma and the monoclonal antibody (FH9) defining this disialosyl structure. *Biochemistry* **25**: 2859-2866.
126. Galili, U., Basbaum, C. B., Shohet, S. B., Buehler, J. and Macher, B. A. (1987) Identification of erythrocyte Galα1-3Gal glycosphingolipids with a mouse monoclonal antibody, Gal-13. *J. Biol. Chem.* **262**: 4683-4688.
127. Hilkens, J., Buijs, F., Hilgers, J., Hageman, P., Calafat, J., Sonnenberg, A. and van der Valk, M. (1984) Monoclonal antibodies against human milk-fat globule membranes detecting differentiation antigens of the mammary gland and its tumors. *Int. J. Cancer* **34**: 197-206.
128. Gooi, H. C., Jones, N. J., Hounsell, E. F., Scudder, P., Hilkens, J., Hilgers, J. and Feizi, T. (1985) Novel antigenic specificity involving the blood group antigen, Lea, in combination with onco-developmental antigen, SSEA-1, recognized by two monoclonal antibodies to human milk-fat globule membranes. *Biochem. Biophys. Res. Commun.* **131**: 543-550.
129. Pettijohn, D. E., Stranahan, P. L., Due, C., Ronne, E., Sorensen, H. R. and Olsson, L. (1987) Glycoproteins distinguishing non-small cell from small cell human lung carcinoma recognized by monoclonal antibody 43-9F. *Cancer Res.* **47**: 1161-1169.
130. Martensson, S., Due, C., Pahlsson; P., Nilsson, B., Eriksson, H., Zopf, D., Olsson, L. and Lundblad, A. (1988) A carbohydrate epitope associated with human squamous lung cancer. *Cancer Res.* **48**: 2125-2131.
131. Nudelman, E. D., Mandel, U., Levery, S. B., Kaizu, T. and Hakomori, S. (1989) A series of disialogangliosides with binary 2→3 sialosyllactosamine structure, defined by monoclonal antibody NUH2, are oncodevelopmentally regulated antigens. *J. Biol. Chem.* **264**: 18719-18725.
132. Watarai, S., Kushi, Y., Shigeto, R., Misawa, N., Eishi, Y., Handa, S. and Yasuda, T. (1995) Production of monoclonal antibodies directed to Hanganutziu-Deicher active gangliosides, *N*-glycolylneuraminic acid-containing gangliosides. *J. Biochem. (Tokyo)* **117**: 1062-1069.
133. Abe, K., Levery, S. B. and Hakomori, S. (1984) The antibody specific to type 1 chain blood group A determinant. *J. Immunol.* **132**: 1951-1954.

134. Furukawa, K., Clausen, H., Hakomori, S., Sakamoto, J., Look, K., Lundblad, A., Mattes, M. J. and Lloyd, K. O. (1985) Analysis of the specificity of five murine anti-blood group A monoclonal antibodies, including one that identifies type 3 and type 4 A determinants. *Biochemistry* **24**: 7820-7826.

135. Clausen, H., Hakomori, S., Graem, N. and Dabelsteen, E. (1986) Incompatible A antigen expressed in tumors of blood group O individuals: immunochemical, immunohistologic, and enzymatic characterization. *J. Immunol.* **136**: 326-330.

136. Voak, D., Sacks, S., Alderson, T., Takei, F., Lennox, E., Jarvis, J., Milstein, C. and Darnborough, J. (1980) Monoclonal anti-A from a hybrid-myeloma: evaluating as a blood grouping reagent. *Vox Sang.* **39**: 134-140.

137. Clausen, H., McKibbin, J. M. and Hakomori, S. (1985) Monoclonal antibodies defining blood group A variants with difucosyl type 1 chain (ALe^b) and difucosyl type 2 chain (ALe^y). *Biochemistry* **24**: 6190-6194.

138. Clausen, H., Levery, S. B., Nudelman, E., Tsuchiya, S. and Hakomori, S. (1985) Repetitive A epitope (type 3 chain A) defined by blood group A_1-specific monoclonal antibody TH-1: chemical basis of qualitative A_1 and A_2 distinction. *Proc. Natl. Acad. Sci. U. S. A.* **82**: 1199-1203.

139. Clausen, H., Stroud, M., Parker, J., Springer, G. and Hakomori, S. (1988) Monoclonal antibodies directed to the blood group A associated structure, galactosyl-A: specificity and relation to the Thomsen-Friedenreich antigen. *Mol. Immunol.* **25**: 199-204.

140. Bundle, D. R., Gidney, M. A., Kassam, N. and Rahman, A. F. (1982) Hybridomas specific for carbohydrates; synthetic human blood group antigens for the production, selection, and characterization of monoclonal typing reagents. *J. Immunol.* **129**: 678-672.

141. Hansson, G. C., Karlsson, K. A., Larson, G., McKibbin, J. M., Blaszczyk, M., Herlyn, M., Steplewski, Z. and Koprowski, H. (1983) Mouse monoclonal antibodies against human cancer cell lines with specificities for blood group and related antigens. Characterization by antibody binding to glycosphingolipids in a chromatogram binding assay. *J. Biol. Chem.* **258**: 4091-4097.

142. Jessell, T. M. and Dodd, J. (1985) Structure and expression of differentiation antigens on functional subclasses of primary sensory neurons. *Philos. Trans. R. Soc. Lond. B. Biol. Sci.* **308**: 271-281.

143. Sueyoshi, S., Nagakura, H., Kato, A., Uetsuki, S., Nakayama, Y. and Adachi, M. (1992) Monoclonal antibody GOM-2 binds to blood group B-Le^y active glycolipid antigens on human gastric cancer cells, KATO-III. *Glycoconj. J.* **9**: 99-108.

144. Nudelman, E., Kannagi, R., Hakomori, S., Parsons, M., Lipinski, M., Wiels, J., Fellous, M. and Tursz, T. (1983) A glycolipid antigen associated with Burkitt lymphoma defined by a monoclonal antibody. *Science* **220**: 509-511.

145. Kojima, H., Tsuchiya, S., Sekiguchi, K., Gelinas, R. and Hakomori, S. (1987) Predefined gene transfer for expression of a glycosphingolipid antigen by transfection with a cosmid genomic library prepared from a cell line in which the specific glycosphingolipid is highly expressed. *Biochem. Biophys. Res. Commun.* **143**: 716-722.

146. Toda, S., Yokobori, Y., Sen, Q., Kannagi, R., Taki, T., Kasama, K., Tanaka, S., Takata, K., and Kimura, H. Globotriaosylceramide (Gb_3) in bull sperm, testis and gonadal fluids. (1997) XVIth International Symposium on Glycoconjugates, Zurich, Switzerland, September 7-12.

147. Kotani, M., Kawashima, I., Ozawa, H., Ogura, K., Ariga, T. and Tai, T. (1994) Generation of one set of murine monoclonal antibodies specific for globo-series glycolipids: evidence for differential distribution of the glycolipids in rat small intestine. *Arch. Biochem. Biophys.* **310**: 89-96.

148. Schwarting, G. A. and Crandall, J. E. (1991) Subsets of olfactory and vomeronasal sensory epithelial cells and axons revealed by monoclonal antibodies to carbohydrate antigens. *Brain Res.* **547**: 239-248.

149. Madassery, J. V., Gillard, B., Marcus, D. M. and Nahm, M. H. (1991) Subpopulations of B cells in germinal centers. III. HJ6, a monoclonal antibody, binds globoside and a subpopulation of germinal center B cells. *J. Immunol.* **147**: 823-829.

150. Brodin, N. T., Thurin, J., Karlsson, K. A., Martensson, S. and Sjogren, H. O. (1989) Monoclonal antibodies to glycolipids of the isoglobo-series detect tumor-associated antigens in rats. *Int. J. Cancer* **43**: 317-326.

151. Kannagi, R., Levery, S. B., Ishigami, S., Hakomori, S., Shevinsky, L. H., Knowles, B. B. and Solter, D. (1983) New globoseries glycosphingolipids in human teratocarcinoma reactive with the monoclonal antibody directed to a developmentally-regulated antigen, SSEA-3. *J. Biol. Chem.* **258**: 8934-8942.

152. Marcus, D. M., Gilbert, S., Sekine, M. and Suzuki, A. (1988) Monoclonal antibodies that bind to galactosylgloboside (SSEA-3 antigen). *Arch. Biochem. Biophys.* **262**: 620-625.

153. Schrump, D. S., Furukawa, K., Yamaguchi, H., Lloyd, K. O. and Old, L. J. (1988) Recognition of galactosylgloboside by monoclonal antibodies derived from patients with primary lung cancer. *Proc. Natl. Acad. Sci. U. S. A.* **85**: 4441-4445.

154. Kannagi, R., Cochran, N. A., Ishigami, F., Hakomori, S., Andrews, P. W., Knowles, B. B. and Solter, D. (1983) Stage-specific embryonic antigens (SSEA-3 and -4) are epitopes of a unique globo-series ganglioside isolated from human teratocarcinoma cells. *Eur. Mol. Biol. Orgn. J.* **2**: 2355-2361.

155. Saito, S., Levery, S. B., Salyan, M. E. K., Goldberg, R. I. and Hakomori, S. (1994) Common tetrasaccharide epitope NeuAcα2→3Galβ1→3(NeuAcα2→6)GalNAc, presented by different carrier glycosylceramides or *O*-linked peptides, is recognized by different antibodies and ligands having distinct specificities. *J. Biol. Chem.* **269**: 5644-5652.

156. Satoh, M., Handa, K., Saito, S., Tokuyama, S., Ito, A., Miyao, N., Orikasa, S. and Hakomori, S. (1996) Disialosyl galactosylgloboside as an adhesion molecule expressed on renal cell carcinoma and its relationship to metastatic potential. *Cancer Res.* **56**: 1932-1938.

157. Stern, P. L., Willison, K. R., Lennox, E., Galfre, G., Milstein, C., Secher, D. and Ziegler, A. (1978) Monoclonal antibodies as probes for differentiation and tumor-associated antigens: a Forssman specificity on teratocarcinoma stem cells. *Cell* **14**: 775-783.

158. Willison, K. R. and Stern, P. L. (1978) Expression of a Forssman antigenic specificity in the preimplantation mouse embryo. *Cell* **14**: 785-793.

159. Nowinski, R., Berglund, C., Lane, J., Lostrom, M., Bernstein, I., Young, W. W.,Jr., Hakomori, S. I., Hill, L. and Cooney, M. (1980) Human monoclonal antibody against Forssman antigen. *Science* **210**: 537-539.

160. Sonnenberg, A., van Balen, P., Hengeveld, T., Kolvenbag, G. J., Van Hoeven, R. P. and Hilgers, J. (1986) Monoclonal antibodies detecting different epitopes on the Forssman glycolipid hapten. *J. Immunol.* **137**: 1264-1269.

161. Gathuru, J. K., Miyoshi, I. and Naiki, M. (1991) In vitro immunization of rat spleen lymphocytes with Forssman glycosphingolipid antigen and the generation of a monoclonal antibody. *J. Immunol. Methods* **137**: 95-102.

162. Bremer, E. G., Levery, S. B., Sonnio, S., Ghidoni, R., Canevari, S., Kannagi, R. and Hakomori, S. (1984) Characterization of a glycosphingolipid antigen defined by the monoclonal antibody MBr1 expressed in normal and neo-plastic epithelial cells of human mammary gland. *J. Biol. Chem.* **259**: 14773-14777.

163. Kannagi, R., Levery, S. B. and Hakomori, S. (1984) Blood group H antigen with globo-series structure: Isolation and characterization from human blood group O erythrocytes. *FEBS Lett.* **175**: 397-401.

164. Shigeta, K., Kirihata, Y., Ito, Y., Ogawa, T., Hakomori, S. and Kannagi, R. (1987) Monoclonal antibody directed to a chemically-synthesized cancer-associated glycosphingolipid antigens. *J. Biol. Chem.* **262**: 1358-1362.

165. Hirabayashi, Y., Fujita, S. C., Kon, K. and Ando, S. (1991) Characterization of a novel Le^x-active ganglioside from chick intestinal tissues recognized by murine monoclonal antibody 188C1. *J. Biol. Chem.* **266**: 10268-10274.

166. Eddy, E. M., Muller, C. H. and Lingwood, C. A. (1985) Preparation of monoclonal antibody to sulfatoxygalactosylglycerolipid by in vitro immunization with a glycolipid-

glass conjugate. *J. Immunol. Methods* **81**: 137-146.

167. Hofstetter, W., Bologa, L., Wetterwald, A., Z'graggen, A., Blaser, K. and Herschkowitz, N. (1984) Production and characterization of monoclonal antibodies to the myelin glycolipid sulfatide. *J. Neurosci. Res.* **11**: 341-350.
168. Miyake, M., Taki, T., Kannagi, R. and Hitomi, S. (1992) First establishment of a human monoclonal antibody directed to sulfated glycosphingolipids S_{M4s}-Gal and S_{M4g}, from a patient with lung cancer. *Cancer Res.* **52**: 2292-2297.
169. Chou, K. H., Ilyas, A. A., Evans, J. E., Quarles, R. H. and Jungalwala, F. B. (1985) Structure of a glycolipid reacting with monoclonal IgM in neuropathy and with HNK-1. *Biochem. Biophys. Res. Commun.* **128**: 383-388.
170. Shashoua, V. E., Daniel, P. F., Moore, M. E. and Jungalwala, F. B. (1986) Demonstration of glucuronic acid on brain glycoproteins which react with HNK-1 antibody. *Biochem. Biophys. Res. Commun.* **138**: 902-909.
171. Chou, D. K., Ilyas, A. A., Evans, J. E., Costello, C., Quarles, R. H. and Jungalwala, F. B. (1986) Structure of sulfated glucuronyl glycolipids in the nervous system reacting with HNK-1 antibody and some IgM paraproteins in neuropathy. *J. Biol. Chem.* **261**: 11717-11725.
172. Borroni, E., Derrington, E. A. and Whittaker, V. P. (1989) Characterization of a monoclonal antibody directed against a sulphoglycolipid that is evolutionarily conserved and developmentally regulated in rat brain. *Cell Tissue Res.* **256**: 373-380.
173. Hiraiwa, N., Iida, N., Ishizuka, I., Itai, S., Shigeta, K., Fukuda, Y., Kannagi, R. and Imura, H. (1988) Monoclonal antibodies directed to a disulfated glycosphingolipid associated with human hepatocellular carcinoma, S_{B1a} ($GgOse_4Cer$- II^3IV^3-*bis*-sulfate). *Cancer Res.* **48**: 6769-6774.
174. Hiraiwa, N., Fukuda, Y., Imura, H., Tadano-Aritomi, K., Nagai, K., Ishizuka, I. and Kannagi, R. (1990) Accumulation of highly acidic sulfated glycosphingolipids in human hepatocellular carcinoma defined by a series of monoclonal antibodies. *Cancer Res.* **50**: 2917-2928.
175. Veerman, E. C. I., Bolscher, J. G. M., Appelmelk, B. J., Bloemena, E., Van den Berg, T. K. and Amerongen, A. V. N. (1997) A monoclonal antibody directed against high M_r salivary mucins recognizes the SO_3-3Galβ1-3GlcNAc moiety of sulfo- Lewis[a]: A histochemical survey of human and rat tissue. *Glycobiology* **7**: 37-43.
176. Izawa, M., Kumamoto, K., Mitsuoka, C., Kanamori, A., Ohmori, K., Ishida, H., Nakamura, S., Kurata-Miura, K., Sasaki, K., Nishi, T. and Kannagi, R. (2000) Expression of sialyl 6-sulfo Lewis x is inversely correlated with conventional sialyl Lewis x expression in human colorectal cancer. *Cancer Res.* in press.
177. Kimura, H., Shinya, N., Nishihara, S., Kaneko, M., Irimura, T. and Narimatsu, H. (1997) Distinct substrate specificities of five human α-1,3-fucosyltransferases for *in vivo* synthesis of the sialyl Lewis x and Lewis x epitopes. *Biochem. Biophys. Res. Commun.* **237**: 131-137.
178. Uchimura, K., Muramatsu, H., Kadomatsu, K., Fan, Q. W., Kurosawa, N., Mitsuoka, C., Kannagi, R., Habuchi, O. and Muramatsu, T. (1998) Molecular cloning and characterization of an *N*-acetylglucosamine-6-*O*-sulfotransferase. *J. Biol. Chem.* **273**: 22577-22583.
179. Mitsuoka, C., Sawada-Kasugai, M., Ando-Furui, K., Izawa, M., Nakanishi, H., Nakamura, S., Ishida, H., Kiso, M. and Kannagi, R. (1998) Identification of a major carbohydrate capping group of the L-selectin ligand on high endothelial venules in human lymph nodes as 6-sulfo sialyl Lewis x. *J. Biol. Chem.* **273**: 11225-11233.
180. Rahman, A. F. and Longenecker, B. M. (1982) A monoclonal antibody specific for the Thomsen-Friedenreich cryptic T antigen. *J. Immunol.* **129**: 2021-2024.
181. Longenecker, B. M., Rahman, A. F., Leigh, J. B., Purser, R. A., Greenberg, A. H., Willans, D. J., Keller, O., Petrik, P. K., Thay, T. Y. and Suresh, M. R. (1984) Monoclonal antibody against a cryptic carbohydrate antigen of murine and human lymphocytes. I. Antigen expression in non-cryptic or unsubstituted form on certain murine lymphomas, on a spontaneous murine mammary carcinoma, and on several human adenocarcinomas. *Int. J. Cancer* **33**: 123-129.
182. Karsten, U., Butschak, G., Cao, Y., Goletz, S. and Hanisch, F. G. (1995) A new monoclonal antibody (A78-G/A7) to the Thomsen-Friedenreich pan-tumor antigen. *Hybridoma* **14**: 37-44.

183. Hanisch, F. G., Stadie, T. and Bosslet, K. (1995) Monoclonal antibody BW835 defines a site-specific Thomsen-Friedenreich disaccharide linked to threonine within the VTSA motif of MUC1 tandem repeats. *Cancer Res.* **55**: 4036-4040.

184. Longenecker, B. M., Willans, D. J., MacLean, G. D., Selvaraj, S., Suresh, M. R. and Noujaim, A. A. (1987) Monoclonal antibodies and synthetic tumor-associated glycoconjugates in the study of the expression of Thomsen-Friedenreich-like and Tn-like antigens on human cancers. *J. Natl. Cancer Inst.* **78**: 489-496.

185. Hirohashi, S., Clausen, H., Yamada, T., Shimosato, Y. and Hakomori, S. (1985) Blood group A cross-reacting epitope defined by monoclonal antibodies NCC-LU-35 and -81 expressed in cancer of blood group O or B individuals: its identification as Tn antigen. *Proc. Natl. Acad. Sci. USA* **82**: 7039-7043.

186. Roxby, D. J., Morley, A. A. and Burpee, M. (1987) Detection of the Tn antigen in leukaemia using monoclonal anti-Tn antibody and immunohistochemistry. *Br. J. Haematol.* **67**: 153-156.

187. King, M. J., Parsons, S. F., Wu, A. M. and Jones, N. (1991) Immunochemical studies on the differential binding properties of two monoclonal antibodies reacting with Tn red cells. *Transfusion* **31**: 142-149.

188. Takahashi, H. K., Metoki, R. and Hakomori, S. (1988) Immunoglobulin G_3 monoclonal antibody directed to Tn antigen (tumor-associated α-*N*-acetylgalactosaminyl epitope) that does not cross-react with blood group A antigen. *Cancer Res.* **48**: 4361-4367.

189. Thurnher, M., Clausen, H., Sharon, N. and Berger, E. G. (1993) Use of O-glycosylation-defective human lymphoid cell lines and flow cytometry to delineate the specificity of *Moluccella laevis* lectin and monoclonal antibody 5F4 for the Tn antigen (GalNAc-α-1-O-Ser/Thr). *Immunol. Lett.* **36**: 239-244.

190. Numata, Y., Nakada, H., Fukui, S., Kitagawa, H., Ozaki, K., Inoue, M., Kawasaki, T., Funakoshi, I. and Yamashina, I. (1990) A monoclonal antibody directed to Tn antigen. *Biochem. Biophys. Res. Commun.* **170**: 981-985.

191. Huang, J., Byrd, J. C., Siddiki, B., Yuan, M., Lau, E. and Kim, Y. S. (1992) Monoclonal antibodies against partially deglycosylated colon cancer mucin that recognize Tn antigen. *Dis. Markers* **10**: 81-94.

192. Kurosaka, A., Fukui, S., Kitagawa, H., Nakada, H., Numata, Y., Funakoshi, I., Kawasaki, T. and Yamashina, I. (1987) Mucin-carbohydrate directed monoclonal antibody. *FEBS Lett.* **215**: 137-139.

193. Kjeldsen, T., Clausen, H., Hirohashi, S., Ogawa, T., Iijima, H. and Hakomori, S. (1988) Preparation and characterization of monoclonal antibodies directed to the tumor-associated O-linked sialosyl-2→6 α-*N*-acetylgalactosaminyl (sialosyl-Tn) epitope. *Cancer Res.* **48**: 2214-2220.

194. Nuti, M., Teramoto, Y. A., Mariani-Costantini, R., Hand, P. H., Colcher, D. and Schlom, J. (1982) A monoclonal antibody (B72.3) defines patterns of distribution of a novel tumor-associated antigen in human mammary carcinoma cell populations. *Int. J. Cancer* **29**: 539-545.

195. Springer, G. F., Desai, P. R., Robinson, M. K., Tegtmeyer, H. and Scanlon, E. F. (1986) The fundamental and diagnostic role of T and Tn antigens in breast carcinoma at the earliest histologic stage and throughout. *Prog. Clin. Biol. Res.* **204**: 47-70.

196. Hanisch, F. G., Uhlenbruck, G., Egge, H. and Peter Katalinic, J. (1989) A B72.3 second-generation-monoclonal antibody (CC49) defines the mucin-carried carbohydrate epitope Galβ(1-3) [NeuAcα(2-6)]GalNAc. *Biol. Chem. Hoppe Seyler* **370**: 21-26.

197. Muroi, K., Amemiya, Y., Sievers, E. L., Miura, Y., Hakamori, S. I. and Loken, M. R. (1997) Expression of sialosyl-T and disialosyl-T antigens in erythroid cells. *Leuk. Lymphoma.* **25**: 403-414.

198. Hanisch, F. G., Koldovsky, U. and Borchard, F. (1993) Monoclonal antibody 2B5 defines a truncated *O*-glycan, GlcNAcβ1-3Galβ1-4GlcNAcβ1-6 (GalNAc), on mucins from deep gastric and duodenal glands as well as metaplasia and neoplasia of gastric differentiation. *Cancer Res.* **53**: 4791-4796.

199. Ishihara, K., Kurihara, M., Goso, Y., Urata, T., Ota, H., Katsuyama, T. and Hotta, K. (1996) Peripheral α-linked N-acetylglucosamine on the carbohydrate moiety of mucin

derived from mammalian gastric gland mucous cells: epitope recognized by a newly characterized monoclonal antibody. *Biochem. J.* **318**: 409-416.

200. Yamashita, Y., Chung, Y. S., Murayama, K., Kannagi, R. and Sowa, M. (1995) Alterations in gastric mucin with malignant transformation: novel pathway of mucin synthesis. *J. Natl. Cancer Inst.* **87**: 441-446.
201. Kumamoto, K., Mitsuoka, C., Izawa, M., Kimura, N., Otsubo, N., Ishida, H., Kiso, M., Yamada, T., Hirohashi, S. and Kannagi, R. (1998) Specific detection of sialyl Lewis x determinant carried on the mucin GlcNAcβ1$\rightarrow$6GalNAcα core structure as tumor-associated antigen. *Biochem. Biophys. Res. Commun.* **247**: 514-517.
202. Turner, J. R., Tartakoff, A. M. and Greenspan, N. S. (1990) Cytologic assessment of nuclear and cytoplasmic *O*-linked *N*-acetylglucosamine distribution by using anti-streptococcal monoclonal antibodies. *Proc. Natl. Acad. Sci. U. S. A.* **87**: 5608-5612.
203. Srikrishna, G., Varki, N. M., Newell, P. C., Varki, A. and Freeze, H. H. (1997) An IgG monoclonal antibody against *Dictyostelium discoideum* glycoproteins specifically recognizes Fucα1,6GlcNAcβ in the core of *N*-linked glycans - Localized expression of core-fucosylated glycoconjugates in human tissues. *J. Biol. Chem.* **272**: 25743-25752.
204. Caterson, B., Christner, J. E. and Baker, J. R. (1983) Identification of a monoclonal antibody that specifically recognizes corneal and skeletal keratan sulfate. Monoclonal antibodies to cartilage proteoglycan. *J. Biol. Chem.* **258**: 8848-8854.
205. Sorrell, J. M., Carrino, D. A., Baber, M. A., Asselineau, D. and Caplan, A. I. (1999) A monoclonal antibody which recognizes a glycosaminoglycan epitope in both dermatan sulfate and chondroitin sulfate proteoglycans of human skin. *Histochem. J.* **31**: 549-558.
206. Kure, S. and Yoshie, O. (1986) A syngeneic monoclonal antibody to murine Meth-A sarcoma (HepSS-1) recognizes heparan sulfate glycosaminoglycan (HS-GAG): cell density and transformation dependent alteration in cell surface HS-GAG defined by HepSS-1. *J. Immunol.* **137**: 3900-3908.
207. Yamagata, M., Kimata, K., Oike, Y., Tani, K., Maeda, N., Yoshida, K., Shimomura, Y., Yoneda, M. and Suzuki, S. (1987) A monoclonal antibody that specifically recognizes a glucuronic acid 2-sulfate-containing determinant in intact chondroitin sulfate chain. *J. Biol. Chem.* **262**: 4146-4152.
208. David, G., Bai, X. M., Van der Schueren, B., Cassiman, J. J. and Van den Berghe, H. (1992) Developmental changes in heparan sulfate expression: in situ detection with mAbs. *J. Cell Biol.* **119**: 961-975.
209. Couchman, J. R., Caterson, B., Christner, J. E. and Baker, J. R. (1984) Mapping by monoclonal antibody detection of glycosaminoglycans in connective tissues. *Nature* **307**: 650-652.
210. Bouchon, B., Levery, S. B., Clausen, H. and Hakomori, S. (1992) Production and characterization of a monoclonal antibody (BBH5) directed to ganglioside lactone. *Glycoconjugate J.* **9**: 27-38.
211. Ding, K., Rosen, A., Ray, A. K. and Magnusson, G. (1992) Anti-GM3-lactam monoclonal antibodies of the IgG type recognize natural GM3-ganglioside lactone but not GM3-ganglioside. *Glycoconjugate J.* **9**: 303-306.
212. Sjoberg, E. R., Chammas, R., Ozawa, H., Kawashima, I., Khoo, K. H., Morris, H. R., Dell, A., Tai, T. and Varki, A. (1995) Expression of de-*N*-acetyl-gangliosides in human melanoma cells is induced by genistein or nocodazole. *J. Biol. Chem.* **270**: 2921-2930.
213. Tai, T., Kawashima, I., Tada, N. and Ikegami, S. (1988) Different reactivities of monoclonal antibodies to ganglioside lactones. *Biochim. Biophys. Acta* **958**: 134-138.
214. Kanamori, A., Inoue, S., Xulei, Z., Zuber, C., Roth, J., Kitajima, K., Ye, J., Troy, F. A. and Inoue, Y. (1994) Monoclonal antibody specific for α2$\rightarrow$8-linked oligo deaminated neuraminic acid (KDN) sequences in glycoproteins. Preparation and characterization of a monoclonal antibody and its application in immunohistochemistry. *Histochemistry* **101**: 333-340.

INDEX

MIX
Papier aus verantwortungsvollen Quellen
Paper from responsible sources
FSC® C105338

If you have any concerns about our products,
you can contact us on
ProductSafety@springernature.com

In case Publisher is established outside the EU,
the EU authorized representative is:
Springer Nature Customer Service Center GmbH
Europaplatz 3, 69115 Heidelberg, Germany

Printed by Libri Plureos GmbH
in Hamburg, Germany